AF443227

Geophysical Monograph 145

Timescales of the Paleomagnetic Field

James E. T. Channell
Dennis V. Kent
William Lowrie
Joseph G. Meert
Editors

American Geophysical Union
Washington, DC

Library of Congress Cataloging-in-Publication Data

Timescales of the paleomagnetic field / James E.T. Channell ... [et al.], editors.
     p. cm. -- (Geophysical monograph ; 145)
  Includes bibliographical references.
  ISBN 0-87590-410-6
  1. Paleomagnetism. 2. Stratigraphic correlation. I. Channell, J. II. Series.

  QE501.4.P35T56 2004
  538'.727--dc22

                          2004057403

  ISBN 87590-410-6
  ISSN 0065-8448

CONTENTS

PREFACE

To mark the 70th birthday of Neil D. Opdyke, a Chapman Conference entitled "Timescales of the Internal Geomagnetic Field" was held at the University of Florida in Gainesville on March 9-11, 2003. This AGU Chapman Conference was sponsored by the U.S. National Science Foundation, University of Florida, Florida Museum of Natural History, and 2G Enterprises. Forty-one talks and twenty-three posters were presented during the three-day meeting. This monograph contains twenty-four of those papers, and is a balanced subset of the papers presented at the conference. The monograph is divided into three parts. Part 1 deals with the geocentric axial dipole (GAD) hypothesis, continental reconstruction, and long-term geomagnetic field behavior. Part 2 comprises papers on magnetic polarity stratigraphy and the acquisition of sedimentary magnetization. Part 3 deals with secular variation, paleointensity, and short-term geomagnetic field behavior. These are all topics that have been substantially impacted by Neil's scientific work.

Some 40 years after the 1960s revolution in the Earth Sciences, most of the practitioners of that time have retired from the scene. One who has not yet retired and maintains his irrepressible zest for our science is Neil Opdyke. A brief synopsis of Neil's research accomplishments (outlined below) does not, in itself, do justice to the inspiration that many of us continue to receive through Neil's good-natured enjoyment of sound scientific debate. He has been a regular participant and mainstay of the GP section at AGU for about half a century!

After graduating in geology from Columbia College in New York, Neil began his career in paleomagnetism following a chance meeting with S.K. Runcorn in the early summer of 1955. Work as a field assistant with Runcorn in Arizona during that summer led to his recruitment to Cambridge as a graduate student. When, in 1956, Runcorn moved to the Department of Physics at King's College, Durham University (later to become the University of Newcastle-upon-Tyne), Neil moved out of the drafty rooms of Gonville and Caius College (Cambridge) to the damp North Sea climes of Newcastle. The seeds of quantification of continental drift were being sown during the late 1950s both at Cambridge and at Newcastle, and Neil's research aimed to establish the case for continental drift from both paleomagnetic data and paleowind directions. In 1958, Neil moved from Newcastle to Rice University marrying Margie Wilson on the way. After a year

Timescales of the Paleomagnetic Field
Geophysical Monograph Series 145
Copyright 2004 by the American Geophysical Union
10.1029/145GM00

at Rice, they traveled under a Fulbright fellowship to Ted Irving's laboratory in Canberra, Australia, and from there to Salisbury (now Harare, Zimbabwe). Neil's paleomagnetic work in Africa in the early 1960s contributed significantly to the eventual documentation of continental drift, and this work still forms an integral part of African apparent polar wander paths.

One of Neil's most important contributions to Earth Science was his pioneering use of magnetic polarity stratigraphy as a means of global correlation. His magnetostratigraphic studies on marine piston cores done at Lamont Doherty Geological (now Earth) Observatory in the mid-1960s remain a model of how biomagnetostratigraphy should be done, and established the importance of magnetic stratigraphy as an integral component of geologic timescales. In 1969, Opdyke and Henry used marine core data for a convincing test of the GAD hypothesis that is central to the use of paleomagnetism in continental reconstruction. Neil's work with N.J. Shackleton in 1973 marked the beginning of the integration of oxygen isotope stratigraphy and magnetostratigraphy that has led to current methods of tuning timescales. Neil pioneered magnetic stratigraphy in terrestrial (non-marine) sediments and produced some of the most impressive records, notably from Pakistan and southwestern USA. These studies led to a vastly improved time frame for vertebrate evolution and allowed the documentation of mammal migration.

Since his move from Lamont to the University of Florida in 1981, Neil has constructed the first polarity timescale for the Carboniferous and Early Permian using magnetostratigraphic data from Colorado, Pennsylvania, NE Canada and Australia. He has been involved in paleomagnetic studies in China that refined our knowledge of the paleogeography and tectonic history of the vast Asian landmass. Neil is presently engrossed in studies of magnetic directions and paleointensities in young (<5 Ma) volcanic rocks along the American Cordillera from Patagonia to Alaska, as well as from Australia. These projects are designed to determine the precise time-averaged structure of the geomagnetic field, central to the refinement of the GAD hypothesis and for models of the geodynamo.

This brings us to the topics covered in this volume.

Part 1 is mainly concerned with the GAD hypothesis, the idea that the time-averaged geomagnetic field closely approximates the field of a geocentric axial dipole, which has served us well for reconstruction of the mosaic of continents and plates through time. At the next level of reconstruction, however, inconsistencies are apparent and have been difficult to reconcile. The GAD hypothesis has been challenged on two counts. First, paleomagnetic data may imply persistent contributions of significant non-axial dipole (NAD) fields, particularly for the Tertiary of central Asia, for the Permo-Triassic of Pangea, and possibly for much of the Precambrian. In addition, analyses of paleomagnetic data for the last 5 Myr indicate small but significant NAD contributions in the time-averaged field. The series of papers in Part 1 take different views on these issues, ranging from evidence for significant NAD fields, to explanations of apparent time-averaged NAD features in terms of artifacts of data distribution or recording process.

Part 2 deals with magnetic polarity stratigraphy, chrono-stratigraphy, and the acquisition of magnetization in sediments. The geomagnetic polarity record is central to the construction of geologic timescales, and provides the principal tool for calibration of marine and terrestrial biozonations. The polarity record continues to evolve with the recognition of brief polarity subchrons, and the limitations to this evolution may lie with the sedimentary recording process. The polarity record of the geomagnetic field indicates changes in reversal frequency on 10^7-10^8 year time-scales that are incompatible with core processes and must, therefore, be attributed to prolonged interactions between the mantle and core.

Part 3 deals with secular variation and short-term field behavior, toward the other end of the variability spectrum. High-sedimentation-rate marine and lake sediments have, in the last few years, revolutionized our understanding of the behavior of the geomagnetic field. The presence of ubiquitous short-lived (~5 kyr duration) polarity subchrons or excursions in the Brunhes and Matuyama Chrons, coupled with high-quality relative paleointensity records, indicate that periods of low paleointensity are frequently accompanied by short-lived but major perturbations of the direction of the geomagnetic field. The study of sub-Milankovitch-scale paleoclimate requires stratigraphic correlation at an appropriate (millennial-scale) resolution. As correlation at this scale is not easily achieved through traditional isotopic methods, geomagnetic paleointensity records and associated directional perturbations will become significant for this purpose. In addition, understanding this short-term geomagnetic behavior is important for constraining models of the geodynamo.

This volume is dedicated to Neil Opdyke on the occasion of his 70th birthday. In recognition of the importance of his work to the Earth Sciences, he has been awarded the George P. Woollard Award of the Geological Society of America (1987), the Fleming Medal of the American Geophysical Union (1996), and has been elected to the National Academy of Sciences (1996) and the American Academy of Arts and Sciences (1998).

James E.T. Channell
Dennis V. Kent
William Lowrie
Joseph G. Meert
Editors

Geocentric Axial Dipole Hypothesis: A Least Squares Perspective

Michael McElhinny

Gondwana Consultants, Port Macquarie, New South Wales, Australia

Departures from the geocentric axial dipole (GAD) model of the time-averaged paleomagnetic field have been proposed both for the time-interval 0–5 Ma and for the Mesozoic and Paleozoic. At present the basic problem is that there are not enough data of sufficient quality to be able to determine second order terms other than a small persistent geocentric axial quadrupole (GAQ). Even for the interval 0–5 Ma using the lava flow database, the majority of the data were derived around 25 years ago when demagnetization analytical procedures were not as robust as those currently in use. There are many ways in which an artificial geocentric axial octupole (GAO) term can arise from data artifacts and these are discussed in some detail using a least squares approach. No differences between the normal and reverse fields are seen in the data. Application of the random paleogeography test suggests that persistent GAO terms are present in the Mesozoic, Paleozoic and Precambrian. However it has now been shown that the method is flawed because the basic assumption of random paleogeography is not valid. It has been proposed that the problem of the reconstruction of Pangea (Pangea A versus Pangea B) during the early Mesozoic and Paleozoic can be resolved if persistent GAO terms are present. However, many of the data used for this conclusion were derived in the 1970s and need to be redone using modern methods. At present the only distinguishable second order feature is a persistent GAQ of about 4% of the GAD field.

1. INTRODUCTION

The Geocentric Axial Dipole (GAD) hypothesis represents the fundamental assumption used when calculating paleomagnetic poles for use in determining apparent polar wander paths. It was first introduced into paleomagnetism by Hospers [1954], who made use of the new statistical methods developed by Fisher [1953]. This was the first demonstration that the average of virtual geomagnetic poles (VGP) over several thousand years in Recent times centered about the geographic pole. Creer et al. [1954] explicitly invoked the GAD

hypothesis when putting forward the concept of the apparent polar wander path for the interpretation of paleomagnetic results from Great Britain. This use of the GAD hypothesis is now standard procedure in paleomagnetism.

It was originally thought that the validity of the GAD hypothesis was demonstrated by the fact that paleomagnetic poles for the past few million years centered about the present geographic pole [*Cox and Doell*, 1960; *Irving*, 1964; *McElhinny*, 1973]. Unfortunately this observation is not by itself sufficient to demonstrate that the time-averaged field is purely that of a geocentric axial dipole (g_1^0). Any geocentric axial field represented by zonal harmonics (g_1^0, g_2^0, g_3^0 etc) will also produce paleomagnetic poles centered about the geographic pole. Opdyke and Henry [1969], using the inclinations observed in 52 deep-sea sediment cores world-wide,

Timescales of the Paleomagnetic Field
Geophysical Monograph Series 145

showed that the plot of mean inclination versus latitude was consistent with the latitude variation of inclination expected from the GAD hypothesis for the past few million years (Figure 1). However, this was only a first-order solution and showed that the GAD was the dominant term. Wilson [1970, 1971, 1972] noted that the paleomagnetic poles for the past few million years tended to plot too far away from the observation site along a great circle joining the site to the geographic pole. Also the poles tended to plot to the right of the geographic pole when viewed from the observation site. These effects were referred to as the far-sided effect and the right-handed effect.

Wilson [1971] introduced the concept of the common-site longitude pole position in which all observation sites are placed at zero longitude. This was a convenient way to analyze the far-sided and right-handed effects. A recent analysis of observations covering the time interval 0–5 Ma using this method by Quidelleur et al. [1994] and McElhinny et al. [1996] is shown in Figure 2 for separated normal and reverse polarity data. The mean of the reverse polarity data is significantly more far-sided and right-handed than that for the normal polarity data. This difference between the means of the normal and reverse polarity data is not significant when only data from igneous rocks are considered, although the far-sided and right-handed effects remain. Wilson [1970, 1971, 1972] modeled the far-sided effect as originating from an axial dipole

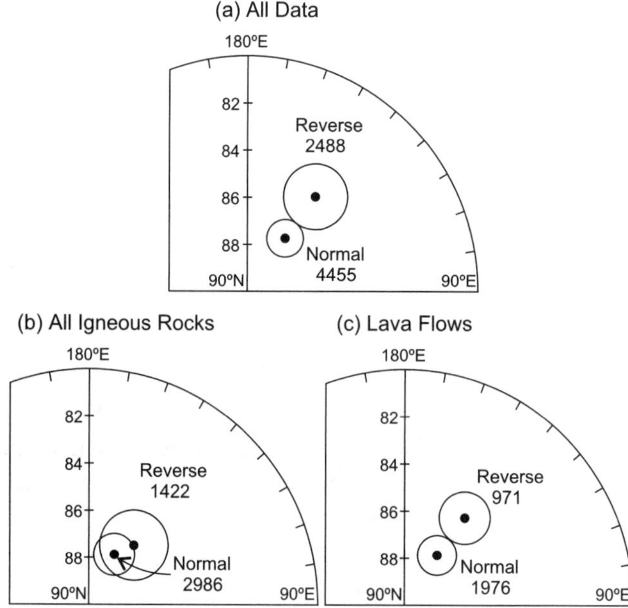

Figure 2. Common site longitude global mean pole positions for 0–5 Ma plotted on a polar stereographic projection (latitudes >80°N) with their 95% circles of confidence. The numbers of sites used in determining each mean are indicated. (a) and (b) are from global data analyzed by McElhinny et al. [1996]. (c) is from a global lava database analyzed by Quidelleur et al. [1994].

source displaced northward along the axis of rotation. However, this is a non-unique solution for modeling geomagnetic field sources and it is more appropriate to use spherical harmonics to describe these effects. In this case the offset dipole model is equivalent to a geocentric axial dipole (g_1^0) plus a geocentric axial quadrupole (g_2^0). Egbert [1992] has shown that there is a natural sampling bias in VGP longitudes in which their distribution peaks 90° away from the sampling longitude. However, the observed right-handed effect is probably too large to be explained in this way. It now appears that the right-handed effect may represent an artifact resulting from inadequate magnetic cleaning or arise from the poor geographical distribution of the data.

The first attempts to carry out spherical harmonic analyses of the time-averaged field over the past few million years were carried out by Wells [1973], Creer et al. [1973] and Georgi [1974]. However, the use of poor quality or unevenly distributed data can result in very inaccurate spherical harmonic descriptions as reflected in the widely different results obtained by these authors. Wells [1973] concluded that only the zonal harmonics were significant after a careful consideration of all the errors involved. However, Creer et al. [1973] and Georgi [1974] concluded that there were significant non-zonal harmonics present, some of them of comparable size to the zonal ones.

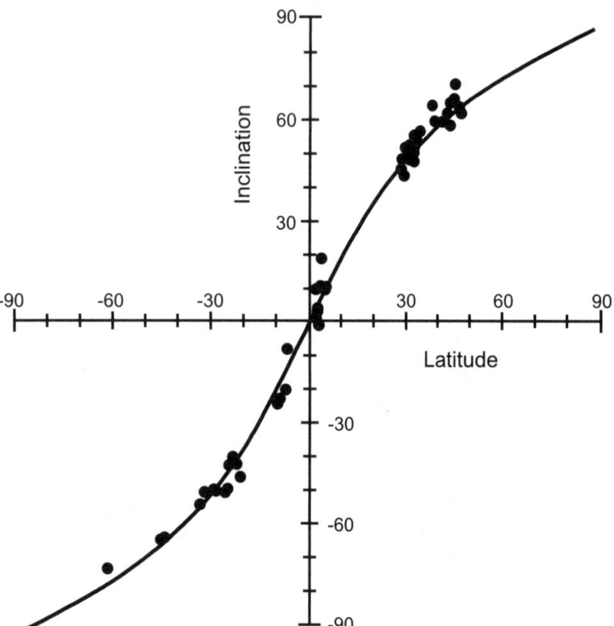

Figure 1. Mean inclinations observed in 52 deep-sea sediment cores covering the past few million years plotted as a function of latitude. The solid curve is the variation expected from a GAD field. Redrawn from Opdyke and Henry [1969].

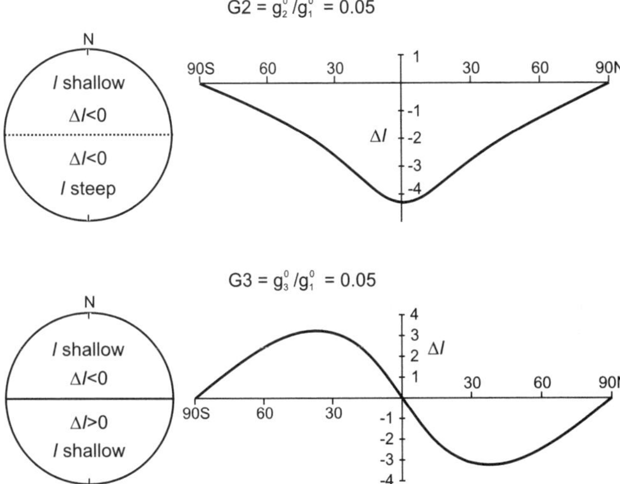

Figure 3. Latitude variation of the inclination anomaly ΔI (departure from a GAD field) expected from a geocentric axial quadrupole (- g_2^0) or geocentric axial octupole (-g_3^0) with magnitude 0.05 of the GAD term, when added to the GAD ($g_1{}^0$) field.

2. INCLINATION ANALYSIS OF DATA 0–5 MA

Cox [1975] showed that the difference of inclination ΔI between that of the observed field and that due to a geocentric axial dipole may be used for determining the zonal harmonics. The method is illustrated in Figure 3 following Merrill and McElhinny [1977] and assumes that the non-zonal har-

monics are negligible. This assumption can be tested by analysing the departures of the mean declinations (D) from the expected zero value (see §4). For a geocentric axial quadrupole (g_2^0), the variation of ΔI with latitude depends upon the ratio G2= g_2^0 / g_1^0. For positive values of G2, ΔI is negative everywhere with its greatest (negative) value occurring at the equator. This corresponds with a shallowing of the inclination in the northern hemisphere and a steepening in the southern hemisphere. For a geocentric axial octupole (g_3^0), the variation of ΔI with latitude depends upon the ratio G3= g_3^0 / g_1^0. For positive values of G3, ΔI is negative in the northern hemisphere, positive in the southern hemisphere, and zero at the equator. This corresponds with a shallowing of the inclination in both hemispheres.

Of course if there is a departure from pure axial symmetry at any given region, then the mean declination (D) will differ from zero. In this case ΔI will differ slightly from the value ΔI_z expected from purely zonal harmonics. McElhinny et al. [1996] have shown that a good approximation for ΔI_z is given by the relation:

$$\sin \Delta I_z = \sin \Delta I \cos D.$$

For small values of D, the correction will thus be small.

McElhinny et al. [1996] analyzed all global paleomagnetic data in the time range 0–5 Ma that conformed with certain minimum selection criteria. This is the only substantive analysis of all types of paleomagnetic data in this time range. The

Table 1. Normalized Gauss coefficients (G2, G3) for the time interval 0-5 Ma calculated for various combinations of paleomagnetic data. Least squares analysis of McElhinny et al. [1996]

Data combination	Polarity	G2	G3	Chi^2
All data combined	N	0.043±0.030	0.017±0.031	0.24
	R	0.063±0.035	0.056±0.037	0.26
Continents all data	N	0.041±0.037	0.031±0.035	0.08
	R	0.054±0.026	0.094±0.056	0.51
Continents igneous rocks	N	0.039±0.048	0.029±0.027	0.15
	R	0.059±0.048	0.058±0.056	0.41
Brunhes igneous rocks	N	0.032±0.025	0.030±0.023	0.63
Matuyama igneous rocks	R	Insufficient data available		
Ocean cores (drift corrected)	N	0.026±0.021	-0.027±0.019	0.28
	R	0.035±0.012	-0.014±0.013	0.69
Continents igneous + ocean cores	N	0.038±0.025	0.013±0.027	0.26
	R	0.049±0.022	0.028±0.024	0.42
Brunhes igneous + ocean cores	**N**	**0.033±0.019**	0.010±0.021	0.70
Matuyama igneous + ocean cores	**R**	**0.042±0.022**	0.012±0.022	0.37
Combined B+M igneous + ocean cores	**N+R**	**0.038±0.012**	0.011±0.012	0.55

purpose was to see if different effects could be seen in the data from sediments, igneous rocks, and ocean cores. This analysis is highlighted in this paper because, unlike other analyses, it was able to show how the quality and type of data can seriously affect the determinations of second-order terms beyond the GAD. A least squares method was used to determine the best fitting G2+G3+G4 combination. The ΔI values were averaged into latitude bands to obtain at least 10 or more averages from pole to pole to create a reasonable distribution of observation points for modeling. This method has the effect of minimizing aliasing problems (see §3.1). Results were subdivided into those from igneous rocks, sediments, and ocean cores for both normal and reverse polarity data as given in Table 1. In all cases it was shown that the G4 term is small and not significant.

Originally McElhinny and Merrill [1977] and other workers thought that there were significant differences between the results obtained from normal and reverse polarity data sets. Reverse polarity data always seem to give larger values for G2 than the normal polarity data. However, Table 1 shows that the G2 values obtained for the normal and reverse fields are not significantly different within error limits regardless of which data combination is used. This conclusion is supported independently by Quidelleur and Courtillot [1996]. On the other hand, wide variations in values for G3 are obtained depending on the data combination used. Artificial G3 values can arise from data artifacts in a number of ways as discussed in §3. McElhinny et al. [1996] therefore concluded that only the G2 values had any sig-

nificance in their analysis. A similar conclusion was arrived at by Quidelleur and Courtillot [1996] and Carlut and Courtillot [1998] from spherical harmonic analyses.

The data combination that appears to minimize the effects arising from the production of artificial G3 values is that from the Brunhes and Matuyama igneous rocks and ocean cores. These combinations produce the lowest values of G3, neither of which is significant. Also the G2 values deduced for the normal and reverse polarity data sets are not significant. Therefore, McElhinny et al. [1996] proposed that the combination of both polarity data sets provides the best estimate of G2 = 0.038 ± 0.012 as is illustrated in Figure 4. In this case ΔI has been plotted against GAD inclination as this spreads the data points more evenly between the poles.

3. ERRORS ASSOCIATED WITH G3 TERMS

There are many ways in which values of G3 can arise from data artifacts as summarized by McElhinny et al. [1996].

3.1. Inclination Errors in Sediments

A shallowing effect on the inclinations observed in sediments can arise from DRM processes [*King and Rees*, 1966] or compaction effects [*Blow and Hamilton*, 1978]. Note that in Table 1 larger G3 values are deduced for the continental data that include sedimentary data when compared with those when only igneous rocks are considered. A curious observation is that the G3 values for the ocean sediment data are always negative. There is no obvious reason why these should differ in this way from the continental data.

3.2. Inclination Errors in Lavas

These can arise from the bulk demagnetizing effect (shape anisotropy) in lavas [*Coe*, 1979]. This is unlikely to be a significant effect except in very thin lavas.

3.3. The Use of Unit Vectors

In a computer simulation and analysis of the paleomagnetic vectors observed in lake sediment paleosecular variation studies, Creer [1983] observed that the lack of inclusion of intensity values in paleomagnetic vectors could cause a shallowing of the observed inclinations. Recently more high quality data from lava flows and sediments have included absolute and relative paleointensity determinations in order to overcome this problem (see discussion by *Johnson et al.* [2003]).

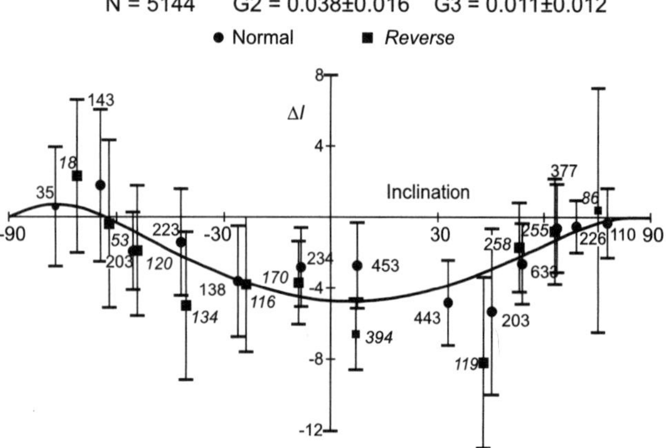

Figure 4. Best fit zonal harmonic model (G2+G3+G4) for the combined Brunhes and Matuyama igneous rock and ocean core sediments data given by McElhinny et al. [1996]. Normal and reverse polarity ΔI values are averaged separately in latitude bands. The mean values of ΔI are shown with 95% confidence limits and are plotted as a function of the dipole inclination corresponding to the mean latitude in each case. N is the total number of sites with the number of sites used in each mean indicated.

3.4. Incomplete Magnetic Cleaning of Brunhes-Age Overprints

McElhinny et al. [1996] have discussed this effect in some detail. A significant artificial G3 can arise in reversely magnetized rocks when a Brunhes-age (normal) overprint has not be completely cleaned through partial demagnetization. Viscous components that would average to the axial dipole field during the Brunhes epoch would have longer relaxation times and would require more careful cleaning procedures.

Suppose that the geomagnetic field during the most recent reverse epochs can be modeled simply as all the coefficients of the present field but with a change of sign. To simulate the time-averaged field, this 'reverse' field can be randomly sampled around lines of latitude to produce an average field within any latitude band. The mean value of ΔI for each of these bands can then be calculated in the same manner as in Figure 4. The resulting plot is shown in Figure 5a. As might be expected, the time-averaged field is dominated by the G2 term, with the G3 and G4 terms being essentially zero. Now add a GAD (normal) field of 5% of the reverse field. This would be the equivalent of what the reversely magnetized rocks would record, but with a 5% longer-tern viscous component that has not been cleaned through incomplete demagnetization procedures.

The mean values of ΔI for each latitude band are then recalculated and are shown in Figure 5b. The result is quite dramatic; an artificial value of +G3 is produced that is even larger than the G2 term, but the G4 term remains unchanged and is still effectively zero. Thus the observation in Table 1 that the G3 terms for reversely magnetized rocks are always greater than those for the corresponding normally magnetized rocks almost certainly arises from incomplete demagnetization of the reversely magnetized rocks.

This observation is stressed here because it relates not only to the use of poor quality data in the 0–5 Ma time interval but also to their use for times older than 5 Ma (see §6). For the 0–5 Ma interval the acquisition of new high quality data (see §5) should minimize this effect.

3.5. Nonorthogonality of Spherical Harmonics

Spherical harmonic functions form an orthogonal set over the surface of a sphere, so it is a simple matter to obtain the individual coefficients with observations uniformly spaced over the sphere. Unfortunately, the mapping of the inclination anomaly curve ΔI leads to nonorthogonality. From Figure 3 it might seem that the ΔI curve for G2 alone is orthogonal to that for G3 alone. However, the G2 curve is not exactly symmetrical about the equator, so the curves are not quite orthogonal between -90° and +90°. The ΔI curves for Gi and

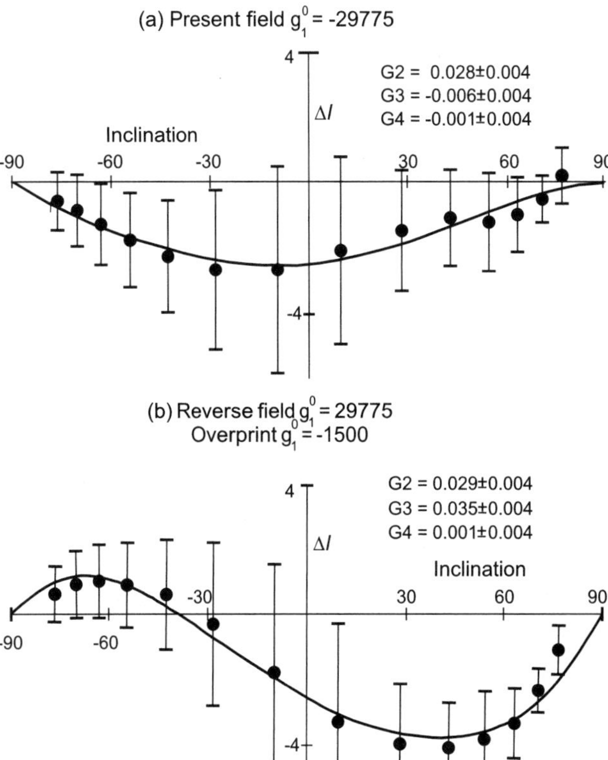

Figure 5. (a) Simulated variation of ΔI with dipole inclination for the reverse polarity field obtained by changing the sign of the coefficients for the present field and then averaging the values of inclination around lines of latitude. (b) Same but with an overprint of g_1^0 = -1500 (5% of the present GAD field). A large artificial value of G3 is produced. From McElhinny et al. [1996].

Gj are distinctly nonorthogonal when i and j are either both even or both odd. Furthermore, the overall ΔI curve for nonzero G2 and G3 is not the sum of the individual ΔI curves for G2 and G3 on their own. For example, at a latitude of 30°S, $\Delta I=-5.54°$ for G2=0.1 and $\Delta I=+6.65°$ for G3=0.1. However, for G2 and G3 combined $\Delta I=0.81°$ when compared with the sum for the separate curves of 1.01°. Therefore the estimates of the coefficients are not entirely independent, but for small values of G2, G3 etc (as is the case here), the effect is small.

3.6. Aliasing

Poor data distribution can cause aliasing between the various Gn terms. As an example, suppose that there is only a G3 term present. From Figure 3 it can be seen that if observations are obtained only north of about 25°N, then this G3 will be aliased into a G2 term. As an example set G3=0.05 and determine the corresponding ΔI values at 25.3°N, 35.5°N, 43.7°N and 63.4°N (the four most northern latitudes for the reverse polarity data in the second line of Table 1). The best fit from modeling these values gives estimates of G2=0.065, G3=0.007 and G4=0.024, so actual G3 term has been aliased almost entirely into a G2 term. A major advantage of the binning used for the analyses of Table 1 is that it creates a reasonable distribution of observation points for modeling, thus minimizing the aliasing problem.

The aliasing problem is exacerbated in the least squares approach where an infinite spherical harmonic representation of the field is truncated to a finite series with only 1,2 or 3 zonal harmonics. Regularized inversions in which one allows more model parameters in such a way that the structure is minimized helps overcome this problem. However, in this case individual coefficient estimates should not be over-interpreted as they could arise purely from the poor quality of the data.

4. DECLINATION ANALYSIS OF DATA 0–5 MA

To test the assumption that the non-zonal harmonics are negligible or at least an order of magnitude lower than G2, Merrill and McElhinny [1977] and McElhinny et al. [1996] analyzed the departure of the declinations (ΔD) from the expected value of zero for a GAD field. For this purpose the reverse polarity data are inverted so that their declinations are northerly about zero. When added to the GAD field, the lowest of the non-zonal harmonics (g_1^1 and h_1^1) will produce a declination anomaly that is a function of latitude. The value of ΔD at each site may thus be normalized to the equator using the relation

$$\tan \Delta D_0 = \tan \Delta D \cos I,$$

where ΔD_0 is the equatorial value of ΔD.

McElhinny et al. [1996] averaged values of ΔD_0 over 45° longitude sectors for both the normal and reverse data sets. These are illustrated in Figure 6 with their 95% confidence limits. Only one of the 16 means is just significantly different from zero. There is some suggestion of a possible equatorial dipole with a peak in the 0–45° longitude sector. Note that this is where the greatest number of sites is located (2050 out of 6443, or 32%). This is a possible explanation for the right-handed effect seen in Figure 1 in which unit weight is given to each site resulting in a bias to data from this longitude sector. What Figure 6 shows is that any non-zonal harmonics are currently below the level of detection with the existing data set.

5. THE LAVA DATABASE AND SPHERICAL HARMONICS 0–5 MA

Paleomagnetic results from lava flows have been used in the study of paleosecular variation. In such studies, the angular dispersion of virtual geomagnetic poles as a function of lat-

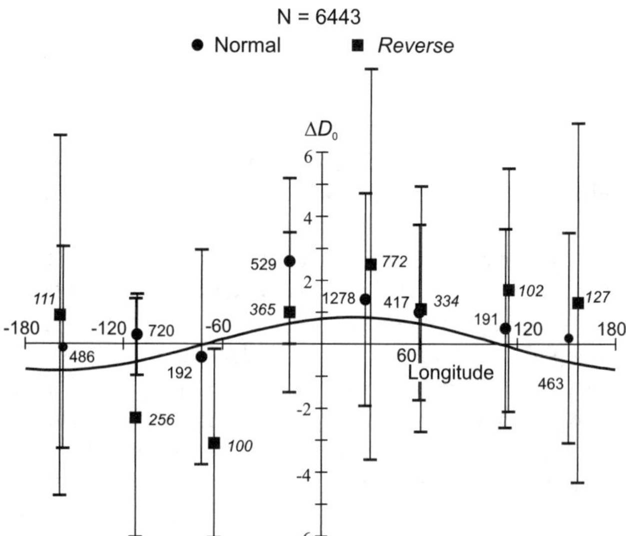

Figure 6. Longitude sector average values of ΔD_0 given by McElhinny et al. [1996] plotted as a function of longitude. Normal and reverse mean values are averaged separately and plotted with their 95% confidence limits. N is the total number of sites with the number of sites used in each maen indicated.

itude is used as a measure of secular variation in the past. McElhinny and Merrill [1975] first set up a database for this purpose and this has been updated more recently for the purpose of attempting full spherical harmonic analyses of the time-averaged field over the time interval 0–5 Ma [*Quidelleur et al.*, 1994; *Johnson and Constable*, 1996]. The selection criteria for these newer data sets are based upon the same principles originally used but with minor differences based upon the authors' own perceptions of what should be included and what should not. As a result McElhinny and McFadden [1997] set up a generalized database in which users could select out data using any desired selection criteria.

When analyzing data from their new database (IAGA paleomagnetic database PSVRL), McElhinny and McFadden [1997] pointed out that many of the data from these young lava flows date from the late 1960s, 1970s and early 1980s when demagnetization procedures had not been formalized through the use of principal component analysis [*Kirschvink*, 1980]. Of the 3917 lava results in the database at that time only 11% had been derived using principal component analysis.

One can see the progression of the demagnetization procedures used by looking at the column *Demagcode* (*DC*) used in the database on the scale 1-4. Table 2 summarizes the various demagnetization procedures used and the corresponding value for the *Demagcode*. Such an historical summary is illustrated in Figure 7. As McElhinny and McFadden [1997] pointed out, the use of such poorly demagnetized data to determine second-order terms in a spherical harmonic analysis is

Table 2. Number of lavas (N) in the PSVRL database [*McElhinny and McFadden*, 1997] according to the demagnetization procedures used. DC is the *Demagcode* in the database according to the procedure described

DC	N	Years	Procedure
1	497	1960–c.1975	NRM only but stability tested on pilot samples
2	2250	1962–c.1985	Bulk cleaning, usually minimum dispersion method
3	532	1970+	Assessment of vector diagrams
4	440	1980+	Principal component analysis
Total	3719	1960–1996	All procedures DC = 1– 4

fraught with danger and they suggested that many of these old studies would need to be redone using modern methods if further progress is to be made. This work is now in progress and some reports relating to this are presented in this volume.

A summary of the least squares analyses of McElhinny et al. [1996] has been given in some detail, because the original purpose of the paper was to show how different data combinations could give quite different values even for the zonal spherical harmonic terms. An obvious conclusion of the analyses is that full spherical harmonic analyses even to fairly low degrees should be viewed with some suspicion because the data are inadequate for this purpose.

However, spherical harmonic analyses of the lava database have been carried out by Gubbins and Kelly [1993], Johnson and Constable [1995, 1997], Quidelleur and Courtillot [1996], Kelly and Gubbins [1997], Carlut and Courtillot [1998], Kokhlov et al. [2001] and Hatakeyama and Kono [2002]. The difference in approach in these spherical harmonic analyses compared with the least-squares approach needs to be emphasized. The least- squares approach uses minimal parameters (zonal harmonics only), whereas the approach used in the spherical harmonic analyses minimizes the non-GAD structure.

Permanent non-zonal aspects of the time-averaged field for the past 5 Myr appear to be present from these spherical harmonic analyses. In most cases these analyses include quantitative tests for whether zonal harmonics fit the data. These tests fail, but in spite of this such analyses still cannot address the fundamental problem of poor data quality. These conclusions may simply arise from data artifacts. Indeed the declination anomalies plotted by Johnson and Constable [1995] in their models for the normal and reverse fields show the same positive anomaly in declination in the 0-45° longitude sector as shown in Figure 6.

Quidelleur and Courtillot [1996] argued that any terms greater than or equal to degree three are not significant. This conclusion was again supported by the analysis of Carlut and Courtillot [1998] using the present data set. There

is no dispute that the dominant term after the geocentric axial dipole is the geocentric axial quadrupole. Table 3 summarizes the values of G2 that have been determined over the past 25 years. Although the least squares and spherical harmonic approaches are different, the values determined for G2 are similar within the error limits. This strongly suggests that any non-GAD terms must be small. The values determined for G2 appear to be converging with the best estimates for the combined normal and reverse fields being in the range 0.037 to 0.042.

One of the outcomes of the analysis by McElhinny and McFadden [1997] of their generalized lava database has been to establish a global program to obtain high quality lava data (including paleointensity determinations) for the 0–5 Ma time interval. This program is now well under way and new high quality data have been published for the Azores [*Johnson et al.*, 1998], Sicily [*Zanella*, 1998], Hawaii [*Laj et al.*, 1999; *Herrero-Bervera and Valet*, 2002], the Caribbean [*Carlut et al.*, 2000], the Canary Islands [*Tauxe et al.*, 2000], Japan [*Morinaga et al.*, 2000], Possession Island in the southern Indian Ocean [*Camps et al.*, 2001], Easter Island [*Miki et al.*, 1998; *Brown*, 2002], western Canada [*Mejia et al.*, 2002], southwestern USA [*Tauxe et al.*, 2003], southern Australia [*Opdyke and Musgrave*, 2004], and Patagonia [*Mejia et al.*, 2004]. Some discussions of these new data sets and global spherical harmonics of the time-averaged field for 0–5 Ma are included in this volume. These new data sets appear to indicate that

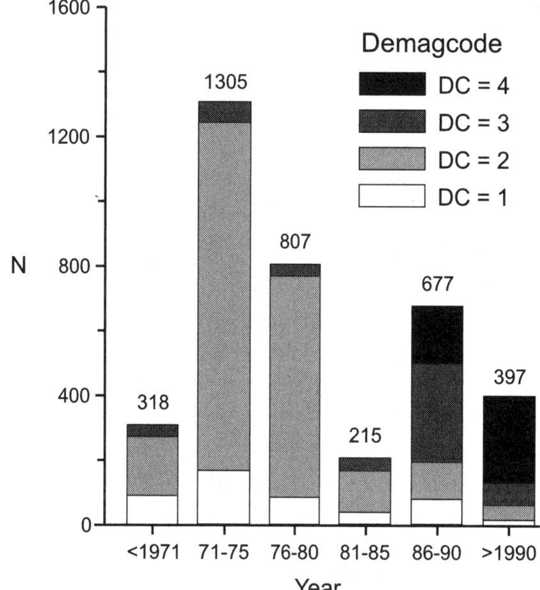

Figure 7. Historical perspective of the demagnetization procedures used in the 0–5 Ma PSVRL database of McElhinny and McFadden [1997]. The DC values between 1 and 4 are described in Table 2. N is the number of lava flows.

Table 3. Values of G2 for the time-averaged field 0-5 Ma as determined by successive authors based on directional data alone. 95% confidence limits are shown where available. M is the method used (LS – least squares; SH – spherical harmonics)

Authors	M	Polarity	G2
Merrill and McElhinny [1977]	LS	N	0.050
	LS	R	0.083
Merrill and McElhinny [1983]	LS	N	0.034±0.006
	LS	R	0.069±0.016
Merrill et al. [1990]	LS	N	0.041±0.006
	LS	R	0.080±0.025
	LS	N+R	0.053
Schneider and Kent [1990]	LS	N	0.026±0.010
	LS	R	0.046±0.014
Johnson and Constable [1995]	SH	N	0.044
	SH	R	0.066
McElhinny et al. [1996]	LS	N	0.033±0.019
	LS	R	0.042±0.022
	LS	N+R	0.038±0.012
Johnson and Constable [1997]	SH	N	0.036
	SH	R	0.038
Kelly and Gubbins [1997]	SH	N+R	0.042
Carlut and Courtillot [1998]	SH	N	0.036±0.006
	SH	R	0.056±0.018
Hatakeyama and Kono [2002]	SH	N	0.043±0.010
	SH	R	0.080±0.025

the time-averaged field is much simpler than previously thought. Most data sets give overall mean values close to the GAD field, confirming that most non-GAD effects previously determined probably result from poor data quality.

6. ZONAL HARMONICS FOR TIMES > 5 MA

6.1. Random Paleogeography Test

Evans [1976] suggested an ingenious way in which to test the GAD hypothesis over geological time. For any given magnetic field at the Earth's surface there is a probability distribution of magnetic inclination (I) for measurements made at randomly distributed sites. This is easily obtained by simply estimating the surface area of the globe corresponding to any set of $|I|$ classes. For example, the dipole field is horizontal at the equator and has $|I|=10°$ at latitude 5.0°. At the poles the field is vertical and has a value of $|I|=80°$ at latitude 70.6°. The surface area of these two zones imply that if sampling is sufficient and geographically random the $0°\leq|I|\leq10°$ band would make up 8.8% of results whereas the $80°\leq|I|\leq90°$ band would make up only 5.7% of results. The present uneven distribution of land and sea on the surface of the Earth and the existence of areas of more intensive study means that in present-day

terms a random geographical sampling of paleomagnetic data has not been undertaken. However, it is argued that over the past 600 Myr continental drift has probably taken place to such an extent that it might be assumed that this has been sufficient to render the paleomagnetic sampling random in a paleogeographical sense.

The dependence of $|I|$ on colatitude θ for axial multipoles can be obtained from the relationship

$$\tan I_l = \frac{-(1+l)P_l}{(\partial P_l / \partial \theta)},$$

where P_l is the Legendre polynomial of degree l and I_l is the inclination arising from the axial multipole of degree l. Evans [1976] compared the observed frequency distribution of $|I|$ from paleomagnetic data over the past 600 Myr with that expected from the first four axial multipoles. He found that, to first order, the observed frequency distribution was most consistent with that expected from a dipole field. Later Piper and Grant [1989] and Kent and Smethurst [1998] updated this analysis using data for the past 3000 Myr. Piper and Grant [1989] concluded that the data were compatible with the GAD hypothesis for the time interval 600–3000 Ma but not for the interval 300–600 Ma. Kent and Smethurst [1998] concluded that the data were compatible with the GAD hypothesis for the Cenozoic and Mesozoic but not for the Paleozoic and Precambrian. However, Meert et al. [2003] have pointed out that there were errors in the application of the χ-square test in both these papers. Consequently the Piper and Grant [1989] data sets for both the Phanerozoic and the Precambrian are not compatible with the GAD hypothesis, and in the Kent and Smethurst [1998] analysis only the Cenozoic data are compatible with the GAD.

McElhinny and McFadden [2000] surmised that, in all of the above cases, the basic assumption of random paleogeographic sampling required by the Evans [1976] test had not been fulfilled. Indeed, Kent and Smethurst [1998] suggested that one possible explanation for the failure of this test in pre-Mesozoic times might be that their results may reflect a tendency for continental lithosphere to have been continuously cycled into the equatorial belt. Meert et al. [2003] have made rigorous analyses of the requirements for random paleogeographic sampling using random walk analysis to test this assumption. They conclude that, on a GAD Earth, sampling over a 600 Myr period will produce a GAD-like distribution of inclinations only 30% of the time. They further show that inadequate sampling can produce false G2 and G3 effects. With the present global paleomagnetic data set, it now appears unlikely that the GAD hypothesis can be tested in this way even using data covering the age of the Earth [*McFadden, 2004*].

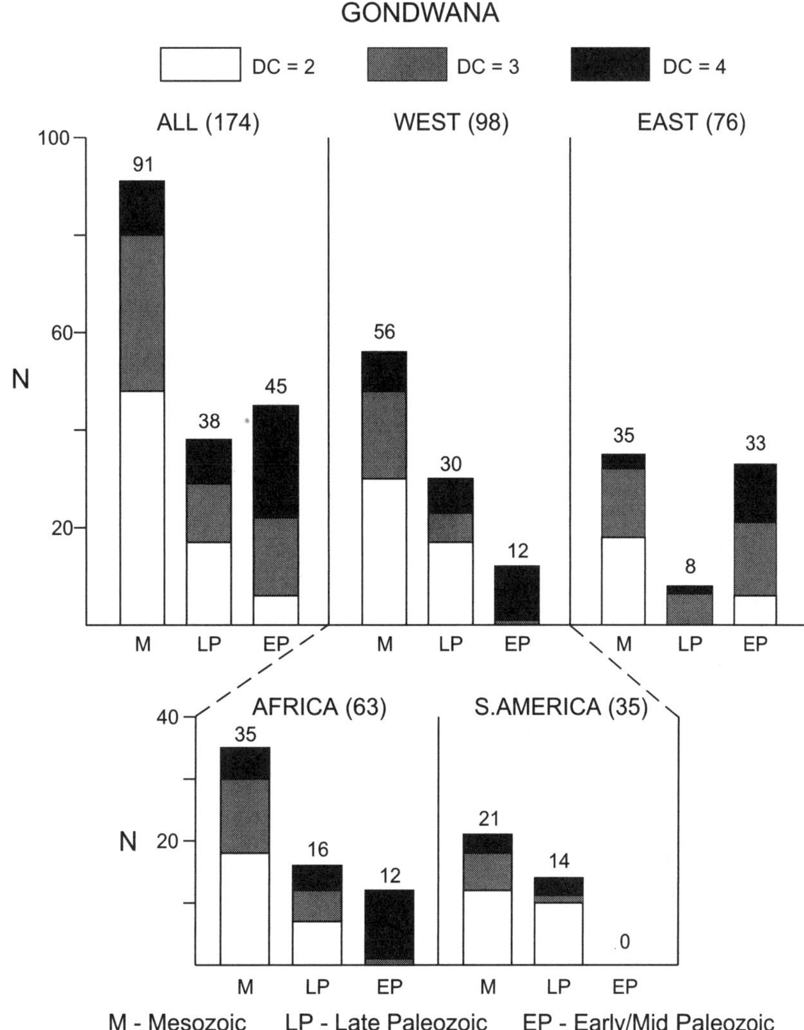

Figure 8. Analysis of the data for Gondwana used by Torsvik and Van der Voo [2002] in terms of the demagnetization procedures used. DC is the *Demagcode* as described in Table 2.

6.2. Mesozoic Reconstructions From Sea-Floor Spreading

Several authors have attempted to determine zonal harmonics for times older than 5 Ma. The method is to reconstruct the relative positions of continents through time back to about 180 Ma using sea-floor spreading magnetic anomalies. Coupland and Van der Voo [1980] were the first to use this procedure and later Livermore et al. [1984] and Lee and Lilley [1986] repeated it. The general conclusions of these studies is that it is only possible to attempt to define the G2 term and that there are insufficient data to attempt a subdivision into normal and reverse components. The problems associated with determining the G3 term over the interval 0–5 Ma alone suggests that estimates of this term would be meaningless. Coupland and Van der Voo [1980] found that G3 was not significant

prior to 50 Ma, whereas Lee and Lilley [1986] found large values of G3 prior to 100 Ma with G2 becoming insignificant. Livermore et al. [1984] found negative G2 values in pre-Cenozoic times, but suggested that those determined for the older ages may not be significantly different from zero.

Schneider and Kent [1990] proposed that values of G3 could be determined in the past by investigating the variation of the field over a large single plate such as Africa during the Tertiary, when it extended over a range of northern and southern latitudes. Their analysis suggests that values of G3 of a few percent only might be possible during the Tertiary.

The varying results discussed above suggest that the data set is inadquate to determine the time variation of G2 even back to 180 Ma. More recently Besse and Courtillot [2002] estimated G2 as 0.03±0.02 when averaged over the past 200 million years.

6.3. Pangea Reconstruction

It is generally agreed from geological evidence that the bulk of Pangea was assembled by the Late Carboniferous and maintained that configuration (Pangea A) until its breakup in the mid-Jurassic. However, paleomagnetic data for the Late Carboniferous to Early Triassic cannot be reconciled with this reconstruction because Gondwana needs to be positioned too far north such that it overlaps with Laurussia. This has led many workers to invoke Pangea B-type reconstructions involving subsequent large dextral transcurrent motions to return to the Pangea A Jurassic fit widely considered to be the starting point for the breakup of Pangea. Van der Voo and Torsvik [2001] and Torsvik and Van der Voo [2002] propose that a possible explanation for this inconsistency would be the existence of persistent non-dipole fields. Torsvik and Van der Voo [2002] compared the Gondwana-Laurussia mean poles with steadily increasing octupole contributions with the following conclusions. With G3≈0.05 the Cretaceous–mid-Jurassic segment is clearly improved. For the Late Triassic–mid-Jurassic G3≈0.1 provides a best fit, whereas for the Permo-Carboniferous G3≈0.15 leads to an almost perfect fit.

There are so many ways in which artificial G3 components can be deduced from data artifacts that these conclusions need to be viewed with some skepticism. The main problem is that many of the data for West Gondwana were determined many years ago before the introduction of Principal Component Analysis as the method for analyzing demagnetization data. Figure 8 analyzes the data set used by Torsvik and Van der Voo [2002] for Gondwana in terms of the *Demagcode* (DC), which describes the demagnetization procedures used in each study (see Table 2). Only a minority of the data has been derived using modern methods (DC=4). The data from South America is dominated by results with DC=2 and are of very poor quality. Until this basic data set is improved, any conclusions regarding the existence of persistent G3 terms must be viewed with considerable suspicion.

6.4. Global Paleointensity Variations

The intensity of the geomagnetic field varies with latitude according to the relation

$$F = F_0 (1 + 3\sin^2 \lambda)^{\frac{1}{2}},$$

where F_0 is the intensity of the GAD field at the equator. The intensity at the poles is thus twice the intensity at the equator. Global paleointensity measurements should conform with the above latitude variation if the GAD hypothesis is valid. Tanaka et al. [1995] and Perrin and Shcherbakov [1997] have summarized paleointensity values for the time intervals 0–10 Ma

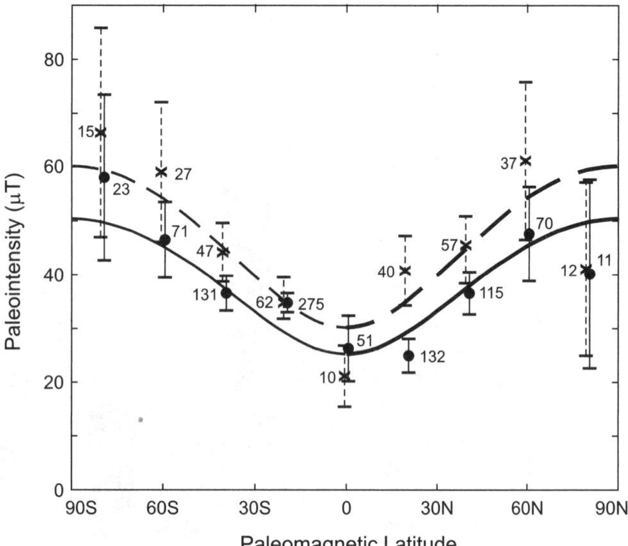

Figure 9. Global paleointensities averaged over 20° latitude bands plotted as a function of paleomagnetic latitude. Vertical bars show the 95% confidence limits for each mean and the numbers of units used in averaging are indicated. Crosses with dashed vertical error bars are for 0–10 Ma (after *Tanaka et al.*, 1995) and solid circles with solid error bars are for 0–400 Ma (after *Perrin and Shcherbakov*, 1997). The curves represent the best fits for a GAD field (dashed curve 0–10 Ma and solid curve for 0–400 Ma).

and 0–400 Ma respectively and calculated mean values averaged over 20? latitude bands. These values are illustrated in Figure 9 and the best-fit curves to each data set for a GAD field are drawn through the data. Perrin and Shcherbakov [1997] argue that where the number of units being averaged is <20, the calculated mean values should be viewed with some caution. A χ-square test indicates that the data are consistent with the latitude variation expected for a GAD field for both data sets. This provides confirmation that the GAD model is valid at least to first order for the past 400 Myr.

7. CONCLUSIONS

The GAD model for the time-averaged field is a good first order approximation at least for the past 400 Myr and probably for the whole of geological time. A persistent geocentric axial quadrupole of about 4% of the axial dipole is present for the past 5 Myr and can be expected to be a permanent feature of the time-averaged field through time. There is no evidence that the normal and reverse mean fields differ from one another. The random paleogeography test of Evans [1976] has been shown by Meert et al. [2003] and McFadden [2004] to be flawed with the present global data set. Attempts at detailed spherical harmonic analyses of the time-averaged field, even for the past 5 Myr, suffer from poor data and their global distribution. There are too many ways in which false axial geocentric octupole terms can arise from data artifacts and poor data quality for these terms to have any reality. The explanation for the problem of Pangea reconstruction during the late Paleozoic and early Mesozoic in terms of persistent geocentric axial octupole terms should be viewed with some skepticism at the present time in view of the poor data quality from the Gondwana continents, especially West Gondwana.

Acknowledgments. I thank the American Geophysical Union for a grant to assist in attending the Chapman conference, and Neil Opdyke for further financial assistance. My thanks to Sergei Pisarevsky and Dennis Kent, who both provided copies of several critical papers, and to Joe Meert and Phil McFadden for permission to cite their papers, which were still in press at the time of writing this paper.

REFERENCES

Besse, J., and V. Courtillot, Apparent and true polar wander and the geometry of the geomagnetic field in the last 200 million years, *J. Geophys. Res., 107,* doi:10.1029/2000JB000050, 2002.

Blow, R. A., and N. Hamilton, Effect of compaction on the acquisition of a detrital remanent magnetization in fine-grained sediments, *Geophys. J. R. Astron. Soc., 52,* 13–23, 1978.

Brown, L., Paleosecular variation from Easter Island revisited: modern demagnetization of a 1970s data set. *Phys. Earth Planet. Int., 133,* 73–81, 2002.

Camps, P., B. Henry, M. Prevot, and L. Faynot, Geomagnetic paleosecular variation recorded in Plio-Pleistocene volcanic rocks from Possession Island (Crozet Archipelago), southern Indian Ocean, *J. Geophys. Res., 106,* 1961–1971, 2001.

Carlut, J., and V. Courtillot, How complex is the time-averaged geomagnetic field over the past 5 million years? *Geophys. J. Int., 134,* 527–544, 1998.

Carlut, J., X. Quidelleur, V. Courtillot, and G. Boudon, Paleomagnetic directions and K-Ar dating of 0 to 1 Ma lava flows from La Guadeloupe Island (French West Indies): implications for time-averaged field models, *J. Geophys. Res., 105,* 835–849, 2000.

Coe, R. S., The effect of shape anisotropy on TRM direction, *Geophys. J. R. Astron. Soc., 56,* 369–383, 1979.

Coupland, D. H., and R. Van der Voo, Long-term non-dipole components in the geomagnetic field during the last 130 Ma, *J. Geophys. Res., 85,* 3529–3548, 1980.

Cox, A. The frequency of geomagnetic reversals and the symmetry of the non-dipole field, *Rev. Geophys., 13,* 35–51, 1975.

Cox, A., and R. R. Doell, Review of paleomagnetism, *Geol. Soc. Amer. Bull., 71,* 645–768, 1960.

Creer, K. M., Computer synthesis of geomagnetic palaeosecular variations, *Nature, 304,* 695–699, 1983.

Creer, K. M., E. Irving, and S. K. Runcorn, The direction of the geomagnetic field in remote epochs in Great Britain, *J. Geomagn. Geoelectr., 6,* 163–168, 1954.

Creer, K. M., D. T. Georgi, and W. Lowrie, On the representation of the Quaternary and late Tertiary geomagnetic field in terms of dipoles and quadrupoles, *Geophys. J. R. Astron. Soc., 33,* 323–345, 1973.

Evans, M. E., Test of the non-dipole nature of the geomagnetic field throughout Phanerozoic time, *Nature, 262,* 676–677, 1976.

Fisher, R. A., Dispersion on a sphere, *Proc. R. Soc. London, A 217,* 295–305, 1953.

Georgi, D. T., Spherical harmonic analysis of palaeomagnetic inclination data, *Geophys. J. R. Astron. Soc., 39,* 71–86, 1974.

Gubbins, D., and P. Kelly, Persistent patterns in the geomagnetic field over the past 2.5 Myr, *Nature, 365,* 829–832, 1993.

Hatakeyama, T., and M. Kono, Geomagnetic field models for the last 5 Myr: time-averaged field and secular variation, *Phys. Earth Planet. Int., 133,* 181–201, 2002.

Herrero-Bervera, E., and J-P. Valet, Paleomagnetic secular variation of the Honolulu Volcanic Series (33–700 ka), Oahu (Hawaii), *Phys. Earth Planet. Int., 133,* 83–97, 2002.

Hospers, J., Rock magnetism and polar wandering, *Nature, 173,* 1183, 1954.

Irving, E., *Paleomagnetism and Its Application to Geological and Geophysical Problems,* 399pp., John Wiley, New York, 1964.

Johnson, C. L., and C. G. Constable, The time-averaged geomagnetic field as recorded in lava flows over the past 5 Myr, *Geophys. J. Int., 122,* 489–519, 1995.

Johnson, C. L., and C. G. Constable, Palaeosecular variation recorded by lavas over the last 5 Myr, *Philos. Trans. R.. Soc. London, A 354,* 89–141, 1996.

Johnson, C. L., and C. G. Constable, The time-averaged geomagnetic field: global and regional biases for 0–5 Ma, *Geophys. J. Int., 131,* 643–666, 1997.

Johnson, C.L., C.G. Constable, and L.Tauxe, Mapping long-term changes in Earth's magnetic field, *Science, 300,* 2044–2045, 2003.

Johnson, C. L., J. R. Wijbrans, C. G. Constable, J. Gee, H. Staudigal, L. Tauxe, V-H. Forjaz, and M. Salgueiro, ^{40}Ar-^{39}Ar ages and paleomagnetism of Sao Miguel lavas, Azores, *Earth Planet. Sci. Lett., 160,* 637–649.

Kelly, P., and D. Gubbins, The geomagnetic field over the past 5 million years, *Geophys. J. Int., 128,* 315–330, 1997.

Kent, D. V., and M. A. Smethurst, Shallow bias of paleomagnetic inclinations in the Paleozoic and Precambrian, *Earth Planet. Sci. Lett., 160,* 391–402, 1998.

King, R. F. and A. I. Rees, Detrital magnetism in sediments: An examination of some theoretical models, *J. Geophys Res., 71,* 561–571, 1966.

Kirschvink, J. L., The least squares line and plane and the analysis of palaeomagnetic data, *Geophys. J. R. Astron. Soc., 62,* 699–718, 1980.

Kokhlov, A., G. Hulot, and J. Carlut, Towards a self-consistent approach to palaeomagnetic field modeling, *Geophys, J. Int., 145,* 157–171, 2001.

Laj, C., H. Guillou, N. Szeremeta, and R. S. Coe, Geomagnetic secular variation at Hawaii around 3 Ma from a sequence of 107 lavas at Kaena Point (Oahu), *Earth Planet. Sci. Lett., 170,* 365–376, 1999.

Lee, S., and F .E. M. Lilley, On paleomagnetic data and dynamo theory, *J. Geomagn. Geoelectr.*, *38*, 797–806, 1986.

Livermore. R. A., F. J. Vine, and A. G. Smith, Plate motions and the geomagnetic field. II. Jurassic to Tertiary, *Geophys. J. R. Astron. Soc.*, *79*, 939–961, 1984.

McElhinny, M. W., *Palaeomagnetism and Plate Tectonics*, 358pp., Cambridge Univ. Press, Cambridge, 1973.

McElhinny, M. W., and P. L. McFadden, Palaeosecular variation over the past 5 Myr based on a new generalized database, *Geophys. J. Int.*, *131*, 240–252, 1997.

McElhinny, M. W., and P. L. McFadden, *Paleomagnetism: Continents and Oceans*, 386pp., Academic Press, San Diego, 2000.

McElhinny, M. W., and R. T. Merrill, Geomagnetic secular variation over the past 5 m.y., *Rev. Geophys. Space Phys.*, *13*, 687–708, 1975.

McElhinny, M. W., P. L. McFadden, and R. T. Merrill, The time-averaged geomagnetic field 0–5 Ma, *J. Geophys. Res.*, *101*, 25007–25027, 1996.

McFadden, P. L., Is 600 Myr long enough for the random palaeo-geographic test of the geomagnetic axial dipole assumption? *Geophys. J. Int.,* (in press) 2004.

Meert, J. G., E. Tamrat, and J. Spearman, Non-dipole fields and inclination bias: Insights from a random walk analysis, *Earth Planet. Sci. Lett.*, *214*, 395–408, 2003.

Mejia, V., R. W. Barendregt, and N. Opdyke, Paleosecular variation of Brunhes age lava flows from British Columbia, Canada, *Geochem. Geophys. Geosyst.*, *3*(12), 8801, doi:10.1029/ 2002 GC000353, 2002..

Mejia, V., Opdyke, N. D., J. F. Vilas, B. S. Singer, and J. S. Stoner, Paleomagnetic secular variation of Plio-Plestocene lavas from southern Patagonia, *Geochem. Geophys. Geosyst., 5,* (in press) 2004.

Merrill, R. T., and McElhinny, M. W., Anomalies in the time-averaged paleomagnetic field and their implications for the lower mantle, *Rev. Geophys. Space Phys.*, *15*, 309–323, 1977.

Miki, M., H. Inokuchi, S. Yamaguchi, J. Matsuda, K. Nagao, N. Isazaki, and K. Yaskawa, Geomagnetic secular variation in Easter Island, southeast Pacific, *Phys. Earth Planet. Int., 106*, 93–101, 1998.

Morinaga, H., T. Matsumoto, Y. Okimura, and T. Matsuda, Paleomagnetism of Pliocene to Pleistocene lava flows in the northern part of Hyogo prefecture, northwest Japan and Brunhes Chron paleosecular variation in Japan, *Earth Planets & Space, 52,* 437–443, 2000.

Opdyke, N. D., and K. W. Henry, A test of the dipole hypothesis, *Earth Planet. Sci. Lett., 6*, 139–151, 1969.

Opdyke, N. D., and R. Musgrave, Paleomagnetic results from the Newer Volcanics of Victoria: contribution to the time averaged field initiative, *Geochem. Geophys. Geosyst., 5,* (in press) 2004.

Piper, J. D. A. and S. Grant, A palaeomagnetic test of the axial dipole assumption and implications for continental distributions throughout Phanerozoic time, *Phys. Earth Planet. Int., 55*, 37–53, 1989.

Perrin, M., and V. Shcherbakov, Paleointensity of the Earth's magnetic field for the past 400 Ma: evidence for a dipole structure during the Mesozoic low, *J. Geomagn. Geoelectr., 49*, 601–614, 1997.

Quidelleur, X., and V. Courtilot, On low-degree spherical harmonic models of paleosecular variation, *Phys. Earth Planet. Int., 95*, 55–77, 1996.

Quidelleur, X., J. P. Valet, V. Courtillot, and G. Hulot, Long-term geometry of the geomagnetic field for the last 5 million years: An updated secular variation database from volcanic sequences, *Geophys. Res. Lett., 21*, 1639–1642, 1994.

Schneider, D. A., and Kent, D. V., Testing models of the Tertiary paleomagnetic field, *Earth Planet. Sci. Lett., 101*, 260–271, 1990.

Tanaka, H., M. Kono, and H. Uchimura, Some global features of paleointensity in geological time, *Geophys. J. Int., 120*, 97–102, 1995.

Tauxe, L., H. Staudigal, and J. R. Wijbrans, Paleomagnetism and ^{40}Ar-^{39}Ar ages from La Palma in the Canary Islands, *Geochem. Geophys. Geosyst., 1*(9), doi:10.1029/2000GC000063, 2000.

Tauxe, L., C. Constable, C. L. Johnson, A. A. P. Koppers, W. R. Miller, and H. Staudigal, Paleomagnetism of the southwestern U.S.A. recorded by 0–5 Ma igneous rocks, *Geochem. Geophys. Geosyst., 4*(4), 8802, doi:10.1029/2002GC000343, 2003.

Torsvik, T. H., and R. Van der Voo, Refining Gondwana and Pangea palaeogeography: Estimates of Phanerozoic non-dipole (octupole) fields, *Geophys. J. Int., 151*, 771–794, 2002.

Van der Voo, R., and T. H. Torsvik, Evidence for late Paleozoic and Mesozoic non-dipole fields provides an explanation for the Pangea reconstruction problem, *Earth Planet. Sci. Lett., 187*, 71–81, 2001.

Wells, J. M., Non-linear spherical harmonic analysis of paleomagnetic data, In *Methods in Computational Geophysics* (ed. B.A. Bolt), pp. 239–269, Academic Press, San Diego, 1973.

Wilson, R. L., Permanent aspects of the Earth's non-dipole magnetic field over Upper Tertiary times, *Geophys. J. R. Astron. Soc., 19*, 417–437, 1970.

Wilson, R. L., Dipole offset: the time-averaged palaeomagnetic field over the past 25 million years, *Geophys. J. R. Astron. Soc., 22*, 491–504, 1971.

Wilson, R. L., Palaeomagnetic differences between normal and reversed field sources, and the problem of far-sided and right-handed pole positions, *Geophys. J. R. Astron. Soc., 28*, 295–304, 1972.

Zanella, E., Paleomagnetism of Pleistocene rocks from Pantelleria Island (Sicily Channell), Italy, *Phys. Earth Planet. Int., 108*, 291–303, 1998.

Michael W. McElhinny, Gondwana Consultants, 31 Laguna Place, Port Macquarie, New South Wales 2444, Australia. (mikemce@midcoast.com.au)

The Case for Pangea B, and the Intra-Pangean Megashear

E. Irving

Pacific Geoscience Centre, Geological Survey of Canada, North Saanich, British Columbia,
Canada

In the mid-1950s we discovered that paleopoles determined from rocks of Late Carboniferous through Triassic age from Europe and North America agreed after closure of the North Atlantic. By contrast, Australian paleopoles from rocks of the same geological systems, although they were brought closer when reassembled into the traditional Pangea configuration (Pangea A), still differed by 15°-20°. This anomaly, the Intra-Pangea Paleomagnetic (IPP) anomaly has now been found throughout Gondwana. It can be explained by placing Gondwana 3-4000 km further east and north with respect to Laurasia, without ocean between them. This is Pangea B. Between the Early Permian and Late Triassic Epochs, Pangea B was transformed into Pangea A by ~3500 km dextral shear (Intra-Pangea Megashear) following the line of the mid-Carboniferous Appalachian-Variscan orogenic belt. The precise timing of the megashear is uncertain.

Several difficulties have been raised. Pangea A has, with minor adjustment, been kept intact, and the IPP anomaly has been explained by invoking one or more of the following: the presence of large, long-term, zonal octupole components in the geomagnetic field during the Late Carboniferous through Triassic Periods, errors in reconstructing continents, errors in inclination or dating, or in isolating original magnetizations. In the past several years, observations, especially from the western Mediterranean region, have, I believe removed each of these possibilities as sole causes of the IPP anomaly. However, this does not mean that, in the future, new difficulties, or old difficulties in combination, will not be raised, but it does mean, that on the basis of present evidence, the ideas of Pangea B and the Intra-Pangean Megashear survive and are germane to discussions of end-Paleozoic geology.

"If ever possible at all, it (that is the reconstruction of pre-Jurassic continental drift) will only be by patiently working backward step by step. And even the first stepback to the Permo-Carboniferous has not yet been taken with world-wide confidence" [*Holmes*, 1965, p. 1245].

Timescales of the Paleomagnetic Field
Geophysical Monograph Series 145
Copyright 2004 by the American Geophysical Union
10.1029/145GM02

1. INTRODUCTION

This essay is about Pangea B, a novel paleogeography for the end of the Paleozoic that, for the past 25 years, has been shunned by almost everybody. My purpose is to state the case for Pangea B, to note its transformation to Pangea A (the Pangea of Wegener), its longevity compared with Pangea A, and some of its potential ramifications. I shall also describe the difficulties that have been raised against the idea and how they have been dealt with. My approach is historical, and my intention is to be brief.

During the first half of the 20th century, two schemes of continental drift were proposed for the latter half of the Phanerozoic. Their essence was that the Earth once possessed either one supercontinent, Pangea [*Wegener*, 1924], or two, Laurasia and Gondwana, with substantially different histories [*Du Toit*, 1937]. Both schemes agree that these supercontinents began to fragment in the Mesozoic and subsequently drifted, as the modern continents, to their present positions.

Wegener closed the Atlantic Ocean, placed Africa to the immediate south of Europe, and South America to the south of North America, a configuration called Pangea A. Köppen and Wegener [1924] developed a theory of past climate zonation. They compiled the global distribution of temperature sensitive deposits from which they estimated the positions of ancient geographic poles (paleopoles), and constructed paleogeographic grids. Figure 1 is their Late Carboniferous map. Pangea A, according to them, was situated mainly in the southern hemisphere, and existed from the Devonian through Jurassic Periods (~200 Ma). Wegener proposed that Pangea broke-up in the Cretaceous and the fragments drifting apart and northwards, so continents are now mainly in the northern hemisphere.

Pangea A has become perhaps the best known icon of tectonics. Wegener's notion of a relatively immobile Pangea, little modified during this long interval, is adopted by almost everybody. The elegance and visual impact of the Early Jurassic reconstruction of Pangea A by Bullard et al. [1965] and Smith and Hallam [1970] has so strongly reinforced that image that there has been a common reluctance to question seriously its applicability to earlier time.

There has, however, remained an undercurrent of alternative ideas, stemming from Du Toit's belief in two substantially independent supercontinents. He described his concept as "radically different" from Wegener's. (Du Toit actually did attempt to place them together for a limited geological interval (his Figure 40) but with little resemblance to Wegener's Pangea, and his discussion was, to say the least, unenthusiastic.) According to Lester King, who knew him, Du Toit tried repeatedly to put them together, but failed to do so to his satisfaction (L. C. King, personal communication to author ca 1956). I suspect the reason was that Du Toit had set himself a high standard of acceptance for juxtaposing now separate continents, based on his detailed studies of the stratigraphies of South America and Africa and their correlation. It was, I suspect, because of his failure to find what he considered acceptable correlations between Laurasia and Gondwana, that he was reluctant to speculate about their relative positions. The repeated similarities in continental Devonian through Triassic strata of Europe and North America (Old and New Red Sandstone and Coal Measures for instance) provided ample grounds for uniting them in that interval, and the Carboniferous through Jurassic strata of the southern continents (Gondwana System) linked India and the southern continents together for a similar interval. However, there is no stratigraphic sequence of comparable scale and duration linking Laurasia and Gondwana accurately.

It was not until marine and other studies in the latter half of the 20th century that it was shown, beyond reasonable doubt that, most continental crust was gathered into a single supercontinent at the beginning of the Jurassic, similar in all essen-

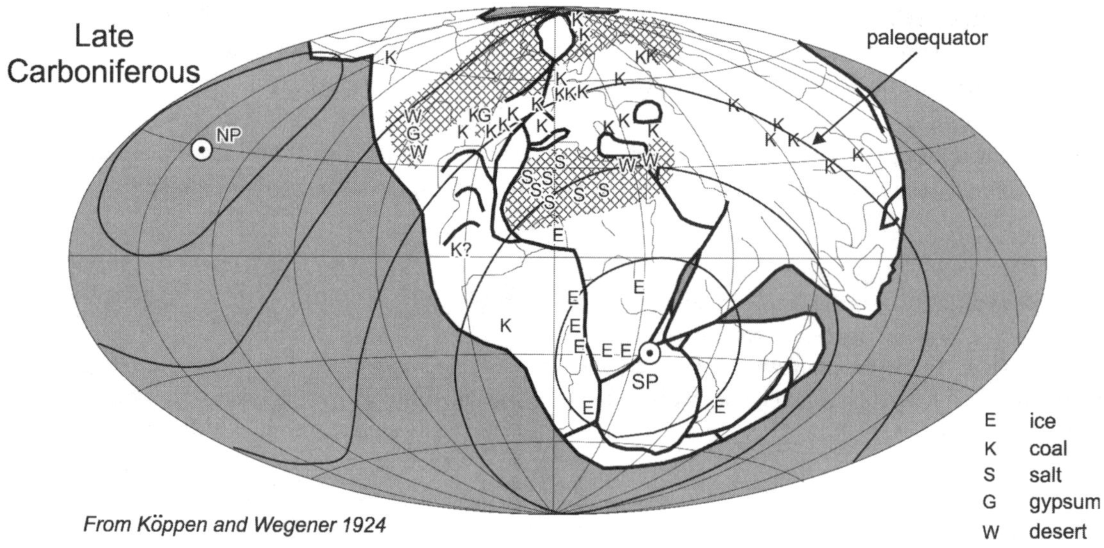

Figure 1. NP and SP are north and south paleogeographic poles with parallels of paleolatitudes constructed around them every 30° in bold lines. Arid regions are cross-hatched. Continents assembled in the Pangea A of Wegener [1922] with Africa fixed relative to the grid.

tials to Wegener's Pangea A. But were Du Toit's doubts about the applicability of Pangea to earlier times justified? How long prior to the Jurassic did Pangea A exist? My purpose is to describe how, after half a century of stratigraphic, paleomagnetic and geochronologic work, correlations of the type sought by Du Toit have been found, and they indicate a Pangea very different from Wegener's.

The story is crucial for the theory of paleomagnetic work because if Pangea was essentially fixed for Late Paleozoic and Triassic times, then, as we shall see, it implies that paleomagnetic pole positions obtained from pre-Jurassic rocks may be subject to systematic errors of "ten or more degrees" [*Briden et al.* 1971, p. 115] and might therefore be of qualitative interest only for most of geological history.

It is appropriate that this review of the paleomagnetism of Carboniferous through Triassic rocks should be included in a volume in honor of Neil Opdyke. His first paper on paleomagnetism recorded early work on the Triassic strata of New Jersey, and was concerned with the perennial problem of inclination error in sedimentary rocks [*Opdyke*, 1961]. He has, over his career, studied extensively Permo-Carboniferous strata. Also his current research is focused on the question of the time-average nature of the geomagnetic field, a topic of central interest herein.

2. 1950S: DISCOVERY OF THE IPP ANOMALY

Testing Wegener's Pangea required an accurate map. Carey provided this by fitting continental shelf edges and reversing strains in oroclines [*Carey*, 1958]. He promptly (in time for the Hobart Symposium on continental drift in March 1956) sent it to me as a large blueprint, and I tested it immediately (Figure 2).

At that time the method of calculating ancient pole positions (or paleopoles) from paleomagnetic data had just been developed [*Creer et al.*, 1954; *Hospers*, 1954], and a sufficient number of paleopoles had been obtained to begin to construct paths of apparent polar wander (APW) for different continents. Corrected to Carey's reconstruction of the northern continents (Laurasia, comprising re-assembled North America, Europe and northern Asia), Carboniferous, Permian and Triassic paleopoles for Europe and North America were in good agreement (Figure 2a). However, Australian paleopoles when corrected to Pangea were systematically different from their northern hemisphere equivalents (Figure 2b). Carey [1958, p. 281] himself also made paleomagnetic tests of his map for the Hobart Symposium. When he repositioned paleopoles for Jurassic rocks on his Pangea, they were well grouped, falling inside a small circle of radius 13.5°, comparable to their individual errors, but for Permian and Carboniferous paleopoles the corresponding enclosing circles

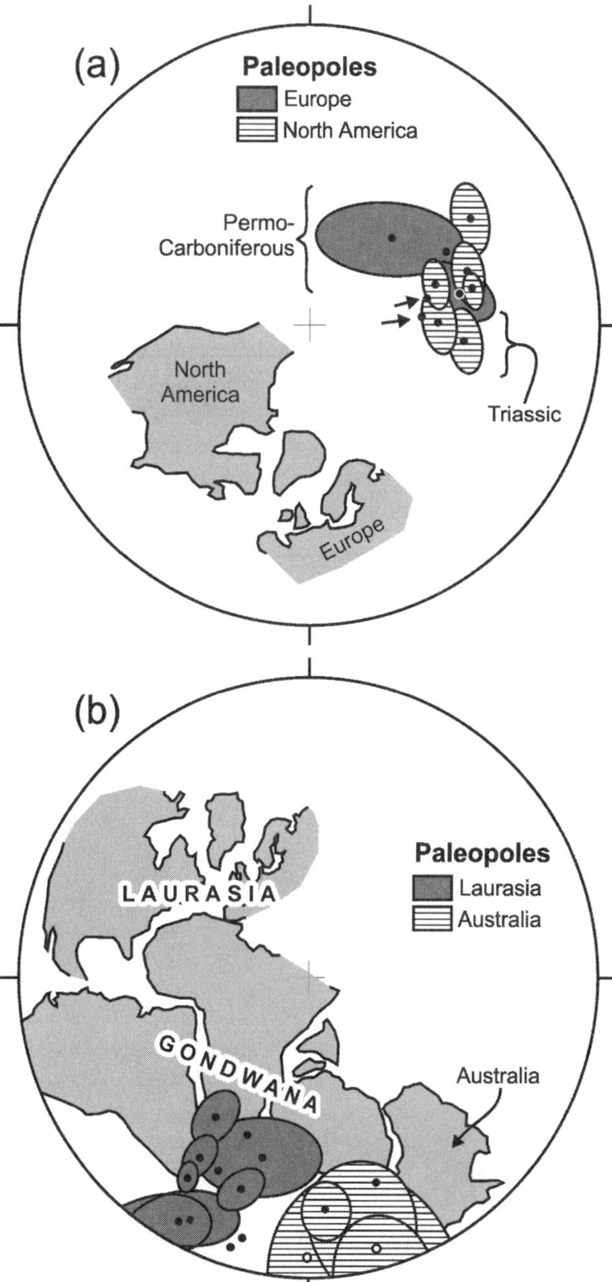

Figure 2. Late Carboniferous, Permian and Triassic paleopoles (P = 0.05) assembled on Carey's 1956 Pangea. The map used by Carey at the Hobart symposium and reproduced in Carey [1958]. (a) North paleopoles from Europe and North America after closing the Atlantic [*Irving*, 1957]. (b) Comparison of Carboniferous, Permian and Triassic south paleopoles from Laurasia and Australia showing the Intra-Pangea paleomagnetic (IPP) anomaly [*Jaeger and Irving*, 1957; republished *Irving*, 1979].

were several times larger, 45° and 35°, respectively. When considering this, Carey hinted at the need to adjust continental positions within Pangea, to correct for what he called "Late Paleozoic orogenesis". Jaeger and Irving [1957] noted that the paleopoles of Figure 2b could be brought into agreement "by moving Australia around Antartica to a position nearer to South America", implying that Carey's Pangea reconstruction was incorrect. We were wrong, and it looks as if Carey was correct; except for the placements of Madagascar and India, which did not affect this particular argument, his reconstruction was close to modern versions.

The anomaly of Figure 2b, even though it was based on very early measurements of natural remanent magnetization without demagnetization, has been confirmed by all subsequent analyses; although details have changed, the grouping of European and North American paleopoles in Figure 2a and the separation of Laurasian and Australian palepoles in Figure 2b are still essentially the same. More recent examples are

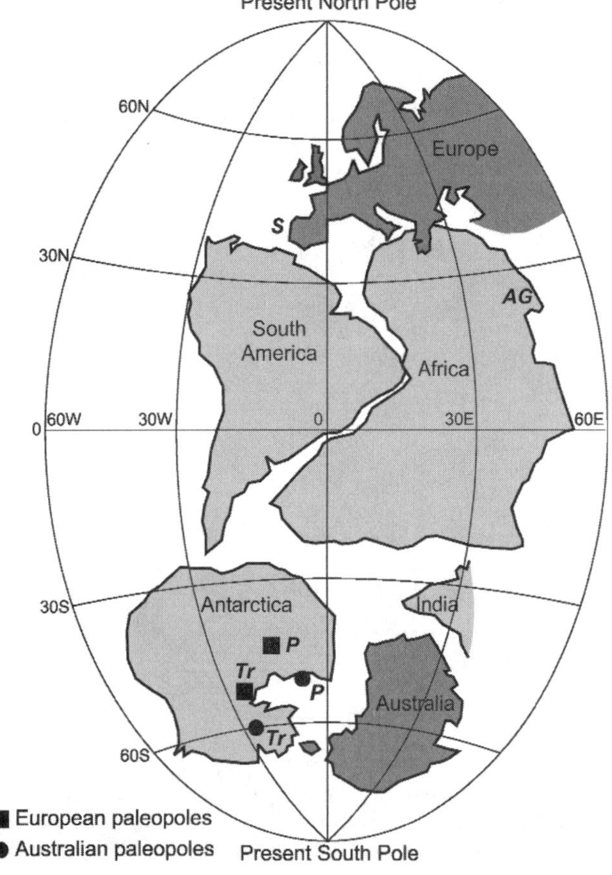

Figure 3. Relative positions of Europe and Australia determined in the early 1960s by overlying paleopoles for Permian and Triassic Periods and moving continents accordingly: small separations reflect statistical uncertainty. Intervening continents arrranged according to *Carey* [1958]. Europe fixed. Spain (S) would be near Algeria (AG) in Pangea A-1. From Irving [1964]. With permission from John Wiley and Son.

shown in Figures 8 and 12. The separation on Figure 2b is referred to as the Intra-Pangean Paleomagnetic (IPP) anomaly, and it is characterized by Late Carboniferous through Jurassic APW segments about 30° long and about 20° apart when repositioned on Pangea A. The segments converge in the Early Jurassic Epoch.

In retrospect, the good agreement between Carey's reconstruction of Laurasia and the paleomagnetic data, and the poor agreement with data from rocks of the same geological periods from Australia, was forceful evidence that Pangea A might have been short-lived. But as just noted, the data had been obtained without benefit of demagnetization, although there were tilt-tests, reversal tests, and repeatability and uniformity, laterally and vertically, had been demonstrated in several instances. Demagnetization techniques were not developed until the late '50s and not applied generally until the early '60s. Also the interpretation of paleomagnetic work as indicative of continental drift was under intense criticism (for example *Graham*, [1956]: *Stehli*, [1957]). In an atmosphere of almost universal opposition to drift, our attempts to pursue mobilistic arguments had little effect, so entrenched was the fixist viewpoint. The immediate need was to redo and expand all paleomagnetic surveys, applying newly developing demagnetization techniques and dating methods. This took several years, and, as a consequence, interest in the IPP anomaly were not revived until the mid-'60s.

3. MID-1960S AND EARLY 1970S: RENEWED DISCUSSION, EULER ROTATIONS

By the mid-60s, there were sufficient data to calculate average paleopoles for geological periods or epochs for each of the major continents. This led to attempts to use Late Carboniferous through Triassic paleopoles to make pre-Jurassic reconstructions, and to determine displacements within orogenic belts. It also led to the re-statement of the IPP anomaly in terms of gross overlaps of continental crust, and to the realization that these overlaps could be reconciled by invoking gross dextral shear between northern and southern continents.

If there are APW segments from different continents that are matched, paleopole by paleopole, then the segments can be brought together and their continents of origin moved accordingly. The minimum requirement is for there to be two pairs of paleopoles from each continent with ages that are different within but similar between continents. At the time, the Permian and Triassic paleopole pairs from Australia and Europe seemed to be appropriately matched, and overlaying them gave Figure 3, which differs from Pangea A. Gondwana occupies more easterly longitudes.

Continents, when placed in their paleomagnetically determined Permian latitudes but in the relative longitudes of

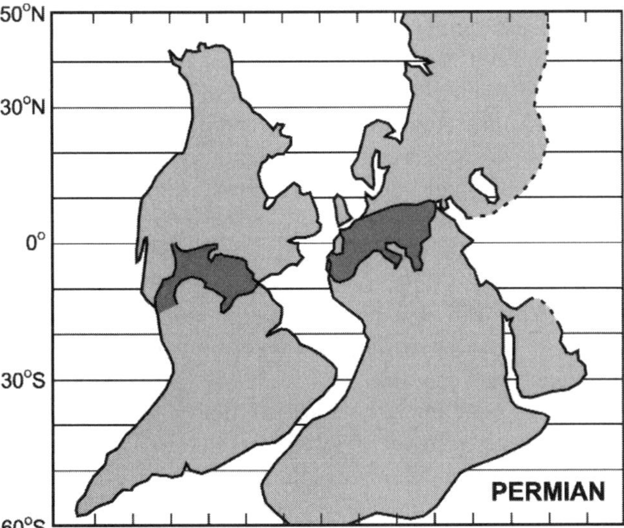

Figure 4. Overlap problem using Permian paleolatitudes as they were known in the mid-1960s. From Irving [1967].

Pangea A, gave very large overlaps (Figure 4). Northern continents were too far south or the southern continents too far north. As Figure 3 shows, paleopoles and latitudes could be made consistent by moving Gondwana eastward. Van Hilten [1964] and de Boer [1965] studied further this possibility.

By the early '60s, good agreement among paleopoles from a score of studies of Permian (mainly Early Permian) rocks from northern and middle Europe and Russia east of the Urals Mountains had been established. These studies were quickly

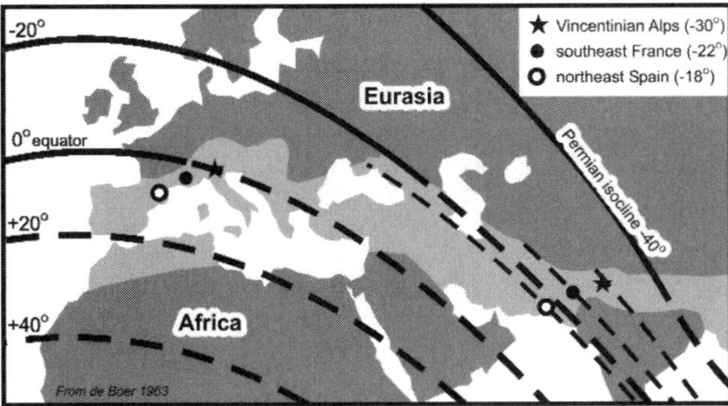

Figure 5. Displacements along the Alpine Orogenic Belt according to de Boer [1965]. Permian isoclines (lines of equal inclination) determined from middle and northern Europe are extended to the south as dashed lines. De Boer did not give errors. His argument regarding the Southern Alps still stand, but later work has called in question his reconstruction of French and Spanish localities. Criticism of de Boer's work has been made by Zijderveld et al. [1970]. The general principle, however, is correct, and this was its first use for estimating large displacements within orogenic belts. With permission from American Geophysical Union.

extended into the adjacent Alpine orogenic belt, and inclinations from the southern Alps of northeastern Italy, southeastern France and northeastern Spain were found to be about 20° higher than expected (Figure 5). Because the inclinations were upwards, and because the rocks had been formed during the Kiaman Reversed Superchron, which had just then been recognised, they must have been in the northern hemisphere. Hence, the likely positions for them within the Alpine orogenic zone lay far to the east, 4500 km to the east according to de Boer [1965]. He suggested that these localities are "outposts of the African continent" and that the discrepancies in inclination reflected displacement of Gondwana with respect to Europe.

Van Hilten [1964] drew maps to describe this displacement. He imposed few constraints on longitude. Placing Gondwana in the Permian very far (>6000 km) to the east of

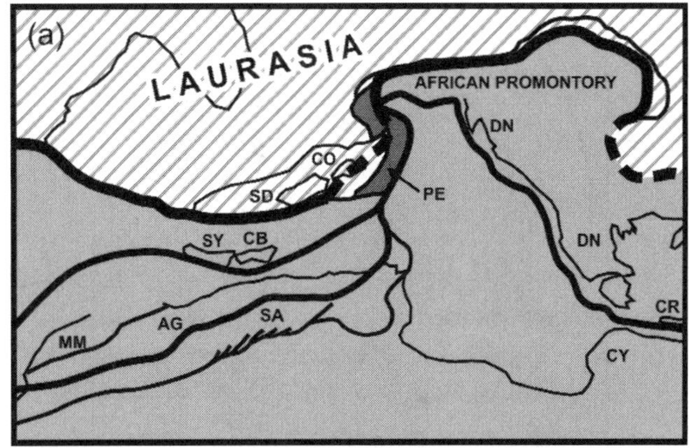

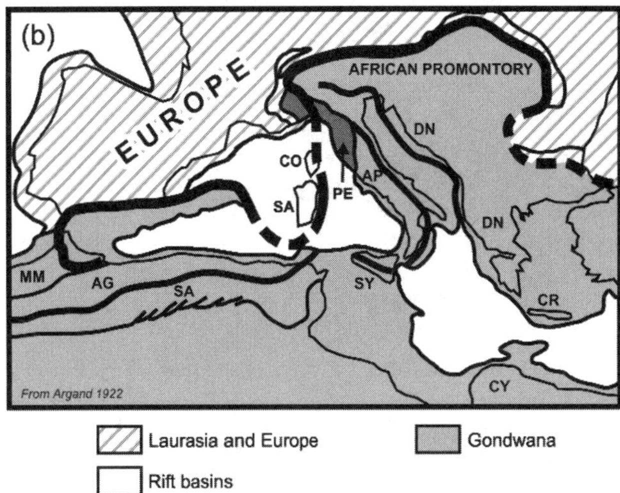

Figure 6. Argand's tectonic reconstructions, (a) before (Cretaceous) and (b) after (later Cenozoic) the Alpine compression, showing his concept of an African promontory or Adria [*Argand*, 1922]. AG = Algerian Atlas, AP = Apennines, CA = Calabria, CO = Corsica, CR = Crete, CY = Cyrenaica, DN = Dinarides, MM = Moroccan Meseta, PE = Pennides, SA = Saharan Atlas, SY = Sicily. Simplified from Argand [1922].

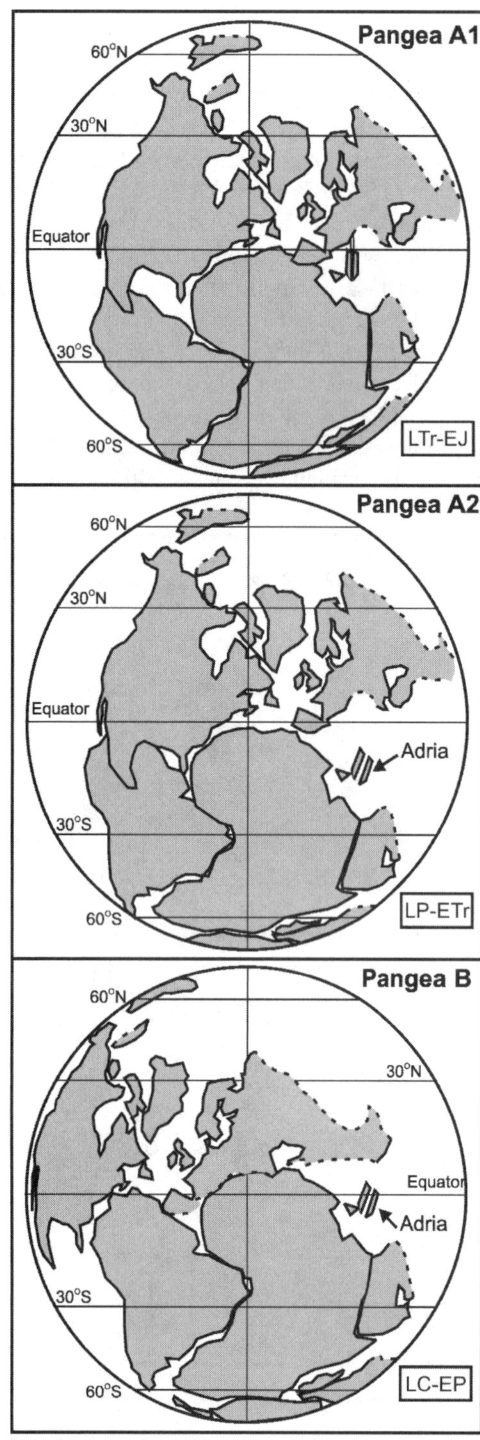

Figure 7. Pangeas. Top is Late Triassic to Early Jurassic Pangea A-1, Bullard et al. [1965], and Smith and Hallam, [1971]. Middle is Pangea A-2, as in (a) with relative position of Gondwana and Laurasia from Van der Voo and French [1974]. Bottom is Pangea B, Late Carboniferous to Early Permian, Irving [1977]; Morel and Irving, [1981]. From Livermore et al. [1986]. With permission from Nature.

Europe, and began moving it westward in the Triassic reaching its present position in the Eocene. He called this the "Tethys Twist", which lasted almost 200 Ma, and which occurred before and during the opening of the modern oceans. At no stage was there the required assembly of continental crust in an Early Jurassic Pangea, and the timing, magnitude and global context of the megashear were not acceptable. Moreover, the conclusions of Van Hilten [1964] were questioned on paleomagnetic grounds by Zijderveld et al. [1970] who gave new results from western Europe and the southern Alps of Italy and argued that there was little difference between them.

However, one notable feature of Van Hilten's maps has endured: his placement of Italy jutting out from what is now Tunisia. This idea of an autochthonous African promontory (now called Adria) is due to Argand [1922] and is of key importance. His maps for the end of the main alpine compression, and after the subsequent extension which opened the ocean basins of the Atlantic and western Mediterranean, are shown in Figure 6. Argand's African promontory pushed deep into central Europe.

Up to this time, continental reconstructions had been made graphically. Computer-based methods employing Euler's point theorem were introduced in the 1960s and led to formalizations

Figure 8. Mean paleopoles from Laurasia and Gondwana showing misfit of Late Paleozoic and Early Triassic south paleopoles when reassembled on Pangea A-1 [Bullard et al., 1965; Smith and Hallam, 1971]. Mean poles are for the P-C Carboniferous and Early Permian, P-Tr Early Permian and Triassic, Tru-Jm Late Triassic to Middle Jurassic [*Schmidt*, 1977].

of Wegener's Pangea. First was Pangea A-1 (Figure 7), the reconstruction of the peri-Atlantic continents by Bullard et al. [1965], and of Gondwana by Smith and Hallam [1970]. Paleopoles for the Late Triassic and Early Jurassic from Gondwana and Laurussia (comprising reconstructed North America and Eurasia east of the Urals) reconstructed to Pangea A-1 were found to be in excellent agreement (Figure 8). A mid-'70s example showed this in more detail, confirming the general correctness of Pangea A-1 for the Early Jurassic Epoch (Figure 9).

Computer-drawn maps were quickly exploited by Briden et al. [1971] in an attempt to study geomagnetic field behaviour. They set aside tectonic explanations, arguing on geological grounds "that no important plate margins cross the reassembly west of the Urals between mid-Lower Permian and Jurassic time, and this Pangea represented relative positions of the major continents in Permo-Triassic time." They found that Permian and Triassic paleopoles, when re-assembled relative to Pangea A-1, became better grouped, but that there remained systematic residual differences among paleopoles from different continental blocks—they had resurrected the IPP anomaly. They found that the distribution of inclinations versus paleolatitude resemble "a combination

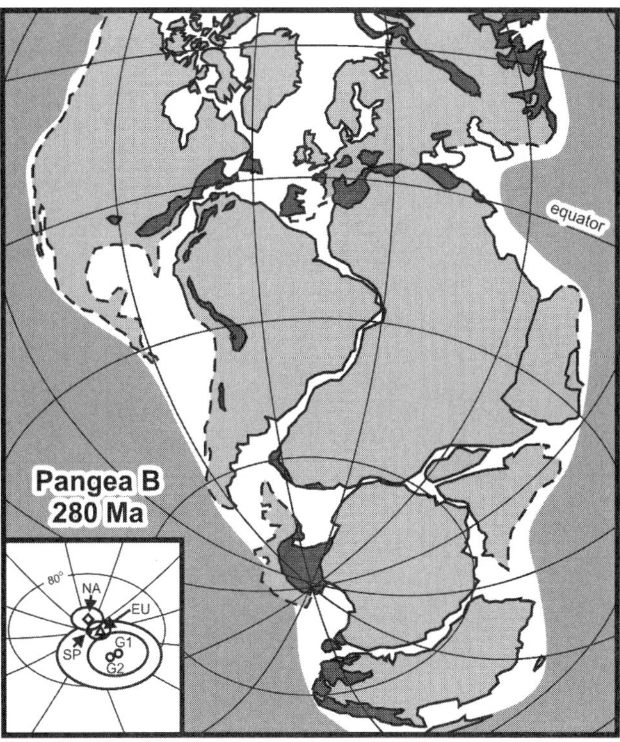

Figure 10. Pangea B. Mean south paleopoles (P = 0.63) in the inset calculated as means within windows of 30 my duration centered on 280 Ma. EU Europe, G1 all and G2 west Gondwana, NA North America, SP south paleopole. Darker shading, Carboniferous orogenic belts. From Morel and Irving [1981]. With permission from American Geophysical Union.

of dipole and octupole fields rather than a purely dipole field". The IPP anomaly, they suggested, was a geomagnetic, not a tectonic phenomenon, an idea that was to re-emerged 30 years later. This difficulty (failure of the geocentric dipole model of the geomagnetic field) was the first of several that would be raised against tectonic interpretations of the IPP anomaly. (*Frankel* [1985] discusses the role that raising difficulties plays in scientific controversies.)

Van der Voo and French [1974] returned to tectonics. They attempted to explain the IPP anomaly by rotating Gondwana 20° about a pole in NW Africa, closing the Gulf of Mexico and creating Pangea A-2 (Figure 7). They "tentatively suggested that this was valid from Late Carboniferous to Late Permian times". The dextral rotation from Pangea A-2 to A-1 occurred in "latest Permian/Early Triassic". Originally, Pangea A-2 was based on only two paleopoles (from two localities in what were then considered to be the Late Carboniferous to Early Permian Middle Paganzo Formation of Argentina [*Embleton*, 1970]) from the whole of Gondwana, although others were available at that time.

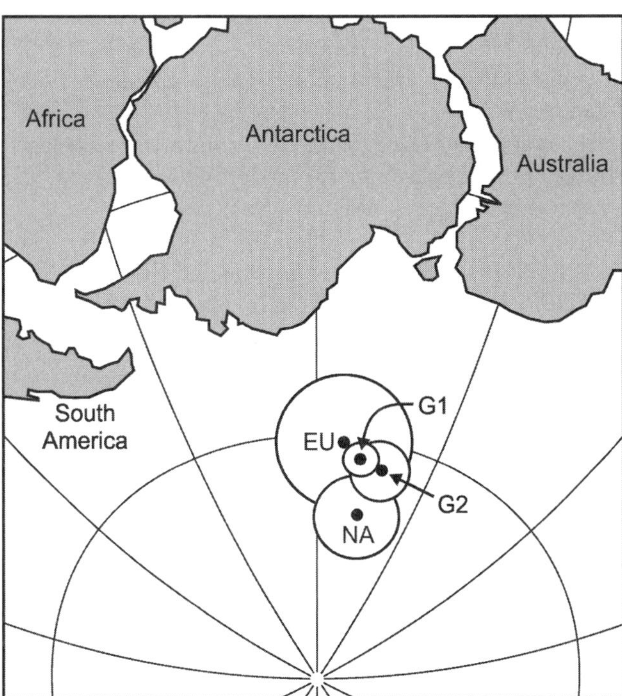

Figure 9. Early Jurassic mean paleopoles, probable errors (P = 0.63) plotted on Pangea A-1, assembled as in Figure 7. G1 all Gondwana, G2 West Gondwana, EU Europe, NA North America. From Morel and Irving [1980]. With permission from Journal of Geomagnetism and Geoelectricity.

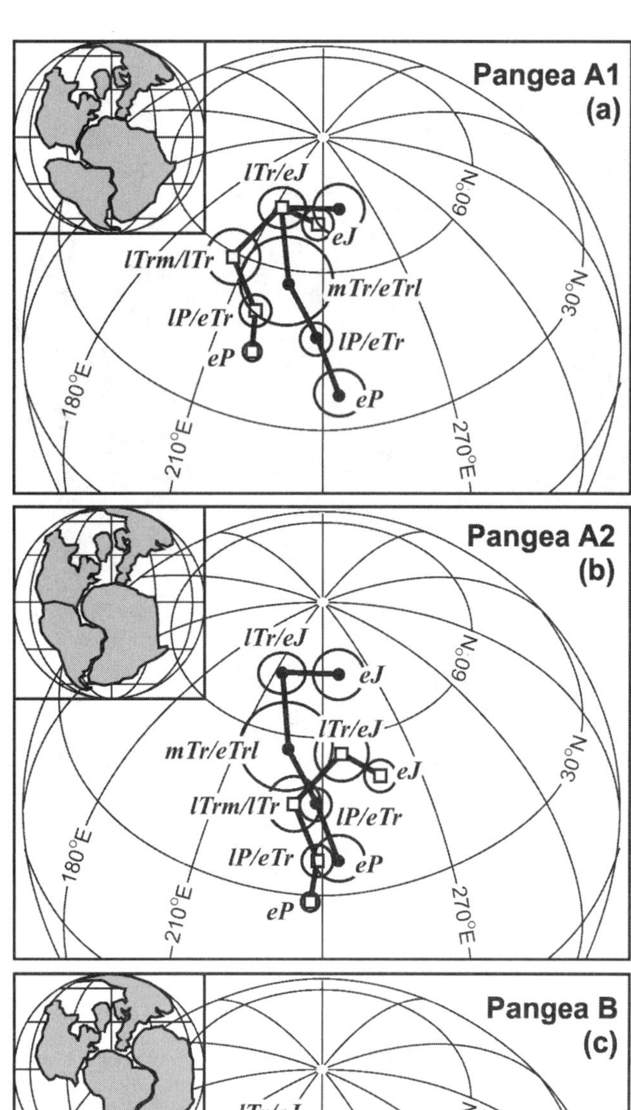

Figure 11. Early Permian through Early Jurassic APW mean paleopole (P = 0.95) by epoch for West Gondwana and Laurussia relative to Pangea (a) A-1, (b) A-2 and (c) B. eP = Early Permian; lP = Late Permian; eTr = Early Triassic; lTrm = late Middle Triassic; mTr = Middle Triassic; eTrl = early Late Triassic; lTr = Late Triassic; eJ = Early Jurassic. From Muttoni et al. [1996]. With permission from Elsevier Science Publishers.

4. LATER 1970S AND EARLY 1980S: PANGEA B INTRODUCED

As data accumulated, it became possible to construct APW paths for Carboniferous and later times by assigning numerical ages to individual paleopoles and averaging them over moving time-windows [*Irving*, 1977]. Initially, windows were 30 or 40 my wide. Now with more data, they are usually 20 my. Two features became apparent. Paleopoles from different parts of Gondwana confirmed the reconstruction of Smith and Hallam (which closely followed Du Toit's Gondwana) justifying the use of data then available from all Gondwana (compare G1 and G2 in Figure 9). Notably, there was good agreement between igneous and sedimentary rocks in the interval latest Carboniferous and Early Permian, indicating no significant inclination error in data available at that time from sedimentary rocks. Second, it was shown, that the data did not fit Pangea A-2 (Figure 11), which led to a search for a conforming reconstruction. This was achieved by unifying paleopoles interval by interval working backwards from Pangea A-1 in the Early Jurassic, by maintaining consistency with paleomagnetically determined paleolatitudes and azimuths, by restricting motions to a minimum between intervals, and by not allowing continents to overlap. The result (Figure 10) was Pangea B for the Early Permian Epoch. (*Westphal* [1977] achieved a reconstruction similar to Pangea B using very different arguments.)

The main features of Pangea B as originally recognised [*Irving*, 1977; *Kanasewich et al.*,1978; *Irving*, 1979; *Morel and Irving*, 1981] are these.

(1) For the interval "latest Carboniferous and Early Permian" Africa was placed below Europe and South America below North America. Pangea B lasted ~50 Ma or more.

(2) In Pangea B the Variscan of Europe is opposite the Mauretainides, and the Appalachians opposite the Late Paleozoic orogenic belt of northwest South America. This provides for the face-to-face, mid-Carboniferous, continental collision of Gondwana and Laurasia along the line of the Appalachian-Variscan orogenic belt. On Pangeas A-1 or A-2, the Variscan orogenic belt faces an open Tethys (Figure 7) in the mid-Carboniferous and neither provides a framework for collision orogeny.

(3) Pangea B transformed to Pangea A-1 by ~ 3500 km dextral megashear along the line of the Appalachian-Variscan orogenic belt. This opened NeoTethys westward.

(4) The megashear occurred between Early Permian and Early Jurassic, mainly, it was thought, during the Triassic. The megashear long predated the Cenozoic Alpine orogeny, and was not the long, drawn-out "Tethys Twist" of Van Hilten.

(5) Pangea A-1 lasted only a short time before it began to disintegrate with the opening of the mid-Atlantic Ocean in the Early Jurassic.

There is no paleomagnetic constraint on longitude and Gondwana could have been either further east than in Figure 10 (for example, harking back to *van Hilten* [1964], *Smith et al.* [1981] proposed Pangea C with Gondwana >6000 km to east), or to the west of South America. Neither possibility meets the constraint of minimum motion, and neither provides matching continental margins for the Appalachian-Variscan collision belt.

5. LATER 1980S AND EARLY 1990S: PANGEA A-2 ASCENDANT

During these years there were several global reviews of paleomagnetic data and attempts made to draw Late Paleozoic maps based on them. Livermore et al. [1985, 1986, p. 164] "considered it unlikely that Pangea ever existed in. . . configuration B". They wrote, (p. 165) "the most probable evolution appeared to be the formation of a supercontinental assembly resembling Pangea A-2 during the late Devonian followed by Permo-Triassic readjustments involving dextral shear". This meant that Pangea was little changed from Late Devonian through Carboniferous, and it transformed from A-2 to A-1 in the Permo-Triassic. Scotese and McKerrow [1990] in their Paleozoic map sequence, retained Pangea A from Namurian to the end of the Permian. Smith and Livermore [1991] re-surveyed global data and concluded that the "most plausible pre-mid-Triassic Pangea is very close to that of Van der Voo and French", that is Pangea A-2. Van der Voo [1993] also reviewed the global data and concluded that Pangea A-2 was consistent with it. He argued that the implied rate of motion of the megashear on the Pangea B model (~10 cm/y) was too great to be plausible and saw no geological evidence that justified it. He was (p. 121) "not in favour of considering Pangea B... seriously". Readers of the literature of these years will be hard pressed to find a good word for Pangea B. Support for Pangea A-1 and A-2 was essential unanimous.

6. LATER 1990S TO PRESENT: PANGEA B REVIVED

In the later '90s, a narrower focus, a western Mediterranean focus, was adopted by Muttoni and colleagues, a focus better suited than global reviews to addressing individually the difficulties that had been raised against Pangea B. From northern Adria, the southern Alps of Italy where Dutch workers had begun work 35 years earlier, Muttoni et al. [1996] described new Triassic results. Then they compiled previous Early Permian to Late Triassic data, and renewed the paleopole comparisons with Africa that others had begun [*Channell et*

al., 1979; *Lowrie,* 1986; *Channell,* 1996]. They noted how well they agreed, including data from igneous and sedimentary rocks. Further work from Libya strengthened their case [*Muttoni et al.*, 2001], confirming Argand's concept of an African promontary (Figure 6) and boosting the African APW path.

To determine the time of transition of Pangea B to A-1, Muttoni et al. [1996] reconstructed, epoch by epoch, Early Permian to Early Jurassic APW segments relative to Pangea A-1. The paths diverge prior to the Late Triassic, indicating that motion continued through the Middle Triassic (Figure 11a). Reconstructed relative to Pangea B, the APW paths agreed only in the Early Permian. The Intra-Pangean Megashear, therefore, was post-Early Permian and pre-Late Triassic, as Morel and Irving [1981] argued. They then assembled paleopoles from Laurussia and Africa, and found (p. 108) that "The reconstruction. . . that satisfies the Late Permian/Early Triassic paleopoles is similar to Pangea A-2. . .". This led them to propose that the major part (~3000 km of shear at ~10 cm/yr) of the transition was complete by the Late Permian/Early Triassic, the remaining ~500 km from Pangea A-2 to A-1 occurring between the Early and Late Triassic. It is worth noting however, that, when reconstructed relative to Pangea A-2, the mean Late Permian/Early Triassic paleopoles for Laurussia and Gondwana remained significantly different (Figure 11b), indicating that, at that time, Gondwana was still too far north to be compatible with Pangea A-2, and that substantial shear, therefore, continued from the Late Permian through the Middle Triassic.

Further examples of the IPP anomaly were provided by Torcq et al. [1997]. They studied strata from Arabia of latest Permian (256 ± 10 Ma) and Early Triassic (244 ± 11 Ma) ages. Rotating their paleopoles into Pangea A-1 and A-2 configurations did not bring them into conformity with APW paths for Laurussia; differences of ~20°, over twice the combined 95% errors, remained. In particular, their review of West Gondwana data (available at that time from Africa, Madagascar, and South America; they did not use data from Adria believing it somewhat allochthonous) yielded an average Early Triassic paleopole. Unifying this with the corresponding Early Triassic paleopole from Eurasia, and swinging Gondwana counter-clockwise, a minimal amount to bring paleolatitudes into agreement with Laurussia, yielded a reconstruction very similar to Pangea B, which they considered (p. 559) "the most likely solution". Because the Early Triassic was divergent, they argued that the megashear was entirely Triassic and occurred at a rate of 9 cm/y, based on 3700 km displacement.

Thus, since the mid-90s, there has been disagreement amongst proponents of Pangea B about the time of transition. Irving [1977], Morel and Irving [1981] and Torcq [1997] argued that the shear was predominantly Triassic, Muttoni et

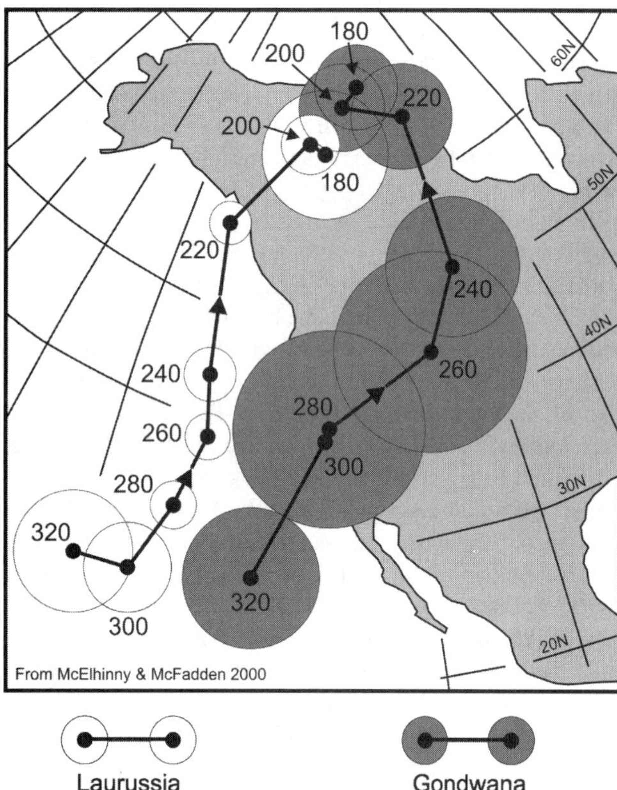

Figure 12. Late Carboniferous through Early Jurassic APW for Gondwana and Laurussia plotted on Pangea A-1 using rotations of Lottes and Rowley [1990]. Paleopole errors P = 0.95. From McElhinny and McFadden [2000]. With permission from Academic Press.

al. [1996] that it was predominantly Permian. All agreed on the configuration of a Permo-Carboniferous Pangea B, and that it changed to Pangea A-1 sometime between Early Permian and Late Triassic.

At the end of the century, old criticisms of Pangea B were once again revived on the basis of global reviews. McElhinny and McFadden [2000], using Pangea A-1 as their base, compiled Late Carboniferous through Early Jurassic data from Laurussia and Gondwana, applied moving windows revealing once again the IPP anomaly (Figure 12). They (p. 299) gave paths for both West and East Gondwana, which at only one time (240 Ma) appeared to differ significantly. Notably, East and West Gondwana paleopoles for the Early Permian (280 Ma), which are central to the Pangea B discussion, are in good agreement. Nevertheless, these authors held out the possibility that errors in reconstruction could have affected mean paleopoles, and they described Pangea B as "implausible" (p. 301).

To test the effect of applying different Euler rotations when assemblying Early Permian Gondwana, four estimates of the 280 Ma paleopole for Gondwana obtained since 1980 are given

in Figure 13. The various Euler rotations used are documented in the caption. Paleopoles are from Gondwana as a whole (MI(1), MM280), from West Gondwana (MI(2)), and from Adria/Africa (MA). The first three are based on data from igneous and sedimentary rocks, the last from igneous rocks only. There are no significant differences, showing how durable this result has been over two decades. The use of different Euler rotations to assemble Gondwana has no significant effect on the determination of the Early Permian paleopole for Gondwana. Furthermore, the absence of significant differences indicates that during the Early Permian Epoch, the time-averaged field measured over 120° of arc across Gondwana was not significantly different from that of a geocentric dipole.

Van der Voo and Torsvik [2001] and Torsvik and Van der Voo [2002] also rejected Pangea B, questioning again the applicability of the time-averaged geocentric dipole model on which it is based. They revived the idea of Briden et al. [1971], that for the Permo-Triassic there had been long-term (much longer than non-dipole components of the field in the past few millon years) zonal octupole components whose magni-

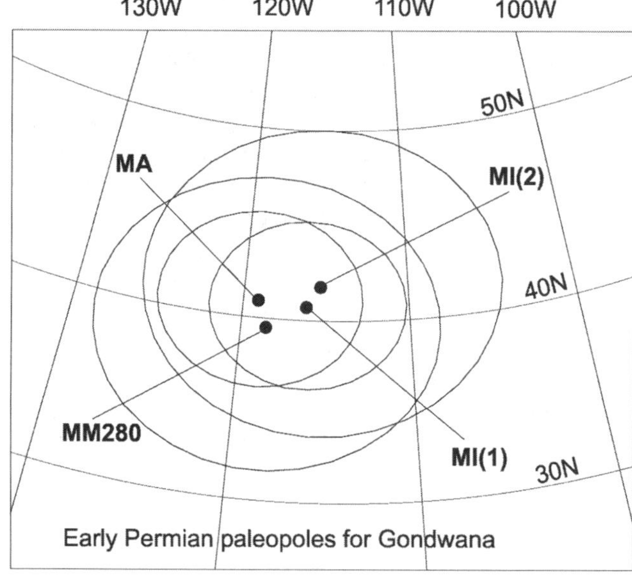

Figure 13. Comparison of four estimates, errors (P = 0.95) of the Early Permian paleopole for Gondwana. MI(1) all Gondwana (N=13) and MI(2) West Gondwana (Africa and South America, N=6) using rotations of Bullard et al. [1965] and Smith and Hallam [1970], 285-255 Ma, Morel and Irving [1981]; MM280 sedimentary and igneous rocks (N=14), 290-270 Ma, all Gondwana assembled using rotations of Lottes and Rowley [1990] analysis of McElhinny and McFadden [2000], this is the 280 Gondwana paleopole of Figure 12; MA igneous rocks only (N=9), 284-276 Ma, from Adria and west Africa (Morocco), Muttoni et al. [2003]. West Africa held fixed. Ages in Ma as given in originals. N the number of studies combined to obtain these mean paleopoles.

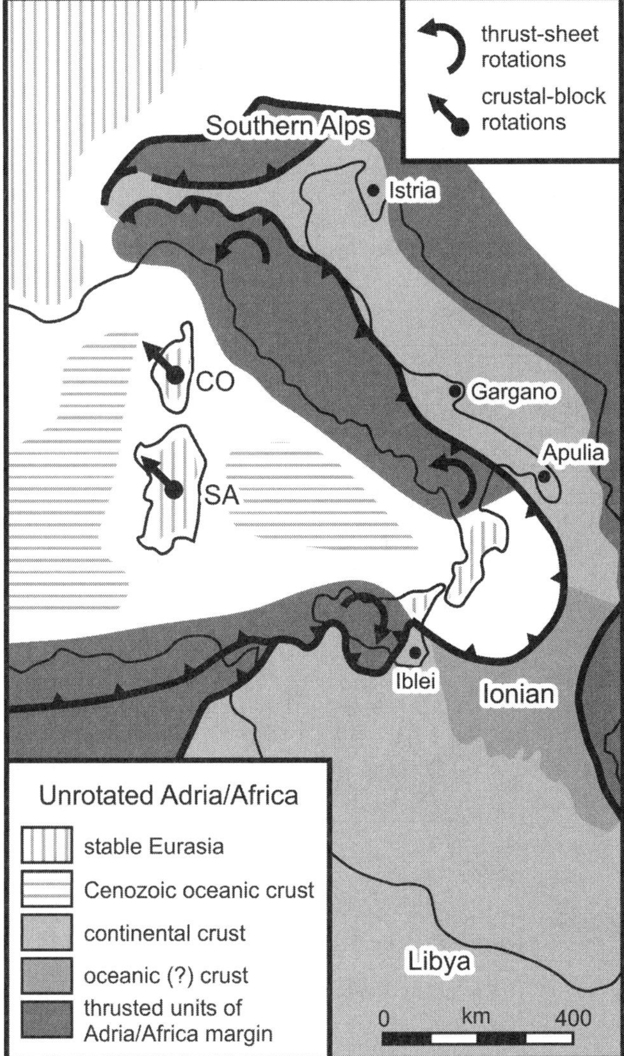

Figure 14. Adria redrawn from Muttoni et al. [2003] showing key paleomagnetic study areas mentioned in text. The rotations of Corsica (CO) and Sardinia (SA) from paleomagnetic observations were originally invoked by Argand (Figure 6). With permission from Elsevier Science Publishers.

tude was 10-20% of the dipole field. Rochette and Vandame [2001] also rejected Pangea B, arguing that it is an artifact of inclination error in sedimentary rocks.

Meanwhile, Muttoni et al. [2003] had been extending their regional western Mediterranean study by resampling Early Permian volcanic units of the southern Alps, essentially the same strata studied earlier by Dutch workers, for example by Zijderveld et al. [1970], who claimed (p. 639) that "paleomagnetic indication for megatectonic translation of the southern Alps is virtually absent". According to Muttoni et al.

[2003] these rocks are coupled to the Variscan basement and are little affected by Alpine thrusting. They gave results in good agreement with data from Early Permian rocks from Morocco, establishing an accurate well dated Early Permian paleopole for Adria/Africa. Then they reviewed younger data from Africa and Adria, both sedimentary and volcanic, from the southern Alps, Istria, the Gargano Peninsula, Apulia and Iblei for the Middle Triassic (9 studies), Late Jurassic to Early Cretaceous (6 studies), Late Cretaceous (25 studies), and Mio-Pliocene (11 studies) (Figure 14). There was good agreement within all five epochs, re-confirming, Argand's African Promontory (Adria) from Permian to the Cenozoic (Figure 6). Muttoni and colleagues then calculated the mean Early Permian paleopole for Europe, using results only from igneous rocks, matched it with that from Adria/Africa, and swung Gondwana east and north a minimal amount until its latitudes were conformable with those of Europe but without continental overlap. They obtained Figure 15a, essentially Pangea B of Figure 10, but with added speculations. They portrayed a world at the end of the Paleozoic in which the Intra-Permian Megashear was a diffuse zone linked to the western extension of NeoTethys and to the Late Paleozoic strike-slip systems of Arthaud and Matte [1977]. There were relatively few mid-ocean ridges. By the end of the Paleozoic a configuration close to Pangea A-2 had been achieved (Figure 15b).

Further, Muttoni et al. [2003] maintained that these Early Permian data are from low paleolatitudes in which octupole effects are "minimized if not virtually excluded". Inclination errors too are negligible, they argued, because their analysis now rested solely on igneous rocks.

7. AGE OF PROPOSED INTRA-PANGEAN MEGASHEAR

Muttoni et al. [1996] argued that the megashear occurred in the ~20 my interval between the deposition of localized Early Permian igneous sequences of the southern Alps and the overlying Late Permian overlap assemblages (Bellerephon Formation, Verrucano Sandstones etc). The latter were deposited slowly, signalling the waning of tectonic activity following the ~3500 km Intra-Permian Megashear. "A similar evolution" according to Muttoni et al. [2003] "occurred in large portions of central Europe" across the width of the Variscan foldbelt. Shear, they argued, occurred at the very high rate of ~15 cm/y. They cited other geological factors supportive of an intra-Permian age for the transition. Recently acquired paleomagnetic data from Late Permian strata from Morocco and Argentina support their interpretation [*Besse et al.*, 2003].

Alternatively, Late Permian and Early Triassic paleopoles from Gondwana may be considered part of the IPP anomaly, as is apparent in the analyses of Morel and Irving [1981] and

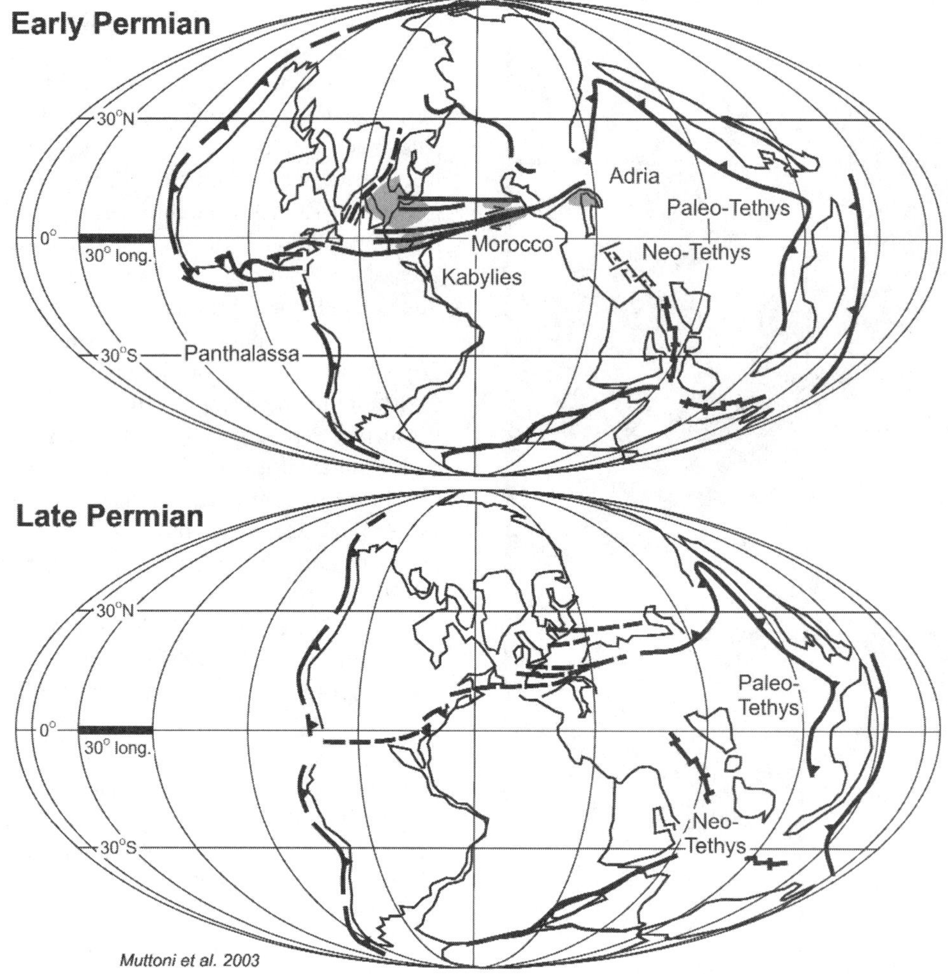

Figure 15. Early and Late Permian reconstructions and the Intra-Permian megashear of Muttoni et al. [2003]. Observations critical to their argument are from the shaded areas. The Intra-Pangean Megashear occurred along a diffuse zone. With permission from Elsevier Science Publishers.

Torcq et al. [1997], who interpreted this to mean that the megashear occurred mainly in the Triassic. What geological evidence is there for this? Little so far as I am aware, but it is an idea that is perhaps worth considering because it prompts interesting questions. There was global regression of oceans in the Late Permian (Figure 16). Can this be related to the possible low volume of ocean ridges at a time when most continental crust was assembled together and stood high, resulting in the progressive loss of shelf habitat and its associated life in the Late Permian (documented in *Erwin* [1993], chapter 3)? Was the extrusion of the Siberian Traps a result of the inital rupture of Pangea B, and was it the cause of the more sudden extinctions at the end of the Paleozoic [c.f. *Courtillot et al.*, 1996]? Was the Early Triassic global transgression a consequence of the development of new ocean ridges resulting from the rupture of Pangea B and the opening westward of NeoTethys? Consideration of the possibility of a mid-Permian age for the Intra-Pangean Megashear would lead to a set of different but equally interesting questions.

8. DISCUSSION AND SUMMARY

Du Toit's idea that Gondwana and Laurasia were largely independent supercontinents during much of the Late Paleozoic has been very much in the shadows. Wegener's Pangea (or later versions of it) little changed from Carboniferous to Jurassic, has held the spotlight. Paleomagnetism in the 1950s provided the first physical evidence of Mesozoic-Cenozoic continental drift, and this evidence, although now fully accepted, was, at the time, strongly criticized and largely ignored. At the same time, paleomagnetism also provided the first physical evidence of Permo-Triassic drift, which, as we have seen, is, after four decades, still largely not accepted. Evidences for late Mesozoic-Cenozoic and Permo-Triassic

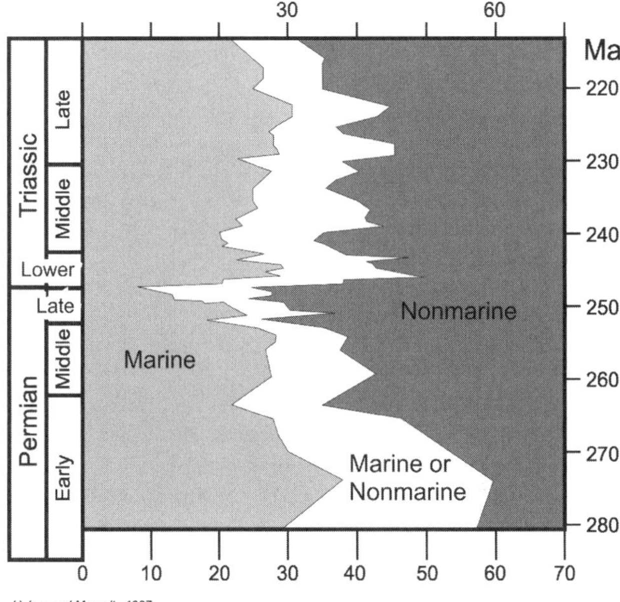

Figure 16. Percent of continent covered by sea. Based on analysis of 68 major sedimentary basins. From Holser and Magaritz [1987]. With permission from Taylor & Francis.

drift were discovered at the same time, being contained in the same observations, so that the IPP anomaly was first described at much the same time as the distinctness of the European and North American APW paths was established [*Runcorn*, 1956; *Irving*, 1956]. This arose because the focus in the '50s was on red beds, mafic igneous rocks, and vulcanigenic sedimentary rocks (good recorders of the geomagnetic field), and they are abundant in Permo-Triassic sequences. Hence, data came to be concentrated in that interval, and those from the northern continents could be fully explained by late Mesozoic-Cenozoic drift, whereas those from Australia could not.

As observations accumulated in the '60s and early '70s, a handful of workers realized that there may have been gross shear between northern and southern continents. In the late '70s, a specific reconstruction (Pangea B) was proposed to explain the IPP anomaly, a reconstruction which placed Gondwana ~3500 km to the east in the Late Carboniferous and Early Permian. Africa collided with Europe and South with North America, resulting in the Appalachian-Variscan orogeny. Pangea B was transformed into Pangea A by megashear, mainly during the Triassic Period, bringing Africa and North America into contact.

During the '70s, '80s and '90s, most workers believed that there were systematic errors in Permo-Triassic paleopole determinations. They restated and elaborated old difficulties, rejecting Pangea B, which, as a consequence, almost everybody ignored. These same workers accepted that the time-averaged

geomagnetic field approximated to a geocentric dipole for post-Triassic time, but believed that during Permo-Triassic time it did not. However, as they worked further back in time, they re-affirmed their belief in the time-averaged geocentric dipole field model, and constructed extensive Early Carboniferous and earlier map sequences accordingly (see for instance *Scotese and McKerrow* [1990], *Van der Voo* [1993]). It was as if these workers singled out the Permo-Triassic geomagnetic field as exceptional, even though there was (and still is) no paleomagnetic evidence that it was structurally different from earlier or later fields. They were willing to reject the model of a time-averaged geocentric dipole field for the Permo-Triassic even though they accepted it and it served well for other geological periods.

Meanwhile, lonely advocates of Pangea B continued to maintain that the data were best satisfied by retaining the time-averaged geocentric dipole model throughout the life of the geomagnetic field. This, in essence, was, and still is, the main issue separating supporters of Pangea B and the Intra-Pangean Megashear from those who reject them.

In the mid-'90s, Muttoni and colleagues developed a new regional strategy centred on the western Mediterranean. By restricting comparisons to West Gondwana, they sidestepped several difficulties, for example, the supposed reconstruction uncertainties for the larger Gondwana. Their main conclusions were, (1) the confirmation of Argand's Africa promontory, which boosted the APW path for West Gondwana as observed from low paleolatitudes, (2) that the exclusive use of well dated igneous rocks discounts systematic inclination or dating errors, and (3) that the megashear occurred mainly during the Permian. They showed that the Adria/Africa link is strong, and answered all individual difficulties as presently stated.

This new regional strategy has greatly strengthened the case for Pangea B and makes it unlikely that any single source of error alone is responsible for the IPP anomaly. However, several sources of error may have operated simultaneously. Also different data-selection criteria (which are, to a considerable degree, subjective) have been applied in the various reviews cited above. There is, therefore, the possibility that two or more of these factors may have interacted synergetically to produce the IPP anomaly as an artifact. This is likely to be the next line of attack on Pangea B. It would, I suggest, require coincidences beyond what can be reasonably assumed. It would be very unlikely that these various errors in data that record events spanning 100 my, have been so finely balanced as to have produced just the observed paleolatitude differences between Laurussia and Gondwana required to create the appearance of the IPP anomaly and Pangea B.

From the foregoing mobilist arguments, three major tectonic events in the later Paleozoic-early Mesozoic are identified: (1) the collision of Gondwana and Laurasia and the creation of the Variscan-Appalachian orogenic belt in the mid-Carboniferous forming Pangea B, which lasted about 50 my, (2) the ~3500 km dextral Trans-Pangean Megashear creating the short-lived Pangea A-1 in the Late Triassic, (3) the fragmenting of Pangea A-1 forming the modern oceans. Pangea B is seen to be the end-point of Paleozoic paleogeography, and its disruption by the Trans-Pangean Megashear as either the culminating event of the Paleozoic or the starting point of Mesozoic-Cenozoic paleogeography; from either viewpoint, the beginning of the current tectonic cycle is not the Early Jurassic but the end of the Paleozoic.

Perhaps the most interesting contribution of the concepts of Pangea B and the Intra-Pangean Megashear is that they offer a fresh framework for speculations about the kinematics of the Appalachian-Variscan orogeny, and the origin of end-Paleozoic extinctions. Certainly their most useful consequence is that they relieve paleomagnetists of the burden, noted by Briden et al. [1971], that serious systematic errors >10° may be ubiquitous in pre-Jurassic paleomagnetic data.

Acknowledgments. Giovanni Muttoni and Dennis Kent kindly gave me a preprint of their paper [*Muttoni et al.*, 2003]. G. Muttoni sent Figure 16 in digital format. Darrel Cowan, Hank Frankel, Randy Enkin, and R. Van der Voo reviewed the paper, and Judith Baker prepared the manuscript. I am most grateful to them all. Geological Survey of Canada publication number 2003122.

REFERENCES

Argand, E., *La tectonique de l'Asie,* C.R. Congr. Geol., Internat., 1. 171–372, 1922. English translation by A. V. Carozzi, Hafner Press, 218 pp., 1977.

Arthaud P and P. R. Matte, Late Paleozoic strike-slip in southern Europe and north Africa: result of right-lateral shear between the Appalachian and the Urals, *Geol. Soc. Amer. Bull., 88,* 1305–1320, 1977.

Besse, J., O. Oufi, A. Rapalini, and V. E. Courtillot, Paleomagnetism of Late Paleozoic series in Morocco and Argentina: implications of GAD hypothesis and Pangea reconstruction. *EOS, Trans. Amer. Geophys. Union, 84,* no 46, Fall Meet. Suppl., Abs. GP410-05, 2003.

Briden, J. C., A. G. Smith, and J. T. Sallomy, The Geomagnetic Field in Permo-Triassic time, *Geophys. J. Roy. Astron. Soc., 23,* 101–118, 1971.

Bullard, E. C., J. E. Everitt and A. G. Smith, The fit of continents around the Atlantic, *Phil. Trans. Roy. Soc., 258,* 41–51, 1965.

Carey, S. W., The tectonic approach to continental drift, in *Continental Drift a symposium, 1956,* University of Tasmania, 177–358, 1958.

Channell J. E. T., Paleomagnetism and paleogeography of Adria, in Paleomagnetism and tectonics of the Mediterranean region, edit. A. Morris and D H Tarling, *Geol. Soc., London, Spec. Publ., 105,* 119–132, 1996.

Channell, J. E. T., B. D'Argenio, and F. Horvath, Adria, the African Promontory in Mesozoic Mediterranean paleogeography, *Earth. Sci. Rev., 15,* 213–292, 1979.

Courtillot, V., J. J. Jeager, Z. Yang, G. Feroud, and C. Hoffman, The influence of continental flood basalts on mass extinctions: where do we stand? Geol. Soc. Amer. Spec. Pap. 307, 513–525, 1996.

Creer, K. M., E. Irving and S. K. Runcorn, The directions of the geomagnetic field in remote epochs in Great Britain, *J. Geomag., Geoelec.,6,* 163–168, 1954.

de Boer. J., Paleomagnetic implications of megatectonic movements in the Tethys, *J. Geophys. Res., 70,* 931–944. 1965.

Du Toit, A. L. *Our Wandering Continents,* Oliver and Boyd, 336 pp., 1937.

Embleton, B. J. J., Paleomagnetic results for the Permian of South America and a comparison with African and Australian data, *Geophys, J, Roy, Astr. Soc. 21,* 105–118, 1970.

Erwin, D. H. *The Great Paleozoic Crisis,* Columbia Univ. Press, 327 pp., 1993.

Frankel, H. "The continental drift debate." In *Resolution of Scientific Controversies Theoretical Perspectives on Closure,* eds. A. Caplan and H. T. Englehaeart. Cambridge Univ. Press, pp 312–373, 1985

Graham, J. W., Paleomagnetism and magnetostriction, *J. Geophys., Res., 61,* 735–739, 1956.

Holmes, A., *Principles of Physical Geology,* Ronald Press, 1288 pp., 1965.

Holser, W. T. and M. Magaritz, Events near the Permo-Triassic boundary, *Modern Geology, 11,* 155–180, 1987.

Hospers, J., Rock magnetism and polar wandering, *Nature, 173,* 1183,1954.

Irving, E., Paleomagnetic and paleoclimate aspects of polar wandering, *Geofis. Pura. Applic., 33,* 23–41, 1956.

Irving, E., Rock Magnetism a new approach to paleogeographic problems, *Phil. Mag. Suppl., 6,* 194–218, 1957.

Irving, E., *Paleomagnetism,* Wiley, 499 pp., 1964.

Irving, E., Paleomagnetic evidence for shear along Tethys, *Systematics Assoc., 7,* 59–76, 1967.

Irving, E., Drift of the major continents, *Nature, 270,* 304–309, 1977.

Irving, E., Pole positions and continental drift since the Devonian, in *The Earth its origins, structure and evolution,* ed. M. W. McElhinny Academic Press, 567–593, 1979.

Jeager, J. C., and E. Irving, Paleomagnetism and reconstruction of Gonwanaland, *C. R. 3rd Congr. Pacific and Indian Ocean Sciences,* 233–242, Imprimerie Officielle, Tananarive, Madagascar, 1957.

Kanasewich, E. R., J. Havskov and M. E. Evans, Plate tectonics in the Phanerozoic, *Can. J. Earth Sci., 15,* 919–955, 1978.

Köppen, W. and Wegener, A., Die klimate der geologischen vorzeit, Borntrager, Berlin, 1924.

Livermore, R. A., A. G. Smith and J. C. Briden, Palaeomagnetic constraints on the distribution of continents in the Late Silurian and early Devonian, *Phil. Trans. Roy. Soc. 309,* 29–56, 1985.

Livermore R. A., A. G. Smith and F. J. Vine, Late Paleozoic to early Mesozoic evolution of Pangaea, *Nature, 322,* 162–165, 1986.

Lottes, A. L. and D. B. Rowley, Reconstruction of Laurasian and Gondwanan segments of Permian Pangea. In *Paleozoic Paleo-*

geography and Biogeography, W. S. McKerrow and C. R. Scotese, (eds) *Mem. Geol. Soc. Amer., 12*, 383–395, 1990.

Lowrie, W., Paleomagnetism and the Adriatic Promontory: a reappraisal, *Tectonics, 5*, 8798–8807, 1986.

McElhinny, M. W. and P. L. McFadden, *Paleomagnetism,* Academic Press, 380pp, 2000.

Morel, P. and E. Irving The concept of a mobile Pangea and the continuity of continental drift. *J. Geomag. Geoelec. 32,* 39–45, 1980.

Morel, P. and E. Irving, Paleomagnetism and evolution of Pangea, *J. Geophys. Res., 86*, 1858–1887, 1981.

Muttoni, G., E. Garzanti, L. Alfonso, C. Simonetta, D. Germani, W. Lowrie, Motion of Africa and Adria since the Permian: paleomagnetic and paleoclimatic constraints from northern Libya, *Earth Planet. Sci. Lett., 192,* 159–174, 2001.

Muttoni, G., D .V. Kent, and J. E. T. Channell, Evolution of Pangea, paleomagnetic constraints from the Southern Alps, *Earth Planet. Sci. Lett., 140*, 97–112, 1996.

Muttoni, G., D. V. Kent, E. Garzanti, P. Brack, N. Abrahamsen and M. Gaetani, The mid-Permian revolution from Pangea 'B' to 'A', *Earth Planet. Sci. Lett., 215*, 379–394**,** 2003.

Opdyke, N. D., The paleomagnetism of the New Jersey Triassic: a field test of the inclination error in red sediments, *J. Geophys. Res., 66*, 1941–1949, 1961.

Rochette, P. and D. Vandamme, Pangea B; an artifact of incorrect paleomagnetic assumptions? *Annali di Geofisica, 44,* 649–658, 2001.

Runcorn, S. K., Paleomagnetic comparisons between Europe and North America, *Proc. Geol. Assoc. Canada, 8,* 77–85, 1956.

Schmidt, P. W., The Late Palaeozoic and Mesozoic palaeomagnetism of Australia, Ph. D. Thesis, Australian National University, 1977.

Scotese, C. R. and W. S. McKerrow, Revised world maps and introduction, *Geol. Soc. Mem. 12,* 1–21, 1990.

Smith, A. G., and A. Hallam, The fit of the southern continents, *Nature, 225,* 139–144, 1970.

Smith, A. G., J. C. Briden and G. E. Drewry, Phanerozoic world maps, *Paleont. Assoc. London, Spec. Pap. 12,* 1–42, 1973.

Smith, A. G. and R. A. Livermore, Pangea in Permian to Jurassic time, *Tectonophysics, 187,* 135–179, 1991.

Smith, A. G., A. M. Hurley, and J. C. Briden, *Phanerozoic paleocontinental world maps*, Cambridge Univ. Press, 102 pp., 1981.

Stehli, F., Permian climate zonation and its implications, *Amer. J. Sci., 255,* 607–618, 1957.

Torcq, F., J. Besse, D. Vaslet, J. Marcoux, L. E. Ricou, M. Halawani and M. Basahel, Paleomagnetic results from Saudi Arabia and the Permo-Triassic Pangea configuration, *Earth Planet. Sci. Lett., 148*, 553–567, 1997.

Torsvik, T. H. and R. Van der Voo, Refined Gondwana and Pangea paleogeography: estimates of Phanerozoic non-dipole (octupole) fields, *Geophys. J. Int., 151*, 771–194, 2002.

Van der Voo, R., *Paleomagnetism of Atlantic, Tethys and Iapetus Oceans,* Cambridge Univ. Press, 411 pp, 1993.

Van der Voo, R. and R. B. French, Apparent polar wandering for the Atlantic bordering continents: Late Carboniferous to Eocene, *Earth Sci. Rev., 10,* 99–119, 1974.

Van der Voo, R. and T. H. Torsvik, Evidence Late Paleozoic and Mesozoic non-dopile fields provide an explanation for Pangea reconstruction problems, *Earth. Planet. Sci. Lett., 187,* 71–81, 2001.

Van Hilten, D., Evaluation of some geotectonic hypotheses by paleomagnetism, *Tectonophysics, 1,* 3–71, 1964.

Wegener, A., *Origin of Continents and Oceans*, English translation of 3rd edit by J. G. A. Skerl, (1924), Methuen, 212 pp; 1922.

Westphal, M., Configuration of the geomagnetic field and reconstruction of Pangea in the Permian Period, *Nature, 267,* 136–137, 1977.

Zijderveld, J. D. A., F. J. A. Hazeu, M. Nardin and R. Van der Voo, Shear in the Tethys and the Permian paleomagnetism in the Southern Alps, including new results, *Tectonophysics, 10,* 639–661, 1970.

E. Irving, Pacific Geoscience Centre, Geological Survey of Canada, 9860 West Saanich Road, North Saanich, BC, Canada, V8L 4B2.

The Quality of the European Permo-Triassic Paleopoles and Its Impact on Pangea Reconstructions

Rob Van der Voo

Department of Geological Sciences, University of Michigan, Ann Arbor, Michigan.

Trond H. Torsvik

VISTA, c/o Geodynamic Centre, NGU, Trondheim, Norway

More than 60 paleopoles for Europe external to the Alpine belt, with ages between 300 and 220 Ma, meet minimum reliability criteria and seem to define one of the best-determined apparent polar wander path segments for any time or any continent. We classified results according to characteristics that may produce a systematic bias in pole positions and which include lithology (volcanics vs. sedimentary rocks), demagnetization code (as provided by the global data-base), and quality of age control. Early Permian (280 ± 10 Ma) mean poles based on different subsets can differ by more than 10 degrees, suggesting that a combination of (1) inclination shallowing in sediments, (2) unrecognized present-day field overprints that have not been adequately removed by demagnetization treatments, and (3) underestimation of rock ages, may introduce a bias. All three causes, as well as possible octupole contributions to the total geomagnetic field, would conspire to yield more southerly paleolatitudes for Europe than warranted. This matters greatly for Pangea reconstructions based on paleomagnetic data. With the highest paleolatitude (23 degrees N at Oslo), a Pangea-A type fit is easily possible without any overlap of the continents, whereas the lowest (12 degrees N) produces a large overlap between Gondwana and Laurussia in a Pangea-A type fit, which could lead to the conclusion that a Pangea-B type reconstruction is preferable for the Early Permian. It appears likely that careful selection of data with the highest quality is the key to avoiding such tectonically problematic issues as Pangea B in the Early Permian.

1. INTRODUCTION

Phanerozoic apparent polar wander path (APWP) segments determined for the major continents can differ enormously in quality; not only can the number of paleopoles per unit of time vary considerably, but the reliability of the individual results—however estimated—can also range widely. Because

Timescales of the Paleomagnetic Field
Geophysical Monograph Series 145
Published in 2004 by the American Geophysical Union
10.1029/145GM03

syntheses of APWPs and continental reconstructions must begin with data selection, this is, given the variable data quality, inherently a rather subjective exercise. Indeed, several classical controversies in paleomagnetism arose from the different selection criteria employed for the construction of APWPs, especially for pre-Mesozoic times for which few other tools exist to aid in continental reconstructions.

There is general agreement, however, that one of the best-determined APWP segments for any period and any of the major continents is that of Europe for Permian time. But one may well ask: "how robust is this APWP"? In this study we analyze the available paleomagnetic data for the interval 300—220 Ma for Europe external to the Alpine belt in order to assess the robustness of this APWP segment in an attempt to determine what factors may impart a bias and thereby compromise data quality. For this interval there are more than 60 individual paleomagnetic pole positions in the global database (GPMDB) [*McElhinny and Lock*, 1996; *McElhinny and Smethurst*, 2001], not including the earliest studies that were carried out without demagnetization. Additions to the database have been made regularly, and include some 10-15 studies made in the last decade; rock types include volcanics (mostly extrusives) as well as sediments (mostly red beds), and the results come from many countries spanning most of western Europe between the Alpine ranges, the Atlantic and Arctic Oceans and the western edge of the Russian Platform. A weakness with the GPMDB (as with many other geo-databases) is that changes from the originally assigned rock ages are not easily uncovered—thus, we have made a lot of efforts in the present compilation to include new age assignments as well as results from radiogenic-isotope work in progress (Table 1).

Paleomagnetists are well accustomed to the necessity of averaging directional data, because of secular variation and non-systematic sampling or measurement errors, and so forth. However, it is important to distinguish between random errors (noise), and systematic errors (bias), because the latter are more likely, of course, to produce erroneous mean poles. Although paleomagnetists are generally cognizant of the possibility of systematic errors that may bias an outcome, few discriminating analyses have been carried out to detect such a bias. For the Early Permian of Europe, there are (at least) four factors that may introduce a bias, and interestingly, all four are likely to have the same effect, namely to assign paleolatitudes to Europe's paleogeographic positions that are lower (more southerly) than warranted. These four factors are: (1) a flattening of the inclination in sedimentary rocks, (2) a contamination by a younger overprint, (3) a contribution of an octupole field to the total field, and (4) an age assignment that is too young (in terms of Ma) because of new developments in the geological time scale for the Permian of Europe or new geochronological determinations that have not yet been taken into account [e.g., *Rochette and Vandamme*, 2001]. These four factors conspire to have the same effect, as will be explained below, not least because the geomagnetic field was reversed throughout the Kiaman interval, while the European continent was steadily moving northward throughout the Permo-Triassic.

Table 1 lists the paleomagnetic results extracted from the GPMDB (plus a few entries not in the database), with columns included that may be used as discriminants for testing three of the four factors listed above (1, 2, and 4). These columns include the demagnetization code of the GPMDB [*McElhinny and Lock*, 1996], rock type (volcanic or sedimentary), and a simple quality factor for the age determination of a study. A separate test will be carried out to see whether octupole fields can be detected as a function of paleo-north-south distance across western Europe.

This study does indeed find evidence of a bias, and we suspect that multiple causes contribute simultaneously. We argue that the effect of this is that published paleogeographic positions of Europe have been systematically too far south by up to 10°. In view of the difficulties in reconciling paleomagnetic data from the Atlantic-bordering continents with conventional Pangea reconstructions in the Permian (see *Hallam* [1983]; *Van der Voo* [1993]; *Van der Voo and Torsvik* [2001]; *Rochette and Vandamme* [2001]; *Muttoni et al.* [2003]; and *Clark and Lackie* [2003], for discussions of the controversial Pangea B reconstruction), this takes on global importance and is not simply a minor technicality that leads to a trivial correction to be applied to the APW path of Europe. We suspect that some APWP segments for other continents and other periods may be similarly affected, and if this is true one could conclude that this places an inherent restriction on the precision with which paleolatitudes can be determined regardless of the statistical parameters such as α_{95}.

2. ANALYSES

Four tests (comparisons) have been devised in order to determine whether a bias may exist in the Permian paleopoles of stable Europe. As already mentioned, these non-random errors may be caused by an inclination error in sedimentary rocks, by the presence of unrecognized overprints, by erroneous age assignments, or by octupole fields. Each of these will be discussed in the following paragraphs, and each test is based on a comparison of mean poles based on certain selection criteria, to be discussed further below. Table 2 lists these mean poles, which have been calculated for 20-million year windows, as a function of age and grouped by selection criteria; for 250 ± 10 and 280 ± 10 Ma, these mean poles are plotted in Figure 1. The paleolatitudes calculated from these mean poles for a specific locality in Europe (Oslo, Norway) are also listed in this table and are illustrated in Figure 2 for 250 and 280 Ma.

2.1. Inclination Error

To test for inclination error, we have compared the mean poles and their paleolatitudes, as derived from sedimentary and igneous rocks, because the latter are unlikely to be afflicted by inclination shallowing. In Figure 2, the mean paleolatitudes extrapolated from all volcanic and all sedimentary results (without any further filtering) can be compared; the paleolatitude from the sedimentary rocks is about the same as that

Table 1. Paleopoles for Europe, External to the Alpine Belt, for the Interval 210 – 310 Ma

QPOL CODE	T	DC	AQ	Age Comment	AGE	α_{95}	Plat	Plon	Reference (GPMDB REFNO)
Triassic									
6M Rhaetian Sediments, Germany, France	S	4	2	205-210	208	8	50	112	Edel & Duringer 1997 (3141)
5M Merci Mudstone, Somerset, UK	S	4	2	Norian (221-210)	215	5.1	50	128	Briden & Daniels 1999 (3311)
5N Sunnhordland dikes, Norway	V	4	3	Ar/Ar (219-223) [*Fossen & Dunlap*, 1999]	221±5	4.6	50	125	Walderhaug 1993 (not in GPMDB)
6M Gipskeuper seds., Germany	S	4	2	Carnian (224-228)	226	6	49	131	Edel & Duringer 1997 (3141)
6N Musschelkalk Carbonates, Poland	S	4	2	Anisian-Ladinian (242-227)	234	12	53	123	Symons et al. 1995 (3253)
6M Heming Limestone, Paris Basin, France	S	4	2	Anisian-Ladinian (242-227)	234	3	54	141	Théveniaut et al. 1992 (2411)
5M Bunter & Musschelk., Germany	S	2	2	Skythian-Ladinian (250-227)	239	15	49	146	Rother 1971 (158)
4N Kingscourt red beds, Ireland	S	2	2	239-245	242	7	59	146	Mulder 1972 (not in GPMDB)
6M Upper Buntsandstein, France,	S	2	2	Olenekian (242-245)	243	5	43	146	Biquand 1977 (1028)
5R Lunner Dikes, Oslo, Norway	V	4	3	Ar-Ar (237-246)	243±5	5.9	53	164	Torsvik et al. 1998 (3188)
Permian									
6M Sudetes Sediments, Poland	S	4	2	Zechstein-Bunts. (245-258)	251	5	50	163	Nawrocki 1997 (3161)
5R Dome de Barrot Red beds, France	S	3	2	Early Thuringian (253-258)	255	2.7	46	147	Van den Ende 1977 (652)
6M Massif des Maures Sediments, France	S	4	2	Thuringian (251-258)	255	4	51	161	Merabet & Daly 1986 (1408)
5R Esterel sediments, France	S	3	2	Saxonian-Thuringian (251-270)	261	5	47	151	Zijderveld 1975 (165)
6R Brive Basin Sediments, France	S	4	2	Saxonian-Thuringian (251-270)	261	4	49	163	Chen et al. 1997 (3144)
3R Upper Lodeve Ss., France	S	3	2	Saxonian (258-270)	264	4.6	47	156	Kruseman 1962 (168)
4R Permian red beds, Lodeve basin, France	S	2	2	Saxonian (258-270)	264	---	53	151	Evans & Maillol 1986 (1207)
5R Esterel extrusives, France	V	3	2	Saxonian (258-270)	264	6.1	52	142	Zijderveld 1975 (165)
5R Lodeve Basin, France	S	3	2	Saxonian (258-270)	264	1.5	49	154	Merabet & Guillaume 1988 (1813)
5R Saxonian Red Ss. Lodeve Basin, France	S	4	2	Saxonian (258-270)	264	4	51	144	Smith et al. 1991 (2361)
4R Bohuslan Dikes Combined, Sweden	V	3	1	250-300	275	8.6	51	165	Thorning & Abrahamsen 1980 (1155)
4R Scania Melaphyre dikes, Sweden	V	3	3	Ar-Ar plagiocl. (Eide in progress)	279±10	11	54	172	Bylund 1974 (2222)
3R Brumunddal Lavas, Norway	V	3	2	Rb/Sr	279±9	0	48	138	Storetvedt & Petersen 1970 (169)
4M Moissey Volcanics., Jura, France	V	4	1	Autunian-Saxonian (258-300)	279	6.7	41	172	Thompson et al. 1986 (1205)
3R Mauchline Lavas, Scotland	V	3	1	Autunian (270-290)	280	14	47	157	Harcombe-Smee et al. 1996 (3093)
5R Bohemian Massif Igneous, Germany	V	3	2	K/Ar, Rb/Sr	280	10	42	166	Soffel & Harzer 1991 (2356)

Table 1. (continued)

QPOL CODE	T	DC	AQ	Age Comment	AGE	α_{95}	Plat	Plon	Reference (GPMDB REFNO)
4R Bohemian Qtz. Porphyry, Germany	V	3	2	As above	280	7	37	161	Thomas et al. 1997 (3145)
4R Ringerike Lavas, Oslo, Norway	V	4	2	Rb/Sr (276-291) [*Torsvik et al.* 1998]	281±4	13	45	157	Douglass 1989 (1830 but Permian pole not listed in GPMDB)
4R Sarna Alkaline Intruion, Sweden	V	3	2	Rb/Sr (recalculated)	281±14	6.9	38	166	Smith & Piper 1979 (1735)
5R Oslo Volcanics, Norway	V	3	2	Rb-Sr (274-294) [*Torsvik et al*, 1998]	281±2	1	47	157	Van Everdingen 1960 (915)
5R Krkonose Basin Oil Shales, Czech R.	S	3	1	Autunian (270-300)	285	2	40	166	Krs et al. 1992 (2444)
4R Krakow Volcanics., Poland	V	2	1	Autunian (270-300)	285	7.9	43	165	Birkenmajer & Nairn 1964 (275)
3R Lower Silesia Volcanics, Poland [a]	V	2	1	Autunian (270-300)	285	13	40	172	Birkenmajer et al. 1968 (465)
4R Lodeve B Comp., France	S	3	1	Autunian (270-300)	285	17	49	162	Cogné et al. 1990 (2454)
5R Lodeve Basin, France	S	4	1	Autunian (270-300)	285	2	42	169	Merabet & Guillaume 1988 (1813)
5R Bohemian Red beds, Czech R.	S	2	1	Autunian-Rotlieg. (260-290)	285	4	41	165	Krs 1968 (167)
3R Lower Lodeve Ss., France	S	3	1	Autunian (270-300)	285	7.7	44	170	Kruseman 1962 (168)
5R Mt. Hunneberg sill, Sweden	V	2	2	K/Ar (recalculated)	285±8	6.3	38	166	Mulder 1971 (2211)
5R North Sudetic Basin Sediments, Poland	S	4	1	Autunian (270-300)	285	5.1	44	184	Nawrocki 1997 (3161)
4R North Sudetic Basin Volcanics, Poland	V	4	1	Autunian (270-300)	285	8.1	42	174	Nawrocki 1997 (3161)
4R Intrasudetic Basin Sediments, Poland	S	4	1	Autunian (270-300)	285	6.8	37	160	Nawrocki 1997 (3161)
5R Intrasudetic Basin Volcanics, Poland	V	4	1	Autunian (270-300)	285	3.2	43	172	Nawrocki 1997 (3161)
5R Exeter Lavas, UK	V	3	3	Ar/Ar; K/Ar (282-291) [*Edwards et al.*, 1997]	286±7	4	50	150	Zijderveld 1967 (165)
4R Exeter Lavas, UK	V	3	3	As above	286±7	10	48	163	Cornwell 1967 (411)
4R Black Forest Volcs. , Germany [c]	V	3	2	Rb-Sr	286±5	5.9	49	176	Konrad & Nairn 1972 (170)
3R Black Forest Rhyolites, Germany	V	4	2	Rb/Sr	286±5	1	42	173	Edel & Schneider 1995 (2941)
4R Thuringer Forest Sediments, Germany	S	3	1	Early Rotliegendes (274-300) [*Menning*, 1995] (age calibration)	287	5.8	42	160	Mauritsch & Rother 1983 (1792)
4R Stabben Sill, Norway,	V	4	2	Rb/Sr (recalculated)	291±8	2.4	32	174	Sturt & Torsvik 1987 (1540)
3R Nahe Volcanics, Germany	V	2	3	Ar-Ar,Rb-Sr [*Lippolt et al.,* 1989]	291±1	13	46	167	Nijenhuis 1961 (940)
3R Saar-Nahe Volcanics, Germany[b]	V	2	3	Ar-Ar,Rb-Sr [*Lippolt et al.,* 1989]	291±1	16	41	169	Berthold et al. 1975 (712)

Table 1. (continued)

QPOL CODE	T	DC	AQ	Age Comment	AGE	α_{95}	Plat	Plon	Reference (GPMDB REFNO)
Carboniferous									
4R Great Whin Sill, UK	V	3	3	Ar-Ar plagioclase (Timmerman pers. com.)	294±2	4.8	44	159	Storetvedt & Gidskehaug 1969 (585)
4R Nideck-Donon Volcanics, France	V	1	3	Ar-Ar	294±1	4	47	168	Roche et al. 1962 (1010)
3R Lower Nideck Volcanics, France	V	2	3	Ar-Ar	294±1	19	42	168	Westphal 1972 (174)
5R Scania dolerite dikes, Sweden,	V	3	3	Ar/Ar plagiocl. (287±2, 300±10) (Timmerman, pers. com.)	294±18	11	37	174	Mulder 1971 (2211)
5R Scania Dolerites, Sweden,	V	3	3	As above	294±18	6.5	38	168	Bylund 1974 (2222)
4R Thuringer Forest Volcanics, Germany,	V	3	1	Early Rotliegendes (290-300)	295	7.1	37	170	Mauritsch & Rother 1983 (1792)
4R Lower Silesia Volcanics, Poland	V	2	2	Westphalian-Stephanian 310-292	296	14	43	174	Birkenmajer et al. 1968 (465)
3R Arendal Diabase Dikes, Norway	V	2	3	Ar/Ar whole rock (Figure 4)	296±10	7.1	43	160	Halvorsen 1972 (175)
4R Ny-Hellesund Diabases, Norway	V	2	3	As above	296±10	2.9	39	161	Halvorsen 1970 (626)
5R Peterhead Dike, Scotland	V	4	3	Ar/Ar plagioclase (Figure 4)	297±12	1.3	41	162	Torsvik 1985 (1535)
5R Mt. Billinger Sill, Sweden	V	2	3	Ar/Ar plagiocl., whole rock 298±2, 300±2) (Timmerman, pers. comm.	299±3	4	31	174	Mulder 1971 (2211)
3R Wackerfield Dike, England, U.K	V	3	2	K/Ar	303±5	3	49	169	Tarling et al. 1973 (180)
5R Queensferry Sill, Midland V., Scotland	V	4	3	Ar-Ar plag. (Torsvik in progress)	305±10	5.2	38	174	Torsvik et al. 1989 (2447)
5R Westph.-Steph. Red beds, Czech R	S	2	2	Westphalian-Stephanian 310-292	305	9	38	163	Krs 1968 (167)

QPOL=Q factor (from Van der Voo [1993]; 7 is best) and magnetic polarity (N=Normal, R=Reverse; M=Mixed); T=Rock Type (V=Volcanics; S=Sediments); DC=Demagnetization Code used in the Global PaleoMagnetic Database (GPMDB) (1=pilot demagnetization, 2=all samples blanket treatment, 3=stereoplots and vector plots and 4=principal component analysis or more); AQ=Age Quality (3=Ar/Ar or U/Pb age; 2=K/Ar, Rb/Sr or stratigraphic age known within ± 10 Ma; 1=no radiogenic isotope age dating available *and* age not known within ± 10Ma); α_{95}=95% confidence circle; Plat/Plon=Pole Latitude/Longitude; GPMDB REFNO=Global Palaeomagnetic Database Reference Number in Version 2002 [*McElhinny & Lock*, 1996]. References for paleopoles listed in the database are not given in the bibliography of this paper, as they can be looked up in the GPMDB by REFNO.

[a] Recalculated because mean direction and pole as listed in the original paper do not agree. Excluded in the new calculation were (1) sediments, and (2) results without any demagnetization. Recalculated mean is based on N=11 sites.

[b] This is the pole that is listed in the paper. While there are indeed normal-polarity results in the study, this pole does not include any. The N-polarity results have much steeper inclinations and are probably post-Permian.

[c] Recalculated mean for volcanics only, N=17 sites (numbers 1,8,9,11,12,2a, 2b,3,4,5,6,26,27,13,14,15,18), mean declination/inclination= 187.8/-15.1; k = 37.3 and α_{95} = 5.9.

Table 2. Mean Paleopoles for Europe

All Poles

Age± 10 Ma	Pole Lat.	Pole Long.	N	A$_{95}$	λ
220	50	128	3	3	31
230	51	133	5	6	30
240	52	144	6	8	28
250	51	154	6	7	24
260	50	153	10	3	24
270	48	157	14	4	21
280	44	165	27	3	16
290	42	167	38	2	14
300	41	168	18	3	12

Age Quality 2 and 3

Age± 10 Ma	Pole Lat.	Pole Long.	N	A$_{95}$	λ
220	50	128	3	3	31
230	51	133	5	6	30
240	52	144	6	8	28
250	51	154	6	7	24
260	50	153	10	3	24
270	48	155	11	5	22
280	45	162	12	5	18
290	42	166	24	3	14
300	41	168	17	3	12

Age Quality 3

Age± 10 Ma	Pole Lat.	Pole Long.	N	A$_{95}$	λ
220	50	125	1	0	32
230	50	125	1	0	32
240	53	164	1	0	25
250	53	164	1	0	25
270	54	172	1	0	25
280	51	161	3	12	23
290	42	165	13	5	14
300	41	167	12	3	12

Demag Code 1 and 2

Age± 10 Ma	Pole Lat.	Pole Long.	N	A$_{95}$	λ
240	51	146	2	36	26
250	51	146	2	36	26
260	53	151	1	0	27
270	53	151	1	0	12
280	41	167	4	4	12
290	41	167	12	3	13
300	41	167	9	4	13

Demag Code 3 and 4

Age± 10 Ma	Pole Lat.	Pole Long.	N	A$_{95}$	λ
220	50	128	3	3	31
230	52	130	4	6	32
240	55	143	3	19	31
250	50	158	4	7	23
260	49	153	9	4	23
270	48	157	13	4	21
280	45	165	23	3	17
290	43	166	26	2	14
300	40	169	9	4	12

Table 2. (continued)

All Sediments

Age± 10 Ma	Pole Lat.	Pole Long.	N	A$_{95}$	λ
220	50	130	2	5	30
230	52	135	4	8	30
240	52	141	5	8	29
250	50	153	5	8	24
260	49	154	9	3	23
270	49	153	6	4	23
280	42	167	8	5	14
290	42	167	8	5	14
300	38	163	1	0	10

All Volcanics

Age± 10 Ma	Pole Lat.	Pole Long.	N	A$_{95}$	λ
220	50	125	1	0	32
230	50	125	1	0	32
240	53	164	1	0	25
250	53	164	1	0	25
260	51	142	1	0	28
270	47	159	8	7	20
280	45	165	19	3	17
290	42	166	30	2	14
300	41	168	17	3	12

Volcanics: Age Quality 3 & Demag Code 3 and 4

Age± 10 Ma	Pole Lat.	Pole Long.	N	A$_{95}$	λ
220	50	125	1	0	32
230	50	125	1	0	32
240	53	164	1	0	25
250	53	164	1	0	25
270	54	172	1	0	25
280	51	161	3	12	23
290	43	163	6	7	16
300	40	168	5	6	11

Mean poles are calculated using a ± 10 million-year window, with selection criteria as indicated. N is the number of poles used, and A$_{95}$ is the radius of the 95% confidence cone around the mean pole. Palaeolatitude (λ) is calculated for a common reference locality (Oslo, Norway: 60°N, 10°E).

for the one volcanic result for 250 Ma, whereas for 280 Ma there is a difference of 3° between the mean poles as well as the extrapolated paleolatitudes. The value for the volcanic results is higher, consistent with a small effect due to inclination shallowing in the sediments. This difference is an average, but it is noteworthy that in at least one case, involving the Estérel results (Table 1, 261–264 Ma), a larger difference of about 10° between extrusives and red beds has clearly been documented [*Zijderveld*, 1975]; this difference for intercalated and therefore contemporaneous rocks can only be explained by inclination shallowing [*Vlag et al.,* 1997].

2.2. Testing With the "Demag Code" for Overprint Bias

The demag code is determined and entered by the compilers of the GMPDB, with higher codes representing progressively more thorough demagnetization techniques [*McElhinny and Lock*, 1996]. In Table 1 we have included these codes (column DQ) and they are explained in the table's legend. Our test is based on the reasoning that most overprints will be of recent vintage (i.e., pdf = present-day field), which in Europe means northerly and steeply down. If unrecognized pdf components contaminate a reversed Early Permian (Kiaman) magnetization, the resulting measured vector sum will be drawn downwards, so that the typical reversed-polarity, southerly and shallow-upwards direction becomes shallower. The corresponding paleolatitude in that case is too low. For normal-polarity Late Permian or Triassic magnetizations, a pdf overprint will steepen the resultant vector sum, and the calculated paleolatitude then becomes too high. For mixed-polarity results, the effects may cancel each other. Our test further relies on the (unproven) assumption that investigators using higher demag-code techniques will have had a better chance to detect, and correct for, the presence of pdf overprints.

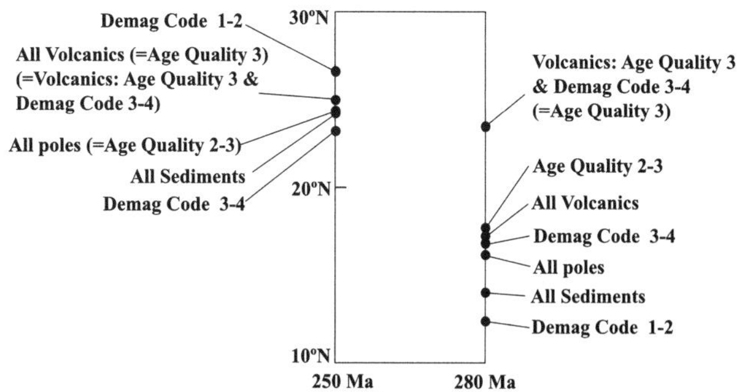

Figure 2. Estimated paleolatitudes for Europe (calculated for the Oslo reference locality at 60°N, 10°E) based on various data selections (see text and Table 2).

Averaged results based on demag codes 1–2 can be compared with means based on codes 3–4 in Table 2. Figure 2 shows these, again as extrapolated paleolatitudes for both 250 and 280 Ma, and these values are different by 3° for 250 Ma and by 5° for 280 Ma. For 280 Ma, the paleolatitudes based on the higher demag codes are higher, just as one would expect for the predominantly reversed polarity results of this part of the Kiaman interval (only two results with ages between 260 and 305 Ma, out of a total of 52 poles, have mixed polarities). In contrast, for the mostly mixed-polarity results included in the 250 Ma mean poles, the lower demag code mean gives the higher paleolatitude. It must be noted that there are not many entries making up the mean poles for the 250 Ma age window, and that the significance of the demag-code test is therefore less clear than for the well-documented 280 Ma mean poles.

2.3. Age Bias

In the last two decades, refinements in geochronological studies have changed the geological time scale calibration (Figure 3). Compared to the time scale of Van Eysinga [1975], the Permian-Carboniferous boundary has been assigned a 12-m.y. older age, so that it is now placed at 292 Ma. Moreover, most recent discussions of the classical western European stage names such as Autunian and Rotliegendes now assign the lowermost parts of these stages to the latest Carboniferous (Gzhelian or Stephanian; see Fig. 3). This means that a typical paleomagnetic result from the Early Permian in France, i.e., classically assigned to the Autunian, is now deemed to be 270–300 Ma, with an estimated age listed as 285 Ma in Table 1. Moreover, the top of the Autunian stage (or Lower Rotliegendes) has in some publications been assigned, at least locally, an age as old as 283 Ma [*Menning*, 1995], resulting in an estimated mean age of 292 Ma for Autunian rocks. In previous

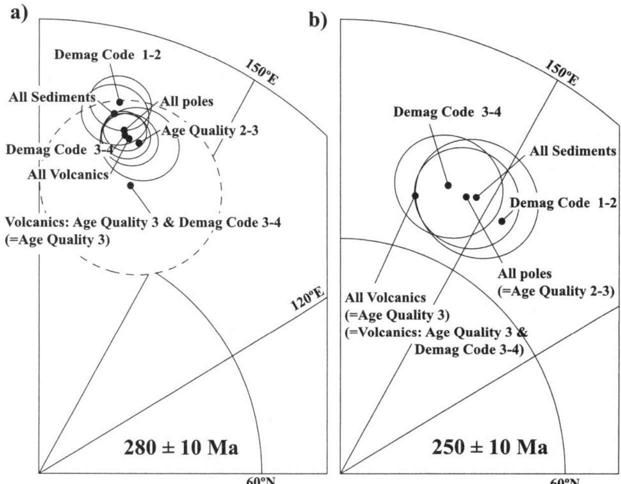

Figure 1. 280 and 250 Ma running mean poles (from Table 2). Note that A$_{95}$'s for two mean poles in (b) are not shown as they are based on only one or two entries.

compilations (e.g., *Van der Voo* [1993] or before), such a pole would have been assigned an age of 277 Ma (268–286 Ma) or even younger. For volcanic results, modern geochronological work (^{40}Ar-^{39}Ar and U-Pb) has also typically found older ages than with the K-Ar (because of argon loss) or Rb-Sr techniques used a few decades ago. Two examples of recent Ar-Ar dating of paleomagnetically important Late Carboniferous results, the Arendal Diabase dikes and the Peterhead Dike are presented in Figure 4. In 1993, these two results were assigned ages, which not only had much larger error margins, but were, in addition, between 7 and 11 million years younger [*Van der Voo*, 1993] than has now been determined. Another example of changes in assigned (numerical) ages can be found in the case of the Estérel volcanics, where previously the age was typically taken to be Late Permian

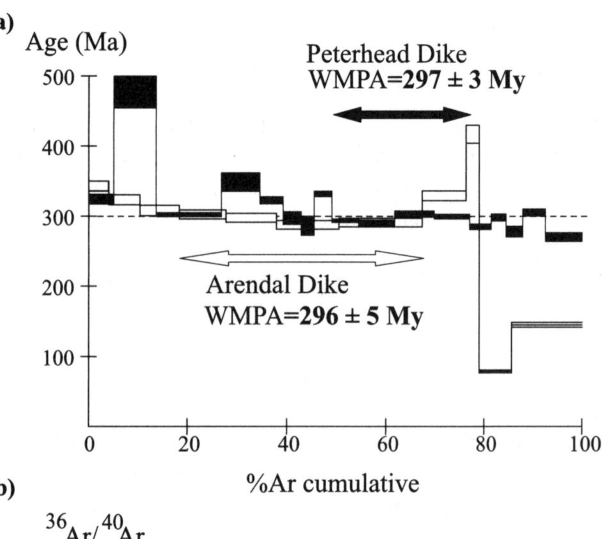

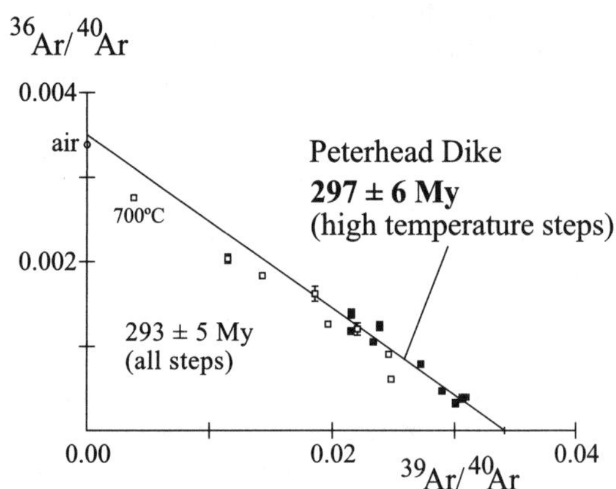

Figure 4. (a) ^{40}Ar/^{39}Ar incremental heating experiment on whole rock (WR) from the Arendal Diabase dikes (S. Norway), conducted at the ^{40}Ar/^{39}Ar laboratory at the Université Blaise Pascal et Centre National de la Recherche Scientifique (France), following the analytical protocol of Arnaud et al. [1993]. A plagioclase sample from the Peterhead dike (NE Scotland) was analyzed at the ^{40}Ar/^{39}Ar Geochronology Laboratory at the Norwegian Geological Survey (cf. *Eide et al.* [2002] for detailed procedures). The Arendal WR experiment yields a ^{40}Ar/^{39}Ar weighted mean plateau age (WMPA) of 296 ± 5 Ma (1σ, including error in J-value) and includes 51% of the released gas (4 steps). The Peterhead plagioclase release spectrum is somewhat irregular but the majority of ages vary between 285–305 Ma. A statistical overlap of ages at the 95% confidence level is achieved for 4 steps (1070–1220°C) but the WMPA age (297 ± 3 Ma) only includes 29% gas. (b) The inverse isochron for the Peterhead plagioclase yields an age of 293 ± 5 Ma (mean squares of weighted deviations (MSWD)=6.4) when all heating steps are included. Removing low temperature steps (T<1070°C; open symbols) yields an age of 297 ± 6 Ma and is better correlated (MSWD=3.3; ^{40}Ar/^{36}Ar=285.9 ± 19.7). In Table 1 we have assigned the Peterhead dike an age of 297 Ma. The Arendal and Ny-Hellesund dike entries have been given a mean age of 296 Ma based on the Arendal dike spectra in (a). Note that all numerical errors in Table 1 are listed as 2σ.

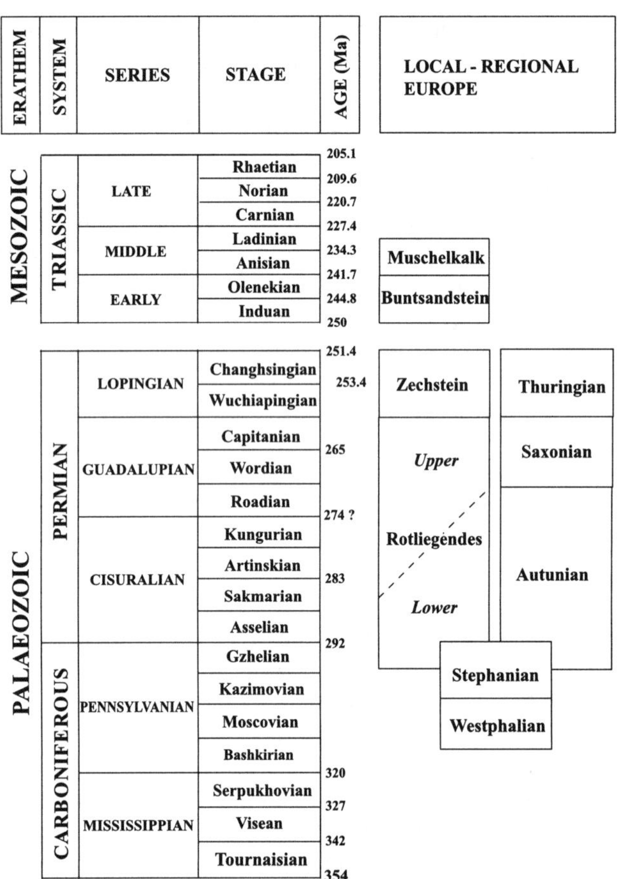

Figure 3. International commission on stratigraphy timescale (http://micropress.org/stratigraphy/) and tentative correlation with local-regional European stage names [*Menning*, 1995]. Note the uncertainty in assigning a numerical boundary (stippled line) for the Lower Rotliegendes (~Autunian) and the Upper Rotliegendes (~Saxonian). We have used the 270–300 Ma range for the duration of the Autunian (Table 1) but 285–300 Ma is often used as well [e.g., *Menning*, 1995].

Table 3. Some Mean Poles Based on Volcanics Only

Area	Age	N	A95	Lat.	Long.	λ
Scandinavia	280	5	10	47	158	20
Scandinavia	290	6	7	40	167	12
Central Europe	280	6	7	45	165	17
Central Europe	290	9	4	44	164	16
S. Alps [a]	280	68[a]	4	45	238	22
S. Alps [b]	280	7	6	43	242	21
Siberian Traps[c]	251	3	6	56	150	30

Age = mean age of $\pm$10 m.y. window, N is the number of entries (poles, or [a]sites), used for comparisons discussed in the text, such as mean Scandinavian and Central European poles for 280 and 290 Ma (with demag code 3-4 and age quality 2-3), poles from the Southern Alps (280 Ma) and the Siberian Traps (251 Ma). The paleolatitude (λ) is calculated for a common reference locality (Oslo, Norway: 60°N, 10°E).
[a] [*Zijderveld*, 1970], [b] from Muttoni et al. [2003])
[c] Pole is the mean of three entries, which are (1) GPMDB-REFNO 2832 (59°N/150°E), (2) a combined pole (including REFNO 3486) at 53°N/154°E, given in Torsvik & Andersen [2002] and (3) from Gurevitch et al., in press (56°N/146°E)

(245–268 Ma; e.g., *Van der Voo* [1993]), but where improved Ar-Ar dating techniques (results reported by *Vlag et al.* [1997]) now yield an upper limit age of 264 $\pm$ 2 Ma for a dike cutting the flows, and an older limit of 278 $\pm$ 2 Ma for the lowermost basalt in the sequence. Because most of the paleomagnetic results come from flows higher in the section, we have used an age of 264 Ma for the Estérel entry in Table 1.

The general outcome of the newer ages is that results with a given paleopole location of, say 38°N, 165°E, are now included in the 290 or 300 Ma age bins, whereas before they were often included in the 260, 270 or 280 Ma age bins (compare Table 1 with *Torsvik et al.* [2001]; their Table 1b). Because Europe is steadily drifting northward during the Permian, the effect on its paleogeographic position is that for the extrapolated paleolatitude (10°N) from this result, the assigned age is now older (i.e., 292 Ma). Vice-versa, for a given younger age (e.g., 270 Ma) the paleolatitude will become more northerly with the removal (reassignment) of several lower-paleolatitude results to older age bins.

We test for this "age bias" in paleopoles and paleolatitudes, by comparing mean poles based on age-determination quality (all vs. AQ $\geq$ 2 vs. AQ = 3; note that the categories 'all results' or 'all volcanics' include AQ = 1). In Figure 2, this comparison results in higher paleolatitudes for the higher-quality age determinations for both 250 and 280 Ma. For 280 Ma, the mean paleopole and mean paleolatitude are shifted by 5° (~ 550 km). Compared to the most recent APW of Torsvik et al. [2001] paleolatitudes (using all poles) are systematically

higher (1–4 degrees, i.e. ~ 110–440 km) which is solely caused by our new age calibration in Table 1.

2.4. Octupole Bias

The effects of octupole fields are maximum at mid-latitudes and zero at the equator and the poles. If a zonal octupole field is of the same sign as the main dipole field, the effect is to lower the observed values at all intermediate paleolatitudes. As already noted, sedimentary inclination error, pdf overprints on reversed Kiaman magnetizations, and erroneously young age assignments all have the same biasing effect on paleolatitudes as such octupole fields. Evidence for octupole fields in the ancient geological past has recently been presented [*Kent and Smethurst*, 1998; *Van der Voo and Torsvik*, 2001; *Torsvik and Van der Voo*, 2002], but their existence has remained controversial [e.g., *Muttoni et al.*, 2003] which illustrates the difficulty of the detection methods. Because the best test for Permian octupole fields would require data from a broad range of paleolatitudes, which is not possible with European data from the limited north-south range of France to Norway, we have opted for a simpler test that compares a lower and a higher-paleolatitude dataset, both based on volcanic rocks only (mean poles are listed in Table 3). The former includes results between the equator and about 15°N, coming from France, Germany, Poland, the Czech Republic and the United Kingdom, collectively designated as "Central Europe". The higher-latitude data come from Scandinavia, i.e., Norway and Sweden. The test would be supportive of the existence of octupole fields if the higher-latitude data set would produce a mean paleopole that is far-sided, as seen from Europe, with respect to the mean paleopole for Central Europe. In other words, in the presence of octupole fields, the mean paleolatitude of the Scandinavian data, as calculated for Oslo, would be lower than that extrapolated to Oslo from the mean Central European paleopole.

In Figure 5 we can see that for 290 Ma, the Scandinavian result is the lower one, supportive of octupole fields. However, in the 280 Ma comparison the opposite is observed, indicating that for this time window the European data do not seem to require the presence of an octupole field.

Lastly, we have also compared the extrapolated paleolatitude from recent and very reliable results from the Siberian Traps (mean age 251 Ma; Table 3) with the mean 250 Ma European paleolatitudes of Figure 2, and found that the extrapolated paleolatitude from the Siberian Traps is 30°N, which is higher than any of the values obtained from western European results (see Table 2 at 250 Ma). This last test is therefore also not consistent with an octupole contribution for this time. Alternatively, suturing between Baltica and North Taimyr and Siberia may not have been complete by 250 Ma as evidenced

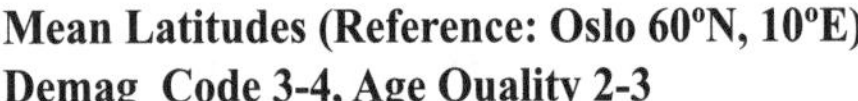

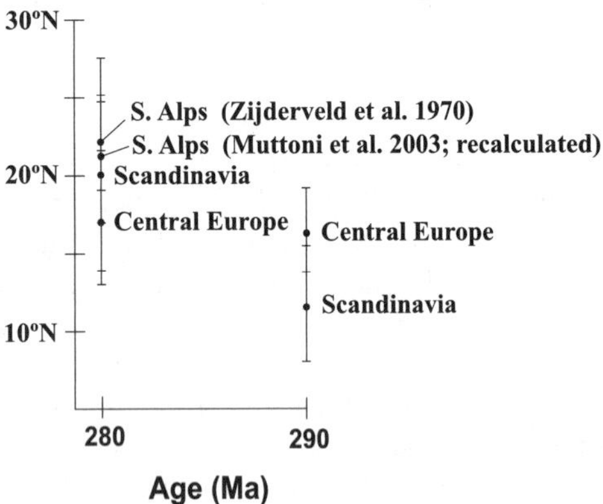

Figure 5. Estimated paleolatitudes for Europe (calculated for the Oslo reference locality at 60°N, 10°E) based on mean paleopoles for Central Europe and Scandinavia (see text and Table 3) and on mean paleopoles for the Southern Alps [*Zijderveld et al.,* 1970; *Muttoni et al.,* 2003], recalculated from the seven volcanic results in the Southern Alps, but excluding results from Morocco.

by post-Permian folding (notably affecting also the volcanics in South and Central Taimyr that are related to the Siberian Traps) along the Arctic margins of Siberia and Baltica [*Torsvik and Andersen* 2002].

3. INTERPRETATION AND IMPLICATIONS FOR PANGEA

The analyses just described illustrate that the Early Permian paleopole database from western Europe appears to contain several biases, many of which conspire to lower the paleolatitudes of the continent from what they should be. It must be noted that multiple causes may underlie an inclination anomaly in a given study; in other words, a too-shallow inclination in a given sedimentary rocks may have been caused by inclination error or by unrecognized overprints, or even by other factors such as erroneous age assignments or octupole fields. Because our analysis is purely statistical, the conclusions derived from our tests assume variables that are not necessarily independent.

Taking the extrapolated ("reference") paleolatitudes of Oslo as our point of departure in this discussion, we see that the values for 280 Ma (Table 2) vary between a low of 12° to a high of 23°N. The low value is not a peculiarity of our particular analysis. A recently published mean paleopole for 280 Ma [*Muttoni et al.,* 2003] predicts a paleolatitude at Oslo of 14°N, which is 9° lower than the high value of 23°N that we obtain using the best-dated, best-demagnetized volcanic results.

To visualize the paleolatitude differences, Figure 6 has been constructed for 250 and 280 Ma. Particularly for 280 Ma, the filtering makes a large difference. Numbers within each reconstruction represent the number of individual paleopoles that make up a particular mean, with the numbers of entries diminishing, of course, as the filtering threshold increases. As already noted the total range between lowest and highest paleolatitudes in the 280 Ma plots is about 10°.

For the 250±10 Ma interval, the paleolatitudes are much less variable, because the corresponding mean poles are all more or less equidistant from Europe (Fig. 1b). Also, it must be noted that far fewer entries are available for 250 Ma than for 280 Ma.

In Figure 6, we have included one additional paleolatitudinal position at 250 Ma for Eurasia, on the assumption that the mean Siberian Trap pole (Table 3) is representative of the entire continent. It can be seen that this results in a position that is about 5° more northerly for Europe. However, that this is still inadequate to solve the overlap created by maintaining the longitudinal positions of the continents in a Pangea A configuration [*Bullard et al.,* 1965], can be seen in Figure 7. Here we have used a mean Gondwanan paleopole for 250±10 Ma from a recently updated compilation [*Torsvik and Van der Voo,* 2002] at 48°N, 252°E (A$_{95}$ = 9, N = 13 poles); this mean pole is based on a conventional Fisherian calculation without any correcting for non-dipole fields, nor any filtering for age or demagnetization quality as done in this study. The reasons that we have refrained from attempting such filtering lie simply in the fact that Gondwana's database for the Late Permian-Early Triassic is very limited and that suspicions about the overall quality of almost all the results are frequently voiced [e.g., *Muttoni et al.,* 1996; *Clark and Lackie,* 2003]. The overlap in Figure 7 (~1000 km) is too large to be called statistically insignificant, and if—as we (and some others, e.g., *Lottes and Rowley* [1990]; *Muttoni et al.* [1996]; *Weil et al.* [2001]) believe—a latest Permian Pangea A configuration is deemed preferable over the Pangea B reconstruction, one must indeed resort to questioning the quality of the Gondwana data [*Clark and Lackie,* 2003] or the applicability of the geomagnetic 100%-dipole field model.

In contrast to the situation at 250 Ma, the paleolatitudinal overlap between Gondwana and Europe in a longitudinally constrained Pangea-A type fit at 280 Ma is significantly diminished if one accepts our filtering according to age and demagnetization quality criteria for results from volcanics only. Using the mean pole for Europe at 280 Ma at 51°N, 161°E (Table 2) and a mean pole at 30°N, 240°E (A$_{95}$ = 4, N = 14 poles, without correction for non-dipole fields or any filtering for quality [*Torsvik and Van der Voo,* 2002]) for Gondwana for the same age, we obtain a very acceptable Pangea A reconstruction, as illustrated in Figure 8. And even if we use a mean Gondwana pole based on the Moroccan results listed by Muttoni et al. [2003] at 37.5°N, 237.5°E, there is insignif-

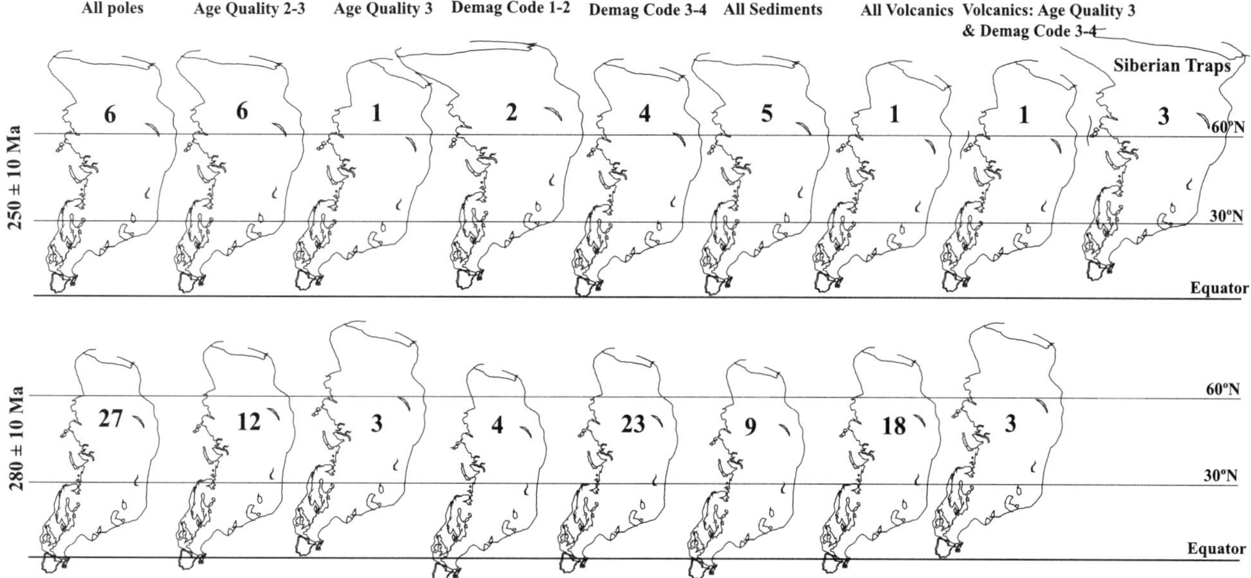

Figure 6. Reconstructions of Eurasia according to poles in Figure 1. Numbers = number of poles. Note that eastern Eurasia (Siberia, Kazakhstan etc.) was probably not fully amalgamated by 280 Ma. At 250 Ma we also show Eurasia reconstructed according to the mean pole from the Siberian Traps (Table 3).

icant overlap between North America and Europe (Laurussia) on the one hand, and Gondwana on the other. Thus, the Early Permian paleomagnetic results of Gondwana and Laurussia apparently can be reconciled with a Pangea-A type reconstruction, provided they are carefully selected from the results for volcanic rocks only using quality criteria for age control and sophistication in demagnetization, without invoking nondipole fields as was done earlier by us [*Van der Voo and Torsvik,* 2001].

Also shown in Figures 7 and 8 are the paleomagnetically determined positions of the Southern Alps, i.e., merely as based on mean paleopoles from Early and latest Permian rocks in northern Italy. The age of the widespread Early Permian volcanics in the Southern Alps is determined by U-Pb techniques as 276–284 Ma, as reported in Muttoni et al. [2003]. Muttoni and colleagues list paleopoles from a variety of sources; giving unit weight to each study, a mean paleopole is obtained as listed in Table 3. Also listed there is a mean pole for the same rocks, giving unit weight to site means [*Zijderveld et al.,* 1970]; the two mean poles are very similar. The paleolatitudes determined by these two Early Permian studies are about 10°N, and are used to locate the Southern Alps in Figure 8 (stars). The area fits comfortably at the western Tethyan margin, but it must be noted that the relative position of the Southern Alps with respect to Africa is different from that of today. Proposals for the position of the Southern Alps with respect to Gondwana are rather varied. Arguments for the Southern Alps as part of an Adria promontory of Africa have most recently been reinforced by Muttoni et al. [2003], but we

note that many other authors [e.g., *Dewey et al.,* 1973; *Dercourt et al.,* 1986; *Stampfli et al.,* 1998; *Stampfli and Borel,* 2002; *Vai* 2003] have argued for a position of the Southern Alps that is uncoupled from Africa. Needless to say that, if Southern Alpine paleopoles are taken as a proxy for Gondwana in the Early Permian, a Pangea-A type reconstruction inevitably produces an unacceptable overlap, even with the "improved" European mean pole of this study.

4. CONCLUSIONS

The main conclusion to come out of our analyses is that a significant bias appears to exist in the Early Permian (~280 Ma) paleopole data from western Europe, which displaces the paleolatitudinal position of this continent southward and thereby causes overlap in Pangea-A type continental reconstructions. Multiple evidence for this bias, generally all of the same sign, has been found through comparisons of (1) results from volcanic and sedimentary rocks, (2) results from formations with good age control vs. those with less well-defined ages, and (3) results based on more modern versus less sophisticated demagnetization techniques. The underlying causes are related to (1) inclination error in sedimentary rocks, (2) a systematic shift towards older ages with more modern geochronology, and (3) the suspected presence of a north and downward directed overprint that has not been satisfactorily eliminated by less sophisticated demagnetization.

The recognition that this bias exists changes the way the mean Early Permian paleopole is computed, because simply

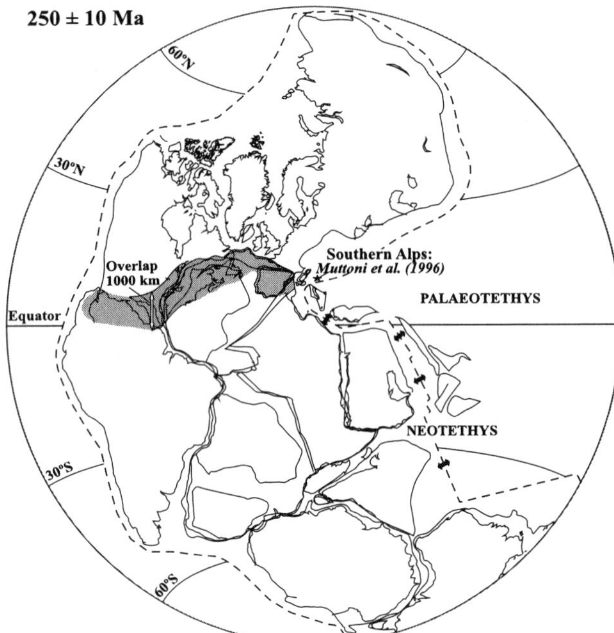

Figure 7. 250 Ma Pangea reconstruction, where Laurussia's position is based on the mean pole from the Siberian Traps (Table 3). Gondwana's position is based on Torsvik and Van der Voo [2002]. The result from the Siberian Traps produces the most northerly position for Eurasia, but still results in an overall continental overlap of 1000 km. The latitudinal position of the Southern Alps near the Permo-Triassic boundary (from *Muttoni et al.* [1996]) is indicated as a star. The Neotethys Ocean between Gondwana's Arabian margin and Cimmerian terranes had opened at about 265 Ma, and was well-developed by 250 Ma [*Stampfli and Borel,* 2002].

taking the average of all results must now clearly be seen as unsatisfactory. Establishing a-priori selection criteria with the three above-mentioned causes of bias in mind, avoids the perception that poles are selected subjectively to support a given argument. Yet, application of these selection criteria can make a difference of up to 10° in paleolatitude for the position of Europe at 280 Ma. On the other hand, we must acknowledge that a high degree of quality-filtering inevitably leads to a smaller dataset from which the continental position is derived. One cannot improve the confidence one has in the quality of a result, without the trade-off of a reduction of the number of entries that make up this mean pole.

The Permian paleolatitude of Europe matters, enormously, for Pangea reconstructions. With our newly calculated and highly filtered Early Permian paleopole for Europe, a Pangea-A type reconstruction has been obtained that is entirely compatible with the Gondwana paleopole (without any filtering or correcting), and which avoids overlap in a north-south sense (Fig. 8). Moreover, for 280 Ma, no octupole field contributions (as previously suggested by Van der Voo and Torsvik [2001]) need to be invoked, nor is there any sign of an octupole bias

internal to the 280 ± 10 Ma paleopole dataset from western Europe. Our Pangea-A reconstruction, in contrast to Pangea-B type reconstructions [e.g. *Muttoni et al.,* 2003], also makes geological sense since the Variscan suturing between Gondwana and Laurussia had terminated by the Early Permian. The main Variscan orogeny (peak metamorphism at ~300 Ma) ended by the Late Carboniferous and metamorphic terranes were unroofed before deposition of latest Carboniferous and Early Permian strata [*Matte,* 2001].

For 250 ± 10 Ma, the situation is somewhat different. Within western Europe, evidence for bias has been found only to a very minor extent in lower-age quality and sedimentary results, whereas for these mostly mixed-polarity paleopoles a lower demagnetization quality does not translate into lower paleo-

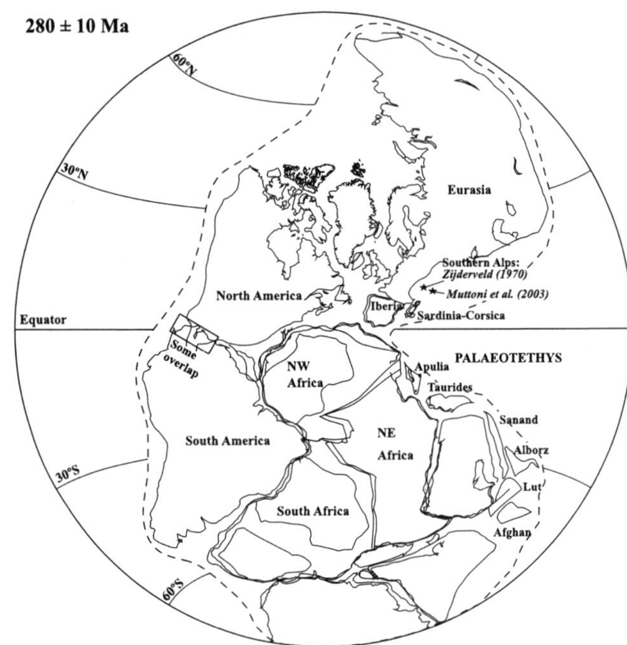

Figure 8. 280 Ma Pangea reconstruction, where Laurussia's position is based on Europe's volcanics-only result with age quality 3 and demag code 3–4. Gondwana's position is based on the running mean pole of Torsvik and Van der Voo [2002]. Note that this allows an almost perfect Pangea A reconstruction except for a minor overlap of South America and Mexican terranes. We show the position of the Southern Alps (stars) based on two different poles (Table 3) that both yield low northerly paleolatitudes. The Southern Alps were probably originally a peri-Gondwanan terrane, but may have rifted off the NW Gondwana margin during the early Devonian together with the Armorica and Iberian massifs and Sardinia-Corsica. The Paleotethys opened behind them [*Torsvik and Cocks,* 2004]. The stippled line shows the approximate size of Pangea. Several terranes in the eastern Paleothethys (e.g. South and North China; not shown in the diagram) were not part of Pangea at this time. In addition eastern Eurasia (Siberia, Kazakhstan etc.) was probably not fully amalgamated by this time.

latitude (as it did for the reversed Kiaman results). This is not surprising, because an overprint on normal-polarity rocks steepens the inclination and increases the paleolatitude. Taking a mean 251 Ma paleopole obtained from the Siberian Traps as representative of western Europe, a 5°-higher paleolatitudinal position is achieved, but this is not sufficient to eliminate an overlap in a longitudinally constrained Pangea-A type reconstruction. To avoid all overlap, one must call into question the quality of the 250±10 Ma paleopole results from all Atlantic-bordering continents, or alternatively, resort once again to octupole contributions to the total geomagnetic field. However, a comparison between Siberian and European 250±10 Ma paleomagnetic poles is not indicative of the presence of octupole fields.

In contrast, at 290±10 Ma, the Scandinavian paleomagnetic directions are too shallow, compared to the results from western Europe. The latter span a paleolatitude band of 0–15°N, and come from France, Germany, the Czech Republic, Poland and the U.K. This comparison between Scandinavian and central-European paleolatitudes could be taken as support for octupole field contributions.

Collectively, however, the evidence for octupole fields from the western European Permo-Triassic dataset is ambiguous. So, while the 'cachet' of an octupole bias may be diminishing, it appears more likely now that careful attention to data quality is the key to avoiding such tectonic conundrums as the paleomagnetic need for Pangea B. The 250±10 Ma dataset is a clear example in need of careful inspection and new investigations in the Gondwana realm; this was recognized already earlier by Muttoni et al. [1996], who argued that no reliable Late Permian—Middle Triassic data were available for West Gondwana.

Acknowledgments. We thank Eric Tohver and Phil McCausland for valuable comments on the manuscript and three anonymous colleagues for their reviews. We also thank Martin Timmerman and Elizabeth Eide for supplying unpublished ^{40}Ar/^{39}Ar ages. This study was supported by grant EAR 9909231 (to RVdV) from the Earth Sciences Division of the National Science Foundation.

REFERENCES

Arnaud, N. O., M. Brunel, J. M. Cantagrel, P. Tapponnier, High cooling and denudation rates at Kongur Shan, eastern Pamir (Xinjiang, China) revealed by ^{40}Ar/^{39}Ar alkali feldspar thermochronology, *Tectonics,* 12, 1335–1346, 1993.

Bullard, E. C., J. E. Everett, and A. G. Smith, A symposium on continental drift IV. The fit of the continents around the Atlantic, *Phil Trans Roy. Soc. London,* A 258, 41–51, 1965.

Clark, D. A., and M.A. Lackie, Palaeomagnetism of the Early Permian Mount Leyshon Intrusive Complex and Tuckers Igneous Complex, North Queensland, Australia, *Geophys. J. Int.,* 153, 523–547, 2003.

Dercourt, J., et al., Geological evolution of the Tethys belt from the Atlantic to the Pamirs since the Lias, *Tectonophysics,* 123, 241–315, 1986.

Dewey, J. F., W.C. Pitman III, W. B. F. Ryan, and J. Bonnin, Plate tec tonics and the evolution of the Alpine system, *Geol. Soc. America Bull.,* 84, 3137–3180, 1973.

Edwards, R. A., G. Warrington, R. C. Scrivener, N. S. Jones, H. W. Haslam, and L. Ault, The Exeter group, south Devon, England: a contribution to the early post-Variscan stratigraphy of northwest Europe, *Geol. Mag.,* 134, 177–197, 1997.

Eide, E. A., P. T. Osmundsen, G.B. Meyer, M.A. Kendrick, and F. Corfu, The Nesna Shear Zone, N-central Norway: an ^{40}Ar/^{39}Ar record of Early Devonian-Early Carboniferous ductile extension and unroofing. *Norw. J. Geol.,* 82, 317–339, 2002.

Fossen, H., and W. J. Dunlap, On the age and tectonic significance of Permo-Triassic dikes in the Bergen-Sunnhordland region, southwestern Norway, *Nor. Geol. Tidskr.,* 79, 169–177, 1999.

Gurevitch, E. L., C. Heunemannn, V. Rad'ko, M. Westphal, V. Bachtadse, J. P. Pozzi, and H. Feinberg, Palaeomagnetism and magnetostratigraphy of the Permian-Triassic Siberian trap basalts, *Tectonophysics* (in press).

Hallam, A., Supposed Permo-Triassic megashear between Laurasia and Gondwana, *Nature,* 301, 499–502, 1983.

Kent, D. V., and M. A. Smethurst, Shallow bias of paleomagnetic inclinations in the Paleozoic and Precambrian, *Earth Planet. Sci. Lett.,* 160, 391–402, 1998.

Lippolt, H. J., J. C. Hess, I. Raczek, and V. Venzlaff, Isotopic evidence for the stratigraphic position of the Saar-Nahe Rotliegendes volcanism, II. Rb-Sr investigations, *N. Jahrb. Geol. Paleontol. Monatsh.,* 9, 539–552, 1989.

Lottes, A. L. and D. B. Rowley, Reconstruction of the Laurasian and Gondwana segments of Permian Pangaea, *in Palaeozoic Palaeogeography and Biogeography,* edited by W. S. McKerrow and C. R. Scotese, Geol. Soc. Lond. Mem., 12, 383–395, 1990.

Matte, P., The Variscan collage and orogeny (480–290 Ma) and the tectonic definition of the Armorica microplate: a review, *Terra Nova,* 13, 122–128, 2001.

McElhinny, M. W. and J. Lock, IAGA paleomagnetic databases with ACCESS, *Surv. Geophysics,* 17, 575–591, 1996.

McElhinny, M. W. and M. A. Smethurst, The global paleomagnetic database: current status, *EOS,* 82 (39), 436, 2001.

Menning, M., A numerical time scale for the Permian and Triassic periods: An integrated time analysis, *in The Permian of northern Pangea,* vol. 1, edited by P. A. Scholle, T. M. Peryt, and D. S. Ulmer-Scholle, pp. 77–97, Springer-Verlag, Berlin, 1995.

Mulder, F. G., Paleomagnetic results from Irish Carboniferous and Triassic rocks, *Scient. Proc. Roy. Dublin Soc.,* A4, 343–349, 1972.

Muttoni, G., D. V. Kent, and J. E. T. Channell, Evolution of Pangea: Paleomagnetic constraints from the Southern Alps, Italy, *Earth Planet. Sci. Lett.,* 140, 97–112, 1996.

Muttoni, G., D. V. Kent, E. Garzanti, P. Brack, N. Abrahamsen, and M. Gaetani, Early Permian Pangea 'B' to Late Permian Pangea 'A', *Earth Planet. Sci. Lett.,* 215, 379–394, 2003.

Rochette, P., and D. Vandamme, Pangea B: An artifact of incorrect paleomagnetic assumptions? *Annali di Geofisica,* 44, 649–658, 2001.

Stampfli, G. M., and G. D. Borel, A plate tectonic model for the Paleozoic and Mesozoic constrained by dynamic plate boundaries and restored synthetic oceanic isochrons, *Earth Planet. Sci. Lett.,* 196, 17–33, 2002.

Stampfli, G. M., J. Mosar, D. Marquer, R. Marchant, T. Baudin, and G. Borel, Subduction and obduction processes in the Swiss Alps, *Tectonophysics,* 296, 159–204, 1998.

Torsvik, T. H., and T. B. Andersen, The Taimyr Fold Belt, Arctic Siberia: Timing of pre-fold remagnetisation and regional tectonics, *Tectonophysics,* 352, 335–348, 2002.

Torsvik, T. H., and L. R. M. Cocks, Earth history 400–250 Ma: A palaeomagnetic, fauna and facies review, *J. Geol. Soc. Lond.* (in press), 2004.

Torsvik, T. H., and R. Van der Voo, Refining Gondwana and Pangea paleogeography: Estimates of Phanerozoic non-dipole (octupole) fields, *Geophys. J. Intern.,* 151, 771–794, 2002.

Torsvik, T. H., E. A. Eide, J. G. Meert, M. A. Smethurst, and H. J. Walderhaug, The Oslo Rift: New palaeomagnetic and [40]Ar/[39]Ar age constraints, *J. Geophys. Int.,* 135, 1045–1059, 1998.

Torsvik, T. H., R. Van der Voo, J. G. Meert, J. Mosar, and H. J. Walderhaug, Reconstructions of the continents around the North Atlantic at about the 60th parallel, *Earth Planet. Sci. Lett.,* 187, 55–69, 2001.

Vai, G. B., Development of the palaeogeography of Pangaea from Late Carboniferous to Early Permian, *Palaeogeogr., Palaeoclimatol., Palaeoecol.,* 196, 125–155, 2003.

Van der Voo, R., *Paleomagnetism of the Atlantic, Tethys and Iapetus Oceans,* 411 pp, Cambridge University Press, Cambridge, 1993.

Van der Voo, R., and T. H. Torsvik, Evidence for late Paleozoic and Mesozoic non-dipole fields provides an explanation for the Pangea reconstruction problems, *Earth Planet. Sci. Lettt.,* 187, 71–81, 2001.

Van Eysinga, F. W. B. (compiler), *Geological time table, 3rd edition,* Elsevier, Amsterdam, 1975.

Vlag, P., D. Vandamme, P. Rochette, and C. Spinelli, Paleomagnetism of the Esterel rocks: A revisit 22 years after the thesis of Hans Zijderveld, *Geologie Mijnb.,* 76, 21–33, 1997.

Walderhaug, H. J., Rock magnetic and magnetic fabric variations across three thin alkaline dikes from Sunnhordland, W. Norway: Influence of initial mineralogy and secondary alterations, *Geophys. J. Intern.,* 115, 97–108, 1993.

Weil, A. B., R. Van der Voo and B. A. van der Pluijm, Oroclinal bending and evidence against the Pangea megashear: The Cantabria-Asturias Arc (northern Spain), *Geology,* 29, 991–994, 2001.

Zijderveld, J. D. A., *Paleomagnetism of the Esterel rocks,* PhD thesis Utrecht University, Utrecht, 199 pp., 1975.

Zijderveld, J. D. A., G. J. A. Hazeu, M. Nardin, and R. Van der Voo, Shear in the Tethys and the Permian paleomagnetism in the Southern Alps, including new results, *Tectonophysics,* 10, 639–661, 1970.

T. H. Torsvik, VISTA, c/o Geodynamic Centre, Leiv Eriksonsvei 39, N-7491 Trondheim, Norway. (trond.torsvik@ngu.no)

R. Van der Voo, Department of Geological Sciences, University of Michigan, Ann Arbor, Michigan 48109. (voo@umich.edu)

On the Origin and Distribution of Magnolias: Tectonics, DNA and Climate Change

R. J. Hebda

Royal British Columbia Museum, and University of Victoria, Victoria, British Columbia, Canada

E. Irving

Pacific Geoscience Center, Geological Survey of Canada, North Saanich, British Columbia, Canada

Extant magnolias have a classic disjunct distribution in southeast Asia and in the Americas between Canada and Brazil, and nowhere in between. Of the 17 sections (about 210 species) in two subgenera, only two, *Tulipastrum* and *Rhytidospermum*, are truly disjunct. Molecular analyses reveal that several North American species are basal forms suggesting that magnolias originated in North America, as indicated by their fossil record. We recognize four elements in their evolution. (1) Ancestral magnolias originated in the Late Cretaceous of North America in high mid-latitudes (45°–60°N) at low altitudes in a greenhouse climate. (2) During the exceptionally warm climate of the Eocene, magnolias spread eastwards, via the Disko Island and Thulean isthmuses, first to Europe, and then across Asia, still at low altitudes and high mid-latitudes. (3) With mid-Cenozoic global cooling, they shifted to lower mid-latitudes (30°–45°N), becoming extinct in Europe and southern Siberia, dividing a once continuous distribution into two, centred in eastern Asia and in North America. (4) In the late Cenozoic, as ice-house conditions developed, magnolias migrated southward from both centres into moist warm temperate upland sites in the newly uplifted mountains ranges of South and Central America, southeast Asia, and the High Archipelago, where they diversified. Thus the late Cenozoic evolution of magnolias is characterized by impoverishment of northern and diversification of southern species, the latter being driven by a combination of high relief and climate oscillations, and neither of the present centers of diversity is the center of origin. Disjunction at the generic level and within section *Tulipastrum* likely occurred as part of the general mid-Cenozoic southward displacement assisted by the development of north-south water barriers, especially the Turgai Strait across western Siberia. Disjunction in section *Rhytidospermum* could be Neogene.

"that grand subject, that almost keystone of the laws of creation, geographical distribution" (Charles Darwin in a letter to Joseph Hooker ca 1860 quoted in *Good* [1974].)

Timescales of the Paleomagnetic Field
Geophysical Monograph Series 145
Copyright 2004 by the American Geophysical Union
10.1029/145GM04

1. INTRODUCTION

For his Ph.D. (1956–59) Neil Opdyke studied the relationships between paleomagnetically determined latitudes and paleoclimates as inferred from temperature-sensitive deposits and from wind directions in eolian sandstones [*Opdyke*, 1961a,b]. By placing such geological deposits at their original latitude, he was able to discuss possible past changes in climatic zones and in the positions of continents. Being able to do so is key to unraveling the origin and paleogeographical dispersal of plant groups and individual species, in a way not available to earlier phytogeographers such as Cain [1944] and Good [1974]. In our contribution to this volume in Neil's honor, we take advantage of advances made since his early work to re-examine the paleogeography of the genus *Magnolia sensu lato*. Our contribution is an example of how the use of the geocentric axial dipole model of the geomagnetic field, which Neil Opdyke helped to establish and continues to study, provides a firm basis for estimating ancient latitude and hence for understanding the paleobiogeography of this important group of flowering plants. We use "magnolia" in the broad sense to include all fossil and extant members of the Magnoliaceae except *Liriodendron* and its direct fossil ancestors.

The genus *Magnolia* was described by Linnaeus in 1737 from material collected in what were then colonies of British North America. The type species is *Magnolia virginiana L.*, the laurel magnolia or sweet bay [*Fernald*, 1959]. Magnolias are members of the Paleogene "boreotropical" flora, as described by many workers [*Tiffney*, 1985 and references therein]. Here we focus on the evolution of a single genus within that larger flora.

Magnolias flourished throughout the Cenozoic, and there are ancestral forms in the Late Cretaceous Epoch connecting them to one of the oldest lineages of flowering plants [*Delevoryas and Mickle*, 1995; *Crane*, 1996]. Viewed as a whole, extant magnolias have a classic disjunct distribution; they occur only in the Americas and in southeastern Asia (Figure 1a). Yet, as a group, there is little doubt that they have a common ancestry, so their past distribution must have been very different. Our purpose is to re-examine old questions,— where did magnolias originate and how did they achieve their modern distribution?

As a consequence of their remarkable geographical distribution, their horticultural value, and their purported, near-basal position in the evolution of flowering plants, magnolias have long been of central interest in taxonomic inquiry and the study of plant geography [for example *Hutchinson*, 1969; *Takhtajan*, 1969; *Tiffney*, 1977; *Stewart and Rothwell*, 1993; *Crane, 1996*; *Manchester*, 1999]. Recent studies of DNA [*Qiu et al.*, 1995; *Azuma et al.*, 2001; *Kim et al.*, 2001], the fossil record, and paleogeography, in which we include paleoclimates, make it possible to examine the distribution of magnolias from the perspective of their long-term history and its relationship to topographic and climatic changes. From this examination we propose a four stage model by which their past and present distribution can be understood.

We acknowledge the important studies and ideas of others [e.g. *Tiffney*, 1977, 1985, 2000; *Figlar*, 1993, *Crane, 1996*;

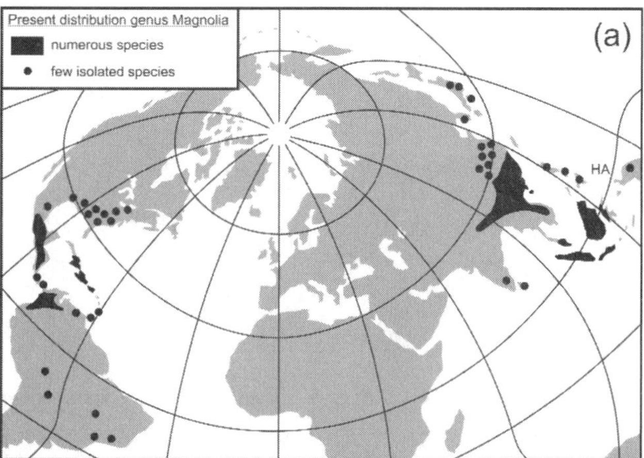

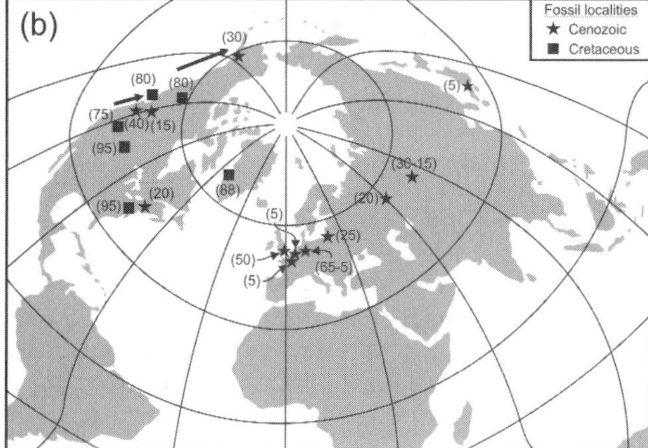

Figure 1. (a) Occurrences of extant magnolias. (b) Well established fossil occurrences of Cenozoic magnolias and Cretaceous ancestral taxa. As discussed in the text, there are other occurrences of fossil leaves and pollen that are not definitively magnolia and are therefore not shown, with the exception of the Oligocene Alaskan occurrence which is included because of its transported character. Numbers in parentheses are approximate ages in millions of years.

Manchester, 1999; *Azuma et al.*, 2001; *Manchester and Tiffney*, 2001; *Tiffney and Manchester*, 2001] and integrate them into our considerations. By bringing together paleogeography and molecular biology in the context of our previous ideas about the origin and evolution of rhododendrons [*Irving and Hebda*, 1993], we believe that advances can be made. In particular, we assume, as we did in our rhododendron study, that, by-and-large, magnolia species share adaptations (morphologic, physiologic and other) to a broad but definable macroclimate. In other words, coded into the genetic material of the majority of magnolia species, there is a common ability to persist and thrive under a broadly similar range of climates. Nowadays they largely inhabit moist, warm temperate habitats, which occur at low altitudes in low mid-latitudes and at higher altitudes in low latitudes. We assume that this was always their prefer-

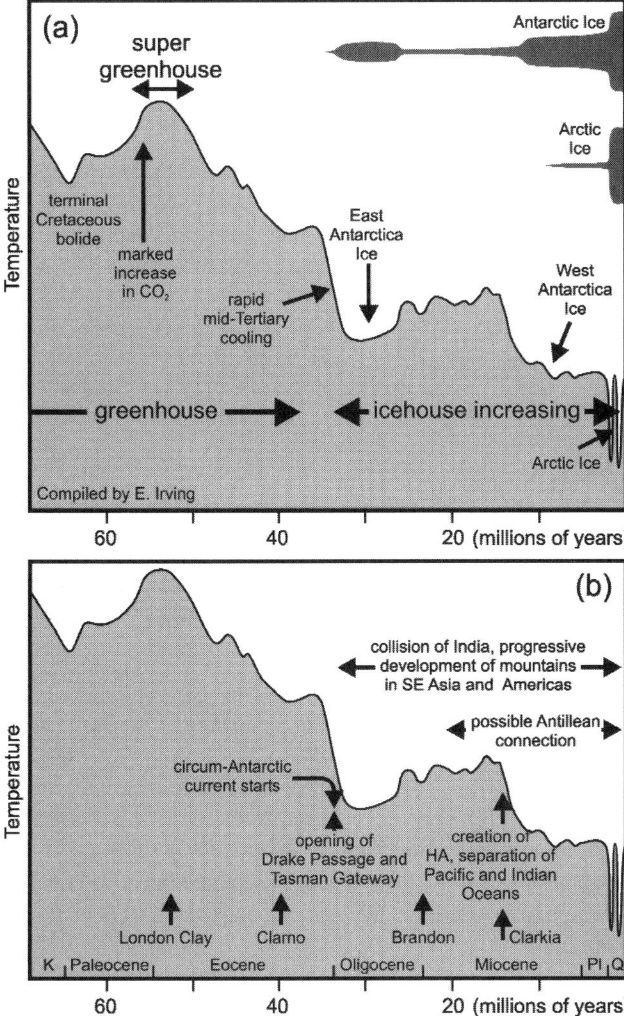

Figure 2. Climatic (a) and paleogeographical (b) events during the past 70 Ma. HA = High Archipelago, Pl = Pliocene, Q = Quaternary. Compiled fby E. I. from various sources, notably Davey et al. [2001] Graham [1999] and Zachos et al. [2001]. London Clay, Clarno, Brandon and Clarkia are notable fossil magnolia localities. Variation in mid-latitudes of mean sea-level temperature much simplified.

ence. Based on this assumption we can then estimate the past distribution, not only from fossil occurrences [*Tiffney*, 1977; *Azuma et al.*, 2001 and references therein], but also, with less precision but greater generality, from the past distribution of congenial habitats as inferred from paleogeography.

Continental drift cannot be the sole factor in determining the distribution of extant magnolias, because reconstructing continents for Late Cretaceous and Paleogene Epochs does not bring together the group's two principal zones of modern occurrence. Thus, we make use of previous ideas concerning large changes [*Savin*, 1977] in the past distribution of moist warm temperate climate, and in the development of upland

habitat (see for instance Myers [2003]). These changes are linked to the mid- and Late Cenozoic evolution of global climate from a non-glacial (greenhouse) to a glacial (ice-house) regime [*Prothero et al.*, 2003], and to the contemporaneous creation of regions of moist, warm temperate climate in the newly uplifted mountains of the Americas, southeast Asia, and the High Archipelago (Philippines, Indonesia and Malaysia) between southeast Asia and Australia (Figure 2b, HA). We use the nomenclature in Figure 3 for describing climatic conditions and latitudes.

2. TAXONOMY OF MODERN MAGNOLIAS

Until recently, magnolias in the broad sense (subfamily Magnolioideae in the family Magnoliaceae) were divided among about half a dozen genera based on morphology and geographic isolation [*Nooteboom*, 1993]. There was much discussion concerning whether all these should be recognized as separate genera, the relationships among them, and the assignment of species to different genera [*Callaway*, 1994; *Nooteboom*, 1996, 2000; *Gardiner*, 2000]. Recent studies of the position of flowers, the presence and absence of floral stipes [*Figlar*, 2000], and especially of chloroplast DNA (cpDNA), have clarified relationships within the genus. Specifically, Figlar's study showed that morphological differences, once thought to separate *Magnolia* and *Michelia* are not valid, and cpDNA studies demonstrated that the variation within the genus *Magnolia*, as it was originally narrowly accepted, is greater than the variation among closely related genera [*Azuma*

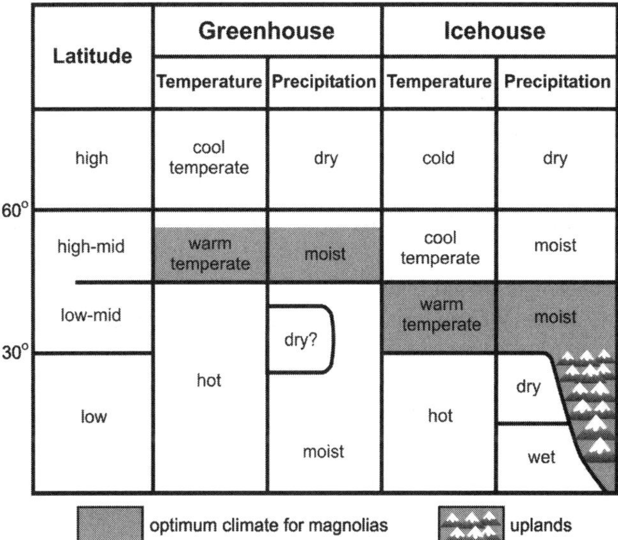

Latitude	Greenhouse		Icehouse	
	Temperature	Precipitation	Temperature	Precipitation
high	cool temperate	dry	cold	dry
high-mid	warm temperate	moist	cool temperate	moist
low-mid	hot	dry?	warm temperate	moist
low		moist	hot	dry / wet

Figure 3. Climate and latitude model used in our discussion. Note that the latitudes of climate zones are different for icehouse and greenhouse regimes, and that the term "tropical" is not used.

Table 1. Sections, numbers (in brackets) and distribution of *Magnolia* species. Compiled with modifications from Magnolia Society website [*Figlar*, 2001].

Section	Distribution
Subgenus *Magnolia* (130)	
Gwillimia (17)	southern China to Indonesia and Philippines
Talauma (31)	Mexico, Caribbean, to Brazil
Magnolia (1)	southeast U.S.A.
Theorhodon (34)	southeast U.S.A., Caribbean, Mexico to Ecuador
Rhytidospermum (6)	Japan, China, southeast U.S.A.
Oyama (3)	Japan, China, Himalayas
Mangletia (29)	southeast and south Asia, Indonesia, Borneo
Kmeria (3)	southeast Asia
Gynopodium (6) [a]	Taiwan, south China
Subgenus *Yulania* (81)	
Yulania (8) [b]	Japan, south China, Himalayas
Cylindricae (1)	east China
Buergeria (4)	Japan, Korea, east China
Tulipastrum (2)	eastern North America, east China
Michelia (51)	Japan, southeast and south Asia, Indonesia, Borneo
Elmerillia (4)	Philippines, Indonesia, New Guinea
Maingola (8)	south Asia, Indonesia, Borneo
Aromadendron (5)	Indonesia, Borneo, Philippines
Genus *Pachylarnax* (2)	south and southeast Asia

[a] includes section *Mangletiastrum* [b] includes *M. liliiflora*

et al., 2001; *Kim et al.*, 2001]. For example, species in the old genus *Michelia* cluster closely with *Magnolia* species [*Azuma et al.*, 2001; *Kim et al.*, 2001]. On the other hand, several species in the old section Rhytidospermum, long considered to be indisputably magnolias, are, in fact, not closely related. Also, from the perspective of cpDNA, western hemisphere, especially North American species in the section *Rhytidospermum*, appear to be the most diverse and basal in the cpDNA tree, suggesting that they could be relicts of the ancestral stock. The distinctness of the genus *Liriodendron* is, however, upheld by all these studies.

Taking into account this new evidence, the Magnolia Society [*Figlar*, 2001 and *Figlar and Nooteboom*, 2004] has recently proposed a new classification, according to which the taxa in the subfamily are now grouped into a single genus *Magnolia*, with two subgenera and 17 sections (Table 1). The 210 or so species are unevenly distributed among the sections, one (*Michelia*) having 51 species, others only one (section *Magnolia* for example).

For our analysis, the level at which the natural entities or taxa are recognized is not especially critical. More important is the recognition of the natural groupings within *Magnolia*, because we can reasonably assume them to be genetic entities with a common place of origin. For our purposes, the recent molecular studies are particularly useful because they identify what appear to be the few true disjunctions, that is, closely related species that are now widely separated geographically. These groups and species are of particular interest in our examination.

3. GEOGRAPHICAL DISTRIBUTION OF MAGNOLIAS BY SECTION

In the Magnolia Society classification [*Figlar*, 2001], there are two main groups, the subgenera *Magnolia* and *Yulania*. These have long been recognized (see Callaway [1994]), though recent studies [*Azuma et al.*, 2001 for example] have changed the composition of groups, by including species once included in other genera such as *Michelia*.

Subgenus *Magnolia* is large, with 9 sections (Table 1), including both evergreen and deciduous species. It is well represented in the Americas, southeast Asia and the High Archipelago (Figures 4a–c).

Sections *Talauma* of the Americas, and *Gwillimia* of Asia consist of evergreen species, living at latitudes of less than 30° and in lowland to mountain sites [*Callaway*, 1994] (Figure 4a). Though these two geographically separated sections were once thought to be related, molecular studies demonstrate that they are not partners of a disjunct pair in the strict sense. Five sections, each occurring in only the Old or New

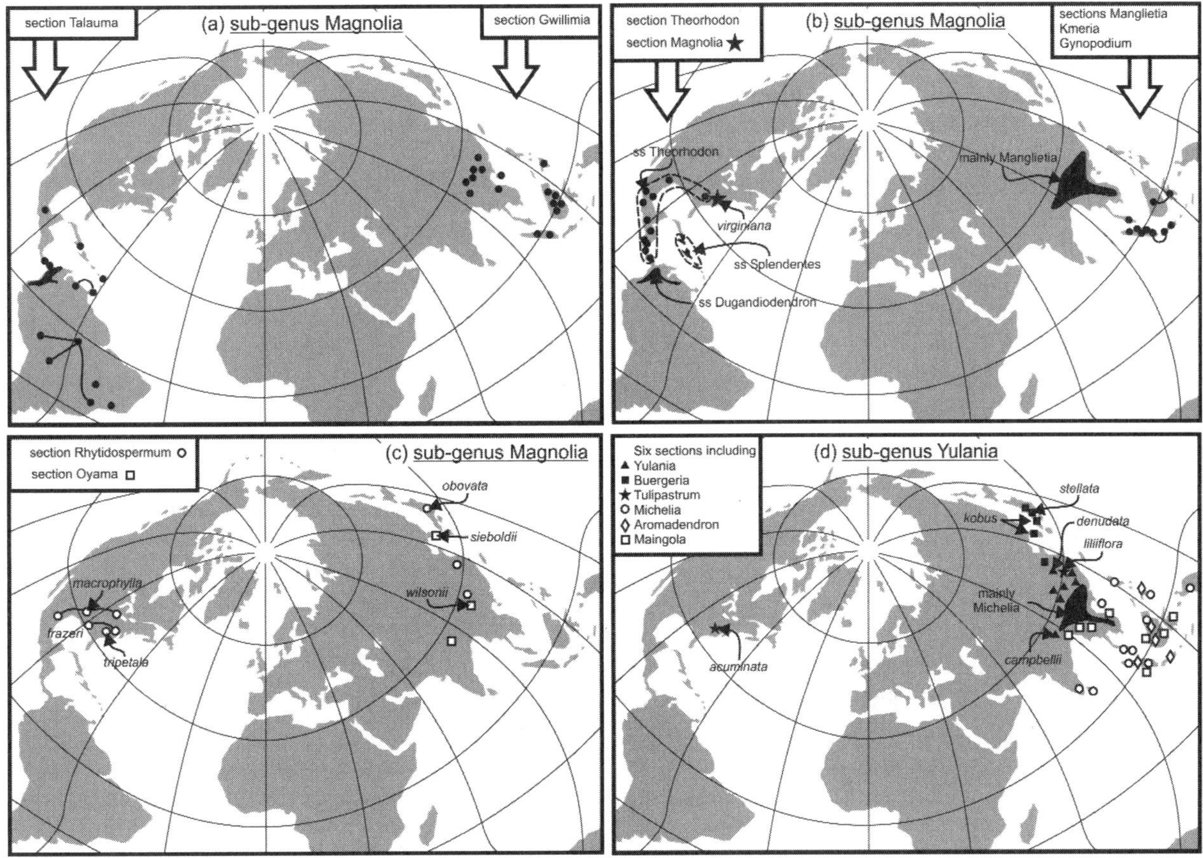

Figure 4. Distribution of extant magnolias by section. Subgenus *Magnolia* (a to c), subgenus *Yulania* (d). Some small sections are omitted. See complete list, Table 1. Where practical, occurrences of individual species are marked. Where there are too many species, the general area is indicated. Widespread occurrences of the same species are linked by lines. Compiled from Figlar [2001].

World, are not disjunct (Figure 4b). Three Asiatic sections, *Manglietia, Kmeria* and *Gynopodium* occur mainly in low-latitude upland habitats; and are mostly evergreen. Section *Theorhodon*, from the Americas, consists mostly of large evergreen trees living in moist mountains settings. In the Magnolia Society classification, the old genus *Dugandiodendron* of Columbia and Venezuela is now considered a sub-section within *Theorhodon*. The other American section (section *Magnolia*) has only one species, *M. virginiana*, which is deciduous to evergreen, and occurs in mid-latitudes mainly at low altitude, and which, according to cpDNA is related to species in section *Theorhodon* [*Azuma et al.*, 2001; *Kim et al.*, 2001].

The small sections *Rhytidospermum* and *Oyama* of the sub-genus *Magnolia*, whose constituent species have been identified through DNA studies [*Qiu et al.*, 1995; *Azuma et al.*, 2001; *Kim et al.*, 2001], are molecularly very diverse. Section *Rhytidospermum*, all members of which are deciduous trees or shrubs, has a truly disjunct distribution with the North American species *M. tripetala* (umbrella tree) closely related to

Asian species (Figure 4c). *M. macrophylla* (great-leaved magnolia), which ranges widely from Ohio to eastern Mexico, is particularly interesting because its cpDNA characteristics place it near the bottom of the magnolia evolutionary tree [*Kim et al.*, 2001]. The strictly Asian and relatively northern (Japan to Nepal) section *Oyama* includes only three, mostly deciduous, species, and is a source of many horticultural forms.

The subgenus *Yulania* has fewer species than subgenus *Magnolia* (Table 1). Except for a single species from eastern North America, it occurs only in Asia, where it ranges from Japan to the Himalayas and the High Archipelago (Figure 4d). Within *Yulania*, there are eight sections falling into two broad groups; a large one of upland, low latitude, evergreen trees, and a small one of northern deciduous shrubs and small trees. The largest section, *Michelia* (which in the new classification includes species previously grouped in the former genus *Michelia*), consists of evergreen trees and ranges from southern China south into the High Archipelago. Species once placed in the genera *Elmerillia* and *Aromadendron* now have their own sections (Table 1). Based on molecular studies [*Kim et al.*, 2001], the largely low latitude section *Maingola* can now be included in the subgenus *Yulania*. There is also a small group of mid-latitude, lowland taxa, mainly from northern China and Japan belonging to the section *Buergeria*. In the small section *Tulipastrum*, the single North American species (*M. acuminata*) is disjunct, being somewhat related to *M. liliiflora* from China [*Azuma et al., 2001*]. Though more closely related to Yulanias than to other magnolias, *M. acuminata* is the most distinct within the group, based on cpDNA analyses [*Kim et al.*, 2001], and it hybridizes with other Yulanias [*Callaway*, 1994]. Parents of the widely grown, spring-flowering hybrids, *Magnolia* x *soulangeana*, come from the subgenus *Yulania*.

In summary, the genus as a whole is disjunctive (Figure 1a), but, of the 17 sections, only two can be said to have "true" continental disjuncts, and both have few species.

4. GEOGRAPHIC DISTRIBUTION OF FOSSILS

The fossil record of a group of plants documents its minimum distribution in time and space. Fossils also provide information concerning the chronological order of the evolution of critical features and lineages, thereby imposing a true-time framework onto molecular or phylogenetic trees [*Azuma et al.*, 2001; *Manchester*, 1999; *Manchester and Tiffney*, 2001; *Tiffney and Manchester*, 2001]. Fossil seeds and fructifications are the most diagnostic and reliably identified. Leaves are less easy to identify confidently, particularly leaves of Cretaceous and Paleogene age. Pollen occurs infrequently, and, as with leaves, can be mistaken for that of other types of plants [*Sun et al.*, 1995]; fossil occurrences of leaves or pollen only are not, therefore, given much weight here.

Compared to many other groups of plants, the fossil record of Magnoliaceae is relatively good [*Mai*, 1971; *Peigler*, 1989, *Tiffney*, 1977; *Crane*, 1996; and *Azuma et al.*, 2001]. In particular, there are unambiguous records of seeds and fructifications [*Azuma et al.*, 2001 and references therein]. Even the early history of the group is accessible through Cretaceous fructifications [*Dilcher and Crane*, 1984; *Delevoryas and Mickle*, 1995; *Crane*, 1996]. Leaves occur widely too, and some have been related closely to extant species [*Figlar*, 1993; *Baranova and Figlar*, 2000].

The two regions where magnolias grow today are crudely linked geographically by a broad belt of well established fossil localities (Figure 1). The belt stretches across the northern hemisphere from western to eastern North America, Greenland, through northern Europe, Russia, southern Siberia to eastern and southeastern Asia, and is broken only by the Labrador Sea and the Atlantic Ocean.

The earliest magnolias, presumed ancestors of Magnoliaceae in the strict sense, are of Cretaceous age and occur exclusively in North America and Greenland. These fossils include fruits, leaves and other remains from the ~95 Ma Dakota Sandstone of Kansas [*Dilcher and Crane*, 1984], a ~90 Ma fruit from western Greenland [*Crane*, 1996], a ~80 Ma fruiting axis from the Nanaimo Group of Vancouver Island, British Columbia [*Delevoryas and Mickle*, 1995], and ~ 90 Ma reproductive structures from the Late Cretaceous of New Jersey [*Crane*, 1996]. The Vancouver Island occurrence may not originally have been located where it is today. Paleomagnetic evidence indicates that Vancouver Island, at the time, was situated 2000 km or more further south (arrow in figure 5(a)) along the western seaboard of North America [*Irving et al.*, 1996; *Enkin et al.*, 2002; 2003; *Johnston*, 2001]. By contrast to the exclusively North American distribution of potentially *Magnolia*-related fossils, *Liriodendron*-related seeds appear to occur more widely in the Cretaceous of the northern hemisphere ranging as far east as Kazakhstan [*Crane*, 1996]. True *Liriodendron* seeds are not known prior to the Oligocene Epoch [*Manchester*, 1999]

Unambiguous Cenozoic *Magnolia* fossils are abundant and widespread [*Azuma et al.*, 2001]. Within a single region, there is often a range of ages, but, including the Cretaceous ancestral forms, maximum ages are oldest in North America (~95 Ma) and youngest in eastern Asia (~5 Ma), with northern Europe (~60-5 Ma) and southern Siberia (~20 Ma) being intermediate. A late Eocene to early Oligocene leaf occurrence in Tibet [*Tao*, 1988a, 1988b], if verified, would place *Magnolia* in east-central Asia slightly earlier than is implied by the pattern just described.

The southern Alaskan leaves reported from the Oligocene Katalla Formation [*Wolfe*, 1977], despite the uncertainty of identifying them, merit consideration, because today they are in a northern location. The Katalla Formation rests on the Yakutat Terrane, which comprises Paleogene to Recent strata, and is very active tectonically. If it had remained fixed the deposit would have had a paleolatitude of about 70° (Figure 5c). However, it is bounded on the east by the currently active dextral (west-side moving north) Fairweather Fault. Seismic and geodetic (GPS) data indicate that it is currently being displaced from the south, and its impingement on the Cordillera of southwest Alaska is responsible for elevating the St. Elias Mountains and for thrusting across the Yukon Territory [*Mazzotti and Hyndman*, 2002]. Cowan [2003], has recently reviewed the evidence comprehensively, and argues "that the Yakutat terrane originated about 1300 km further south of its present position". How much of this motion was accomplished after the Katalla Formation was laid down is unknown, but clearly it was then well south of its present position relative to North America.

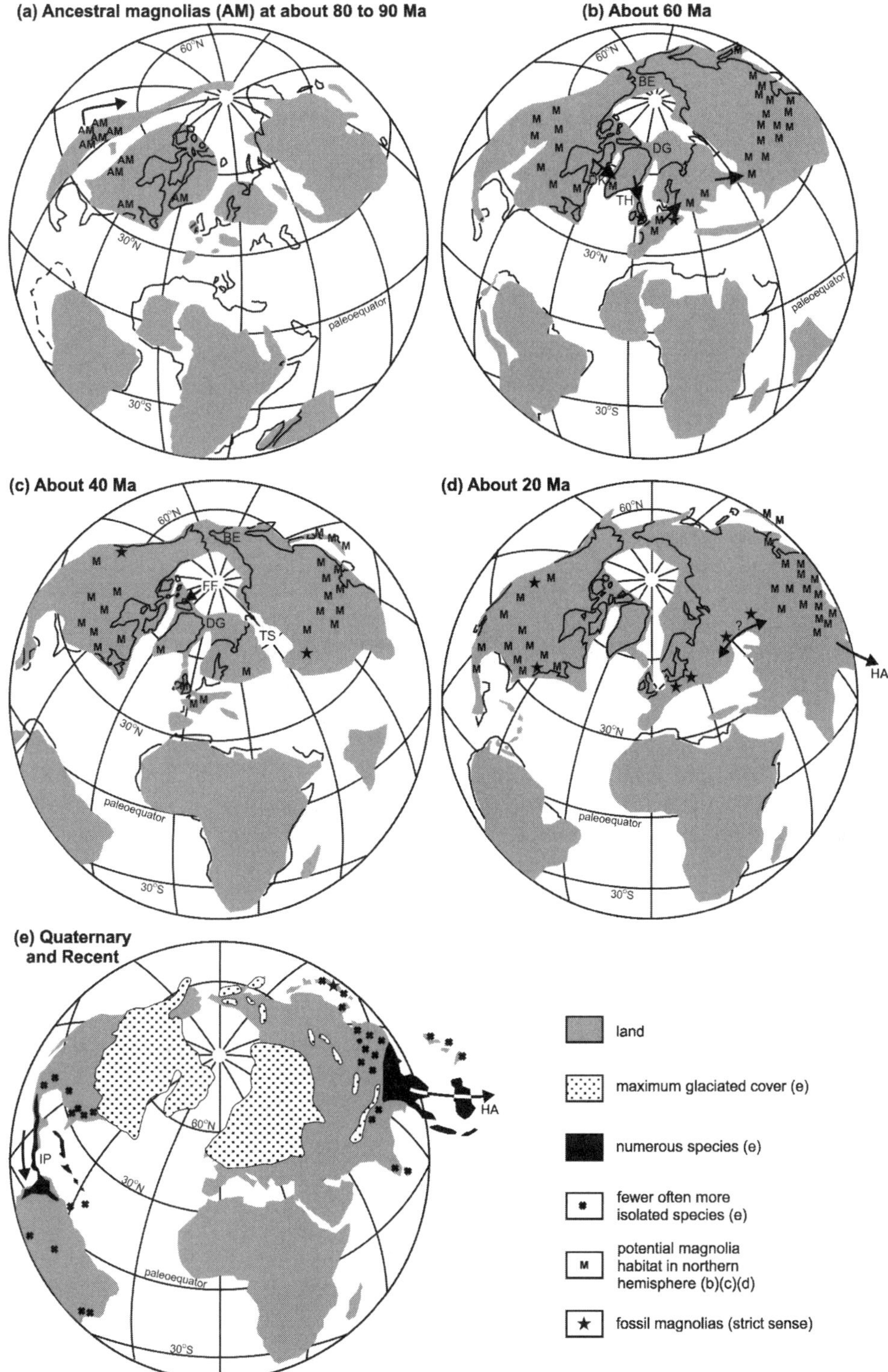

Figure 5. Proposed past distribution. (a) Possible latest Cretaceous northward migration of Vancouver Island shown. (b to d) Climate and habitat congenial to magnolias based on Figure 3 may not apply to the section Rhytidospermum. (e) Quaternary and Recent shows southward migration during the later Cenozoic along Isthmus of Panama (IP) and into High Archipelago (HA). Land-bridges are BE = Beringia, DG = De Geer, DK = Disko Island volcanoes, TH = Thulean (Iceland-Faroes).TS = Turgai Strait. FF = Fossil forest Axel Heiberg Island.

Critical is the absence of well established fossil magnolias in today's regions of highest diversity, southern southeast Asia and Central America/northern South America. The fossil floras of southern Mexico and Central America are moderately (pollen) to poorly (megafossils) known [*Graham*, 1999]. In Graham's opinion (personal communication February 2002) magnolias were not likely in the region until at least the Pliocene. Notably, no magnolia has been reported in the well studied Plio-Pleistocene basins of Colombia [*Hooghiemstra*, 1984; *Wijninga*, 1996]. There is some knowledge of fossil floras of parts of southern southeast Asia especially China [*Li et al.*, 1995(a)], yet no unambiguous magnolia fossils earlier than mid-late Cenozoic have, to the best of our knowledge, been confirmed [*Liu and Zheng, Li and Zheng* 1995; *Li et al.*, 1995(b)].

The quality of the fossil record permits some consideration at the sectional level. Talauma-type seeds have been recorded from the Eocene London Clay [*Chandler,* 1964; *Tiffney* 1977]. Also the Eocene Clarno beds of Oregon may contain remains of a representative of section *Talauma* [*Manchester*, 1994]. There is a good record of the section *Theorodon* in North America, including *M. septentrionalis* (similar to *M. grandiflora*) in the Oligocene Brandon Lignite of New England [*Tiffney*, 1977], and leaves of *M. latahensis* (also similar to *M. grandiflora*) in the Miocene Clarkia beds of Idaho [*Baranova and Figlar,* 2000]. The cpDNA sequence obtained from fossil magnolia leaves from the Clarkia beds confirms the presence of the subgenus *Magnolia* by the Miocene Epoch in western North America, well outside today's range [*Golenberg et al.,* [1990]. *Mai* [1971] and *Tiffney* [1977] places section *Manglietia* in Europe with certainty in the Oligocene and possibly as early as Eocene. Section *Rhytidospermum* may be present in the Oligocene Brandon Lignite (as *M. waltonii*), although this fossil may belong to the section *Tulipastrum* in the subgenus *Yulania* [*Tiffney*, 1977].

The subgenus *Yulania* has a scattered record that is strongly developed in Europe and adjacent areas of Russia [*Azuma et al.*, 2001] ranging from the Eocene (*Michelia*-type fossils) into the early Pleistocene Epoch. Section *Tulipastrum* is apparently represented by leaves in the Miocene Clarkia flora [*Figlar,* 1993].

Thus, both magnolia subgenera were present well back into the Cenozoic, perhaps as early as Eocene under a greenhouse climate, and almost certainly by the end of the Oligocene [*Azuma et al.*, 2001] after global cooling had begun (Figure 2). Within each subgenus, several sections were also probably established by the Oligocene. It is notable that subgenus *Yulania*, except for *M. acuminata*, is strictly Asian from the perspective of the fossil record and extant species.

5. PALEOGEOGRAPHY AND PALEOCLIMATE

Understanding the present and fossil distribution of an ancient land-based group such as magnolias requires knowledge of changes in the distribution of land and sea, in the latitudinal disposition of land, and in global climate.

5.1. The Distribution of Land and Sea

The history of land masses is shown in Figure 5 based on plate tectonic (for positions of continental lithosphere), paleomagnetic evidence (for geographical grid), and the distribution of terrestrial and marine sediments (for coastlines). The base maps are those of Irving [1982] with coastlines added from Smith et al. [2001]. Although old, these base maps display well magnolia distribution, and, in the past 20 years, estimates, in the period of interest, of the positions of continents and their latitudes have changed little. There has been added detail and accuracies (now better than 5°) have improved, but we are interested here in generalities.

In the Late Cretaceous Epoch, North America and most of Eurasia belonged to the same continental lithospheric block, but, owing to high sealevel, shallow epicontinental seas covered much of it (Figure 5a). Rising above sea level were four, very large, low-lying islands: the western Cordillera of North America, eastern North America and Greenland, the Baltic Shield, and an extensive area of northeastern Asia. India was in mid-southern latitudes; it had, along with Madagascar, disengaged from Africa.

In the Eocene Epoch, land became more extensive as sea level fell. North America became connected with eastern Asia by the Beringian land-bridge (Figure 5b, BE). The Labrador Sea and North Atlantic Ocean began to open, but volcanic isthmuses or chains of islands (Thulean and Disko Island volcanic edifices, TH and DK) spanned them, at least semi-continuously. The Turgai Gulf partially separated the Baltic Shield and the Asian land area, but land was continuous at its southern end. India began to drift north.

By the late Eocene Epoch, North America was almost entirely emergent. However, the belt of land that had encircled the northern hemisphere in the Early Eocene was now breached by the Turgai Strait in the east (TS, Figure 5c), and in the west by the northward extension of the Labrador Sea. Between them, Greenland and the Baltic Shield remained connected by the De Geer land-bridge, DG (Figure 5c).

By the Miocene Epoch the Atlantic had become a substantial ocean, but the Thulean land-bridge remained mainly emergent (Figure 5d). The Turgai Strait had ceased to exist and, once again, land extended from western Europe to eastern Asia. India had now collided with Asia and the Alpine-Himalayan orogenic belt began to form. South America remained isolated, although there is the possibility of intermittent island connection via an early Antillean arc [*Iturralde-Vinent and MacPhee*, 1999], but we are aware of no evidence that it was used by magnolias at this time. The modern continents were emergent by the Pliocene Epoch, and the High Archipelago had begun to develop as Australia approached southeast Asia. The Himalayas and mountain ranges of southeast Asia were uplifted (see for example *Harrison*, 1992). As the Atlantic opened, the Americas moved westward, over-riding oceanic lithosphere to the west. Vast volcanic arcs developed along the western margins of the Americas, margins that were uplifted in the Pliocene [see for instance *Parrish*, 1981; *van der Hammen* 1974], culminating with the completion of the Panama isthmus.

5.2. Latitudinal Disposition of Land and Sea

Throughout the period of interest, northern land areas moved not only relative to each other but also relative to the geographic pole, and these motions affected their climate. In the Late Cretaceous the north geographic pole was near what was to become the Beringian land-bridge, and subsequently moved from there to its present position (note the migration across latitude parallels in Figure 5). Thus, during the first half of the history of magnolias, the latitudes of northwestern North America and northeast Asia, were about 20° higher than now, and the latitudes of western Europe about 10° lower. Notably, throughout its existence, the Beringian connection was never far south of 70°N.

5.3. Global Climate Change

During the Late Cretaceous through Eocene Epochs, Earth was under a greenhouse regime (Figure 2), marked by high CO_2 atmospheric content [*Savin*, 1977; *Barron and Washington*, 1985; *Barron*, 1987; *Prothero et al.*, 2003; *Zachos et al.*, 2001]. Mean temperatures were markedly higher, and the equator to pole temperature gradient apparently less steep than today (Figure 3), so that the warm temperate belt suitable for magnolias was about 10 to 15° north of its position today. The absence or incomplete development of the Cordilleran and the Alpine-Himalayan mountain systems, and the presence of inland seaways, meant that interiors of northern continents did not have today's severe winters and hot summers, and extensive drought-causing rain-shadows were probably limited or absent.

Marked cooling occurred in the early Oligocene Epoch when the Antarctic ice cap first appeared (Figure 2). It's development might be linked to the opening of the Drake Passage between Antarctica and South America, and of the Tasman Gateway between Antarctica and Australia [*Exon et al.*, 2002].

These waterways allowed the development of the circum-Antarctic current, which limited southern penetration of warm equatorial water (Figure 2).

Followed by an interval of warming in the late Oligocene and early Miocene, cooler climates returned in the later Miocene, the cooling possibly related to the development of the High Archipelago as a barrier to flow of warm Pacific water into the Indian Ocean.

The Pliocene Epoch remained generally cool, but in the mid-Pliocene there was a warm interval about 3.5° C warmer than present [*Graham*, 1999]. Full ice-house conditions developed in the Quaternary, with sea-level ice of fluctuating extent repeatedly covering large areas in both north and south polar latitudes. This pattern may have been triggered by the uplift of the Isthmus of Panama, blocking flow from Atlantic to Pacific. The Alpine-Himalayan and Cordilleran orogenic belts were uplifted, creating vast rain-shadow regions. Arid continental conditions inimical to magnolias were established in the centres of large land-masses.

5.4. Summary

Paleogeographic features important for the history of magnolias are as follows: (1) Late Cretaceous isolation of North America under equable greenhouse conditions; (2) establishment in the early Eocene Epoch under super-greenhouse conditions of continuous land from western North America to east Asia via Beringia at high latitudes, and of volcanic land bridges connecting North America with Eurasia at mid-latitude; (3) isolation of Europe from east Asia by the Turgai Straight in the late Eocene Epoch; (4) separation of Europe from North America in the mid-Cenozoic as global cooling began, but with climates warmer than at present persisting into the Pliocene Epoch; (5) uplift of the Alpine-Himalayan and Cordilleran mountain belts, and the connection of North and South America; (6) the development of full ice-house conditions and extensive continental climates in the latest Cenozoic.

6. PROPOSED PALEOGEOGRAPHIC ORIGIN AND SUBSEQUENT HISTORY OF MAGNOLIAS

An outline is given in Figures 5 and 6. We have tried to incorporate the new understanding of the groups within the genus [*Figlar*, 2001], and of the fossil record, and to keep in mind that migration would be restricted by the distribution of seas, and facilitated by the presence of land with moist warm-temperate climates. We have sought to explain the widespread fossil distribution, the present restricted bimodal distribution, and the peculiarities of the distribution of modern sections and their respective diversities.

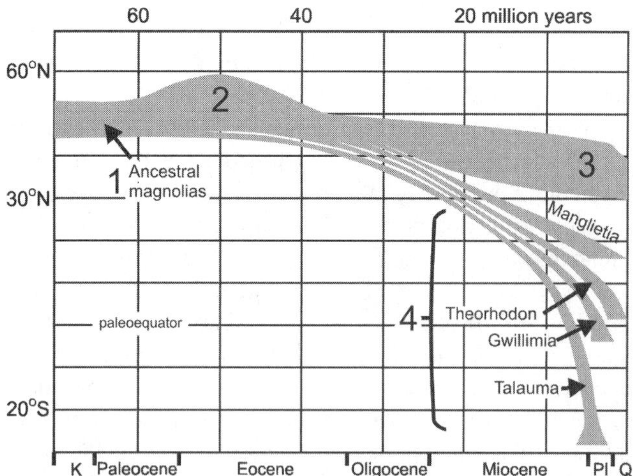

Figure 6. The four elements of magnolia evolution and their latitudinal distribution. Width of the trends indicates approximate latitudinal spread, not species diversity. (1) Ancestral magnolias in North America in high mid-latitudes. (2) Rapid spread across Eurasia during the Eocene climatic maximum still in high mid-latitudes. (3) Southward migration of remnant forms of (2) to low mid-latitudes during later Cenozoic cooling. (4) Southward migration into the newly formed mountains and uplands of the Americas and southeast Asia; divergence of sections very speculative.

6.1. Place of Origin

Ancestral and early magnolias have been documented in North America. There are, to the best of our knowledge, no comparable occurrences elsewhere. There is no extensive fossil record of Late Cretaceous ancestral magnolias or their Early Cenozoic descendants in Central America, South America, in eastern or southeastern Asia, places where they are now most abundant and diverse. This pattern contrasts with that of *Liriodendron*, whose distinctive fossil leaves are preserved in strata of Maastrichtian age in Japan and North America [*Manchester*, 1999]. Of course, a record could yet be discovered. Significantly, in China, although plant remains generally are well known from the Paleogene of northern China [*Li and Zheng*, 1995] there is no solid evidence for the occurrence there of early magnolias. Furthermore, cpDNA studies point to a North American "cradle" because several of the apparently "basal" magnolias still live there. Thus, North America seems their most likely place of origin.

Under the greenhouse climate of Late Cretaceous through Eocene time, magnolias and ancestral magnolias occurred in high mid-latitudes of 45° to 60°N (Figure 5a, b). This interpretation is consistent with a concept of the widespread ancestral "boreotropical" flora [*Tiffney*, 1985 and references therein] and Manchester's analysis of reliably identified fossil plant taxa [*Manchester*, 1999] although, as our analysis shows, the climatological term "boreotemperate" flora would perhaps

be more appropriate. We also note that it is species that evolve and disperse, not floras. Each species and in general the genus, has its own unique history. The Cretaceous ancestors of magnolias seem to have been confined to North America for many tens of millions of years, presumably because broad epicontinental seas then separated North America from Eurasia, and birds, a possible means of dispersal, may not then have been capable of long distance flight [*Hope*, 2002; *Zhou and Hou*, 2002]. Northern high mid-latitudes and low altitudes seem to have been preferred by magnolias until the Oligocene Epoch, when global cooling commenced (Figure 2).

6.2. Migration out of North America

Following the recession of the Late Cretaceous epicontinental seas a super-greenhouse climate developed between 55 and 50 Ma (Figure 2). The Labrador Sea first and then the North Atlantic Ocean began to open, spanned initially by volcanic chains or isthmuses situated in mid-latitudes. Their latitude and proximity to moderating ocean water indicates that, under greenhouse conditions, these North Atlantic intercontinental connections would have had climates suitable for magnolias (Figure 3, 5b, 6) [*Tiffney*, 1985]. We propose that magnolias migrated eastward from North America and Greenland via this connection, first to Europe, and then across southern Siberia to eastern Asia. Southern Siberia, at a paleolatitude of about 50°, probably had a mild climate at that time because of wide expanses of ocean to the south, not mountains as nowadays (Figure 5b). As magnolias spread and filled the zone of climatic potential, they left behind a trail of populations and fossils from North America through northern Europe and central Asia (Figure 1b).

Besides the link via the Thulean isthmus or island chain, there are two other possible migration routes, westward via the Beringian land-bridge, and northeastward via the De Geer land-bridge through Spitzbergen and across the Barents Shelf, which was exposed in the Paleogene (Figure 5b and c). Both are situated, during the early and mid-Cenozoic, at latitudes about 70°N, over 20° higher than the Thulean route. So far as we know, no unambiguous occurrences of Cenozoic fossil magnolias have been found at such high paleolatitudes, even under greenhouse conditions, although possible magnolia pollen has been reported by McIntyre [1991] from the Eocene Fossil Forest of Axel Heiburg Island at latitude 75°N [*Irving and Wynne*, 1991], but no definitive fossils such as fruiting bodies. Possible Alaskan Oligocene leaf fossils [*Wolfe*, 1977] probably were transported tectonically from southern latitudes, as already described. Therefore, we consider the Beringian and De Geer land-bridges less likely routes for the early Tertiary migration of magnolias out of North America. Tiffney [1985] and Manchester [1999] have made similar climatic arguments for the boreotropical (our boreotemperate) flora as a whole. Presumably, it is at this time that the stage is now set for the genus-level disjunction within *Magnolia* genus through the southward displacement of the various constituents of this flora [*Tiffney*, 1985 and references therein].

6.3. Isolation and Creation of Disjuncts

The Atlantic continued to open, and towards the end of the Oligocene Epoch, the Turgai Strait extended southward, restricting further, or even prohibiting migration into Asia from the west (Figure 5c). Land connections via Europe become more tenuous as the Atlantic Ocean widened and latitude increased, but the Thulean land-bridge, according to Smith et al. [2001], remained almost continuous until the Miocene, and the remnant Disko Island volcanoes may have reduced the obstacle of the Labrador Sea. Cross-Atlantic migration routes may, therefore, have been open but were less hospitable than earlier.

As a result of these changes and of the cooling which began in the Oligocene, disjunction at the genus level commenced, and the 15 sections without modern disjunctions began evolving separately in North America and the Old World. Also it was, we suggest, at this time that the disjunction within section *Tulipastrum* may have developed.

The situation in *section Rhytidospermum* (*M. tripetala* and its close relatives in Asia) is different; DNA evidence suggests that disjunction occurred later than in *Tulipastrum* [*Qiu et al.*, 1995], at about 20 to 27 Ma (Miocene) according to Azuma et al. [2001], after the commencement of global cooling. At that time there was continuous land between North America and eastern Asia via Beringia, but at a latitude (70°) greater than that of any securely identified magnolia fossils of this or any other age (Figures 1 and 5). To invoke migration via Beringia in the Neogene therefore has, at present, no supporting fossil evidence and does not appear to meet the broad climatic criteria we have assumed for the genus. Nevertheless we shall consider it.

Species in section *Rhytidospermum* are relatively hardy [*Callaway*, 1994; *Gardiner*, 2000]. Molecular studies suggest that the species became disjunct long after the commencement of cooling in the Oligocene. Suppose, therefore, that despite the absence of fossil evidence, species of this exceptional section once did grow at latitude 70°. If so, *M. tripetala* may be an end-member of an atypical hardy magnolia group that, during the Neogene (Figure 5d), ranged continuously across Beringia from east Asia to North America, the range only becoming severed late in the Cenozoic by the onset of full ice-house climate (Figure 5e). Perhaps an analogy can be drawn with the genus *Rhododendron*, which has examples of such atypical hardy species, *Rhododenron lapponicum* (the

type species described by Linnaeus from Lapland) and *Rhododendron groenlandicum* (Labrador Tea). These two species have a high latitude distributions today, circumpolar in the former case. Perhaps a climate capable of supporting an atypical hardy magnolia may have persisted as late as the middle Pliocene in Alaska [*Graham*, 1999]. Another possibility is that of a long-distance dispersal event that somehow brought the ancestor of *M. tripetala* or the Asian taxa across the Pacific.

Our conclusion regarding the generally dominant role of the trans-Atlantic (Thulean) route and the small or negligible role of Beringia in magnolia migrations is in general accord with that of Manchester [1999] based of fossil records of very many other mesophytic taxa.

As the core climatic zone of magnolias shifted south (Figures 3, 5d, 6), cooling, aridification, and the associated development across central Eurasia and North America of deserts and grasslands inimicable to magnolias, broke the connection between the populations of east Asia and Europe, and in North America, eventually restricted them to the southeast of the continent and Mexico. Contact between European and North American magnolias was severed as the North Atlantic widened, and the Thulean and Disko Island land-bridges sank.

6.4. Southern Migration and Speciation

Before the mid-Cenozoic, magnolias, according to the fossil record, had predominantly lived at relatively low altitudes (London Clay for example; in Chandler [1964]) often on or near a coastal plain. With the development of icehouse climate in the later Cenozoic, ice-caps developed at both poles (Figure 2) and the wide zone of lowland habitat once available to magnolias shrank. It was at this time that magnolias expanded their range southward into the moist, warm-temperate, high valleys of newly uplifted mountains. In the Americas, species of section *Theorhodon* dispersed along the recently formed ranges of Mexico and Central America, along the Isthmus of Panama and perhaps also the Antillean Arc into the Andes of northern South America where they diversified [*Azuma et al.*, 2001]. Likewise, in Asia, species of sections *Gwillimia* and *Michelia* migrated into the mountains of southwest China, Indochina and the High Archipelago, where they too diversified.

The absence of known Cenozoic magnolia fossils from South America (for example there are none recorded in the Plio-Pleistocene Bogota Basin [*Hooghiemstra*, 1984; *Wijninga*, 1996]) or southeast Asia, indicates recent immigration. The numerous closely related extant species that occur there suggest youthful taxa currently undergoing speciation. After all, the habitats in which the southerly diaspora occur are themselves young, geologically speaking, because the uplift of mountains and the erosion of valleys inhabited by the southward diasporas have been comparatively recent.

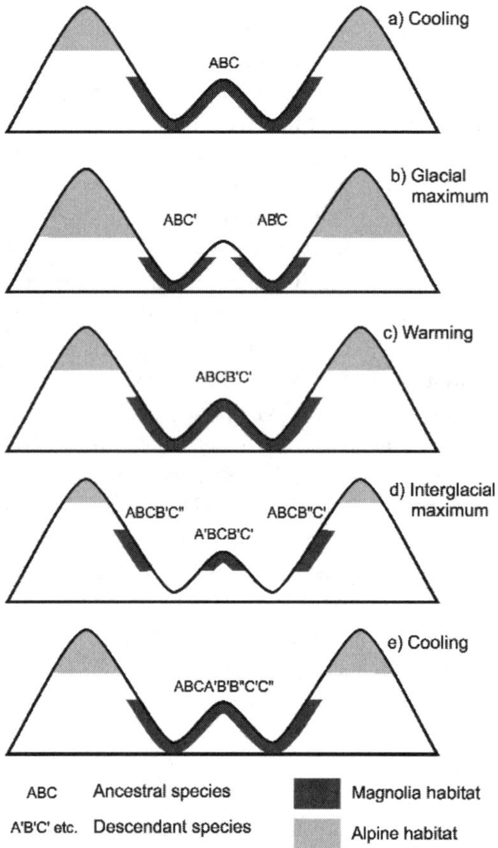

Figure 7. The High Relief Climate Oscillation (HIRCO) speciation model illustrated. During late Cenozoic, glacial-interglacial cycling climatic zones moved up and down the steep valley sides eroded as a result of high precipitation in recently formed mountains. (a) During cooling after an interglacial interval, populations of species A, B, and C are distributed across lower valley slopes and the low divide. (b) During maximum cooling, populations disperse downslope becoming separated by the low divide, and species B in one valley and C are replaced by B' and C' through genetic drift and selection. (c) Warming allows populations to ascend and five species are now everywhere present. (d) Warming continues and these mixed populations move up-slope, are split into three by the topography, and genetic drift and selection produces three new species, C" replaces C' on the left-hand slope, A' replaces A on the central ridge and B" evolves from B' on the right-hand slope. (e) Cooling causes populations to descend and remix. During the entire cycle, species increase from three to eight. Cycles are assumed to vary with periodicities of 20,000 to 100,000 years driven by orbital variations. Modified from Irving and Hebda [1993].

To explain this recent speciation in these regions of young high relief, we invoke the high relief, climate oscillation (HIRCO) model that we proposed earlier to account for the comparable speciation in the genus *Rhododendron* [*Irving and Hebda*, 1993]. In these newly uplifted areas, magnolias encountered climates that were zoned according to altitude.

Climatic oscillations with periodicities of 20 to 100 thousand years resulting from orbital variations, moved the moist, warm-temperate climate zone favorable to magnolias, up and down the flanks of valleys (Figure 7). Magnolia species shifted their range accordingly. Populations underwent a cycle of repeated fragmentation, coalescencing and intermingling, each time with varying constituents, thus promoting speciation. The HIRCO speciation model could be effective to some degree in all high mountains in mid- and low latitudes, but it is likely to be most effective in mountains of extreme relief and very high rainfall in low latitudes,—in the east Himalaya/southeast China region, in the High Archipelago, and in the tropical cordillera of Central America and northern South America, regions where the diversity of extant magnolias is highest.

The majority of extant species of magnolias are, we suggest members of these two geologically recent southward dispersal events. It was, we contend, in this way and at this comparatively recent time, that the numerous species of the modern non-disjunct sections evolved. By contrast, the northern magnolias of Japan, northern China, Korea, and southeastern North America have few species, apparently residuals from what was once a more extensive and diverse magnolia flora. It is from these more northerly, better known, relatively hardy, and mostly older taxa, that magnolias of temperate gardens have been derived, bringing much widespread horticultural interest to the genus.

7. SUMMARY

The history of magnolias has been controlled by interrelationships between global climate change, changes in distribution of land and sea, changes in the latitude of land areas, and tectonic uplift. Magnolias most likely originated in North America in high mid-latitudes under a greenhouse climate. They probably spread eastward through Europe to western Asia in the Eocene and subsequently to eastern Asia, remaining in high mid-latitudes, still under greenhouse climate. As a consequence of global cooling in the Oligocene, magnolias began to migrate to lower mid-latitudes and became isolated, first into three (North America, Europe, eastern Asia) or more and then into two centres, (North America and eastern Asia). In the later Neogene they migrated south to upland valleys in the newly-uplifted mountain belts of southeast Asia and Central and South America. Here they rapidly diversified as a result of the vertical oscillation of climate zones under variable icehouse conditions (HIRCO speciation model). The Cenozoic evolution and dispersal of magnolias, as we have argued for rhododendrons, was characterized by the impoverishment of species in the north and by diversification of species in the south.

Note added in proof: For the most current classification, consult Figlar and Nooteboom (2004). Among other things, they have now split section *Theorhodon* between sections *Magnolia* and *Talauma*, elevated each of *Magnolia fraseri* and relatives and *Magnolia macrophylla* and relatives into sections, and elevated section *Gynopodium* into a new subgenus. These changes affect the details of Figure 4 but do not affect our arguments or conclusions.

Acknowledgments. We began to compile the material for this paper for a lecture delivered before the Magnolia Society International in Bellevue, Washington, USA, in March 2002, and we are very grateful to the program organizer, Diane Thompson, for inviting us; it was the stimulus of her invitation that caused us to undertake this task. Richard Figlar and Alan Graham have been very helpful with providing information, and listening to and commenting on some of our ideas. Geraldine Allen, Darrel Cowan, Laurie Donovan, Karin Helmens, Gary Kaiser, B.H. Tiffney, Qi-bin Zhang and Ken Marr provided key information. S.R. Manchester C.R. Barnes and D. Christophel gave very helpful reviews. Richard Franklin drew the figures and Judith Baker prepared the manuscript. We are very grateful to everybody for their help. Geological Survey of Canada publication number 2003200.

REFERENCES

Azuma, H., J. C. Garci-Franco, V. Rico-Gray, and L. B. Thien, Molecular phylogeny of the Magnoliaceae: the biogeography of tropical and temperate disjunctions, *Amer. J. Bot., 88,* 2275–2285, 2001.

Baranova, N. A., and Figlar, R. B., Leaf Features of (fossil) Miocene *Magnolia latahensis* show affinity to extant *Magnolia grandiflora. Proc. Internat. Symp. on the Family Magnoliaceae,* Science Press, Beijing, pp. 58–64, 2000.

Barron, E. J., and W. M. Washington, Cretaceous climate: high atmosphere CO_2 as a plausible mechanism, (editors), Carbon Cycles and Atmospheric CO_2, natural variations, Archean to present. *Geophys. Mon. Amer. Geophys. Union,* 32, 546–553, 1985.

Barron, E. J., Eocene equator-to-pole surface ocean temperatures; a significant climate problem? *Paleooceanography,* 2, 729–739, 1987.

Cain, S. A., *Foundations of Plant Geography.* Harper, New York. 556 p., 1944.

Callaway, D. J., The World of Magnolias. Timber Press. Portland, Oregon. 260 p., 1994.

Chandler, M. E. J., *The Lower Tertiary floras of southern England, IV. A summary and survey of findings in light of recent botanical observations.* London: British Museum, 1964.

Cowan, D. S., Revisiting the Baranof-Leech River hypothesis for early Tertiary coastwise transport of the Chugach-Prince William terrane, *Earth Planet. Sci. Lett.,* 213, 463–475, 2003.

Crane P. R., The phylogenetic position and fossil history of the Magnoliaceae. In: Hunt, D. (ed.) *Magnolias and their Allies.* Proceedings of an International Symposium, Royal Holloway,

University of London, Egham, Surrey, U.K. 12–13 April 1996. pp. 1–38, 1996.

Davey et al., Drilling for Antarctic Cenozoic climate and tectonic history at Cape Roberts, southwestern Ross Sea, *Eos, Trans. Amer. Geophys. Union, 82,* 585–590, 2001.

Delevoryas, T., and J. E. Mickle, Upper Cretaceous Magnoliaceous fruit from British Columbia, *Amer. J. Bot., 82,* 763–768, 1995.

Dilcher, D. L., and P. R. Crane, Archaeanthus; an early angiosperm from the Cenomanian of the western interior of North America. *Ann. Missouri Bot. Gard, 71,* 351–383, 1984.

Enkin, R. J., J. Baker, and P. S. Mustard, Paleomagnetism of the Late Cretaceous Nanaimo Group, southwestern British Columbia, *Can. J. Earth Sci., 38,* 1403–1422, 2002.

Enkin, R. J., J. B. Mahoney, J. Baker, J. Riesterer, and M. L. Haskin, Deciphering shallow inclinations: implications for Late Cretaceous strata overlapping the Insular/Intermontane. Superterrane boundary in the southern Canadian Cordillera, *J. Geophys. Res.,* 108, B4, 2186, doi:10.1029/2002JB001983,2003.

Exon, N., and 28 co-authors, Drilling reveals climate consequences of Tasman Gateway opening, *Eos, Trans. Amer. Geophys. Union, 83,* 253–259, 2002.

Fernald, M. L., *Gray's Manual of Botany, 8th Centennial Edition.* American Book Co. New York. 1632 p., 1959.

Figlar, R. B., Stone Magnolias. *Arnoldia, 53,* 2–9, 1993.

Figlar, R. B., Proleptic branch initiation in *Michelia* and *Magnolia* subgenus Yulania provides basis for combinations in subfamily Magnolioideae. In Liu,Y., H. Fan, Z. Chen, Q. Wu and Q. Zeng, (eds.) *Proc. Internat. Symp. on the Family Magnoliaceae,* Science Press, Beijing, pp. 26–37, 2000.

Figlar, R. B., Classification of Magnoliaceae in Magnolia Society Website, www.magnoliasociety.org. December 2001.

Figlar, R. B. and H. P. Nooteboom. Notes of new Magnoliaceae IV. A new system of Magnolioideae (Magnoliaceae), new combinations in Malesian Magnolia and some corrections in previously proposed names. *Blumea 49 (in press), 2004.*

Gardiner, J., *Magnolias: A Gardeners Guide,* Timber Press. Portland Oregon. 249 p., 2000.

Golenberg, E. M., D. E. Giannasi, M. T. Clegg, C. J. Smiley, M. Durbin, D. Henderson, and G. Zurawski., Chloroplast DNA sequence from a Miocene *Magnolia* species. *Nature, 344,* 656–658, 1990.

Graham, A., *Late Cretaceous History of North American Vegetation North of Mexico.* Oxford Univ. Press, 1999.

Good, R., *The Geography of the Flowering Plants. 3rd edition,* Longman, London, 1974.

Harrison, T. M., P. Copeland, W. S. F. Kidd and A. Yin, Raising Tibet, *Science, 255,* 1663–1670, 1992.

Hooghiemstra, H., *Vegetational and climatic history of the high plains of Bogota, Columbia, Dissertationes Botanicae, 79.* Cramer, Vaduz, 368 pages, 1984.

Hope, S., The Mesozoic radiation of Neornithes. In: L.M. Chiappe and L.M. Witmer (eds). *Mesozoic Birds.* Univ. Calif. Press, Berkley. pp. 339–388, 2002.

Hutchinson, J., *Evolution and Phylogeny of Flowering Plants.* Academic Press, London, 717 p., 1969.

Irving, E., Fragmentation and assembly of the continents, *Geophysical Surveys, 5,* 299–332, 1982.

Irving, E., and R. J. Hebda, Concerning the origin and distribution of Rhododendrons. *J. Amer. Rhododendron Soc., 48,* 139–162, 1993.

Irving, E., P. J. Wynne, D. J. Thorkelson, and P. Schiarizza, Large (1000 to 4000 km) northward movements of tectonic domains in the northern Cordillera, 83 to 45 Ma, *J. Geophys. Res., 101,* 17,901–17,916, 1996.

Irving, E., and P. J. Wynne, The paleolatitude of the Eocene fossil forest of Arctic Canada. In: R.L Christie and N,J. McMillan (eds.) Tertiary Fossil Forest of the Geodetic Hills. Axel Heiberg Island, Arctic Archipelago. *Geol. Survey Canada, Bull., 403,* 209–211, 1991.

Iturralde-Vinent, M. A., and R. D. E. MacPhee, Paleogeography of the Caribbean region: implications for Cenozoic biogeography, *Amer. Museum Nat. Hist. Bull., 238,* 1–95, 1999.

Johnston, S. T., The great Alaskan train wreck; reconciliation of paleomagnetic and geological data, *Earth Planet. Sci. Lett., 193,* 259–272, 2001.

Kim, S., C.W. Park, Y-D. Kim, and Y. Suh, Phylogenetic relationships in family Magnoliaceae inferred from *NDHF* sequences. *Amer. J. Bot., 88,* 717–728, 2001.

Li, X., Z. Zhou, C. Cai, G. Sun, S. Ouyang, and L. Deng (editors), *Fossil Floras of China Through the Geological Ages (English Edition).* Guangdong Science and Technology Press, Guangzhou. China, 694 p., 1995(a).

Li, H., W. Li, and J. Liu, Quaternary floras. Chapter 12. In: Li, X., Z. Zhou, C. Cai, G. Sun, S. Ouyang, and L. Deng (eds.). *Fossil Floras of China Through the Geological Ages (English Edition).* Guangdong Science and Technology Press, Guangzhou, China. pp. 552–588, 1995(b).

Li, H. and Y. Zheng, Paleogene floras. Chapter 10 In: Li, X., Z. Zhou, C. Cai, G. Sun, S. Ouyang, and L. Deng (eds.). *Fossil Floras of China Through the Geological Ages (English Edition).* Guangdong Science and Technology Press, Guangzhou, China. pp. 455–505, 1995.

Liu, Y, and Y. Zheng, Neogene floras. Chapter 11. In: Li, X., Z. Zhou, C. Cai, G. Sun, S. Ouyang, and L. Deng (eds.). *Fossil Floras of China Through the Geological Ages (English Edition).* Guangdong Science and Technology Press, Guangzhou, China. pp. 506–551, 1995.

Mai, D. H., Fossile Funde von Manglietia BLUME (Magnoliaceae) *Feddes Repert, 82,* 441–448, 1971.

Manchester, S. R., Fruits and seeds of the Middle Eocene Nut Beds flora, Clarno Formation, North Central Oregon, *Palaeontogr. Amer., 58,* 1–205, 1994.

Manchester, S.R., Biogeographical relationships of North American Tertiary Floras, *Ann. Missouri Bot. Gard.,86,* 472–522, 1999.

Manchester, S. R., and B. H. Tiffney, Integration of paleobotanical and neobotanical data in the assessment of phytogeographic history of Holarctic Angiosperm clades, *Int. J. Plant Sci., 162,* S19–S27, 2001.

Mazzotti, S. and R. D. Hyndman, Yakutat collision and strain transfer across the northern cordillera, *Geology, 30,* 495–498, 2002.

McIntyre, D. J., Pollen and spore flora of an Eocene forest, Eastern Axel Heiberg Island, N.W.T.. In: R.L Christie, and N.J. McMillan (eds). Tertiary Fossil Forest of the Geodetic Hills, Axel Heiburg Island, Arctic Archipelago, *Geol. Survey Canada, Bull., 403.* 83–97, 1991.

Myers, J. A., Terrestrial Eocene-Oligocene vegetation in the Pacific Northwest. In: *From Greenhouse to Icehouse*: Prothero, D.R., Ivany, L.C. and E.A. Nesbitt (eds.) Columbia University Press, New York, pp.172–185, 2003.

Nooteboom, H. P. Magnoliaceae. In Kubitzki, K. (ed.) *The Families and Genera of Vascular Plants, 2,* 391–401, Springer Verlag, Berlin, 1993.

Nooteboom, H. P., The tropical Magnoliaceae and their classification. In: Hunt, D. (ed.) *Magnolias and their Allies.* Proceedings of an International Symposium, Royal Holloway, University of London, Egham, Surrey, U.K. 12–13 April 1996. pp. 71–80, 1996.

Nooteboom, H. P., Different look at the classification of the Magnoliaceae, In: Liu,Y., H. Fan, Z. Chen, Q. Wu and Q. Zeng, (eds.) *Proceedings of the International symposium on the Family Magnoliaceae,* pp 26–37. Science Press, Beijing, China, 2000.

Opdyke, N. D., The impact of paleomagnetism on paleoclimatic studies. *Intern. J. Bioclim. and Biometr, 3,* 1–6, 1961a.

Opdyke, N. D., The climatological significance of desert sandstones. In *Descriptive Palaeoclimatology* edited by A. E. M. Nairn, Interscience, pp 45–60, 1961b.

Parrish, R. R., Cenozoic thermal and tectonic history of the Coast Mountains of British Columbia as revealed by fission track and geological data and quantitative thermal models; Ph.D. thesis, University of British Columbia, Vancouver, 116 p., 1981.

Peigler, R. S., Fossil Magnoliaceae: review of the literature, *J. Magnolia Soc., 25,* 1–11, 1989.

Prothero, D. R., L. C. Ivany, and E. A. Nesbitt, *From Greenhouse to Icehouse: The Marine Eocene-Oligocene Transition.* Columbia University Press, New York, 2003.

Qiu, Y-L., C. R. Parks, and M. W. Chase, Molecular divergence in the eastern Asian-eastern North American disjunct section *Rhytidospermum* of *Magnolia* (Magnoliaceae). *Amer. J. Botany. 82,* 1589–1598, 1995.

Savin, S. M., The History of the Earth's surface temperature during the past 100 million years, *Ann. Rev. Earth Planet. Sci.,5,* 319–355, 1977.

Smith A. G., D. G. Smith, and B. M. Funnel. *Atlas of Mesozoic and Cenozoic Coastlines,* Cambridge Univ. Press. 60p. 2001.

Stewart, W. N. and G. W. Rothwell, *Paleobotany and the Evolution of Plants, 2nd Ed.* Cambridge Univ. Press, 521 p., 1993.

Sun, G., Z. Cao, H. Li, and X. Wang, Cretaceous Floras. Chapter 9 In: Li, X., Z. Zhou, C. Cai, G. Sun, S. Ouyang, and L. Deng (editors). *Fossil Floras of China Through the Geological Ages (English Edition).* Guangdong Science and Technology Press Guangzhou, China. pp 411–452, 1995.

Takhtajan, A., *Flowering Plants Origin and Dispersal.* Oliver and Boyd, Edinburgh. 310 p., 1969.

Tao, J., Fossil flora and paleoclimatic significance of Liuqu Formation, Lazi County, Tibet. *Mem. Inst. Geol., Acad. Sinica, 3,* 223–238, 1988a

Tao, J., The Paleocene flora and paleoclimate of Liuqu Formation in Xizang (Tibet), In: Whyte, R.O. et al. (eds.) The *Palaeoenvironment of East Asia from the mid-Tertiary, Vol. 1. Geology, Sea level changes, palaeoclimatology, palaeobotany.* Hong Kong: Centre of Asian Studies, University of Hong Kong, pp 520–522, 1988b.

Tiffney, B. H., Fruits and seeds of the Brandon Lignite: Magnoliaceae *Bot. J. Linn. Soc., 75,* 299–323, 1977.

Tiffney, B. H., The Eocene North Atlantic land bridge: its importance in Tertiary and modern phytogeography of the northern hemisphere. *J. Arnold Arboretum, 66,* 243–273, 1985.

Tiffney, B. H., Geographic and climatic influence on the Cretaceous and Tertiary history of Euramerican floristic similarity. *Acta Universitatis Carolinae – geologica, 44,* 5–16, 2000.

Tiffney, B. H. and S. R. Manchester, The use of geological and paleontological evidence in evaluating plant phytogeographic hypotheses in the Northern Hemisphere Tertiary, *Int. J. Plant Sci., 162,* S3–S17, 2001.

Van der Hammen, T., The Pleistocene changes in vegetation and climate in tropical South America, *J. Biogeography, 1,* 3–26, 1974.

Wijninga, V .M., Neogene evolution of the north Andean flora . In: V.M. Wijninga (ed.), *Paleobotany and palynology of Neogene sediments from the high plain of Bogota (Colombia),* 117–135, chapter 8, 1996.

Wolfe, J. A., Paleogene floras from the Gulf of Alaska region. *U.S. Geol. Surv. Prof. Paper, 997,* 1–108, 1977.

Zachos, J., M. Pagani, L. Sloan, J. T. Thomas, K. Billups, Trends, rhythms, and aberrations in global climate 65 Ma to present. *Science, 292,* 686–693, 2001.

Zhou, Z., and L. Hou, The discovery and study of Mesozoic birds in China. In L.M. Chiappe and L.M. Witmer (eds). *Mesozoic Birds.* Univ. Calif. Press, Berkeley. pp 160–208, 2002.

R. Hebda, Royal British Columbia Museum, 675 Belleville St., Victoria, B.C., Canada, V8W 9W2, and Biology Department and School of Earth and Ocean Sciences, University of Victoria, P.O. Box 1700, Victoria, B.C., Canada, V8W 2Y2.

E. Irving, Pacific Geoscience Center, Geological Survey of Canada, North Saanich, B.C., Canada, V8L 4B2.

A Long-term Octupolar Component in the Geomagnetic Field? (0–200 Million Years B.P.)

Vincent Courtillot and Jean Besse

Laboratoire de Paléomagnétisme, Institut de Physique du Globe, Paris, France

The hypothesis of a geocentric axial dipole (GAD), which is fundamental to pale-omagnetism and plate reconstructions, has recently been challenged by suggestions that significant long-term octupolar contributions of between 6–25% of the GAD may have existed for Precambrian through early Tertiary time. For instance, Si and Van der Voo [2001] propose that a value of 6% would account for the low inclina-tions observed in central Asia in the Cretaceous and early Tertiary. Following and updating our previous analysis of the global paleomagnetic data base [*Besse and Cour-tillot*, 2002], we attempt to find evidence for octupolar contributions in the 0–200 Ma period. An important component of our analysis is the inclusion of data from sites believed to have possibly undergone a tectonic rotation about a local vertical axis (con-tributing 174 out of 465 data). We analyze the positions of mean poles in 20-Ma win-dows in common-site longitude, respectively for the northern mid-latitudes, equatorial and southern mid-latitudes, searching for the distinctive antisymmetrical pattern expected for a dipole plus octupole (without quadrupole) field. We next analyze the distribution of "latitude anomaly" (derived from the inclination anomaly, i.e. observed minus expected in case of a pure dipole) versus dipole latitude. Based on these various data manipulations, we find no robust evidence for an octupole and esti-mate that values on the order of 5% are unlikely to have been exceeded in the last 200 Ma. A preliminary 200 Ma overall mean field has a quadrupole component on the order of 3 ± 2% (i.e. significant) and an octupole of 3 ± 8% (i.e. not signifi-cant). A by-product of the analysis using poles from "free to rotate" sites is clear con-firmation of the amplitude of rotations undergone by for instance parts of the Adriatic promontory of Africa or the Colorado plateau.

1. INTRODUCTION

Paleomagnetism has been successful in the last half century in constraining and in cases solving geophysical problems of global importance, such as dynamo generation in the Earth's core and plate tectonic reconstructions. Most of the models developed from the '60s to the '80s assumed that, when averaged over a sufficient amount of time, in excess of a few thousand years, the

Timescales of the Paleomagnetic Field
Geophysical Monograph Series 145
Copyright 2004 by the American Geophysical Union
10.1029/145GM05

Earth's magnetic field could be described accurately by a (geo-) centered axial dipole (GAD) (going back at least to Hospers and Runcorn in the early to mid-50s; see e.g. *Courtillot et al* [1992]). Slight departures from a purely centered dipole field were noted as early as 1970, when Wilson argued for a far-sided and right-handed distribution of virtual geomagnetic poles during the Cenozoic. Briden et al [1970] also questioned a purely GAD model. Assuming a given fit of the continents at the 1000-m-depth contour to have held for Permo-Triassic time, these authors find systematic departures of the then available paleomagnetic data from a GAD. They eliminate the possibil-ities of major errors in reconstruction or Earth contraction and

"/>

note that a significant octupolar component would improve the fit to the data. Since it is well known that the instantaneous field does depart significantly from a GAD, the question has been to find the time scales over which non-dipole terms could (or could not?) be considered as negligible. Carlut et al [1999] argued that mean values need be taken over only a few thousand years in order for significant, non-dipole low-degree terms to emerge from noise.

As noted by Kent and Smethurst [1998], empirical support for the GAD hypothesis has mostly come from analyses of the time-averaged field in the last 5 Ma, when data are abundant and plate tectonic motions can in general be neglected. Numerous recent analyses have provided an improving view of the long-term geometry of the field and its secular variation in these last 5 Ma [e.g. *Constable and Parker*, 1988; *Gubbins and Kelly*, 1993; *Quidelleur et al*, 1994; *Johnson and Constable*, 1995; *McElhinny et al*, 1996; *Johnson and Constable*, 1997; *Kelly and Gubbins*, 1997; *Carlut and Courtillot*, 1998]. Differences between these studies correspond to differences in databases and inversion techniques and probably reflect the confidence one can have on individual harmonic terms. Although there is argument over the significance of higher degree zonal and non-zonal terms, all models agree that the mean field over the last 5 Ma is dominantly GAD (g_1^0, denoted simply g_1 in the following), with a contributing axial quadrupole (g_2^0, denoted g_2 in the following) on the order of 5% of g_1. The axial octupole (g_3^0, denoted g_3 in the following) in these models is always less than 3% of g_1 (between 1 and 1.6% for *Johnson and Constable* [1997], and *Carlut and Courtillot* [1998]; 2.9% for *Kelly and Gubbins* [1997]); McElhinny et al [1996] find values on the order of 1 to 3% but with larger uncertainties and conclude that "g_3 is not large enough to be discernible from the currently available data") and many other terms are less than 1% of g_1 (below, we use the notation $G2 = g_2/g_1$, $G3 = g_3/g_1$). Carlut and Courtillot considered that all terms below a threshold of about 300nT (i.e. 1% of g_1) could not be considered as significant though this view was not shared by Kelly and Gubbins [1997], and to a lesser extent Johnson and Constable [1997].

Other authors have attempted to extract non GAD terms from global paleomagnetic databases extending further back in the past, when models are increasingly less well constrained and a number of uncertainties come into play (such as tectonic mobility, or inclination shallowing). Particular attention has been given to the last 200 to 250 Ma, for which plate kinematic reconstructions are reasonably well constrained thanks to oceanic data. Evans [1976] introduced an observational test, based on the frequency distribution of paleomagnetic inclinations, to discriminate between various hypotheses and concluded, based on data available up to 1971 (1271 |I| values binned to a total of 430) and ranging back to the Cambrian,

"that the geomagnetic field has been essentially dipolar throughout Phanerozoic time". The rationale of this approach was that for any given zonal multipole field component, such frequency distributions are characteristic if data are available from a large enough set of randomly distributed sites for a given time window. Evans' [1976] analysis of course only allows to show whether, based on observed and predicted distributions, the field can (or cannot) be distinguished from a GAD field. A number of subsequent analyses have detected an axial quadrupole component with an amplitude on the order of 3 to 6% of the axial dipole [e.g. *Coupland and Van der Voo*, 1980; *Merrill and McElhinny*, 1983], when data are averaged over the last 100 Ma or more, thus confirming some of Wilson's early ideas. Livermore et al [1983, 1984] tried to extract such a quadrupolar term in a worldwide paleomagnetic database going back 200 Ma. Using the method of Evans [1976] with ten times more data ranging back to 3000 Ma, Piper and Grant [1989] concluded that "the dipole model is a more efficient description of the geomagnetic field than quadrupolar and higher order configurations over most of geological time" (with the marginal exception of the 1000–600 Ma window). However, Meert et al [2003] have very recently shown that the assumption underlying the "inclination-only" studies ("that considerable continental drift has occurred over a sufficiently long period to render paleomagnetic sampling random in a paleogeographic sense") is not valid, and that "time-averaged inclination-only studies using the extant paleomagnetic database should be viewed with extreme caution".

In their 1990 review, Schneider and Kent concluded, from an analysis of 185 deep-sea sediment piston cores from low to middle latitudes from the last 2.5Ma, in favor of an axial quadrupole/dipole contribution of about 2.5 to 5%, and a standing octupole/dipole of 2 to 3% magnitude. They estimated "that pre-Pliocene values could have been 2 or 3 times larger". Besse and Courtillot [1991] argued that, for the period prior to 5 Ma going to 200 Ma, a significant quadrupolar term could not be extracted unequivocally from the data available at that time. Based on the analysis of 9490 spot readings of the field from 930 groups of I data covering the 0–5 Ma period, McElhinny et al [1996] found an acceptable zonal fit and best estimates of g_2/g_1 on the order of 3 to 4 ± 2%, and as recalled above could not determine a significant g_3 ("We do not believe it is possible to obtain a G3 estimate that can reliably be attributed to the time-averaged field rather than to some contamination"). They noted that "paleomagnetic data are often incorrect at the second order level required" for such studies "because of problems associated with the acquisition and processing of the remanent magnetization" (dip corrections, undetected local rotations, undetected minor overprints, inclination shallowing mimicking a g_3 effect, problems with space and time averaging, aliasing due to small errors,. . .).

In a recent update of our 1991 study [*Besse and Courtillot*, 2002, hereafter referred to as BC02], we again addressed the question of dipolarity of the past geomagnetic field. All sites were rotated to a common longitude (taken by reference to be 0) for each 20 Ma time window, and all such frames were stacked from 0 to 200 Ma. When a mean pole was computed in each time window, a jagged path resulted. Poles tended to follow an erratic course, yet remaining most of the time in the hemisphere opposite to the one centered on the reference meridian. The angular distance from the pole was always small (on the order of 2° on average) and the uncertainties such that most 95% confidence intervals included the geographical pole. When a grand 200 Ma average was computed, the mean pole was found to lie on the far side of the reference meridian, 1.4±1.2° away from the geographical pole. If taken at face value, this would imply a far-sided effect due to a persistent quadrupole on the order of 3% (±2%) of the dipole. We concluded that there were indications in favor of a persistent quadrupole on the 200 Ma time scale, with an amplitude on the order of half the one obtained for the last 5 Ma. However, these values are not distinguishable from zero in any individual 10 or 20 Ma time frame (except at 52 and 59 Ma), and the effect is on the order of contributions from other sources of paleomagnetic uncertainty. In conclusion, the GAD model (g_1, sometimes including, but most often neglecting a modest quadrupolar component g_2) has not been questioned much in most recent paleomagnetic studies.

2. RECENT STUDIES REVEAL OCTUPOLAR FIELDS

However, a paper by Kent and Smethurst [1998], followed by papers by Van der Voo and Torsvik [2001], Si and Van der Voo [2001] and Torsvik and Van der Voo [2002], have generated new interest in the potential existence of significant higher degree non-dipolar components to the long-term average geomagnetic field. Using various characteristics of inclinations in the available databases, these authors argue in favor of a significant axial octupole g_3, at least for certain time periods.

Using version GPMDB v. 3.1 [*McElhinny and Lock*, 1976] of the global paleomagnetic database, Kent and Smethurst [1998] follow the approach of Evans [1976] and analyze the frequency distributions of the absolute value of inclination $|I|$ for various subsets of the data. Kent and Smethurst [1998] test the null hypothesis of a purely GAD mean field against observed distributions, and compare them with distributions expected for combinations of GAD with quadrupole and octupole components. They analyze 6419 mean inclinations from rocks dating back to 3500 Ma, excluding only results considered as secondary magnetizations by the original authors. In order to have sufficient numbers of data for statistical significance of the analysis, they select four time windows cov-

ering the Cenozoic, Mesozoic, Paleozoic and Precambrian. They find that the binned data for the Cenozoic and Mesozoic cannot be told apart from a GAD model, whereas for the Paleozoic and Precambrian the GAD model can be rejected. They next analyze reasons for the latter observation, which Piper and Grant [1989] had attributed to remagnetization in the late Paleozoic. They reject this hypothesis, which they show should have led to GAD-like distributions; they also reject the hypothesis of inclination shallowing due to sediment compaction, by separating sedimentary and igneous data in the database and showing that similar distributions of $|I|$ are arrived at. Kent and Smethurst [1998] recognize that the distributions for pre-Mesozoic times could be due to preferential cycling of the continents by mantle convection in the equatorial belt, resulting in an anomalous number of low $|I|$ values. But they choose to explore the alternative possibility that these distributions could indeed be due to non-GAD terms, and find good fits to the observations when 10% and 25% contributions of the axial quadrupole and octupole (respectively) relative to the axial dipole are included. This would imply a long-term decrease of the G3 term, which Kent and Smethurst suggest might be linked to growth of the inner core. Indeed a field with a 25% G3 contribution would be quite different from the recent field. At the core-mantle boundary, the octupolar component would almost be of the same order as the dipole (G3 at the Earth's surface must be multiplied by the ratio of Earth's to core radius squared, i.e. 3.3 x 0.25 = 0.84), implying disappearance of the dipolar dominance of the field at the CMB, which has apparently prevailed since the Mesozoic. Meert et al's [2003] rejection of the assumption of random paleogeographic sampling also applies to the analysis of Kent and Smethurst [1998].

The impetus for the studies of Van der Voo and Torsvik [2001] and Si and Van der Voo [2001] comes from the fact that certain large-scale paleomagnetic observations, which are in some cases interpreted in plate tectonic terms, could simply have arisen from non-GAD field geometries. They cite three important examples: anomalously low inclinations in central Asia [e.g. *Westphal*, 1993; *Chauvin et al*, 1996; *Cogné et al*, 1999], late Permian reconstructions of Pangea (A vs. B; e.g. *Van der Voo et al*, 1976; *Irving*, 1977; *Smith and Livermore*, 1991; *Van der Voo*, 1993; *Torcq et al*, 1997], and discrepancies between poles from northern vs. central areas of Laurentia and Europe [*Kent and Clemmensen*, 1996; *Nawrocki*, 1999; *Torsvik et al*, 2001]. These authors use a database of 300 to 40 Ma poles from Europe and North America constructed by Torsvik et al [2001], from which they exclude results with low reliability (Q < 3, after *Van der Voo* [1993]) and from displaced terranes. The authors rotate the poles at 1 Ma increments with a 20 Ma moving window, determine a mean paleopole and compare each individual paleolatitude observation with the paleolatitude predicted from the mean paleopole which is closest in time (at a 1 Ma resolution). They then stack the data in three large time intervals corresponding to 200–300, 120–200 and 40–120 Ma, after having "centered" each distribution. Van der Voo and Torsvik argue that if there is no systematic departure from a GAD field for this large continental mass, data points representing observed vs. predicted inclinations should fall along the diagonal line. They find that such is the case for the 120–200 Ma window, but not for the 200–300 and 40–120 Ma windows. They then argue that overlooked secondary magnetizations or inclination errors cannot be the cause of the anomalous slopes and conclude that the most likely source is a contribution from a long-term non-dipole field. Van der Voo and Torsvik note that a G2 contribution is observed neither in the late Paleozoic nor in the early Mesozoic paleomagnetic data and reconstructions, particularly for continents bordering the Central and North Atlantic, and would not yield the anomalously low slopes they find at low paleolatitudes. On the opposite, a G3 contribution of about 0.1 would have the same effect as observed, and such a contribution would remove the need for alternate Pangea reconstructions: more precisely, a 0.1 G3 would bring continents in a Pangea A fit in the Late Carboniferous (if Australian poles are removed; otherwise, a 0.2 value is required), whereas a value of 0.2 is required for the Late Permian-Early Triassic. Van der Voo and Torsvik rightly note that the time around 250 Ma is the one where the largest deviations between Gondwana and Laurussia are found if a GAD hypothesis is used. In a reconstruction of North Atlantic bordering continents with particular reference to high latitude sites, Torsvik et al [2001] find that a G3 of about 0.08 reconciles the hot spot locations in reconstructions of the Atlantic and Indian oceans in the Cretaceous and early Tertiary. Si and Van der Voo [2001] analyze paleolatitude discrepancies found in Central Asia for a period from about 20 to 80 Ma: they compile two datasets from the year 2000 version of the GPMDB, using only data with Q (or demagnetization code) > 2, $k >$, 10 and $\alpha_{95} < 15°$, one to determine a reference APWP for the stable parts of the North Atlantic bordering continents, the other for Asian sites between 20–50° in latitude and 40–120° in longitude. They find that a G3 ratio of 0.06 or greater would significantly reduce the paleolatitude discrepancies between the two sets. However, they recognize in that case that contributions from inclination shallowing and NS convergence between central Asian blocks and Siberia are likely contributors, though unlikely to explain the discrepancies fully. Torsvik and Van der Voo [2002] complement their previous analysis [*Van der Voo and Torsvik*, 2001], using the pole list of Van der Voo [1993] with newer poles, revised ages and Q > 2. When trying to fit the separate APWPs of Gondwana and Laurussia for the period from 65 back to 350 Ma, they find that a G3 of 0.05 improves

the Cretaceous-Mid-Jurassic fit over a GAD, a "G3 of 0.1 provides a best fit for the Late Triassic- Mid-Jurassic", and G3 of 0.15 "leads to an almost perfect match in the Permo-Carboniferous segments". They note however that GAD gives the best match for the 130 Ma poles. They conclude that a Pangea-A fit from 310 to 170 Ma requires significant G3 contributions, and propose a global non-dipole field corrected APWP from 0 to 330 Ma which they consider "better than any GAD-based APW path". Torsvik and Van der Voo emphasize though that the inclination error [see *Rochette and Vandamme*, 2001] is a possible alternate mechanism that could have led to excessively strong estimates of G3 for the Paleozoic, (as found by *Kent and Smethurst* [1998]). Indeed, a flattening F of 0.75 reconciles magmatic and sedimentary data for both Gondwana and Laurussia back to 210 Ma with no further need for a G3 contribution (their figure 12d), but a systematic offset and the possible need for G3 remains prior to 230 Ma.

In summary, some recent studies argue for rather significant contributions of a long-term octupole (G3 up to 0.25) in the field, but the amplitudes and periods for which this is thought to apply differ somewhat (as do the data and techniques used to determine these contributions). One problem is that uncertainties are seldom given and are likely often on the same order of magnitude or larger than the values themselves. Kent and Smethurst [1998] find no need for non-GAD components in the Mesozoic and Cenozoic, but suggest a G3 up to 0.25 prior to 250 Ma. We have pointed out that this is not a "modest" contribution, as it brings the amplitude of the octupolar component close to that of the dipole at the CMB. Van der Voo and Torsvik [2001] note that no G2 is required in the Late Paleozoic-Early Mesozoic. Besse and Courtillot [2002] find a persistent G2 on the order of only 0.03 ± 0.02 averaged over the last 200 Ma; values computed in 10 Ma windows are seldom more than 4° away from the mean pole (their figure 7), implying G2 averages on these durations no larger than 0.08, with uncertainties on the same order of magnitude or larger. Schneider and Kent [1990] find values of G3 on the order of 0.05 to 0.06 for the Oligocene, while Torsvik et al [2001] and Si and Van der Voo [2001] suggest it may be 0.06 up to 0.1 for the Cretaceous and Early Tertiary. Finally, Torsvik and Van der Voo [2002] find evidence for a G3 of order 0.05 for 70–180 Ma (but 0 around 130 Ma), 0.1 for 180–220 Ma and 0.15 for 250–330 Ma. These discrepancies are of course due to the state and use of the database, with (more often than not) not enough data in each time window to compute statistically significant averages. In our BC02 paper, we spent some time discussing the presence of a small long-term quadrupole, but did not discuss how the database could be used to further check for the presence of higher order terms. The present paper undertakes this, i.e. a determination of maximum bounds on G3 contributions to the field in the last 200 Ma, using 10

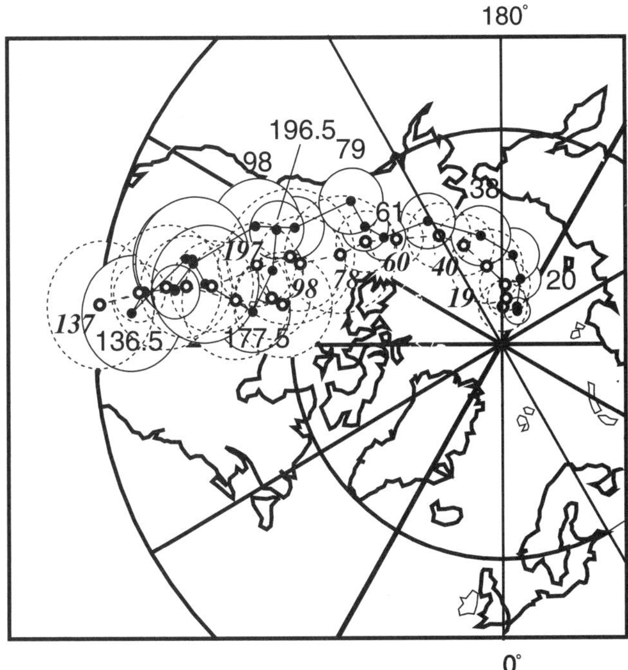

Figure 1. A comparison between the new synthetic apparent polar wander path derived in this paper (full dots and solid confidence intervals) and the one derived by Besse and Courtillot (2002) (open dots and dashed confidence intervals). Both in South African coordinates (see Table 1). Mean ages (in Ma) corresponding to some poles are indicated.

or 20 Ma long running time-windows. The data presentation is given in the Appendix.

3. DISCUSSION OF THE DATA AND THE RESULTING SYNTHETIC APWP

The data were combined in the same manner described in BC02. This resulted in a new estimate of the synthetic APWP for all these plates (our third since BC91 and BC02), which was computed in 20 Ma windows every 10 Ma back to 210 Ma (CB03). This is not the central point of the present paper and we only discuss this new synthetic APWP cursorily (Figure 1). The 95% confidence intervals of individual 20 Ma mean poles are in general less than 4°, except at 80, 100 and 140 to 180 Ma (in one case, at 150 Ma, it reaches 6°). In *all* cases, mean poles for the BC02 and CB03 curves are within the joint intersection of their 95% confidence intervals. The two synthetic paths, shown in South African coordinates in Figure 1, are essentially identical back to 70 Ma. The main difference occurs as a small jump at 80 Ma; this is largely due to inclusion of 11 data from Apulia and may therefore be considered with some caution. Between 80 and 110 Ma the two paths are offset by about 3 to 4°. The older parts are almost indistinguishable.

3.1. Analysis of Outliers

We have carefully analyzed all data points in each individual time window, in order to check for anomalous points or outliers. Outliers are defined as those paleopoles more than 20° away from the 10 Ma running mean. Out of 470 data, we found only 17 outliers. Three corresponded to data that should clearly be removed from our selection: they were 1) a 1978 pole from Italy, dated 29 Ma, in which it is indicated that no tectonic correction was applied, and which is contradicted by other concordant results from Italy from the same period (REFNO 1940/RESULTNO 1784); a 1984 pole from Finland, dated 77.5 Ma and linked to impact melt rocks for which no structural study is available (REFNO 2225/RESULTNO 5829); a 1993 pole from East Antarctica, dated 176.5 Ma, regarded by the original authors as transitional (REFNO 2721/RESULTNO 7080). Six corresponded with data we had first considered as autochthonous, and which were clearly displaced but easily reconciled with the mean if allowed to rotate about a vertical axis. They were transferred from "o" to "m"-type category (see Appendix for definition of these codes). They were: 1) a 89.5 Ma pole from Sudan; 2) 77.5 and 89.0 Ma poles from Egypt, close to the Gulf of Suez and an area of transition from Red Sea rifting to Dead Sea transform faulting prone to local rotations (REFNO 3260/RESULTNO 8394, 8396), 3) two 155.5 Ma poles from Patagonia, on the edge of the southern Andes and therefore clearly prone to local rotations (REFNO 3302/RESULTNO 8583, 8584); 4) a 170 Ma pole from the Chon Aike intrusives (REFNO 2558/RESULTNO 2182). This left 8 outliers for which we found no reason to change the status and which were therefore retained in our selection (4 data between 0.5 and 8.5 Ma from Chaîne des Puys, Nigeria, Guinea and Silesia, the 19.5 Ma Cairo basalts, a 57.5 Ma pole from the Sydney basin, 107.5 Ma intrusives from Spitsbergen, and a 192 Ma pole from Yorkshire (barely an outlier). We then redid the entire calculation of the synthetic APWP with our final selected sub-database consisting of 465 data. The 95 confidence intervals from 4 time windows improved by up to 1° (at 80 Ma), with now only five above 4° and none reaching 6° (100 Ma, and 140 to 180 Ma). This is actually the APWP given in Table 1.

3.2. Some Examples of Rotated Areas

It is particularly interesting to see how the "m"-type data (see Appendix) contribute to the individual means, and whether they are always compatible with the means based on autochthonous data only. We will give one illustration. Rotations are clearly seen to be required by some data from Italy and the Colorado Plateau at 50–70 Ma: we see in Figure 2 that when a number of poles are left free to rotate about local vertical axes

Table 1. Master apparent polar wander paths in South African coordinates for the past 200 Ma (20 Ma window)

Window	BC02					CB03				
	Age	N	λ (°N)	ϕ (°E)	A95	Age	N	λ (°N)	ϕ (°E)	A95
0	3.1	30	86.2	176.9	2.6	2.6	146	86.2	152.5	1.4
10	8.3	54	85.3	173.5	2.0	5.0	181	85.8	156.0	1.2
20	18.9	38	83.9	175.9	2.7	19.6	66	83.0	162.6	2.2
30	29.5	23	81.8	190.7	3.9	28.5	43	80.7	172.0	2.8
40	40.0	24	79.0	201.1	3.2	38.2	34	78.5	190.8	3.1
50	52.2	31	76.9	210.3	3.4	51.8	38	75.0	211.3	2.9
60	59.7	45	74.3	225.7	2.9	60.8	54	73.2	228.5	2.4
70	67.3	34	71.9	233.7	3.2	68.5	52	70.9	230.0	2.5
80	77.9	14	70.3	241.6	6.1	78.9	34	67.9	227.2	3.4
90	90.0	13	66.7	248.7	4.9	89.5	23	64.5	241.2	3.2
100	97.6	12	65.4	248.0	7.0	97.8	15	60.6	245.0	4.9
110	113.6	17	57.9	259.0	4.6	110.9	19	55.3	255.7	3.9
120	119.1	20	54.0	261.1	2.7	119.7	26	53.8	260.7	2.0
130	126.4	14	50.0	262.1	2.9	127.1	23	50.6	261.8	2.1
140	136.8	7	45.7	264.5	6.2	136.7	13	49.4	265.3	5.7
150	151.6	10	52.9	260.6	6.2	152.4	18	54.4	255.2	5.9
160	162.3	15	55.1	259.9	5.1	161.4	28	56.6	256.0	4.3
170	173.4	21	60.8	260.9	6.0	171.5	32	58.6	259.9	4.3
180	178.8	18	66.0	260.1	5.4	177.7	29	62.8	262.9	4.1
190	189.7	23	64.7	259.0	4.2	191.4	38	64.1	252.6	3.3
200	196.7	19	62.2	252.4	4.3	196.5	35	62.9	243.6	3.0
210						208.1	16	61.1	232.7	3.8

Window: center of time window in Ma; Age: actual mean age in Ma; N: number of data in window; λ and ϕ: latitude and longitude of mean pole in degrees; A95: 95% confidence interval of pole.

(m-type data), they can be brought in full agreement with the mean pole from stable sites. Two poles from the Colorado plateau display a clear clockwise rotation of about 20°, whereas a Sicilian pole shows the large (about 110°) clockwise rotation of allochthonous parts of the Apulian promontory of Africa. Lebanon is also a nice example at 100–120 Ma, and the Falkland Islands at 180–200 Ma.

4. AN ATTEMPT TO UNCOVER OCTUPOLAR COMPONENTS IN THE SELECTED DATABASE

Since in BC02 we addressed the problem of a long-term G2 (quadrupolar) component, in this paper we focus on trying to extract any evidence for a significant G3 (octupolar) component, as proposed in the recent studies summarized in section 2 of this paper.

We rotated each mean pole in each 20 Ma time window (obtained in the previous section) in order to bring it at the rotation axis (geographic North pole), and then rotated all sites contributing to the selected sub-database to a common reference longitude (the so-called "common site longitude"). We then analyzed the transformed data in two different ways. As in previous papers, the idea was to try and extract from noise a latitude dependence of observed inclinations, in order to try and see if they departed in a systematic way from what is expected for a GAD (plus a small G2 as found for instance in paper BC02). The difference between the inclinations I at a given site expected from a purely GAD field and a GAD plus G3 only field geometry is a simple antisymmetric function of latitude obeying the equation:

$$\tan(I) = \cot\theta \, (1 + G3(5\cos^2\theta - 3))/(2 + 3.G3(5\cos^2\theta - 1))$$

θ being colatitude; see e.g. McElhinny et al [1996]). The differences in corresponding VGP latitudes follow a similar curve (Figure 3). We see from Figure 3 that one way to try and uncover the octupolar component in the case of few and noisy data could be to simply compute three averages: one over equatorial latitudes (say 25°S to 25°N), and two in opposite mid- to high-latitude bands (say 25° to 70°N, and 25 to 70°S). This should lead to negligible values for the equatorial mean (formally 0) and significant and opposite values for the mid- to high-latitude bands, directly proportional to G3 (in the case of Figure 3, with a G3 of 0.03, these values are approximately

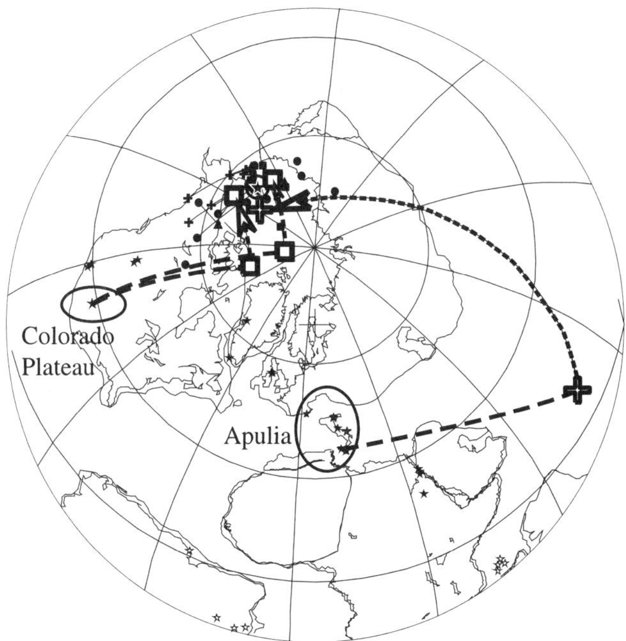

Figure 2. An example of data from a 20 Ma time-window (50–70 Ma). Site locations are indicated by stars and paleomagnetic poles from autochtonous sites by solid dots. Poles from allochthonous sites which have likely suffered significant rotations about locally vertical axes are shown: poles from the Colorado plateau (large open squares) and Apulian promontory of Africa (large open cross) are left free to rotate and the positions they attain are also shown, with arrows indicating the rotations. Other poles from allochthonous sites are shown as small crosses before and after rotation. The overall mean pole for this time window is shown as a larger star in a circle.

+ and – 2.2°, roughly proportional to G3 for these small values; Figure 3 also shows the latitude anomalies for a 0.03 quadrupole, and a model with both G2 = 0.03 and G3 = 0.03). With this in mind, we first computed mean poles for each time window in the three latitude bands suggested above. Results are shown in Table 2a and are illustrated in Figure 4. A significant G3 component should result in the three mean poles in each window being aligned on the common meridian, with the equatorial-band mean pole not significantly different from the North pole, the northern mid- to high-latitude band mean pole being far-sided, and the southern mid- to high-latitude band pole being near-sided by the same amount, as already suggested by Figure 3 (Figure 4a).

In Table 2a, we can first check the overall significance of departures of each individual mean pole from the North pole (i.e. distance from the North pole less than the corresponding 95% confidence interval). Out of 58 cases (in 2 cases there are not enough data to compute a mean), 33 cannot be distinguished from the North pole, 7 are less than 1° above that limit, and 18 are more than 1° above that limit, yet never more than 5° (only four—70N, 80N, 100S and 160E—are more

than 3° above that limit; the 70N notation has 70 standing for the mean age of the window (70 Ma) and N for the northern mid- to high latitude band; the two other bands are S for southern mid- to high latitudes, and E for equatorial latitudes). Out of these, 9 values have less than 8 contributing paleopoles, and hence carry less confidence, reducing the number of possibly significant results from 25 to 16 out of 60. So whatever the long-term non-GAD and G2 field components, they are bound to be small. Let us first have a look at the mean poles for the equatorial latitudes: 9 are insignificantly different from the North pole, and all others but one are no more than 2° farther away than their 95% confidence radius. The exception, at 160 Ma (where the difference is 5°), has only 6 contributing paleopoles. We would need to have a detailed look at the geographical distribution of contributing sites, since the hypothesis that the values for the equatorial band should lie on the North pole implies a homogeneous distribution of sites, which is of course often likely not to hold. If we now compare N and S estimates, whose antisymmetry about the

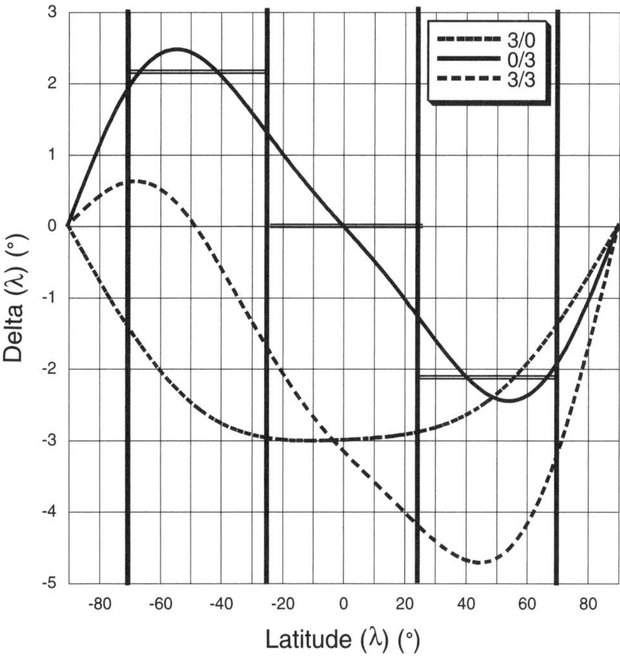

Figure 3. The "latitude anomaly" (actual latitude minus latitude derived from inclination under the assumption of a purely dipole field) as a function of latitude in the cases of three field models, all with an axial dipole contribution $g_{10} = g_1$ and a) an axial quadrupole contribution $g_2 = 0.03g_1$ and no octupole (short dashed line), b) an axial octupole contribution $g_3 = 0.03g_1$ and no quadrupole (solid line) and c) an axial quadrupole contribution $g_2 = 0.03g_1$ and an axial octupole contribution $g_3 = 0.03g_1$ (longer dashed line). Mean latitude anomalies are shown as double lines for the pure octupolar case when averaged in the three latitude bands 70–25°N, 25°N–25°S, 25–70°S.

Table 2a. Mean poles for three latitude bands for the past 200 Ma (20 Ma window)

Window	Latitude band	N	λ (°N)	φ(°E)	A95	(90°-λ)	δ
10	S	21	88.7	348.3	3.4	1.3	-2.1
	E	47	87.5	227.4	2.2	2.5	0.3
	N	112	86.9	146.9	1.6	3.1	1.5
20	S	15	88.8	95.2	4.6	1.2	-3.4
	E	20	89.1	44.4	3.3	0.9	-2.4
	N	32	85.4	104.6	2.7	4.6	1.9
30	S	8	85.8	229.3	5.3	4.2	-1.1
	E	10	88.4	339	4.5	1.6	-2.9
	N	24	84.9	112.3	3.0	5.1	2.1
40	S	4	85.2	132.6	14	4.8	-9.2
	E	11	88.9	10.7	4.6	1.1	-3.5
	N	19	84.9	137.7	4.2	5.1	0.9
50	S	6	84.0	205	11.4	6,0	-5.4
	E	4	87.7	91.2	11.8	2.3	-9.5
	N	28	85.2	135.7	3	4.8	1.8
60	S	12	84.6	221	4.3	5.4	1.1
	E	14	83.7	234.4	4.5	6.3	1.8
	N	28	85.4	139.8	2.4	4.6	2.2
70	S	11	87.1	241	4	2.9	-1.1
	E	20	82.5	223.4	5.3	7.5	2.2
	N	20	83.1	121.5	3	6.9	3.9
80	S	8	86.6	70.7	4.2	3.4	-0.8
		9	64.1	260.1	14.3	26.9	12.6
	N	17	82.7	109.7	3.5	7.3	3.8
90	S	7	86.0	106.1	3.8	4,0	0.2
	E	7	83.0	293.5	9.4	7,0	-2.4
	N	9	87.1	203.6	3.1	2.9	-0.2
100	S	3	82.4	138.8	3.5	7.6	4.1
	E	5	86.9	52.8	7.3	3.1	-4.2
	N	7	84.8	183.3	6.3	5.2	-1.1
110	S	7	87.8	28.3	5.1	2.2	-2.9
	E	6	82.0	258.5	6.2	8,0	1.8
	N	6	85.3	185.6	7.4	4.7	-2.7
120	S	12	88.8	36	3.2	1.2	-2,0
	E	6	82.1	260.4	6.1	7.9	1.8
	N	8	89.6	102.1	1.8	0.4	-1.4
130	S	6	88.8	195	4.6	1.2	-3.4
	E	7	85.4	243.7	5	4.6	-0.4
	N	8	89.0	291.6	2.7	1,0	-1.7
140	S	3	85.7	61.7	1.9	4.3	2.4
	E	6	88.1	236.1	4.7	1.9	-2.8
	N	4	82.2	267.7	12.3	7.8	-4.5
150	S	1					
	E	7	79.9	132.4	8	10.1	2.1
	N	10	84.3	253.7	6.5	5.7	-0.8
160	S	4	82.1	312.7	7.1	7.9	0.8
	E	6	78.5	139.1	6.5	11.5	5,0
	N	16	87.3	221	5.9	2.7	-3.2

δ = difference between colatitude (90°-λ) and 95% uncertainty A95.

Table 2a. (continued)

Window	Latitude band	N	λ (°N)	φ(°E)	A95	(90°-λ)	δ
170	S	15	87.2	349.4	4.3	2.8	-1.5
	E	3	83.4	156.1	12.7	6.6	-6.1
	N	11	85.3	228.2	8.7	4.7	-4,0
180	S	16	85.2	9.3	4.8	4.8	0,0
	E	4	82.8	158.3	5.9	7.2	1.3
	N	6	85.6	249.3	10.3	4.4	-5.9
190	S	8	82.4	350.8	7.5	7.6	0.1
	E	21	85.8	159.4	3.5	4.2	0.7
	N	9	83.2	304.6	5.3	6.8	1.5
200	S	3	85.4	256.1	17.2	4.6	-12.6
	E	25	86.7	160.3	3.2	3.3	0.1
	N	7	83.2	277.3	8	6.8	-1.2

pole on the common meridian would be diagnostic of a G3 effect, we find that these are seldom significant. The N and S means are actually both far-sided by a similar (often not statistically significant) amount in most time windows between 20 and 110 Ma. The difference between the N and S mean is not statistically different from zero at 20, 40, 50, and most means from 100 to 200 Ma (except 150 Ma), often but not always (e.g. 130 Ma) because one of the 95% confidence intervals is rather large (and the corresponding number of contributing paleopoles small). Because the S means often have smaller numbers of data and larger 95% confidence intervals, we now have a look at the better determined means, i.e. the ones from the N-latitude band. The only systematic feature we find is that N means from 10 to 110 Ma are (often significantly) far-sided. Values are on the order of 4° (from 3 to 5°), with 95% confidence intervals on the order of 2 to 3°. If we assume that the S poles are anti-symmetrical (this is often allowed by their larger 95% confidence intervals), in order to maximize the potential G3 contribution, the total S to N-mean distance would be on the order of 8°, with a (very roughly estimated) uncertainty no less than 6°. Taken at face value, this would correspond to a G3 of 8 ± 6%. However, inspection of Table 2a and Figure 4 shows that the effect is not systematic: 1) it is not seen prior to 80 Ma, 2) it does not really apply between 20 and 70 Ma, 3) the departures from zero are often not much larger for the N and S-poles than for the E-poles, which should not happen. . . So there really is no solid evidence there for a significant G3 term, for instance reaching 10% for long periods of geological time, such as the late Mesozoic and Cenozoic as suggested by Si and Van der Voo [2001]. We show in Table 2b the grand means for the 0–65, 65–210 and 0–210 Ma 200 Ma periods (Figure 5 a,b,c). In the 0–65 Ma case, only the N-pole is significantly offset from the North pole, by 3.5 ± 1.2° (N = 168 data). The angular distance between the N- and S-mean is 4.3 ± 2.7° but S is not near-sided and the offset along the common site meridian

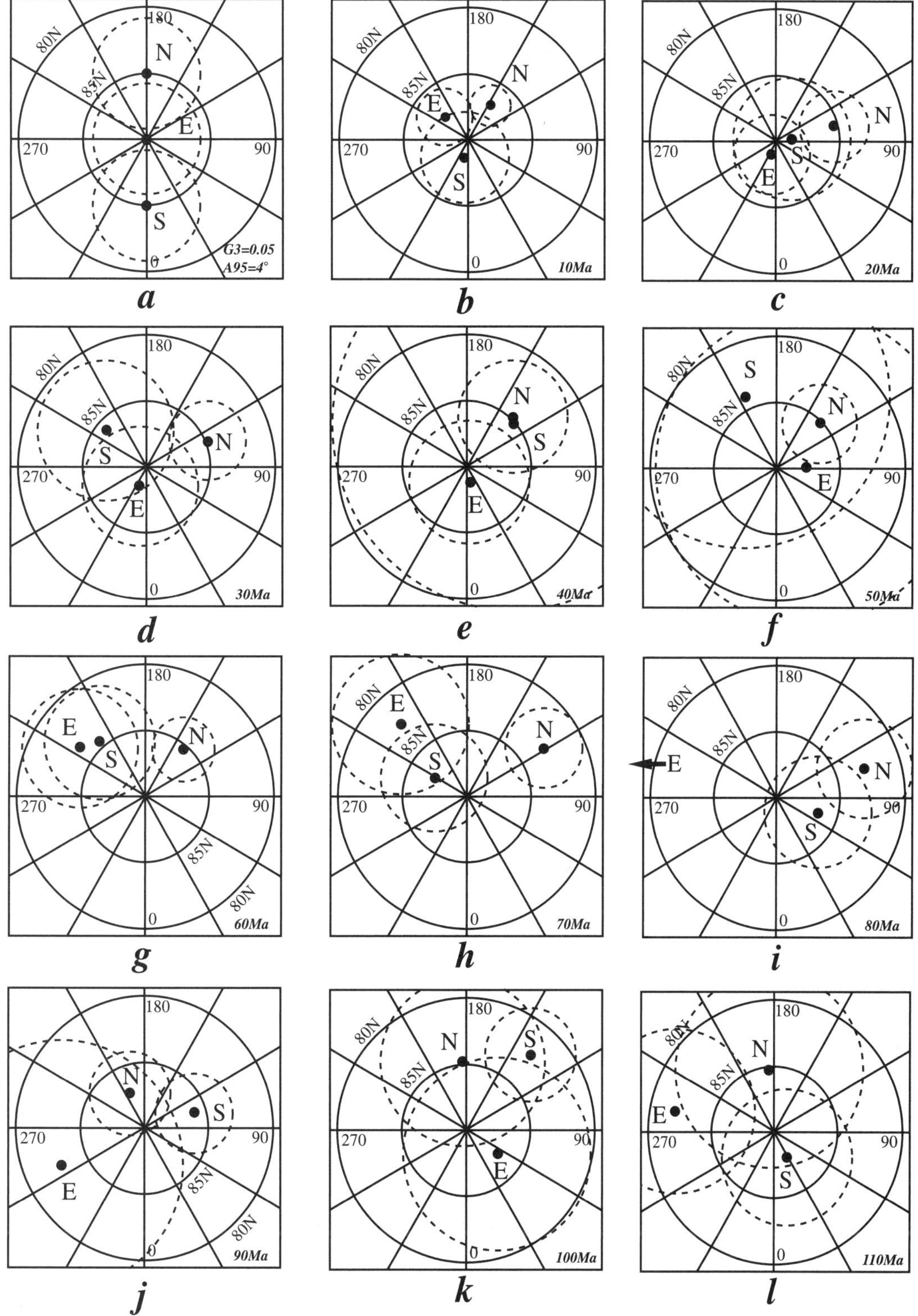

Figure 4. Mean poles (dots) and 95% confidence intervals (dashed circles) in three latitude bands (70–25°N: pole marked N; 25°N–25°S: pole marked E; 25–70°S: pole marked S), in successive 20 Ma time windows from 0 to 210 Ma, centered on the indicated age (in Ma). The data have been brought to common-site longitude as indicated in the text. Cartoon a is for a theoretical distribution with a field consisting of a dipole plus an axial octupole contribution $g_3 = 0.03g_1$ and no quadrupole.

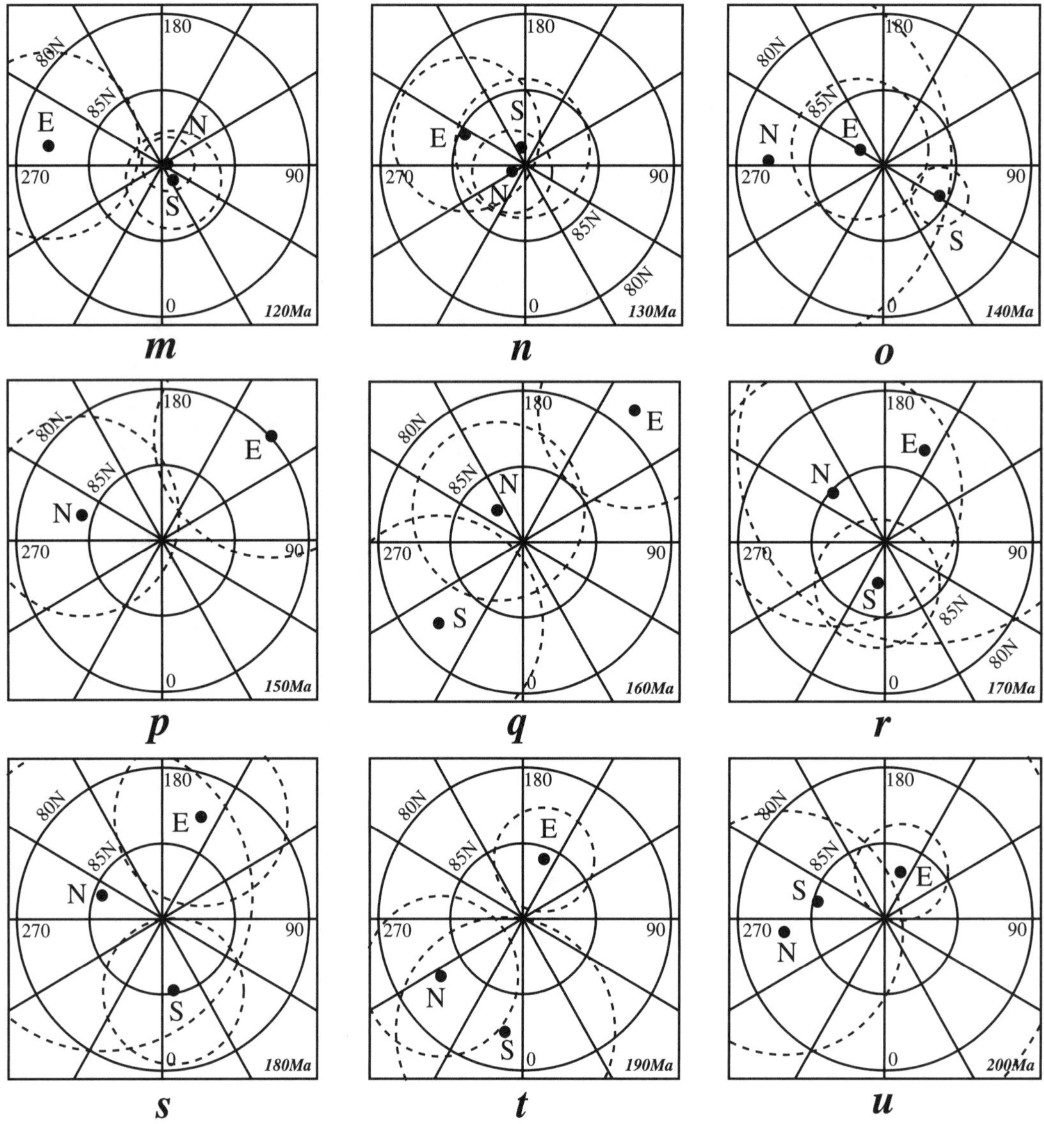

Figure 4. (continued)

is insignificant. For the 65–210 Ma mean, the N and E poles are undistinguishable and there is a slight indication of N/S antisymmetry, with an angular distance of only 4.5 ± 3.5°. There is also some indication of a N/S antisymmetry in the overall mean for the whole 200 Ma period, with an angular N/S-mean difference reaching about 3.6 ± 1.9°. But again, the S-mean is not significantly different from the North pole, whereas the E-pole is, being about as far-sided than the N-pole.

We have finally separated from the database poles corresponding to extrusive (N = 247), intrusive (N = 106) and sedimentary (N = 115) rocks. These new sub-databases are now too restricted in number to attempt an analysis in 20 Ma windows, but we can resume the analysis for the entire set. This is shown in Table 2c (and Figure 5d,e,f). The three mean poles for the

extrusives are essentially identical to those previously found for the entire database, though they have slightly larger α_{95}, by 0.5°. The α_{95} for the intrusives are doubled, and in that case the three means are indistinguishable from either the North pole or each other, so that no non-GAD contribution is detectable. The α_{95} for the sediments are similar, though the value is much larger (at 5°) for the S-mean which has only 15 contributing poles. The two well determined means, N- and E- are almost identical, with the N-mean 3.8 ± 2.1° away from the North pole. It may be significant to underline the fact that the N-means for all data between either 0 and 65 Ma or 0 and 210 Ma, for data restricted either to extrusives or to sediments are similarly offset from the North pole, being far sided by about 3°. Only the extrusives show a significantly right-handed mean.

Table 2b. Overall mean poles for three latitude bands for 0-65, 65-210 and 0-210 Ma periods

Window	Latitude band	N	λ (°N)	φ (°E)	A95	(90°-λ)	δ
0-65 all	S	39	88.1	239.4	2.4	1.9	-0.5
	E	75	87.8	229.9	1.8	2.2	0.4
	N	168	86.5	139.7	1.2	3.5	2.3
0-210 all	S	89	89.2	339.9	1.6	0.8	-0.8
	E	139	87.6	211.9	1.3	2.4	1.1
	N	234	87.2	153.3	1.1	2.8	1.7
65-210 all	S	54	88.4	0.0	1.9	1.6	-0.3
	E	75	86.2	192.2	2.1	3.8	1.7
	N	65	86.9	212.0	2.9	3.1	0.2

Table 2c. Overall mean poles for three latitude bands for 0-200 Ma, separately for intrusive, extrusive and sedimentary rocks

Window	Latitude band	N	λ (°N)	φ (°E)	A95	(90°-λ)	δ
0-210 int	S	50	88.8	328.3	2.0	1.2	-0.8
	E	81	87.5	231.5	1.7	2.5	0.8
	N	115	86.3	139.6	1.8	3.7	1.9
0-210 ext	S	23	88.2	32.1	3.4	1.8	-1.6
	E	26	88.2	163.2	2.9	1.8	-1.1
	N	56	88.9	164.8	2.3	1.1	-1.2
0-210 sed	S	15	88.3	219.4	5.1	1.7	-3.4
	E	32	86.6	176.6	2.6	3.4	0.8
	N	64	86.2	176.3	2.1	3.8	1.7

δ = difference between colatitude (90°-λ) and 95% uncertainty A95.

The fact that means based separately on extrusives and sediments are compatible (but for the right handed effect which is not analyzed further here) eliminates the need to resort to a systematic effect of syn-sedimentary inclination shallowing (as concluded in their respective studies by Kent and Smethurst [1998] and Van der Voo and Torsvik [2001]).

So any evidence for a long-term G3 component is at best rather limited. Amplitudes would not be in excess of 6 ± 4% for the various means (all data, extrusives only, sediments only) for the last 200 Ma. And all departures from the expected geometry (N/S-mean asymmetry and E-mean centered on the North pole) show that the G3 hypothesis is generally not validated, and that the data are still either too few or too noisy to allow a clear identification.

We have next had a look at the detailed distributions of "latitude anomaly" vs. latitude. We have constructed this distribution for our selected data in the 0–210 Ma window (N = 465 data). It forms a very dispersed set, with a rather high noise level and few data points for which the uncertainty

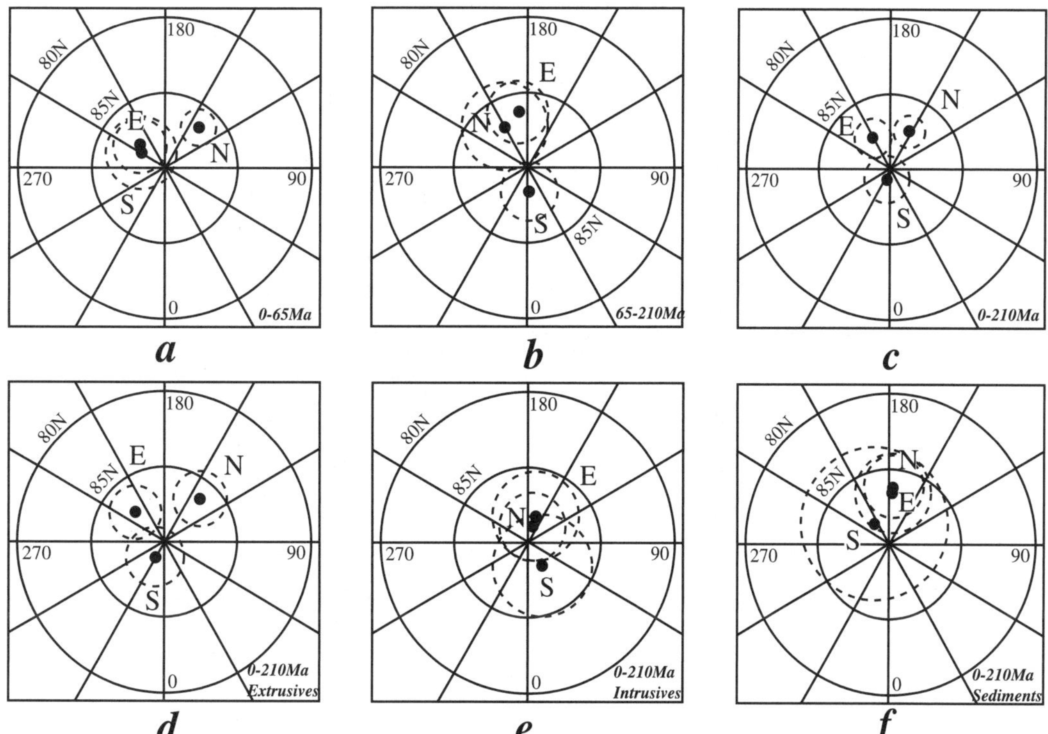

Figure 5. Same as Figure 4, a) for the Cenozoic data (0–65 Ma); b) for the 65–210 Ma period; c) for the full data base (0–210 Ma); d) for the whole 0–210 Ma period but with data restricted to extrusive rocks; e) same restricted to intrusives; f) same restricted to sediments.

does not include 0. Following McElhinny et al [1996], we average values within 13 latitude bands (90–50°S, then in 10° bands, finally 60–90°N) and estimate both the standard deviation and the 95% confidence limit on each mean (Figure 6). We see that the distribution is rather scattered with large uncertainties, and some significant jumps from one band of latitude to the next. Fitting a zonal model to the data would clearly require rather significant terms beyond order 3. In Figure 6, we also display the three expected distributions from Figure 3, that is with (G2 = 0, G3 = 3%), (G2 = 3%, G3 = 0), and (G2 = 3%, G3 = 3%). None of these simple models satisfy the data. The "quadrupole-only" curve gives a rather good fit overall, except in three latitude bands in the southern hemisphere centered on 5°S, 35°S and 45°S. The "octupole-only" curve does not fit the data at 25°S and 15°N, and is an equally

good or bad fit. Combining both effects gives a similar result. All that can be said with these data are that mean components are unlikely to exceed these values on the order of 3%. The latitude anomaly tends to be systematically negative in the entire northern hemisphere, and zero on average in the southern hemisphere, implying a break in hemispherical symmetry (values in the S- and N- mid- to high-latitude bands as defined above are respectively 0.9 ± 6.5° and –2.8 ± 5.9°). Taken at face value, the amplitude difference between N- and S- mid latitudes is about 4° (actually a non-significant 3.7 ± 8.8°), corresponding to a non-significant G3 of 2.5%. The same analysis was performed separately for the 0–5 Ma and 5–210 Ma data sets (figure not shown here). It shows the same signature, implying that there is no systematic bias due to the more abundant 0–5 Ma data set. For this recent time period (N = 123

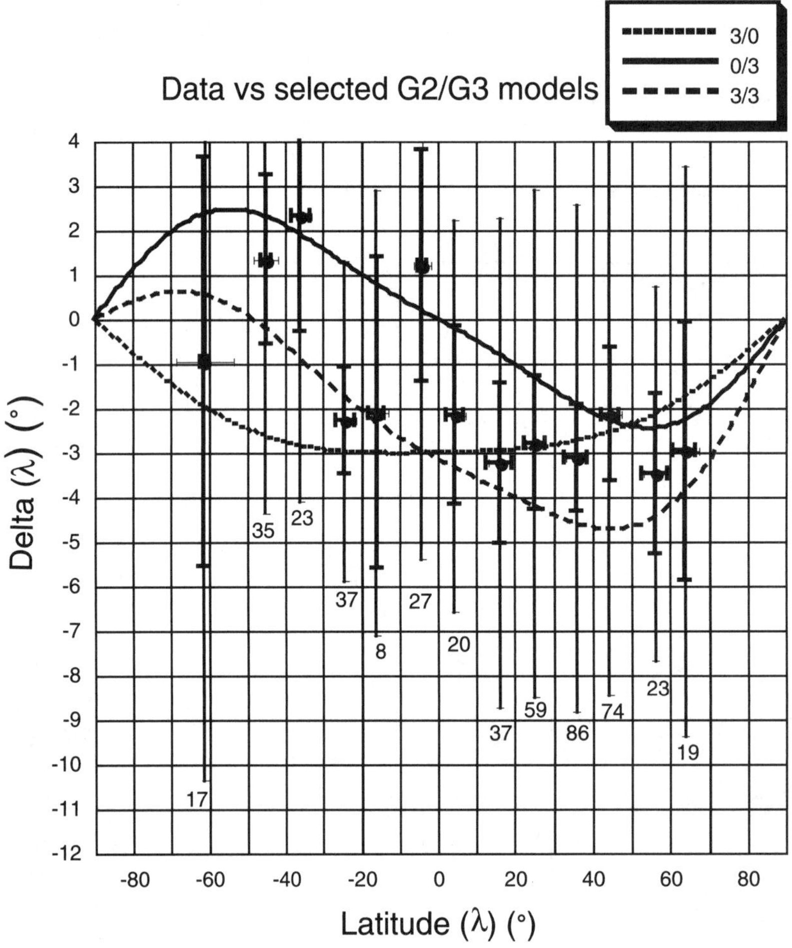

Figure 6. "Latitude anomalies" as a function of latitude (as in Figure 3). Data in common site longitude from our selected database are averaged 13 latitude windows as in McElhinny et al (1996), with standard deviation, 95% confidence interval and number of data in band indicated. Three theoretical curves are shown for comparison, for the same three models as shown in Figure 3. No model fits all data. An octupolar component much larger than 3% of the dipole would contradict the data.

data), our results can be compared with many other, more elaborate studies. The southern hemisphere data are not numerous enough and are very noisy. The northern hemisphere data are more numerous and consistent and once again indicate a significant negative anomaly of the same amplitude as the overall mean (the N-mean is $-2.6 \pm 5.6°$, and its difference with respect to the 200 Ma mean is $0.2 \pm 7.5°$). More specifically, there is an aberrant point due to two data near 52°S (the only data South of 50°S!). Otherwise, all data are consistent with a G2 of 3%, with a particularly good fit for the northern hemisphere data. The southern hemisphere data are slightly better fitted with a combined (G2 = 3%, G3 = 3%) but a purely octupolar model is inconsistent with most data. Separating extrusive and sedimentary data does not reveal significant differences. The sedimentary data are too few in the southern hemisphere for a meaningful analysis. They may appear to have slightly larger negative values in the northern latitude band than the extrusives, but only by a non-significant 10% ($-4.0 \pm 6.4°$ vs. $-2.8 \pm 5.5°$, i.e. a difference of $0.2 \pm 8.4°$).

5. DISCUSSION AND CONCLUSION

The analyses presented in this paper for the last 210 Ma indicate that there is no compelling evidence of a significant octupolar contribution to the field. One might be concerned that the outliers that we removed could be those influenced by non-dipole fields. We have found only 17 outliers in 470 data (i.e. less than 4% of the total). Out of these 17, we retained 8, changed the status of 6 for serious tectonic reasons, and suppressed only 3 for reasons we believe are rather readily acceptable. This cannot be the reason why non-dipole terms actually present in the mean field would have become erased. Our analysis reveals a consistent pattern at any time scale from tens to 200 Ma and for any rock type, in which the total amplitude of the latitude anomaly is on the order of 3 to 4° with an uncertainty on the order of 8°, so again a potential G3 on the order of 3 to 4% ($\pm 8\%$) is allowed but not required by the data. Yet, many other sources of dispersion remain and the detailed geometry of the data distributions does not reduce to the simple one expected for an octupolar contribution to the field. The presence of a small but significant mean quadrupole has been elaborated on previously (e.g. *Besse and Courtillot* [2000]). It remains more reasonable, in our view, at the present stage of development of the database, to search for explanations with well known physical background (even if not always easy to model or justify) such as remagnetization, syn- or post-sedimentary inclination shallowing or small- or large-scale tectonic effects in order to try and account for what appear to be anomalous regional observations, such as those pertaining to the Cenozoic Asian inclination problem or even the Pangea A vs. B debate: even when sampling is restricted to sites close to the paleo-equator of large supercontinents, the APWPs for North Africa and Laurussia are significantly different since about the middle Triassic (J. Besse et al., work in progress). Poles and reconstructions appear to be consistent with an A-type Pangea back to 260 Ma, but a B-Pangea is favored prior to 260 Ma. In that case also, there is apparently no need to resort to (and no strong evidence in favor of) a significant octupolar component of the field, back to at least 320 Ma. We note that problems occur around 250 Ma, i.e. the time of the Permo-Triassic boundary, for which clearly more data are required for a robust analysis. Detailed analysis of the available database is a rewarding and useful exercise. For instance, the overall Cenozoic rotations of data from the Apulian/Italian peninsula or from the Colorado plateau appear to be well supported by comparison of data from these areas with the rest of the global database. Increasing and improving significantly the database remains a very significant goal for paleomagnetists, and hopefully we should be able to start constraining more significantly the percentage of quadrupolar and octupolar contributions to the GAD, thereby improving both the quality of paleogeographic reconstructions and constraints to geodynamo modelers. For the time being, our best estimates for mean quadrupole and octupole in the last 200 Ma are respectively on the order of $3 \pm 2\%$ and $3 \pm 8\%$.

Acknowledgments. We are happy to contribute to this volume in the honor of Neil Opdyke, one of the founding fathers and most important figures of modern paleomagnetism. We thank the organizers of the Chapman conference organized in 2003 in Gainesville, Florida, to celebrate this great friend and colleague. We acknowledge reviews by Trond Torsvik and an anonymous reviewer. We wish to point the reader to the paper by M. McElhinny in the same volume, which the author kindly made available to us after this chapter was written, but prior to publication. We are also grateful to M. Flanigan and J. Meert for discussion. IPGP Contribution NS 1956.

APPENDIX: PRESENTATION OF THE DATA

We recall the data used in our previous paper (BC02), which this paper largely follows, and indicate updates to these data.

In BC02, we used the June 1999 version of the GPMDB (v3.3), which includes all paleomagnetic data published up to the end of 1997. We used the data from most plates: NAM, EUR, AFR, IND, AUS, ANT, SAM and GRE (Greenland). We retained data from both igneous and sedimentary rocks. The selection criteria applied to sift the base and extract (hopefully robust and high-quality) subsets were the following (most being similar to those used by other authors; see *McElhinny and Lock* [1995], and also *Van der Voo* [1993]): 1) at least 6 sites and 36 samples per study; 2) a 95% confidence interval less than 10° in the Cenozoic and 15° in the Mesozoic; 3) evidence for successful alternating field and/or thermal demagnetization (i.e. Demagnetization Code equal to or larger than 2); 4) dating uncertainties less than 15 Ma (this is one of the main nagging problems in updates of the database); and 5) absence of remagnetization ensured by the fact that differences between magnetization and stratigraphic ages should be less than 5 Ma and by rejecting poles with a negative fold test or negative reversal test. One of the key aspects of our selection process was an attempt to identify mobile zones from which data should be excluded from what should in principle be rigid plates. Our task was then to determine whether the extent of deformation required that the data be rejected or not. A major source of error is linked to local tectonic rotations about vertical axes. Detailed analysis of individual poles and comparison with synthetic APWPs allow one to spot remaining outliers with a previously undetected tectonic origin. In some cases, we have used another method: some poles that where suspected to have undergone rotations without significant poleward motion have been integrated using a method derived from McFadden and McElhinny [1988]. We treated these poles using only inclination data, i.e. as small circle constraints. Another problem was to determine to which extent large plates could be considered as rigid. The ocean-based kinematic syntheses of Müller et al [1993], which imply that Africa and South America need to be subdivided into a small number of rigid subplates, were used. Asian data were excluded from the compilation, which comprised 221 poles.

Here, contrary to BC02, we do not include data from ODP/DSDP drilling sites or skewness poles coming from analyses of marine magnetic anomaly profiles (this results in the loss of only 21 data compared to BC02). We have been working with version v.4.3 of the GPMDB as available in July 2002, which includes all published data up to March 2002. Our time interval is from 0 to 220 Ma. For the same reasons discussed in BC02, we have excluded all data from central and eastern Asia, except India, and the Pacific Ocean, which could not reliably be reconstructed to the other plates. This results in 1850 data. We have next excluded data for which 1) the number of samples was less than 15; 2) the number of sites was less than 6; 3) the 95% confidence interval was larger than 15°; 4) the demagnetization code was less than 2; 5) the dating uncertainties were larger than 15 Ma; 6) there was evidence of remagnetization (differences between magnetization and stratigraphic ages larger than 5 Ma and/or a negative fold test). All superceded data, and data combined elsewhere in the GPMDB, were removed (at least those we were able to identify; there may well be cases that escaped our scrutiny). This set of filters reduced the data to 1022. A number of studies (124) for which there was no indication of number of sites, often implying magnetostratigraphic stud-

ies, were set aside. The remaining 898 data were next separated in three categories, based on our understanding of their tectonic environment. Since this is a step that is not indicated in the GPMDB, we spend some time explaining it (the complete data table is published in www.ipgp.jussieu.fr/~besse/pubs/c&b2003). Based on information available in the GPMDB, and on location of sites relative to major tectonic belts, we considered that the data either represented sites autochthonous to a given plate at a given age (code o), or, given the complexity and/or amplitude of tectonic movements, could not be quantitatively and independently reconstructed to the plate they now belonged to (code n). But we considered that a number of sites, which were located close to a major autochthonous plate, could not have suffered major lateral movement and were only likely to have rotated about vertical axes (code m). Examples of "n"-type sites are found for example in Iberia, central America, the Alaskan peninsula, the Caribbean, western North America, Chile, the Caucasus and parts of eastern Europe, Greece and parts of the eastern Mediterranean, parts of Antarctica, . . . We retained as members of the "m"-type category sites from the Adriatic promontory of Africa (Albania, Croatia, Sicily . . .), Bohemia, basins in Argentina, the Colorado plateau and a thin elongated belt on the eastern edge of stable north America (including sites in Alberta), Tasmania and parts of eastern Australia, parts of East Antarctica. The final working database is that consisting of remaining data of both "o" and "m"- types. When constructing mean paleopoles for any given time window, the "m"-type data provide "inclination-only" information, since the sites could have rotated about vertical axes since acquisition of original magnetization. These can be included as small circle constraints following the method of McFadden and McElhinny [1988]. This final database contains 470 entries, 296 autochthonous and 174 having possibly undergone a rotation. This is a large drop from the original 1850 data but we believe these are minimum criteria to produce significant results. Note that this figure can be compared to the 221 poles of BC02, so that there is a further significant increase over our previous attempt, implying that the analysis presented here cannot be considered as simply a repeat of earlier papers.

REFERENCES

Besse, J., and V. Courtillot, Revised and synthetic polar wander paths of the African, Eurasian, North American and Indian plates, and true polar wander since 200 Ma, *J. Geophys. Res.*, 96, 4029–50, 1991.

Besse, J., and V. Courtillot, Apparent and true polar wander and the geometry of the geomagnetic field over the last 200 Myr, *J. Geophys. Res.*, 107, no. B11, 2300, doi:10.1029/2000JB000050, 2002.

Besse, J., and V. Courtillot, Correction to "Apparent and true polar wander and the geometry of the geomagnetic field over the last 200 Myr, *J. Geophys. Res.*, 107, no. B11, 2300, doi:10.1029/2000JB000050, 2002", *J. Geophys. Res.*, 108, no. B10, 2469, doi:10.1029/2003JB002684, 2003.

Briden, J. C., A. G. Smith, and J. T. Sallomy, The geomagnetic field in Permo-Triassic time, *Geophys. J. Roy. Astron. Soc.*, 23, 101–117, 1970.

Carlut, J., and V. Courtillot, How complex is the time-averaged geomagnetic field over the past 5 Myr? *Geophys. J. Int.*, 134, 527–544, 1998.

Carlut, J., V. Courtillot, and G. Hulot, Over how much time should the geomagnetic field be averaged to obtain the mean-paleomagnetic field ? *Terra Nova*, 11, 239–243, 1999.

Chauvin, A., H. Perroud, and M. L. Bazhenov, Anomalous low palaeomagnetic inclinations from Oligocene-Lower Miocene red beds of the south-west Tien Shan, Central Asia, *Geophys. J. Int.*, 126, 303–313, 1996.

Channell, J. E. T., Catalano, R., D'Argenio, B., Palaeomagnetism and deformation of the Mesozoic continental margin in Sicily, *Tectonophysics*, 61, 391–407, 1980.

Cogné, J. P., N. Halim, Y. Chen and V. Courtillot. Resolving the problem of shallow magnetizations of Tertiary age in Asia: insights from paleomagnetic data from the Qiangtang, Kunlun, and Qaidam blocks (Tibet, China), and a new hypothesis. *J. Geophys Res.*, 104, 17 715–17 734, 1999.

Constable, C. G., and R. L. Parker, Statistics of the geomagnetic secular variation for the past 5 Myr., *J. Geophys. Res.*, 93, 11569–81, 1988.

Coupland, D. H., and R. Van der Voo, Long-term nondipole components in the geomagnetic field during the last 130 Ma, *J. Geophys. Res.*, 85, 3529–3548, 1980.

Courtillot, V., J. P. Valet, G. Hulot, and J. L. Le Mouël, The Earth's magnetic field: which geometry?, *Eos Transactions Am. Geophys. U.*, 73, 337–342, 1992.

Evans, M. E., Test of the dipolar nature of the geomagnetic field throughout Phanerozoic time, *Nature*, 262, 676–677, 1976.

Gubbins, D., and Kelly, P., Persistent patterns in the geomagnetic field over the past 2.5 Myr. *Nature*, 365, 829–832, 1993.

Hagstrum, J. T., and Lipman, P. W., Late Cretaceous paleomagnetism of the Tucson Mountains: implications for vertical axis rotations in south central Arizona, *J. Geophys. Res.*, 96, 16069–16081, 1991.

Irving, E., Drift of the major continents since the Devonian, *Nature*, 270, 304–309, 1977.

Johnson, C. L., and C. G. Constable, The time-averaged field as recorded by lava flo ws over the past 5 Myrs, *Geophys. J. Int.*, 122, 489–519, 1995.

Johnson, C., and C. G. Constable, The time-averaged geomagnetic field: Global and regional databases for 0–5 Ma, *Geophys. J. Int.*, 131, 643–666,. 1997.

Kelly, P., and D. Gubbins, The geomagnetic field over the past 5 million years. *Geophys. J. Int.*, 128, 1–16, 1997.

Kent, D. V., and L. B. Clemmensen, Paleomagnetism and cycle stratigraphy of the Triassic Fleming Fjord and Gipsdalen Formations of East Greenland, *Bull. Geol. Soc. Denmark*, 42, 121–136, 1996.

Kent, D. V., and M. A. Smethurst, Shallow bias of paleomagnetic inclinations in the Paleozoic and Precambrian, *Earth Planet. Sci. Lett.*, 160, 391–402, 1998.

Livermore, R. A., F. J. Vine and A. G. Smith, Plate motions and the geomagnetic field—I. Quaternary and late Tertiary, *Geophys. J.R. Astr. Soc.*, 73, 153–171, 1983.

Livermore, R. A., F. J. Vine and A. G. Smith, Plate motions and the geomagnetic field—II. Jurassic to Tertiary, *Geophys. J.R. Astr. Soc.*, 79, 939–961, 1984.

McElhinny, M. W., and J. Lock, Four IAGA databases released in one package, *EOS, Trans. Amer. Geophys. Union*, 76, 266, 1995.

McElhinny, M. W., P. L. McFadden, and R. T. Merrill, The time averaged paleomagnetic field 0–5 Ma, *J. Geophys. Res.*, 101, 25007–25027, 1996.

McFadden, P. L., and M. W. McElhinny, The combined analysis of remagnetization circle and direct observation in paleomagnetism, *Earth Planet. Sci. Lett.*, 87, 161–72, 1988.

Meert, J. G., E. Tamrat, and J. Spearman, Non dipole fields and inclination bias: insights from a random walk analysis, *Earth Planet. Sci. Lett.*, 214, 395–408, 2003.

Merrill, R. T., M. W. McElhinny, and P. L. McFadden, *The Magnetic Field of the Earth; Paleomagnetism, the Core and the Deep Mantle*, 527 pp., Academic Press, London, 1998.

Nawrocki, J., Paleomagnetism of Permian through Early Triassic sequences in central Spitsbergen: implications for paleogeography, *Earth Planet. Sci. Lett.*, 169, 59–70, 1999.

Piper, J. D. A., and S. Grant, A paleomagnetic test of the axial dipole assumption and implications for continental distributions through geological time, *Phys. Earth Planet. Int.*, 55, 37–53, 1989.

Quidelleur, X., J. P. Valet, V. Courtillot, and G. Hulot, Long-term geometry of the geomagnetic field for the last five million years: An updated secular variation database, *Geophys. Res. Lett.*, 21, 1639–1642, 1994.

Rochette, P., and D. Vandamme, Pangea B: an artifact of incorrect paleomagnetic assumptions?, *Ann. Geofis.*, 44, 649–658, 2001.

Schneider, D. A., and D. V. Kent, The time-averaged paleomagnetic field, *Rev. Geophys.*, 28, 71–96, 1990.

Si, J., and R. Van der Voo, Too-low magnetic inclinations in central Asia: an indication of a long-term Tertiary non-dipole field?, *Terra Nova*, 13, 471–478, 2001.

Torcq, F., J. Besse, D. Vaslet, J. Marcoux, L. E. Ricou, M. Halawani, and M. Basahel, Paleomagnetic results from Saudi Arabia and the Permo-Triassic Pangea configuration, *Earth Planet. Sci. Lett.*, 148, 553–567, 1997.

Torsvik, T. H., and R. Van der Voo, Refining Gondwana and Pangea paleogeography: estimates of Phanerozoic non-dipole (octupole) fields, *Geophys. J. Int.*, 151, 771–794, 2002.

Torsvik, T. H., J. Mosar, and E. A. Eide, Cretaceous-Tertiary geodynamics: a North Atlantic exercise, *Geophys. J. Int.*, 146, 850–867, 2001.

Van der Voo, R., *Paleomagnetism of the Atlantic, Tethys and Iapetus Oceans*, Cambridge University Press, Cambridge, 411pp., 1993.

Van der Voo, R., and T. H. Torsvik, Evidence for late Paleozoic and Mesozoic non-dipole fields provides an explanation for the Pangea reconstruction problems, *Earth Planet. Sci. Lett.*, 187, 71–81, 2001.

Westphal, M., Did a large departure from the geocentric axial dipole hypothesis occur during the Eocene? Evidence from the magnetic polar wander path of Eurasia, *Earth Planet. Sci. Lett.*, 117, 15–28, 1993.

Wilson, R. L., Permanent Aspects of the Earth's Non-dipole Magnetic Field over Upper Tertiary Times, *Geophys. J. R. Astr. Soc.*, 19, 417–37, 1970.

Jean Besse, Institut de Physique du Globe de Paris, 4 place Jussieu, 75232 Paris Cedex 05, France. (besse@ipgp.jussieu.fr)

Vincent Courtillot, Institut de Physique du Globe de Paris, 4 place Jussieu, 75232 Paris Cedex 05, France. (courtil@ipgp.jussieu.fr)

The Paradox of Low Field Values and the Long-Term History of the Geodynamo

John A. Tarduno and Alexei V. Smirnov

Department of Earth and Environmental Sciences, University of Rochester, Rochester, New York

The long-term strength of the geomagnetic field has been defined on the basis of Thellier-Thellier paleointensity experiments using igneous whole-rock samples and submarine basaltic glass. Yet many of the virtual dipole moments derived from these analyses are comparable to values that characterize geomagnetic excursions and reversal transitions. The low field stability implied by these values represents a paradox: it is in stark contrast to what is known from the history of paleosecular variation and the frequency of geomagnetic reversals. Taken at face value, the paleointensity data imply the field has been extraordinarily energetic during the last 10 million years. But factors that could change the magnetic field energy are characterized by time scales orders of magnitude greater than those needed to account for this apparent signal. We suggest instead that processes that lead to field underestimates (natural and experimental alteration) are far more common than is usually supposed. In particular, we focus here on the roles of low-temperature oxidation (maghemitization) and hydrothermal alteration.

We conclude that the mean strength of the field since Mesozoic times was probably similar to that of the last 10 million years (7 to 8×10^{22} Am^2), except during periods of very low (superchron) and very high (e.g. the Late Jurassic) reversal frequency. Paleointensity estimates for the latter interval are anomalously low relative to all available data and, together with Thellier paleointensity estimates based on analyses of single plagioclase crystals, suggest that reversal rate and field intensity are inversely related.

1. INTRODUCTION

We start with a consideration of the rapidly expanding number of high resolution geomagnetic records derived from the Ocean Drilling Program. These data have lent considerable support to the idea that times of low geomagnetic field intensity are also often characterized by excursional field directions. For these sedimentary data the field value is a relative paleointensity derived by normalizing the natural remanent magnetization (NRM) by a rock magnetic parameter to account

Timescales of the Paleomagnetic Field
Geophysical Monograph Series 145
Copyright 2004 by the American Geophysical Union
10.1029/145GM06

for concentration variations, and a field excursion is defined by low latitude virtual geomagnetic poles [*Valet and Meynadier*, 1993, *Channell et al.*, 2002]. Excursions, which may be manifestations of extreme secular variation, appear to represent short (~500 year) reversals in the outer core [*Gubbins*, 1999].

To utilize the sedimentary records of relative paleointensity further to understand the geodynamo it is necessary to calibrate the data with an absolute paleointensity measure. This is done using the Thellier-Thellier method [*Thellier and Thellier*, 1959] as applied to young, relatively unweathered volcanic rocks. The Thellier-Thellier method is accepted as the most reliable approach to absolute paleointensity determination because it relies on thermoremanent magnetization

(TRM), the best understood magnetization process. After calibration, the sediment data suggests a mean time-averaged field strength between 6.0 and 8.0×10^{22} Am^2 for the Brunhes chron [*Valet*, 2003].

A comparison of the calibrated sedimentary paleointensity values with known excursions has led to the concept of a critical field strength; once the field drops below this value excursions are typically observed. Guyodo and Valet [1999] proposed a critical dipole moment of approximately 4×10^{22} Am^2, noting that a reduction of the present day dipole to this value would result in the emergence of large non-dipole fields at many locations. This concept is further supported by continued analyses of the recent field, in which the rapid decrease of the dipole appears to be associated with the growth of localized reversed flux patches [*Hulot et al.*, 2002].

The definition of a critical value based on calibrated sedimentary records is supported by analyses of virtual dipole moments (VDMs) from lavas spanning the interval 0.03 to 10 Ma by Tanaka et al. [1995]. Lavas yielding paleointensity values of 4×10^{22} Am^2 or less were found to be associated with virtual geomagnetic poles (VGPs) having latitudes less than 45°; the latter is a commonly employed definition of excursional field behavior. The critical field value also is consistent with field values defined by high-resolution Thellier analyses of volcanic rocks formed during excursions [*Chauvin et al.*, 1989; *Goguitchaichvili et al.*, 2001] and field reversals [*Prévot et al.*, 1985].

Because the field is typically not in an excursional or reversing state, virtual dipole moments at or less than the critical value should be relatively rare. Yet databases of paleointensity estimates derived from analyses of basalts and submarine basaltic glass [*Selkin and Tauxe*, 2000] formed prior to the last 10 million years contain many such values. This represents a paradox because the attendant increase in relative non-dipole contributions to the field that should occur when the total field intensity drops to strengths near the critical value, predicted by Brunhes sediment data and VDM/VGP datasets for 0.03–10 Ma, conflicts with what is known about the trends of secular variation with time. Below we use paleointensity databases to define this paradox, examine the implications for the geodynamo of assuming the paleointensity database values are accurate, and offer alternative explanations based on natural geologic processes and rock magnetism.

2. PALEOINTENSITY AND PALEOSECULAR VARIATION

We consider first the recently updated paleointensity database maintained by Perrin et al. [1998]. We select only results based on Thellier analyses, using the virtual dipole moments (VDMs) from the database. When no VDM is reported, we calculate the value (if possible) from the field intensity. We supplement these data with values reported in references compiled by Heller et al. [2002]. No transitional data were selected, and the standard deviation of the paleointensity estimates was limited to ≤25%; this yielded 702 data points.

The data fall naturally into 11 age bins, separated by intervals with few or no data (Figure 1). Mean dipole moments calculated from these bins suggests that for the last 10 million years the field has generally been at an intensity higher than that of the last 260 million years (we note that the nominal mean for 50–70 Ma is higher, but seems drawn to this high value by only 3 VDMs). Only during the interval from 270–320 Ma are the mean values similar to the most recent field. There are also interesting characteristics of the range of values within each bin. The relatively densely populated 0–10 Ma interval shows a range that extends to over 14×10^{22} Am^2. Values this high are not common in any of the older age bins.

This seeming deficiency of high field values in the data from whole rocks older than 10 million years is also clear in histograms of VDM values (Figure 2a-b). We note that Heller et al. [2002] produced similar histograms and further interpreted these only in terms of field history. We also compare the histogram values with Thellier results available from submarine basaltic glass as reported by Selkin and Tauxe [2000]. The VDM's for Thellier analyses of submarine basaltic glass older than 10 million years show an even greater lack of high field values (Figure 2c).

Next, we compiled the percentage of the whole rock VDM data that were less, or equal, to the critical value defined by Brunhes data versus time. If the Thellier whole rock results are accurate, this analysis (Figure 3) suggests that during the period 10 to 320 Ma, the field was below the critical field strength ~44% of the time. We also consider the paleointensity database based on Thellier analyses of submarine basaltic glass as reported by Selkin and Tauxe [2000]. A similar analysis suggests the field was below the critical strength threshold for nearly 60% of the 10 to 170 Ma interval.

These numbers are remarkable because taken at face value they imply that the field remained at near transitional intensity for a considerable portion of its history. We assess this conclusion by looking at the variation of paleosecular variation with time. Paleosecular variation (PSV) is typically defined by the angular dispersion of the lava mean directions, S:

$$S^2 = \frac{1}{N-1} \sum_{i=1}^{N} \Delta_i^2 \qquad (1)$$

where N is the number of virtual geomagnetic poles (VPGs) and Δ_i is the angle between the ith VGP and the mean VGP. McFadden et al. [1991] updated and expanded upon early PSV analyses, proposing a model (Model G) where the field

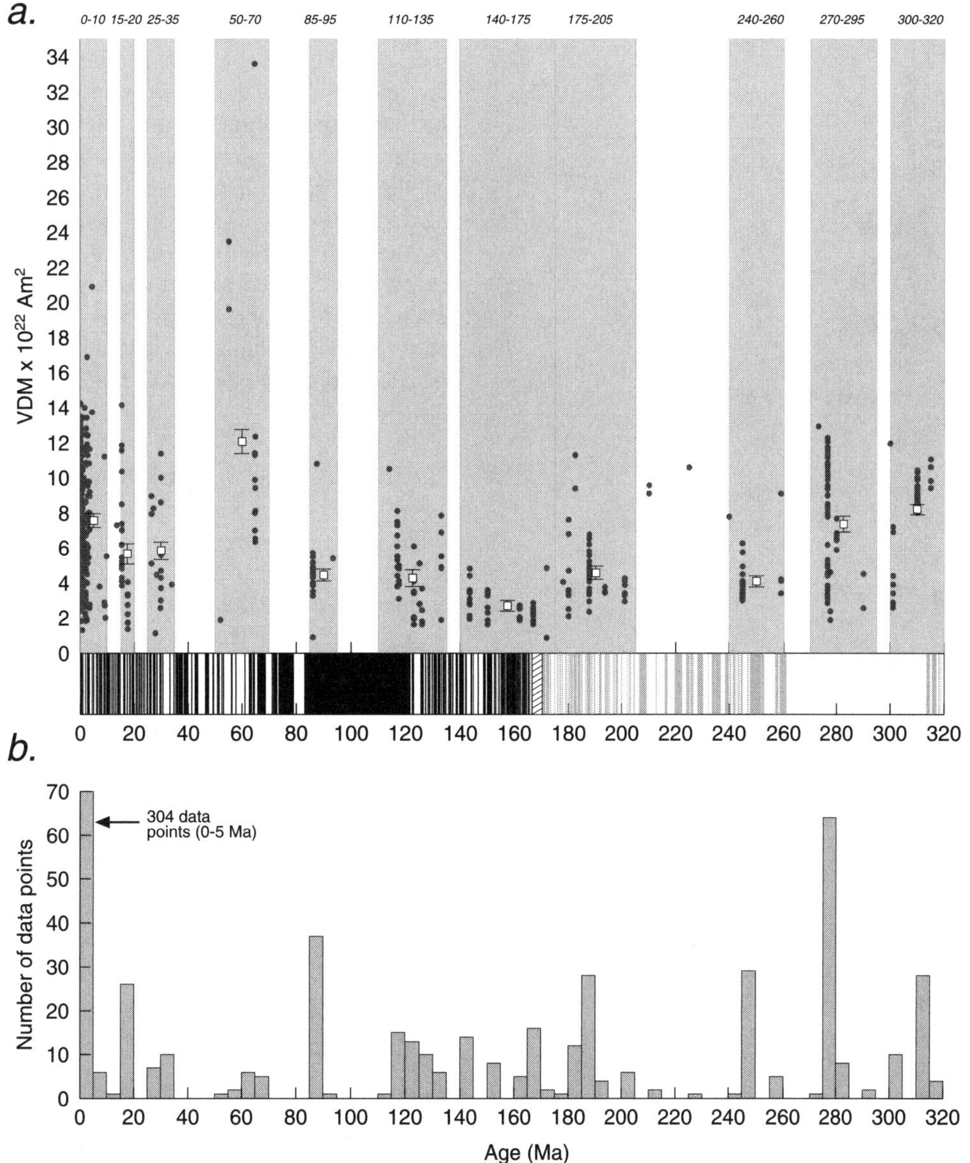

Figure 1. (a) Paleointensity data (virtual dipole moments, VDM) based on Thellier analyses of whole rocks (basalts and baked contacts) shown as gray circles. These data fall into 11 age bins (gray boxes) separated by ages for which no or very little data are available. Nominal means (and 1 errors) are also shown (boxes). (b) Number of data versus time.

is composed of two independent families: a dipole (or anti-symmetric) family, which is linear with respect to latitude ($S_p = b\lambda$) and a quadrupole (or symmetric) family that is constant with latitude ($S_s = a$). The total VGP angular dispersion (S_λ) is given by:

$$S_\lambda = \sqrt{(S_s)^2 + (S_p)^2} \qquad (2)$$

Although complete independence of the two families is unlikely [*Johnson and Constable*, 1996], the model is nonetheless useful as an approximation. McFadden et al. [1991] pro-

posed that the families varied systematically with respect to reversal rate during the last 160 million years. In particular, the smallest contribution of the quadrupole family was noted for the lowest rate of reversals, which was represented by an interval centered on the Cretaceous Normal Polarity Superchron (K-N). A reassessment of this model for the K-N Superchron, using new data, and a more stringent data selection of older data, has supported this suggestion [*Tarduno et al.*, 2002]. Secular variation studies are not sufficiently abundant to allow a continuous analysis for times older than 160 Ma. However, Rochette et al. [1997] discuss data supporting a lower quadru-

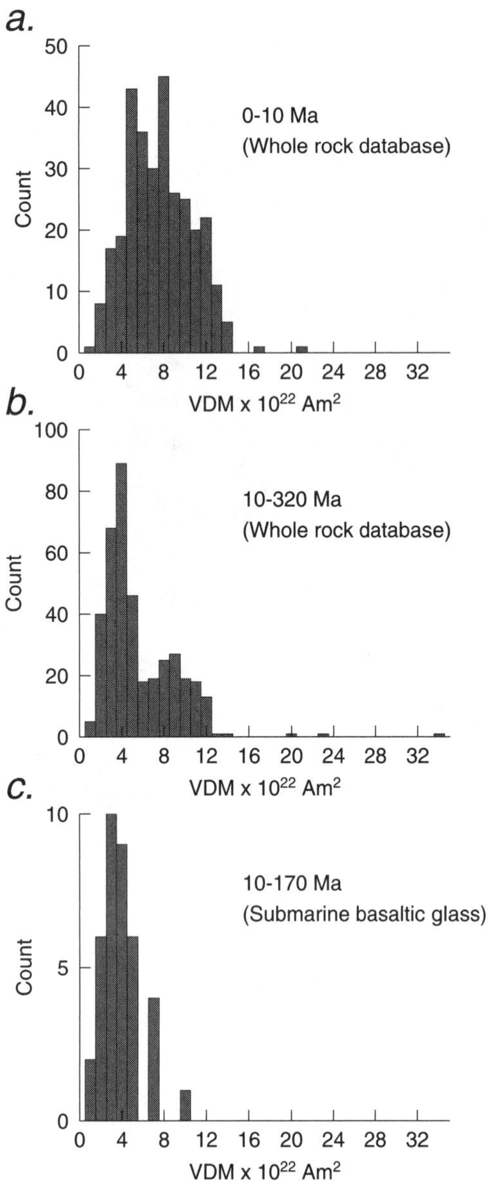

a.

*0-10 Ma
(Whole rock database)*

b.

*10-320 Ma
(Whole rock database)*

c.

*10-170 Ma
(Submarine basaltic glass)*

Figure 2. Histograms of virtual dipole moments derived from Thellier analyses of whole rocks formed between 0 and 10 Ma (a), whole rocks formed between 10 and 320 Ma (b) and submarine basaltic glass formed between 10 and 170 Ma (c).

pole family contribution for the Kiaman Reversed Polarity Superchron.

True excursional directions are typically excluded from these analyses; for example Johnson and Constable [1996] exclude VGP's with latitudes <55°. If the field were actually in an excursional state over 40% of the time (as suggested by the whole rock VDM data), large percentages of the data would have been rejected from nearly all PSV studies; this has clearly not been the case for select high resolution studies [e.g., *Tarduno et al.*, 2001; 2002].

Some direct comparisons of the paleosecular variation estimates and the age-binned estimates of the percentage of time the field fell below the critical value (Figure 3a-b) bounding the Cretaceous Normal Polarity Superchron further illustrate the paradox. The paleosecular secular variation data suggest that the quadrupole family contributions to the field should be near a minimum, yet the whole rock basalt values suggest that the field was below the critical value ~37% (85–95 Ma) and ~47% (110–135 Ma) of the time; these percentages are much higher than those for the last 10 million years (~12%) when the field is thought to have a significantly higher contributions from the quadrupole family.

3. GEOLOGIC AND ROCK MAGNETIC SOLUTIONS

The analyses above define an apparent inconsistency between paleointensity and directional databases. Either the Brunhes field (and the field for the last 10 million years) has been very much unlike that of the last 320 million years (an issue we address later), or there is a bias toward low field values in the paleointensity data from older rocks. The most probable and prevalent process that could be responsible for low field bias in the Thellier database of VDM's derived from whole rocks is probably the simplest: weathering. Paleomagnetists go to great lengths to sample the freshest material available in the field. Nevertheless, the production of iron rich clays through weathering (which can occur on geological time-scales) can essentially prime whole rocks for experimental alteration. During Thellier experiments fine-grained magnetic particles can form from these clays. The onset of this alteration can be subtle, eluding detection by standard reliability criteria. The formation of these fine-grained particles can ultimately result in a bias toward low field estimates [*Cottrell and Tarduno*, 2000]. Here we examine the more direct effects of weathering on the *in situ* magnetic mineralogy of basalts. We also discuss the effects of hydrothermal fluids that can accompany the formation of baked contacts. We note that evidence for experimental alteration and low-field bias in submarine basaltic glass is presented in Smirnov and Tarduno [2003].

Most early Cenozoic and older lavas, which are claimed to be "fresh" in the literature, actually contain titanomagnetites that have undergone low-temperature oxidation. The effect of this process on paleointensity analyses, especially those derived from paleointensity databases [*Heller et al.*, 2002; *Biggin and Thomas*, 2003], is usually overlooked. Titanomaghemite can be represented by:

$$Fe^{3+}_{[2-2x+z(1+x)]R}\,Fe^{2+}_{(1+x)(1-z)R}\,Ti^{4+}_{xR}\square_{3(1-R)}O^{2-}_4$$

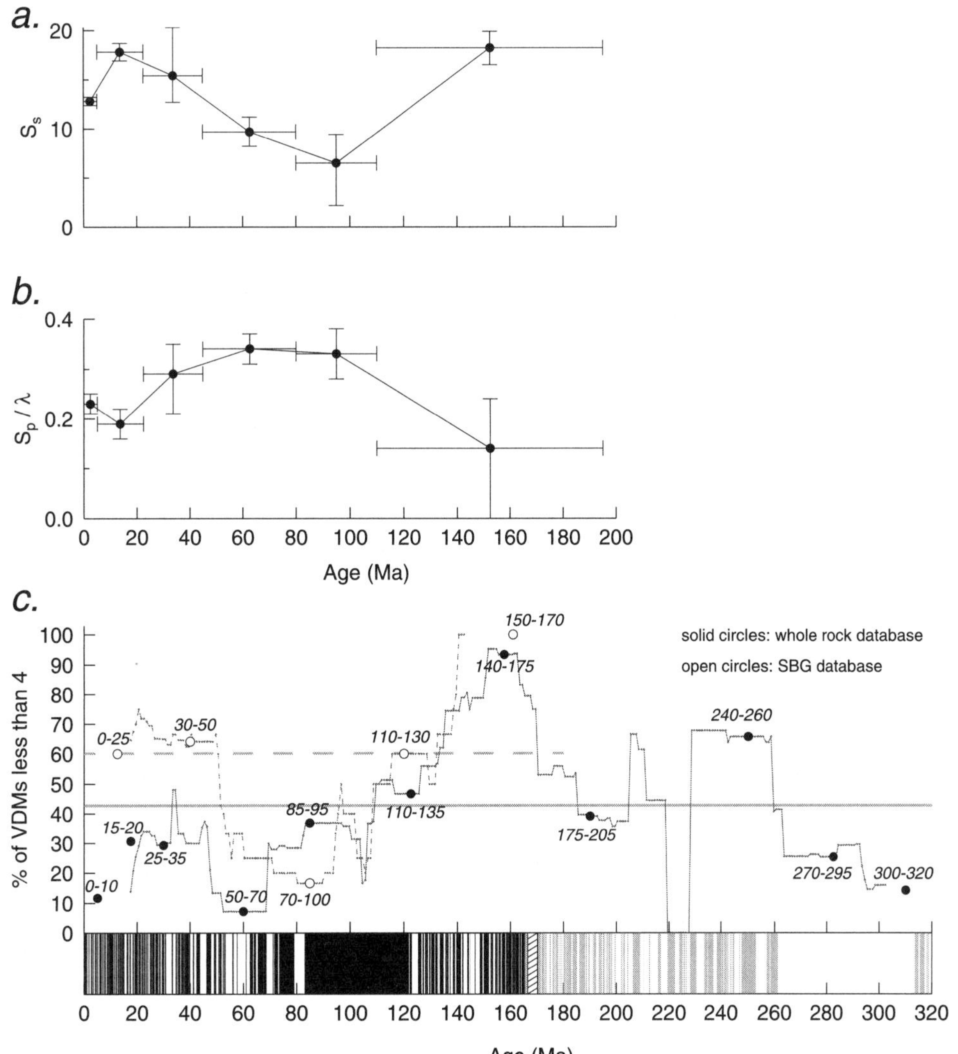

Figure 3. (a) Variation of secondary (or quadrupole family) field contribution versus time from McFadden et al. [1991]. (b) Variation of primary (dipole family) contribution versus time. (c) Percentage of virtual dipole moment (VDM) data less or equal to the critical field value defined by Brunhes-age data (4×10^{22} Am^2, see text). Data derived from Thellier analyses of whole rocks (solids) and submarine basaltic glass (open circles) are shown, together with mean values (solid and dashed heavy horizontal lines, respectively). Also shown is a running mean, based on a window 35-million-years long, of each data set (solid light line for whole rock data, dashed light line for submarine basaltic glass).

where x is Ti content, $R=8/[8+z(1+x)]$, represents a lattice vacancy, and z is the oxidation parameter [*O'Reilly*, 1984]. Because of low temperature oxidation, typical lavas will contain a NRM that consists of an original TRM, and a chemical remanent magnetization (CRM) carried by titanomaghemite. The proportion of CRM will vary with increasing z.

In terms of the ultimate effect on paleointensity data, there are two mechanisms to consider. First, there will be a decrease in intensity associated with the lower saturation magnetization (M_s) [e.g., *Özdemir and O'Reilly*, 1982; *Bleil and Peterson*, 1983]. Second, the direction of CRM relative to the original TRM must be considered. Experimental data, as well as the

very existence of a marine magnetic anomaly pattern, suggests that the CRM is broadly parallel to the original TRM. Here we extend the model of Özdemir and Dunlop [1985] who considered phase coupled TRM and CRM in titanomaghemite. These authors noted that the process should not be completely efficient. Below we examine a simple end-member model.

3.1. Low Temperature Oxidation of Titanomagnetite

In the case of pseudo-single and multidomain domain grains, a titanomaghemite rim can form from low temperature oxidation (Figure 4a). We assume that grains forming this rim have single domain (SD) behavior. It is unlikely that the CRM carried by these rim grains will be positively coupled to the NRM; instead the weak exchange interaction will favor magnetization in the ambient field (which could be of opposite polarity from that of the original NRM). We attribute a 50% efficiency to the CRM, reflecting an equal probability of being magnetized in a normal field (assumed to be approximately parallel to the original NRM) or a reversed polarity field.

The overall magnetization (NRM + CRM) decreases due to the decrease in net M_s accompanying oxidation and the less-than-complete positive coupling of the CRM and NRM (Figure 4b). However, during a Thellier experiment, the oxidized grains will be completely magnetized parallel to the applied field (Figure 4c). Because of their SD character, the "rim" grains will contribute more to the pTRM than the host grains. These effects result in a substantial shallowing of the NRM-TRM slope during a Thellier paleointensity experiment and a bias toward low paleointensity estimates. Considering moderate oxidation states ($z = 0.3–0.5$) the low-field bias is large (30–40%) (Figure 4d).

Real rocks typical of those used in paleointensity experiments contain mixtures of single, pseudo-single and multidomain grains, oxidized to various levels. Single domain grains will also be oxidized under low temperature conditions, but through cracking this rim will contain distinct superparamagnetic grains [*Özdemir et al.*, 1993; *Smirnov and Tarduno*, 2000]. In some larger single domain grains, it is possible that an oxidized rim could form similar to the case described above. The effects on paleointensity in such a case, however, are more difficult to predict, and will depend of the balance between effects of the local exchange interaction fields, magnetostatic interaction and the microcoercivity of the rim grains.

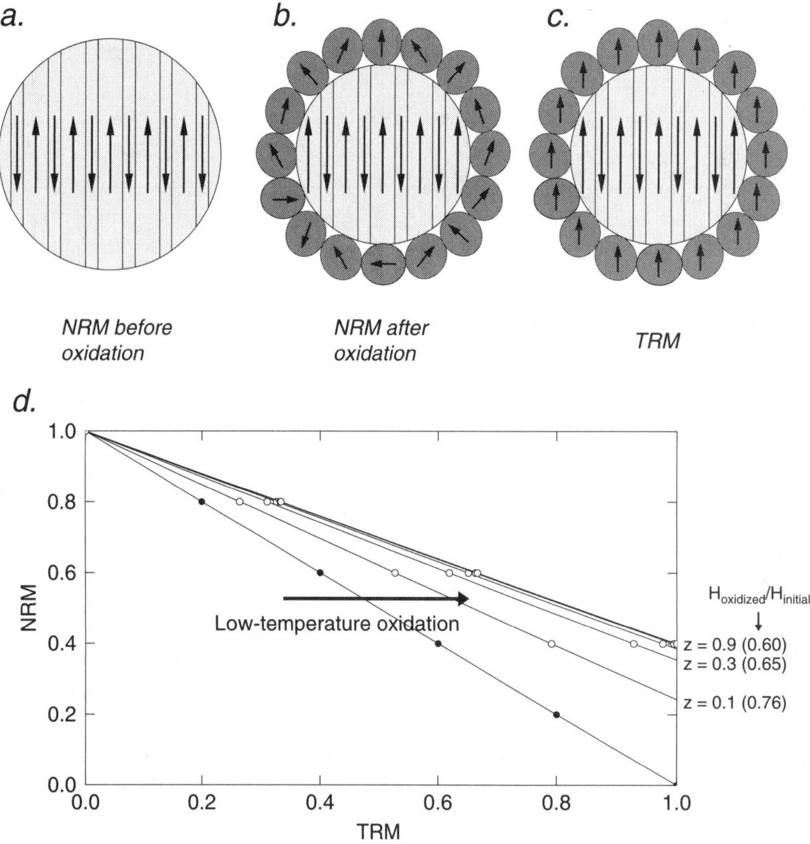

Figure 4. Effects on paleointensity data given low temperature oxidation of pseudo-single and multi-domain titanomagnetite grains. (a) Grain before oxidation. (b) Grain after oxidation; gray represents oxidized rim grains. (c) Oxidized grain after the application of a laboratory thermoremanent magnetization (TRM). (d). Effect of progressive oxidation on paleointensity determinations, where z is oxidation parameter, and $H_{oxidized}/H_{initial}$ is the ratio of the final field determination in a Thellier experiment using a rock characterized by the oxidized grains relative to that derived from an experiment characterized by a rock having unoxidized grains.

The simple model of pseudo-single/multidomain oxidation described above is applicable only in the unblocking temperature range below the inversion of titanomaghemite [*Readman and O'Reilly*, 1970; *Özdemir*, 1987] that accompanies the heating. This temperature range is a function of Ti content, and ranges between 250°C for maghemite, to ~475°C for Ti-rich particles [*Dunlop and Özdemir*, 1997]. Inversion results in the creation of magnetite capable of very high experimental TRM acquisition, and thus is the source of an even greater low field bias. A massive inversion should be easily detectable in the slopes of NRM-TRM plots, or through other rock magnetic tests. However, it does not seem to have been universally accepted that when such effects are observed, data obtained from unblocking temperatures lower than that of obvious inversion effects should be questioned [e.g., *Tanaka and Kono*, 2002]. It is not possible to retrieve this type of information from current generations of the paleointensity database.

Furthermore, because titanomaghemite may comprise only part of the total magnetic grain assemblage, the effect of inversion may be subtle, highlighting the need for partial TRM checks [*Coe*, 1967]. Such checks are a basic reliability requirement in many [e.g., *Selkin and Tauxe*, 2000; *Tarduno et al.*, 2001; *Goguitchaichvili et al.*, 2002] but not all [e.g., *Heller et al.*, 2002; *Biggin and Thomas*, 2003] studies. Techniques for paleointensity analysis which do not rely on heating, such as the Shaw method, would of course not detect inversion, but would suffer similar biases as described here (under the controversial assumption that in rocks typical of those used in paleointensity studies, which contain mixtures of composition and grain sizes, ARM accurately models TRM).

3.2. Hydrothermal Formation of Magnetite

Another natural alteration process which could be important in paleointensity studies is the effect of hydrothermal fluids that can accompany the emplacement of igneous rocks. The importance of these fluids lies in their capacity to facilitate the precipitation of new magnetite grains. If the new grains grow larger than the SP/SD threshold, they are able to acquire a CRM which may be characterized by a broad spectrum of unblocking temperatures and coercive forces [e.g., *Dunlop and Özdemir*, 1997]. The intensity of grain-growth CRM is thought to be weaker than the intensity of TRM imparted in the same magnetic field [*Stacey and Banerjee*, 1974]. While the exact CRM/TRM is difficult to predict, some experimental data for magnetite have shown that the CRM intensity could be only 50% that of the TRM strength [*Hoye and Evans*, 1975].

Because both CRM and TRM are carried by the same distribution of magnetic grains, they should have similar stabilities with respect to thermal treatment [*Dunlop and Özdemir*, 1997]. Therefore, during a Thellier experiment, the replacement of a CRM related to hydrothermal processes in nature with a laboratory TRM will result in a bias in paleointensity estimates toward low values.

The effects of hydrothermal process described might be particularly important for what is sometimes considered to be one of the best paleointensity recorders: baked contacts. Many baked contacts involve the shallow subsurface emplacement of igneous bodies, where extended hydrothermal effects can be expected.

4. DISCUSSION AND CONCLUSIONS

At any location on Earth extraordinarily high or low intensity values can occur due to the natural variation of the magnetic field. We have followed the assumption here, employed by many authors [e.g., *Prévot et al.*, 1985; *Heller et al.*, 2002; *Biggin and Thomas*, 2003], that by sampling enough VDM's through time we can recover an accurate long-term signal of the geomagnetic field, disregarding temporal averaging at any given site. We note that an alternative, more conservative method requires time-averaged paleomagnetic dipole moments before proceeding with conclusions on the geodynamo. The latter approach indicates that only a handful of datasets are available to constrain rigorously the field [e.g., *Tarduno et al.*, 2001; 2002].

The VDM data are spaced unevenly in time, and some intervals are devoid of data. Nevertheless, proceeding with standard assumptions and the databases of whole rocks (basalts and baked contacts) and submarine basaltic glass VDM values, the data indicate that the field has been below the critical strength defined by Brunhes-age data, below which excursions are typically observed, for large percentages of the interval 10 to 320 million years. This is a paradox because an increase in the quadrupole family contributions to the field, as would be expected during these times of very low field strengths, is not observed.

Could the field have been in a lower energy state, such that the low apparent paleointensity values would not be accompanied by enhanced secular variation? Some simulations of the geodynamo call for alternations between high and low total kinetic and magnetic energies, with reversals occurring at high energy states [*Li et al.*, 2002]. However, applying such models to the long-term history of the geodynamo carries with it the suggestion that the Brunhes chron (and during the last 10 million years) is somehow unusually energetic, and that the geodynamo rarely had as much total energy during the prior 320 million years.

Processes that could result in changes in the energy state of the geodynamo include large scale flow in the mantle altering the pattern of heat flux across the core mantle boundary [*Glatzmaier et al.*, 1999], and growth of the inner core [*Smirnov et al.*, 2003]. Mantle flow of the scale needed to change core-mantle heat flux requires time scale on the order of 100 million years, and hence cannot provide an explanation for the differences between the paleointensity data of the last 10 million years and older time intervals. Similarly, inner core growth should not be a factor for the time interval considered here. Thus we find no reason to believe the field during the last 10 millions years should have more energy than other intervals in the Cenozoic and early Cretaceous. And given the plethora of processes that can result in low paleointensity estimates as rock age, this explanation seems especially unlikely.

Low field bias can be caused by subtle experimental alteration of basalt [*Cottrell and Tarduno*, 2000] and basaltic glass [*Smirnov and Tarduno*, 2003]. Moreover, we show here that a low field bias is associated with weathering of titanomagnetites. An evaluation and correction for the effect in any given study necessitates information on magnetic grain sizes and oxidation states, as well as further work on the efficiency of CRM-TRM coupling. Detailed magnetic grain size and oxidation state information is rarely available and is at present decoupled from databases used by some authors to summarize paleointensity history. Furthermore, the potential role of hydrothermal process should be investigated, especially in studies of baked contacts, the results from which compose substantial portions of some paleointensity syntheses for Mesozoic rocks [e.g., *Prévot et al.*, 1990].

An important corollary of our analysis of the present paleointensity database is that statistical treatments of the bulk data are unlikely to recover the true field value. Instead rigorous rock magnetic selection criteria, coupled with attempts to examine more closely the prospects of paleointensity measurements using basic rock forming minerals [e.g., *Cottrell and Tarduno*, 1999] must continue to be developed to expand upon the presently available small data set of reliable paleointensity measurements.

At present, the best estimate of the long-term intensity should be tied to Thellier analyses of basalt formed during past ten million years that have seen the least weathering. These data suggest a mean value between 7 and 8×10^{22} Am^2. Field stability, as measured by average reversal rate and secular variation, suggests the field was at or above this level for most of the last 320 million years.

In closing, we note further that there is only one feature that stands out as an exception in the basalt VDM data (at 95% confidence) portrayed in Figure 3: the exceedingly low values between 150 and 170 Ma. As noted by McElhinny and McFadden [2000] this interval also corresponds to the reduced marine magnetic anomaly amplitudes, recognized by

Larson and Hilde [1975]. If the anomalies defined in Jurassic crust [*Handschumacher et al.*, 1988] are true reversals, this interval may have been characterized by the highest reversal frequency of the last 320 million years. Although the paleointensity bias makes the absolute value impossible to determine, Figure 3c hints at relative changes. In this view, the real, and hence reduced, "Mesozoic Dipole Low" [*Prévot et al.*, 1990; *Goguitchaichvili et al.*, 2002] may be an interval corresponding to high reversal frequency [*McElhinny and Larson*, 2003], further supporting the idea of links between field strength and reversal frequency [*Cox*, 1968; *Tarduno et al.*, 2001; 2002].

Acknowledgments. This work was supported by the National Science Foundation. We thank the conveners for organizing the conference honoring the achievements of Neil Opdyke, Rob Coe for helpful discussions, and the anonymous reviewers for their comments. We see the long-term field as being more robust and stable than is implied by current paleointensity data from whole rocks, which may be compromised by natural and experimental alteration. In our opinion this interpretation is more consistent with Neil's seminal contributions establishing the dipole dominance of the field, than other analyses which take the paleointensity data at face value.

REFERENCES

Biggin, A. J., and D. N. Thomas, Analysis of long-term variations in the geomagnetic poloidal field intensity and evaluation of their relationship with global geodynamics, *Geophys. J. Int., 152,* 392–415, 2003.

Bleil, U., and N. Petersen, Variations in magnetization intensity and low-temperature oxidation of ocean floor basalt, *Nature, 301,* 384–388, 1983.

Channell, J. E. T., A. Mazaud, P. Sullivan, S. Turner, and M. E. Raymo, Geomagnetic excursions and paleointensities in the Matuyama Chron at Ocean Drilling Program Sites 983 and 984 (Iceland Basin), *J. Geophys. Research, 107,* 10.1029, 2002.

Chauvin, A., R. A. Duncan, N. Bonhommet, and S. Levi, Paleointensity of the Earth's magnetic field and K-Ar dating of the Louchadiere volcanic flow (Central France): New evidence for the Laschamp excursion, *Geophys. Res. Lett., 16,* 1189–1192, 1989.

Coe, R. S., The determination of paleointensities of the earth's magnetic field with emphasis on mechanisms which could cause non-ideal behaviour in Thelliers method, *J. Geomagn. Geoelectr., 19,* 157–179, 1967.

Cottrell, R. D., and J. A. Tarduno, Geomagnetic paleointensity derived from single plagioclase crystals, *Earth Planet. Sci. Lett., 169,* 1–5, 1999.

Cottrell, R. D., and J. A. Tarduno, In search of high fidelity geomagnetic paleointensities: A comparison of single crystal and whole rock Thellier-Thellier analyses, *J. Geophys. Res., 105,* 23,579–23,594, 2000.

Cox, A., Lengths of geomagnetic polarity intervals, *J. Geophys. Res.,* *73,* 3247–3260, 1968.

Dunlop, D. J., and Ö. Özdemir, *Rock Magnetism: Fundamentals and Frontiers,* 573 pp., Cambridge University Press, New York, 1997.

Glatzmaier, G. A., R. S. Coe, L. Hongre, and P. H. Roberts, The role of the Earth's mantle in controlling the frequency of geomagnetic reversals, *Nature, 401,* 885–890, 1999.

Goguitchaichvili, A., P. Camps, and J. Urrutia-Fucugauchi, On the features of the geodynamo following reversals or excursions by absolute geomagnetic paleointensity data, *Phys. Earth Planet. Inter., 124,* 81–93, 2001.

Goguitchaichvili, A., L. M. Alva-Valdivia, J. Urrutia, and J. Morales, On the reliability of Mesozoic Dipole Low: New absolute paleointensity results from Paraná flood basalts (Brazil), *Geophys. Res. Lett., 29,* 10.1029, 2002.

Gubbins, D., The distinction between geomagnetic excursions and reversals, *Geophys. J. Int., 137,* F1–F3, 1999.

Guyodo, Y. and J.-P. Valet, Global changes in intensity of the Earth's magnetic field during the past 800 kyr, *Nature, 399,* 249–252, 1999.

Handschumacher, D. W., W. W. Sager, T. W. C. Hilde, and D. R. Bracey, Pre-Cretaceous tectonic evolution of the Pacific plate and extension of the geomagnetic polarity reversal time scale with implications for the origin of the Jurassic "quiet zone", *Tectonophysics, 155,* 365–380, 1988.

Heller, R., R. T. Merrill, and P. L. McFadden, The variation of intensity of earth's magnetic field with time, *Phys. Earth Planet. Inter., 131,* 237–249, 2002.

Hoye, G. S., and M. E. Evans, Remanent magnetization in oxidized olivine, *Geophys. J. Roy. Astron. Soc., 41,* 139–151, 1975.

Hulot, G., C. Eymin, B. Langlais, M. Mandea, and N. Olsen, Small-scale structure of the geodynamo inferred from Oersted and Magsat satellite data, *Nature, 416,* 620–623, 2002.

Johnson, C. L., and C. G. Constable, Palaeosecular variation recorded by lava flows over the past five million years, *Phil. Trans. Roy. Soc. London A, 354,* 89–141, 1996.

Larson, R. L., and T. W. C. Hilde, A revised time scale of magnetic reversals for Early Cretaceous and Late Jurassic, *J. Geophys. Res., 80,* 2586–2594, 1975.

Li, J., T. Sato, and A. Kageyama, Repeated and sudden reversals of the dipole field generated by a spherical dynamo action, *Science, 295,* 1887–1890, 2002.

McElhinny, M. W., and R. L. Larson, Jurassic dipole defined from land and sea, *Eos, Transactions American Geophysical Union, 84,* 362, 366, 2003.

McElhinny, M. W., and P. L. McFadden, *Paleomagnetism, Continents and Oceans,* 385 pp., Academic Press, San Diego, 2000.

McFadden, P. L., R. T. Merrill, M. W. McElhinny, and S. Lee, Reversals of the Earth's magnetic field and temporal variations of the dynamo families, *J. Geophys. Res., 96,* 3923–3922, 1991.

O'Reilly, W., *Rock and Mineral Magnetism,* 220 pp., Blackie, Glasgow and London, and Chapman and Hill, New York, 1984.

Özdemir, Ö., Inversion of titanomaghemite, *Phys. Earth Planet. Inter., 46,* 184–196, 1987.

Özdemir, Ö. and D. J. Dunlop, An experimental study of chemical remanent magnetization of synthetic monodomain titanomaghemites with initial thermoremanent magnetization, *J. Geophys. Res., 90,* 11,513–11,523, 1985.

Özdemir, Ö. D. J. Dunlop, and B. M. Moskowitz, The effect of oxidation on the Verwey transition in magnetite, *Geophys. Res. Lett., 20,* 1671–1674, 1993.

Özdemir, Ö. and W. O'Reilly, Magnetic hysteresis properties of synthetic monodomain titanomaghemites, *Earth Planet. Sci. Lett., 57,* 437–447, 1982.

Perrin, M., E. Schnepp, and V. Shcherbakov, Paleointensity database update, *EOS, Transactions of the American Geophysical Union, 79,* 198, 1998.

Prévot, M., M. E. Derder, M. O. McWilliams, and J. Thompson, Intensity of the Earth's magnetic field: Evidence for a Mesozoic dipole low, *Earth Planet. Sci. Lett., 97,* 129–139, 1990.

Prévot, M., E. A. Mankinen, R. S. Coe, and S. Grommé, The Steens Mountain (Oregon) geomagnetic polarity transition 2. Field intensity variations and discussion of reversal models, *J. Geophys. Res., 90,* 10,417–10,448, 1985.

Readman, P. W. and W. O'Reilly, The synthesis and inversion of non-stoichiometric titanomagnetites, *Phys. Earth Planet. Inter., 4,* 121–128, 1970.

Rochette, P, F. Ben Atig, H. Collombat, D. Vandamme, and P. Vlag, Low paleosecular variation at the equator: a paleomagnetic pilgrimage from Galapagos to Esterel with Allan Cox and Hans Zijderveld, *Geologie en Mijnbouw, 76,* 9–19, 1997.

Selkin, P. A., and L. Tauxe, Long-term variations in palaeointensity, *Phil. Trans. Roy. Soc. London A, 358,* 1065–1088, 2000.

Smirnov, A. V., and J. A. Tarduno, Low-temperature magnetic properties of pelagic sediments (Ocean Drilling Program site 805C): Tracers of maghemitization and magnetic mineral reduction, *J. Geophys. Res., 105,* 16,457–16471, 2000.

Smirnov, A. V., and J. A. Tarduno, Magnetic hysteresis monitoring of Cretaceous submarine basaltic glass during Thellier paleointensity experiments: Evidence for alteration and attendant low field bias, *Earth Planet. Sci. Lett., 206,* 571–585, 2003.

Smirnov, A. V., J. A. Tarduno, and B. N. Pisakin, Paleointensity of the early geodynamo (2.45 Ga) as recorded in Karelia: A single crystal approach, *Geology, 31,* 415–418, 2003.

Stacey, F. D., and S. K. Banerjee, *The Physical Principles of Rock Magnetism,* 195 pp., Elsevier, Amsterdam, 1974.

Tanaka, H., and M. Kono, Paleointensities from a Cretaceous basalt platform in Inner Mongolia, northeastern China, *Phys. Earth Planet. Inter., 133,* 147–157, 2002.

Tanaka, H., M. Kono and H. Uchimura, Some global features of palaeointensity in geological time, *Geophys. J. Int., 120,* 97–102, 1995.

Tarduno, J. A., R. D. Cottrell, and A. V. Smirnov, High geomagnetic field intensity during the mid-Cretaceous from Thellier analyses of single plagioclase crystals, *Science, 291,* 1779–1783, 2001.

Tarduno, J. A., R. D. Cottrell, and A. V. Smirnov, The Cretaceous superchron geodynamo: Observations near the tangent cylinder,

Proceedings of the National Academy of Sciences of the United States of America, 99, 14020–14025, 2002.

Thellier, E. and O. Thellier, Sur l'intensité du champ magnétique terrestre dans le passé historique et géologique, *Annales Géophysique, 15,* 285–376, 1959.

Valet, J.-P., Time variations in geomagnetic intensity, *Rev. Geophys., 41,* 10.1029/2001RG000104, 2003.

Valet, J. P. and L. Meynadier, Geomagnetic field intensity and reversals during the past four million years, *Nature, 366,* 234–238, 1993.

J. A. Tarduno and A. V. Smirnov, Department of Earth and Environmental Sciences, University of Rochester, 227 Hutchison Hall, Rochester, New York 14627.

Intensity and Polarity of the Geomagnetic Field During Precambrian Time

David J. Dunlop and Yongjae Yu[1]

Geophysics, Department of Physics, University of Toronto, Ontario, Canada

There are only 24 Thellier-type paleointensity estimates for Precambrian rocks. Because orogenesis is episodic, these cluster in a few time intervals: 7 between 820 and 1240 Ma, 7 between 1850 and 2215 Ma, 8 between 2450 and 2765 Ma, and 2 earlier Archean results (3470 Ma). Most late Archean-early Proterozoic results are from dikes. Two results are from large intrusions with multiple phases and slow cooling. Late Precambrian results from the slowly uplifted Grenville Province have an even longer cooling history but partial thermoremanent magnetizations (TRMs) can be dated fairly accurately by $^{40}Ar/^{39}Ar$ geochronology. The geographic distribution is uneven. Most results are from Canada, Greenland and Baltica, with only one study each from Africa and Australia. The African and Australian results are thermal overprints rather than primary TRMs. Numbers of acceptable results are often small and standard deviations large. Despite these limitations, the virtual dipole moments for almost all studies lie within a range 0.5–1.5 times the published 0.3–300 Ma average, and only two results are conspicuously high ($>10^{23}$ Am^2). There is no obvious record of onset and growth of a dynamo field in the Archean or early Proterozoic. A detailed record of a Middle Proterozoic polarity transition has all the expected features: 180° reversal along a great circle path, precursor excursion on the same path, a major drop in intensity during the reversal, and reduced intensity before and after. Reversal frequency may have been lower than in recent times, with several chrons of length 20–50 myr around 1460–1400 Ma, 1150–1050 Ma, and 1050–820 Ma, but there are also records of 3 myr and shorter chrons in continuous stratigraphic sections. The dipolar nature of the field is debatable. Roughly contemporaneous results confirm the expected dipole dependence of field strength and paleosecular variation on paleolatitude. On the other hand, there is an apparent excess of directional results with low inclinations and paleolatitudes compared to the dipole prediction but this may be due to bias built into the method.

1. INTRODUCTION

[1]Currently at Geosciences Research Division, Scripps Institution of Oceanography, La Jolla, California.

Timescales of the Paleomagnetic Field
Geophysical Monograph Series 145

Interest in Precambrian paleomagnetic field intensity was sparked by Hale's [1987a] proposal, on the basis of very limited data, that the Earth's dipole moment had increased markedly around the Archean-Proterozoic boundary (ca. 2.5 Ga), perhaps signaling nucleation/growth of the inner core. Theoretical estimates of the time of inner core formation [*Stevenson et al.*, 1983; *Buffett et al.*, 1992; *Labrosse et al.*,

1997, 2001] vary so widely that any firm observational constraint would be extremely valuable. Although there may be additional energy sources [*Buffett et al.*, 2000], growth of the inner core is believed to be essential for maintaining convection and driving the geodynamo [*Stevenson*, 2003].

The inner core may also influence field excursions and reversals through its stabilizing influence on fluid flow patterns in the outer core [*Gubbins*, 1999]. We lack data for paleosecular variation in the Precambrian, and given the time resolution required, we may never have this data. The polarity record has recently been much enhanced by two magnetostratigraphic studies of long, essentially continuous sedimentary sections in the middle Proterozoic [*Gallet et al.*, 2000; *Elston et al.*, 2002]. Together with earlier more fragmentary evidence, they suggest that reversals were less frequent than at present, with some polarity intervals approaching the length of superchrons.

In a recent comprehensive review of time variations of geomagnetic field intensity [*Valet*, 2003], a single page was devoted to the Precambrian, which represents almost 9/10 of Earth history, and the data presented comprised only 9 studies. Even with the appearance in the past two years of several high-quality studies, there remains a paucity of data, as well as some serious contradictions between early and more recent studies of the same rock formations. We choose in this review to include all Thellier-type results [*Thellier and Thellier*, 1959] for the Precambrian based on a reasonable number (more than 3–4) of determinations but to present them in a hierarchy based on reliability.

We first present data from rapidly cooled rocks, usually with positive baked contact tests demonstrating that the natural remanent magnetization (NRM) is a primary thermoremanent magnetization (TRM), for which partial TRM checks were carried out, and with 10 or more determinations. Only 6 results are in this A category. If one of the criteria is not fulfilled—the rocks cooled slowly or partial TRM checks were not performed or there are <10 determinations—we classify the results as category B. An additional 10 results are in this category. If two or more of the criteria are not fulfilled, the results are in category C. The 8 results in this category are of doubtful reliability but to exclude them entirely would either depopulate an entire age interval or would conceal a discrepancy between results of similar age from the same area.

2. MULTIVECTORIAL PALEOINTENSITY DETERMINATION

Precambrian rocks are often remagnetized. The time interval between primary magnetization and overprinting can be hundreds of millions of years (myr). If the overprinting is thermal, resulting in a partial TRM, it may be possible to cleanly separate the NRM into non-overlapping vectors using thermal demagnetization. This is always possible for single-domain grains and is approximately the case also for small pseudo-single-domain grains. It has become routine practice to determine two paleomagnetic poles from the two NRM vectors of different ages, but only recently have two paleofield intensities been recovered from multivectorial NRMs [*Yu and Dunlop*, 2002].

When the primary and overprinting NRMs are at a large angle to each other, conventional Arai plots give too low a paleointensity for the overprint. The reason for this underestimation is that the percentage change in the resultant of two near-perpendicular vectors during demagnetization is much less than the percentage change in the vector that demagnetizes first. For example, if the overprint and primary vectors have equal magnitudes and are at 90° to each other, their resultant changes from 1.414 to 1, a decrease of 30%, as the overprint decreases by 100%, from 1 to 0.

The cure for this problem is to work directly from vector plots, which are always determined in the course of a Thellier experiment. Then changes in the full length of the overprint vector are correctly measured. There is no problem in analyzing the primary TRM vector because it only begins to thermally demagnetize after the overprint has been completely removed. Thus a necessary condition for multivectorial paleointensity determination is vectors with sharp intersections on a Zijderveld vector diagram.

Figure 1 illustrates Arai plots for high- and low-blocking-temperature vectors of equal magnitudes at angles of 60°, 90° and 180° to each other. These cases represent remagnetization in nature either after a long interval, with a large lati-

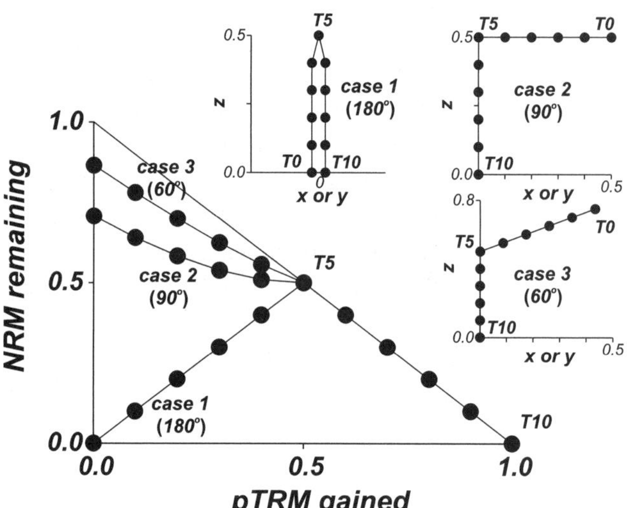

Figure 1. Simulated Arai plots for antiparallel, perpendicular, and oblique (60°) overprinting of TRM by partial TRM at temperature T_5. Equal overprinting and primary fields are assumed.

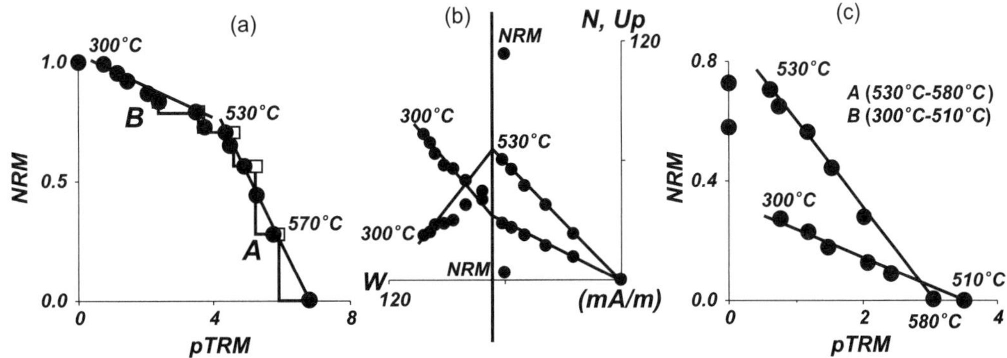

Figure 2. Vector diagram and conventional and separated Arai plots for a Cordova Gabbro sample with A and B NRM vectors. Open squares are partial TRM checks.

tudinal component of plate motion, or following a field reversal. Antiparallel remagnetization following a reversal results in a correct paleointensity estimate, but for 60° and 90° overprints, the paleofield is underestimated by 25–30% and 50–60%, respectively. As well as having average slopes <1, the Arai plot segments for these overprints are curved, but the curvature is easy to miss with real data, particularly in the 60° case.

An actual example of multivectorial paleointensity determination appears in Figure 2. The Cordova Gabbro of the late Precambrian Grenville Province, a medium- to high-grade orogen in Ontario, was thermally remagnetized around 1000 Ma and then partially remagnetized around 850 Ma [*Yu and Dunlop*, 2002]. The two NRMs (A and B) are partial TRMs which separate cleanly in thermal demagnetization and have an angular separation of about 60° (vector diagram, Figure 2). The 300–510°C segment of the conventional Arai plot is curved (the partial TRM checks at 510°C and below reproduce this curvature perfectly) but it is nevertheless possible to fit the data by a straight line (Figure 2, left). However the loss in total (vector resultant) NRM over this interval is only 0.20, whereas the actual change in the B vector between 300 and 510°C is 0.28 (separated A and B Arai plots: Figure 2, right). The conventional Arai plot underestimates the B paleofield by 30%.

3. THE PRECAMBRIAN PALEOINTENSITY RECORD

Table 1 lists all currently available Thellier-type paleointensity data for the Precambrian based on at least 3–4 determinations. Only one formation (Grenville dike, Gananoque, Ont.) is included from the reconnaissance study by Schwarz and Symons [1969], all others being based on 1 or 2 determinations, and none are included from Kobayashi [1968]. Some category C results published only in thesis form [*Hale*, 1985] have been included because of their special interest, as discussed later.

There are only 19 Precambrian formations for which paleointensity data are available but Table 1 has 24 entries. The Cordova Gabbro has two distinct partial TRMs, each dated, which yield two paleofield intensities of different ages (see section 2). Keweenawan igneous rocks record a highly asymmetric reversal, the normal (N) and reverse (R) mean inclinations being +34° and –70°, respectively. One possible explanation of the asymmetry is a substantial difference between the times of magnetization of N and R rocks, which are therefore listed separately. The Biscotasing and Matachewan dikes have each been studied twice, with rather different results. Both sets of data are listed. Finally, basaltic and peridotitic komatiites of the Komati Formation yield very different results and are listed separately because of the possibility that they were magnetized at different times. Conversely, some results that were treated separately in previous compilations are recalculated here as a single entry. These are the overprinted Hamersley Basin lavas and the Yellowknife 8a dikes. Samples from different sites were averaged separately in the original studies, but there is no reason to believe they have very different ages of magnetization.

As well as being sparse, the Precambrian paleointensity data are very unevenly spread geographically. All but two of the results are from the Laurentian-Baltic Shield, with a single result each from the South African and Australian Shields. They are also unevenly spread temporally. Because orogenesis is episodic, the data cluster in a few time intervals: 6 (formations) between 820 and 1240 Ma, 6 between 1850 and 2215 Ma, 7 between 2450 and 2765 Ma, and a single earlier Archean result (3470 Ma). There are major gaps between 1240 and 1850 Ma and pre-2765 Ma.

3.1. Category A Results

There are only 6 category A results, from rapidly cooled rocks for which the NRM has been shown (or can be presumed) to be primary TRM, incorporating partial TRM checks,

Table 1. Thellier-type paleointensity determinations for the Precambrian

Formation	Location	Age, Ma	M	Ref.	VADM, 10^{22} Am^2	N	M	Ref.	Quality
Grenville dike	Gananoque, Ont., Canada	817 ± 70	K/Ar	1	5.77 ± 1.39	4	T	2	C
Cordova Gabbro B	Marmora, Ontario	850 ± 50	Ar/Ar	3	1.82 ± 0.38	15	T+	4	B
Cordova Gabbro A	Marmora, Ontario	1000 ± 50	Ar/Ar	3	3.12 ± 0.36	18	T+	4	B
Keweenawan igneous N	L. Superior, Ontario	1092 ± 5	U/Pb	5	11.56 ± 3.63	20	T	6	B
Keweenawan igneous R	L. Superior, Ontario	1102 ± 5	U/Pb	5	10.92 ± 2.99	34	T	6	B
Abitibi dikes	N. Ontario	1141 ± 1	U/Pb	7	1.36 ± 0.45	23	T+	8	A
Tudor Gabbro	Madoc, Ontario	1240 ± 2	U/Pb	9	4.58 ± 0.84	45	T+	10	B
Sudbury IC	Sudbury, Ontario	1850 ± 1	U/Pb	11	3.77 ± 1.42	47	T	12	C
Hamersley lavas overprint	W. Australia	2000 ± 100	uplift	13	2.39 ± 0.91	22	T+	13	B
Fort Frances dikes	NW Ontario	2076 ± 5	U/Pb	14	1.00 ± 0.58	7	T+	8	B
Marathon dikes	L. Superior, Ontario	2121 ± 14	U/Pb	14	1.06 ± 0.38	12	T+	8	A
Biscotasing dike	Munro Twp., N. Ontario	2150 ± 20	Ar/Ar	15	18.4 ± 2.7	4	T	16	C
Biscotasing dikes	Abitibi Belt, N. Ontario	2167 ± 2	U/Pb	17	0.77 ± 0.19	6	T+	8	B
Seneterre dikes	Abitibi Belt, N. Ontario	2216 ± 8	U/Pb	17	1.26 ± 0.67	4	T+	8	B
Burakova IC border dikes	Karelia, Russia	2449 ± 1	U/Pb	18	8.43 ± 2.11	15	T+	19	A
Matachewan dikes	Munro Twp., N. Ontario	2465 ± 20	U/Pb	20	5.71 ± 0.38	4	T	16	C
Matachewan dikes	N. Ontario	2465 ± 20	U/Pb	20	2.80 ± 0.87	26	T+	8	A
Yellowknife 8a dikes	N.W.T., Canada	2631 ± 11	U/Pb	21	6.15 ± 1.25	15	T+	22	A
Stillwater IC	Montana, USA	2703 ± 50	Sm/Nd	23	7.20 ± 2.20	59	T+	24	B
Abitibi gabbros	Munro Twp., N. Ontario	2703 ± 3	U/Pb	25	3.28 ± 1.38	4	T	16	C
Greenland dike	W. Greenland	2752 ± 63	K/Ar	26	1.92 ± 0.62	12	T+	26	A
Abitibi komatiites	Munro Twp., N. Ontario	2765 ± 42	Sm/Nd	27	3.42 ± 1.18	4	T	16	C
Komati mafic komatiites	South Africa	3470 ± 20	Ar/Ar	28	2.54 ± 0.33	4	T	29	C
Komati ultramafic komatiites	South Africa	3470 ± 20	Ar/Ar	28	4.78 ± 0.17	3	T	29	C

M is method; Ref. is reference; VADM is virtual axial dipole moment; N is the number of Thellier determinations. Quality is judged on the basis of 3 criteria: N = 10 or more vs. <10; partial TRM checks (T+) vs. no checks (T); fast cooling, usually with positive contact test demonstrating primary TRM, vs. slow cooling. References: 1, Wanless et al. [1966]; 2, Schwarz & Symons [1969]; 3, Cosca et al. [1991, 1992], Lopez-Martinez & York [1983]; 4, Yu & Dunlop [2002]; 5, Palmer & Davis [1987]; 6, Pesonen & Halls [1983]; 7, Krogh et al. [1987]; 8, Macouin et al. [2003]; 9, Mezger et al. [1993]; 10, Yu & Dunlop [2001]; 11, Krogh et al. [1984]; 12, Schwarz & Symons [1970]; 13, Sumita et al. [2001]; 14, Buchan et al. [1996]; 15, Hanes & York [1979]; 16, Hale [1985, 1987a]; 17, Buchan et al. [1993]; 18, Amelin et al. [1995]; 19, Smirnov et al. [2003]; 20, Heaman [1997]; 21, MacLachlan & Helmstaedt [1995]; 22, Yoshihara & Hamano [2000]; 23, Premo et al. [1990]; 24, Selkin [2003]; 25, Nunes & Jensen [1980]; 26, Morimoto et al. [1997]; 27, Zindler et al. [1978]; 28, Lopez-Martinez et al. [1984]; 29, Hale [1987b], Hale & Dunlop [1984].

and with 10 or more independent determinations. These are plotted in Figure 3a. At first glance, the data suggest a strong increase in the Earth's virtual axial dipole moment (VADM) toward the close of the Archean, as suggested by Hale [1987a].

However, there are two problems. First, the VADMs drop back to very low values at 2120 Ma and 1140 Ma. Second, there is a conflict between high and low values at nominally the same time, 2450–2465 Ma. The Burakovka igneous complex of

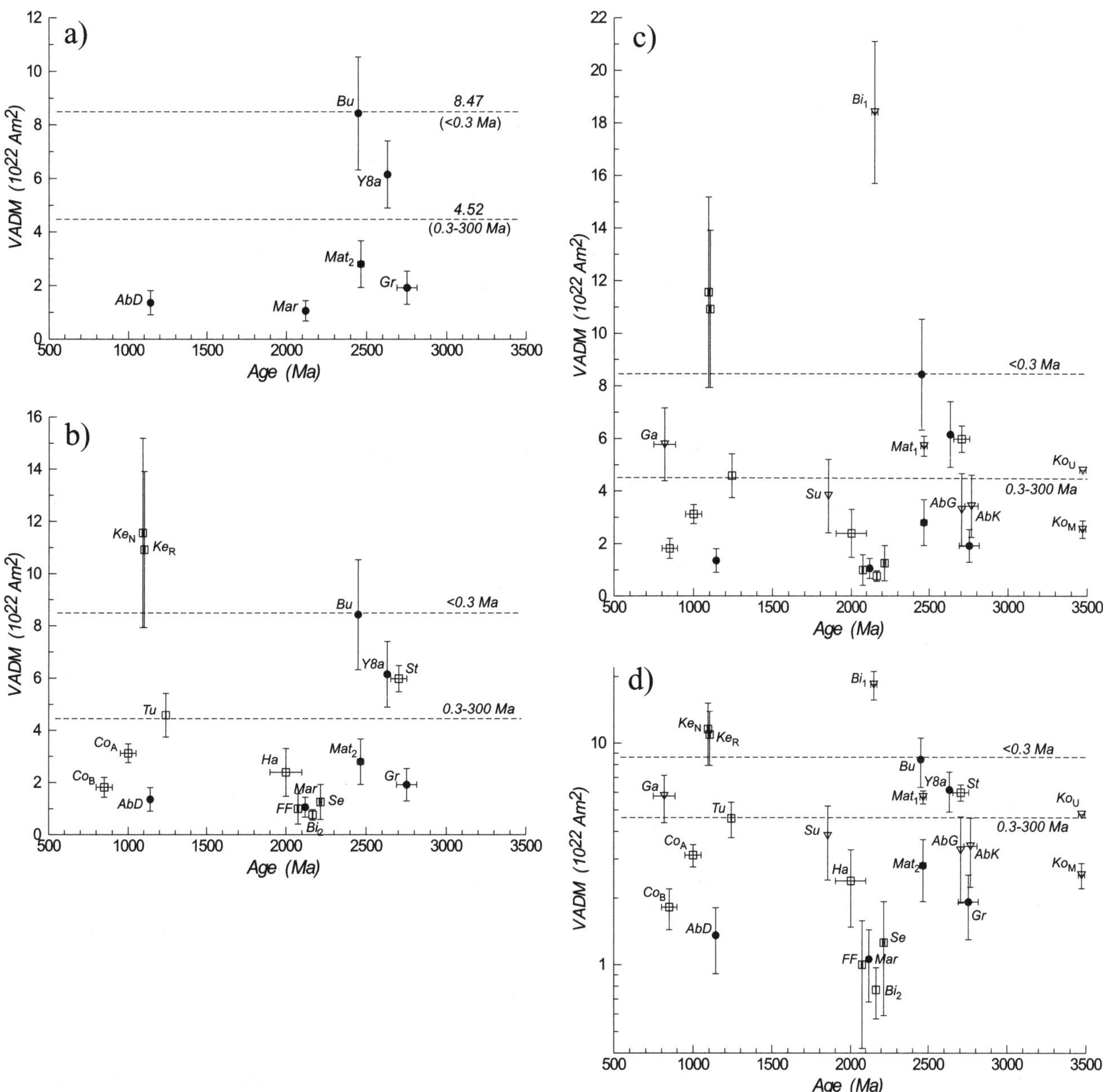

Figure 3. Virtual axial dipole moment (VADM) as a function of age for Precambrian rocks. (a) Category A data. (b) Category A (circles) and B (squares) data. (c) Category A, B, and C (triangles) data. (d) same as (c) but with a logarithmic VADM scale. Labels key to entries in Table 1. A data: AbD, Abitibi dikes; Mar, Marathon dikes; Mat$_2$, Matachewan dikes; Bu, Burakovka dikes; Y8a, Yellowknife 8a dikes; Gr, Greenland dike. B data: Co$_A$, Co$_B$, Cordova Gabbro A & B; Ke$_N$, Ke$_R$, Keweenawan normal & reverse; Tu, Tudor Gabbro; Ha, Hamersley lavas; FF, Fort Frances dikes; Bi$_2$, Biscotasing dikes; Se, Seneterre dikes; St, Stillwater IC. C data: Ga, Grenville (Gananoque) dike; Su, Sudbury IC; Bi$_1$, Biscotasing dike (Munro); Mat$_1$, Matachewan dike (Munro); AbG, Abitibi gabbros; AbK, Abitibi komatiites; Ko$_M$, Ko$_U$, Komati mafic & ultramafic komatiites.

Karelia is tightly dated at 2449 ± 1 Ma [*Amelin et al.*, 1995] and the border dikes which fed the intrusion cannot be greatly different in age. The range of ages for the Matachewan dikes of Ontario includes 2450 Ma; indeed, Heaman [1997] argues for the existence of a 2450 Ma large igneous province including both areas on the basis of the U/Pb dates. The contrast between the Burakovka and Matachewan VADMs is about a factor 3 [*Smirnov et al.*, 2003; *Macouin et al.*, 2003], which is not beyond the bounds of possibility but is at the extreme limit of paleointensity variation over the past few thousand years. At any rate the data demonstrate that high VADMs, similar to Phanerozoic and recent values, are reasonably common in the late Archean and earliest Proterozoic.

Although almost all of the 6 dikes shown in Figure 3a have a positive baked contact test, the test was often performed on a dike of the same swarm from a different geographic area. This can be a risky inference. In Ontario, what was originally named the Abitibi dike swarm has been shown to comprise three swarms, Abitibi, Biscotasing and Seneterre, with similar pale-omagnetic directions but very different ages. Most satisfying is a contact test performed at one of the sites used in the paleointensity work. Even better is a contact test based on identical paleointensities as well as directions in intrusion and baked country rock. Such a test has rarely been performed because of the difficulty of finding well-exposed and magnetically well-behaved contact rocks.

3.2. Category B Results

Ten additional results are in category B. They are plotted along with the category A data in Figure 3b. With one exception, the Keweenawan, these VADMs fall within a range of 0.5 to 1.5 times the Phanerozoic average (excluding the last 0.3 Ma) of about 4.5×10^{22} Am^2 [*Juárez et al.*, 1998; *Selkin and Tauxe*, 2000]. This average relies heavily on submarine basaltic glass results, which are systematically lower than other data. There are possible physical reasons: a CRM rather than TRM origin for the NRM [*Heller et al.*, 2002]; or production of new magnetite during heating in the Thellier experiment for the older glasses [*Smirnov and Tarduno*, 2003]. Either mechanism would produce a low paleofield estimate.

The case for higher paleofields from the late Archean onward is weakened by most of the category B data. But how much trust should we place in them?

Three dike swarms, the Fort Frances, Biscotasing and Seneterre, are in category B because their numbers of determinations are <10 (7, 6, and 4, respectively). Nevertheless their VADMs are in good agreement with that of the Marathon dikes, the fourth swarm in the 2100–2200 Ma age interval, and all were studied with equal care [*Macouin et al.*, 2003]. They are probably reliable.

The Keweenawan results are supported by positive contact tests, and are based on large sample sets (20 N, 34 R). They lack only partial TRM checks. Although Valet [2003] considers them poorly determined, because of their dispersion and the asymmetry of N and R directions, we do not share that opinion. The standard deviation of N and R data, taken separately or collectively, is about ±30%, quite comparable to that of most other data and less than many of the weaker VADMs. Furthermore the average VADMs of N and R samples are almost identical, as one would expect if the paleofield was dipolar and asymmetry was produced by a rapid latitudinal shift during the ~10 myr between R and N magnetization ages (1102 ± 5 and 1092 ± 5 Ma, respectively).

The remaining category B results are for studies with partial TRM checks and adequate numbers of samples but from units that cooled slowly. The present Stillwater igneous complex (IC) data set [*Selkin et al.*, 2000] is supplemented by a large number of soon-to-be-published results [*Selkin*, 2003]. (Early results by Bergh [1970] similar to these are not listed.) The 1850 Ma (*not* 1700 Ma) Sudbury and the 2700 Ma Stillwater layered ICs have multiple phases and a long intrusion and cooling history. Different phases may possibly record different paleofield intensities. The Sudbury results are for 47 specimens distributed through a 300-m section. There is considerable variation in directions as well as intensities, suggesting differences in magnetization age, although the authors make a single average of all results [*Schwarz and Symons*, 1970].

Layered intrusions have a pronounced fabric anisotropy, which has seldom been taken into account. Selkin et al. [2000], using the anisotropy of anhysteretic remanent magnetization to correct their results, estimated that the paleofield had about one-half the intensity indicated by raw Thellier data. Much smaller corrections were calculated by Yu and Dunlop [2001, 2002] for the Tudor and Cordova Gabbros, which are small intrusions lacking obvious fabric.

A major concern with slowly cooled units is the difficulty of determining a precise age of magnetization. This problem becomes acute in the case of slowly uplifted metamorphic terrains with a cooling history of tens or even hundreds of myr. Three of the results in category B are from slowly cooled units in the Grenville Province (Ontario). The Tudor Gabbro was not reheated above 500°C during burial metamorphism and the highest blocking-temperature fraction of its primary TRM survived [*Yu and Dunlop*, 2001]. In this case, the cooling time was no more than a few thousand years and the magnetization age is well determined. On the other hand, the Cordova Gabbro was reheated above its highest blocking temperatures and both the A and B NRMs were acquired during very slow cooling over the general time interval 1050–800 Ma. By ^{40}Ar/^{39}Ar dating of minerals like hornblende, phlogopite, biotite, muscovite and tremolite, which have different closure temperatures to Ar diffusion, it is possible to establish a regional cooling curve to which NRM blocking temperatures can be matched [*Lopez-Martinez and York*, 1983, 1984; *Cosca et al.*, 1991, 1992]. In this way, magnetization ages of 1000 ± 50 Ma and 850 ± 50 Ma were estimated for the Cordova A and B, respectively [*Yu and Dunlop*, 2002]. For the thermally overprinted lavas of the Hamersley Basin (Western Australia) [*Sumita et al.*, 2001], Ar/Ar constraints are lacking and the age of magnetization is correspondingly less certain (estimated to be 2000 ± 100 Ma).

Another problem with slowly cooled units is that TRM and partial TRM intensities depend on cooling rate. There are several theoretical estimates of this effect for single-domain grains [*Dunlop and Özdemir*, 1997, Chap. 16], of which the best known is by Halgedahl et al. [1980], but they all deal with total TRM intensity, not with partial TRMs in different blocking temperature ranges. Furthermore, whereas TRM intensity is predicted and observed to increase with increasing cooling time for single-domain grains, there is some experimental evidence that TRM intensity *decreases* with slower cooling for multidomain grains [*Perrin*, 1998].

The cooling-rate effect on paleointensity is potentially important. Hale [1987b] and Selkin et al. [2000] calculate, using Halgedahl et al.'s model, that their paleointensity values should be reduced by factors of 1.55 and 1.45, respectively, to allow for slow cooling. However, a correction has seldom been applied and its theoretical and experimental basis remains inadequate. Therefore we report in Table 1 values uncorrected for cooling rate. Since most paleointensities, even for coarse-grained rocks, are based on NRMs with single-domain character, the uncorrected values are probably in general too high.

3.3. Category C Results

The 8 results in category C are added to the A and B data in Figure 3c. Most of these studies have marginal numbers of determinations and none used partial TRM checks. They cannot be considered reliable, yet they should not be disregarded entirely. The youngest (817 Ma) and oldest (3470 Ma) Precambrian data are in this category. So are results for 4 units from Munro Township in the Abitibi greenstone belt (N. Ontario) which are mainly of historical interest. Hale [1987a] used these VADMs to make his case for onset of a vigorously convecting dynamo around 2500 Ma, but the supporting Arai plots and sample results are mainly unpublished [*Hale*, 1985].

Munro Township is a classic locality. Here Irving and Naldrett [1977] demonstrated a double positive contact test for a large "Abitibi" dike cutting Matachewan dikes and Archean gabbro. This was expanded and used to estimate burial depth by Buchan and Schwarz [1981]. Schutts and Dunlop [1981] extended studies more broadly in the Abitibi belt and meas-

ured "Abitibi" and Matachewan overprints of gabbros, basalts and komatiites as well as characteristic NRMs of some of these country rocks. The stage was thus set for Hale's [1985] paleointensity studies of 2765 Ma komatiites, 2703 Ma gabbros and their baked contact basalts, 2465 Ma Matachewan dikes and baked contacts, and the "Abitibi" dike, which had been dated at 2150 ± 20 Ma (^{40}Ar/^{39}Ar) by Hanes and York [1979], together with its baked contact rocks (Matachewan and basalts). (This dating showed that the "Abitibi" dike was of quite a different age from the 1140 Ma Abitibi swarm, and it was subsequently named Preissac. It is so referenced by *Yu and Dunlop* [2001], but is properly named Biscotasing [*Buchan et al.*, 1993].)

What is unique about Hale's [1985] study are the matching paleointensity determinations on gabbro and diabase intrusions and their baked contact rocks. The most important of these is the result for the Munro Twp. Biscotasing dike (1 determination), baked Matachewan contact (2 determinations), and baked Archean basalt (1 determination), all of which agree within $\pm15\%$ and are very high: $18.4 \pm 2.7 \times 10^{22}$ Am^2 (Figure 3c). More puzzling is the fact that the recent Biscotasing dike study by Macouin et al. [2003] yields an average VADM that is 20 times lower: $0.77 \pm 0.20 \times 10^{22}$ Am^2 for 6 samples. Hale [1985] does not show all his Arai plots, but the baked Archean basalt sample utilizes data over the entire range 200–550°C, above which no NRM remains, and one baked Matachewan diabase shows linearity from 200–525°C, utilizing about one-third of the NRM. Macouin et al. [2003] do not show an example of their Biscotasing data, but most of their data for other dike swarms utilize higher temperature ranges, typically 500–550°C. It may be that Hale's lower temperature data really reflect a viscous overprint, but if so it is curious that it affected all three rocks types identically and that his other three units do not show similar high-intensity overprinting.

The original Biscotasing result, although high even by Phanerozoic standards, is not beyond the realm of possibility. It is slightly more than twice the present-day VADM, and about 50% higher than the VADMs of the Cretaceous Rajmahal Traps [*Tarduno et al.*, 2001] and Keweenawan rocks. Munro Township deserves restudy, given the pivotal role played by the Biscotasing VADM in Hale's [1987a] model and the conflict between it and more recent studies in the Abitibi Belt.

3.4. Dispersion and Quality of Precambrian VADMs

Figure 3c gives the impression that higher mean VADM intensity is accompanied by larger scatter of individual determinations. This is not so in fact. When the VADM data are plotted on a logarithmic scale (Figure 3d), so that equal percentage errors appear as equal-length bars at any VADM level, it is clear that the high VADMs have average or below-average standard deviations.

An important factor is the number N of individual determinations. Not surprisingly, when N is small, standard deviation tends to be larger, although there are some exceptions. For example, the Sudbury Complex VADM (47 determinations) has a ±38% standard deviation. The sampling covered 300 m of section with various lithologies and likely there are real variations in time of magnetization, although the authors chose to combine all results. Combining or separating results can greatly affect apparent scatter. We have chosen to average the Komati Fm. basaltic and peridotitic komatiite data separately, on the grounds that they may have different magnetization ages. Even though N is small (3 or 4), the separate means have very small scatter. However, if the data for both were combined, the standard deviation would increase greatly because the two means differ by about a factor 2. This is the reason for the relatively large errors for results from the Hamersley Basin lavas and the Yellowknife 8a dikes in Figure 3. The authors averaged their results separately by sites, but we see no justification for this. The Hamersley thermal overprint is regional, while both 8a dikes are from the same swarm, sampled within <2 km of each other. Therefore we have combined results. The between-site scatter in both cases is much greater than within-site scatter, and the error bars increase accordingly.

None of these problems would arise if we had the luxury of selecting data by really stringent criteria: rapidly cooled units, with primary TRM demonstrated at the sites used for paleointensity work and dated radiometrically at the same sites or in the immediate vicinity. But the Precambrian does not provide us with many such ideal situations. Lavas, the mainstay of the Phanerozoic paleointensity database, are rarely usable because even low greenschist grade metamorphism alters the magnetic minerals. The only Precambrian determinations for nearly pristine lava flows are the Keweenawan, results that have been distrusted in previous compilations, and possibly some sites of Abitibi komatiitic basalts sampled sparsely by Hale [1985]. The Komati Fm., which was at first believed to carry a primary TRM, was later reassessed by Hale [1987b] as thermally overprinted.

Diabase dikes are the best Precambrian targets because they resist later metamorphism better than lavas and because many swarms seem to have been intruded late in the tectonic history of structural provinces into cool crust at relatively shallow levels. It is no accident that all 6 category-A results in Table 1 and Figure 3a are from diabase dikes. Nevertheless there is reason to be cautious. These dikes are notoriously difficult to date by U/Pb. Often only one or two dikes in a swarm have been dated and these may be at a large distance from the area

sampled paleomagnetically. The Matachewan swarm gives individual dates that are tightly determined but the results from different areas differ by almost 30 Ma, 3–10 times the error of individual measurements [*Heaman*, 1997]. Thus contemporaneity cannot be taken for granted over broad regions, either because propagation of the dikes took a long time, or because the crust has been eroded to different depths in different areas, or both. We must also beware of swarms that group dikes of similar trend which turn out to have quite different ages. The Abitibi/Preissac/Biscotasing/Seneterre dikes are a classic example. The 2217 Ma Nipissing dikes/ sills of Ontario, which are worthy future paleointensity candidates [*Donadini et al.*, 2003], have three different paleomagnetic signatures in different areas. The final problem with dikes and sills is that they are widely separated time lines in an extremely long geologic record.

By contrast, the Precambrian abounds in slowly cooled rocks, from small stocks to large, originally deep-seated igneous complexes like the Sudbury, Burakovka and Stillwater, to entire volcanic-gneiss terrains that have been exhumed after burial at depth. Some of these rocks contain ideal magnetic carriers in the form of single-domain magnetite inclusions in silicate minerals, which have buffered them against alteration. The problem is in dating the time of magnetization within acceptably narrow bounds and, when this becomes feasible, in applying a suitable cooling-rate correction to the paleointensity. Much progress has been made in the dating [*Lopez-Martinez and York*, 1983; *Cosca et al.*, 1991, 1992] but the cooling-rate question is almost unconstrained by experimental data. These are the next frontiers in Precambrian paleointensity determination.

4. WAS THE FIELD DIPOLAR IN PRECAMBRIAN TIME?

We have only the slenderest of evidence for or against a Precambrian dipole field. Three units or pairs of units have results with sufficiently different paleolatitudes to test the expected latitude dependence of a dipole field (Figure 4). These are the Cordova Gabbro A and B, with respective paleointensities and paleolatitudes $B_p = 18.1$ µT, $\lambda_p = -40.3°$ and $B_p = 7.1$ µT, $\lambda_p = -4.6°$, the Keweenawan N and R, with $B_p = 51$ µT, $\lambda_p = +18.5°$ and $B_p = 73$ µT, $\lambda_p = -54.5°$, and the Abitibi gabbros/ baked basalts and komatiites, with $B_p = 14.3$ µT, $\lambda_p = -17.5°$ and $B_p = 26.0$ µT, $\lambda_p = -78.0°$. These follow the dipole curves most convincingly for the Keweenawan results, fairly well for the Abitibi pair, and least convincingly for the Cordova A and B. This is in the same order as increasing age differences between results in a pair. The Keweenawan N and R magnetizations are ~10 myr different in age, the Abitibi gabbro and komatiite magnetization ages are ~60 myr differ-

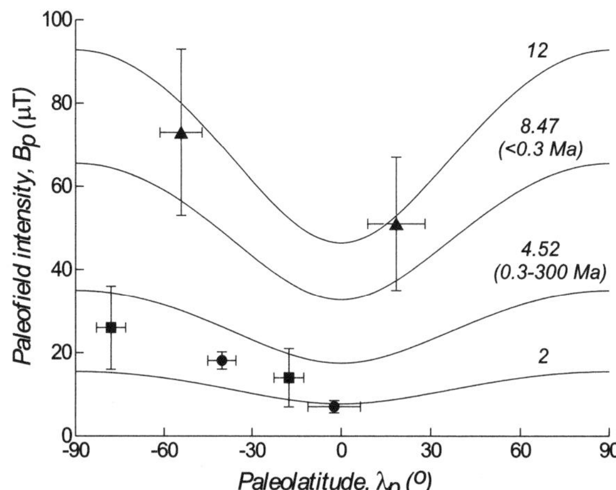

Figure 4. Dependence of paleointensity B_p on paleolatitude λ_p for Keweenawan N and R (triangles), Abitibi gabbros and komatiites (squares), and Cordova A and B (circles) compared with the expected dependence for a geocentric axial dipole (GAD) field.

ent, and the Cordova A and B magnetization ages are ~150 myr different.

A more convincing comparison would be between different formations with the same age but widely different paleolatitudes. There is a single such comparison possible from Table 1, between the 2703 ± 50 Ma Stillwater and 2703 ± 2 Abitibi gabbro/basalt results, with respective paleolatitudes $\lambda_p = -82°$ and $-17.5°$. The VADMs (which incorporate latitude correction) of 7.20 and 3.28×10^{22} Am^2 do not agree closely. However, this does not necessarily mean the field was not dipolar. The Earth's field has varied over the last few thousand years by at least a factor 2 and it is certainly not possible to demonstrate contemporaneity to within a thousand years for any pair of Precambrian results. The Stillwater and Abitibi magnetization ages, ostensibly the same, are in reality probably different by 10–20 Ma or more. The two formations are located <30° apart in the Laurentian Shield, yet their paleolatitudes differ by ~60°.

The only practically achievable test of dipolarity may be the one carried out by Evans [1976] for Phanerozoic time and updated and extended to the Precambrian by Kent and Smethurst [1998]. Grouping data within broad time intervals, they examined the frequency distribution of paleomagnetic inclination I, which is distinctively different for dipole, quadrupole and octupole fields. For both unweighted and spatiotemporally binned Paleozoic and Precambrian data, Kent and Smethurst noted a surplus (relative to that expected for a geocentric axial dipole or GAD field) of low I values and a deficiency of high I values. A different statement of the same result for the Precambrian was given earlier by Lapointe et al. [1978]: "What happened to the high-latitude palaeomagnetic poles?" To carry out this test, one must know the distri-

bution of the continents, which is increasingly uncertain from the Paleozoic back into the Precambrian. Kent and Smethurst [1998] point to geodynamically triggered cycling of continents through low latitudes as a possible source for low-I bias. Selective rejection of results from the Canadian Shield with high positive I similar to the present Earth's field on the grounds of presumed viscous overprinting [Lapointe et al., 1978] is also a plausible source of bias.

Although not definitive, an early test of dipolarity was carried out by Schwarz and Symons [1969] in their reconnaissance paleointensity study of (mainly) Canadian Shield rocks. They compared the cumulative distributions of paleointensities (reduced to equatorial intensity, equivalent to calculating a VADM) for samples with λ_p either <25° or >25°. The agreement was excellent. Although the authors disclaimed any conclusive verification of the GAD hypothesis, the positive result of their test is in interesting conflict with later studies. Of their 41 data, 32 are Precambrian, and 25° is approximately the crossover point between surplus (low-I) and deficiency (high-I) of data in Kent and Smethurst's study, which admittedly uses much more data (1277 results).

A final line of evidence is the analysis of between-site dispersion of remanence directions in Keweenawan rocks as a function of λ_p by Halls and Pesonen [1982]. The variation of the circular standard deviation with λ_p (covering the range $15° \leq \lambda_p \leq 60°$) follows quite well the corresponding curve for the paleosecular variation (PSV) of the present and Tertiary fields for all 9 R studies and 5 of 6 N studies. The R and N averages fall essentially on the curve. This, together with the concordance of Keweenawan R and N paleointensities (Figure 4, Table 1), argues against a special nondipolar state of the field as the cause of R-N reversal asymmetry. Halls and Pesonen's study is the most detailed available of Precambrian PSV.

5. THE PRECAMBRIAN POLARITY REVERSAL RECORD

There are only three continuous or nearly continuous magnetostratigraphic records for the Precambrian. The best controlled of these is for the Belt-Purcell Supergroup of Montana and Alberta, covering the time interval 1460–1400 Ma approximately [Elston et al., 2002]. The others are for the Malgina and Linok Formations of the Siberian craton (~1100–1050 Ma; [Gallet et al., 2000]) and the upper Grand Canyon Supergroup (~1150–1050 Ma) together with Keweenawan data of the same general age [Elston et al., 2002]. The Siberian and North American records are contradictory, the Siberian record showing 16 symmetric (180°) reversals between 1100 and 1050 Ma and the North American formations recording a single asymmetric reversal in this interval.

Figure 5 compares the Belt-Purcell polarity zonation to the Jurassic to present polarity record on matching timescales. Between ~1450 Ma and 1400 Ma, only 3 major reversals are recorded, with short intervals of mixed polarities near the top of the record. The upper N chron is 5–10 myr long. The middle R chron (~30 myr) is almost as long as the Cretaceous N superchron (126–85 Ma).

The Belt-Purcell polarity sequence is indicated on the apparent polar wander path (APWP) of Figure 6. Paleopoles for the Grand Canyon Supergroup (G) and the Keweenawan (K) rocks of Lake Superior and some alkaline complexes immediately to the east are also plotted. Their polarities are not indicated, but an N to R reversal occurs around G4/K3 and an R to N reversal at G7–G8/K8–K12. The latter reversal is highly asymmetric in the Keweenawan igneous rocks, as noted earlier, and also in the Grand Canyon sedimentary rocks, but not in contemporaneous alkaline rocks immediately east of Lake Superior (e.g., K13 and also results not plotted for the Nemogosenda and Lackner Lake carbonatites and the Shenango alkaline complex [*Costanzo-Alvarez et al.*, 1993]). The almost perfectly 180° reversals recorded by the alkaline rocks imply that the pronounced reversal asymmetry ($I \sim +34°$ for N samples, $I \sim -70°$ for R samples) in Keweenawan rocks is due neither to rapid APW nor to non-dipole fields but to an undetected secondary magnetization added to the primary N

Figure 6. Laurentian APWP from 1500 to 1000 Ma. Poles B1–B8, Belt-Purcell Supergroup (closed, N; open, R; half-open, mixed or transitional). Poles G1–G14, Grand Canyon Supergroup (solid lines, continuous section; dashed lines, unconformities). Poles K1–K21, Keweenawan and nearby carbonatite results (after *Elston et al.* [2002]). Reprinted by permission of the Geological Society of America.

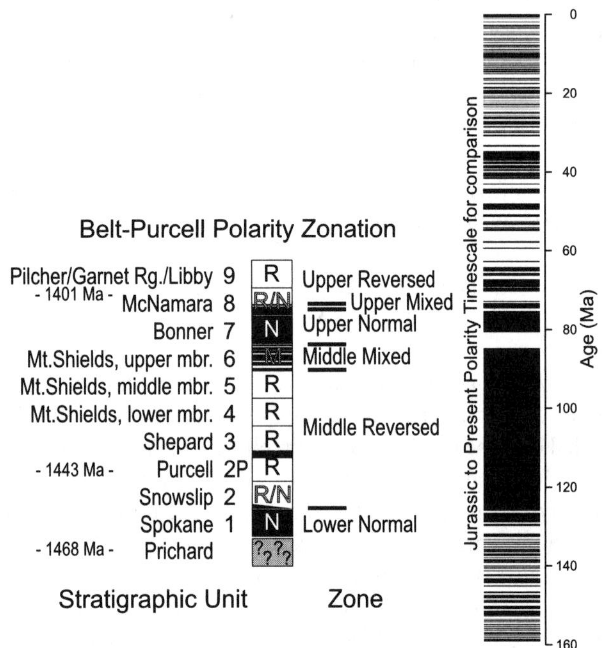

Figure 5. Polarity zonation in the Belt-Purcell Supergroup compared with the Jurassic to present polarity timescale (after *Elston et al.* [2002]). Reprinted by permission of the Geological Society of America.

and R NRMs. Halls and Pesonen [1982] considered this possibility but discounted it.

Figure 7 illustrates the polarity zonation measured for the Malgina Formation from the SE margin of the Siberian Shield. In this time interval, 1100–1050 Ma, Gallet et al. [2000] measured 16 reversals, correlated across 4 sections 50–100 km apart. In the Linok Formation of NW Siberia, the magnetostratigraphy is less complete but at least 6 reversals are recorded. In the same time interval in North America, the G and K sequences record only a single reversal. Although there could be gaps in the Keweenawan record, the Grand Canyon section from G7 to G13 is continuous (indicated by solid tie lines in Figure 6; dashed lines cross unconformities). The disagreement is therefore puzzling.

Another point of disagreement between the North American and Siberian records is that the reversals in the latter are symmetric. Figure 8a shows mean directions of both polarities from all Malgina sections after bedding correction and Figure 8b combines both polarities from the different sections. The R and N directions are exactly 180° apart. This result and the North American carbonatite data make it unlikely that there was a persistent non-reversing non-dipole component to the geomagnetic field around 1100–1050 Ma.

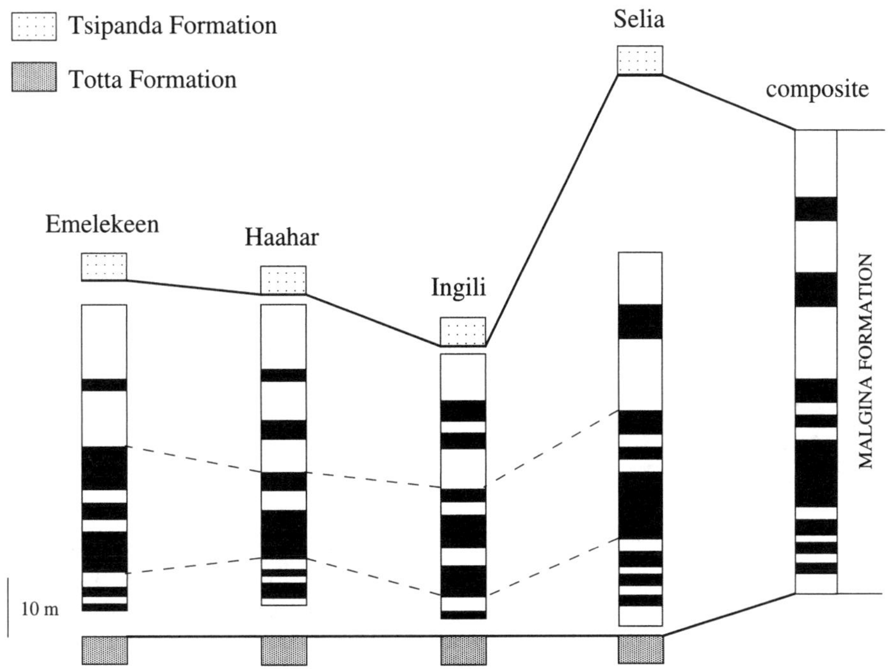

Figure 7. Magnetostratigraphy of the 1100–1050 Ma Malgina Formation, Siberia (after *Gallet et al.* [2000]).

For the time interval 1020–820 Ma, immediately follow-ing Keweenawan time, there are no magnetostratigraphic records but the paleomagnetism has been studied in unusual detail, mainly for the slowly uplifted Grenville orogen. Few of the paleopoles are primary (the Tudor Gabbro is an excep-tion) but the general age succession has been established by ^{40}Ar/^{39}Ar thermochronometry of key formations [*Lopez-Mar-tinez and York*, 1983; *Cosca et al.*, 1991, 1992; *Warnock et al.*, 2000]. There is still debate about the sequence of poles, one school favoring a counterclockwise linkage [*Weil et al.*, 1998] and the other a clockwise sequence [*Costanzo-Alvarez and Dunlop*, 1998]. In the clockwise version shown in Figure 9, the oldest results (MD, TH, MM) have N polarity and link with the youngest (~1050 Ma) Keweenawan N poles of Figure 6. There are then a cluster of 12 R results in the interval ~1000–950 Ma, two mixed polarity results, 8 N results around 900 Ma and three overprints of R polarity (one of them the Cordova B) around 850 Ma. Because all these results average over con-siderable cooling times, they may conceal a finer polarity zonation but in the simplest picture, there are only 4 dominant polarity chrons in this ~200 myr interval. As with the Belt, Grand Canyon and Keweenawan sequences (but not the Siber-ian ones), the Grenville seems to have had long periods of a single polarity.

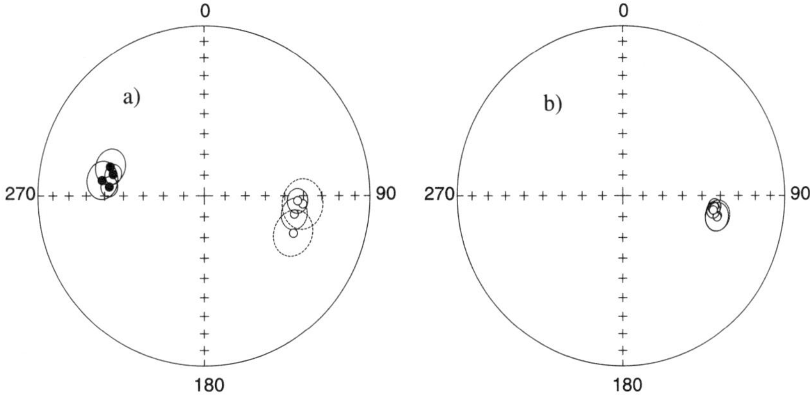

Figure 8. (a) N and R site mean directions for the Malgina Formation after bedding correction. (b) Malgina N and R results combined, showing 180° reversals (after *Gallet et al.* [2000]).

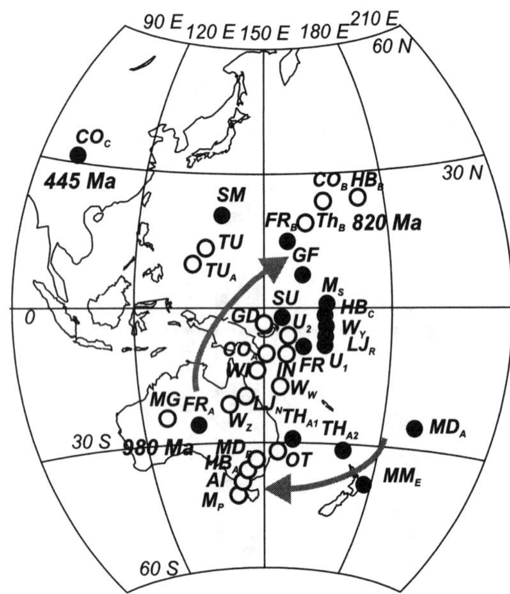

Figure 9. The Grenville (1020–820 Ma) APWP, clockwise version. The polarity zonation is N-R-mixed-N-R (closed symbols, N; open, R; half-open, mixed).

The oldest reversal for which the sense of polarity change is known is recorded by the 2465 ± 20 Ma Matachewan dike swarm [*Halls*, 1991]. Most Matachewan dikes have R polarity (relative to the present field: the true polarity is unknown because of gaps in the APWP) but N polarity dikes are also fairly common, particularly in crustal blocks uplifted from greater original depths. Since the later-cooled dikes record a normal field, the polarity reversal must have been R to N. This is confirmed at 4 sites in northern Ontario flanking the Kapuskasing Uplift, just east of Lake Superior, where an N dike cuts an older R dike. The opposite relationship is never observed, implying that only one reversal occurred during the duration of Matachewan dike intrusion.

6. DETAILED RECORDING OF A PROTEROZOIC POLARITY TRANSITION

Bingham and Evans [1975] reported a full polarity transition, preceded by a field excursion along the same path, recorded in a continuous 600-m section of the 1800–1650 Ma Stark Formation of the Great Slave Supergroup, N.W.T., Canada. The transition occurs across ~25 m of section and is recorded at 6 sites, each with 2–4 samples. The reversal is a full 180° and the path of virtual geomagnetic poles (VGPs) is a great circle passing close to the study location. The mean NRM intensity decreases by about a factor 3 during the transition. The drop in intensity covers the same time interval as the reversal in directions, with periods of somewhat reduced

intensity immediately above and below the transition zone. The excursion, with an ~80° directional change and recorded at only one site (5 m of section), coincides with the earlier intensity reduction and seems to be a precursor to the full field reversal.

The only feature that is somewhat different from modern polarity transition records is the duration of the reversal, which is estimated to be 10^4–10^5 yr on the basis of standard sedimentation rates. If this calibration is correct, the interval over which the entire 600 m section was deposited and magnetized would be 0.25–2.5 myr, during which time a single reversal occurred.

7. DISCUSSION AND EVALUATION

Precambrian paleointensity data have many limitations. We are often forced to work with unfavorable rocks that have been partially or totally remagnetized. Demonstrating that the NRM components are purely thermal is no easy task. The timing of remagnetization(s) may or may not be clear, and if the overprinting occurred during slow uplift, a regional thermochronometric cooling curve is required, not a single radiometric age.

Since volcanic rocks are generally unusable because of metamorphic alteration, dikes and sills are the targets of choice. However, these may not have cooled as rapidly as is usually assumed. At the time of the emplacement (2167 Ma) of the Munro Township Biscotasing dike, the present erosion surface was at a temperature of 180–220°C [*Buchan and Schwarz*, 1981]. At 1250 Ma, when dikes that cut the Sudbury IC were intruded, the presently exposed area was at a temperature of 260–280°C [*Schwarz and Buchan*, 1982]. Even in these relatively low-grade areas of the Superior and Southern Provinces, uplift of 7–10 km has occurred and the dikes were emplaced in warm rather than cold crust. In the Grenville Province, burial was to depths more like 10–25 km [*Anovitz and Essene*, 1990] and only rare bodies, such as the low-grade Tudor and Cordova Gabbros, are usable.

Country rocks with 200–270°C temperatures should not unduly retard fixing of the NRM fraction used for paleointensity determination, which normally has higher blocking temperatures. However, they do promote viscous remagnetization much beyond the usual 250–300°C limit of magnetite laboratory unblocking temperatures [*Pullaiah et al.*, 1975]. If uplift is slow and reversals infrequent, this viscous overprinting may continue for a long time. A crude estimate of the minimum acceptable temperature for Thellier data would be the sum of the limit for room temperature viscous overprinting (250–300°C) plus the ambient temperature at the time of intrusion.

Cooling-rate correction of partial TRM intensities may also be sensitive to burial and uplift. For this, one must have a realistic idea of the original depth of burial and the rate of regional uplift. For dike swarms, TRM intensity is likely not much affected by erosional uplift because cooling to regional ambient temperature through the blocking temperature range used in paleointensity work occurred before uplift began. For deep-seated intrusions, on the other hand, uplift cooling may be all important. The effective cooling history of the Tudor Gabbro NRM (all blocking temperatures >500°C) is only a few thousand years, while that of the nearby, more deeply buried Cordova Gabbro spans tens of myr.

Although we have no well-tested theory of cooling-rate correction, almost any model predicts a dependence as the logarithm of time. Thus differences in cooling rate between chilled margins and slowly cooled interiors of intrusions, or between narrow and very thick dikes, could have as significant an effect on their TRM intensities and derived paleointensities as the difference in cooling rates between the Tudor and Cordova Gabbro NRMs. In support of this speculation, Halls et al. [2003] report that 20-m wide Matachewan dikes have 50% greater paleointensity values than 4-m wide dikes, and 17-m wide 2.1 Ga dikes have 60% higher paleointensities than 3-m wide dikes (4–6 dikes in each average). Clearly even dikes do not always cool quickly enough to give paleointensity and VADM values that can be compared directly with Phanerozoic volcanic data.

How does the Precambrian field compare with the Phanerozoic field? Despite the limitations of the Precambrian data—unfavorable rock types, long cooling histories and unknown cooling-rate corrections, sometimes uncertain magnetization ages, small numbers of acceptable samples, and lack of partial TRM checks in early studies—all but a very few VADMs lie within a range 0.5–1.5 times the 0.3–300 Ma average. The higher and lower VADMs have similar percentage errors, which are generally similar to those of Phanerozoic data, suggesting that the scatter is due to PSV.

There is no obvious record of onset or growth of a dynamo field in the Archean or early Proterozoic. Category-A data (Figure 3a) do hint at a peak around the close of the Archean, much as hypothesized by Hale [1987a]. However, there is a serious disagreement between the high VADM recorded by the Burakovka dikes [Smirnov et al., 2003] (results from separated plagioclase crystals, which protect the magnetite they contain from alteration) and the low VADM for the Matachewan dikes [Macouin et al., 2003]. (We note in passing that intermediate to high values of 5.7 and ~9 × 10^{22} Am^2 were reported for Matachewan dikes in early studies [Hale, 1985; Kobayashi, 1968] and that a very high, non-Thellier determination for the late Archean of South Africa of about 12 × 10^{22} Am^2 was reported by McElhinny and Evans [1968].) Some authors believe that the average field remained low throughout the Precambrian and increased only during the

Phanerozoic [Macouin et al., 2003]. All models of field evolution are based on smaller subsets of the full data presented in Table 1 and Figures 3a–c. None is definitive at our present stage of knowledge (see Prévot and Perrin [1992]).

In addition to VADMs, the Precambrian field has other points of similarity to the modern field. The one Precambrian polarity transition recorded in detail [Bingham and Evans, 1975] has the hallmarks of a typical transition. The field strength drops about a factor 3 during the directional reversal, with smaller decreases in strength before and after. The VGP path is a great circle, with a brief excursion along the same path shortly before the successful reversal. The duration of the reversal is atypical but is calibrated against an assumed sedimentation rate and could be in error.

There may have been a low reversal frequency in the Precambrian although it is possible that some short reversals have been missed because of lower time resolution of the Precambrian recorders. A single reversal occurred during Matachewan dike intrusion [Halls, 1991]. In the highly continuous Belt-Purcell Supergroup [Elston et al., 2002] there are only two short intervals of mixed polarities in a 60 myr long record. Two N and two R chrons dominate, one R chron being about 30 myr long. In the 1150–1050 Ma interval, the Grand Canyon Supergroup records only two reversals, and so (in a less continuous record) do the Keweenawan rocks of Lake Superior [Elston et al., 2002]. However, a well-resolved record from Siberia between 1100 and 1050 Ma [Gallet et al., 2000] shows 16 reversals, all symmetric in contrast to the asymmetric reversals in the Grand Canyon and Keweenawan data. Finally, in Grenville time (1050–820 Ma), 4 main polarity chrons are recorded. The rocks are all uplifted and slowly cooled, and could have averaged over shorter chrons. Nevertheless there must have been exceptionally long intervals of one polarity or shorter chrons could not be concealed.

How close was the Precambrian field to a GAD? The evidence is fragmentary and contradictory. Asymmetric reversals in the Keweenawan have been cited as evidence for long-lived non-dipole features that did not reverse with the main dipole field [Nevanlinna and Pesonen, 1983]. The bulk of the evidence is against such a model. The N and R Keweenawan data, despite their very different average I values, give indistinguishable VADMs ([Halls and Pesonen, 1982]; Figures 3b, 4), supporting the axial dipole assumption used to calculate them. There are also numerous 180° reversals recorded elsewhere for the same time period [Gallet et al., 2000]. Other data (Cordova, Abitibi) approximately support a GAD variation of paleointensity with paleolatitude λ_p, but they are less convincing because of substantial time differences between results in each pair (Figure 4). Another line of evidence is recorded PSV, in the form of between-site dispersion of paleomagnetic directions. For all published

Keweenawan studies, Halls and Pesonen [1982] showed that the circular standard deviation follows closely the dependence on λ_p expected for a GAD field.

The main evidence against a GAD field is the excess of low-I or low-λ_p results in a random sampling in time and space of Precambrian paleomagnetic results [*Kent and Smethurst*, 1998]. Although the data set tested is large, there are possible sources of bias: discrimination against high-λ_p results in the Canadian Shield data [*Lapointe et al.*, 1978], and possible geodynamic controls favoring low latitudes of the continents [*Kent and Smethurst*, 1998]. Although the jury is still out, we prefer the null hypothesis, that the Precambrian field was not markedly less dipolar than the recent field, until a compelling case to the contrary is made.

Acknowledgments. We thank Mélina Macouin, Jean-Pierre Valet, Alexei Smirnov and John Tarduno for giving us access to their new results before publication. Constructive reviews by John Tarduno and Peter Selkin improved the paper. This research was supported by the Natural Sciences and Engineering Research Council of Canada through grant A7709 to D.J.D.

REFERENCES

Amelin, Yu. V., L. M. Heaman, and V. S. Semenov, U-Pb geochronology of layered mafic intrusions of the eastern Baltic shield: Implications for the timing and duration of Paleoproterozoic continental rifting, *Precambrian Res., 75*, 31–46, 1995.

Anovitz, L. M., and E. J. Essene, Thermobarometry and pressure-temperature paths in the Grenville Province of Ontario, *J. Petrol., 31*, 197–241, 1990.

Bergh, H. W., Paleomagnetism of the Stillwater Complex, Montana, In *Palaeogeophysics*, edited by S. K. Runcorn, Academic Press, London, 143–158, 1970.

Bingham, D. K., and M. E. Evans, Precambrian geomagnetic field reversal, *Nature, 253*, 332–333, 1975.

Buchan, K. L., and E. J. Schwarz, Uplift estimated from remanent magnetization: Munro area of Superior Province since 2150 Ma ago, *Can. J. Earth Sci., 18*, 1164–1173, 1981.

Buchan, K. L., J. K. Mortensen, and K. D. Card, Northeast-trending early Proterozoic dykes of southern Superior Province: Multiple episodes of emplacement recognized from integrated paleomagnetism and U-Pb geochronology, *Can. J. Earth Sci., 30*, 1286–1296, 1993.

Buchan, K. L., H. C. Halls, and J. K. Mortensen, Paleomagnetism, U-Pb geochronology, and geochemistry of Marathon Dykes, Superior Province, and comparison with the Fort Frances swarm, *Can. J. Earth Sci., 33*, 1583–1595, 1996.

Buffett, B. A., E. J. Garnero, and R. Jeanloz, Sediments at the top of Earth's core, *Science, 290*, 1338–1342, 2000.

Buffett, B. A., H. E. Huppert, J. R. Lister, and A. W. Woods, Analytical model for solidification of the Earth's core, *Nature, 356*, 329–331, 1992.

Cosca, M. A., J. F. Sutter, and E. J. Essene, Cooling and inferred uplift/erosion history of the Grenville Orogen, Ontario: Constraints from ^{40}Ar/^{39}Ar thermochronology, *Tectonics, 10*, 959–977, 1991.

Cosca, M. A., E. J. Essene, M. J. Kunk, and J. F. Sutter, Differential unroofing within the Central Metasedimentary Belt of the Grenville Orogen: Constraints from ^{40}Ar/^{39}Ar thermochronology, *Contrib. Mineral. Petrol., 110*, 211–225, 1992.

Costanzo-Alvarez, V., and D. J. Dunlop, A regional paleomagnetic study of lithotectonic domains in the Central Gneiss Belt, Grenville Province, *Earth Planet. Sci. Lett., 157*, 89–103, 1998.

Costanzo-Alvarez, V., D. J. Dunlop, and L. J. Pesonen, Paleomagnetism of alkaline complexes and remagnetization in the Kapuskasing Structural Zone, Ontario, Canada, *J. Geophys. Res., 98*, 4063–4079, 1993.

Donadini, F., L. J. Pesonen, and S. Harlan, Preliminary Thellier paleointensity measurements on Precambrian rocks (abstract), *Eur. Geophys. Soc., Geophys. Res. Abstr., 5*, 00560, 2003.

Dunlop, D. J., and Ö. Özdemir, *Rock Magnetism: Fundamentals and Frontiers*, Cambridge Univ. Press, New York, 573 pp., 1997.

Elston, D. P., R. J. Enkin, J. Baker, and D. K. Kisilevsky, Tightening the Belt: Paleomagnetic-stratigraphic constraints on deposition, correlation, and deformation of the Middle Proterozoic (ca. 1.4 Ga) Belt-Purcell Supergroup, United States and Canada, *Geol. Soc. Am. Bull., 114*, 619–638, 2002.

Evans, M. E., Test of the dipolar nature of the geomagnetic field throughout Phanerozoic time, *Nature, 262*, 676–677, 1976.

Gallet, Y., V. E. Pavlov, M. A. Semikhatov, and P. Yu. Petrov, Late Mesoproterozoic magnetostratigraphic results from Siberia: Paleogeographic implications and magnetic field behavior, *J. Geophys. Res., 105*, 16,481–16,499, 2000.

Gubbins, D., The distinction between geomagnetic excursions and reversals, *Geophys. J. Int., 137*, F1–F3, 1999.

Hale, C. J., Evidence of the Archean geomagnetic field, Ph.D. thesis, 188 pp., Univ. of Toronto, Toronto, Ont., Canada, 1985 (available from University Microfilms, Ann Arbor, MI).

Hale, C. J., Paleomagnetic data suggest link between the Archean-Proterozoic boundary and inner-core nucleation, *Nature, 329*, 233–237, 1987a.

Hale, C. J., The intensity of the geomagnetic field at 3.5 Ga: Paleointensity results from the Komati Formation, Barberton Mountain Land, South Africa, *Earth Planet. Sci. Lett., 86*, 354–364, 1987b.

Hale, C. J., and D. J. Dunlop, Evidence for an early Archean geomagnetic field: A paleomagnetic study of the Komati Formation, Barberton Greenstone Belt, South Africa, *Geophys. Res. Lett., 11*, 97–100, 1984.

Halgedahl, S. L., R. Day, and M. Fuller, The effect of cooling rate on the intensity of weak-field TRM in single-domain magnetite, *J. Geophys. Res., 85*, 3690–3698, 1980.

Halls, H. C., The Matachewan dyke swarm, Canada: An early Proterozoic magnetic field reversal, *Earth Planet. Sci. Lett., 105*, 279–292, 1991.

Halls, H. C., and L. J. Pesonen, Paleomagnetism of Keweenawan rocks, *Geol. Soc. Am. Memoir, 156*, 173–201, 1982.

Halls, H., N. McArdle, M. Gratton, M. Hill, and J. Shaw, Microwave paleointensities from dyke chilled margins: A way to obtain long term variations in geodynamo intensity for the last three billion years? (abstract), *Eur. Geophys. Soc., Geophys. Res. Abstr., 5*, 13837, 2003.

Hanes, J. A., and D. York, A detailed ^{40}Ar/^{39}Ar study of an Abitibi dike from the Canadian Superior Province, *Can. J. Earth Sci., 16*, 1060–1070, 1979.

Heaman, L. M., Global mafic magmatism at 2.45 Ga: Remnants of an ancient large igneous province? *Geology, 25*, 299–302, 1997.

Heller, R., R. T. Merrill, and P. L. McFadden, The variation of intensity of Earth's magnetic field with time, *Phys. Earth Planet. Inter., 131*, 237–249, 2002.

Irving, E., and A. J. Naldrett, Paleomagnetism in Abitibi greenstone belt and Abitibi and Matachewan diabase dikes: Evidence of the Archean geomagnetic field, *J. Geol., 85*, 157–176, 1977.

Juárez, M. T., L. Tauxe, J. S. Gee, and T. Pick, The intensity of the Earth's magnetic field over the past 160 million years, *Nature, 394*, 878–881, 1998.

Kent, D. V., and M. A. Smethurst, Shallow bias of paleomagnetic inclinations in the Paleozoic and Precambrian, *Earth Planet. Sci. Lett., 160*, 391–402, 1998.

Kobayashi, K., Paleomagnetic determination of the intensity of the geomagnetic field in the Precambrian period, *Phys. Earth Planet. Inter., 1*, 387–395, 1968.

Krogh, T. E., D. W. Davis, and F. Corfu, Precise U-Pb zircon and baddeleyite ages for the Sudbury area, In *The Geology and Ore Deposits of the Sudbury Structure*, edited by E. G. Pye, A. J. Naldrett, and P. E. Giblin, *Ont. Geol. Surv. Spec. Vol., 1*, 431–446, 1984.

Krogh, T. E., and 8 others, Precise U-Pb isotopic ages of diabase dykes and mafic to ultramafic rocks using trace amounts of baddeleyite and zircon, In *Mafic Dyke Swarms*, edited by H. C. Halls and W. F. Fahrig, *Geol. Assoc. Can. Spec. Paper, 34*, 147–152, 1987.

Labrosse, S., J-P. Poirier, and J-L. LeMouël, On cooling of the Earth's core, *Phys. Earth Planet. Inter., 99*, 1–17, 1997.

Labrosse, S., J-P. Poirier, and J-L. LeMouël, The age of the inner core, *Earth Planet. Sci. Lett., 190*, 111–123, 2001.

Lapointe, P. L., J. L. Roy, and W. A. Morris, What happened to the high-latitude palaeomagnetic poles? *Nature, 273*, 655–657, 1978.

Lopez-Martinez, M., and D. York, Further thermochronometric unraveling of the age and paleomagnetic record of the Grenville Province, *Can. J. Earth Sci., 20*, 953–960, 1983.

Lopez-Martinez, M., D. York, C. M. Hall, and J. A. Hanes, 3.46 Ga ^{40}Ar/^{39}Ar ages for terrestrial rocks: Barberton Mountain komatiites, *Nature, 207*, 352–355, 1984.

MacLachlan, K., and H. Helmstaedt, Geology and geochemistry of an Archean mafic dike complex in the Chan formation: Basis for

revised plate-tectonic model of the Yellowknife greenstone belt, *Can. J. Earth Sci., 32*, 614–630, 1995.

Macouin, M., J-P. Valet, J. Besse, K. Buchan, R. Ernst, M. Le Goff, and U. Scharer, Low paleointensities recorded in 1 to 2.4 Ga Proterozoic dykes, Superior Province, Canada, *Earth Planet. Sci. Lett., 213*, 79–95, 2003.

McElhinny, M. W., and M. E. Evans, An investigation of the strength of the geomagnetic field in the early Precambrian, *Phys. Earth Planet. Inter., 1*, 485–497, 1968.

Mezger, K., E. J. Essene, B. A. van der Pluijm, and A. N. Halliday, U-Pb geochronology of the Grenville orogen of Ontario and New York: Constraints on ancient crustal tectonics, *Contrib. Miner. Petrol., 114*, 13–26, 1993.

Morimoto, E., Y. Otofuji, M. Miki, H. Tanaka, and T. Itaya, Preliminary paleomagnetic results of an Archean dolerite dyke of west Greenland: Geomagnetic field intensity at 2.8 Ga, *Geophys. J. Int., 128*, 585–593, 1997.

Nevanlinna, H., and L. J. Pesonen, Late Precambrian Keweenawan asymmetric polarities as analyzed by axial offset dipole geomagnetic models, *J. Geophys. Res., 88*, 645–658, 1983.

Nunes, P. D., and L. S. Jensen, Geochronology of the Abitibi metavolcanic belt, Kirkland Lake area—Progress report, In *Summary of Geochronology Studies 1977–1979*, edited by E. G. Pye, *Ont. Geol. Surv. Misc. Paper, 92*, 56 pp., 1980.

Palmer, H. C., and D. W. Davis, Paleomagnetism and U-Pb geochronology of volcanic rocks from Michipicoten Island, Lake Superior, Canada: Precise calibration of the Keweenawan polar wander track, *Precambrian Res., 37*, 157–171, 1987.

Perrin, M., Paleointensity determination, magnetic domain structure, and selection criteria, *J. Geophys. Res., 103*, 30, 591–30,600, 1998.

Pesonen, L. J., and H. C. Halls, Geomagnetic field intensity and reversal asymmetry in late Precambrian Keweenawan rocks, *Geophys. J. Roy. Astron. Soc., 73*, 241–270, 1983.

Premo, W., R. Helz, M. Zientek, and R. Langston, U/Pb and Sm/Nd ages for the Stillwater Complex and its associated sills and dikes, Beartooth Mountains, Montana: Identification of a parent magma, *Geology, 18*, 1065–1068, 1990.

Prévot, M., and M. Perrin, Intensity of the Earth's magnetic field since Precambrian from Thellier-type palaeointensity data and inferences on the thermal history of the core, *Geophys. J. Int., 108*, 613–620, 1992.

Pullaiah, G., E. Irving, K. L. Buchan, and D. J. Dunlop, Magnetization changes caused by burial and uplift, *Earth Planet. Sci. Lett., 28*, 133–143, 1975.

Schutts, L. D., and D. J. Dunlop, Proterozoic magnetic overprinting of Archaean rocks in the Canadian Superior Province, *Nature, 291*, 642–645, 1981.

Schwarz, E. J., and K. L. Buchan, Uplift deduced from remanent magnetization: Sudbury area since 1250 Ma ago, *Earth Planet. Sci. Lett., 58*, 65–74, 1982.

Schwarz, E. J., and D. T. A. Symons, Geomagnetic intensity between 100 million and 2500 million years ago, *Phys. Earth Planet. Inter., 2*, 11–18, 1969.

Schwarz, E. J., and D. T. A. Symons, Paleomagnetic field intensity during the cooling of the Sudbury irruptive 1700 million years ago, *J. Geophys. Res., 75*, 6631–6640, 1970.

Selkin, P. A., and L. Tauxe, Long-term variations in paleointensity, *Phil. Trans. Roy. Soc. London, A358*, 1065–1088, 2000.

Selkin, P. A., J. S. Gee, L. Tauxe, W. P. Meurer, and A. J. Newell, The effect of remanence anisotropy on paleointensity estimates: A case study from the Archean Stillwater Complex, *Earth Planet. Sci. Lett., 183*, 403–416, 2000.

Selkin, P. A., Archean paleointensity from layered intrusions, Ph.D. thesis, Univ. Calif. San Diego, 306 pp., 2003.

Smirnov, A. V., and J. A. Tarduno, Magnetic hysteresis monitoring of Cretaceous submarine basaltic glass during Thellier paleointensity experiments: evidence for alteration and attendant low field bias, *Earth Planet. Sci. Lett., 206*, 571–585, 2003.

Smirnov, A. V., J. A. Tarduno, and B. N. Pisakin, Paleointensity of the early geodynamo (2.45 Ga) as recorded in Karelia: A single-crystal approach, *Geology, 31*, 415–418, 2003.

Stevenson, D. J., Planetary magnetic fields, *Earth Planet. Sci. Lett., 208*, 1–11, 2003.

Stevenson, D. J., T. Spohn, and G. Schubert, Magnetism and thermal evolution of the terrestrial planets, *Icarus, 54*, 466–489, 1983.

Sumita, I., T. Hatakeyama, A. Yoshihara, and Y. Hamano, Paleomagnetism of late Archean rocks of Hamersley basin, Western Australia and the paleointensity at early Proterozoic, *Phys. Earth Planet. Inter., 128*, 223–241, 2001.

Tarduno, J. A., R. D. Cottrell, and A. V. Smirnov, High geomagnetic intensity during the mid-Cretaceous from Thellier analyses of plagioclase crystals, *Science, 291*, 1779–1783, 2001.

Thellier, E., and O. Thellier, Sur l'intensité du champ magnétique terrestre dans le passé historique et géologique, *Ann. Géophys., 15*, 285–376, 1959.

Valet, J-P., Time variations in geomagnetic intensity, *Rev. Geophys., 41 (1)*, 1004, doi:10.1029/ 2001RG000104, 2003.

Wanless, R. K., R. D. Stevens, G. R. Lachance, and J. Y. H. Rimsaite, Age determinations and geological studies: K/Ar isotope ages, *Pap. Geol. Surv. Can., 65–17*, 101 pp., 1966.

Warnock, A. C., K. P. Kodama, and P. K. Zeitler, Using thermochronometry and low-temperature demagnetization to accurately date Precambrian paleomagnetic poles, *J. Geophys. Res., 105*, 19,435–19,453, 2000.

Weil, A. B., R. Van der Voo, C. MacNiocaill, and J. G. Meert, The Proterozoic supercontinent Rodinia: Paleomagnetically derived reconstructions for 1100 to 800 Ma, *Earth Planet. Sci. Lett., 154*, 13–24, 1998.

Yoshihara, A., and Y. Hamano, Intensity of the Earth's magnetic field in late Archean obtained from diabase dikes of the Slave Province, Canada, *Phys. Earth Planet. Inter., 117*, 295–307, 2000.

Yu, Y., and D. J. Dunlop, Paleointensity determination on the Late Precambrian Tudor Gabbro, Ontario, *J. Geophys. Res., 106*, 26,331–26,343, 2001.

Yu, Y., and D. J. Dunlop, Multivectorial paleointensity determination from the Cordova Gabbro, southern Ontario, *Earth Planet. Sci. Lett., 203*, 983–998, 2002.

Zindler, A., C. Brooks, N. T. Arndt, and S. R. Hart, Nd and Sr isotope data from komatiitic and tholeiitic rocks of Munro Township, Ontario, *U. S. Geol. Surv. Open File Rep. 78–701*, 469–471, 1978.

David J. Dunlop and Yongjae Yu, Geophysics, Department of Physics, University of Toronto, Toronto, Ontario, M5S 1A7, Canada. (dunlop@physics.utoronto.ca; yjyu@ucsd.edu)

A Simplified Statistical Model for the Geomagnetic Field and the Detection of Shallow Bias in Paleomagnetic Inclinations: Was the Ancient Magnetic Field Dipolar?

Lisa Tauxe

Scripps Institution of Oceanography, La Jolla, California

Dennis V. Kent

Department of Geological Sciences, Rutgers University, Piscataway, New Jersey, and
Lamont-Doherty Earth Observatory, Palisades, New York

The assumption that the time-averaged geomagnetic field closely approximates that of a geocentric axial dipole (GAD) is valid for at least the last 5 million years and most paleomagnetic studies make this implicit assumption. Inclination anomalies observed in several recent studies have called the essential GAD nature of the ancient geomagnetic field into question, calling on large (up to 20%) contributions of the axial octupolar term to the geocentric axial dipole in the spherical harmonic expansion to explain shallow inclinations for even the Miocene. In this paper, we develop a simplified statistical model for paleosecular variation (PSV) of the geomagnetic field that can be used to predict paleomagnetic observables. The model predicts that virtual geomagnetic pole (VGP) distributions are circularly symmetric, implying that the associated directions are not, particularly at lower latitudes. Elongation of directions is North-South and varies smoothly as a function of latitude (and inclination). We use the model to characterize distributions expected from PSV to distinguish between directional anomalies resulting from sedimentary inclination error and from non-zero non-dipole terms, in particular a persistent axial octupole term. We develop methodologies to correct the shallow bias resulting from sedimentary inclination error. Application to a study of Oligo-Miocene redbeds in central Asia confirms that the reported discrepancies from a GAD field in this region are most probably due to sedimentary inclination error rather than a non-GAD field geometry or undetected crustal shortening. Although non-GAD fields can be imagined that explain the data equally well, the principle of least astonishment requires us to consider plausible mechanisms such as sedimentary inclination error as the cause of persistent shallow bias before resorting to the very "expensive" option of throwing out the GAD hypothesis.

Timescales of the Paleomagnetic Field
Geophysical Monograph Series 145
Copyright 2004 by the American Geophysical Union
10.1029/145GM08

1. INTRODUCTION

One of the most useful assumptions in paleomagnetism is that the geomagnetic field is on average close to that of a geocentric axial dipole (GAD). The GAD model is a specific, testable hypothesis, which for the past few million years provides an excellent fit to global data such that the largest non-GAD contribution is generally found to be no more than about 5% of GAD (e.g. *Opdyke and Henry* [1969]; *Merrill and McElhinny* [1977]; *Schneider and Kent* [1988]; *Gubbins and Kelly* [1993]; *Kelly and Gubbins* [1997]; *Quidelleur et al.* [1994]; *Johnson and Constable* [1995]; *Johnson and Constable* [1998]; *Carlut et al.* [2000]).

It is difficult to test the GAD hypothesis in ancient times owing to plate movements, rock deformation and remagnetization. Nevertheless, the inescapable conclusion from a variety of paleomagnetic data and analysis techniques is that there is often a strong bias toward shallow inclinations that appears inconsistent with a GAD model for the ancient time-averaged geomagnetic field (e.g., *Kent and Smethurst* [1998], *Westphal* [1993], *Si and Van der Voo* [2001]). There are many potential causes for mean directions and distributions that are biased shallow. Given the fundamental utility of the GAD assumption in paleomagnetism, alternative mechanisms deserve a closer look. In this paper we will consider depositional inclination error, especially in detrital hematite, as a better explanation for many of the discrepant observations.

In this paper, we will examine the evidence for inclination anomalies in the Central Asian redbed sediments and explore the possible explanations. Then we will develop a simple statistical model for paleosecular variation which predicts distributions of geomagnetic vectors in agreement with the currently available data sets. The model can be modified to include arbitrary non-zero gauss terms and we investigate the effect of adding an arbitrary amount of non-zero axial octupole on the predicted distributions. We then consider the consequences of sedimentary inclination error on distributions of directions and propose two methods for detecting and correcting for the resulting shallow bias.

2. INCLINATION ANOMALIES IN CENTRAL ASIAN RED BEDS

Paleomagnetic poles obtained from globally distributed locations will tend to average out non-dipole field contributions; hence globally averaged paleopoles should reflect mainly the GAD field. These poles are often used to predict directions at specific locations. The difference between the predicted and observed directions could be caused by local non-dipole field effects, local artifacts caused by rock deformation, or by magnetic recording biases. Paleomagnetists typically assume that the geomagnetic field has been essentially GAD in order to estimate a reference pole for a given plate at the desired time (e.g., *Irving* [1964]). If the rotation parameters linking various plates are known, then these reference poles can be used to predict directions expected from a GAD field at any place on any linked plate. An updated compilation of paleopoles for the Atlantic-bordering continents for 0-200 Ma shows very good agreement with the GAD model, with only a small (~3%) axial quadrupolar contribution [*Besse and Courtillot*, 2002]. Despite general agreement, comparison of predicted directions with those observed reveals a persistent shallow bias in Cenozoic paleomagnetic directions from sediments of Central Asia (see, e.g., *Thomas et al.* [1993]; *Chauvin et al.* [1996]; *Cogné et al.* [1999]; *Si and Van der Voo* [2001]; *Dupont-Nivet et al.* [2002]; *Gilder et al.* [2003]).

Non-GAD geomagnetic fields, in particular axial octupolar fields ($\overline{g}_3^0$), have been called on to explain the Central Asian inclination anomalies (e.g., *Thomas et al.* [1993]; *Chauvin et al.* [1996]; *Van der Voo and Torsvik* [2001]; *Si and Van der Voo* [2001]; *Dupont-Nivet et al.* [2002]). The logic according to, for example, Si and Van der Voo [2001], is that the reference poles are largely based on results from the UK and North America. A non-zero axial octupolar contribution with the same sign as the dipole makes directions in mid-northern latitudes shallower than expected from a GAD field. These directions, when converted to paleomagnetic poles will be "far-sided." If this reference pole is then used to predict directions in Asia, the predicted directions will be too steep. The effect is amplified by the fact that the actual directions observed in Asia in the same octupolar field will be shallower than expected from GAD. Typical contributions of $\overline{g}_3^0$ called for are between 10 and 20% of the average axial dipole.

While most studies attribute the observed inclination shallowing to non-GAD geomagnetic fields, there are alternative interpretations. Cogné et al. [1999] attributed the effect to a large degree of tectonic shortening. Recently, sedimentary inclination error (either by compaction or by initial depositional processes) has gained favor as a possible explanation for the effect (e.g., *Gilder et al.* [2001]; *Dupont-Nivet et al.* [2002]; *Tan et al.* [2003]; *Gilder et al.* [2003]).

The overwhelming majority of the paleomagnetic data from Central Asia come from red beds whose remanence is attributed to detrital hematite by the authors. Tauxe and Kent [1984] studied natural (modern) and laboratory redeposited sediments with detrital hematite. Results from their redeposition experiments are shown in Figure 1a. These data demonstrate a pronounced bias toward shallow inclinations that follow the "inclination error formula" of King [1955]:

$$\tan(I_o) = f \tan(I_f) \tag{1}$$

where I_o and I_f are the observed and applied field inclinations, respectively, and f is an empirical coefficient (here called the "flattening factor"), estimated to be about 0.55 (dashed line in Figure 1a) in these particular sediments.

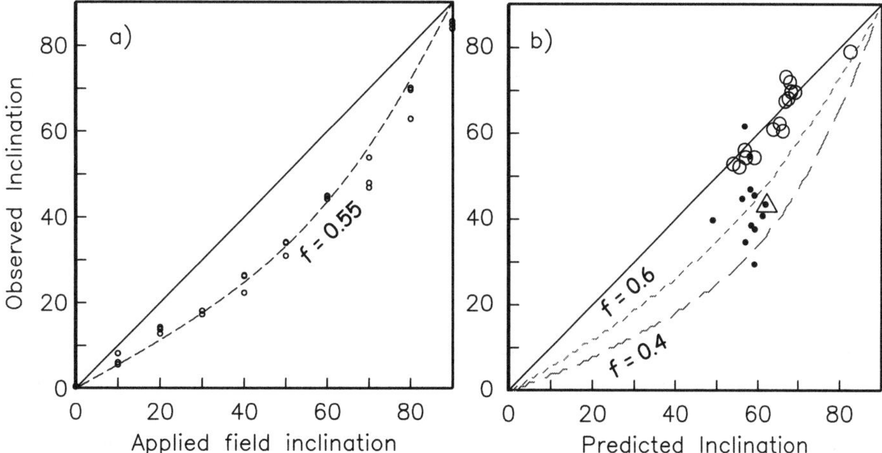

Figure 1. a) Observed inclination versus applied field determined for natural sediments with detrital hematite (data of *Tauxe and Kent* [1984]). Dashed line is for $f = 0.55$. b) Filled (sediments) and open (basalts) circles are observed inclination versus predicted inclination from the APWP for Europe of Besse and Courtillot [2002] for data compiled by Gilder et al. [2003]. Triangle is the magnetostratigraphic study of Gilder et al. [2001] to be discussed later. Dashed lines are the function $\tan(I_o) = f \tan(I_f)$ for $f = 0.4$ and $f = 0.6$ (as labelled).

Gilder et al. [2003] compiled paleomagnetic data for Cenozoic and Mesozoic sediments and basalts from Central Asia north of the Tibetan plateau. We replot their red bed data compilation as observed inclinations versus predicted inclinations from Besse and Courtillot [2002] as solid dots in Figure 1. Data from igneous rocks are shown as open circles. Theoretical curves for inclination error with $f = 0.4$ and 0.6 are shown on Figure 1b as dashed lines. In general, the observed inclinations from the sediments fall well below the expected values. Data from basaltic units in the same compilation do not display a shallow bias as shown also by Bazhenov and Mikolaichuk [2002]. These data show conclusively that the sedimentary units are shallower than the basaltic data; hence Gilder et al. [2003] strongly argue for inclination error as a cause of the inclination anomaly.

Based on the predicted and observed inclinations shown in Figure 1b, it is reasonable to interpret the shallow inclinations from Central Asian sediments as resulting from sedimentary inclination error. However, non-GAD field geometry or crustal motion inconsistent with geological observations have been called upon as plausible explanations for shallow inclinations in many tectonic studies. Some means for discriminating among the various possibilities would be of general use.

We suggest in this paper that when inclination error occurs, it distorts the original distribution of directions in ways that should be distinguishable from the other mechanisms of inclination shallowing if the characterisics of the data set as a whole are considered. Before we begin to explore the effect of inclination error on distributions of directions, we need to understand what distributions we might expect from secular variation of the geomagnetic field itself. We therefore will first consider statistical paleosecular variation models capable of predicting directions as a function of position on the surface of the Earth. We then will investigate the effect of non-zero average octupolar components and finally we will characterize the effect of sedimentary inclination error on directions.

3. PALEOMAGNETIC CONSTRAINTS AND STATISTICAL MODELS OF THE GEOMAGNETIC FIELD

3.1. The Giant Gaussian Process

To predict the distribution of directions produced by paleosecular variation of the geomagnetic field, we require a statistical model to generate plausible sets of geomagnetic field vectors. A good starting point is the model of Constable and Parker [1988] (hereafter CP88). This models the time varying geomagnetic field as a "Giant Gaussian Process" (GGP) whereby the gauss coefficients g_l^m, h_l^m (except for the axial dipolar term, g_1^0 and in some models also the axial quadrupole term g_2^0) have zero mean and standard deviations that are a function of degree l. For $l \geq 2$ these standard deviations are given by

$$\sigma_l^2 = \frac{(c/a)^2 l \alpha^2}{(l+1)(2l+1)} \quad (2)$$

where c/a is the ratio of the core radius to that of Earth and α is a fitted parameter. The parameters used in the model of CP88 are listed in Table 1.

Table 1. Parameters for various PSV models

Parameter	CP88	CJ98	QC96	TK03.GAD
$\bar{g}_1^0$	$-30\mu\mathrm{T}$	$-30\mu\mathrm{T}$	$-30\mu\mathrm{T}$	$-18\mu\mathrm{T}$
$\bar{g}_2^0$	$-1.8\mu\mathrm{T}$	$-1.5\ \mu\mathrm{T}$	$-1.2\mu\mathrm{T}$	0
α	$27.7\ \mu\mathrm{T}$	$15\mu\mathrm{T}$	$27.7\mu\mathrm{T}$	$7.5\mu\mathrm{T}$
β				3.8
σ_1^0	$0.5\sigma_l = 3\mu\mathrm{T}$	$3.5\sigma_l = 11.72\mu\mathrm{T}$	$3\mu\mathrm{T}$	$\beta\sigma_l = 6.4\mu\mathrm{T}$
σ_1^1	$0.5\sigma_l = 3\mu\mathrm{T}$	$0.5\sigma_l = 1.67\mu\mathrm{T}$	$3.0\mu\mathrm{T}$	$\sigma_l = 1.7\mu\mathrm{T}$
σ_2^0, σ_2^2	$\sigma_l = 2.14\mu\mathrm{T}$	$\sigma_l = 1.16\mu\mathrm{T}$	$1.3\mu\mathrm{T}$	$\sigma_l = 0.6\mu\mathrm{T}$
σ_2^1	$\sigma_l = 2.14\mu\mathrm{T}$	$3.5\sigma_l = 4.06\mu\mathrm{T}$	$4.3\mu\mathrm{T}$	$\beta\sigma_l = 2.2\mu\mathrm{T}$
$l - m$ odd	σ_l	σ_l	σ_l	$\beta\sigma_l$
$l - m$ even	σ_l	σ_l	σ_l	σ_l
$c/a{=}0.547$	$(\sigma_l)^2 =$	$(c/a)^{2l}\alpha^2/[(l + 1)(2l + 1)]$		

Many data sets show a persistent offset in equatorial inclinations at least in reverse polarity data sets, consistent with a small non-zero mean axial quadrupolar term ($\bar{g}_2^0$). We are ignoring this effect in the present paper because the bias introduced by $\bar{g}_2^0$ is negligible for the latitude of the Asian studies and has not been considered as a possible explanation for the inclination anomalies observed there. Hence the version of CP88 and other models discussed here are the "GAD" versions in which the axial quadrupolar term has zero mean (e.g., CP88.GAD).

The advantage of using a statistical model like CP88 is that distributions of directions with various non-zero gauss coefficients (such as the axial quadrupole or octupole term) can be generated and compared with the paleomagnetic observations and with other model predictions. One simply draws coefficients for a field model from gaussian distributions with the specified means and standard deviations and calculates the geomagnetic elements at a given position using the usual formulae (see *Constable and Parker* [1988] for details). The main disadvantage of the CP88 model is that it fails to account for the observed variations in dispersion of the virtual geo-

magnetic poles (VGPs) calculated from directions as a function of latitude (see e.g., *McFadden et al.* [1988]; *Kono and Tanaka* [1995]; *Constable and Johnson* [1999]).

3.2. VGP Scatter as a Function of Latitude

McElhinny and McFadden [1997] (hereafter MM97) compiled an updated paleosecular variation of recent lavas (PSVRL) database of directions from lava flows from the last 5 million years that met their minimum acceptance criteria. They also estimated angular standard deviation of the scatter S of the VGPs with latitude. S (e.g., *Cox* [1969]) is defined as

$$S^2 = (N-1)^{-1}\sum_{i=1}^{N}(\Delta_i)^2$$

where N is the number of observations and Δ is the angle between the ith VGP and the spin axis. In Figure 2 we show the variation of VGP scatter as a function of latitude from the compilation of MM97 as dots. One criterion for the PSVRL database is that VGPs are rejected if they are more than $45°$ from the spin axis in order to avoid over-representation of transitional data. The MM97 estimates of S also used a variable VGP colatitude cutoff as suggested by Vandamme [1994], which is an iterative process whereby the cutoff is found by $\theta = 1.85S + 5°$. Cutoffs range from $25°$ at the equator to $\sim 42°$ near the pole. Values of S based on trimmed data sets are here termed S'. The predicted behavior of S' from the CP88.GAD model is shown as a dashed line in Figure 2.

The fact that VGP scatter increases with latitude has been known for decades (e.g., *Cox* [1962]). As pointed out by McFadden et al. [1988] among others, gauss coefficients that are asymmetric about the equator (those with $l - m$ odd) contribute more strongly to the scatter in VGPs at high latitude than those that are symmetric about the equator (those with $l - m$ even). In order to improve the fit of the statistical paleosecular variation model to their compilation of paleomagnetic observations, Quidelleur and Courtillot [1996] proposed

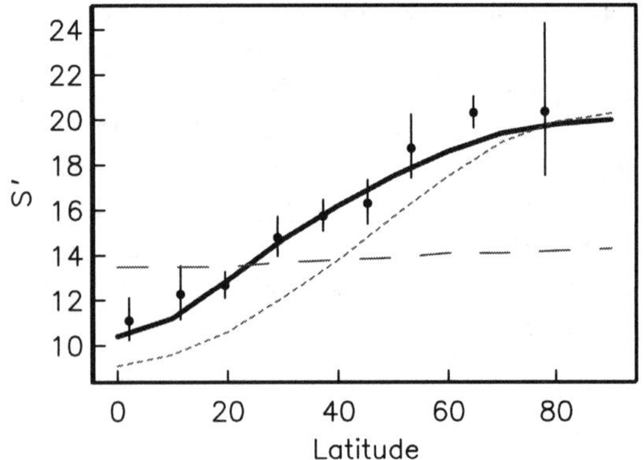

Figure 2. Estimated behavior of S' from the data compilation of MM97 (circles). The dashed line is the predicted behavior from CP88.GAD, the dotted line is from CJ98.GAD and the heavy solid line is from TK03.GAD.

Table 2. Predicted observables from the TK03.GAD field model from 10,000 realizations

λ	$\overline{B}\mu\mathrm{T}$	$\overline{I}$	$<I>$	κ	E	S_l^u	$S_l^{'u}$
0	18.6 ± 7.4	-0.1	0	15.0	2.9	$14.8^{15.3}_{14.4}$	$10.4^{10.6}_{10.3}$
10	19.5 ± 7.7	17.5	19.4	15.1	2.6	$15.6^{15.1}_{16.0}$	$11.2^{11.4}_{11.0}$
20	21.6 ± 8.6	33.3	36.1	15.5	2.1	$16.9^{17.3}_{16.5}$	$12.9^{13.0}_{12.7}$
30	24.5 ± 9.7	46.4	49.1	16.9	1.8	$18.3^{18.7}_{17.9}$	$14.7^{14.8}_{14.5}$
40	27.5 ± 10.7	56.9	59.2	19.4	1.5	$19.6^{19.9}_{19.2}$	$16.2^{16.4}_{16.1}$
50	30.6 ± 11.6	65.4	67.2	22.6	1.3	$20.7^{21.1}_{20.4}$	$17.5^{17.7}_{17.4}$
60	33.2 ± 12.2	72.6	73.9	26.1	1.2	$21.7^{22.1}_{21.3}$	$18.6^{18.8}_{18.4}$
70	35.2 ± 12.7	78.9	79.7	29.1	1.1	$22.4^{22.8}_{22.1}$	$19.4^{19.6}_{19.2}$
80	36.4 ± 13.0	84.6	84.9	31.2	1.0	$22.9^{23.2}_{22.5}$	$19.8^{20.0}_{19.6}$
90	36.8 ± 13.1	89.8	90.0	31.8	1.0	$23.0^{23.4}_{22.7}$	$20.0^{20.2}_{19.8}$

λ: Latitude ($^\circ$), $\overline{B}$: average field intensity, $\overline{I}$: average field inclination, $<I>$ inclination expected from dipole formula; κ: Fisher [1953] precision parameter, E: elongation, S: VGP scatter, untrimmed, S': VGP scatter, trimmed using Vandamme [1994] criterion. A best-fit polynomial to the inclination-elongation data is $E = 2.88 - 0.0087I - .0005I^2$.

a variation on the CP88 model (hereafter QC96; see Table 1). QC96 improves the fit by decreasing the variance in the σ_2^0 (symmetric) term and increasing the variance in the σ_2^1 (asymmetric) terms relative to the CP88 model.

The most recent of the GGP type models is that of Constable and Johnson [1999] (hereafter CJ98). CJ98 attempts to fit a compilation of lava flow data for the last 5 million years [*Johnson and Constable*, 1996]. The variant of their model with zero average for the non-dipole terms is here called CJ98.GAD. Like QC96, CJ98 achieves an increase in high latitude VGP scatter by adding power to the asymmetric terms and decreasing power in the symmetric terms relative to CP88. The prediction of the behavior of S' with latitude of CJ98.GAD (CJ98 as in Table 1 but with $\overline{g}_2^0 = 0$) is shown as a dotted line in Figure 2. It does a good job of predicting the VGP scatter observed at high latitude, but under predicts scatter at lower latitudes. [We note that CJ98 was designed to fit a different data compilation with more stringent VGP co-latitude cut-offs than MM97.]

3.3. A Simplified Giant Gausssian Process Paleosecular Variation Model

As discussed in the previous section and seen in Table 1, both QC96 and CJ98 manipulate the variance for each gauss term separately to achieve a fit to the paleomagnetic observations. We propose here a return to the simplicity of CP88, but modify it to properly account for the observed dependence of S on latitude. The latitudinal dependence of S can be simulated by having larger variance in the asymmetric gauss coefficients than in the symmetric ones. We therefore fit the first order properties of the paleosecular variation data base (average intensity and dispersion of VGPs with lati-

tude) with three parameters: the average axial dipole term $\overline{g}_1^0$, α as in Equation 2 and β defined as the ratio $\frac{\sigma_l^m (l-m:\mathrm{odd})}{\sigma_l^m (l-m:\mathrm{even})}$ for a given degree l.

Because our understanding of the average field intensity has changed recently (e.g., *Selkin and Tauxe* [2000]), none of the statistical models fit the observed average intensity of the magnetic field very well, having an average field approximately equal to the present field. Selkin and Tauxe [2000] arrived at an average of approximately half that value, that is, the virtual axial dipole moment (VADM) of the present field is approximately 80 ZAm^2 [NB: Z = 10^{21}] whereas the average VADM for the last five million years is approximately 46 ZAm. We therefore set the value of the average axial dipole term ($\overline{g}_1^0$) to give the correct average VADM (see Table 1).

Once the value for the axial dipole term has been set, α has a strong effect on the scatter of VGPs observed at equatorial latitudes. Having changed $\overline{g}_1^0$, then, we must also change α if we are to fit the data at least as well as prior models. As noted previously, β has the strongest effect at high latitudes. Therefore we chose α and β to give the best fit to the VGP scatter as estimated by MM97. Our preferred values are listed in Table 1 and the fit of predicted S' from 10,000 realizations to the dataset of MM97 is shown as the heavy solid line in Figure 2. (Both trimmed (S') and untrimmed (S) estimates are listed in Table 2; trimming typically reduces S by about 2°.) As it was designed to do, the TK03.GAD model fits the data of MM97 very well.

A departure from previous models in the TK03.GAD model is that the axial dipole term (g_1^0; asymmetric degree one term) is no longer treated specially apart from its non-zero mean value; its variation is treated identically to other terms. We plot the values for σ_l^m for the two families (symmetric and asymmetric) in Figure 3a.

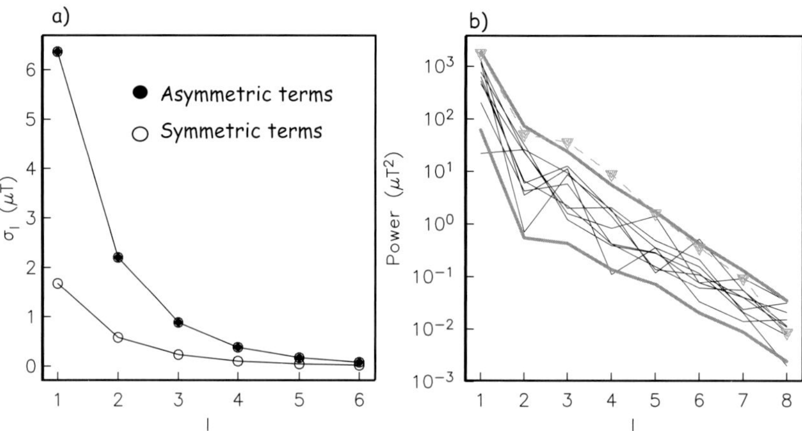

Figure 3. a) Variation of σ_l as a function of degree l for the symmetric ($l-m$ even) and asymmetric ($l-m$ odd) gauss terms for model TK03.GAD (see Table 1). b) Power evaluated for representative realizations of the TK03.GAD (thin lines). 95% confidence bounds derived from 10,000 realizations (heavy lines). Power spectrum of the IGRF for 1995 (dashed line with triangles).

3.4. Lowes Spectrum

Constable and Parker [1988] used the present (1980 International Geomagnetic Reference Field) as a guide for constructing CP88. TK03 is constructed to fit the paleomagnetic data for the last 5 million years instead. To compare the statistical behavior of model TK03.GAD with the present geomagnetic field, we can calculate the power R_i in each degree (see, e.g., CP88) using the formula of Lowes [1974]:

$$R_i = \sum_{m=0}^{l} (l+1)[(g_l^m)^2 + (h_l^m)^2]$$

for realizations of the model. We plot 25 so-called "Lowes-spectra" as thin dashed lines in Figure 3b. Averaging 10,000 such realizations gives 95% confidence bounds on the model which are plotted as heavy lines in Figure 3b. We also plot the Lowes spectrum of the 1995 International Geomagnetic Reference Field as triangles. The fact that the spectrum of the present field lies at the very upper bound of realizations from our statistical field model supports the contention of Hulot and Gallet [1996] that the present field is not a good guide for the time-averaged field. In fact, it appears to be quite an unusual field state.

3.5. Predicted Distributions of Geomagnetic Vectors

Geomagnetic field vectors evaluated at various latitudes (λ) from 1000 realizations of model TK03.GAD are shown in Plate 1a-d. We have plotted each realization as a vector end-point in three dimensions where the RGB color value reflects the contributions of North (red), East (green) and Down (blue). The same realizations are plotted along the principal directions of each cloud of points in Plate 1e–h. This projection is in many ways similar to an equal area projection of unit vectors centered on the principal direction, but we have included the intensity information as well.

The fact that we can simulate the full geomagnetic field vector allows us to predict the behavior of various parameters in frequent use in paleomagnetic studies, for example VGPs and virtual dipole moments (VDMs). We convert intensity values from 1000 realizations of TK03.GAD to Virtual Axial Dipole Moments (VADM) and plot them against the VGP latitude of the associated direction (Figure 4). This plot exhibits the well known pattern from the geomagnetic field (e.g., *Tanaka et al.* [1995]) of low VADMs associated with low VGP latitudes. We note that while low paleointensities fre-

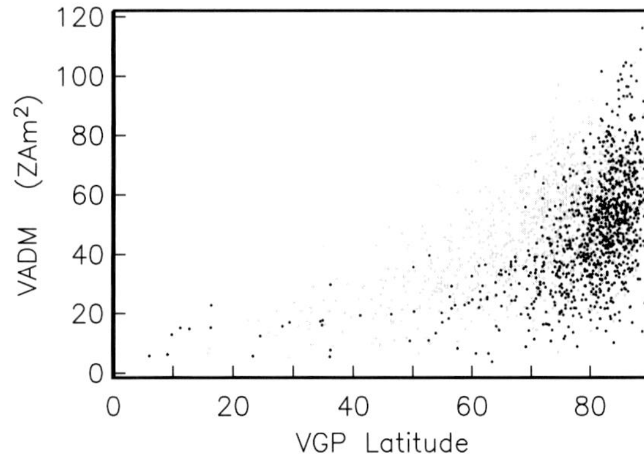

Figure 4. Vectors from realizations of TK03.GAD converted to VGP latitude and VADM for the equator (black) and the polar (lighter) observation sites.

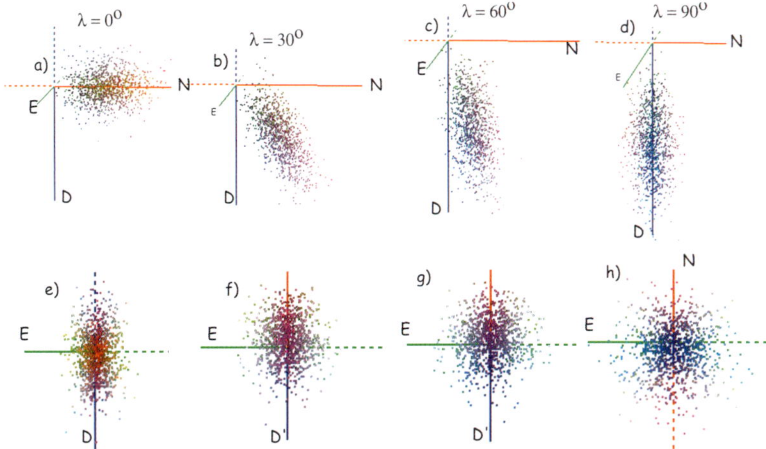

Plate 1. 1000 realizations of TK03.GAD projected as North (red), East (green) and Down (blue) components. Each dot is assigned the RGB color corresponding to the contributions from each component. a–d) All North axes are 40 μT long. (South, East and Up axis are the dashed lines. a) Equator, b) 30°N, c) 60°N, d) 90°N. e–h) Same data as a–d) but projected along the principal axis for each data cloud. All East axes are 20 μT. Axes labelled D' are projections in the N–S plane looking along the expected direction at that latitude. e) Equator, f) 30°N, g) 60°N, h) 90°N.

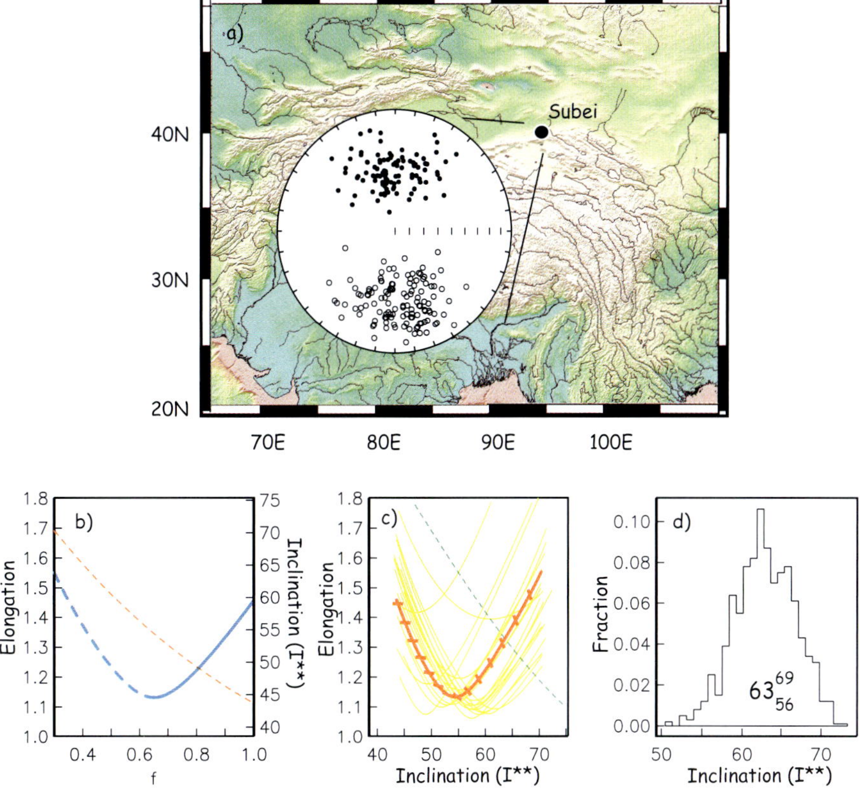

Plate 2. a) Paleomagnetic directions of Oligo-Miocene redbeds from Asia [*Gilder et al.* 2001] in equal area projection (stratigraphic coordinates). b) Plot of elongation (heavy solid and dashed line) and inclination (dashed) as a function of unflattening by the parameter f in Equation 2. Elongation is E–W (N–S) when heavy line is solid (dashed) c) Plot of elongation versus inclination for the data in b) (solid) and for the TK03.GAD model (dashed). Also shown are results from 20 bootstrapped datasets. The crossing points represents the inclination/elongation pair most consistent with the TK03.GAD model. d) Histogram of crossing points from 1000 bootstrapped datasets. The most frequent inclination (63°) is exactly that predicted from the Besse and Courtillot [2001] European APWP. The 95% confidence bounds on this estimate are 56–69°.

quently occur with no directional deviations, all highly divergent directions are associated with low paleointensity. It is therefore perhaps inadvisable to identify "excursions" on the basis of intensity records alone as excursions are by definition intervals of deviant observation sites directions. The lighter points in Figure 4 are from observations sites at the pole, while the darker (black) points are evaluated at the equator. There are many more divergent VGP latitudes in the polar simulations than at the equator from the same field models. This model would predict therefore that excursions would only rarely be observed globally, as deviant directions (defined as > 45° from the pole) are much more prevalent

at high latitude observation sites than at low latitude observation sites in the model. Furthermore, our model suggests that the initial selection procedure of MM97 would exclude many observations from high latitude sites while including the comparable observations from the same field state observed at low latitudes.

Because of the comparative dearth of intensity information in routine paleomagnetic data sets, paleomagnetists rarely consider both direction and intensity in a single plot. Plots similar to those shown in Plate 1 cannot be constructed from the current data base with enough data points to fully characterize the vector distribution of the paleomagnetic field as

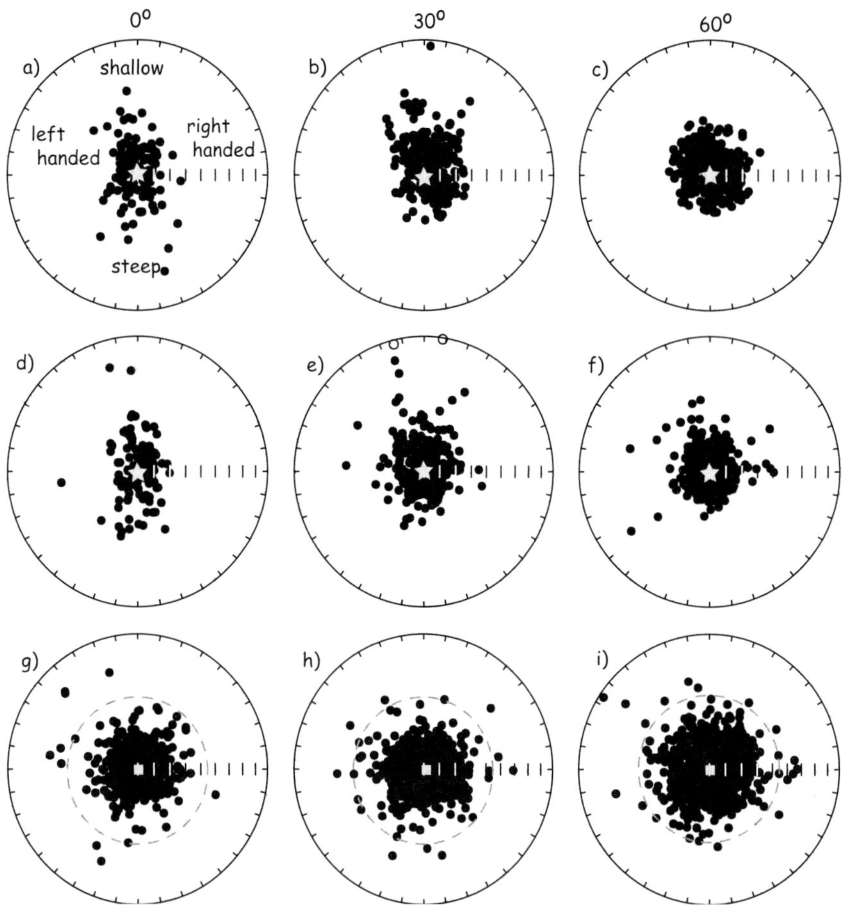

Figure 5. a) Paleomagnetic directions from the PSVRL database (see *McElhinny and McFadden* [1997]) compiled for latitude band 0–5° (N&S). Antipodes of reverse directions are used. The expected direction is at the star at the center of the equal area projection. Directions in the upper (lower) half are shallower (steeper) than expected and those to the right (left) are right-handed (left-handed). b) Same as a) but for 25–35° (N&S) latitude band. c) Same as a) but for 55–65° (N&S) latitude band. d) Same as a) but directions are from realizations of the TK03.GAD model evaluated at 0° latitude. There are the same number of directions as in a). e) Same as b) but for TK03.GAD model at 30° latitude. f) Same as c) but for the TK03.GAD model at 60° latitude. g–i) The associated VGP positions of the model realizations of d–f) plotted in polar projection (squares are the poles). The dashed circle is the 45° cutoff used as an initial cutoff for entry into the PSVRL database. All VGPs outside of this circle would have been eliminated as "transitional" or "excursional". Calculations of S' eliminate additional VGPs based on the variable cutoff criterion (see text).

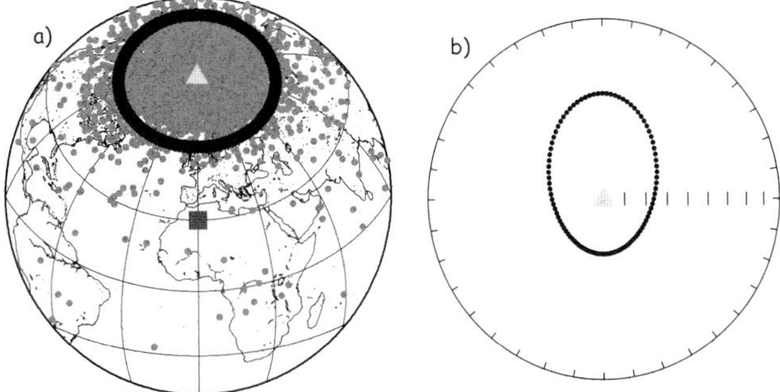

Figure 6. a) VGPs from geomagnetic vectors evaluated at 30°N (site of observation shown as square). The geographic pole is shown as a triangle. A set of VGP positions at 60°N are shown at the site of observation [squares in (a)] converted from the black ring. b) Directions observed at the site of observation square in a) converted from black ring of VGPs in a) which correspond to the VGP positions at 60°N. These directions have been projected along expected direction at site of observation (triangle). Note that a circularly symmetric ring about the geographic pole gives an asymmetric distribution of directions with a shallow bias.

a function of latitude. Instead, paleomagnetists generally plot directional data as unit vectors in equal area projection.

To illustrate how directions behave as a function of latitude, we plot directional data selected from the PSVRL database (downloaded in January 2002 from the NGDC website) for 10° latitude bands in Figures 5a–c. The directions are plotted (taking the antipodes of the reverse directions) projected in equal area projections along the expected direction at each latitude from a GAD field (*Hoffman* [1984]). In addition to the criteria of MM97 for inclusion in the database, we selected data with demagnetization codes of 2 or better from sites with at least 3 specimens and a κ of at least 100. We show realizations of the same number of directions drawn from TK03.GAD (Figures 5d–f) and the associated VGPs (Figures 5g–i). Note that no VGPs generated from TK03.GAD were trimmed in these plots.

One observation from model TK03.GAD is that the simulated distributions of VGPs are circularly symmetric at all latitudes. [NB: The VGP distributions are not Fisherian *sensu strictu* as the distribution of latitudes is not exponential, having a low latitude tail.] Circular symmetry of VGPs implies that the corresponding distributions of directions cannot be symmetric everywhere. In fact there is no essentially dipolar field structure that can give rise to Fisher distributed directions everywhere, so it is generally true that data sets of geomagnetic field directions would not be expected to be Fisher distributed. Although this has long been suspected (e.g., *Creer* [1959]), it has been largely ignored in routine paleomagnetic studies (but see important exceptions by *Baag and Helsley* [1974], *Kono* [1997], *Beck* [1999], and *Tanaka* [1999]).

An immediate consequence of circularly symmetric VGPs is illustrated in Figure 6a in which we plot as small dots the VGP positions from field vectors drawn from TK03.GAD evaluated at the sampling site (30°N; square). The geographic pole is indicated by the triangle. We also show a black ring of VGP positions at 60°N. This ring is converted to the expected direction at the sampling site in Figure 6b with the expected direction at the center of the diagram. The ring of VGPs maps into an ellipse that is asymmetrical with a significant shallow bias. Because the most shallow directions are associated with the low intensities (see e.g., Plate 1b and Figure 4), they do not bias the vector mean significantly. They do, however, bias the average inclination derived from unit vectors (see Table 2). This inclination anomaly varies from zero at the equator to a maximum of about 3° at mid-latitudes and was predicted by Creer [1983] from a secular variation model based on migrating radial dipoles. The essential feature of Creer's model was that VGP distributions are circularly symmetric which is also a key feature of the types of models considered here. Also noted by Creer [1983], the inclination anomaly has the same form as a non-zero contribution of the g_3^0 term. The magnitude of the effect is not large enough, however, to explain inclination anomalies of ~20° under consideration here.

To characterize the elongation of the distribution of directions derived from Fisher distributed VGPs as a function of latitude, Tanaka [1999] used the ratio of the 95% confidence radii α_{31}/α_{32} from Bingham statistics [*Bingham*, 1964; *Bingham*, 1974]. The radii of the Bingham ellipses are ultimately based on the eigenvalues of the "orientation matrix" **T** [*Scheidegger*, 1965] which is defined as:

$$\mathbf{T} = \begin{pmatrix} \sum x_{1j}x_{1i} & \sum x_{1j}x_{2i} & \sum x_{1j}x_{3j} \\ \sum x_{1j}x_{2j} & \sum x_{2j}x_{2j} & \sum x_{2j}x_{3j} \\ \sum x_{1j}x_{3j} & \sum x_{2j}x_{3j} & \sum x_{3j}x_{3j} \end{pmatrix}$$

where x_{ij} are the ith component of the jth unit vector. The eigenvalues τ_i and eigenvectors $\mathbf{V}_i$ reflect the shape and orientation of the distribution of directions, respectively. For Fisher distributions, the eigenvector $\mathbf{V}_1$ associated with the maximum eigenvalue τ_1 is coincident with the Fisher mean direction. $\mathbf{V}_2$ and $\mathbf{V}_3$ are in the directions of the major and minor axes of the Bingham confidence ellipse whose radii are related through a non-linear transformation to the eigenvalues. Here we use the eigenvalues themselves and follow Tauxe [1998] who defined an elongation parameter E as the ratio τ_2/τ_3 to quantify the asymmetry in the distributions of directions seen in both the PSVRL dataset and the TK03.GAD model (Figure 5a–f). (Note that this is different from the elongation defined later by Beck [1999]). The elongation direction is the declination of $\mathbf{V}_2$ or $D_{\mathbf{V}_2}$.

We plot the behavior of elongation for the PSVRL data compilation in approximately 20° latitude bands in Figure 7 as solid dots. The dots are placed at the average latitude of the data set and the horizontal bars indicate the latitude window from which the data were drawn. Also shown is the variation of E predicted from TK03.GAD (triangles in Figure 7). E varies in the TK03.GAD model from ~3 (rather elongate) at the equator to unity (approximately symmetric) near the poles (see also Table 2). D_{V_2} remains essentially zero for all distributions that have significant elongation. In other words, the distributions of field directions tend to be elongated in the North-South direction. The distributions of VGPs, however, remain highly symmetric (see Figure 5g–i). We also show the variation of elongation with latitude from the CJ98.GAD model (circles) for comparison. Even with quite different statistical behavior of the field, the variation of E with latitude is rather similar. We also plot the inclination variation with latitude (λ) predicted from the dipole formula $\tan I = 2 \tan \lambda$ as squares in Figure 7.

As an aside, given the expectation for elliptical distributions of directions derived from inherently GAD fields, it is likely to be inappropriate to use Fisher statistics on directional data sets. Love and Constable [2003] offer a means for incorporating intensity information into the averaging process, but as yet have only dealt with the isotropic case. A glance at Plate 1 suggests that distributions of paleomagnetic vectors are unlikely to be isotropic (which would have data clouds that are "round" as opposed to the triaxial distributions observed here) and there is a need for anisotropic statistical methods for dealing with geomagnetic vector data. Until the theory is more developed, a non-parametric bootstrap (see *Tauxe* [1998]) is probably the least biased way to get confidence intervals for distributions of directions or their components.

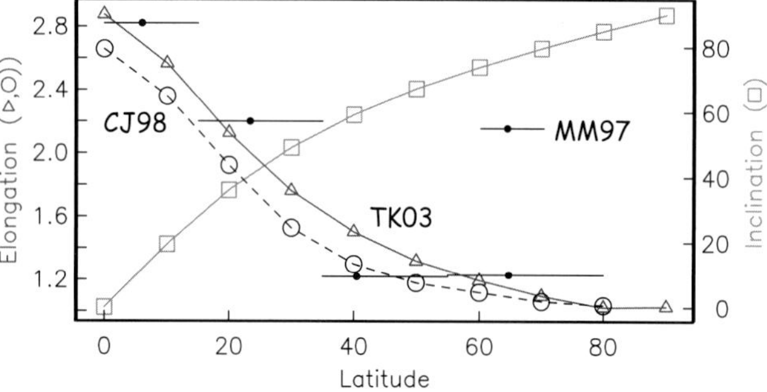

Figure 7. Variation of elongation (triangles) and average inclination (squares) versus latitude for the TK03.GAD model. Also shown is elongation from CJ98.GAD (circles) and the selected directions from the PSVRL database (see text). By about 60°N latitude, distributions of directions are virtually circularly symmetric.

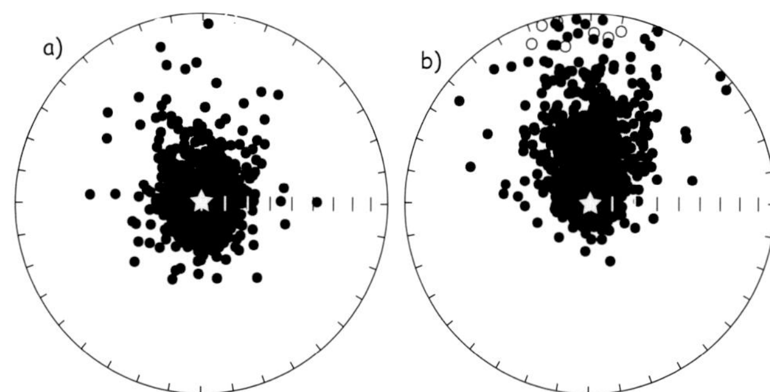

Figure 8. Equal area projections as in Figure 5. Sets of directions evaluated for 30° latitude, projected along direction expected from a GAD field. a) Directions drawn from TK03.GAD. b) Directions drawn from TK03.g30 ($\overline{g}_3^0$ set to 20% $\overline{g}_1^0$).

3.6. Contribution of Non-Zero Mean Octupolar Term

We are interested in this paper in the difference between directional dispersion that results from non-GAD contributions (in particular the octupole) and dispersion that comes from sedimentary processes. Therefore, it is worth considering what effect the axial octupolar contribution ($\overline{g}_3^0$, frequently called upon to explain the inclination anomalies in the ancient field) would have on directions observed in the paleomagnetic field. In Figure 8 we illustrate the effect of non-zero octupolar components on the distribution of directions observed at 30° latitude. Figure 8a shows the distribution of directions drawn from TK03.GAD as viewed down the expected direction from a GAD field. Figure 8b shows TK03, but with the $\overline{g}_3^0$ term set to 20% of $\overline{g}_1^0$ (TK03.g30). The average inclination of this set of directions is 30.4°, compared to 49° expected from the dipole formula (see Table 3). In general, the addition of a non-zero axial octupolar component of the same sign as $\overline{g}_1^0$ at mid latitude tends to increase the elongation in the N–S direction and decrease the average inclination. As noted earlier, this has an identical form to the bias that results from neglecting the intensity information. However, the inclination anomaly of Central Asian red beds is ~ 20° at 40°N, far larger than can be achieved by ignoring intensity; one requires a non-zero mean contribution of 10–20% for the g_3^0 term to explain the observation.

Table 3. Predicted observables from the TK03.g30 field model with non-zero g_3^0 mean evaluated at 30° N. See caption for Table 2.

% g_3^0	$\bar{I}$	E
5	42.6	2.0
10	38.4	2.3
15	34.1	2.6
20	29.6	3.0
30	20.3	3.9

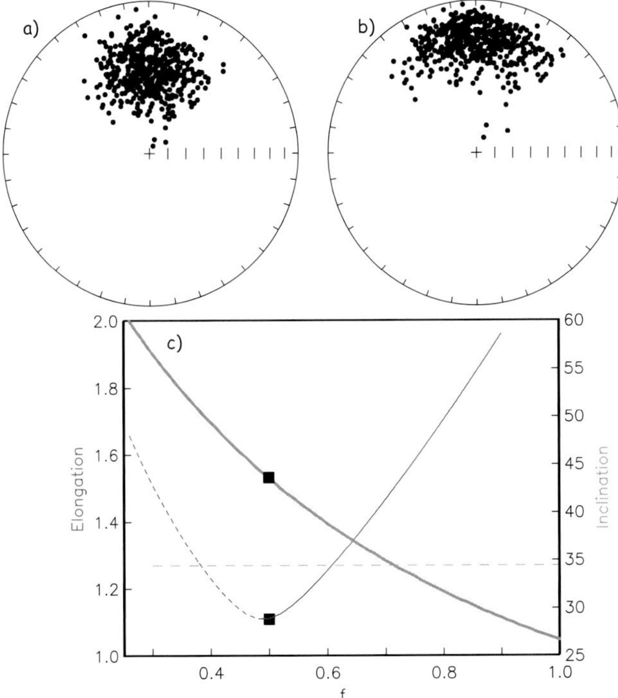

Figure 9. a) Equal area projections of 500 directions drawn from a Fisher distribution with $\kappa = 20, \bar{D} = 0°, \bar{I} = 45°$. Center of diagram is the vertical. b) Directions from a) with inclination distorted by function $\tan(I_o) = f \tan(I_f)$, setting $f = 0.5$. c) Elongation (solid line with dashed extension) and inclination (heavy solid line) as a function of the transformation to "undo" the inclination error (see text). The elongation changes from East-West (solid portion of elongation curve) to North-South (dashed portion) at about $f = 0.5$. 95% of data sets drawn from Fisher distributions with $N = 500$, $\kappa = 20$ have elongations below the horizontal dashed line ($E = 1.27$). The original elongation (inclination) values, 1.11 (43°), are plotted as black squares.

3.7. Sedimentary Inclination Error

We are now in a position to examine the effect of sedimentary flattening "inclination error" of, e.g., King [1955] on various distributions of directions. To investigate the effect of inclination error on a set of directions, we draw 500 directions from a Fisher distribution [Fisher, 1953] with a precision parameter κ of 20, a true mean declination $\hat{D} = 0°$ and an inclination of $\hat{I} = 45°$. [We use the program **fishrot** in the PMAG1.7 software distribution available at http://sorcerer.ucsd.edu/software/.] The calculated mean direction of the data set is $\bar{D} = 0.1°, \bar{I} = 43.6°, \alpha_{95} = 1.4°$ (see Figure 9a).

We transformed each inclination (I) of this data set to a new inclination (I^*) by the "inclination error" formula (Equation 1) with $f = 0.5$. The transformed directions (D, which remains the same) and I^* are shown in Figure 9b. The new distribution has a flattened mean inclination of $\bar{I}^* = 26.7°$, and

is clearly distorted from a Fisher distribution with a pronounced East-West elongation.

To assess the degree of asymmetry in the directions, we use the eigenanalysis of the orientation tensor as before. In a Fisher distribution, eigenvalues τ_2 and τ_3 are statistically indistinguishable making the distribution of data symmetric about the principal direction (E is close to unity). [Monte Carlo simulation of 1000 Fisher distributions with $N = 500$, $\kappa = 20$ have E < 1.1 95% of the time.] If we suppose that the asymmetry in a given data set was caused by "inclination error" acting on an initially symmetric distribution, we could invert the data by:

$$\tan(I^{**}) = (1/f)\, \tan(I^*). \qquad (3)$$

Calculating the eigenparameters for a variety of values of f would allow us to determine the value of f that brings the data to minimum elongation.

Results of such an inversion on the distorted data of Figure 9b are shown in Figure 9c in which we plot the elongation (dashed and solid line) and mean inclination (heavy solid line) as a function of f. The value of f that achieves minimum elongation is $f = 0.5$. The mean direction of the inverted data set using $f = 0.5$ is of course identical to the original in this example.

4. DETECTION/CORRECTION OF INCLINATION ERROR

4.1. "Correction-by-site" Method

While the distribution of directions derived from the geomagnetic field is unlikely to be Fisher distributed except at high latitude, the individual sample directions from each site are in fact expected to be Fisher distributed. Random perturbations in the recording and orienting processes will predominate over field variations in the short time span represented by the site. Therefore, if one had enough samples per site, one could seek the f that minimizes E at a site level, find D, I^{**} (using Equation 3) and recalculate the site means based on the D, I^{**} sample directions.

We illustrate the so-called "correction-by-site" (CBS) procedure for a hypothetical study in Figure 10. In Figure 10a, we show the set of 100 directions drawn from TK03.GAD evaluated at 30°N (the large dots; drawn from those shown in Figure 5e). The average of these is $\bar{D} = 358.7°, \bar{I} = 46.3°$. For each of these "sites", we draw 20 "sample" directions from a Fisher distribution with $\kappa = 100$, shown as small dots. We transformed each sample direction using the inclination error formula with a flattening factor f of 0.5. The transformed D, I^* are shown as small dots in Figure 10b. The average of the "flattened" site means (shown as large dots) is $\bar{D} = 358.8°, \bar{I}^* = 29.1°$.

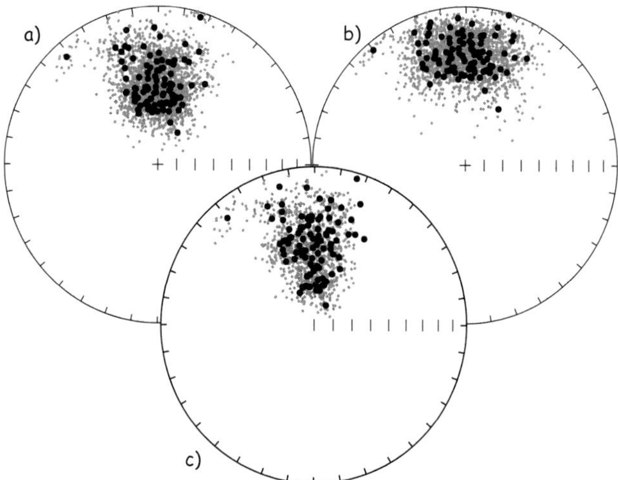

Figure 10. a) Hypothetical Fisher distributed sample directions (small dots) for each site mean (large dots) simulated for 100 hypothetical sites whose directions were drawn from TK03.GAD at 30°N. There are 20 samples at each site. b) Data from a) after transforming to I^{**} using $f = 0.5$. c) Data from b) after seeking the value of f that minimizes E within a site, inverting for I^{**} using that f in Equation 3. Each "site" mean was recalculated with the D, I^{**} for each sample.

The data from each site were treated as in Figure 9c to find the value of f ($1.0 > f > 0.3$) which minimized elongation. After finding the optimum f at each site, we inverted the sample directions using Equation 3. These D, I^{**} are shown as small dots in Figure 10c. New average values for each site are shown as large dots and the mean of these sites is $\bar{D} = 358.7°, \bar{I} = 46.3°$, virtually identical to the original value.

Our CBS method relies on a few essential assumptions. First, we assume that sample directions at a site level are Fisher distributed and that sufficient samples were obtained to adequately represent that distribution. We assume that every sample at a given site was affected by the identical flattening factor. We do not, however, need to assume any *a priori* distribution of the original geomagnetic field directions. Success of the method will depend on taking enough samples at the site level and sampling uniform enough lithology that the single value of f assumption is reasonable. Based on Monte Carlo simulations, we estimate that perhaps as many as 20 samples are necessary for a robust estimate of f. Similar arguments at the study level by Tauxe et al. [2003] suggest that at least 100 sites are necessary to represent the distribution of directions drawn from plausible models of the geomagnetic field.

4.2. "Elongation/Inclination Method"

Unfortunately, the generally available databases do not yet retain data at the sample level, nor do most studies have both large numbers of sites (≥100) and large numbers of samples per site (≥20). However, one can seek the value of f at a study level that yields an elongation/inclination pair consistent with some geomagnetic field model. We illustrate this "elongation/inclination" (E/I) method in the following.

The E/I method of inclination correction requires a data set large enough to have sampled secular variation of the geomagnetic field and one in which an average value of f can reasonably be estimated for the entire study. Most studies aimed at producing paleomagnetic poles are too small, typically having a few dozen sites. Fortunately, the magnetostratigraphic data set of Gilder et al. [2001] is unusually large, having 222 sites. [There are only ~2 samples per site, however, so we are unable to test the CBS method with this data set.] Directions from this study are shown in Plate 2a. These have a mean of $\bar{D} = 356.1°, \bar{I} = 43.7°$. The initial distribution is elongated E–W, which immediately suggests that the anomalously shallow mean inclination is unlikely to be due to a geomagnetic field with a significant axial octupolar contribution because that always produces N–S elongation.

Assuming that the location of the study (presently located at 39.5°N, 94.7°E) has been fixed to the European coordinate system and taking the 20 Myr pole for Europe from Besse and Courtillot [2002] (81.4°N, 149.7°E), the inclination is predicted to be 63° (see triangle in Figure 1b). These sediments are typical of Asian sedimentary units in having an inclination relative to the predicted values that is some 20° too shallow.

To find the average value of f appropriate for the study using the elongation/inclination method we apply Equation 3 to the data shown in Plate 2a (taking the antipodes of the reverse directions) for a range of values of f (Plate 2b). In Plate 2c, we plot the elongation versus inclination for each set directions transformed using a given value of f. These are plotted along with the elongation/inclination behavior predicted by TK03.GAD. The orientation of D_{V_2} is shown as hatchures on curve for the data (heavy line) in Plate 2c, with vertical lines being N–S and horizontal lines being E–W. A best-fit polynomial to the model inclination-elongation data in Table 2 is: $E = 2.88 - 0.0087I - 0.0005I^2$ and is plotted as a dashed line in Plate 2c. The model (dashed line) and observed elongation/inclination (heavy hatched) curves cross at an inclination of ~64°.

To obtain confidence bounds on the "corrected" inclination, we perform a bootstrap in which 222 randomly chosen sites from the original data set are analyzed in the same fashion. Results from twenty such bootstrapped data sets are shown as thin lines in Plate 2c. A histogram of 1000 crossing points of bootstrap curves with the model elongation-inclination line are plotted in Plate 2d. The mode of the bootstrapped crossings is at an inclination of 63° with 95% of the crossings falling between 56° and 69°. Other paleosecular variation models (e.g., CJ98.GAD) will give different results in detail. However, the estimates are all within a few degrees of each other because the largest differences among models occur in the low inclination regions and all are unity at the pole. The region most sensitive to inclination error is at inclinations of near 45° where the various models are relatively consistent.

The results of the elongation-inclination method virtually rule out a significant role for axial octupolar fields as the cause for the inclination bias observed in the Asian sedimentary rocks and strongly support the sedimentary flattening hypothesis of Gilder et al. [2001], Tan et al. [2003] and Gilder et al. [2003].

5. CONCLUSIONS

We have created a simple statistical field model based on the Giant Gaussian Process approach pioneered by Constable and Parker [1988]. The model was designed to fit currently available estimates for average field intensity and VGP scatter as functions of latitude while retaining the elegant simplicity of the Constable-Parker model. Our model fits the average field intensity found by Selkin and Tauxe [2000] and the VGP scatter as a function of latitude of McElhinny and McFadden [1997]. Realizations of the TK03.GAD model lead to the following observations:

1. Our model fits the paleomagnetic data quite well; it suggests however, that the Lowes spectrum of the present field is at the upper bounds of behavior for the geomagnetic field.

2. In general, directions representing paleosecular variation of the geomagnetic field are not expected to be Fisher distributed, while VGP distributions resulting from those directional data sets are likely to be at least circularly symmetric (although not, in fact, Fisher distributed). The direction of elongation in GAD fields is North-South with the maximum elongation at the equator. Statistical treatment of directional data sets should use a bootstrap technique that assumes no *a priori* distribution. Furthermore, mapping of circularly symmetric VGP distributions results in elliptical directional distributions with a shallow bias in the mean inclination with respect to the expected direction at mid- latitude sites.

3. Recent PSV models are based on data sets that have attempted to eliminate transitional directions from the analysis of distributions of directions and VGPs by using various VGP colatitude cutoff angles. Our statistical field model has no reversals built into it (in fact the g_1^0 term changed sign only 26 times in 10,000 simulations), yet has many VGPs that exceed these arbitrary cutoffs, particularly from high latitude sites of observation. The resulting statistical parameters (e.g., VGP scatter) will therefore underestimate the true variability of the non-transitional geomagnetic field.

4. While large deviations from the geocentric axial dipole axis are always associated with low intensities, low intensities are not always associated with deviant field directions, especially for low latitude sites of observation. Hence "excursions", which by definition are large deviations in direction, cannot be reliably identified by low paleointensity values alone and will only rarely be observed globally.

The principal advantage of using a statistical paleosecular variation model is that we can evaluate various processes that have been called upon to explain anomalous inclinations observed in several data sets of late. In particular, we have varied the contribution of the axial octupolar gauss coefficient and evaluated its effect on the distribution of directions generated from that particular field model. We compared realizations of the octupolar field model with the distribution of directions derived from our TK03.GAD model after "flattening" using the well known inclination error formula of King [1955] [$\tan(I_o) = f \tan (I_f)$] where f is the "flattening factor". Our analyses suggest the following:

1. The contribution of non-zero non-GAD terms to the geomagnetic field changes the distribution of directions. The contribution of a non-zero average axial octupole of the same sign as the axial dipole enhances N–S elongation of the observed directions as well as creating a shallow bias. The predicted distributions are distinctly different from those expected from sedimentary inclination error, which are elongate East-West.

2. We develop two procedures for "correcting" inclinations that have suffered from sedimentary flattening. The first is the correction-by-site (CBS) method. The CBS method requires no *a priori* assumption about the distribution of paleofield directions. It relies instead on the assumption that at a site level, variations in direction are largely due to random errors during sampling and measurement; these are routinely expected to result in Fisher distributed data. If a sufficient number of samples (~20) are available for each site, the value of f can be found that minimizes elongation, returning the data to their original (by assumption) circularly symmetric state. Site means from these adjusted directions can then be used to calculate the mean direction of the entire study. We stress that the sampling strategy must be designed to sample an instant in time and not average out secular variation. Furthermore, each site must sample a homogeneous lithology to ensure a uniform value of f for all samples from the same site.

3. A second method of inclination error correction relies on the behavior of the elongation versus inclination of the statistical field model TK03.GAD which has the best-fitting polynomial function of $E = 2.88 - 0.0087I - 0.005I^2$. Directions are inverted with a range of values of the flattening fac-

tor using the equation $\tan(I^{**}) = (1/f)\,\tan(I_o)$ where I_o is the observed inclination, I^{**} is the transformed inclination and f is the assumed flattening factor. Elongation and mean inclination are calculated for each set of transformed data and plotted in an elongation-inclination plot. The inclination at which the transformed data cross the model is the inclination/elongation pair consistent with the field model. 95% confidence bounds can be found using a bootstrap.

4. Performing the elongation/inclination procedure on the large Oligo-Miocene data set of Gilder et al. [2001] results in an estimate of 63_{56}^{69} for the inclination, precisely that predicted from the apparent polar wander path for Eurasia of Besse and Courtillot [2002]. The initial distribution of data is elongate E-W, which precludes an axial octupolar field as the cause of the inclination anomaly. Depositional inclination error is therefore the likely cause for inclination bias in the Asian red beds.

5. We suspect that inclination error is prevalent in ancient redbeds that carry a detrital magnetization. This will contribute to a shallow bias in statistical distribution of inclinations, as has been observed in pre-Cenozoic data (e.g., *Kent and Smethurst* [1998]). The ability to diagnose sedimentary inclination error by the methods described here should be strong motivation for adequate sampling and for reporting results at the sample level. The fact that data from crystalline rocks may also show a shallow bias (e.g., *Kent and Smethurst* [1998]) could mean that these crystalline data may suffer from some other artifacts, such as undetected tilting. In the meantime, paleopoles for tectonic plates based on sedimentary data, particularly with detrital hematite as the carrier of remanence, should be used with caution.

Acknowledgments. We thank Rob Van der Voo, Wout Krijgsman, Yves Gallet, and Vincent Courtillot for stimulating conversations. Ken Kodama and Stuart Gilder provided helpful reviews and Cathy Constable and Catherine Johnson significantly improved the manuscript with thoughtful suggestions. We thank Julie Bowles for careful proof reading. Stuart Gilder kindly sent us his data. We are grateful to Daniel Staudigel who wrote the program "CloudView" which projects paleomagnetic vectors in three dimensional, color coded plots. This work was partially supported by NSF Grant EAR0003395. Lamont-Doherty Earth Observatory contribution #6551.

REFERENCES

Baag, C., and C. Helsley, Shape analysis of paleosecular variation data, *J. Geophys. Res*, 4923–4932, 1974.

Bazhenov, M., and A. Mikolaichuk, Paleomagnetism of Paleogene basalts from the Tien Shan Kyrgyzstan: rigid Eurasia and dipole geomagnetic field, *Earth Planet. Sci. Lett.*, 155–166, 2002.

Beck, M., On the shape of paleomagnetic data sets, *J. Geophys. Res.*, *104*, 25,427–25,441, 1999.

Besse, J., and V. Courtillot, Apparent and true polar wander and the geometry of the geomagnetic field over the last 200 Myr, *J. Geophys. Res*, *107*, doi:10.1029/2000JB000/050, 2002.

Bingham, C., Distributions on the sphere and on the projective plane, Ph.d. thesis, Yale University, 1964.

Bingham, C., An antipodally symmetric distribution on the sphere, *Ann. Statist., 2*, 1201–1225, 1974.

Carlut, J., and V. Courtillot, How complex is the time-averaged geomagnetic field over the past 5 Myr?, *Geophys. J. Int.,134*, 527–544, 1998.

Carlut, J., X. Quidelleur, V. Courtillot, and G. Boudon, Paleomagnetic directions and K/Ar dating of 0 to 1 Ma lava flows from La Guadeloupe Island (French West Indies): Implications for time-averaged field models, *J. Geophys. Res.,105*, 835–849, 2000.

Chauvin, A., H. Perroud, and M. Bazhenov, Anomalous low paleomagnetic inclinations from Oligocene-Lower Miocene red beds of the south-west Tien Shan, Central Asia, *Geophys. J. Int., 126*, 303–313, 1996.

Cogné, J., N. Halim, Y. Chen, and V. Courtillot, Resolving the problem of shallow magnetizations of Tertiary age in Asia: insights from paleomagnetic data from the Qiangtang, Kunlun, and Qaidam blocks (Tibet, China), and a new hypothesis, *J. Geophys. Res*, *104*, 17,715–17,734, 1999.

Constable, C., and C. Johnson, Anisotropic paleosecular variation models: implications for geomagnetic field observables, *Phys. Earth Planet. Int., 104*, 35–51, 1999.

Constable, C., and R. L. Parker, Statistics of the geomagnetic secular variation for the past 5 m.y., *J. Geophys. Res., 115*, 11,569–11,581, 1988.

Cox, A., Analysis of present geomagnetic field for comparison with paleomagnetic results, *J. Geomag. Geoelectr., 13*, 101–112, 1962.

Cox, A., Research note: Confidence limits for the precision parameter, K, *Geophys. J. Roy. Astron. Soc, 17*, 545–549, 1969.

Creer, K., E. Irving, and Nairn, Paleomagnetism of the Great Whin Sill, *Geophys. J. Int., 17,* 306–323, 1959.

Creer, K. M., Computer synthesis of geomagnetic paleosecular variations, *Nature, 2*, 695–699, 1983.

Dupont-Nivet, G., Z. Guo, R. Butler, and C. Jia, Discordant paleomagnetic direction in Miocene rocks from the central Tarim Basin: evidence for local deformation and inclination shallowing, *Earth Planet. Sci. Lett., 199*, 473–482, 2002.

Fisher, R. A., Dispersion on a sphere, *Proc. Roy. Soc. London, Ser. A, 217*, 295–305, 1953.

Gilder, S., Y. Chen, and S. Sen, Oligo-Miocene magnetostratigraphy and rock magnetism of the Xishuigou section, Subei (Gansu Province, western China) and implications for shallow inclinations in central Asia, *J. Geophys. Res, 106*, 30,505–30,521, 2001.

Gilder, S., Y. Chen, J. Cogné, X. Tan, V. Courtillot, D. Sun, and Y. Li, Paleomagnetism of Upper Jurassic to Lower Cretaceous volcanic and sedimentary rocks from the western Tarim Basin and implications for inclination shallowing and absolute dating of the M-O (ISEA?) chron, *Earth Planet. Sci. Lett., 206*, 587–600, 2003.

Gubbins, D., and P. Kelly, Persistent patterns in the geomagnetic field over the past 2.5 Myr, *Nature, 365*, 829–832, 1993.

Hoffman, K., A method for the display and analysis of transitional paleomagnetic data, *J. Geophys. Res., 89,* 6285–6292, 1984.

Irving, E., *Paleomagnetism and Its Application to Geological and Geophysical Problems,* John Wiley and Sons, Inc., 1964.

Johnson, C., and C. Constable, The time averaged geomagneitc field: global and regional biases for 0-5 Ma, *Geophys. J. Int., 131,* 643–666, 1997.

Johnson, C. L., and C. G. Constable, The time-averaged geomagnetic field as recorded by lava flows over the last 5 Myr, *Geophys. J. Int., 122,* 489–519, 1995.

Johnson, C. L., and C. G. Constable, Palaeosecular variation recorded by lava flows over the past five million years, *Phil. Trans. R. Soc. Lond. A., 354,* 89–141, 1996.

Johnson, C. L., and C. G. Constable, Persistently anomalous Pacific geomagnetic fields, *Geophys. Res. Lett., 25,* 1011–1014, 1998.

Kelly, P., and D. Gubbins, The geomagnetic field over the past 5 million years, *Geophys. J. Int., 128,* 315–330, 1997.

Kent, D. V., and M. Smethurst, Shallow bias of paleomagnetic inclinations in the Paleozoic and Precambrian, *Earth Planet. Sci. Lett., 160,* 391–402, 1998.

King, R. F., The remanent magnetism of artificially deposited sediments, *Mon. Nat. Roy. astr. Soc., Geophys. Suppl., 7,* 115–134, 1955.

Kono, M., Distributions of paleomagnetic directions and poles, *Phys. Earth Planet. Int., 103,* 313–327, 1997.

Kono, M., and H. Tanaka, Mapping the Gauss coefficients to the pole and the models of paleosecular variation, *J. Geomag. Geoelectr., 47,* 115–130, 1995.

Love, J., and C. G. Constable, Gaussian statistics for paleomagnetic vectors, *Geophys. J. Int., 152,* 515–565, 2003.

Lowes, F., Spatial power spectum of the main geomagnetic field and extrapolation to the core, *Geophys. J. R. Astron. Soc., 36,* 717–730, 1974.

McElhinny, M. W., and P. L. McFadden, Palaeosecular variation over the past 5 Myr based on a new generalized database, *Geophys. J. Int., 131,* 240–252, 1997.

McElhinny, M. W., P. L. McFadden, and R. T. Merrill, The time-averaged paleomagnetic field 0-5 Ma, *J. Geophys. Res., 101,* 25,007–25,027, 1996.

McFadden, P. L., R. T. Merrill, and M. W. McElhinny, Dipole/Quadrupole family modeling of paleosecular variation, *J. Geophys. Res, 93,* 11,583–11,588, 1988.

Merrill, R. T., and M. W. McElhinny, Anomalies in the time-averaged paleomagnetic field and their implications for the lower mantle, *Rev. Geophys. Space Phys., 15,* 309–322, 1977.

Opdyke, N. D., and K. W. Henry, A test of the Dipole Hypothesis, *Earth Planet. Sci. Lett., 6,* 139–151, 1969.

Quidelleur, X., J. P. Valet, V. Courtillo, and G. Hulot, Long-term geometry of the geomagnetic field for the last five million years: An updated secular variation database, *Geophys. Res. Lett., 21,* 1639–1642, 1994.

Scheidegger, A. E., On the statistics of the orientation of bedding planes, grain axes, and similar sedimentological data, *U.S. Geo. Surv. Prof. Pap., 525-C,* 164–167, 1965.

Schneider, D., and D. V. Kent, The time-averaged paleomagnetic field, *Rev. Geophys., 18,* 71–96, 1990.

Schneider, D. A., and D. V. Kent, The paleomagnetic field from equatorial deep-sea sediments: axial symmetry and polarity asymmetry, *Science, 242,* 252–256, 1988.

Selkin, P., and L. Tauxe, Long-term variations in paleointensity, *Phil. Trans. Roy. Astron. Soc., 358,* 1065–1088, 2000.

Si, J., and R. Van der Voo, Too-low magnetic inclinations in central Asia: an indication of a long-term Tertiary non-dipole field?, *Terra Nova, 13,* 471–478, 2001.

Tan, X., K. Kodama, H. Chen, D. Fang, D. Sun, and Y. Li, Paleomagnetism and magnetic anisotropy of Cretaceous red beds from the Tarim basin northwest China: Evidence for a rock magnetic cause of anomalously shallow paleomagnetic inclinations from central Asia, *J. Geophys. Res, 108,* doi:10.1029/2001JB001608, 2003.

Tanaka, H., Circular asymmetry of the paleomagnetic directions observed at low latitude volcanic sites, *Earth Planets Space, 51,* 1279–1286, 1999.

Tanaka, H., M. Kono, and H. Uchimura, Some global features of paleointensity in geological time, *Geophys. J. Int., 120,* 97–102, 1995.

Tauxe, L., *Paleomagnetic Principles and Practice,* Kluwer Academic Publishers, 1998.

Tauxe, L., and D. V. Kent, Properties of a detrital remanence carried by hematite from study of modern river deposits and laboratory redeposition experiments, *Geophys. Jour. Roy. astr. Soc., 77,* 543–561, 1984.

Tauxe, L., C. Constable, C. Johnson, W. Miller, and H. Staudigel, Paleomagnetism of the Southwestern U.S.A. recorded by 0-5 Ma igneous rocks, *Geochem. Geophys. Geosyst.,* doi:10.1029/2002GC000343, 2003.

Thomas, J.-C., H. Perroud, P. Cobbold, M. Bazhenov, V. Burtman, A. Chauvin, and E. Sdybakasov, A paleomagnetic study of Tertiary formations from the Kyrgys Tien-Shan and its tectonic implications, *J. Geophys. Res, 98,* 9571–9589, 1993.

Vandamme, D., A new method to determine paleosecular variation, *Phys. Earth Planet. Int., 85,* 131–142, 1994.

Van der Voo, R., and T. H. Torsvik, Evidence for late Paleozoic and Mesozoic non-dipole fields provides an explanation for the Pangea reconstruction problems, *Earth Planet Sci. Lett., 187,* 71–81, 2001.

Westphal, M., Did a large departure from the geocentric axial dipole hypothesis occur during the Eocene? Evidence from the magnetic polar wander path of Eurasia, *Earth Planet. Sci. Lett., 117,* 15–28, 1993.

Dennis V. Kent, Lamont Doherty Geological Observatory, Palisades, New York 10964. (dvk@ldeo.columbia.edu)

L. Tauxe, Scripps Institution of Oceanography, La Jolla, California 92093-0220. (ltauxe@ucsd.edu)

Geomagnetic Polarity Timescales and Reversal Frequency Regimes

William Lowrie

Institute of Geophysics, ETH Hönggerberg, Zürich, Switzerland

Dennis V. Kent

*Department of Geological Sciences, Rutgers University, Piscataway, New Jersey
and Lamont-Doherty Earth Observatory, Palisades, New York*

An analysis of geomagnetic reversal history is made for the most reliable polarity timescales covering the last 160 Myr. The timescale of Cande and Kent [1995] (CK95) is the optimum representation of Cenozoic and Late Cretaceous polarity history, and the timescale of Channell et al. [1995] (CENT94) best represents the Early Cretaceous and Late Jurassic. The CK95 timescale can be divided into two nearly linear segments at Chron C12r. The lengths of chrons in the younger segment have no systematic trend, and so this part of the polarity sequence is considered to be stationary for statistical analysis. The mean chron length is 0.248 Myr and the gamma index, k, for the distribution of chron lengths is 1.6 ± 0.4; inserting just 8 additional short subchrons that have been verified from magnetostratigraphic studies as polarity reversals reduces the mean chron length to 0.219 Myr and k to 1.3 ± 0.3. The older segment is stationary if the two long polarity chrons C33n and C33r adjacent to the Cretaceous Normal Polarity Superchron are omitted; in this case the mean chron length is 0.749 Myr and k is 1.2 ± 0.4. The chrons in the CENT94 timescale are stationary with mean length 0.415 Myr and k is 1.3 ± 0.3. The gamma indices of the chron distributions are not significantly different from a Poisson distribution ($k = 1$), which implies that the reversal process is essentially free of long-term memory. However, if the mean chron duration is an indicator of stability of the reversal process, it appears that long lasting episodes of stable behavior may be followed by abrupt change to another stable regime with a markedly different reversal frequency. There is no significant change of the gamma index from one regime to another although the mean polarity chron length changes by more than a factor of three. This would imply that the probability of a reversal may be constant within each regime but varies inversely with mean interval length from regime to regime.

1. GEOMAGNETIC POLARITY TIMESCALES

Timescales of the Paleomagnetic Field
Geophysical Monograph Series 145
Copyright 2004 by the American Geophysical Union
10.1029/145GM09

The history of polarity reversals of Earth's magnetic field is well known for the last 160 Myr from the record of oceanic magnetic anomalies. Three distinct episodes are defined (Figure 1). The youngest corresponds to the magnetic anomalies

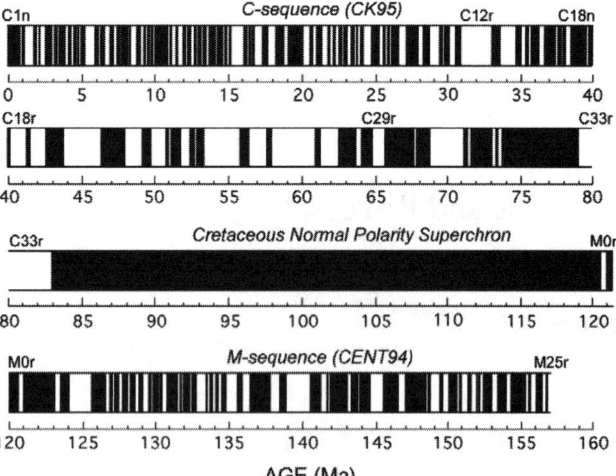

Figure 1. Composite timescale incorporating the current optimum timescales for the C-sequence and M-sequence magnetic anomalies.

formed in the Cenozoic and Late Cretaceous, which are here referred to as the C-sequence. This is separated in the oceanic record from the older M-sequence of anomalies, formed in Late Jurassic and early Cretaceous times, by the Cretaceous Quiet Zone in which correlated magnetic anomalies are absent. The corresponding Cretaceous Normal Polarity Superchron (CNPS) represents a time interval in which Earth's magnetic field evidently did not reverse polarity. Magnetic polarity stratigraphy studies in continental exposures of marine limestones have confirmed the main features of the C- and M-sequences as reflecting reversals of the geomagnetic field and correlated most stage boundaries from the Late Jurassic to the present to the corresponding marine magnetic polarity records [*Alvarez et al.*, 1977; *Channell and Erba*, 1992; *Channell et al.*, 2000; *Channell et al.*, 1979; *Channell and Medizza*, 1981; *Lowrie and Alvarez*, 1981; *Lowrie and Alvarez*, 1984; *Lowrie et al.*, 1982; *Lowrie and Channell*, 1984; *Ogg et al.*, 1984].

Several magnetic polarity timescales have been developed for each of the two reversal sequences. In each case the magnetic anomalies are first interpreted as a block model of oceanic crustal magnetizations of alternating polarity. Multiple profiles are combined to minimize local minor differences in spreading rate on any given profile. The spacings of block boundaries in the resultant composite block model then define the relative lengths of polarity intervals in the timescale. Magnetic stratigraphy has played an important role in dating the polarity sequences by correlating key biostratigraphic stage boundaries or other radiometrically dated datum-levels to the C- and M-sequence polarity record. The rest of the timescale is dated by conversion of polarity boundary locations to numeric ages by interpolation between the tie-levels. In order to avoid

sudden changes in apparent spreading rate at tie-points a best-fit straight line or smooth curve may be fitted to the correlation points. The fundamental assumption here is that sea-floor spreading in selected corridors was constant or at least smoothly varying over long intervals of time, which is justified at least to first order, but is nonetheless a potential source of error in determining block widths and chron durations. The 'absolute ages' of the tie-levels are much less exactly known than the relative lengths of the polarity intervals. In addition to the errors of absolute dating there are stratigraphic errors in relating the radiometrically dated rocks to the biostratigraphically located stage boundary or other datum-levels.

The polarity timescale for each reversal sequence has evolved, accompanying improvements in the resolution of magnetic anomalies, definition of oceanic block models, magnetostratigraphic correlation and the dating of calibration points. The pioneering timescale of Heirtzler et al. [1968], hereafter referred to as HDHPL68, was derived from comparison of marine magnetic profiles in different oceans, from which a block model of crustal magnetization based on a highly levered extrapolation of sea-floor spreading history in the South Atlantic was chosen as best representative for anomalies from the present to the Late Cretaceous (anomaly 32). The Late Cretaceous anomaly sequence was refined and extended to anomaly 33r by analyses of North Pacific and North Atlantic marine profiles [*Cande and Kristoffersen*, 1977]. Magnetostratigraphic correlation of the Cretaceous-Tertiary boundary to anomaly 29r [*Lowrie and Alvarez*, 1977] provided better dating of the older end of the sequence. This resulted in a more accurate timescale [*LaBrecque et al.*, 1977], hereafter referred to as LKC77. These reversal sequences formed the basis of several subsequent timescales [*Mead*, 1996].

A detailed re-evaluation of the C-sequence marine magnetic anomalies and the corresponding block models led to the development of an improved timescale [*Cande and Kent*, 1992a]. Age calibration was achieved by fitting a smooth (cubic spline) curve to nine calibration levels for South Atlantic spreading history. An updated version [*Cande and Kent*, 1995], hereafter referred to as CK95, incorporates improved ages for the calibration levels. It is used as reference timescale for the Late Cretaceous and Cenozoic (C-sequence) in the present paper.

Investigations of magnetic anomalies on the Hawaiian, Japanese and Phoenix lineations in the North Pacific and the Keathley lineations in the North Atlantic were integrated into a polarity sequence and timescale covering marine magnetic anomalies M0 to M25 [*Larson and Hilde*, 1975], which formed the basis for subsequent polarity timescales. Oceanic crust older than the sequence M0–M25 was at first thought to be devoid of lineated magnetic anomalies and was called the Jurassic Quiet Zone by analogy to its Cretaceous counterpart.

Magnetic lineations with low amplitude were described in the youngest part of the Jurassic Quiet Zone and modeled by a block model of polarity characterized by decreasing magnetization with increasing age [*Cande et al.*, 1978]. The M-sequence was extended thereby to M29.

The dating of M-sequence polarity reversals relied initially on estimated bottom ages in DSDP holes drilled on or near the lineations. Subsequent timescales have been formed by attaching better calibration ages to the sequence [*Harland et al.*, 1990; *Kent and Gradstein*, 1985]. These usually consist of a group of ages at each end, with linear interpolation and extrapolation serving to date reversal boundaries between and beyond the tie-points. Channell et al. [1995] compared block models for the anomalies on the Hawaiian, Japanese, Phoenix and Keathley lineations and decided on a new Hawaiian block model with optimum approximation to constant spreading rate, covering magnetic polarity chrons CM0 to CM29. The calibration ages gave favorable comparison with stage boundary ages from magnetostratigraphic correlations. This timescale, hereafter CENT94, is chosen as reference for the M-sequence in this paper. The combined polarity timescales represented by CK95 and CENT94 delineate a Cretaceous Normal Polarity Superchron (CNPS) that began at 120.6 Ma and lasted until 83.0 Ma (Figure 1).

Later airborne and marine deep-tow magnetometer surveys have suggested that the young end of the Jurassic Quiet Zone is characterized by low amplitude, short wavelength anomalies that are lineated [*Handschumacher et al.*, 1988; *Sager et al.*, 1998]. If they are due to short polarity chrons, they extend the polarity history associated with the M-sequence of anomalies from CM29 to CM41, and suggest that the Jurassic Quiet Zone may have formed during an interval of rapidly varying magnetic polarity [*Handschumacher et al.*, 1988; *Sager et al.*, 1998], rather than constant polarity, as inferred for the Cretaceous Quiet Zone. However, the pre-M29 anomalies resemble "tiny wiggles" observed within the C-sequence and many of the features may thus be due to paleointensity fluctuations [*Cande and Kent*, 1992b]. Magnetostratigraphic investigations of Middle to Late Oxfordian limestone sections [*Steiner et al.*, 1985] have described magnetozones with normal and reverse polarity, but these have not yet been correlated satisfactorily to the marine polarity record. In the absence of such confirmation, the possible chrons CM29–CM41 are not discussed further here.

2. REVERSAL FREQUENCY SINCE THE LATE JURASSIC

The variation of reversal behavior in the composite timescale CK95+CENT94 is conveniently displayed by plotting the age of each reversal against the order of its occur-

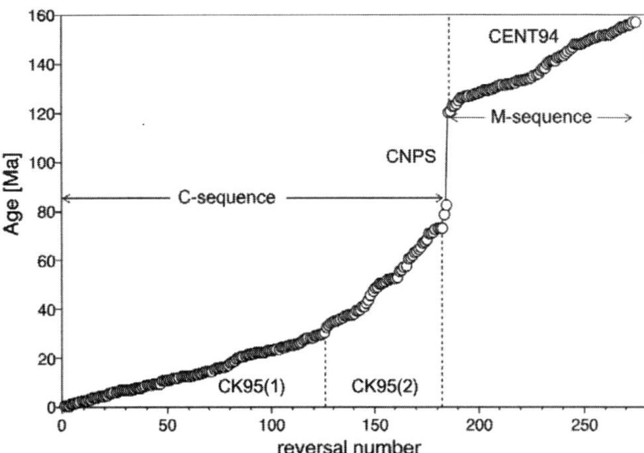

Figure 2. Ages of magnetic reversals in the Cenozoic and Mesozoic oceanic anomaly sequences, plotted against their age order.

rence (Figure 2). The CENT94 data closely define a straight line. The C-sequence can be divided into two nearly linear segments that intersect near Chron C12r at about 31–33 Ma. We subdivide the polarity sequence at Chron C12r because this 2.1 Myr chron lasts about an order of magnitude longer than the average subsequent chron. Two polarity intervals, C33n and C33r, immediately following the CNPS are several times longer than any of the following chrons. They are the second and third longest chrons in the entire 160 Ma sequence and may have closer affinity to the CNPS than to the rest of the polarity sequence. For the purposes of further analysis we define the segment younger than C12r as CK95(1) and the older segment (including C12r but without C33n and C33r) as CK95(2). This choice is somewhat arbitrary but it allows us to subdivide the polarity sequence into the longest possible linear segments without excluding any chrons other than C33n and C33r. Linear segments imply that the reversal process is stationary within the segment and allow calculation of representative statistical parameters. Mean polarity interval lengths are 0.248 Myr for CK95(1), 0.749 Myr for CK95(2), and 0.415 Myr for CENT94. The CNPS, the 37.6 Myr Cretaceous interval of constant normal polarity, is evident as an abrupt discontinuity between the M- and C-sequences and may represent a prolonged disruption of the reversal process.

3. STATISTICAL ANALYSIS OF CHRON LENGTHS

3.1. Statistical Models of Polarity Chron Durations

The observed durations of polarity chrons (10^4–10^7 yr) are typically much longer than the characteristic time constants of magnetohydrodynamic processes in Earth's liquid core (10^2–10^4 yr). To account for the discrepancy, Cox [1968] proposed a stochastic model in which the instantaneous probability of a reversal is constant. For a low probability of occurrence the lengths (x) of polarity chrons conform to an exponential (Poisson) distribution with probability density function $p(x)$ given by

$$p(x) = \frac{1}{\mu}\exp\left(-\frac{x}{\mu}\right)$$

in which μ is the mean chron length.

This model was not, however, obviously supported by the known geomagnetic polarity records, which were notably lacking in short polarity chrons. Instead, the observed distribution of chron durations usually agreed better with a gamma distribution [*Naidu*, 1970], which has the probability density function

$$p(x) = \left(\frac{k}{\mu}\right)^k \frac{x^{k-1}}{\Gamma(k)}\exp\left(-k\frac{x}{\mu}\right)$$

in which k is the gamma index of the distribution and $\Gamma(k)$ is the gamma function of k, defined as

$$\Gamma(k) = \int_0^\infty x^{k-1}e^{-x}dx$$

The exponential distribution corresponds to $k = 1$. A gamma distribution of chrons implies that the reversal process has a memory; the probability of a reversal is not constant with time. Immediately after the occurrence of a reversal the probability of a new one is at first very low and increases with

time. This would suggest that the reversal process has a recovery time during which the probability of a reversal is gradually restored.

Estimation of the gamma index k usually employs the maximum likelihood method, first applied by Naidu [1970] to fixed windows and by Phillips [1977] to a moving window. In this method, the best estimate of k is the solution of the equation

$$\ln k - \Psi(k) = \ln\mu - \frac{1}{N}\sum_N \ln x$$

where $\Psi(k) = d\{\ln\Gamma(k)\}/dk$ is the digamma function. For small numbers of chrons, McFadden [1984] recommends the substitution of $(N-1)$ for N. The variance (σ^2) of k is given by the relationship [*Phillips*, 1977]

$$\sigma^2(k) = \frac{1}{N\left(\Psi'(k) - \dfrac{1}{k}\right)}$$

where $\Psi'(k) = d\Psi(k)/dk$ is the trigamma function. The $\pm 2\sigma$ confidence limits of k may be estimated from this equation.

3.2. Results of Earlier Statistical Analyses

Non-stationarity of the reversal sequences complicates statistical analyses of the chron durations. Naidu [1970] analyzed the statistical properties of the HDHPL68 time-scale

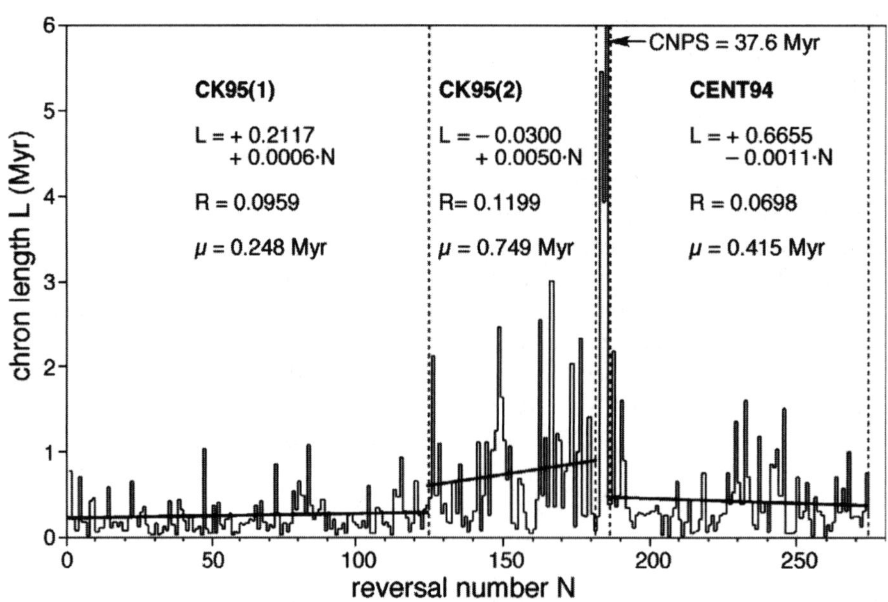

Figure 3. Polarity chron lengths (Myr) in three segments of the CK95 and CENT94 timescales, in which linear fits have non-significant slopes.

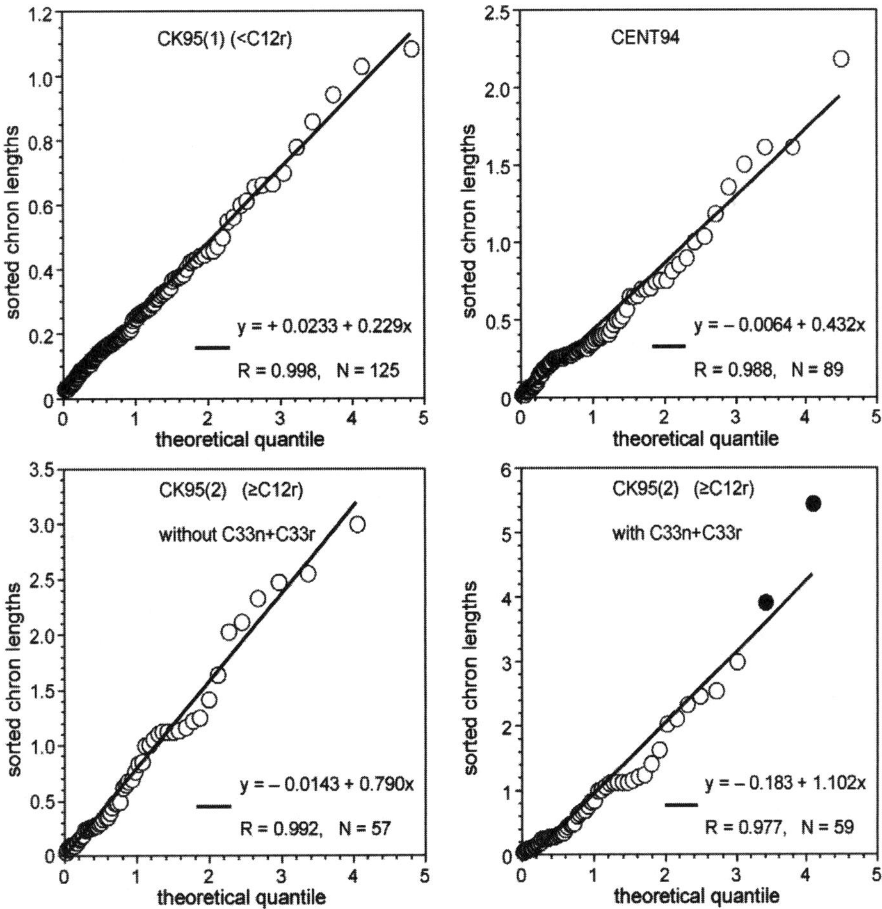

Figure 4. Quantile-quantile plots for the stationary timescale segments CK95(1), CENT94, and CK95(2) show visual agreement of the distributions of chron lengths (Myr) with a Poisson (exponential) distribution; addition of chrons C33n and C33r to CK95(2) gives a poorer fit.

within successive short windows of 8 Myr width. The assumption was made that the sequence was stationary in each short window. The gamma index was estimated to be $k = 2$ for C-sequence polarity chrons younger than 48 Ma and $k = 3.6$ for chrons older than 56 Ma.

A disadvantage of using fixed-width time windows is that the number of chrons in each window is different, which for small numbers might influence interpretation of the statistical properties. To obviate this problem Phillips [1977] analyzed the HDHPL68 timescale with a moving window containing a constant number of polarity intervals. This method has the disadvantage that each window represents a variable length of time at different points in the sequence. Moreover, the window typically includes a rather small number of chrons for robust statistical analysis. For chrons younger than the discontinuity at 45 Ma, the gamma index was $k = 1.55$, but the analysis gave different values of $k = 2.28$ and $k = 1.19$ for the distributions of normal and

reverse chrons, respectively. This was interpreted as indicating that the normal and reverse polarity states of Earth's magnetic field have different stabilities.

Unfortunately, the pioneering HDHPL68 timescale was incomplete and contained some erroneously interpreted polarity intervals, which gave rise to an artificial discontinuity in reversal rate at about 45 Ma. Thus the statistical analyses of this timescale have only historical significance.

Lowrie and Kent [1983] analyzed the improved LKC77 timescale using a moving window of constant 8 Myr width. They found values of $k = 1.52$ for the stationary part of LKC77 younger than 40 Ma; values of $k = 1.90$ and $k = 1.28$ were found for the normal and reverse chrons, respectively, in this interval. McFadden and Merrill [1984] analyzed the LKC77 timescale using 25-interval wide moving windows and obtained an optimum value of $k = 1.25$, with a 95% confidence range of 1.02 to 1.55. The difference in k between the normal and reverse polarity states was attributed to over-sen-

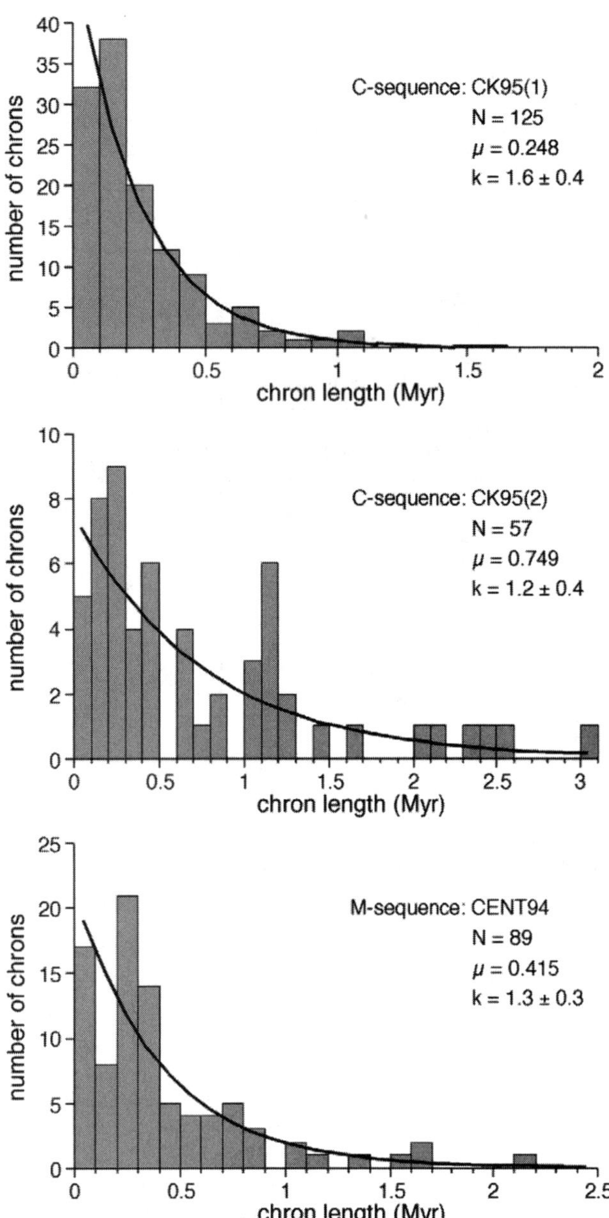

Figure 5. Histograms of chron lengths (Myr) in stationary timescale segments CK95(1), CK95(2), and CENT94 are uneven but show a lack of short chrons. The solid lines are the best fit exponential curves.

sitivity of the analytical method rather than being a real feature of the geomagnetic field.

A moving window analysis creates a false visual impression that the data set is larger than it really is; e.g., using a 25-point moving window, only every 25th analysis is independent. Ideally the entire timescale can be detrended by fitting a suitable time-varying function for the mean value. Lutz and Watson [1988] and Gaffin [1989] modeled reversal rates

with aperiodic functions; Lowrie [1997] proposed using low-order polynomials to represent the non-stationary mean chron length. Marzocchi [1997] detrended the CK95 timescale by fitting an exponential to the reversal rate and obtained $k = 1.4 \pm 0.4$ for the full timescale. He also found the part of CK95 younger than 30 Ma to be stationary without detrending and obtained values of $k = 1.6 \pm 0.4$ for this segment.

4. GAMMA INDEX ESTIMATION FOR C-SEQUENCE AND M-SEQUENCE CHRONS

A plot of the lengths of individual polarity chrons in the composite timescale against their occurrence [*Gallet and Courtillot*, 1995] is shown in Figure 3. Straight lines were fitted by least squares to the chron lengths and the Student t-statistic was used to evaluate the confidence levels of the slopes. The best-fit line to CENT94, with 89 polarity intervals, has slope 0.0011 ± 0.0033 and is not significant. This conclusion that there is no significant trend in CENT94 was also reached by Hulot and Gallet [2003] using a more sophisticated analysis of stationarity proposed by McFadden and Merrill [2000]. The slope of the linear segment CK95(1), with 125 polarity intervals, is 0.0006 ± 0.0011 and is also not significant. These segments may be considered as stationary. If the two very long polarity chrons C33n and C33r are included in the analysis of segment CK95(2), the linear regression has slope 0.0172 ± 0.0145 and is marginally significant. However, if the two chrons are omitted, the best-fit line to the remaining 57 polarity intervals has slope 0.0050 ± 0.0112 and is not significant.

The segments CK95(1), CK95(2) without C32r and C33r, and CENT94 are stationary. A quantile–quantile (Q–Q) test [*Fisher et al.*, 1987] can be applied to the stationary segments to test visually how well the distributions of chron lengths conform to a theoretical distribution, in this case the Poisson (exponential) distribution. To prepare this diagram, the observed chrons are ranked in order of length and plotted against the theoretical quantile (Figure 4). The linear fits indicate good agreement with the Poisson distribution, except for CK95(2) when the long chrons C33n and C33r are included. This disagreement with the rest of the population is further justification for omitting these two anomalously long chrons, which last 5.5 Myr and 3.9 Myr, respectively and together represent an appreciable fraction of the 83 Myr C-sequence. Their durations and location at the end of the CNPS suggest that C33n and C33r might be more closely associated with the kind of field behavior in the CNPS, which lasted 37.6 Myr, rather than the more rapidly reversing later C-sequence.

The maximum likelihood method was employed to determine the optimum gamma indices k for the distributions of chron lengths in each segment. Histograms of the chron lengths

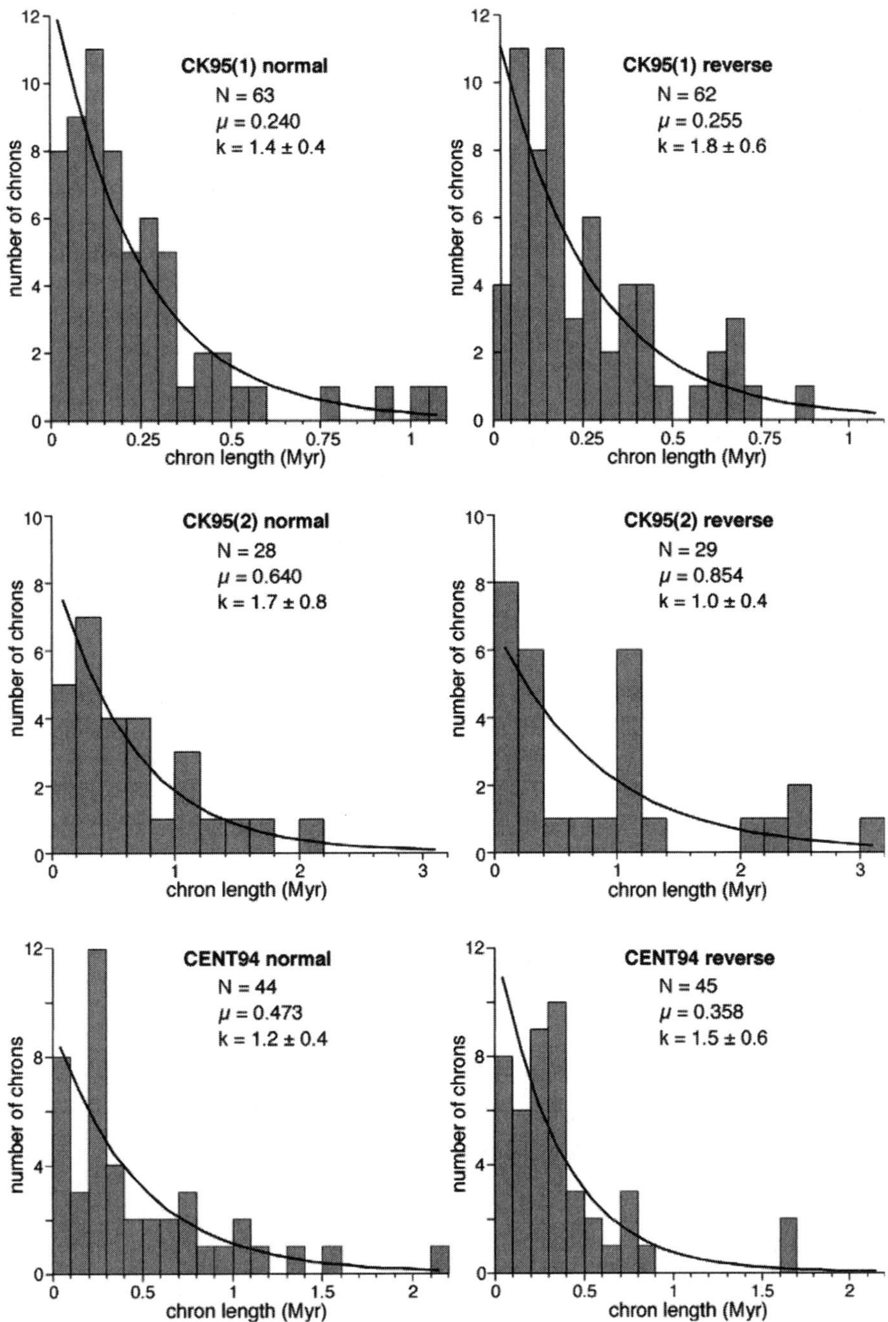

Figure 6. Distributions of normal and reverse polarity chron lengths (Myr) in the stationary timescale segments CK95(1), CK95(2), and CENT94. Solid lines are best fit exponential curves.

are shown in Figure 5 and for the separate distributions of normal and reverse chrons in Figure 6. The values of k obtained for both polarity sets are in the range $1.2 \leq k \leq 1.6$ (Table 1). For CK95(1) the 2σ limits do not include $k = 1$, but for CK95(2) and CENT94 they do (Figure 7). When separated by polarity, only the reverse polarity set for CK95(1) does not include $k = 1$.

5. THE CRETACEOUS NORMAL POLARITY SUPERCHRON

The uniform polarity of the CNPS has been challenged by reports of reversed polarity zones in the Aptian and Albian. Reversely magnetized Aptian marls were reported at a site near Valdorbia in the Umbrian Apennines [*VandenBerg and*

Table 1. Number of chrons (N), range of ages (Ma), mean chron duration (μ, Myr), maximum likelihood estimate of the gamma index (k), and $\pm 2\sigma$ range of k in segments of the composite timescale used in this paper. CK95(1)+ includes 8 additional subchrons listed in Table 2

Timescale	N	Age range	μ	k	$\pm 2\sigma(k)$
CK95(1), N&R	125	0–30.9	0.248	1.56	± 0.36
CK95(1), N	63		0.240	1.41	± 0.45
CK95(1), R	62		0.255	1.75	± 0.58
CK95(1)+, N&R	141	0–30.9	0.219	1.33	± 0.29
CK95(1)+, N	71		0.216	1.13	± 0.34
CK95(1)+, R	70		0.222	1.63	± 0.50
CK95(2), N&R	57	30.9–73.6	0.749	1.18	± 0.40
CK95(2), N	28		0.640	1.65	± 0.81
CK95(2), R	29		0.854	0.96	± 0.45
CENT94, N&R	89	120.6–157.5	0.415	1.26	± 0.34
CENT94, N	44		0.473	1.15	± 0.44
CENT94, R	45		0.358	1.45	± 0.56

Wonders, 1976]. The magnetozone occurred in strongly hematized redbeds and was not found in other coeval Tethyan magnetostratigraphic sections. However, Tarduno [1990] found reversely magnetized samples of Aptian age in DSDP Site 463. Although reversely magnetized beds were found in the Albian Fucoid Marls in the Umbrian Contessa section, their significance as polarity chrons was cast in doubt by rock magnetic evidence that the magnetozone could be remagnetized [*Tarduno et al.*, 1992]. Possible correlative short magnetozones in cores recovered from the Deep Sea Drilling Program and Ocean Drilling Program may be due to accidentally inverted core segments. There is presently no convincing evidence that refutes the interpretation of the CNPS as an uninterrupted lengthy period of constant normal polarity, but there are strong indications that it may be characterized by paleointensity fluctuations [*Cronin et al.*, 2001].

6. EFFECTS OF CRYPTOCHRONS ON POLARITY CHRON DISTRIBUTIONS

Many short wavelength, low amplitude magnetic anomalies ("tiny wiggles") are interspersed in the marine magnetic record. It is uncertain if these represent very short polarity chrons or geomagnetic intensity fluctuations. Cande and LaBrecque [1974] were able to model small scale magnetic anomalies between Anomalies 5 and 5A equally satisfactorily as either short chrons or intensity fluctuations. Cande and Kent [1992b] made a detailed analysis of some of these anom-

alies and favored the latter explanation, but to accommodate the ambiguity they proposed designating them as cryptochrons. Of the 54 cryptochrons catalogued by Cande and Kent [1992b], 46 are potentially of normal polarity and only 8 of reverse polarity. They are unevenly distributed in the polarity record: 46 occur in the Oligocene and Paleocene, and thus would affect primarily the statistical properties of segment CK95(2) if found to be real polarity subchrons. For example, adding all the cryptochrons listed for the optimum timescale of Cande and Kent [1995] reduces the gamma index to $k = 1.3 \pm 0.3$ in CK95(1) and to $k = 0.6 \pm 0.1$ in CK95(2) (Figure 8). However, not all, and perhaps only a few, are likely to be real polarity subchrons. CK95 used a 30 kyr cut-off to separate cryptochrons of uncertain origin from the longer, unambiguous polarity chrons. It is noteworthy that one-half (27) of the 54 cryptochrons identified in CK95 have durations of 10 kyr or less and only 3 have durations longer than 20 kyr (but less than 30 kyr). We suggest that relatively few cryptochrons, most likely the longer ones, represent polarity reversals.

Detailed magnetostratigraphic investigations have generally failed to identify short polarity chrons in sections where CK95 indicates cryptochrons are abundant. A short core, the Massicore, drilled in Eocene-Oligocene marls from the Italian Marches region, contained a complete magnetostratigraphy from Chron C16n to Chron C12r but there were no additional short polarity chrons where CK95 identified cryptochrons in this interval [*Lanci et al.*, 1996]. A rock magnetic

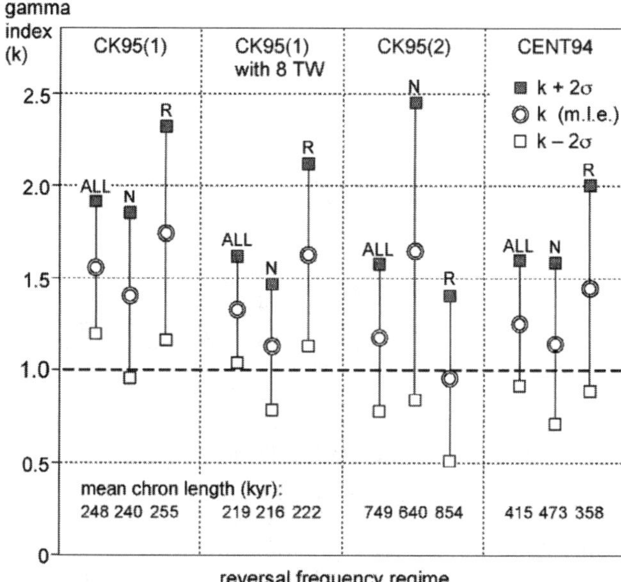

Figure 7. Maximum likelihood estimates of the gamma index k and its $\pm 2\sigma$ range for the timescale segments CK95(1), CK95(1) with 8 short chrons added, CK95(2) and CENT94. The numbers below each range of k give the mean chron length for the segment in kyr.

study showed paleointensity fluctuations at the expected locations of the cryptochrons [*Lanci and Lowrie*, 1997]. Closely sampled Oligocene pelagic sediments from DSDP Site 522 showed consistent decreases in paleointensity that could correspond to cryptochrons [*Tauxe and Hartl*, 1997]. It might be postulated that the discrete sample spacing was too wide to find these short events. However, an almost continuous high-resolution magnetostratigraphy in ODP sites 1218 and 1219, obtained with the on-ship pass-through magnetometer and

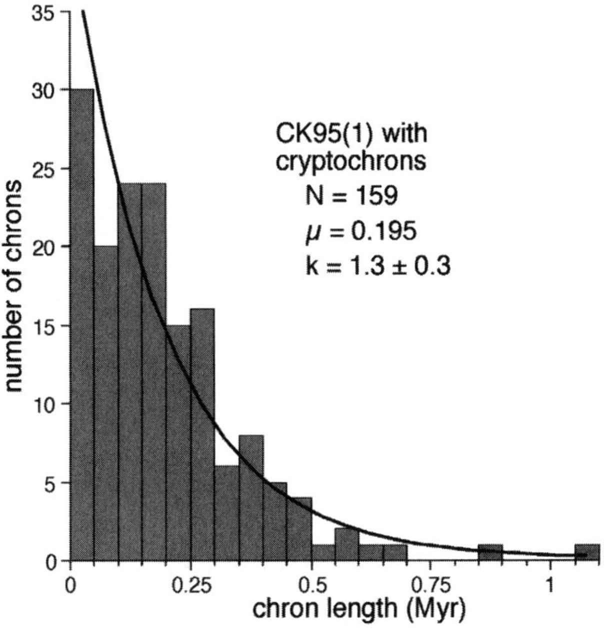

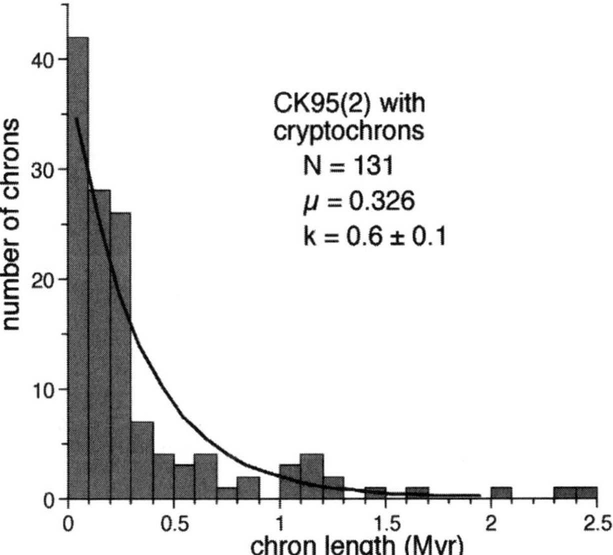

Figure 8. Distributions of chron lengths (Myr) when all cryptochrons predicted in CK95 are added to the two stationary segments CK95(1) and CK95(2). Solid lines are best fit exponential curves.

Table 2. Ages (Ma) of short magnetozones found in magneto-stratigraphic studies that can be considered as real polarity sub-chrons. Ages are calculated with respect to CK95

Subchron	Young end	Old end	Reference
C1r.2r-1n	1.201	1.211	Berggren et al. [1995]
C4r.2r-1n	8.606	8.664	Krijgsman & Kent [2004]
C4Ar.1r-1n	9.097	9.117	Krijgsman & Kent [2004]
C5r.2r-1n	11.167	11.193	Krijgsman & Kent [2004]
C5r.2r-2n	11.352	11.363	Krijgsman & Kent [2004]
C5r.3r-1n	11.555	11.584	Krijgsman & Kent [2004]
C5Dr-1n	17.793	17.854	Channell et al. [2003]
C7Ar-1n	25.678	25.705	Channell et al. [2003]

subsequent u-channel sampling, also showed no short polarity chrons where they were predicted by cryptochrons [*Lanci et al.*, 2002]. Detailed magnetostratigraphic sections in Umbrian carbonate rocks showed variations in paleointensity within the CNPS but no evidence of cryptochrons [*Cronin et al.*, 2001].

There are, however, indications that short polarity chrons not listed in CK95 do exist. For example, the Cobb Mountain Subchron is included in the Late Neogene chronology of Berggren et al. [1995]. Five additional short polarity intervals, each longer than 10 kyr, were identified in a detailed magnetostratigraphic study of Middle to Late Miocene sediments from DSDP Site 608 [*Krijgsman and Kent*, this volume]. Their proposed status as real polarity subchrons was verified on the basis of correlative short polarity intervals reported in ODP Site 854 [*Schneider*, 1995], ODP Site 1092 [*Evans and Channell*, 2003], and/or the Orera section in Spain [*Abdul Aziz and Langereis*, this volume]. Four other features that were also identified in Site 608 were considered by Krijgsman and Kent [this volume] to represent very short (<10 kyr) geomagnetic excursions associated with fluctuations in field intensity, rather than polarity subchrons. Three of these excursions occur in Chron C5n.2n. In an important test of whether "tiny wiggles" represent polarity chrons or paleointensity fluctuations, Bowles et al. [2003] searched explicitly for short polarity features and measured paleointensity variations in Chron C5n.2n in sediments from ODP Site 887. They found no polarity subchrons in this interval; instead, paleointensity variations gave a good match to the low-amplitude oceanic magnetic anomalies from which cryptochrons in this interval had been postulated. Thus, even though directional excursions corresponding to the cryptochrons in Chron C5n.2n may sometimes be found (e.g., *Evans and Channell* [2003]), their very short duration and

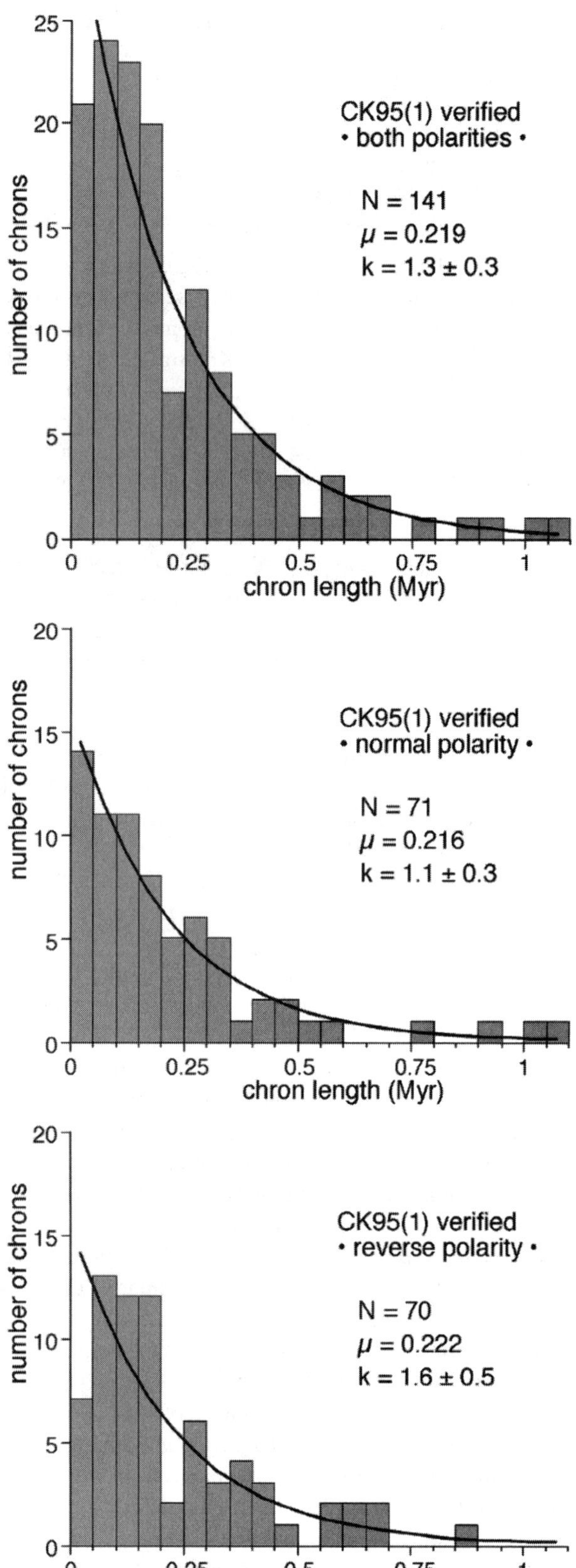

ephemeral character suggest that they do not correspond to global polarity subchrons.

We have incorporated the five additional subchrons identified by Krijgsman and Kent [this volume] and used similar criteria (e.g., antipodal directions to the host polarity interval and durations equal to or greater than 10 kyr as estimated from local sedimentation rates) to assess potential new subchrons in recent studies that were made with very close sampling or continuous measurement, and employing modern paleomagnetic techniques. In this way we identified a total of 8 additional short chrons (Table 2), which all occur in the youngest segment of the Cenozoic timescale, CK95(1). When they are included in CK95(1) the distributions of chrons of both polarities and those of only normal polarity are close to exponential although there is a deficiency in short chrons with reverse polarity (Figure 9). The recomputed mean interval length becomes 0.219 Myr and the gamma index is reduced to $k = 1.3$, with 95% fiducial limits that now include $k = 1$ (Table 2). The difficulty of finding corresponding short polarity intervals in magnetostratigraphic records suggests that most of the oceanic "tiny wiggles" identified as cryptochrons are due to paleointensity variations and are not part of the reversal sequence. We believe that it is unlikely that forthcoming detailed magnetic stratigraphy will reveal many more additional short chrons than have already been reported, but even a few more would further reduce k to a value even closer to unity.

7. DISCUSSION

In preparing geomagnetic polarity timescales no consideration is usually given to the finite duration of a polarity transition, which is thought to last about 4–6 kyr (e.g., *Clement and Kent* [1984]). This time is included in the lengths of polarity chrons, which are measured between the midpoints of polarity transitions, even though the transitional time can be an appreciable fraction of the durations of some short polarity chrons. McFadden and Merrill [1993] recognized that transitions are not instantaneous and introduced the idea of a dead time of 5 kyr following a reversal during which no new reversal can take place. The result is to increase the effective value of k for a Poisson distribution to slightly greater than 1.

McFadden and Merrill [1997] characterized reversal behavior by changes in reversal rate instead of chron length. They interpreted the changes of reversal rate as a gradual decrease prior to the CNPS, when it became zero, followed by a gradual increase subsequently. Our analysis differs fundamentally

Figure 9. Distributions of chron lengths (Myr) when the few short polarity intervals found in magnetostratigraphic studies are added to the two stationary segments CK95(1) and CK95(2). Solid lines are best fit exponential curves.

from theirs because we see evidence for stationary reversal behavior prior to and after the CNPS. In this respect, our point of view is more similar to that of Gallet and Hulot [1997], who regarded the CNPS as representing either an abrupt perturbation of the reversal process or a separate (non)reversal regime rather than being part of a continuous, long-term evolution of reversal rate. The stationarity prior to the CNPS is in agreement with the findings of Hulot and Gallet [2003] that there is little evidence of precursory field behavior that heralded this superchron, contrary to the earlier suggestion of McFadden and Merrill [2000].

Although our analysis reaches similar conclusions, it differs in several ways from that of Gallet and Hulot [1997]. They subdivide the polarity sequence into three segments on the basis of age, at points that differ from ours and for different criteria. Their segment A, covering 130–160 Ma, omits the youngest M-chrons; it is stationary, which we find to be the case for all of CENT94 from 120–160 Ma, including the youngest M-chrons. Segment B of Gallet and Hulot [1997], covering 25–130 Ma, includes the CNPS and is not stationary. Merrill and McFadden [1994] found it inappropriate to consider the CNPS as part of the reversal sequence in statistical analyses; we also omit it from our analysis. In addition, we omit chrons C33n and C33r, which in our view have a similar status to the CNPS. These omissions result in a stationary segment CK95(2). Segment A of Gallet and Hulot [1997], covering 0–25 Ma, is stationary. Our CK95(1) embraces 0–31 Ma and is stationary, a conclusion also reached by Marzocchi [1997].

We note here that the pre-CM29 cryptochrons in the marine anomaly record have the same character as the "tiny wiggles" in the Cenozoic record and, moreover, have not been correlated satisfactorily in magnetostratigraphic studies. We have excluded the pre-CM29 features because the reversal sequence they might represent has not been verified and the anomalies may well be due to paleointensity variations. If they should be found to represent reversals, they should then be regarded as part of a (higher frequency) reversal regime distinct from CENT94, as CK95(1) is distinct from CK95(2).

The increasing reversal rate following the CNPS in the analyses of McFadden and Merrill [1997] and McFadden and Merrill [2000] arises because they consider the C-sequence in its entirety to be non-stationary, whereas we regard it as composed of two reversal regimes, each of which is stationary. Chrons C33r and C33n immediately following the CNPS, if included in CK95(2), would make this segment barely non-stationary. However, they are exceptionally long and might well be classified with the CNPS; together, these chrons can even be considered as a different reversal regime with an exceptionally long mean interval length of 15.7 Myr.

In some previous analyses of reversal statistics, values of k substantially greater than 1 were regarded as due to either an incomplete record or some process that inhibits reversals. McFadden and Merrill [1984] regarded a value of k greater than 1 as an indicator for the fraction of missed short polarity chrons that were not resolved in the polarity timescales because of concatenation with other chrons. They predicted that about 46% of the polarity intervals in the C-sequence contained one or more unresolved short chrons. This prediction has not stood the test of observation. The most detailed analysis of the Late Cretaceous and Cenozoic reversal record, CK95, identified 54 locations where short polarity chrons might occur. Yet few of these cryptochrons have been confirmed as polarity intervals, as summarized above. Moreover, the addition of only 8 verified subchrons in CK95(1) has reduced k to be statistically indistinguishable from 1 (Poissonian, taking into account finite transition time), whereas CENT94 and CK95(2) already have gamma indices statistically indistinguishable from 1. Should more subchrons be confirmed, they would further reduce k to a value even closer to unity in each regime. Interestingly, there is remarkably little difference in the k-values for CK95(1) if only 8 selected subchrons from magnetostratigraphy or all 17 possible cryptochrons from CK95 are added. This is an indication that the statistical properties are becoming very robust. The general absence of short polarity intervals in the CNPS may indicate that the occurrence of cryptochrons is a function of reversal rate, i.e., more might be expected to occur when the geodynamo is already reversing frequently, as in segment CK95(1).

In another scenario, the reversal itself is treated as a special event, following which there is an 'inhibition' period of 40–50 kyr in which the probability of another reversal is reduced [*McFadden and Merrill*, 1993]. Such an inhibition period, whose physical origin is unclear, was introduced in order to explain elevated values of k when the mean interval length was short. However, given our observation that there is no significant change in gamma index k with mean interval length, and that k is close to unity in each reversal regime, the concept of an inhibition period is unnecessary.

Our analysis leads to the following conclusions.

(1) The record of reversal history over the past 160 Myr is composed of distinct segments in which the reversal process was stationary and characterized by a gamma index not distinctly different from unity, i.e. a Poisson process.

(2) Each segment may be looked on as a different regime of reversal behavior, either intrinsic to the dynamo process or triggered externally, e.g., see Gallet and Hulot [1997]. Further progress in understanding the significance of the reversal regimes will require integration of other relevant types of data, such as paleosecular variation [*McFadden et al.*, 1991].

(3) From one regime to another there is no significant change in gamma index k, but the average reversal rates shift abruptly and markedly, for example by a factor 3 in the Cenozoic.

(4) The Poisson model, with k indistinguishable from 1, appears to be a fundamental feature of geomagnetic field reversals. It is not only characteristic of the last 160 Myr as documented here for CK95(1), CK95(2), and CENT94. It is also observed for a 30 Myr interval in the Late Triassic, for which an astronomically calibrated geomagnetic polarity timescale has been developed [*Kent and Olsen*, 1999].

Acknowledgments. This is contribution 6552 of Lamont Doherty Earth Observatory and contribution 1323 of the Institute of Geophysics, ETH Zürich. DVK is grateful to the ETH Zürich for partial support during a sabbatical visit. We thank Ron Merrill and Yves Gallet for helpful reviews.

REFERENCES

Abdul Aziz, H., and C. G. Langereis, Astronomical tuning and duration of three new subchrons (C5r.2r-1n, C5r.2r-2n and (C5r.3r-1n) recorded in a Middle Miocene continental sequence from NE Spain, this volume, pp. 141–160.

Alvarez, W., M. A. Arthur, A. G. Fischer, W. Lowrie, G. Napoleone, I. Premoli Silva, and W. M. Roggenthen, Upper Cretaceous-Paleocene magnetic stratigraphy at Gubbio, Italy. V. Type section for the Late Cretaceous-Paleocene geomagnetic reversal time scale, *Geol. Soc. Amer. Bull.*, *88*, 383–389, 1977.

Berggren, W. A., F. J. Hilgen, C. G. Langereis, D. V. Kent, J. D. Obradovich, I. Raffi, M.E. Raymo, and N. J. Shackleton, Late Neogene chronology: New perspectives in high-resolution stratigraphy, *Geol. Soc. Amer. Bull.*, *107*, 1272–1287, 1995.

Bowles, J., L. Tauxe, J. Gee, D. McMillan, and S. Cande, Source of tiny wiggles in chron C5A: A comparison of sedimentary relative intensity and marine magnetic anomalies, *Geochem., Geophys., Geosys.*, *4*, doi:10.1029/2002GC000489, 2003.

Cande, S., R. L. Larson, and J. L. LaBrecque, Magnetic lineations in the Pacific Jurassic quiet zone, *Earth Planet. Sci. Lett.*, *41*, 434–440, 1978.

Cande, S. C., and D. V. Kent, A new geomagnetic polarity time scale for the Late Cretaceous and Cenozoic, *J. Geophys. Res.*, *97*, 13,917–13,951, 1992a.

Cande, S. C., and D. V. Kent, Ultra-high resolution marine magnetic anomaly-profiles: a record of continuous paleointensity variations?, *J. Geophys. Res.*, *97*, 15,075–15,083, 1992b.

Cande, S. C., and D. V. Kent, Revised calibration of the geomagnetic polarity timescale for the Late Cretaceous and Cenozoic, *J. Geophys. Res.*, *100*, 6093–6095, 1995.

Cande, S. C., and Y. Kristoffersen, Late Cretaceous magnetic anomalies in the North Atlantic, *Earth Planet. Sci. Lett.*, *35*, 215–224, 1977.

Cande, S. C., and J. L. LaBrecque, Behaviour of the earth's palaeomagnetic field from small scale marine magnetic anomalies, *Nature*, *247*, 26–28, 1974.

Channell, J. E. T., and E. Erba, Early Cretaceous polarity chrons CM0 to CM11 recorded in northern Italian land sections near Brescia, *Earth Planet. Sci. Lett.*, *108*, 161–179, 1992.

Channell, J. E. T., E. Erba, G. Muttoni, and F. Tremolada, Early Cretaceous magnetic stratigraphy in the APTICORE drill core and adjacent outcrop at Cismon (Southern Alps, Italy), and correlation to the proposed Barremian-Aptian boundary stratotype, *Geol. Soc. Amer. Bull.*, *112*, 1430–1443, 2000.

Channell, J. E. T., E. Erba, M. Nakanishi, and K. Tamaki, Late Jurassic-Early Cretaceous time scales and oceanic magnetic anomaly block models, in *Geochronology, Timescales, and Stratigraphic Correlation, Special Publication, SEPM*, edited by W.A. Berggren, D. V. Kent, M. Aubry, and J. Hardenbol, pp. 51–64, 1995.

Channell, J. E. T., W. Lowrie, and F. Medizza, Middle and Early Cretaceous magnetic stratigraphy from the Cismon section, northern Italy, *Earth Planet. Sci. Lett.*, *42*, 153–166, 1979.

Channell, J. E. T., and F. Medizza, Upper Cretaceous and Paleogene magnetic stratigraphy and biostratigraphy from the Venetian (Southern) Alps, *Earth Planet. Sci. Lett.*, *55*, 419–432, 1981.

Clement, B. M., and D. V. Kent, Latitudinal dependency of geomagnetic polarity transition durations, *Nature*, *310*, 488–491, 1984.

Cox, A., Lengths of geomagnetic polarity intervals, *J. Geophys. Res.*, *73*, 3247–3260, 1968.

Cronin, M., L. Tauxe, C. Constable, P. Selkin, and T. Pick, Noise in the quiet zone, *Earth Planet. Sci. Lett.*, *190*, 13–30, 2001.

Evans, H. F., and J. E. T. Channell, Upper Miocene magnetic stratigraphy at ODP site 1092 (sub-Antarctic South Atlantic): recognition of 'cryptochrons' in C5n.2n, *Geophysical J. Intl.*, *153*, 483–496, 2003.

Fisher, N. I., T. Lewis, and B. J. Embleton, *Statistical Analysis of Spherical Data*, Cambridge University Press, 1987.

Gaffin, S., Analysis of scaling in the geomagnetic polarity reversal record, *Phys. Earth Planet. Inter.*, *57*, 284–290, 1989.

Gallet, Y., and V. Courtillot, Geomagnetic reversal behaviour since 100 Ma, *Phys. Earth Planet. Inter.*, *92*, 235–244, 1995.

Gallet, Y., and G. Hulot, Stationary and nonstationary behaviour within the geomagnetic polarity time scale, *Geophys. Res. Lett.*, *24*, 1875–1878, 1997.

Handschumacher, D. W., W. W. Sager, T. W. C. Hilde, and D. R. Bracey, Pre-Cretaceous evolution of the Pacific plate and extension of the geomagnetic polarity reversal time scale with implications for the origin of the Jurassic "Quiet Zone", *Tectonophysics*, *155*, 365–380, 1988.

Harland, W. B., R. L. Armstrong, A. V. Cox, L. E. Craig, A. G. Smith, and D.G. Smith, *A Geologic Time Scale 1989*, 263 pp., Cambridge University Press, Cambridge, 1990.

Heirtzler, J. R., G. O. Dickson, E. M. Herron, W. C. Pitman, III, and X. Le Pichon, Marine magnetic anomalies, geomagnetic field reversals and motions of the ocean floor and continents, *J. Geophys. Res.*, *73*, 2119–2136, 1968.

Hulot, G., and Y. Gallet, Do superchrons occur without any palaeomagnetic warning?, *Earth Planet. Sci. Lett.*, *210*, 191–201, 2003.

Kent, D. C., and P. E. Olsen, Astronomically tuned geomagnetic polarity timescale for the Late Triassic, *J. Geophys. Res.*, *104*, 12,831–12,841, 1999.

Kent, D. V., and F. M. Gradstein, A Cretaceous and Jurassic geochronology, *Geol. Soc. Am. Bull.*, *96*, 1419–1427, 1985.

Krijgsman, W., and D. V. Kent, Non-uniform occurrence of short-term polarity fluctuations in the geomagnetic field? New results from Middle to Late Miocene sediments of the North Atlantic (DSDP Site 608), this volume, pp. 161–174.

LaBrecque, J. L., D. V. Kent, and S. C. Cande, Revised magnetic polarity time scale for Late Cretaceous and Cenozoic time, *Geology*, *5*, 330–335, 1977.

Lanci, L., J. M. Pares, and J. E. T. Channell, Miocene-Oligocene magnetostratigraphy from Equatorial Pacific sediments (ODP Site 1218, Leg 199), *Eos Trans. AGU, 83* (47), Fall Meeting Suppl., Abstract PP21D-09, 2002.

Lanci, L., and W. Lowrie, Magnetostratigraphic evidence that "tiny wiggles" in the marine magnetic anomaly record represent geomagnetic paleointensity variations, *Earth Planet. Sci. Lett.*, *148*, 581–592, 1997.

Lanci, L., W. Lowrie, and A. Montanari, Magnetostratigraphy of the Eocene-Oligocene boundary in a short continental drill-core, *Earth Planet. Sci. Lett.*, *143*, 37–48, 1996.

Larson, R. L., and T. W. C. Hilde, A revised time scale of magnetic reversals for the Early Cretaceous and Late Jurassic, *J. Geophys. Res*, *80*, 2586–2594, 1975.

Lowrie, W., Polynomial representation of geomagnetic polarity reversal sequences, *EoS Trans. AGU*, *78*, F193, 1997.

Lowrie, W., and W. Alvarez, Upper Cretaceous-Paleocene magnetic stratigraphy at Gubbio, Italy. III. Upper Cretaceous magnetic stratigraphy, *Geol. Soc. Amer. Bull.*, *88*, 374–377, 1977.

Lowrie, W., and W. Alvarez, One hundred million years of geomagnetic polarity history, *Geology*, *9*, 392–397, 1981.

Lowrie, W., and W. Alvarez, Lower Cretaceous magnetic stratigraphy in Umbrian pelagic limestone sections, *Earth Planet. Sci. Lett.*, *71*, 315–328, 1984.

Lowrie, W., W. Alvarez, G. Napoleone, K. Perch-Nielsen, I. Premoli Silva, and M. Toumarkine, Paleogene magnetic stratigraphy in Umbrian pelagic carbonate rocks: the Contessa sections, Gubbio, *Geol. Soc. Amer. Bull.*, *93*, 414–432, 1982.

Lowrie, W., and J. E. T. Channell, Magnetostratigraphy of the Jurassic-Cretaceous boundary in the Maiolica limestone (Umbria, Italy), *Geology*, *12*, 44–47, 1984.

Lowrie, W., and D. V. Kent, Geomagnetic reversal frequency since the Late Cretaceous, *Earth Planet. Sci. Lett.*, *62*, 305–313, 1983.

Lutz, T. M., and G. S. Watson, Effects of long-term variation on the frequency spectrum of the geomagnetic reversal record, *Nature*, *334*, 240–242, 1988.

Marzocchi, W., Missing reversals in the geomagnetic polarity timescale: their influence on the analysis and in constraining the process that generates geomagnetic reversals, *J. Geophys. Res.*, *102*, 5157–5171, 1997.

McFadden, P. L., Statistical tools for the analysis of geomagnetic reversal sequences, *J. Geophys. Res.*, *89*, 3363–3372, 1984.

McFadden, P. L., and R. T. Merrill, Lower mantle convection and geomagnetism, *J. Geophys. Res.*, *89*, 3354–3362, 1984.

McFadden, P. L., and R. T. Merrill, Inhibition and geomagnetic field reversals, *J. Geophys. Res.*, *98*, 6189–6199, 1993.

McFadden, P. L., and R. T. Merrill, Asymmetry in the reversal rate before and after the Cretaceous Normal Polarity Superchron, *Earth Planet. Sci. Lett.*, *149*, 43–47, 1997.

McFadden, P. L., and R. T. Merrill, Evolution of the geomagnetic reversal rate since 160 Ma: Is the process continuous?, *J. Geophys. Res.*, *105*, 28,455–28,460, 2000.

McFadden, P. L., R. T. Merrill, M. W. McElhinny, and S. Lee, Reversals of the earth's magnetic field and temporal variations of the dynamo families, *J. Geophys. Res.*, *96*, 3923–3933, 1991.

Mead, G. A., Correlation of Cenozoic-Late Cretaceous geomagnetic polarity time scales: An Internet archive, *J. Geophys. Res.*, *101* (8107–8109), 1996.

Merrill, R. T., and P. L. McFadden, Geomagnetic field stability: Reversal events and excursions, *Earth Planet. Sci. Lett.*, *121*, 57–69, 1994.

Naidu, P. S., Statistical structure of geomagnetic field reversals, *J. Geophys. Res.*, *76*, 2649–2662, 1970.

Ogg, J. G., M. B. Steiner, F. Oloriz, and J. M. Tavera, Jurassic magnetostratigraphy, 1. Kimmeridgian-Tithonian of Sierra Gorda and Carcabuey, southern Spain, *Earth Planet. Sci. Lett.*, *71*, 147–162, 1984.

Phillips, J. D., Time variation and asymmetry in the statistics of geomagnetic reversal sequences, *J. Geophys. Res.*, *82*, 835–843, 1977.

Sager, W. W., C. J. Weiss, M. A. Tivey, and H. P. Johnson, Geomagnetic polarity reversal model of deep-tow profiles from the Pacific Jurassic Quiet Zone, *J. Geophys. Res.*, *103*, 5269–5286, 1998.

Schneider, D. A., Paleomagnetism of some Leg 138 sediments: Detailing Miocene magnetostratigraphy, *Proceedings of the Ocean Drilling Program Scientific Results*, *138*, 59–72, 1995.

Steiner, M. B., J. G. Ogg, G. Melendez, and L. Sequeiros, Jurassic magnetostratigraphy, 2. Middle-Late Oxfordian of Aguilon, Iberian Cordillera, northern Spain, *Earth Planet. Sci. Lett.*, *76*, 151–166, 1985.

Tarduno, J., W. Lowrie, W. V. Sliter, T. J. Bralower, and F. Heller, Reversed polarity characteristic magnetizations in the Albian Contessa section, Umbrian Apennines, Italy: Implications for the existence of a mid-Cretaceous mixed polarity interval, *J. Geophys. Res.*, *97*, 241–271, 1992.

Tarduno, J. A., Brief reversed polarity interval during the Cretaceous Normal Polarity Superchron, *Geology*, *18*, 683–686, 1990.

Tauxe, L., and P. Hartl, 11 million years of Oligocene geomagnetic field behavior, *Geophys. J. Intl.*, *128*, 217–229, 1997.

VandenBerg, J., and A. A. H. Wonders, Paleomagnetic evidence of large fault displacement around the Po-Basin, *Tectonophysics*, *33*, 301–320, 1976.

Dennis V. Kent, Lamont-Doherty Earth Observatory, Palisades, New York 10964. (dvk@ldeo.columbia.edu)

William Lowrie, Institute of Geophysics, ETH Hönggerberg, 8093 Zürich, Switzerland. (lowrie@mag.ig.erdw.ethz.ch)

A Middle Eocene–Early Miocene Magnetic Polarity Stratigraphy in Equatorial Pacific Sediments (ODP Site 1220)

Josep M. Parés

Department of Geological Sciences, University of Michigan, Ann Arbor, Michigan

Luca Lanci

*Istituto di Dinamica Ambientale, Universita' di Urbino, Localita Crocicchia, Urbino, Italy, and
Department of Geological Sciences, Rutgers University, Piscataway, New Jersey*

In this paper we report new paleomagnetic and rock-magnetic results from Equatorial Pacific sediments obtained during Ocean Drilling Program Leg 199 (Site 1220; 10.17°N, 142.75°W). ODP Leg 199 was designated to collect sediments along a latitudinal transect in the Pacific Ocean to better understand Paleogene sedimentation patterns and the system of equatorial currents and thus magnetic chronology was an essential part. Continuous paleomagnetic measurements of u-channel samples were complemented by analysis of discrete samples. The results support the overall polarity pattern obtained on the shipboard pass-through magnetometer. Both rock-magnetism and paleomagnetic data show the presence of magnetite as the main carrier of the remanence. The recovered magnetostratigraphy spans from Chron C20r (mid-Middle Eocene) through Chron 6An.1n (Early Miocene) and constitutes an unprecedented record for the Pacific Ocean. Mean sediment accumulation rates range from 8 m/Ma in the middle Eocene to 3 m/Ma in the Late Oligocene. We do not observe short polarity events that might account for tiny wiggles in Chrons C12r or C13r, favoring the interpretation that they represent paleointensity variations of the geomagnetic field. Alternatively, short polarity events might have been smoothed or wiped out by delayed remanence acquisition. When plotted against time, paleomagnetic inclinations show a polarity-dependent anomaly, which is larger during times of normal polarity. Possible origins of such inclination anomaly include partial overprint and a non-dipole field contribution.

1. INTRODUCTION

After the realization that deep-sea sediment cores can be used for establishing a reversal record of the Earth's paleomagnetic field [*Harrison and Funnell*, 1964; *Opdyke et al.*,

Timescales of the Paleomagnetic Field
Geophysical Monograph Series 145
Copyright 2004 by the American Geophysical Union
10.1029/145GM10

1966; *Foster and Opdyke*, 1970], magnetic stratigraphy has become central to the calibration of geologic time. Since the pioneering Vema cores from the Pacific Ocean [*Opdyke*, 1968], the geomagnetic reversal time scale based on deep-sea sediment cores has been substantially improved and it has become fundamental in bridging biozonations and absolute ages [see *Opdyke and Channell*, 1996].

Reversal stratigraphy has been particularly fundamental in the correlation and age assignment of Paleogene sedimentary sequences. The Earth in the Paleogene experienced a variety of relevant events (mammal turnover, mass extinctions, Antarctic ice sheet formation, Eocene impact events) that overall make this time period a primary subject of chronostratigraphic studies based on magnetic reversals. The correlation between Paleogene magnetostratigraphy and biostratigraphy of marine sediments is mostly due to the pelagic sediments sections in northern Italy—Gubbio and Contessa [*Alvarez et al.*, 1977; *Lowrie et al.*, 1982; *Napoleone et al.*, 1983]. When combined, these two classic Italian sections span from the Paleocene (Chron C28n) to the Oligocene/Miocene boundary (Chron C6Cn) and all the magnetic polarity chrons have been confirmed independently in deep-sea sediment cores. Offshore magnetic stratigraphies for the Paleogene come mostly from the South Atlantic [DSDP Sites 522, 523, 524, *Tauxe et al.*, 1983a; ODP Leg 113, Sites 689 and 690, *Speiss*, 1990; and ODP Leg 114, *Clement and Hailwood*, 1991]. In particular Hole 689B produced excellent results for Late Oligocene to Middle Eocene [*Speiss*, 1990], whereas Site 522, in the South Atlantic, allowed Tauxe et al. [1984] to identify Chron 6B (lowermost Early Miocene) through Chron C13n (lowermost Early Eocene) in a single hole. The bulk of these studies have culminated in a well-defined magnetostratigraphic record and related faunal events for the Atlantic Ocean.

Data are lacking though for the Pacific and Indian Oceans impeding correlations of biostratigraphic, climatic, faunistic and impact events among oceans. Ocean Drilling Program Leg 199 offered an unprecedented opportunity to rectify this situation. Leg 199 was designed primarily to obtain new data to define the climatic and oceanographic processes that caused early Paleogene warming and the subsequent transition to cooler climate. Sediments drilled during Leg 199 are likely to document the evolution of the equatorial circulation and productivity in the Paleogene and Neogene. Dating a continuous sequence of Paleogene sediments by magnetic stratigraphy and tying it to biostratigraphy was thus a primary goal of Leg 199. In this paper we report paleomagnetic and rock magnetic results from Site 1220 and we combine them with the biostratigraphic data to produce an integrated magnetostratigraphic record for Paleogene equatorial sediments.

2. LITHOSTRATIGRAPHY AND BIOSTRATIGRAPHY

Site 1220, one of the eight sites drilled during Leg 199, is located in between the Clarion and the Clipperton fracture zones (10.17°N), at magnetic Anomaly 25, of ~ 56 Ma [*Cande et al.*, 1989] (Figure 1). The site is at a water depth of 5218 m and is in an abyssal hill. Site 1220 comprises three different holes and when combined they produce a 200 m thick composite lithologic column of Paleogene and Neogene sediments. On average, sediment accumulation rates range from

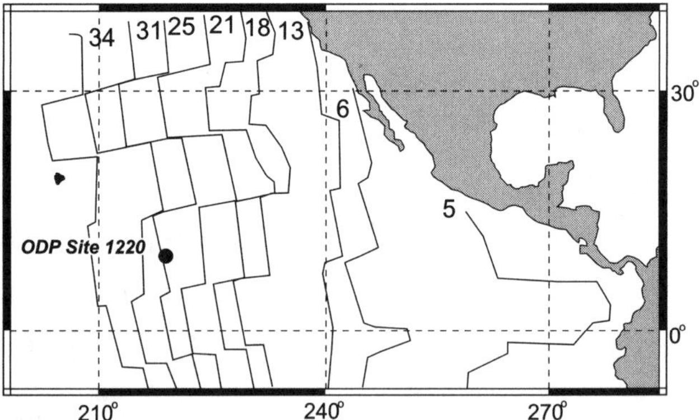

Figure 1. Map of the central tropical Pacific showing location of ODP Site 1220 and main magnetic anomalies of the Oceanic Plate. Site coordinates are 10.17°N 142.75°W at a depth of 5,218 meters below sea level. Magnetic anomalies from Müller et al. [1997].

8 m/Ma in the middle Eocene to 3 m/Ma in the Late Oligocene [*Shipboard Scientific Party*, 2002a].

Seismic reflection profiles collected during the survey cruise (EW9709) revealed four well-defined reflectors below a relatively thin layer of clay in the studied area [*Shipboard Scientific Party*, 2002a]. Lithostratigraphic observations made on board during Leg 199 confirmed the presence of five major sedimentary units (Figure 2). The basement of the stratigraphic pile is marked by fine-grained phaneritic basalt of the oceanic crust, corresponding to Anomaly 25. Directly above the oceanic crust there is a thin layer (~15 m) of partially dolomitized nannofossil ooze, radiolarian nannofossil ooze, calcareous chalk and black clay (Unit V). Unit IV is the thickest unit at Site 1220, consisting of 115 m of radiolarian oozes with clay and chert. Units III and II comprise overall ~60 m of radiolarian and nannofossil oozes with varying clay content. Unit III contains an important component of diatoms (15–45%), especially at the very bottom. Magnetic susceptibility, among other physical properties, mimics the major lithological units, being much lower in the carbonate sediments than in the clay or radiolarian ooze (Figure 2).

The biostratigraphic record at Site 1220 indicates a nearly complete sequence spanning from the uppermost Paleocene to the Lower Miocene. Paleogene and Neogene biostratigraphic zonation is based on radiolarians, planktonic foraminifers and calcareous nannofossils [*Shipboard Scientific Party*, 2002a]. The Paleocene/Eocene boundary was recovered in Hole 1220B within Unit V (not shown in Figure 2) and is marked by the extinction of Paleocene benthic foraminifers, the presence of excursion fauna of planktonic foraminifers and the appearance of the nannofossil genus *Rhomboaster* and the extinction of the nannofossil genus *Fasciculithus*, although none of these events are synchronous with the benthic foraminifer extinction marker used

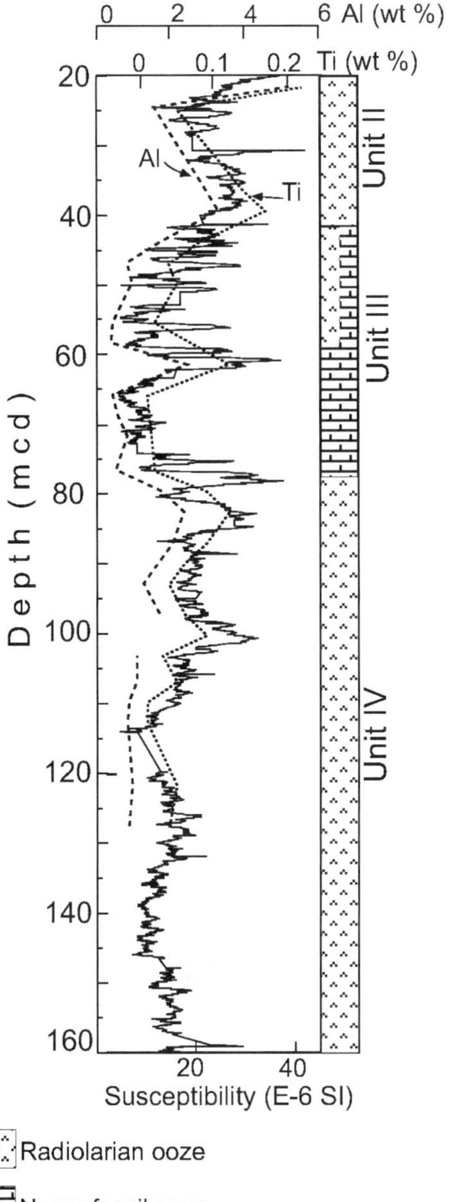

Figure 2. Lithostratigraphic units at Site 1220, including bulk susceptibility and changes of Ti and Al content (Ti and Al concentrations were determined by inductively coupled plasma-atomic emission spectroscopy, *Shipboard Scientific Party*, 2002a).

and planktonic foramminifera. The Oligocene biostratigraphy is based on planktonic foraminifera. Lowermost Oligocene begins with Zone CP16a+b (NP21) and from this zone and stratigraphically upwards, calcareous zones CP16c through CN1 have been identified. Except for the first two cores in Hole 1220A, radiolarians are abundant and well-preserved through the drilled sections. Diagnostic forms have been found which document zones RP7 (Early Eocene) through RN1 (Miocene). Overall, radiolarians and calcareous nannofossils provide robust anchoring points for the correlation of the observed magnetic stratigraphy with the Geomagnetic Polarity Time Scale of Cande and Kent [1995], as will be seen later in this paper.

3. METHODS

U-channel samples of 2 x 2 cm^2 cross-section and 1.5 m in length [*Tauxe et al.*, 1983b] were collected post-cruise from the archive halves in the ODP core Repository at Texas A&M. Natural Remanent Magnetization (NRM) measurements and progressive stepwise alternating field (AF) demagnetization were performed on a 2G Enterprises pass-through cryogenic magnetometer at the University of California, Davis. The u-channel samples were measured at 1-cm intervals and demagnetized at 5 or 10 mT steps up to peak fields of 70 mT or until the magnetic intensity was either too low or unstable. The response function of the magnetometer pick-up coils has a width of ~4.5 cm, implying that only every fourth measurement is independent. Characteristic Remanent Magnetization (ChRM) directions and associated maximum angular deviation (MAD) [*Kirschvink*, 1980] were calculated at each 1 cm interval using the 'pca' routine of Tauxe [1998]. For each interval, the corresponding Virtual Geomagnetic Pole was then computed and used to establish the magnetic polarity. Discrete samples from working halves were also collected for rock-magnetic analysis and additional AF demagnetization. The NRM of discrete samples was measured on a three-axis 2G cryogenic magnetometer housed in a shielded room with a residual field of less than 200 nT at the University of Michigan. Rock magnetic analyses include bulk magnetic susceptibility measured with a KLY-2 (Kappabridge) and AF demagnetization (SI4B, Sapphire Instruments). Hysteresis measurements were made with a Princeton Vibrating Sample Magnetometer (Micro-VSM) at the Institute for Rock Magnetism (Minneapolis, MN).

In low latitude sites with near-horizontal remanence, such as that studied here, the determination of magnetic polarity cannot be based solely upon magnetic inclination, as done for medium to high latitudes. Both declination and inclination are required for reversal stratigraphy in equatorial regions to unambiguously distinguish normal from reverse magneti-

for the P/E boundary at this site. Unfortunately, the sediments that include the P/E were recovered with extended core barrel (XCB) which produced mostly core fragments by the drilling process and thus no magnetic stratigraphic data are available. The Lower Eocene is represented by a calcareous nannofossil assemblage CP9b/NP11. The remaining Eocene sediments are barren of calcareous nannofossils

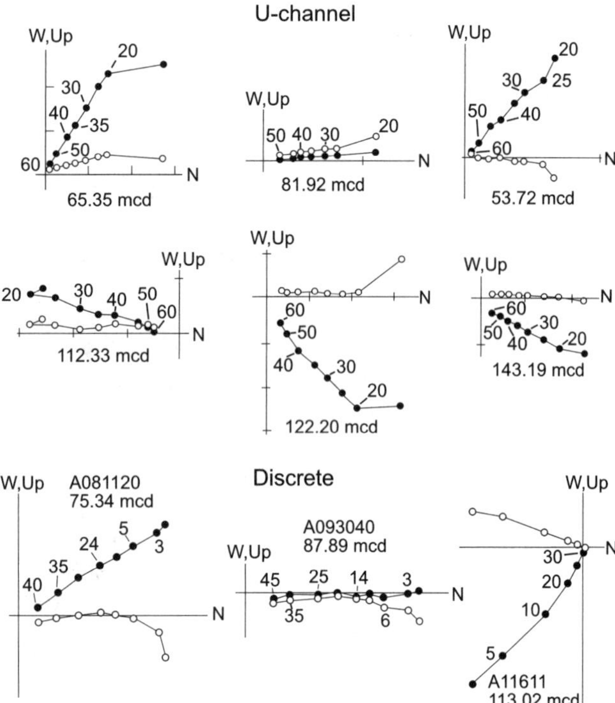

Figure 3. Representative demagnetization diagrams in orthogonal (Zijderveld-type) plots for u-channel and discrete samples. Closed (open) symbols represent projection of the vector end points onto the horizontal (vertical) plane. Demagnetization steps are shown in milli-Tesla.

zation directions. For this study, the orientation of cores recovered by the Advanced Hydraulic Piston corer (APC) has been recorded using the Tensor Tool (Tensor Inc., Austin, TX). The instrument has a three-axis fluxgate magnetometer that records the declination of a reference line on the Core with respect to magnetic North. The Tensor tool readings are recorded continuously at 30 seconds intervals, downloaded to a computer and analyzed once the tool is back on

the ship deck. The accuracy of the Tensor Tool is of the order of ±30°, which is precise enough for distinguishing north from south seeking directions.

4. PALEOMAGNETIC RESULTS

Although shipboard results based on the NRM measurement after a blanket AF demagnetization at 20 mT yielded a robust magnetostratigraphy [*Shipboard Scientific Party, 2002a*], shore-based measurements hold promise for providing a higher resolution paleomagnetic record based on progressive stepwise demagnetization and vector analysis of the magnetization components. AF progressive demagnetization was performed on u-channels up to ~70 mT or when the appearance of spurious magnetization prevented further demagnetization levels, often above 50–60 mT. Typically, a soft component is removed at around 20 mT and at higher fields the ChRM component is defined up to ~60 mT (Figure 3). If spurious magnetization develops it often deflects the demagnetization path slightly away from the origin on the orthogonal diagram. The presence of spurious magnetization has been noticed in a number of deep-sea sediments [e.g., *Tauxe et al.,* 1995] and it can sometimes be cancelled out in discrete samples by applying double demagnetization.

Measured Mr/Ms ratios from representative samples ranged between 0.26 and 0.33 whereas Hcr/Hc ratios were between 1.78 and 2.03. The average value of coercivity Hc is 15.9 mT, ranging from 14.5 to 17.3 mT. When these ratios are plotted on a Day diagram [*Day et al.,* 1977], values for Mr/Ms and Hcr/Hc lie in the pseudosingle-domain (PSD) field (Figure 4). The clustering of the data points in the Day diagram also suggests uniform magnetic properties of the magnetite grains. These results agree with the median destructive field observed in the progressive AF demagnetization diagrams and overall are indicative of magnetite as the main carrier of remanence in the sediments.

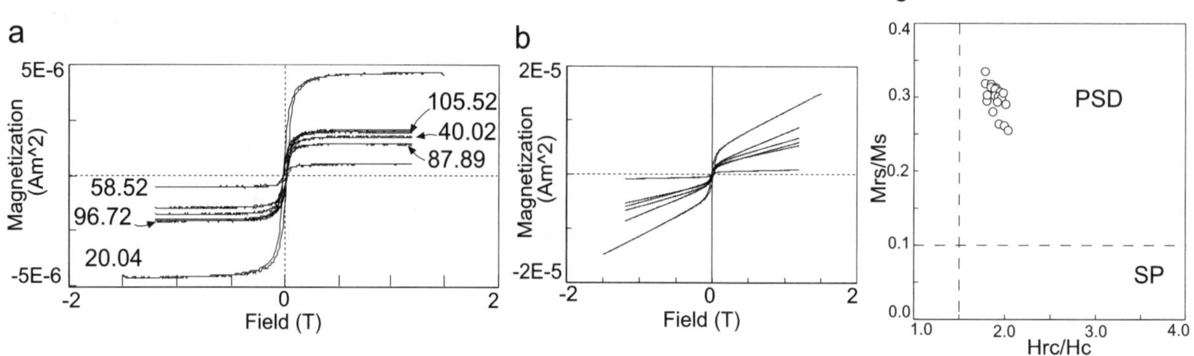

Figure 4. (a) Typical examples of slope corrected hysteresis loops from representative samples at various depth of Site 1220. Numbers show depth of the samples in mcd. (b) Uncorrected hysteresis loops. (c) Hysteresis ratios plotted on a Day et al. [1977] diagram.

We can speculate on the origin of the magnetization using circumstantial evidence based on the concentration changes of two elements through the stratigraphic section, Aluminum and Titanium. The presence of these two elements is typically regarded as a detrital input indicator. In Site 1220, both Ti and Al follow the same pattern as the magnetic susceptibility (Figure 2), suggesting a detrital origin for the magnetite grains carrying remanence in the pelagic sediments.

We have identified a total of 59 reversals, which correspond to polarity chrons C20r (lower Middle Eocene) through C6r (Early Miocene) (Figure 5 and Table 1). Biostratigraphic zones, including radiolarians and calcareous nannofossils, were used for anchoring the magnetic reversals to the geomagnetic reversal time scale providing a solid correlation of the polarity record for the entire section (Figure 6). The magnetostratigraphic record shows a continuous and high-resolution record of the polarity of the Earth's magnetic field from Middle Eocene to Early Miocene. The paleomagnetic record begins in Chron C20r (mid-Middle Eocene) within the radi-

olarian ooze with clay. The lower part of Chron C18n1.n shows some noise in the VGP Latitude, probably due to a partially unremoved barrel imprint. Chrons C13r and C12r, in Late Eocene and Early Oligocene respectively, show no evidence for short polarity intervals, though there are some low VGP latitudes at around 62 and 64 mcd (top Chron C12r). Additional sister cores are being processed to determine the origin of such features. A few more intervals in the mid part of the section show VGP Latitude values that are lower than expected (90° > Lat > 65°) (e.g., chrons C10n.1n through C9n). We noticed that such low values in latitude correspond to dispersion of declination within sections of a given core and not to changes in inclination. Twisting of the core liners before they are sliced into sections seems to be the origin of such deviations in declination.

Overall, paleomagnetism of Site 1220 offers an unprecedented high-resolution magnetostratigraphic record of Neogene and Paleogene equatorial sediments, as shown in Figure 6. Whereas most deep-sea Paleogene magnetostratigraphic

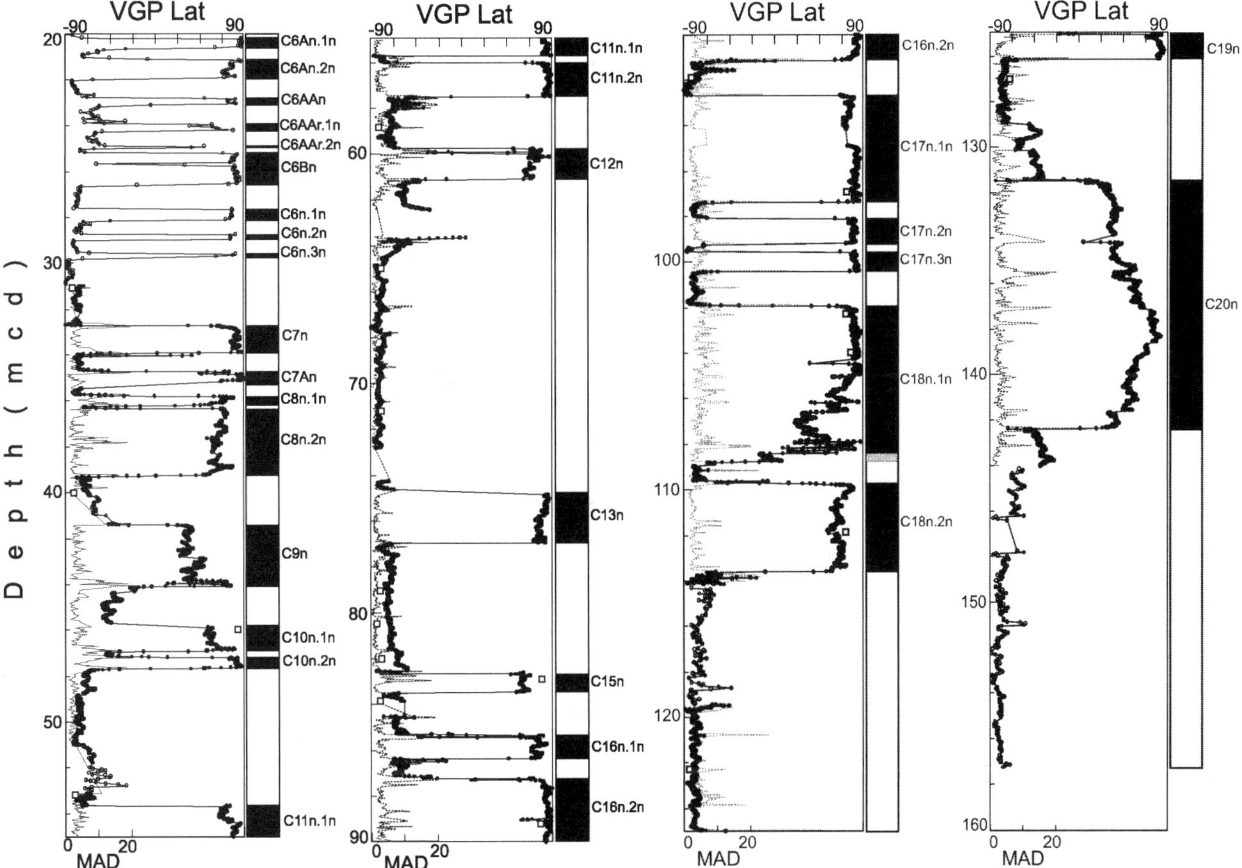

Figure 5. Plot of the virtual geomagnetic pole (VGP) latitudes versus depth at ODP Site 1220. Maximum Angular Deviation (MAD) shown with dashed line. Dots correspond to data obtained from progressive AF demagnetization of u-channels. Open circles indicate blanket AF demagnetization at 20 mT. Open squares show results from discrete samples.

Table 1. Depths of polarity zones in meters composite depth (mcd) at Site 1220 and corresponding chron labels.

Depth	Chron	Age (Ma)	Epoch
20.22	T_C6An.1n	20.518	
20.67	B_C6An.1n	20.725	
21.19	T_C6An.2n	20.996	
22.09	B_C6An.2n	21.320	
22.84	T_C6AAn	21.768	
23.14	B_C6AAn	21.859	
24.18	T_C6AAr.1n	22.151	
24.48	B_C6AAr.1n	22.248	
24.93	T_C6AAr.2n	22.459	
25.04	B_C6AAr.2n	22.493	Early Miocene
25.23	T_C6Bn.1n	22.588	
25.64	B_C6Bn.1n	22.750	
25.75	T_C6Bn.2n	22.804	
26.65	B_C6Bn.2n	23.069	
27.70	T_C6Cn.1n	23.353	
28.14	B_C6Cn.1n	23.535	
28.82	T_C6Cn.2n	23.677	
28.97	B_C6Cn.2n	23.800	
29.57	T_C6Cn.3n	23.999	
29.79	B_C6Cn.3n	24.118	
32.71	T_C7n.1n	24.730	
	B_C7n.1n	24.781	
	T_C7n.2n	24.835	
33.83	B_C7n.2n	25.183	
35.70	T_C8n.1n	25.823	Late Oligocene
36.22	B_C8n.1n	25.951	
36.29	T_C8n.2n	25.992	
39.21	B_C8n.2n	26.554	
41.41	T_C9n	27.027	
44.09	T_C9n	27.972	
45.79	T_C10n.1n	28.283	
46.84	B_C10n.1n	28.512	
47.14	T_C10n.2n	28.578	
47.66	B_C10n.2n	28.745	
53.57	T_C11n.1n	29.401	
55.73	B_C11n.1n	29.662	
56.02	T_C11n.2n	29.765	
57.45	B_C11n.2n	30.098	Early Oligocene
59.70	T_C12n	30.479	
61.19	B_C12n	30.939	
74.80	T_C13n	33.058	
76.97	B_C13n	33.545	
82.64	T_C15n	34.655	
83.47	B_C15n	34.940	
85.33	T_C16n.1n	35.343	
86.36	B_C16n.1n	35.526	
87.21	T_C16n.2n	35.685	Late Eocene
91.11	B_C16n.2n	36.341	
92.67	T_C17n.1n	36.618	
97.38	B_C17n.1n	37.473	
98.05	T_C17n.2n	37.604	

Table 1. (continued)

Depth	Chron	Age (Ma)	Epoch
99.22	B_C17n.2n	37.848	
99.55	T_C17n.3n	37.920	
100.40	B_C17n.3n	38.113	
101.90	T_C18n.1n	38.426	
	B_C18n.1n	39.552	
109.72	T_C18n.2n	39.631	
113.60	B_C18n.2n	40.130	Middle Eocene
125.04	T_C19n	41.257	
126.11	B_C19n	41.521	
131.45	T_C20n	42.536	
142.36	B_C20n	43.789	

records have been established in the Atlantic Ocean or in the Indian Ocean, our results offer a new integrated magneto-biostratigraphy for Pacific equatorial sediments, which will allow new correlations among oceans.

5. IMPLICATIONS FOR PALEOGENE PALEOLATITUDES

Long, continuous and well-dated deep-sea sedimentary records are prime candidates for determining the time-averaged geomagnetic field, including inclination fluctuations. This is particularly important if changes in inclination can provide a detailed paleolatitudinal record of the studied location and contribute to the long-standing question of the Pacific Plate motion in the Paleogene. Bioturbation and relatively low sediment accumulation rates average secular variation, making Site 1220 sediments particularly appropriate for such a goal.

We have grouped inclination data by polarity chrons (age) and computed the corresponding mean inclination to monitor its changes with time (Figure 7). We averaged the inclination of twenty-eight chrons identified in the magnetic stratigraphic record. Polarity chrons used in our analysis include C20r (Middle Eocene) to C6Cr (uppermost Late Oligocene). Given that the u-channel data set has not been deconvolved, we computed mean inclinations in two different ways, first, by using every fourth measurement and, secondly, by including the entire data set. The difference in results is always less than one degree, i.e., well below the resolution. The averages are calculated using the inclination-only method of McFadden and Reid [1982]. The associated circles of 95% confidence range from 0.2° to 2.2°. Data points at the top and bottom of the polarity chrons have been omitted from the calculations in order to avoid possible bias due to transitional fields. In addition, the top and bottom ten centimeters of the cores have also been excluded to allow for possible sediment disruption.

The most conspicuous feature of the inclination data set is that the distribution of normal and reverse inclinations (I_N,

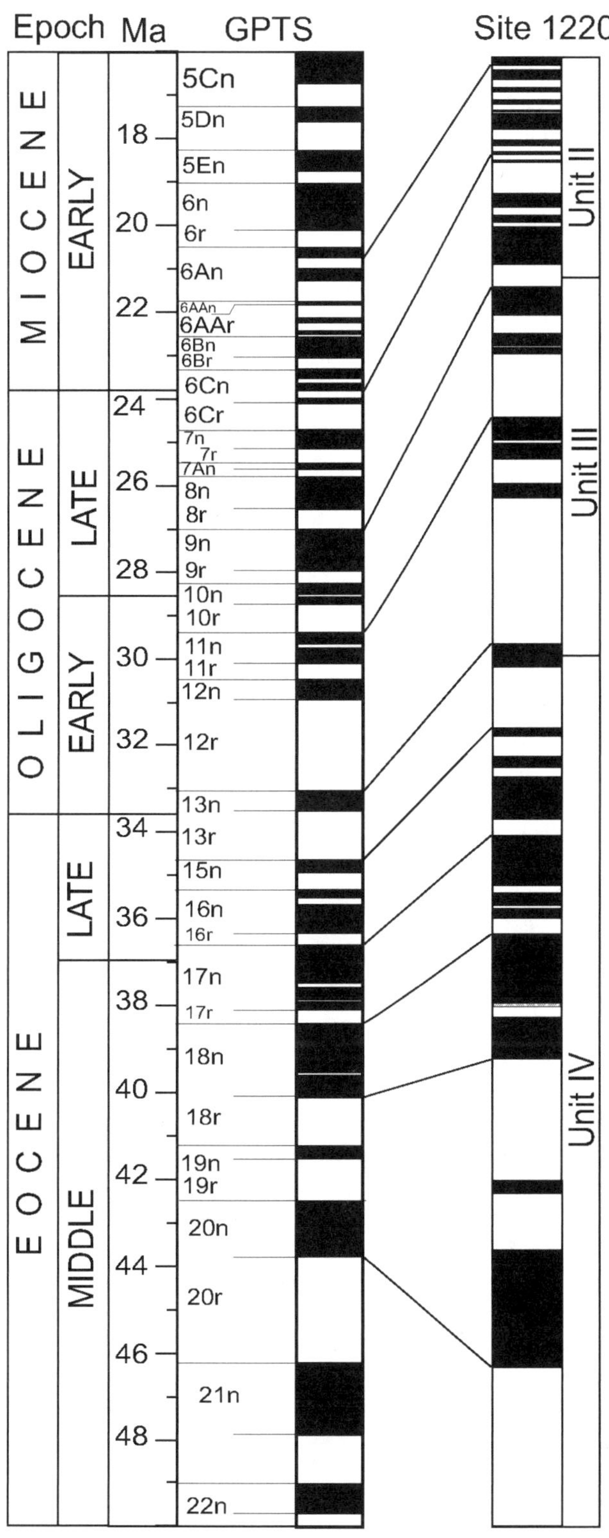

Figure 6. Correlation of the magnetic polarity stratigraphy at ODP Site 1220 with the Geomagnetic Polarity Timescale of Cande and Kent [1992, 1995].

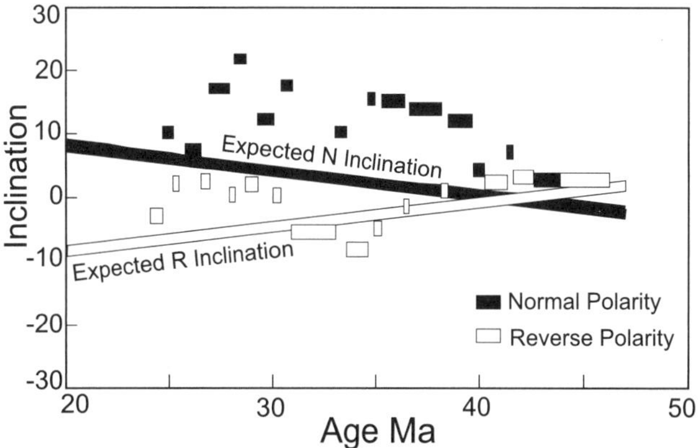

Figure 7. Inclination evolution of Site 1220 sediments (10.17°N, 217.25°E). Solid (open) boxes represent normal (reverse) polarity chrons. Solid (open) heavy lines show the predicted path of inclinations based on relative plate motions [*Engebretson et al.*, 1985].

I_R) is not antipodal. Along the stratigraphic column typically I_N is steeper than I_R. The fact that normal and reverse directions in land-based magnetostratigraphic studies are not always antipodal has been known for a number of years and does not preclude an unambiguous correlation of magnetic polarity sequences to the geomagnetic reversal time scale. Whereas its occurrence has hardly been studied in detail, there is a popular belief that shallower (steeper) reversed (normal) inclinations reflect contamination of the data set by an unremoved present day field overprint. That is, of course, in the northern hemisphere and assuming that the study rocks are located northward of their original position when they formed. This is precisely the case of Site 1220 sediments, which experienced a progressive northward translation during the Cenozoic. Hence, a possible explanation for the observed non-antipodal inclinations is the contribution of an unremoved present day field (PDF) component on the ChRM directions.

An alternative source of inclination anomalies is shallowing of inclination due to sediment compaction and/or depositional processes. It is unlikely though that inclination shallowing alone can explain the observed anomaly: At equatorial regions, paleomagnetic inclinations are close to horizontal and thus least susceptible to shallowing due to depositional processes or compaction. Moreover, while reverse inclinations are shallower, normal inclinations are steeper. Thus, the polarity-dependent pattern that we observe cannot be the result of the systematic shallowing of inclination in sediments as already noted by Schneider and Kent [1988] because the inclination error should be similar in both normal and reverse polarity intervals.

A third possibility to explain the observed inclination anomaly is a non-dipole contribution of the geomagnetic field. A persistent quadrupole component would have, at the equator, a similar effect as a present day field overprint on the paleomagnetic inclinations.

The above possibilities, or a combination of them, can explain the non-antipodal distribution of inclination data along the studied sedimentary column. In addition to the asymmetry between normal and reverse inclinations, we can also compare the observed versus expected values. For this purpose, we must compare our data set with a framework that is not based on paleomagnetic data, because this reference frame might not be exempt from the non-dipole contributions that we are trying to evaluate. Engebretson et al. [1985] produced a model for the displacement history of the oceanic plates in the Pacific Basin based on hotspots. The use of Pacific hotspots as a reference frame is highly questionable though [e.g., *Tarduno and Gee*, 1995; *Tarduno and Cottrell*, 1997; *Cottrell and Tarduno*, 2003]. Recent paleomagnetic results from Leg 197 indicate for example that the Hawaii hotspot moved southward from 81 to 43 Ma with respect to the rotation axis at rates of 3 to 5 cm/y [*Shipboard Scientific Party, 2002b*]. While being aware of the uncertainty of the paleolatitudes for the Pacific Plate in the Paleogene, we will tentatively use the reference frame from Engebretson et al. [1985] for the time interval younger than 45 Ma, although as we will see, the choice of the paleolatitude reference frame does not change the main conclusion of our result. Of note, the reference inclination for ~45 Ma, as predicted by the model that we use, is very close to inclination that we observe for that time interval, when both I_N and I_R have similar values. Because both I_N and I_R have similar values and they coincide with the expected inclination, it is reasonable to assume that there is no contamination, whatever its origin, in the Middle Eocene data set.

Comparing the expected versus observed values of inclination, it follows that on average the anomaly of I_N is higher than that of I_R, or that there is a polarity dependence of the inclination anomaly. Using the concept of "inclination anomaly" introduced by Cox [1975] ($\Delta I = |I_{observed}| - |I_{expected}|$), our data show that $\Delta I_N > \Delta I_R$ and specifically $\Delta I_N > 0$, and that mostly $\Delta I_R < 0$, except for polarity Chron C13r (34.7 to 33.5 Ma). There are two implications of the ΔI that are significant and might shed a hint on the source of the data set distribution. First, an unremoved present day field overprint in the paleomagnetic data set would produce $\Delta I_N < \Delta I_R$ at low latitudes in the northern hemisphere, just opposite to what it is observed here. Second, the value of ΔI varies with time, which would imply a fluctuating contribution of the present day field component in the sediments. There is no correspondence though between lithology and ΔI, undermining a link between PDF contribution and sediment changes.

Previous results on younger sediments [e.g., *Livermore et al.*, 1984, *Schneider and Kent,* 1988, 1990] have shown that non-dipole contributions (quadrupole, octupole) on the paleomagnetic record can produce a bias of the paleomagnetic inclinations. The inclination anomaly for an axial octupole contribution is mostly antisymmetric about the equator; it has no effect at the equator and poles and is greatest at mid-latitudes. For an axial quadrupole the contribution is largely symmetric about the equator, with its greatest effect at the equator and diminishing towards the poles. In the studied location (10° N) the only likely non-dipole effect to explain an inclination bias is a quadrupole contribution (g_2/g_1 ~13%). We note though that a non-dipole contribution such as a quadrupole field does not explain by itself the polarity dependence that we observe ($\Delta I_N > \Delta I_R$). In this respect, there is evidence suggesting that for the past five million years the reverse polarity field has had a larger non-dipole content than the normal field [*Opdyke and Henry*, 1969; *Wilson*, 1970, 1972; *Merrill and McElhinny*, 1977; *Schneider and Kent*, 1988]. Such studies show evidence for an inclination anomaly, $\Delta I_N < \Delta I_R$, in the Plio-Pleistocene time interval. Our observed inclination data set for the Middle Eocene through the uppermost Oligocene could hence be explained by a fluctuating quadrupole field contribution. In contrast to some existing results though [e.g., *Schneider and Kent*, 1988, 1990], the Eocene-Oligocene data would suggest a larger non-dipole contribution during normal polarity states. Complementary data from other Leg 199 sites will shed more light on the origin of the inclination anomaly.

6. CONCLUSIONS

Paleomagnetic data from sediments at ODP Site 1220 provide a high-resolution and continuous magnetostratigraphic record spanning Chrons C20r (Middle Eocene) to C6An.1n (Early Miocene). Such a magnetic stratigraphy provides the first complete Middle Eocene through Early Miocene equatorial record for the Pacific Ocean, which allows detailed correlations of biozones and impact events among oceanic basins and provides paleontologists to develop a detailed biostratigraphic framework for the Paleogene and Neogene.

The absence of short polarity intervals where cryptochrons have been observed in the geomagnetic reversal time scale (e.g., C9n, C12r, C13r) suggests that those are most likely related to fluctuations of the geomagnetic field intensity or that they are extremely fast (<10 ka) events, below the resolution of the studied sediments. Alternatively, these short polarity events might have been completely smoothed or wiped out by early diagenesis [e.g., *Bleil and von Dobeneck*, 1999; *Parés et al.*, this volume]. Further relative paleointensity determinations from Leg 199 sediments are needed in order to shed more light on to the origin of tiny wiggles in the Paleogene.

The inclination anomaly ($|I_N| > |I_R|$) in Site 1220 sediments can be a priori explained by either an unremoved present day field component (or a barrel overprint for that matter) or alternatively by the contribution of a non-dipole (quadrupole) component of the geomagnetic field. Whereas a present day field contribution alone does not explain the observed polarity dependence of ΔI, the non-dipole hypothesis does not agree with previous determinations for the last five million years. In this regard, from the existing literature it follows that there is a greater anomaly during times of reverse polarity in the Pliocene and Pleistocene, as opposed to our own observations in the Paleogene where the anomaly seems to be larger during times of normal polarity. There is a continuing debate about the significance of non-dipole fields on the paleomagnetic data set [e.g., *Kent and Smethurst*, 1998; *Van der Voo and Torsvik*, 2001; *Torsvik and Van der Voo*, 2002]. It is desirable to gather additional paleomagnetic records from the Southern hemisphere and from antipodal longitudes to the studied site in order to constrain the origin of the inclination anomaly in equatorial sediments.

Acknowledgments. P. McCausland, E. Tohver and R. Van der Voo provided constructive comments that helped to improve the manuscript. We thank Ken Verosub at the University of California, Davis, for use of the pass-through cryogenic magnetometer to measure the u-channel samples. We thank Daming Wang for measuring hysteresis parameters and Tony T. Goodman for helping with data management. We are grateful to all our colleagues and the staff of the ODP drilling vessel *Joides Resolution* for the delightful and highly successful Leg 199. Three anonymous referees provided helpful comments and suggestions. The personnel at the Gulf Coast Repository (College Station, TAMU), John Firth, Phil Rumford and Bruce Horan, were very helpful and patient with core sampling. The ODP is sponsored by the U.S. National Science Foundation (NSF) and participating countries under the management of Joint Oceanographic Institutions (JOI) Inc. This research was supported by grants from JOI/USSSP to both authors.

REFERENCES

Alvarez, W., Arthur, M. A., Fischer, A. G., Lowrie, W., Napoleone, G., Premoli-Silva, I. and Roggenthen, W. M, Upper Cretaceous-Paleocene magnetic stratigraphy at Gubbio, Italy. V. Type section for the Late Cretaceous-Paleocene geomagnetic reversal time scale, *Geol. Soc. Am. Bull.*, 88, 383–389, 1977.

Bleil, U. and von Dobeneck, T., Geomagnetic events and relative paleointensity records—Clues to high-resolution paleomagnetic chronostratigraphies of Late Quaternary marine sediments?, In *Use of Proxies in Paleoceanography: Examples from the South Atlantic*, edited by G. Fisher and G. Wefer, pp. 635–654. Springer-Verlag, Berlin, 1999.

Cande, S. C., LaBrecque, J. L., Larson, R. L., Pitman, W. C., III, Golovchenko, X. and Haxby, W. F., Magnetic lineations of the world's ocean basins, *AAPG*, 13:1, 1989.

Cande, S. C. and Kent, D. V., A new geomagnetic polarity timescale for the late Cretaceous and Cenozoic. *J. Geophys. Res.*, 97, 13917–13951, 1992.

Cande, S. C. and Kent, D. V., Revised calibration of the geomagnetic polarity timescale for the Late Cretaceous and Cenozoic. *J. Geophys. Res.*, 100, 6093–6095, 1995.

Clement, B. M. and Hailwood, E. A., 1991. Magnetostratigraphy of sediments from sites 701 and 702, *In Proceedings of the Ocean Drilling Program, Scientific Results*, edited by P. F. Ciesielski, Y. Kristofferson et al., 114, pp. 359–365, ODP, College Station, TX.

Cottrell, R. D. and Tarduno, J. A., A Late Cretaceous pole for the Pacific plate: implications for apparent and true polar wander and the drift of hotspots, *Tectonophysics*, 362, 321–333, 2003.

Cox, A., The frequency of geomagnetic reversals and the symmetry of the nondipole field. *Rev. Geophys.*, 13, 35–51, 1975.

Day, R., Fuller, M. D. and Schmidt, V. A., Hysteresis properties of titanomagnetites: grain size and composition dependence. *Phys. Earth Planet. Int.*, 13, 260–266, 1977.

Engebretson, D. C., Cox, A. and Gordon, R. G., Relative motions between oceanic and continental plates in the Pacific Basin, *Spec. Pap., Geol. Soc. Am.*, 206, 1985.

Foster, J. H. and Opdyke, N. D., Upper Miocene to Recent magnetic stratigraphy in deep-sea sediments, *J. Geophys. Res.*, 75, 4465–4473, 1970.

Harrison, C. G. A. and Funnell, B. M., Relationship of paleomagnetic reversals and micropaleontology in two late Cenozoic cores from the Pacific Ocean, *Nature*, 204, 566, 1964.

Kent, D. V. and Smethurst, M. A., Shallow bias of paleomagnetic inclination in the Paleozoic and Precambrian, *Earth Planet. Sci. Lett.*, 160, 391–402, 1998.

Kirschvink, J. L., The least-squares line and plane and the analysis of paleomagnetic data. Geophys, *J. Roy. Astron. Soc.*, 62, 699–718, 1980.

Livermore, R. A., Vine, F. J. and Smith, A. G., Plate motions and the geomagnetic field II: Jurassic to Tertiary, *Geophys. J. R. Astron. Soc.*, 73, 153–171, 1984.

Lowrie, W., Alvarez, W., Napoleone, G., Perch-Nielson, K., Premoli-Silva, I. and Toumarkine, M., Paleogene magnetic stratigraphy in Umbrian pelagic carbonate rocks: The Contessa sections, Gubbio, *Geol. Soc. Am. Bull.*, 93, 414–432, 1982.

McFadden, P. L. and Reid, A., Analysis of paleomagnetic inclination data, *Geophys. J.R. Astron. Soc.*, 69, 307–319, 1982.

Merrill, R. T. and McElhinny, M. W., Anomalies in the time-averaged paleomagnetic field and their implications for the lower mantle, *Rev. Geophys.*, 15, 309–323, 1977.

Müller, R. D., Roest, W. R., Royer, J.-Y., Gahagan, L. M. and Sclater, J. G., Digital isochrones of the world's ocean floor, *J. Geophys. Res.*, 102, 3211–3214, 1997.

Napoleone, G., Premoli-Silva, F., Heller, F., Cheli, P., Corezzi, S. and Fischer, A. G., Eocene magnetic stratigraphy at Gubbio, Italy, and its implications for Paleogene geochronology, *Geol. Soc. Am. Bull.*, 94, 181–191, 1983.

Opdyke, N. D., The paleomagnetism of oceanic cores. In *History of the Earth's Crust,* edited by (R.A. Phinney, pp. 67–72. Princeton University Press, Princeton, NJ, 1968.

Opdyke, N.D. and Henry, K.W., A test of the dipole hypothesis, *Earth Planet. Sci. Lett.*, 6, 139–151, 1969.

Opdyke, N. D., Glass, B., Hays, J. P. and Foster, J., Paleomagnetic study of Antarctica deep-sea cores, *Science*, 154, 349–357, 1966.

Opdyke, N. D. and Channell, J. E. T., *Magnetic Stratigraphy*, Academic Press, 1996.

Parés, J. M., Van der Voo, R., Yan. M. and Fang, X., After the dust settles: Why is the Blake Event imperfectly recorded in Chinese Loess?, this volume, pp 191–204.

Schneider, D. A. and Kent, D. V., Inclination anomalies from Indian Ocean sediments and the possibility of a standing nondipole field, *J. Geophys. Res.*, 93, 11621–11630, 1988.

Schneider, D. A. and Kent, D. V., The time-averaged paleomagnetic field, *Rev. Geophys.*, 28, 71–96, 1990.

Shipboard Scientific Party, Leg 199 summary, In *Proceedings of the Ocean Drilling Program, Initial Reports, 199,* edited by Lyle, M. W., Wilson, P. A., Janecek, T. R., et al., College Station TX (Ocean Drilling Program), 2002a.

Shipboard Scientific Party, Leg 197 summary, In *Proceedings of the Ocean Drilling Program, Initial Reports, 197,* edited by Tarduno, J. A., Duncan, R. A., Scholl, D. W., et al., College Station TX (Ocean Drilling Program), 2002b.

Speiss, V., Cenozoic magnetostratigraphy of Leg 113 drill sites, Maud Rise, Weddell Sea, Antarctica, In *Proceedings of the Ocean Drilling Program, Scientific Results, 113*, edited by P. F. Barker, J. P. Kennett et al., 113, pp. 261–290, College Station, TX, 1990.

Tarduno, J. A. and Gee, J., Large-scale motion between Pacific and Atlantic hotspots, *Nature*, 378, 477–480, 1995.

Tarduno, J. A. and Cottrell, R. D., Paleomagnetic evidence for motion of the Hawaiian hotspot during formation of the Emperor seamounts, *Earth Planet. Sci. Lett.*, 153, 171–180, 1997.

Tauxe, L., *Paleomagnetic Principles and Practice*, Kluwer Academic Publishers, 299 pp, 1998.

Tauxe, L., Tucker, P., Petersen, N. P. and LaBrecque, J. L., The magnetostratigraphy of Leg 73 sediments,. *Palaeogeogr., Palaeoclimatol. Palaeoecol.*, 42, 65–90, 1983a.

Tauxe, L., LaBrecque, J. L., Dodson, R. and Fuller, M., U-channels: a new technique for paleomagnetic analysis of hydraulic piston cores, *EOS, Trans. Am. Geophys. Union*, 64, 219, 1983.

Tauxe, L., Tucker, P., Petersen, N. P. and LaBrecque, J. L., Magnetostratigraphy of Leg 73 sediments, In *Initial Reports of the Deep Sea Drilling Project*, edited by K. J. Hsu, J. L. LaBrecque et al., 73, pp. 609–621, U.S. Govt. Printing Office, Washington, D.C., 1984.

Tauxe, L., Pick, T. and Kok, Y., Relative paleointensity in sediments: a pseudo-Thellier approach, *Geophys. Res. Lett.*, 22, 2885–2888, 1995.

Torsvik, T. H. and Van der Voo, R., Refining Gondwana and Pangea palaeogeography: estimates of Phanerozoic non-dipole (octupole) fields, *Geophys. J. Int.*, 151, 1–24, 2002.

Van der Voo, R. and Torsvik, T. H., Evidence for Late Paleozoic and Mesozoic non-dipole fields provides an explanation for the Pangea reconstruction problems, *Earth Planet. Sci. Lett.*, 187, 71–81, 2001.

Wilson, R. L., Permanent aspects of the earth's nondipole magnetic field over Upper Tertiary times, *Geophys. J.R. Astron. Soc.*, 19, 417–437, 1970.

Wilson, R. L., Paleomagnetic differences between normal and reversed field sources, and the problem of far-sided and right-handed pole positions, *Geophys. J.R. Astron. Soc.*, 28, 295–304, 1974.

L. Lanci, Istituto di Dinamica Ambientale, Universita' di Urbino, Localita Crocicchia, Urbino, PU 61029, Italy. (llanci@uniurb.it)

J. M. Parés, Department of Geological Sciences, University of Michigan, 2534 C.C. Little Building, Ann Arbor, Michigan. (jmpares@umich.edu)

Astronomical Tuning and Duration of Three New Subchrons (C5r.2r-1n, C5r.2r-2n and C5r.3r-1n) Recorded in a Middle Miocene Continental Sequence From NE Spain

Hayfaa Abdul Aziz[1] and Cor G. Langereis

Paleomagnetic Laboratory 'Fort Hoofddijk', Utrecht, The Netherlands

In the astronomically tuned Miocene continental Orera Composite Section (OCS) of NE Spain, one short interval and two single sample levels of normal polarity were found. The low sampling resolution inhibited determination of whether these normal polarity intervals represent remagnetization, reversal excursions or even short subchrons. We have sampled these intervals with high resolution, to determine their authenticity, and their astronomical duration. Rock magnetic analysis indicates that the main magnetic minerals in the studied rocks are magnetite and (fine-grained) hematite. Furthermore, a positive correlation is observed between rock magnetic parameters and lithology. Geochemical analysis also reveals a strong relation between concentrations of the main elements Al, Ca and Mg and lithologic changes. To characterize the sedimentary environment and possibly identify diagenetic influence on acquisition of the natural remanent magnetization (NRM), multivariate classification techniques were performed using both geochemical and rock magnetic variables. We conclude that the short polarity intervals—denoted as OCS-1, OCS-2 and OCS-3—are not overprints or artefacts of diagenetic processes, but represent real features of the geomagnetic field. The normal polarity interval OCS-1 has an estimated duration of 26 kyr, and corresponds to cryptochron C5r.2r-1 identified at ODP Site 845. We argue that it must now be termed as subchron C5r.2r-1n. OCS-2 has a duration of 15 kyr, and although it does not clearly show a pair of reversals, it must be denoted as subchron C5r.2r-2n. Finally, OCS-3 has a duration of 9 kyr, a full pair of reversals and must be considered as subchron C5r.3r-1n.

1. INTRODUCTION

Recently, an integrated stratigraphic study was performed for middle to late Miocene continental deposits from the Calatayud basin in north-eastern Spain [*Abdul Aziz et al.,*

[1] Faculty of Geosciences, Department of Earth Sciences, Utrecht University, Utrecht, The Netherlands

Timescales of the Paleomagnetic Field
Geophysical Monograph Series 145

2000]. The aim of this study was to establish a high-resolution cyclostratigraphic and magnetostratigraphic framework for cyclically bedded distal alluvial fan-floodplain to lacustrine deposits in the Orera area. The magnetostratigraphy was used to confirm cyclostratigraphic correlations between the different subsections in the study area. Subsequently, a composite magnetostratigraphic and cyclostratigraphic record was established, the so-called Orera Composite Section (OCS). The composite magnetostratigraphic record revealed the presence of at least one new short normal polarity interval [*Abdul Aziz et al.,* 2000; 2003b]. In addition, at two different levels, a single sample with normal polarity was found, so far not

"

recorded in magnetostratigraphic records from the marine realm. However, in many studies of marine magnetic anomaly profiles from sufficiently fast spreading ridges, small magnetic anomalies that are related to geomagnetic field behaviour have been observed. They have been referred to as what they look like: tiny wiggles. Their nature and origin, however, are often not clear. They may be considered to represent very short polarity intervals [*Blakely*, 1974; *Rea and Blakely*, 1975], or intensity fluctuations of the field [e.g. *Cande and Labreque*, 1974; *Cande and Kent*, 1992b; *Gee et al.*, 1996]. In their geomagnetic polarity time scale (GPTS), Cande and Kent [1992a], hereafter referred to as CK92, denoted tiny wiggles as cryptochrons, in view of their uncertain origin. They have used a 30-kyr duration cut-off as a realistic estimate of the typical maximum resolution of marine magnetic anomaly profiles from relatively fast spreading ridges. The cryptochrons have a designation (-1, -2, etc.) following the primary (sub)chron nomenclature, e.g. C1r.2r-1 is the uppermost cryptochron (-1) in the second reversed subchron (.2r) of Chron C1r (the Matuyama Chron). Cande and Kent [1992a] proposed that a cryptochron can be elevated to the status of subchron if it corresponds to a magnetostratigraphically identified pair of reversals. Such a cryptochron will then acquire a polarity suffix like the Cobb Mt. subchron (C1r.2r-1n).

The origin of many cryptochrons in CK92 have not (yet) been confirmed by magnetostratigraphic studies. On the other hand, there is firm evidence for excursions and "reversal excursions". Here, the term excursion is used for virtual geomagnetic poles (VGP) deviating more than 45° from geographical north [*Verosub and Banerjee*, 1977], while it is termed a reversal excursion for VGPs that deviate in excess of 90° from geographical north [*Merrill and McFadden*, 1994] and possibly reaching (near) opposite polarity. Many reversal excursions have been well established in the Brunhes Chron [*Langereis et al.*, 1997; *Lund et al.*, 1998], and there is increasing evidence for them in the Matuyama Chron [*Quidelleur et al.*, 2001]. Invariably, reversal excursions are associated with low paleointensities of the geomagnetic field, both in the Brunhes [*Guyodo and Valet*, 1999] and in the Matuyama [*Guyodo et al.*, 1999; *Channell et al.*, 2002] Chrons, but also for older cryptochrons [e.g., *Roberts and Lewin-Harris*, 2000]. This renders the discussion about the cause for tiny wiggles—either a short period of (near) opposite polarity or a period of low field intensity—rather superficial: it would seem that both phenomena occur at the same time and are linked together.

Several mechanisms have been postulated to explain the occurrence of (reversal) excursions. Gubbins [1999] proposed that excursions are events in which the geomagnetic field reverses in the liquid outer core—which has timescales of 500 yr or less—but not in the solid inner core. The influence of the inner core—with a diffusion time of approximately 3–5 kyr—delays full reversal during which time the original polarity may re-establish itself in the outer core. The typical duration of 3–6 kyr of reversal excursions [*Langereis et al.*, 1997] is consistent with the diffusion time scale. Evidently, this duration is usually much too short to be detected in marine magnetic anomaly profiles, and explains why so few excursions have been detected, even as tiny wiggles.

A clear distinction between reversal excursion and subchron is at best difficult and usually impossible to make. A subchron should at least have a pair of reversals [*Cande and Kent*, 1992a], and reach an opposite polarity during at least a 'measurable' amount of time [cf. *Gubbins*, 1999]. This would require a minimum duration of approximately 10 kyr for the entire interval of excursional (or transitional) directions and opposite polarity. In practice, however, such a definition will meet some serious problems. First of all, a good estimate of excursion duration requires highly accurate dating as well as high-resolution sampling. Many excursions are not observed simply because of their short duration and too low sampling density. Secondly, smoothing and even complete removal of the paleomagnetic signal is the rule rather than the exception, e.g., through post-depositional diagenesis. In the majority of records, even with high resolution sampling and high sedimentation rates, reversal excursions or short subchrons are simply not recorded, or they are incompletely recorded, making the event seem shorter; a subchron may thus appear as an excursion only. A good example is the Cobb Mountain subchron which—if it is recorded at all—has reported durations varying from 10 kyr [*Hsü et al.*, 1990] to 25 kyr [*Clement and Kent*, 1986/1987] and even 33–43 kyr [*Biswas et al.*, 1999; *Channell et al.*, 2002, 2003].

Reversal excursions have been identified in many sites world-wide, but most frequently in relatively young sediments (i.e., of Brunhes and late Matuyama age) simply because of the numerous sedimentary cores of limited depth range. Hence, although excursions are sometimes identified in older rocks [*Hartl et al.*, 1993; *Rösler and Appel*, 1998], they are still rare. Necessary comparisons with other coeval records are usually lacking.

In this paper, we present the detailed study of the short polarity intervals in the Orera Composite Section (Figure 1) in which *Abdul Aziz et al.* [2000, 2003b] reported two levels represented by single normal polarity samples and one normal polarity interval defined by multiple samples in Chron C5r. We present the results of a detailed paleomagnetic, rock magnetic and geochemical study of these intervals to determine their nature and origin and to attempt to assess authenticity, age and duration of the intervals. The astronomical calibration of the OCS enables the required accuracy to date the observed short polarity intervals and to determine their duration. Finally,

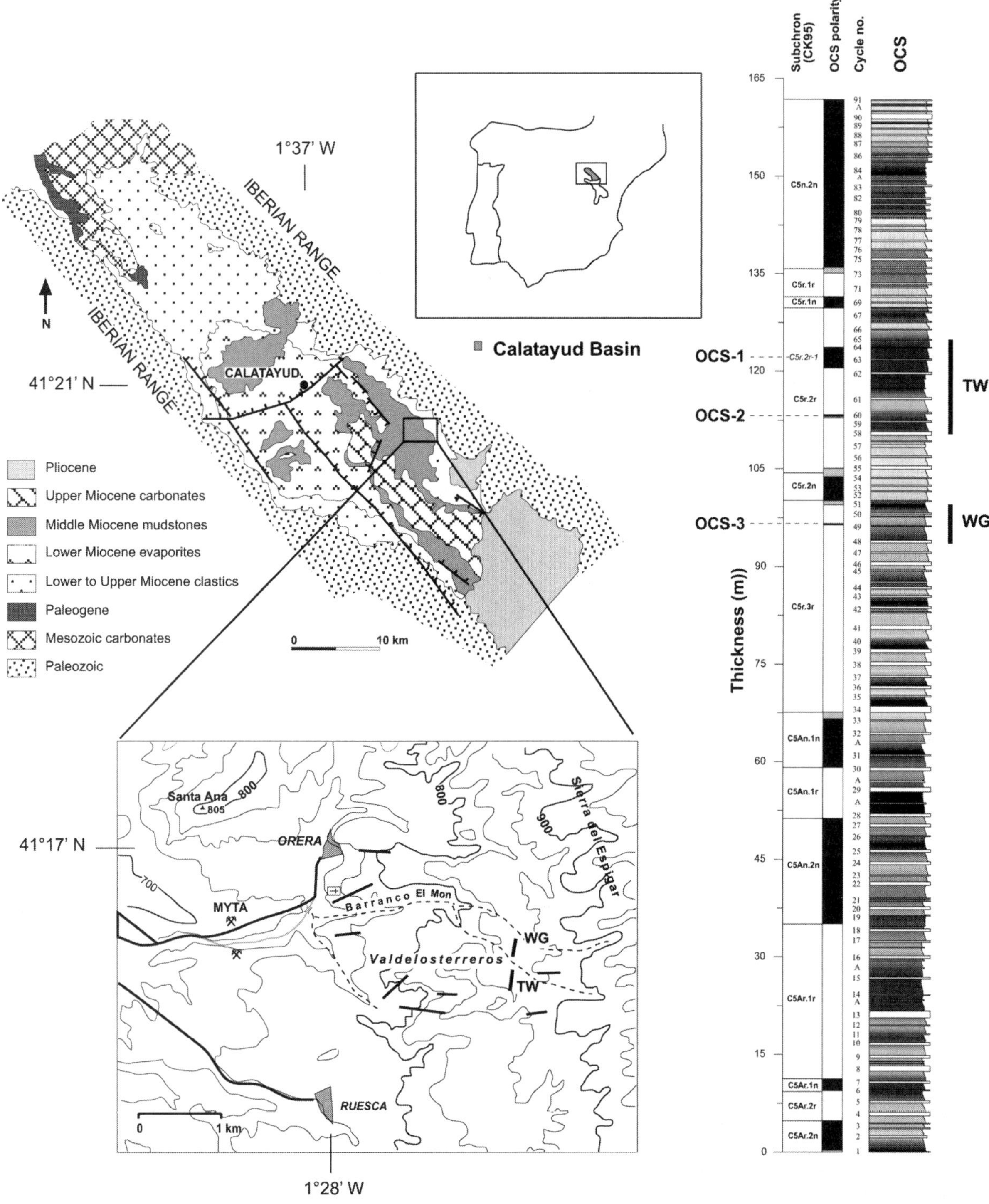

Figure 1. Geologic sketch map of the Calatayud basin and location map of the studied TW and WG sections. The corresponding subchrons in the magnetostratigraphy of the Orera Composite Section (OCS) are according to the correlation with the geomagnetic polarity timescale (GPTS) of Cande and Kent [1995; CK95]. Dark shades in the lithostratigraphic column of the OCS indicate red colored mudstones, lighter shades indicate gray mudstones while white resistant horizons represent the carbonate rocks. The cycle numbers of the OCS are shown left of the lithology column and the positions of the studied sections are indicated with bars.

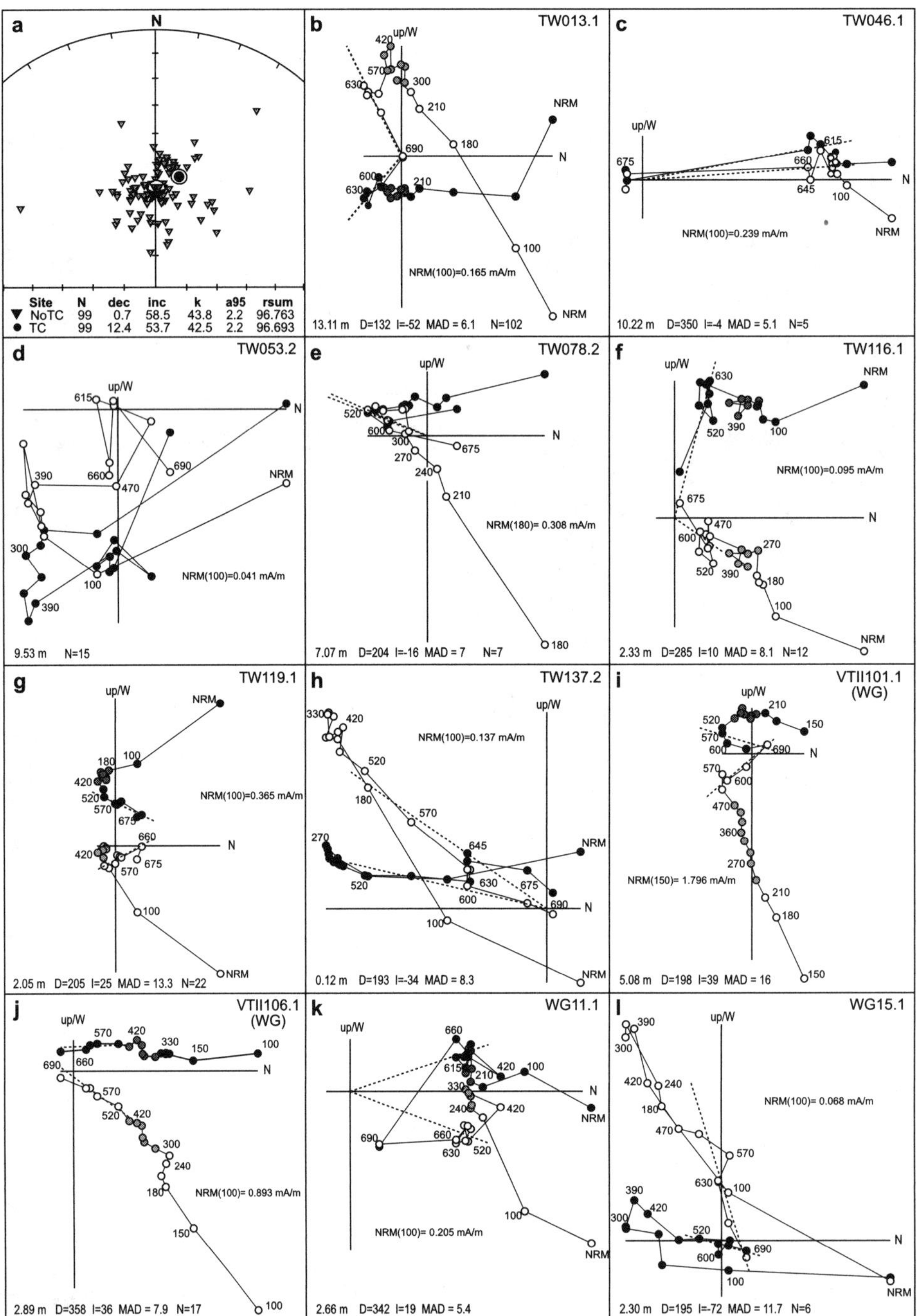

Figure 2. (a) Equal area projection and Fisher parameters of the low temperature (LT) magnetization component in the TW section with and without tectonic correction (TC and NoTC), respectively. The average NoTC direction with 95% confidence interval closely corresponds to the geocentric axial dipole field (GAD, asterisk), indicating a present-day overprint. (b-l) Thermal demagnetization diagrams for selected specimens from the studied TW and WG sections. Black (white) and dark (light) gray shaded circles denote the projection onto the horizontal (vertical) plane. The gray circles illustrate the intermediate temperature component (MT). Dashed lines indicate principal component analysis (PCA) fits to the ChRM, also shown for samples with no clear ChRM directions (e, g, i and l). The numbers indicate temperature steps in °C. Stratigraphic position (in meters), declination (D), inclination (I) and maximum angular deviation (MAD) for each measured specimen are shown. Also the number of specimens (N) with similar demagnetization behavior is indicated. Note that b and h, f and k, g and i show similar behavior.

we discuss whether a distinction can be made concerning their origin in terms of subchrons or reversal excursions.

2.GEOLOGICAL SETTING AND MAGNETOSTRATIGRAPHY

The study area lies near the village Orera, approximately 20 km south-east of Calatayud (Figure 1) where late to middle Miocene sediments crop out as part of the Paleogene and Neogene deposits of the NW–SE trending Calatayud basin [*Abdul Aziz et al.*, 2000]. The mountains of the Iberian Range adjacent to the basin consist of Paleozoic and Mesozoic rocks. Along the basin margin, coarse-grained alluvial fan conglomerates occur, laterally grading to more distal, finer grained alluvial fan-floodplain deposits and palustrine and lacustrine sediments. The central part of the Calatayud basin is dominated by evaporites. The lithological succession in the study area is characterized by a distinct, precession-controlled, cyclic alternation of mudstone and dolomitic carbonate, which are interpreted as floodplain/mudflat and shallow lake deposits, respectively, that accumulated in a shallow lake basin [*Abdul Aziz et al.*, 2003a]. More than 91 small-scale, mudstone-carbonate cycles were recognised, which in turn are hierarchically arranged in intermediate- and large-scale cycles. Cyclostratigraphic correlations between (overlapping) subsections, confirmed by magnetostratigraphy, resulted in the construction of the OCS. Correlation of the magnetic polarity record of the OCS to the GPTS of Cande and Kent [1995] indicated that the succession ranges in age between 10.7 and 12.8 Ma [*Abdul Aziz et al.*, 2000]. Furthermore, the OCS magnetostratigraphic record revealed the presence of three short intervals of normal polarity: two in C5.2r and one in C5.3r (Figure 1). To determine the magnetic component and to distinguish between primary and secondary components, IRM analysis were performed for samples from different types of lithology throughout the OCS [*Abdul Aziz et al.,* 2000]. The results of that study indicated that hematite and magnetite carry the magnetization components in these rocks.

High-resolution carbonate and color records were used to demonstrate the astronomical origin of the sedimentary cyclicity in the OCS and to establish the phase relation with astronomical parameters. Subsequently, each individual mudstone-carbonate cycle was calibrated to the astronomical curves of Laskar [1993] which resulted in the astronomical dating of reversal boundaries in the OCS [*Abdul Aziz et al.*, 2003b], including the short normal polarity interval (cryptochron) C5r.2r-1 first identified by Schneider [1995]. The duration of this interval was calculated to be 31 kyr [*Abdul Aziz et al.*, 2003b]. Because the other two short normal polarity intervals were defined by one sample level only, no age was calculated.

The astronomical calibration provides accurate age information for both dating and constraining duration of the short normal polarity intervals. However, to determine whether these normal polarity intervals are truly reversal excursions or subchrons, it is important to first ensure that the magnetic signal throughout the studied sections has a geomagnetic origin.

3. MATERIALS AND METHODS

We re-sampled the following intervals of subsection VT-II [see Figure 5 in *Abdul Aziz et al.*, 2000]: between cycles 58 and 65 (TW section) and between 48 and 51 (WG section) (Figure 1). These intervals are located within subchrons C5r.2r and C5r.3r, respectively. At each sampling level, two oriented and two non-oriented standard paleomagnetic cores were drilled at an average sampling interval of 10.5 cm, which corresponds to ~1550 kyr. In the short polarity intervals, however, the sample spacing was at an average distance of 9 cm (~1300 kyr). The re-sampling of these intervals has therefore increased the resolution by almost an order of magnitude relative to the original study by Abdul Aziz et al. in 2000. The color of the sediments was measured in the field at an average stratigraphic distance of 8.5 cm using a portable photospectrometer (Minolta CM508i).

To determine the characteristic remanent magnetization (ChRM) directions, at least one specimen per level was thermally demagnetized in a laboratory-built shielded furnace using stepwise temperature increments of 30°C up to 420°C, of 50°C between 420 and 600°C and steps of 15°C between 600 and 690°C. The natural remanent magnetization (NRM) was measured on a horizontal 2G Enterprises DC SQUID cryogenic magnetometer. The initial susceptibility (χ_{in}) was measured on a Kappabridge KLY-2 using a sister sample. An anhysteretic remanent magnetization (ARM) was then imparted to this sample in alternating fields up to a maximum of 250 mT using incremental steps of 20–50 mT, with a DC field of 29 µT.

Acquisition curves of the isothermal remanent magnetization (IRM) were determined for specimens around polarity interval OCS-2. Prior to acquisition, the specimens were demagnetized in a 300 mT alternating field. Subsequently, 28–30 acquisition steps were applied at fields up to 2.5 T. The peak fields were applied with a PM4 pulse magnetizer and the IRM intensity was measured on the cryogenic magnetometer. IRM acquisition curves were analyzed by decomposing the acquired curves into magnetic components using the IRM component fitting of Kruiver et al. [2001]. IRM acquisition curves are assumed to have a cumulative log-normal form with respect to the applied field, and the components add linearly in the acquisition curves provided that no magnetic interactions occur. Each magnetic component can be characterized by the saturation IRM (SIRM), the peak field at which half of the SIRM is reached (B1/2) and the dispersion of its corresponding cumulative log-normal distribution (DP) [*Kruiver et al.*, 2001]. According to the IRM-fitting method of Kruiver et al. [2001], the error in the total IRM estimation for samples with low intensities and noisy IRM acquisition curves will be highest for measurements at low applied fields.

For the geochemical analysis, the non-oriented core specimens were freeze dried up to –40°C and were subsequently powdered using a mortar mill. These specimens were completely destroyed in HF, HNO_3 and $HClO_4$ and element concentrations were obtained with ICP-OES analysis (inductively coupled plasma optical emission spectrometer, Perkin Elmer-type Optima 3000). Accuracy and precision of the analyses was checked by replicate samples and by laboratory standards and was found to be less than 3% for the major elements Al, Ca, Mg, K, Fe, Na and Mn.

To determine relationships between rock magnetic and geochemical parameters and to discern groups with similar characteristics, the multivariate fuzzy *c*-means cluster analysis technique [FCM; *Bezdek et al.*, 1984] was applied. This technique is based on a partitioning method in which the similarity of a case is calculated with respect to all clusters. This similarity is expressed by a membership value that varies between 0 (no similarity between sample and cluster) and 1 (identical) for each case. The memberships to the clusters for one sample add up to 1. An intermediate case is recognised by significant memberships of more than one cluster. The weight of each sample to a cluster centre is related to its membership value so that intermediate samples have less influence on the position of cluster centres.

FCM cluster models were run for 2 to 7 clusters. Because the most appropriate number of clusters is not known, the best solution can be determined from the statistical properties of the FCM model; the partition coefficient F′ should have the highest value and the classification entropy H′ should have the lowest value [*Bezdek et al.*, 1984]. Another method to determine the optimal number of clusters is by displaying the cluster assignments in a non-linear mapping (NLM) plot [*Sammon*, 1969]. In this 2-dimensional image of a multi-dimensional data cloud, clusters are possibly meaningful when the cases of the same clusters coherently group in the NLM plot. If this is not the case, the best solution should then be sought in a model with less clusters [see, e.g., *Dekkers et al.*, 1994].

4. RESULTS

4.1. Paleomagnetic Results

Overall, good paleomagnetic results were obtained for specimens from both the TW and WG sections (Figure 2). The NRM intensities range between 10 and 4500 µA/m, with some exceptions showing high NRM intensities up to 11000 µA/m, mostly associated with the red (silty) mudstones. Conversely, low NRM-intensity values occur in carbonate-rich lithologies. The Zijderveld diagrams of the measured specimens show that the NRM is composed of at least three components. A small, viscous and randomly oriented component, which is removed at temperatures of 100°C, likely represents a laboratory-induced magnetization related to storage. A low temperature component (LT) is removed at 210–240°C. The overall mean direction of this LT component corresponds within error to the geocentric axial dipole (GAD) inclination for the location (41.5°N; inclination of 60.5°), suggesting that LT represents a present-day overprint (Figure 2a).

At higher temperatures two components can be distinguished; an intermediate component (MT), which is completely removed between temperatures 520 and 600°C and is thought to reside in magnetite (or possibly maghemite); and a high temperature component (HT) which is removed between 615 and 690°C, indicative of hematite (Figures 2b, f, h, j, k and l). The HT component, which is found in the bulk of the samples (>95%), is interpreted as the characteristic remanent magnetization (ChRM) and usually its direction can be reliably determined. Some specimens consist of a 'hard' hematite component, which is only removed at temperatures exceeding 660°C (Figures 2c and k). Interpretation of the ChRM directions was sometimes hampered because of poor and noisy demagnetization behavior, mostly associated with low intensities (Figure 2d). Occasionally, the ChRM does not decay towards the origin (Figures 2e, g and l). In some specimens, differences are observed between the directions of the MT and HT components (Figures 2f, g and k) suggesting a slight delay in acquisition. However, this delay is not consistent between the two components, nor does it occur throughout the studied sections. Concerning the origin of the MT and HT components, there are certainly some—still poorly understood—mechanisms that cannot be sufficiently explained and, therefore, we cannot precisely estimate the seemingly inconsistent differences in timing between the two components. In fact, Krijgsman et al. [1997] already noted that in Spanish red beds two different (magnetite and hematite) components had conflicting acquisition times (yet not differing more than a few thousand years, as is the case in this study, since they are within a precession cycle). At the older polarity transition, the magnetite component reverses first (i.e. has a longer delay in acquisition), while at the younger polarity transition the reversal of the hematite component is earlier. This mechanism can possibly in part be explained by different origins of hematite depending on the changing environment within a climatic (precessional) cycle, e.g. detrital vs. pigmentary hematite. The fact that magnetite may have a longer lock-in

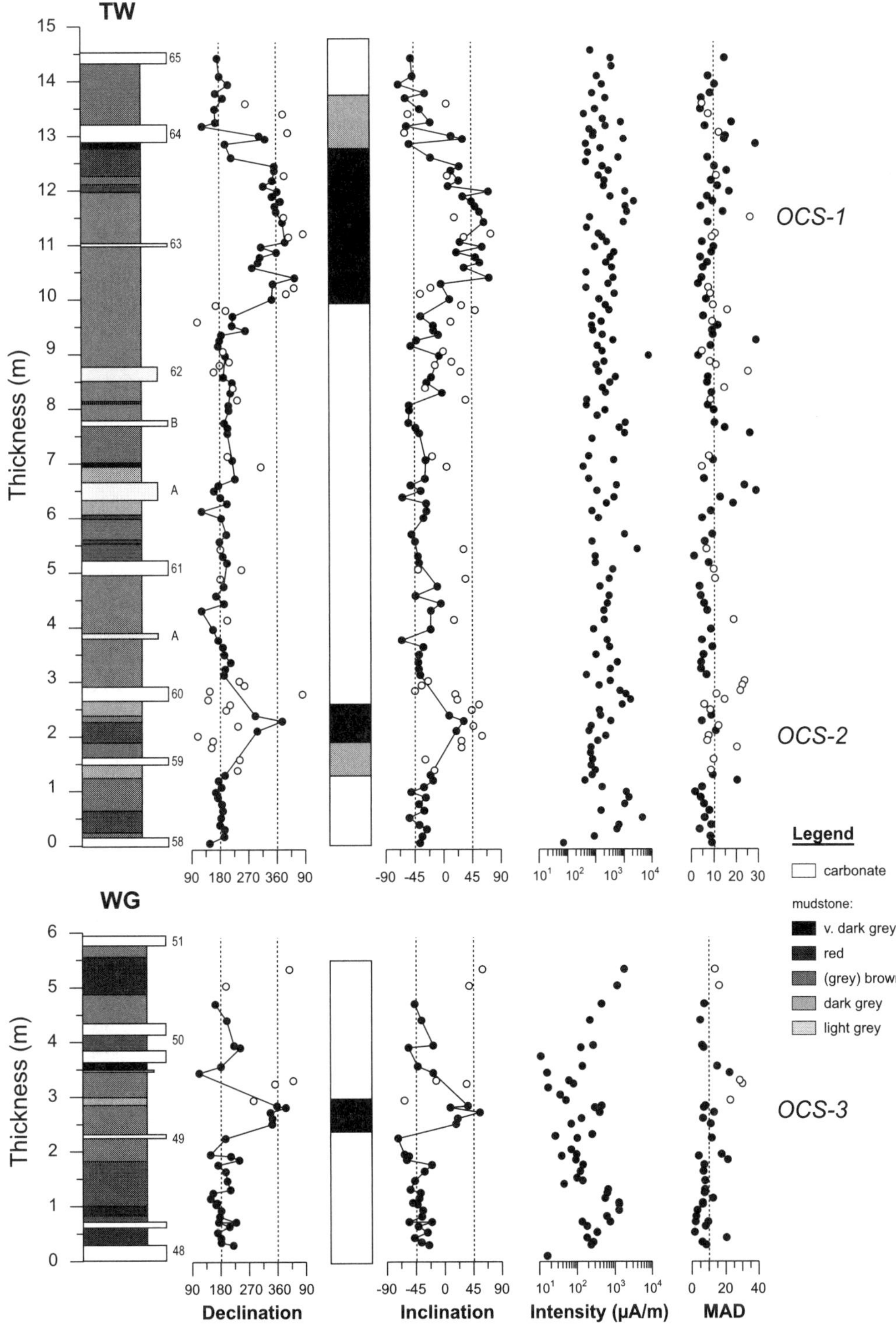

Figure 3. Lithology, magnetostratigraphy, NRM intensity and maximum angular deviation (MAD) for the TW and WG sections. The shadings in the lithology column indicate the colors of the lithology as shown in the legend; the numbers to the right of the column indicate the cycle number according to the cyclostratigraphy of the Orera Composite Section [OCS: *Abdul Aziz et al.,* 2003b]. In the declination, inclination and MAD diagrams, solid dots represent the ChRM directions while open circles show the unclear directions. In the polarity column, black (white) denotes normal (reversed) polarity zones, gray shaded intervals denote zones of unclear polarity.

Table 1. Coercivity values of samples from different lithographies around polarity interval OCS-2

Polarity & Lithography Column	Sample	Lithology	IRM Component Analysis Resuls[a]					
			2 components			3 components		
			SIRM	*B1/2*	*DP*	*SIRM*	*B1/2*	*DP*
	TW 97	d. grey mudstone	240	42	0.40			
			269	204	0.46			
	TW101	d. grey mudstone	95	42	0.36			
			231	204	0.34			
	TW105	d. grey mudstone	535	30	0.24	536	31	0.31
			400	257	0.39	270	162	0.39
						172	912	0.39
	TW108		127	96	0.38	41	45	0.31
			55	407	0.38	88	145	0.30
						51	363	0.42
	TW113	carbonate rich mudstone	131	132	0.47			
			9	1023	0.57			
	TW116	red brown mudstone	337	229	0.47	312	251	0.42
			10	1778	0.80	13	1698	1.06
						20	33	0.35
	TW119	red mudstone	520	33	0.33	520	33	0.32
			1430	812	0.51	1170	631	0.47
						210	1906	0.31
	TW124	Org.-rich grey brown mudstone with carbonate rooting				31	31	0.40
						100	427	0.56
						107	2570	0.22
	TW128	d. grey-brown mudstone				19.5	30	0.45
						138	159	0.32
						8	1000	0.4
	TW131	d. grey mudstone	227	110	0.48	23	26	0.45
			424	407	0.46	503	195	0.42
						147	1479	0.36
	TW134	red mudstone	5680	25	0.28			
			1710	794	0.41			
	TW137	carbonate				11.5	27	0.45
						92	174	0.45
						12.7	562	0.42

[a] Induced remanent magnetization (IRM) component analysis results show the number of resolved components, their saturation IRM (SIRM), their peak field at half of the SIRM (B½) and their dispersion of the cumulative log-normal distribution (DP). The results have been obtained using the Kruiver et al. [2001] method.

time may be related to a NRM acquisition process similar to that observed in the Chinese loess, which shows a considerable lock-in delay. Soil formation processes complicate the matter further and give rise to complex magnetization acquisition processes [*Spassov et al.*, 2003].

By plotting the ChRM directions and NRM intensity in stratigraphic order for the TW section, two intervals of opposite (normal) polarity are revealed (Figure 3): between 10 and 13 m (termed OCS-1) and between 1 and 3 m (termed OCS-2). In the WG section, one distinct normal polarity interval is distinguished between 2 and 4 m, which we designate as OCS-3 (Figure 3). Hence, we confirm the presence of the polarity intervals earlier recorded by Abdul Aziz et al. [2000].

4.2. Rock Magnetic Results

IRM acquisition curves were obtained for twelve samples (from representative lithologies) from polarity interval OCS-2, between 0 and 4.5 m. IRM curves from the same lithology show similar results, therefore we used the results with a well-defined (and minimum) number of magnetization components as a reference (Table 1). Combined with the results in Table 1, the IRM curves indicate that either two or three magnetic components can be distinguished, depending on lithology (Figure 4). In almost all samples, a low-coercivity component is recorded with B1/2 values in the range of 25 to

40 mT, which suggests the presence of magnetite. The second component can be divided into two groups: coercivity values ranging between 130 and 240 mT, or occasionally higher. In general, the lower coercivities are typical of gray mudstones and carbonates while the higher coercivities are characteristic of red mudstones [see also *Abdul Aziz et al.*, 2000]. We

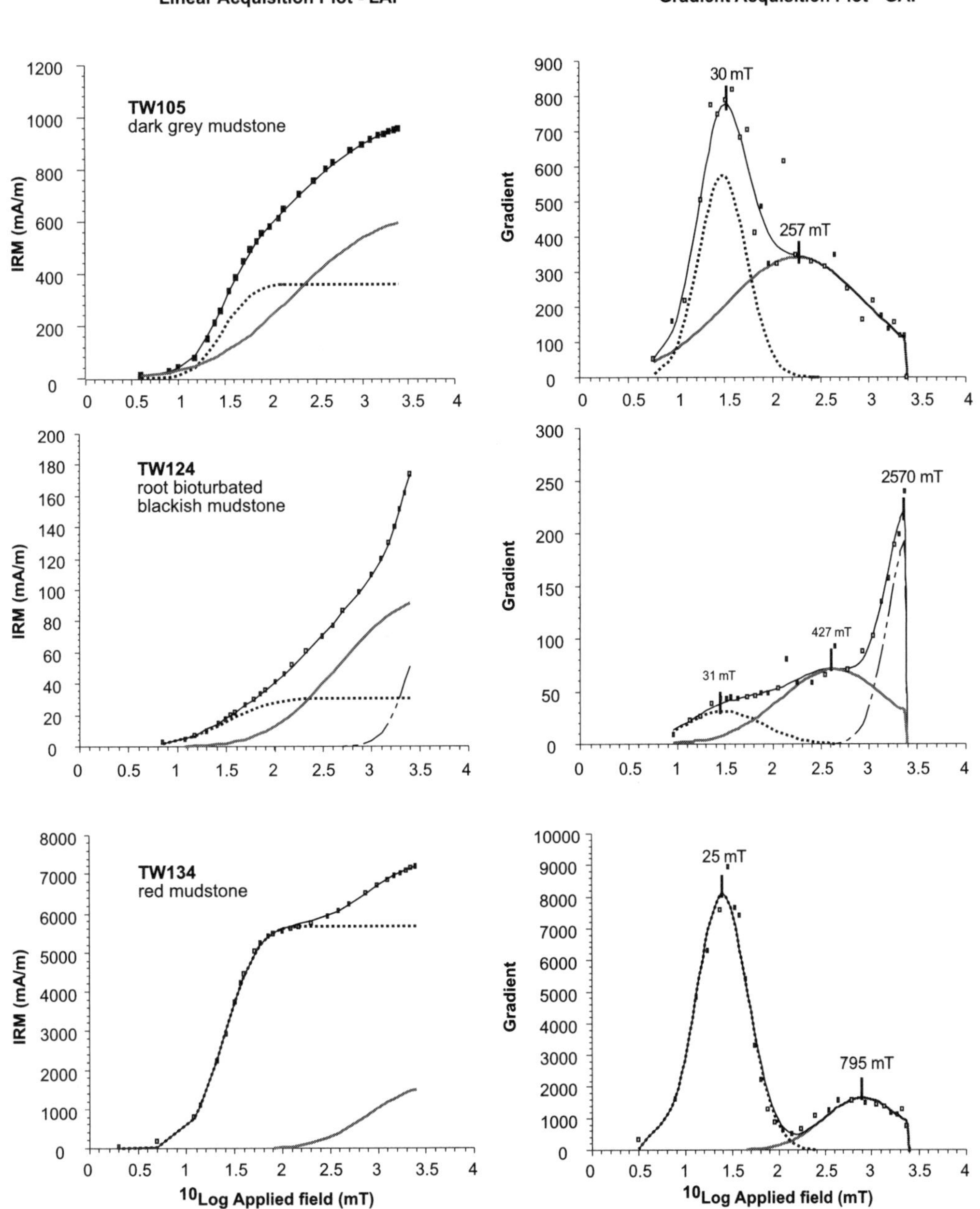

Figure 4. Examples of IRM component analysis for the linear and gradient acquisition plots [*Kruiver et al.*, 2001]. Data points are represented by squares; dotted, shaded and long-dashed lines represent the best fit component 1, 2 and 3, respectively. The solid line is the best fitting sum of components. Component 1 corresponds to magnetite, component 2 to (fine-grained) hematite and component 3 to goethite (see text for details).

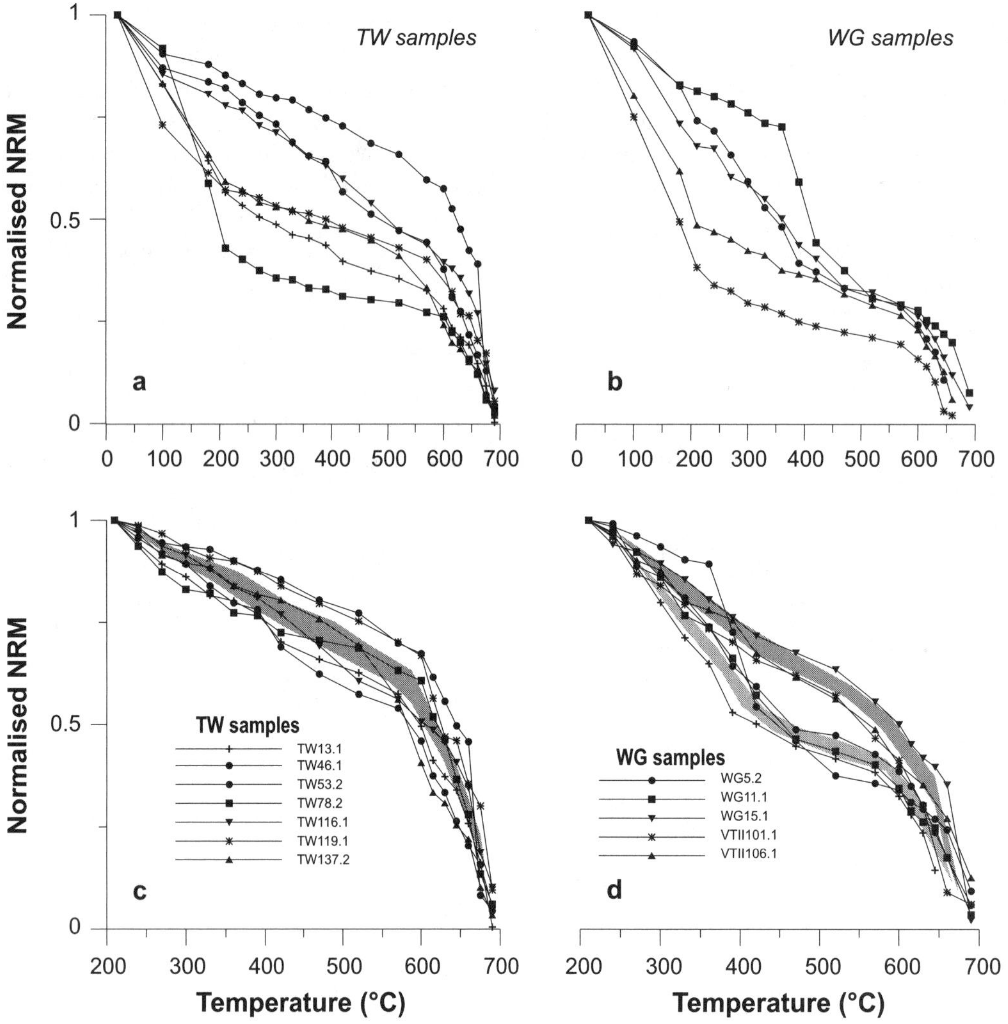

Figure 5. Thermal decay curves of selected samples from the TW and WG sections. The first temperature steps up to 210°C suggest large differences in the thermal decay for various samples (a and b). When excluding these first steps (c and d), a more consistent decay is observed whereby at least 50% of the decay occurs in the temperature interval 600–690°C. Samples from WG indicate that another group of samples with a decay in the temperature interval 400–500°C is also present.

assume that both ranges of values correspond to hematite, where the low coercivities are indicative of (very) fine-grained hematite [*Dekkers and Linssen*, 1989] and the high coercivities indicate less fine-grained hematite. In several samples there is a third component with very high coercivities (900–2500 mT), e.g., in black, root-bioturbated mudstones (Table 1, Figure 4). These coercivity values are typical of fine-grained goethite [*Dekkers*, 1989].

An indication of the magnetic mineral components can also be deduced from the thermal (TH) decay of the NRM. In the normalized TH-decay curves for the TW and WG specimens, a rapid decay is observed between room temperature and 210–240°C; the decrease in intensity varies from 20 to 60% and may be associated with goethite (Fig-

ure 5a, b). For a better comparison of the TH-decay curves between specimens, the thermal steps up to 210°C were excluded. It appears that the TH-decay curves display a similar trend characterized by a slow decay up to approximately 600°C followed by a rapid decrease of the NRM in the temperature range 600 to 690°C (Figure 5c, d). The magnetic mineral associated with the first decay up to 600°C is probably magnetite, while the sharp decay between 600 and 690°C and the unblocking at 690°C corresponds to hematite. The TH-decay curves for specimens from the WG section show more variations than those from TW: some WG specimens show a more rapid TH-decay at lower temperatures not seen in TW specimens (Figure 5d) and could indicate higher concentrations of goethite.

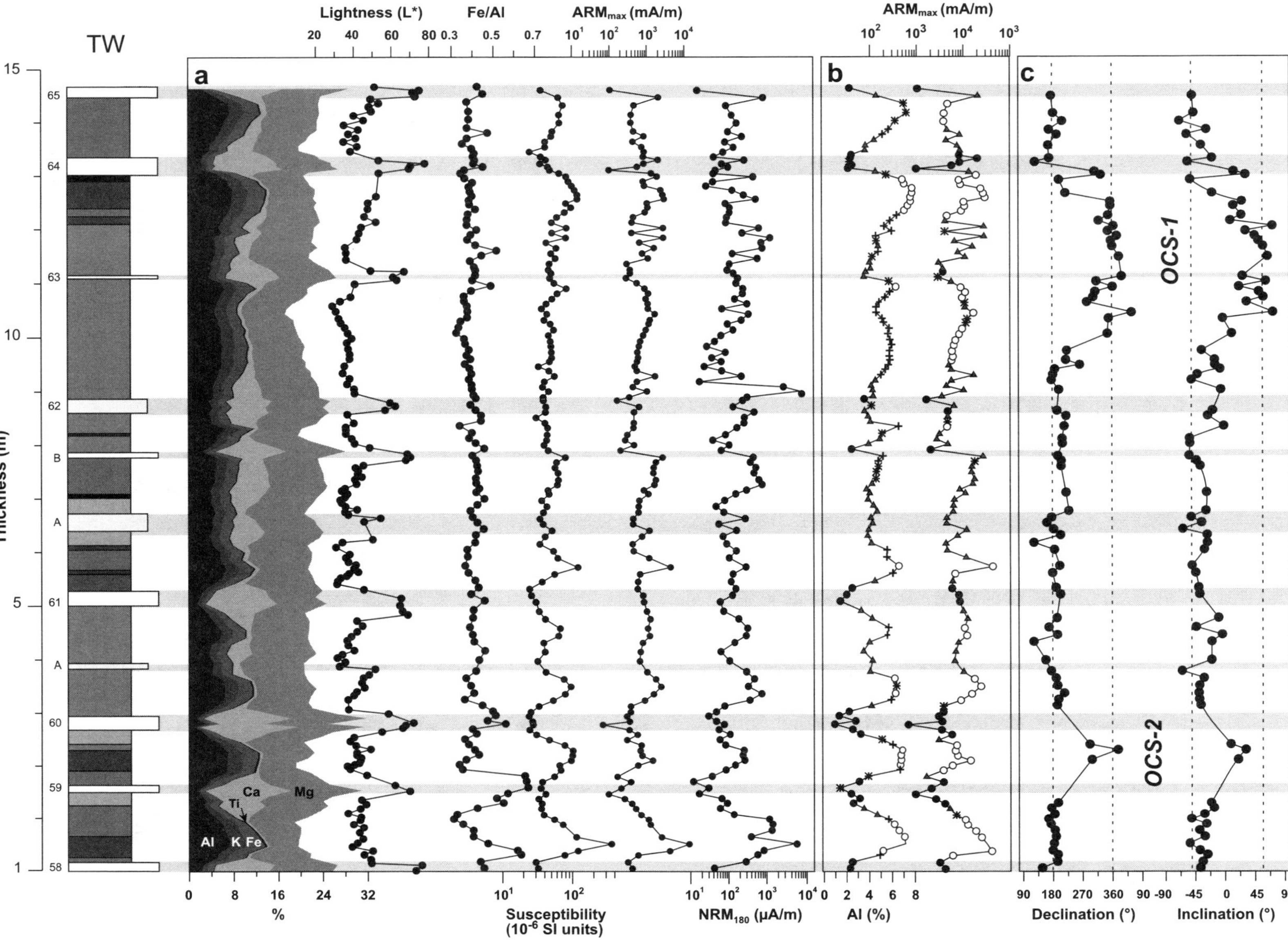

Figure 6. (a) Geochemical variables (aluminium, potassium, iron, titanium, calcium and magnesium) and rock magnetic variables (susceptibility, anhysteretic remanent (ARM_{max}) and natural remanent magnetization at 180°C (NRM_{180})) versus depth; (b) Example of fuzzy cluster assignments for aluminium and ARM_{max} for the 4 and 3 cluster model, respectively. Cluster 1 is represented by black dots, cluster 2 by triangles, cluster 3 by open circles and cluster 4 by crosses (see also Figure 8). Samples intermediate between clusters are denoted with asterisks; (c) Declination and inclination results for the TW section. Note: The unreliable ChRM directions have been excluded.

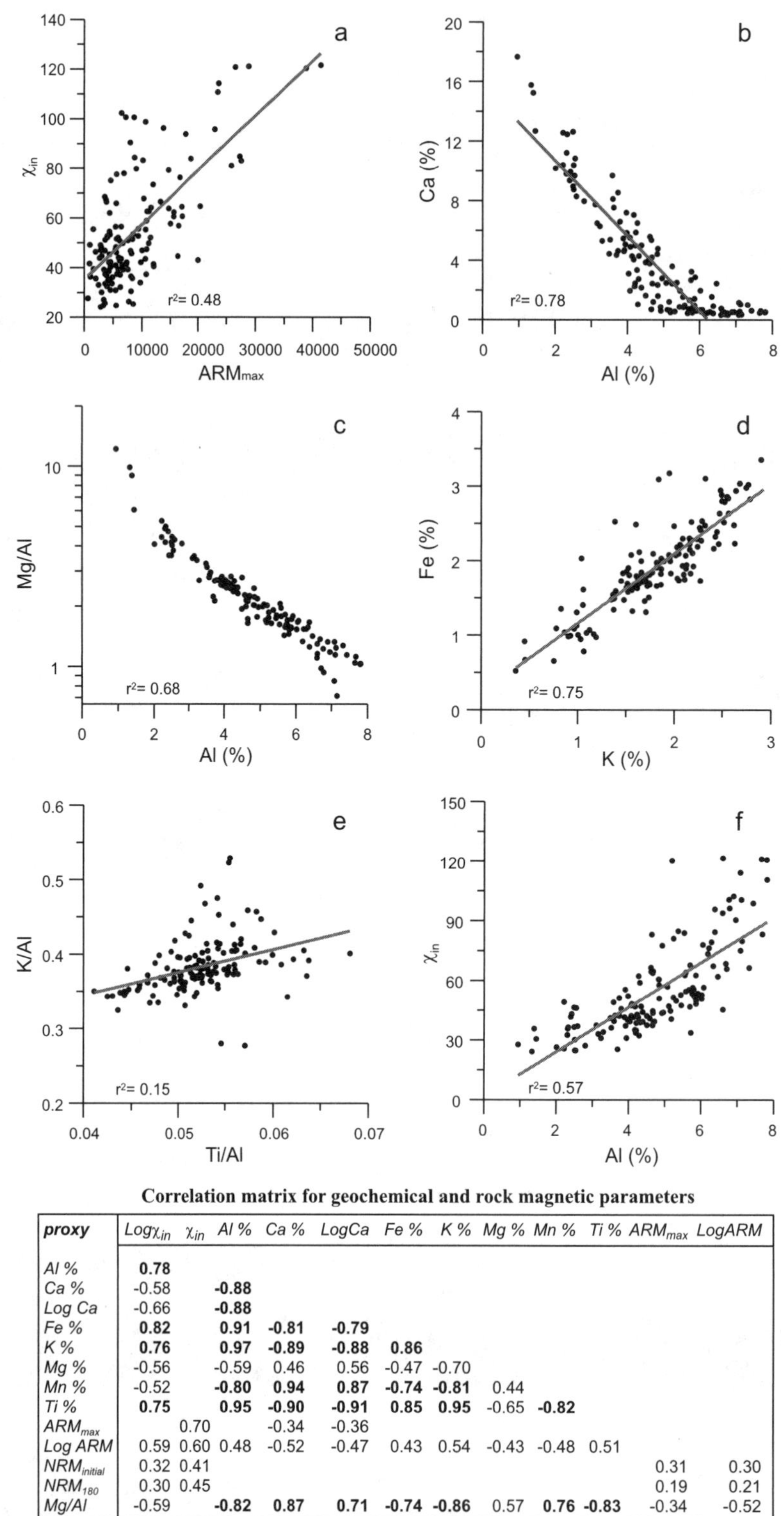

Correlation matrix for geochemical and rock magnetic parameters

proxy	$Log\chi_{in}$	χ_{in}	Al %	Ca %	LogCa	Fe %	K %	Mg %	Mn %	Ti %	ARM_{max}	LogARM
Al %	**0.78**											
Ca %	-0.58		**-0.88**									
Log Ca	-0.66		**-0.88**									
Fe %	**0.82**		**0.91**	-0.81	-0.79							
K %	**0.76**		**0.97**	-0.89	-0.88	0.86						
Mg %	-0.56	-0.59	0.46	0.56	-0.47	-0.70						
Mn %	-0.52		**-0.80**	**0.94**	**0.87**	**-0.74**	**-0.81**	0.44				
Ti %	**0.75**		**0.95**	**-0.90**	**-0.91**	0.85	**0.95**	-0.65	**-0.82**			
ARM_{max}		0.70		-0.34	-0.36							
Log ARM	0.59	0.60	0.48	-0.52	-0.47	0.43	0.54	-0.43	-0.48	0.51		
$NRM_{initial}$	0.32	0.41									0.31	0.30
NRM_{180}	0.30	0.45									0.19	0.21
Mg/Al	-0.59		**-0.82**	**0.87**	**0.71**	**-0.74**	**-0.86**	0.57	**0.76**	**-0.83**	-0.34	-0.52

Figure 7. Scatter diagrams of: (a) susceptibility (χ_{in}) versus anhysteretic remanent magnetization imparted at 250 mT (ARM_{max}); (b) Ca versus Al ; (c) Mg/Al versus Al; (d) Fe versus K; (e) the weathering index proxy K/Al versus Ti/Al; and (f) χ_{in} versus Al. The inserted table shows the correlation coefficients for the geochemical and rock magnetic parameters analyzed in this study.

The initial susceptibility (χ_{in}) strongly depends on the concentration of paramagnetic (clay) minerals, on the concentration of ferrimagnetic minerals and on grain size. Susceptibility data from the TW section show that variations in χ_{in} are influenced by lithology (Figure 6a). The intervals 0–8.5 and 11–14.5 m consist of obvious lithological alternations of carbonates and mudstones, with low χ_{in} values for carbonates and high values for mudstones; the highest χ_{in} values are recorded in red and brown mudstones. The interval 8.5–11 m lacks well-developed carbonate beds which results in more uniform intermediate χ_{in} values (Figure 6a).

ARM acquisition data show that most specimens are not saturated at 250 mT, which is consistent with the high coercivities indicated by the IRM results. ARM imparted up to 250 mT (ARM_{max}) (Figure 6a) show that he analyzed specimens can be divided into three groups depending on their rate of ARM acquisition: a high, an intermediate and a low acquisition rate of the ARM. Fast acquisition is typical for red and brown mudstones while lower acquisition rates are characteristic for gray mudstones and carbonates, with the slowest rates observed for carbonates.

Contrary to χ_{in}, the ARM is usually enhanced in fine-grained magnetic minerals [*Banerjee et al.*, 1981], but it more strongly depends on ferrimagnetic mineral concentration [*Sugiura*, 1979; *King et al.*, 1983; and *Tauxe*, 1991]. In this study, ARM_{max} and χ_{in} do not always show a positive correlation, which suggests that χ_{in} is also determined by changes in (paramagnetic) clay mineral concentration (Figure 7a).

Finally, variations in the NRM intensity at 180°C are related to lithology; low intensities coincide with carbonate beds, while high values correspond to red mudstones (Figure 6a). The peak in NRM_{180} at 9 m seems to be related to formation of magnetic minerals as it occurs just above a conspicuous thin (~1.5 cm) red layer.

Summarising, all rocks in TW contain both magnetite and hematite, with very fine-grained hematite also occurring in carbonates and gray mudstones. Goethite is also present in most of the different types of lithology. Furthermore, the concentration of magnetic minerals is enhanced in the red mudstones and is reduced in the carbonates. Closer investigation reveals that in the lower, well-developed cyclic part (cycles 58–61A) and in the top part (cycles 63–65) of the TW section, the magnetic parameters NRM, χ_{in}, ARM_{max} and IRM appear to be mainly influenced by lithologic changes. In the middle part, however, carbonate beds are poorly developed resulting in a more homogeneous lithology in which the magnetic parameters have more constant values.

4.3. Geochemical Results

X-ray diffraction analysis on samples from the OCS indicated that the mudstones are dominated by clay minerals, which consist mainly of illite, illite-smectite mixed layers and di- and tri-octahedral smectite, with minor amounts of kaolinite [*Abdul Aziz et al.*, 2003a]. The carbonate content in the mudstones is usually negligible, whereas the percentage of silt- to sand-sized quartz and feldspar grains varies from 3 to 35%. The carbonate beds are mainly dolomitic, with carbonate contents varying between 65 and 93%. Bulk ICP analysis on samples from the TW section revealed high concentrations of Al, Ca and Mg. High Al percentages are characteristic of the mudstones and are related to clay mineral content (Figure 6a). In contrast, high Ca and Mg contents are related to carbonate minerals. Fe, K and Ti follow the same trend as Al, while Mn follows Ca (Figure 6a). The lithologic variations are also clearly illustrated by the lightness value L*, where high (i.e., lighter) values correspond to the carbonate beds (Figure 6a).

Ca (%) and Al (%) inversely correlate, especially for Al percentages less than 7% (Figure 7b and inserted table). At higher Al values, Ca percentages are lower than 1% and no clear correlation with Al is observed. Variations in Al (%) are likely related to variations in clay mineral content. Mn concentrations are low in the mudstones but may be 10 to 30 times higher in the carbonates (~1200 ppm). The distribution of Mn is strongly related to carbonate content and is probably determined by manganese absorption and overgrowth on the calcite/dolomite surfaces [*Middelburg et al.*, 1987; *Van Hoof et al.*, 1993]. Mg has relatively high concentrations (500–1100 ppm) in mudstones and in carbonates (800–1300 ppm). A slight anti-correlation with Al (%) is observed, but after correction for the Mg contribution to the clay minerals, by taking the ratio Mg/Al, this anti-correlation becomes more obvious (Figure 7c). This in turn also implies a positive correlation between Mg/Al and Ca (%). Finally, Fe, K and Ti show positive correlations with each other and with Al (%), which indicates that the concentrations of these elements are strongly associated with clay minerals (Figure 7d).

5. DISCUSSION

5.1. Fuzzy Clusters

Interpretation of geochemical elements or ratios in marine environments has been proven to be a sensitive tool to recognize diagenesis. In addition, a clear correlation with magnetic properties has been demonstrated using fuzzy *c*-means cluster analysis techniques [*Dekkers et al.*, 1994; *van Santvoort et al.*, 1997]. Similar studies have been performed in continental settings. In the palustrine-alluvial environment of the Librilla section in southeastern Spain, clusters enriched in iron—mainly related to the high coercivity IRM component of goethite—indicated a clear relationship with samples with overprinted directions [*Kruiver et al.,* 2002]. However, not all samples within this iron-enriched cluster were overprinted. In addition, magnetite overprints did not show any relation with a particular cluster [see for details *Kruiver et al.,* 2002].

Generally, chemical weathering in continental settings is often indicated by the K/Al ratio, which in turn is also an indicator for detrital input, along with the Ti/Al ratio. Therefore, good correlation between these two ratios would reflect changes in clastic input. In this study, however, the correlation between both ratios is poor (r^2=0.38; Figure 7e) and we therefore refrain from using this proxy. Furthermore, considering the closed nature of the Calatayud basin, the parent rock (i.e., Paleozoic slates) and the clay mineralogy (illite, mixed-layer illite-smectite and Mg- smectite), it is clear that complex transformation processes played a role in the clay mineral composition of the detritus in the TW section. This can be illustrated by the good correlation between Al, Fe, K, and Ti (see table in Figure 7). Similarly, Mg is enriched in the carbonate as well as in the clay minerals and therefore the Mg/Al ratio is probably not useful as an indicator for chemical weathering. Nevertheless, relationships between the geochemical elements and the rock magnetic parameters show that the elements dominant in the mudstones (Al, Fe, K and Ti) correlate well with χ_{in} (Figure 7f) while the elements dominant in the carbonates (Ca and Mg) correlate less well (see table to Figure 7). Finally, raised Fe/Al ratios in marine sediments are taken to indicate diagenetic enrichment of Fe. In this study, the enrichment or depletion of Fe is determined relative to the "average" detrital input, as contained in the Fe/Al ratio. In the TW section, Fe-enhancement occurs just above and within carbonate beds of cycles 58, 59 and 60 (Figure 6a). This enhancement seems to be related to high concentrations of carbonate minerals and relatively low ARM_{max} values.

To distinguish subtle patterns or compositional overlaps and to discriminate possible groupings between geochemical parameters and magnetic behaviour we thus used multivariate classification techniques. This statistical technique has been successfully applied in geochemistry [e.g., *Vriend et al.,* 1988] and has also been used to link rock-magnetic behavior to geochemical environment and to climate change in marine settings [*Dekkers et al.,* 1994; *Kruiver et al.,* 2002]. The magnetic and geochemical variables we used in the multivariate fuzzy *c*-means (FCM) and non-linear mapping (NLM) cluster analyses were χ_{in}, ARM_{max} and NRM_{180} and Al (%), Ca (%), Fe (%), K (%), Mg (%), Mn (%), Ti (%), and $CaCO_3$ and $MgCO_3$ both from the carbonate fraction [see *Abdul Aziz et al.,* 2003b], respectively. The color records, represented by the lightness value L* and the redness value a*, were also used because they clearly show the lithologic changes caused by mudstone-carbonate alternations and by red-coloring. The fuzzy cluster analysis was tested by repeating the analysis using different

(random) starting conditions and by running all and selected variables. All the results proved to be stable and consistent. For an illustrative FCM and NLM analysis we have chosen the geochemical variables Al, Ca, Fe, K, Mg and Ti and the rock magnetic variables χ_{in} and ARM_{max}. Prior to the cluster analyses, log-normal distributed variables were logarithmically transformed and then standardized to ascertain equal weight for all variables.

The results for the 3, 4 and 5 cluster models are shown in NLM plots (Figure 8). The cluster centres have been sorted by increasing Al concentrations. In all three cluster models, two dominant groups can be distinguished. One group is characterized by high carbonate mineral contents (Ca and Mg) and the other group by high clay mineral related contents (Al, K, Fe, and Ti). Between these two major clusters, intermediate clusters occur with increasing concentrations of carbonate and decreasing concentrations of clay mineral related elements. In all the different runs of the fuzzy cluster analyses, the geochemical parameters contribute significantly to the cluster partitioning. Furthermore, the magnetic parameters ARM_{max} and χ_{in} also increase with increasing concentrations of clay mineral elements. An exception is the 5 cluster model where a cluster of high ARM_{max} and relatively high χ_{in} values is found with intermediate percentages for both Ca and Al, which may indicate enhanced concentrations of magnetic minerals (Figure 8). The influence of lithological variations on cluster assignments is illustrated in Figure 6b. The carbonate and clastic and rock magnetic mineralogy clearly follows the cyclicity, however, this is not the case in the interval 61A to 63, where the lithological cyclicity is poorly-developed. In the 6 and 7 cluster models (not shown), the ARM_{max} parameter mainly contributes to the cluster partitioning in this interval. Yet, no relation can be established between ARM_{max} and the occurrence of reversals. Finally, Fe-enhancement around cycle 59 and 60, which is strongly confined to carbonate rich lithologies, show a relation with low ARM_{max} and low NRM values. Most samples from these two levels show poor demagnetization behaviour (Figure 6). In contrast, Fe-enrichment just above the carbonate of cycle 58 is related to intermediate ARM_{max} and NRM values and to intermediate carbonate mineral concentrations. The samples from this interval show good and reliable demagnetization behaviour. Hence, we do not find any link between Fe-enrichment and a particular cluster.

Because lithologic changes strongly influence both the geochemical and rock magnetic proxy records, there is a clear correlation between both types of records (Figure 6c). We see no clear correlations, however, between the records and the occurrence of normal polarity samples that determine OCS-1 and -2. Similar clusters, i.e., similar lithologic and rock magnetic properties, show both normal and reversed directions. The normal polarities of OCS-1 and -2 are not related to a particular clus-

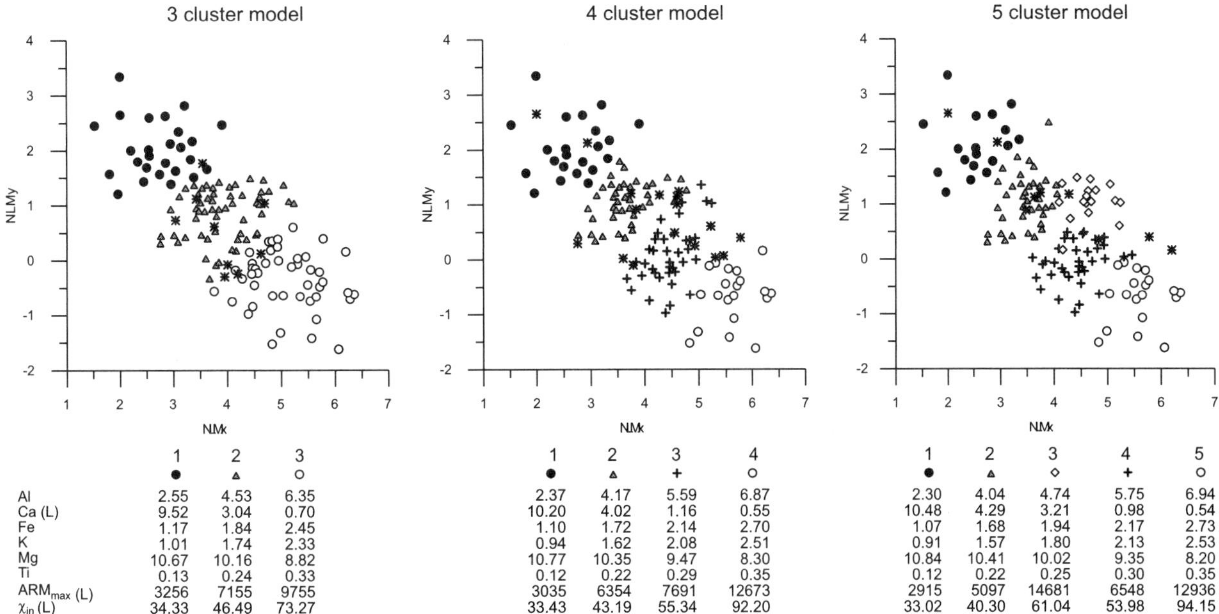

3 cluster model				4 cluster model					5 cluster model				
	1	2	3		1	2	3	4		1	2	3	4	5
	●	▲	○		●	▲	+	○		●	▲	◇	+	○
Al	2.55	4.53	6.35		2.37	4.17	5.59	6.87		2.30	4.04	4.74	5.75	6.94
Ca (L)	9.52	3.04	0.70		10.20	4.02	1.16	0.55		10.48	4.29	3.21	0.98	0.54
Fe	1.17	1.84	2.45		1.10	1.72	2.14	2.70		1.07	1.68	1.94	2.17	2.73
K	1.01	1.74	2.33		0.94	1.62	2.08	2.51		0.91	1.57	1.80	2.13	2.53
Mg	10.67	10.16	8.82		10.77	10.35	9.47	8.30		10.84	10.41	10.02	9.35	8.20
Ti	0.13	0.24	0.33		0.12	0.22	0.29	0.35		0.12	0.22	0.25	0.30	0.35
ARM_{max} (L)	3256	7155	9755		3035	6354	7691	12673		2915	5097	14681	6548	12936
χ_{in} (L)	34.33	46.49	73.27		33.43	43.19	55.34	92.20		33.02	40.30	61.04	53.98	94.15

Figure 8. Non-linear mapping (NLM) plots and cluster centers for the 3, 4 and 5 cluster model. The x and y axes have no physical meaning. The elements are in%, while susceptibility (χ_{in}) and anhysteretic remanent magnetization (ARM_{max}) are in 10^{-6} SI units and mA/m, respectively. Log-transformed parameters are indicated by L. Asterisks indicate samples that are intermediate between clusters.

ter. This is especially evident for the normal directions of OCS-1, which belong to several different clusters (Figure 6b). The short normal polarity interval OCS-3 in the WG section is recorded in dark gray mudstones similar to the mudstones in the TW section. The range of rock types, and their chemical and magnetic mineral composition, records similar ChRM directions. We therefore see no clear evidence for diagenetic changes. However, we realize that small diagenetic changes leading to remagnetization of the primary magnetic component may easily remain undetected in whole rock geochemistry. Hence, the observations described above do not allow us to conclude that no diagenesis has taken place. Perhaps the best argument for true normal polarity directions in OCS-1, -2 and -3 is to confirm these polarity intervals in other independent records, which will be discussed in the next section.

5.2. Evidence For New Subchrons

Since tiny wiggles, or cryptochrons, indicate geomagnetic field behavior of uncertain origin in marine magnetic anomaly profiles, their origin and nature require confirmation from (detailed) magnetostratigraphic studies. The existence of (reversal) excursions, in particular in the Brunhes and Matuyama Chrons, is now widely recognised and many excursions have been confirmed in globally distributed records [*Langereis et al.,* 1997; *Singer et al.,* 1999; *Channell et al.,*

2002]. Duration may be used to distinguish between (reversal) excursions and subchrons, as one could argue that subchrons must have a minimum duration of >6 kyr (i.e. minimally twice the inner core diffusion time of 3–5 kyr). We follow Gubbins [1999] in that, in addition to a pair of reversals, a subchron must have a short duration of the altered polarity state, which is sufficiently long to establish a time-averaged stable field. In practice, this requires a subchron to have a duration >10 kyr. Whether short duration subchrons can be detected in marine magnetic anomaly profiles presents a practical limitation, but it is not fundamental to the question of distinguishing between reversal excursions and subchrons.

Another sound criterion is that a subchron will be characterized by low intensities during the reversals that bound the subchrons, but that the intensities will recover to higher values within the polarity interval, while an excursion will only have a single interval of low intensity [cf. *Roberts and Lewin-Harris,* 2000]. Reversal excursions during the Brunhes and late Matuyama Chrons invariably are associated with single paleointensity minima [*Guyodo and Valet,* 1999; *Guyodo et al.,* 1999]. However, to establish a paleointensity minimum and possible recovery within the subchron requires 'suitable' sediments which is not the case in the present study, considering the variable lithology and the uncertainty regarding normalization of the remanence in (hematite bearing) continental sediments.

Because the short normal polarity intervals in our studied sections most likely represent geomagnetic field behavior, the next step is to determine whether the short normal polarity intervals OCS-1, OCS-2 and OCS-3 represent subchrons or reversal excursions. Therefore, we have investigated the duration of the three polarity intervals and whether they consist of a pair of geomagnetic reversals. If so, then these cryptochrons may be elevated to the status of subchrons [see Appendix on Nomenclature of *Cande and Kent*, 1992a].

The duration of OCS-1, 2 and 3 was determined from the ages of the astronomically calibrated mudstone-carbonate cycles in both the TW and WG sections, whereby the midpoints of carbonate beds were correlated to precession minima (Figure 9) [*Abdul Aziz et al.*, 2003b]. To resolve whether each of these short polarity events consist of a pair of full reversals, the latitudinal variation of the VGP paths and Fisher parameters were compiled from the ChRM directions (Figure 10). No correction was applied for inclination errors.

The VGP paths calculated for the short normal polarity interval OCS-3 in the WG section cluster at southern high latitudes (Figure 10). There are no intermediate VGP positions during the polarity transition and the VGPs abruptly jump to northern high latitudes and stay there for a short period of time, estimated to be longer than 5 kyr. Finally, the VGPs move back to the southern high latitudes. These results and the equal area plot of ChRM directions show that the short OCS-3 interval consists of a pair of reversals, with a short time of fully opposite (normal) polarity. The total duration of OCS-3 is approximately half a precession cycle, or 9 kyr (Table 2). The VGPs indicate opposite polarity direction and, more importantly, this short normal polarity event has been recently confirmed in other independent records, namely in DSDP Site 608 [29 kyr; *Krijgsman and Kent*, this volume] and in the revised calibrated ODP Site 1092 [6 kyr; *Krijgsman and Kent*, this volume; *Evans and Channell*, 2003]. Based on these confirmations, this short polarity interval should be considered as a subchron and, according to the nomenclature of CK92, it acquires the notation of subchron C5.3r-1n.

Figure 9. Astronomical calibration and dating of cycles and reversals in the TW and WG sections [see for details of tuning *Abdul Aziz et al.*, 2003b]. In the polarity columns for both sections, black (white) denotes normal (reversed) polarity; the gray shading represents uncertain polarity. The astronomical precession and eccentricity curves are from Laskar et al. [1993]. The astronomical polarity timescale (APTS) of the Orera Composite Section (OCS) is compared to the geomagnetic polarity timescale (GPTS) of Cande and Kent [1995; CK95]. The *relative* position of short normal polarity intervals OCS-2 and OCS-3 in the GPTS is indicated with gray shaded blocks (see also Table 2). Note that cryptochron C5r.2r-1*n* was identified by Schneider [1995] at ODP Site 845. Its age in the GPTS has been adjusted according to CK95.

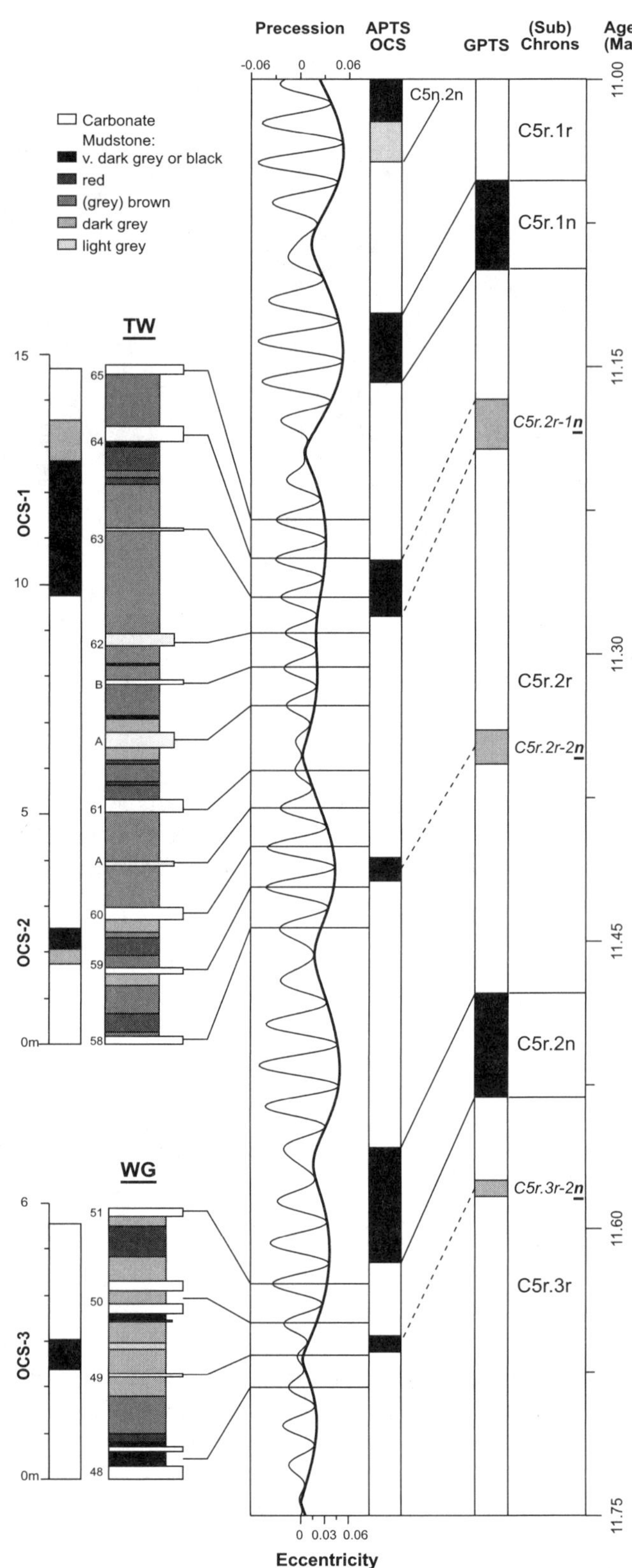

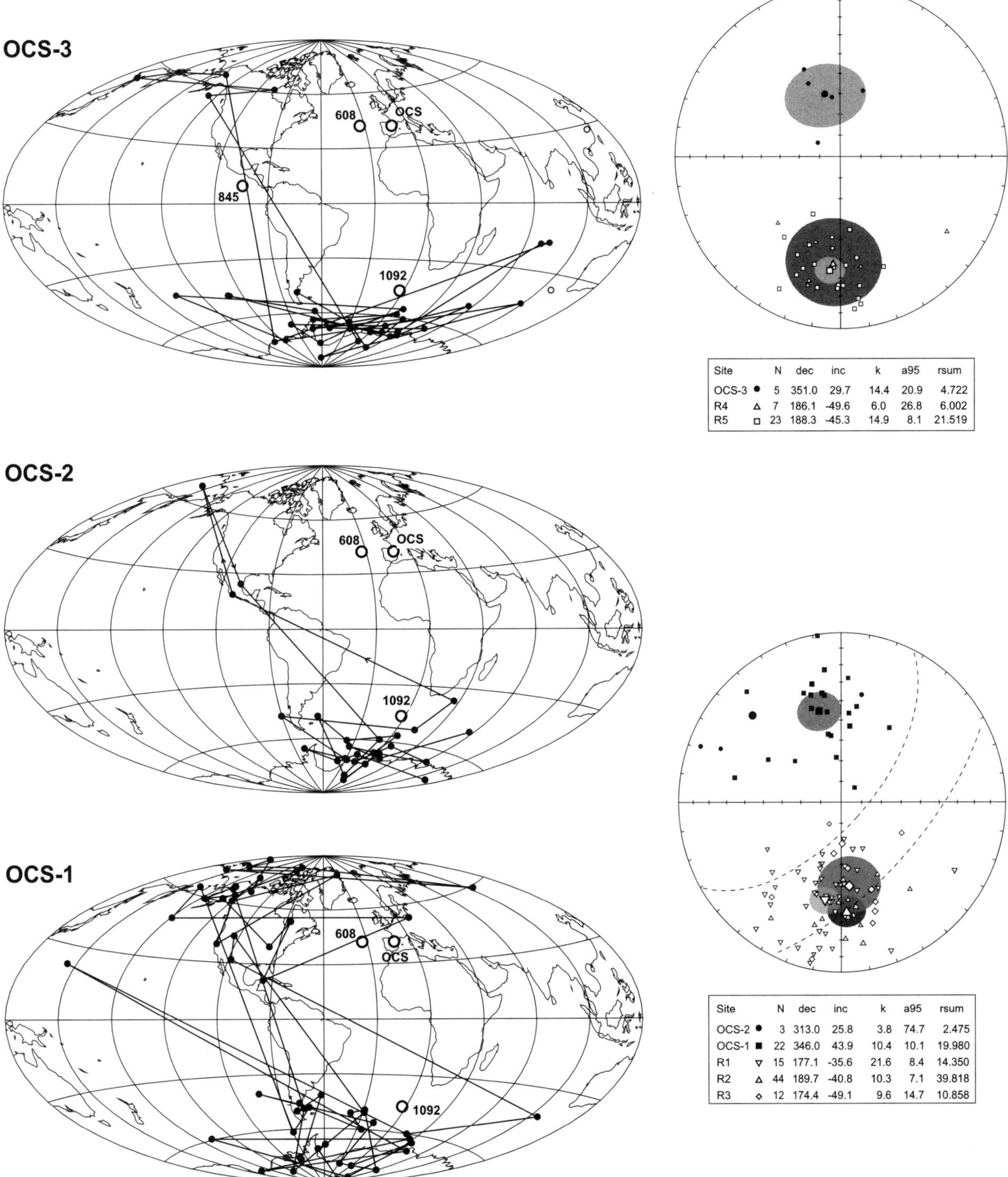

Site		N	dec	inc	k	a95	rsum
OCS-3	●	5	351.0	29.7	14.4	20.9	4.722
R4	△	7	186.1	-49.6	6.0	26.8	6.002
R5	□	23	188.3	-45.3	14.9	8.1	21.519

Site		N	dec	inc	k	a95	rsum
OCS-2	●	3	313.0	25.8	3.8	74.7	2.475
OCS-1	■	22	346.0	43.9	10.4	10.1	19.980
R1	▽	15	177.1	-35.6	21.6	8.4	14.350
R2	△	44	189.7	-40.8	10.3	7.1	39.818
R3	◇	12	174.4	-49.1	9.6	14.7	10.858

Figure 10. Virtual geomagnetic poles (VGPs) and equal area plots with Fisher parameters determined from the characteristic remanent magnetization (ChRM) components of polarity transition intervals in the TW and WG sections. The locations of the studied sections and of the sites discussed in the text are also indicated with large open circles. In the equal area plots, shaded circles show the 95% confidence intervals. R1–R5 indicate the reversed polarity intervals in the TW and WG sections, with R1 the youngest and R5 the oldest reversed interval.

Table 2. Astronomical ages and duration of short polarity intervals OCS-1, 2 and 3

Polarity Interval	Subchron	Stratigraphic level (m)		Age APTS[a] (Ma)	Duration (kyr)	Age GPTS (Ma)
OCS-1	C5r.2r-1**n**	*Top:*	12.77 ± 0.28	11.254 ± 0.004	26 kyr	11.167
		Bottom:	9.78	11.280		11.193
OCS-2	C5r.2r-2 **n**	*Top:*	2.60	11.405	15 kyr	11.320
		Bottom:	1.85 ± 0.15	11.420 ± 0.002		11.335
OCS-3	C5r.3r-1 **n**	*Top:*	2.94	11.658	9 kyr	11.573
		Bottom:	2.33	11.667		11.582

[a] The astronomical ages and duration of OCS-1, 2 and 3 are based on the astronomical polarity time scale of the Orera composite section [*Abdul Aziz et al.*, 2003b]. Note that the ages in the GPTS column are *relative* ages.

The VGPs for the short polarity interval OCS-2 in the TW section shows a slight shift toward the southern Indian Ocean and then an abrupt shift to high northern latitudes (Figure 10). No full normal polarity VGPs are reached during any significant amount of time; the VGPs shift rapidly via central America back to high southern latitudes. Altogether, the VGP paths and the equal area plot results do not show a convincing pair of reversals for OCS-2, nor do they indicate a significant interval with opposite polarity. Yet, the duration of the interval is approximately 15 kyr (Table 2). The VGP behavior argues for a reversal excursion, whereas its relatively long duration supports an origin as subchron. Notably, the short normal polarity interval of OCS-2 has also been confirmed in DSDP Site 608 [*Krijgsman and Kent*, this volume] and in the revised calibrated ODP Site 1092 [*Evans and Channell*, 2003]. Therefore, we conclude that OCS-2 must be denoted as subchron C5r.2r-2n.

The normal polarity interval OCS-1 shows clear antipodal directions, with VGPs apparently confined to a longitudinal band over the Americas (Figure 10). This VGP confinement has been observed in many other records and has been attributed to core-mantle interactions [*Laj et al.*, 1991; *Gubbins and Coe*, 1993], or, alternatively, to sedimentary artefacts [*Rochette*, 1990; *Langereis et al.*, 1992]. Here, we focus on the clear pair of reversals between stratigraphic levels 7.5 and 14.5 m (Figure 10). The VGPs in the southern high to mid latitudes pass the equator to northern mid latitudes before remaining at high northern latitudes. The subsequent change back to reversed polarity is irregular, showing fluctuations between northern and southern latitudes before reaching the reversed polarity direction. The overall character of the VGP path and the equal area plot clearly shows that OCS-1 not only has a full pair of reversals, but it also remains in a fully normal polarity state for at least 26 kyr (Table 2). This implies that OCS-1 should be regarded as a true normal polarity zone, and thus as a subchron. Abdul Aziz et al. [2003b] identified OCS-1 as cryptochron C5r.2r-1, which was earlier recog-

nised by Schneider [1995] in sediments from ODP Site 845. Its presence was confirmed by the marine magnetic anomaly profile of the conjugate Nazca and Pacific plates [*Abdul Aziz et al.*, 2003b; *Cande and Labreque*, 1974]. Furthermore, the duration of C5r.2r-1n in our study (26 kyr) agrees with the estimated duration of 26 kyr by Schneider [1995], which was based on an estimate of the average sedimentation rate. Schneider [1995] also noted that that cryptochron C5r.2r-1 reflected a full normal polarity zone with distinct intensity recovery between the pair of reversals. He refrained from denoting this polarity interval as a normal polarity subchron because it was found in only one core at ODP Site 845. Recently, this polarity event has also been identified in DSDP Site 608 (26 kyr) as well as in ODP Site 1092 (38 kyr) [*Krijgsman and Kent*, this volume; *Evans and Channell*, 2003]. We, thus, confirm that OCS-1, i.e. cryptochron C5r.2r-1, should be elevated to the status of subchron. According to the proposed nomenclature of Cande and Kent [1992a], the notation cryptochron C5r.2r-1 changes into subchron C5r.2r-1n.

6. CONCLUSIONS

We have carried out a detailed re-sampling of three short normal polarity intervals in the continental Orera Composite Section from the Calatayud Basin in NE Spain [*Abdul Aziz et al.*, 2000]. Paleomagnetic, rock magnetic and geochemical analyses of the three short intervals, informally labelled as OCS-1, OCS-2 and OCS-3, indicate no relationship with diagenetic or other sedimentary artefacts and suggest that these features have a geomagnetic origin. The duration of these short normal polarity intervals and the nature of the VGPs, together with their confirmation in DSDP/ODP sites 845, 608 and 1092 (at a wide range of latitudes) indicate that all three intervals represent true subchrons. Following the nomenclature of Cande and Kent [1992a], they should now be denoted as full subchrons C5r.2r-1n (OCS-1), C5r.2r-2n (OCS-2) and C5r.3r-1n (OCS-3) rather than as cryptochrons.

Acknowledgments. We thank Pauline Kruiver for her help with sampling and for advice on the IRM fitting method, Mark Dekkers for introducing the fuzzy analysis techniques, Helen de Waard for performing the ICP measurements and three anonymous reviewers for their critical and constructive comments. This research was supported by the Earth and Life Sciences Foundation (ALW) with financial aid from the Netherlands Organization for Scientific Research (NWO) and by the Netherlands Research Centre for Integrated Solid Earth Science (ISES). The work was carried out under the program of the Vening Meinesz Research School of Geodynamics (VMSG).

REFERENCES

Abdul Aziz, H., E. Sanz-Rubio, J.P. Calvo, F. J. Hilgen, F .J. and W. Krijgsman, Paleoenvironmental reconstruction of a middle Miocene alluvial fan to cyclic shallow lacustrine depositional system in the Calatayud Basin (NE Spain), *Sedimentology,* 50 , 211–236, 2003a.

Abdul Aziz, H., F. J. Hilgen, D. S. Wilson, W. Krijgsman, and J. P. Calvo, An astronomical polarity time scale for the late middle Miocene based on cyclic continental sequences, *J. Geophys. Res.,* 108 (B3), 2159, doi: 10.1029/2002JB001818, 2003b.

Abdul Aziz, H., F. J. Hilgen, W. Krijgsman, E. Sanz-Rubio and J. P. Calvo, Astronomical forcing of sedimentary cycles in the middle to late Miocene continental Calatayud Basin (NE Spain), *Earth Planet. Sci. Lett.,* 177, 9–22, 2000.

Banerjee, S. K., J. W. King, and J. A. Marvin, A rapid method for magnetic granulometry with applications to environmental studies, *Geophys. Res. Lett.,* 8, 333–336, 1981.

Bezdek, C. J., R. Ehrlich and W. Full, FCM: the fuzzy c-means clustering algorithm, *Comp. Geosci.,* 10, 191–203, 1984.

Biswas, D. K., M. Hyodo, Y. Taniguchi, M. Kaneko, S. Katoh, H. Sato, Y. Kinugasa and K. Mizuno, Magnetostratigaphy of Plio-Pleistocene sediments in a 1700-m core from Osaka Bay, southwestern Japan and short geomagnetic events in the middle Matuyama and Brunhes Chrons, *Palaeogeogr. Palaeoclimatol. Palaeoecol.,* 148, 233–248, 1999.

Blakely, R. J., Geomagnetic reversals and crustal spreading rates during the Miocene, *J. Geophys. Res.* 79, 2979–2985, 1974.

Cande, S. C. and D. V. Kent, Revised calibration of the geomagnetic polarity time scale for the late Cretaceous and Cenozoic, *J. Geophys. Res.,* 100, 6093–6095, 1995.

Cande, S. C. and D. V. Kent, A new geomagnetic polarity time-scale for the Late Cretaceous and Cenozoic, *J. Geophys. Res.,* 97, 13,917–13,951, 1992a.

Cande, S. C. and D. V. Kent, Ultrahigh resolution marine magnetic anomaly profiles: a record of continuous paleointensity variations?, *J. Geophys. Res.,* 97, 15,075–15,083, 1992b.

Cande, S. C. and J. L. LaBrecque, Behaviour of the Earth's palaeomagnetic field from small scale marine magnetic anomalies, *Nature,* 247, 26–28, 1974.

Clement, B. M. and D. V. Kent, Short polarity intervals within the Matuyama: transitional field records from the hydraulic piston cored sediments from the North Atlantic, *Earth Planet. Sci. Lett.,* 81, 253–264, 1986/1987.

Channell, J. E. T. and M. E. Raymo, Paleomagnetic record at ODP Site 980 (Feni Drift, Rockall) for the past 1.2 Myrs, *Geochem. Geophys. Geosyst.,* 4 (4), 1033, doi:10.1029/2002GC000440, 2003.

Channell, J. E. T., A. Mazaud, P. Sullivan, S. Turner and M. E. Raymo, Geomagnetic excursions and paleointensities in the Matuyama Chron at Ocean Drilling Program Sites 983 and 984 (Iceland Basin), *J. Geophys. Res.,* 107 (B6), doi:10.1029/2001JB000491, 2002.

Dekkers, M. J., C. G. Langereis, S. P. Vriend, P. J. M. van Santvoort and G.J. de Lange, Fuzzy c-means cluster analysis of early diagentic effect on natural remanent magnetisation acquisition in a 1.1 Myr piston core from the Central Mediterranean, *Phys. Earth Planet. Inter.,* 85, 155–171, 1994.

Dekkers, M. J., Magnetic properties of natural goethite—I. Grain-size dependence of some low- and high-field related rock magnetic parameters measured at room temperature, *Geophys. J. Int.,* 97, 323–340, 1989.

Dekkers, M. J. and J. H. Linssen, Rock magnetic properties of fine-grained natural low-temperature haematite with reference to remanence acquisition mechanisms in red beds, *Geophys. J. Int.,* 99, 1–18. 1989.

Evans, H. F. and J. E. T. Channell, Upper Miocene magnetic stratigraphy at ODP Site 1092 (sub-Antarctic South Atlantic): recognition of 'cryptochrons' in C5n.2n, *Geophys. J. Int.,* 153, 483–496, 2003.

Gee, J., D. A. Schneider and D. V. Kent, Marine magnetic anomalies as recorders of geomagnetic intensity variations, *Earth Planet. Sci. Lett.,* 144, 327–335, 1996.

Gubbins, D. and R. S. Coe, Longitudinally confined geomagnetic reversal paths from non-dipolar transition fields, *Nature,* 362, 51–53, 1993.

Gubbins, D., 1999. The distinction between geomagnetic excursions and reversals, *Geophys. J. Int.,* 137, F1–F3.

Guyodo, Y. and J. P. Valet, Global changes in geomagnetic intensity during the past 800 thousand years. *Nature,* 399, 249–252, 1999.

Guyodo, Y., C. Richter and J.P. Valet, Paleointensity record from Pleistocene sediments (1.4–0 Ma) off the California Margin, *J. Geophys. Res.,* 104, 22,953–22,964, 1999.

Hartl, P., L. Tauxe, and C. Constable, Early Oligocene geomagnetic field behavior from DSDP Site 522, *J. Geophys. Res.,* 98, 19,649–19,666, 1993.

Heslop, D., M. J. Dekkers, P. P. Kruiver and I. H. M. Van Oorschot, Analysis of isothermal remanent magnetisation acquisition curves using the expectation-maximisation algorithm, *Geophys. J. Int.,* 148, 58–64, 2002.

Hsü, V., D. L. Merrill and H. Shibuya, Paleomagnetic transition records of the Cobb Mountain event from sediments of the Celebes and Sulu Seas, *Geophys. Res. Lett.,* 17, 2069–2072, 1990.

King, J. W., S. K. Banerjee, J. A. Marvin and S. P. Lund, Use of small amplitude paleomagnetic fluctuations for correlation and dating of continental climate changes, *Palaeogeogr. Palaeoclimatol. Palaeoecol.,* 42, 167–183, 1983.

Krijgsman, W., W. Delahaije, and C. G. Langereis, Cyclicity and NRM acquisition in the Armantes section (Miocene, Spain): Potential for an astronomical polarity time scale for the continental record, *Geophys. Res. Lett.*, 24, 1027–1030, 1997.

Krijgsman, W. and D. V. Kent, Non-uniform occurrence of small-term fluctuations in the geomagnetic field? New results from Middle to Late Miocene sediments of the North Atlantic (DSDP Site 608), pp 161–174, this volume.

Kruiver, P. P., W. Krijgsman, C. G. Langereis and M. J. Dekkers, Cyclostratigraphy and rock-magnetic investigation of the NRM signal in late Miocene palustrine-alluvial deposits of the Librilla section (SE Spain), *J. Geophys. Res.*, 107, doi: 10.1029/2001JB000945, 2002.

Kruiver, P. P., M. J. Dekkers and D. Heslop, Quantification of magnetic coercivity components by the analysis of acquisition curves of isothermal remanent magnetisation, *Earth Planet. Sci. Lett.*, 189, 269–276, 2001.

Kruiver, P. P., Y. S. Kok, M. J. Dekkers, C. G. Langereis and C. Laj, A pseudo-Thellier relative palaeointensity record, and rock magnetic and geochemical parameters in relation to climate during the last 276 kyr in the Azores region., *Geophys. J. Int.*, 136, 757–770, 1999.

Laj, C., A. Mazaud, R. Weeks, M. Fuller and E. Herrero-Bevera, Geomagnetic reversal paths, *Nature*, 351, 447, 1991.

Langereis, C. G., M. J. Dekkers, G. J. de Lange, M. Paterne, and P. van Santvoort, Magnetostratigraphy and astronomical calibration of the last 1.1 Myr from an eastern Mediterranean piston core and dating of short events in the Brunhes, *Geophys. J. Int.*, 129, 75–94, 1997.

Langereis, C. G., A. A. M. Van Hoof and P. Rochette, Longitudinal confinement of geomagnetic reversal paths as a possible sedimentary artefact, *Nature*, 358, 226–230, 1992.

Laskar, J., F. Joutel and F. Boudin, Orbital, precessional, and insolation quantities for the Earth from −20 Myr to +10 Myr, *Astron. Astrophys.*, 270, 522–533, 1993.

Lund, S. P., G. Acton, B. Clement, M. Hastedt, M. Okada and R. Williams, Geomagnetic field excursions occurred often during the last million years, *EOS, Trans. AGU*, 79, 178–179, 1998.

Merrill, R.T. and P.L. McFadden, Geomagnetic field stability: reversal events and excursions, *Earth Planet. Sci. Lett.*, 121, 57–69, 1994.

Quidelleur, X., J. Carlut, P.-Y. Gillot and V. Soler, Evolution of the geomagnetic field prior to the Matuyama-Brunhes transition: radiometric dating of a 820 ka excursion at La Palma, *Geophys. J. Int.*, 150, F1–F5, 2002.

Rea, D. K. and R. J. Blakely, Short-wavelength anomalies in a region of rapid seafloor spreading, *Nature 255*, 126–128, 1975.

Roberts, A. P. and J. C. Lewin-Harris, Marine magnetic anomalies: evidence that 'tiny wiggles' represent short-period geomagnetic polarity intervals, *Earth Planet. Sci. Lett.*, 183, 375–388, 2000.

Rochette, P., Rationale of geomagnetic reversals versus remanence recording processes in rocks: a critical review, *Earth Planet. Sci. Lett.*, 98, 33–39, 1990.

Rösler, W. and E. Appel, Fidelity and time resolution of the magnetostratigraphic record in Siwalik sediments: high resolution study of a complete polarity transition and evidence for cryptochrons in a Miocene fluviatile section, *Geophys. J. Int.*, 135, 861–875, 1998.

Sammon, J. W., A non-linear mapping for data structure analysis, *IEEE Trans. Comput.*, C18, 401–409, 1969.

Spassov, S., Heller, F., Evans, M. E., Yue, L. P. and von Dobeneck, T., A lock-in model for the complex Matuyama-Brunhes boundary record of the loess/palaeosol sequence at Lingtai (Central Chinese Loess Plateau), *Geophys. J. Int.*, 155, 350–366, 2003.

Sugiura, N., ARM, TRM, and magnetic interactions: concentration dependence, *Earth Planet. Sci. Lett.*, 42, 451–455, 1979.

Singer, B. S., K. A. Hoffman, A. Chauvin, R. S. Coe and M. S. Pringle, Dating transitionally magnetized lavas of the late Matuyama Chron: toward a new ^{40}Ar/^{39}Ar timescale of reversals and events, *J. Geophys. Res.*, 104, 679–693, 1999.

Tauxe, L., Sedimentary records of relative paleointensity of the geomagnetic field: theory and practice, *Rev. Geophys.*, 31, 319–354, 1993.

Van Santvoort, P. J. M., G. J. de Lange, C. G. Langereis and M. J. Dekkers, Geochemical and paleomagnetic evidence for the occurrence of "missing" sapropels in eastern Mediterranean sediments, *Paleoceanography*, 12, 773–786, 1997.

Verosub, K. L. and S. K. Banerjee, Geomagnetic excursions and their paleomagnetic record, *Rev. Geophys. Space Phys.*, 15, 145–155, 1977.

Vriend, S. P., P. F. M. van Gaans, J. Middelburg and A. de Nijs, The application of fuzzy c-means analysis and non-linear mapping to geochemical datasets: examples from Portugal, *Appl. Geochem.*, 3, 213–224, 1988.

H. Abdul Aziz and C. G. Langereis, Paleomagnetic Laboratory 'Fort Hoofddijk', Budapestlaan 17, 3584 CD, Utrecht, Netherlands. (haziz@geo.uu.nl; langer@geo.uu.nl)

Non-Uniform Occurrence of Short-Term Polarity Fluctuations in the Geomagnetic Field? New Results From Middle to Late Miocene Sediments of the North Atlantic (DSDP Site 608)

Wout Krijgsman

Paleomagnetic laboratory "Fort Hoofddijk", Utrecht University, The Netherlands

Dennis V. Kent

*Department of Geological Sciences, Rutgers University, Piscataway, New Jersey, and
Lamont-Doherty Earth Observatory, Palisades, New York*

New magnetostratigraphic results from DSDP Site 608 in the North Atlantic reveal the presence of nine short-term polarity fluctuations that do not correspond to subchrons in the most recent GPTS. Comparison with published results from ODP Sites 845 and 1092, and from the continental Orera Composite Section, shows that all these polarity fluctuations are observed in more than one record. At DSDP Site 608, five of the intervals are ascertained by more than two samples and have an estimated duration that is larger than 10 kyr. We regard these as polarity subchrons, which define five new short polarity intervals in the Miocene: C4r.2r-1n, C4Ar.1r-1n, C5r.2r-1n, C5r.2r-2n, and C5r.3r-1n. Three polarity fluctuations within C5n and one in C5An.2n are each only determined by one sample and are hence not unambiguous. They have an estimated duration < 10 kyr and an inconsistent expression. Consequently, we believe they qualify better as directional excursions and that they are most likely associated with decreases in paleointensity (DIPs) of the geomagnetic field. The identification of these nine new magnetic events suggests that the occurrence of short-term polarity fluctuations in the geomagnetic field is indeed non-uniformly distributed through time. For example, they are relatively more common in the early Late Miocene and the Pleistocene but virtually absent in the latest Miocene and Pliocene. It remains uncertain, however, if this is related to real behavior of the geodynamo. The detection of several new short polarity subchrons in the Middle to Late Miocene time interval may be explained by earlier registration problems in the magnetic anomaly patterns, while the absence of such features in the latest Miocene-early Pliocene could be related to paleomagnetic records of inferior data quality.

Timescales of the Paleomagnetic Field
Geophysical Monograph Series 145
Copyright 2004 by the American Geophysical Union
10.1029/145GM12

1. INTRODUCTION

One of the most fascinating characteristics of the Earth's magnetic field is that the dipole undergoes complete polarity reversals a few times every million years on average. The intervals between reversals have a stable normal or reversed polarity and are called subchrons, chrons, and superchrons, depending on their duration. They define a characteristic pattern of polarity zones through time and, as such, form a fundamental tool for dating of the geological record: the geomagnetic polarity time scale (GPTS). The reliability and completeness of this GPTS is crucial for geochronology but also for understanding the long-term statistical properties of the geomagnetic field. The shortest polarity intervals in the most recent GPTS (CK95 of Cande and Kent [1995]) are typically on the order of 30 kyr in duration, but the magnetic anomaly patterns of fast spreading oceanic plates indicate that smaller-scale variations, so-called 'tiny wiggles', exist as well [*Cande and Kent*, 1992a, 1992b; *Bowers et al.*, 2001]. The origin of 'tiny wiggles', whether they represent short polarity subchrons or mainly decreases in paleointensity is unclear, and consequently they have been referred to as cryptochrons [*Cande and Kent*, 1992a, 1992b]. Detailed and numerous paleomagnetic investigations of the youngest period of the geological record also revealed the presence of many shorter periods of anomalous field behavior referred to as excursions, relatively brief deviations in directions that are larger than what might be expected from normal secular variation and that are usually accompanied by substantial decreases in paleointensity [for a review see *Langereis et al.*, 1997; *Lund et al.*, 1998]. An ongoing debate, parallel to that for cryptochrons, is which excursions qualify as records of short polarity subchrons.

Numerous short-term polarity fluctuations are considered to be an intrinsic part of Earth's magnetic field behavior according to the most recent studies on the geodynamo [*Gubbins*, 1999]. In contrast to this, the occurrence of cryptochrons or excursions in the geological record seems rather non-uniformly distributed through time. They are common in latest Pliocene to Pleistocene (2–0 Ma), middle to early Late Miocene (11–8 Ma), Oligocene (30–24 Ma), and Paleocene (61–53 Ma) records, but seem to be largely absent in records of the latest Miocene and most of the Pliocene, Early Miocene, and Eocene [*Cande and Kent*, 1992a]. The main question associated with these intervals when cryptochrons or excursions are apparently absent is whether they are related to different behavior of the geodynamo or to registration problems of the paleomagnetic signal. Variation in data quality may thus contribute to the apparent irregularity in the temporal distribution of cryptochrons and excursions.

The registration and characterization of short polarity subchrons and excursions in the geological record obviously require a high-resolution sampling and an accurate determination of the paleomagnetic signal, which are not always available. Especially in the Brunhes (C1n) and Matuyama (C1r)

chrons, numerous high-resolution and high-quality paleomagnetic records from complete stratigraphic sequences and detailed volcanic successions are available and many of them reveal the existence of excursions [*Langereis et al.*, 1997; *Lund et al.*, 1998; *Singer et al.*, 1999; *Channell et al.*, 2002]. Detailed comparison of these records, however, commonly reveals significant inconsistencies in the age, duration and the number of registered events. In addition, numerous Mio-Pliocene sections—both in continental and marine environments—have been sampled for the construction of the Mediterranean-based astronomical polarity time scale (APTS) with a sufficiently high resolution to detect cryptochrons [*Langereis and Hilgen*, 1991; *Krijgsman et al.*, 1995; *Van Vugt et al.*, 1998; *Garcés et al.*, 2001]. So far, no solid evidence has been provided for the existence of excursions in records for this time span. This seems to be in agreement with the apparent lack of cryptochrons in the Late Miocene to Early Pliocene ocean floor anomaly patterns [*Cande and Kent*, 1992a]. Recent attempts to extend the Mediterranean APTS downward in time revealed again the presence of several excursions in the Middle to Late Miocene interval [*Abdul Aziz et al.*, 2000, 2003]. This suggests that the occurrence of short-term polarity fluctuations of the geomagnetic field might indeed be restricted to specific intervals in geological history.

The validity of a new subchron at a certain age and location obviously requires confirmation by other records from different parts of the world. For the Middle to Late Miocene interval, the existence of features that may correlate with magnetic anomaly cryptochrons has been reported from continental records [*Tauxe and Opdyke*, 1982; *Garcés et al.*, 1996; *Li et al.*, 1997a, 1997b; *Rösler and Appel*, 1998; *Roperch et al.*, 1999], and also from detailed magnetostratigraphic studies on cores of the Deep Sea Drilling Project (DSDP) or its successor, the Ocean Drilling Program (ODP) [*Schneider et al.*, 1997; *Roberts and Lewin-Harris*, 2000; *Bowles et al.*, 2003; *Evans and Channell*, 2003]. In this study, we present new paleomagnetic results from a detailed re-sampling of the middle to early Late Miocene interval of DSDP Site 608, which contains one of the best and most complete magnetostratigraphic records of this interval in the world [*Clement and Robinson*, 1986].

2. DSDP LEG 94; SITE 608

During DSDP Leg 94, sediment was recovered from 22 holes at six sites in the North Atlantic. The primary objective of this leg was to obtain a continuous high-resolution paleoclimatic record along a transect from 35°N to 55°N in the North Atlantic [*Ruddiman et al.*, 1987]. Paleomagnetic, stable isotope, calcareous nannofossil, foraminiferal, diatom, radiolarian, and dinocyst stratigraphic studies have been performed and made these records a reference section for Miocene isotope studies [*Clement and Robinson*, 1986; *Miller et al.*, 1991; *Gartner, 1992*]. The reliable magnetostratigraphic results, the near-complete recovery, and the abundant fossil content of the sediment provided a time framework in which the paleoclimatic, paleoceanographic, tectonic, and other studies of these sediments could be integrated and together correlated with the GPTS. Site 608 was drilled on the southern flank of the King's Trough tectonic complex, a series of roughly parallel basins and ridges situated 700 km northeast of the Azores (Figure 1). Hole 608 was continuously cored with the variable-length hydraulic piston corer and extended core barrel to basement at 530.9 meters below sea-floor (mbsf) [*Ruddiman et al.*, 1987]. Core recovery was generally over 90%, and an almost continuous stratigraphic sequence was recovered from the mid-upper Oligocene to the Pleistocene. Sediment accumulation rates averaged between 3 and 5 cm/kyr, which allowed the delineation of a detailed magnetic polarity stratigraphy [*Clement and Robinson*, 1986]. Subsequent shore-based additional sampling and laboratory work were done to refine the polarity records [*Miller et al.*, 1991].

The magnetostratigraphic records of the Plio-Pleistocene interval of the Leg 94 sites could be correlated straightforwardly to the GPTS. Moreover, several excursions were observed in the high-resolution records of several holes, which correlated to the Cobb Mountain (1.20–1.21 Ma), 'Gilsa' (~1.58 Ma), and Réunion (2.14–2.15 Ma) events [*Clement and Robinson*, 1986; *Clement and Kent*, 1987]. Correlation of the Miocene polarity log of Site 608 with the GPTS also seems unambiguous, especially with the presence of a 40 meter-thick normal polarity interval in cores 23 through 27 that represents chron C5n.2n, and is furthermore supported by

biostratigraphic data. In several intervals, the Miocene polarity pattern of Site 608 shows even more detail than the standard GPTS. This could indicate additional polarity subchrons that were previously not registered in the polarity time scale.

3. NEW MAGNETOSTRATIGRAPHIC RESULTS FROM DSDP SITE 608

3.1. Methods

The original magnetostratigraphy of Site 608 was measured with a fluxgate spinner magnetometer with shorebased work on a cryogenic [*Clement and Robinson*, 1986]. Several pilot samples were subjected to progressive partial alternating-field (AF) demagnetization, which showed that unstable or secondary components were readily removed by 10 mT [*Clement and Robinson*, 1986]. In addition, a stable component was identified by a linear trajectory that decayed toward the origin at higher fields. On the basis of these results, most samples were demagnetized in only one step of 10 mT [*Clement and Robinson*, 1986].

For our study, the working halves of cores 19 to 34—stored in the ODP East Coast Repository at Lamont-Doherty Earth Observatory (LDEO)—were initially sampled with a resolution of 25 cm, which should correspond to a temporal resolution of approximately 5–10 kyr. All intervals containing polarity reversals and potential subchrons were then re-sampled in higher detail to pinpoint the directional changes as exactly as possible. The resulting 585 samples were stepwise AF-demagnetized with standard fields of 5, 10, 15, 20, 30, 40 mT, and measured on a 2G Model 760 DC-Squid 3-axis cryogenic magnetometer housed in a shielded room at the pale-

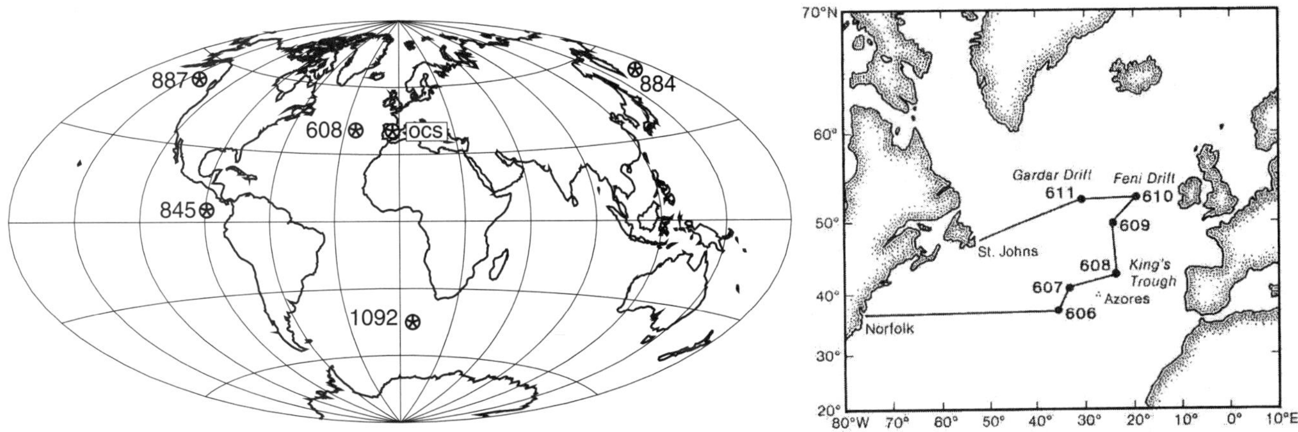

Figure 1. Location map of DSDP Leg 94, with Site 608 located at the southern flank of the King's Trough tectonic complex [after *Ruddiman et al.*, 1987]. On world map are also the locations of the other sites and sections that are discussed in this paper.

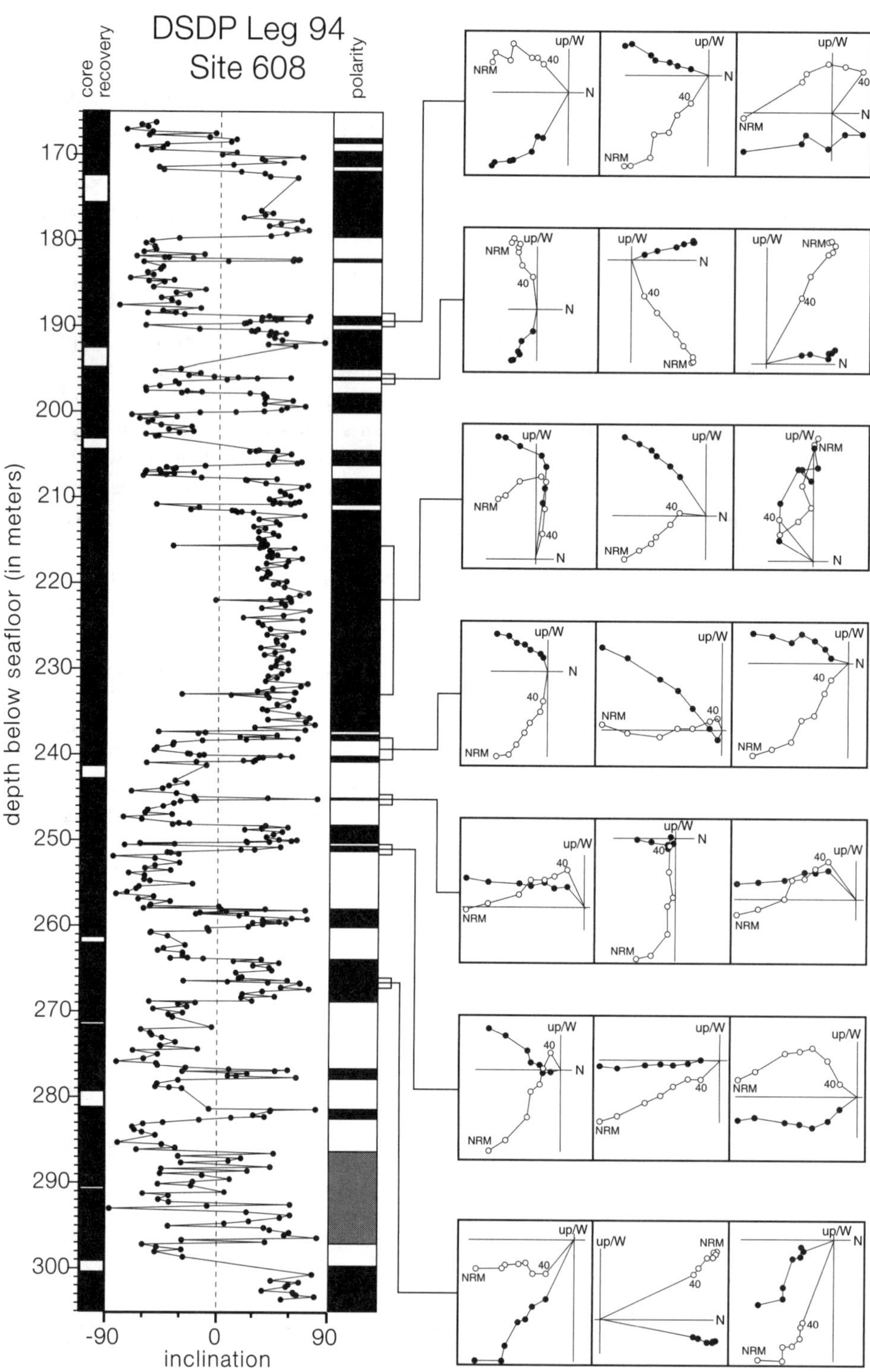

Figure 2. Paleomagnetic results from DSDP Site 608. On the left is the stratigraphic record of the vector end-point component of the inclination record after stepwise AF-demagnetization. In the polarity column black (white) denotes normal (reverse) polarity; gray shading indicates an interval of uncertain polarity. Core recovery is indicated by black colors. On the right are Zijderveld diagrams of all the intervals that contain evidence for short-term polarity excursions. Closed (open) symbols represent projections of vector end-points on the horizontal (vertical) plane after AF-demagnetization with steps of 5, 10, 15, 20, 30, and 40 mT

omagnetic laboratory of LDEO (Figure 2). Principal component analysis was applied to determine the component directions of the NRM, chosen by inspection of vector end-point demagnetization diagrams. Especially the lower part of the sequence revealed demagnetization diagrams where the secondary component was only removed at fields significantly higher than 10 mT, which showed that the extra effort of stepwise demagnetization was crucial to obtain reliable data (Figure 2). The sediments of cores 29 through 34 were frequently brecciated and intervals suitable for sampling were often difficult to find [*Clement and Robinson*, 1986]. Consequently, the data from this part of the section were generally of poorer quality.

Table 1. Position in meters below seafloor (mbsf) of the magnetic reversals in DSDP Site 608

Magnetic reversal	Site 608 depth (mbsf)	GPTS age	APTS age
C3Br.2n (y)	168.19-167.96	7.341	7.455
C3Br.2n (o)	168.74-168.51	7.375	7.492
C4n.1n (y)	169.72-169.45	7.432	7.532
C4n.1n (o)	171.48-171.24	7.562	7.644
C4n.2n (y)	172.01-171.80	7.650	7.697
C4n.2n (o)	179.64-179.61	8.072	8.109
C4r.1n (y)	182.18-182.09	8.225	8.257
C4r.1n (o)	182.49-182.48	8.257	8.303
C4An (y)	190.44-190.38	8.699	8.750
C4An (o)	195.02-192.42*	9.025	9.075
C4Ar.1n (y)	197.83-197.80	9.230	9.280
C4Ar.1n (o)	200.07-200.05	9.308	9.377
C4Ar.2n (y)	204.53-202.89*	9.580	9.629
C4Ar.2n (o)	206.34-206.32	9.642	9.679
C5n.1n (y)	207.85-207.74	9.740	-------
C5n.1n (o)	210.81-210.80	9.880	-------
C5n.2n (y)	211.55-211.45	9.920	-------
C5n.2n (o)	237.38-237.24	10.949	11.053
C5r.1n (y)	237.75-237.60	11.052	11.122
C5r.1n (o)	238.39-238.28	11.099	11.158
C5r.2n (y)	248.32-248.23	11.476	11.558
C5r.2n (o)	250.37-250.31	11.531	11.629
C5An.1n (y)	258.05-257.95	11.935	12.028
C5An.1n (o)	260.25-260.07	12.078	12.127
C5An.2n (y)	263.94-263.89	12.184	12.214
C5An.2n (o)	268.85-268.80	12.401	12.447
C5Ar.1n (y)	276.75-276.59	12.678	12.757
C5Ar.1n (o)	277.82-277.76	12.708	12.780
C5Ar.2n (y)	281.46-281.39*	12.775	12.816
C5Ar.2n (o)	282.66-282.42	12.819	12.881

Ages are in Ma according to the GPTS [*Cande and Kent*, 1995] and APTS [*Hilgen et al.*, 1995; *Abdul Aziz et al.*, 2003]. (o) corresponds to older and (y) corresponds to younger reversal boundary. Asterisks represent reversals that correspond to core breaks and which are hence less accurately determined.

Relative orientation between cores was not available but the magnetic inclinations predicted by an axial dipole field are 60° at Site 608, and thus steep enough to allow an unambiguous polarity determination, and the detection of directional excursions by changes in sign, regardless of the lack of declination control (Figure 3). The average inclination of the samples from Site 608 is 45.3 ± 1.7° for the N = 293 normal polarity set and −43.2 ± 2.0° for the N = 228 reverse polarity set, which are perfectly antipodal within statistical errors but both too shallow for the sampling location presumably because of sedimentary inclination error [*Celaya and Clement*, 1988]. Nevertheless, the results suggest that coring deformation, storage, and sampling problems have not degraded the reliability of the magnetic signal to any appreciable extent.

3.2. Chronology

Plotting the inclination results against stratigraphic level in depth below seafloor shows that a large number (47) of polarity reversals have been registered (Figure 3). The conspicuously long normal magnetozone corresponding to Chron C5n.2n is again easily identified in the middle part of the sequence. From this magnetozone, the correlation to the GPTS [*Cande and Kent*, 1992a, 1995] can be straightforwardly extended upwards and downwards, resulting in the recovery of the complete polarity sequence from C5Ar to C3Br (Figure 3). This implies that we have obtained a rather continuous (apart from several core breaks) polarity record from about 13 to 7.5 Ma. In addition, nine excursions have been identified which do not correspond to subchrons in the GPTS (Figure 3). None of these intervals are associated with core section breaks or biscuited material. Four of them are determined by only one or two samples and are hence not unambiguous, but the other five are ascertained by more than two samples. Following the cryptochron terminology of Cande and Kent [1992a, 1995], these candidate polarity zones are named C4r.2r-1, C4Ar.1r-1, C5n.2n-1, C5n.2n-2, C5n.2n-3, C5r.2r-1, C5r.2r-2, C5r.3r-1 and C5An.2n-1 (Table 2).

The unequivocal correlation of the magnetic polarity stratigraphy of Site 608 to the GPTS establishes a more accurate chronology for the Middle to Late Miocene sediments and stratigraphy of the North Atlantic. The ages for all reversals of Site 608 (Table 1) are given according to the most recent GPTS [*Cande and Kent*, 1995] and to the Mediterranean based astronomical polarity time scale (APTS; [*Hilgen et al.*, 1995; *Abdul Aziz et al.*, 2003]). This allows an accurate estimation of the sediment accumulation rate at Site 608 (Figure 4). This sedimentation rate appears to be fairly constant throughout the Middle Miocene (13–10 Ma) with an average value of 2.3 cm/kyr, but decreases in the Late Miocene (10–7 Ma) to an average of 1.7 cm/kyr. Consequently, the position and dura-

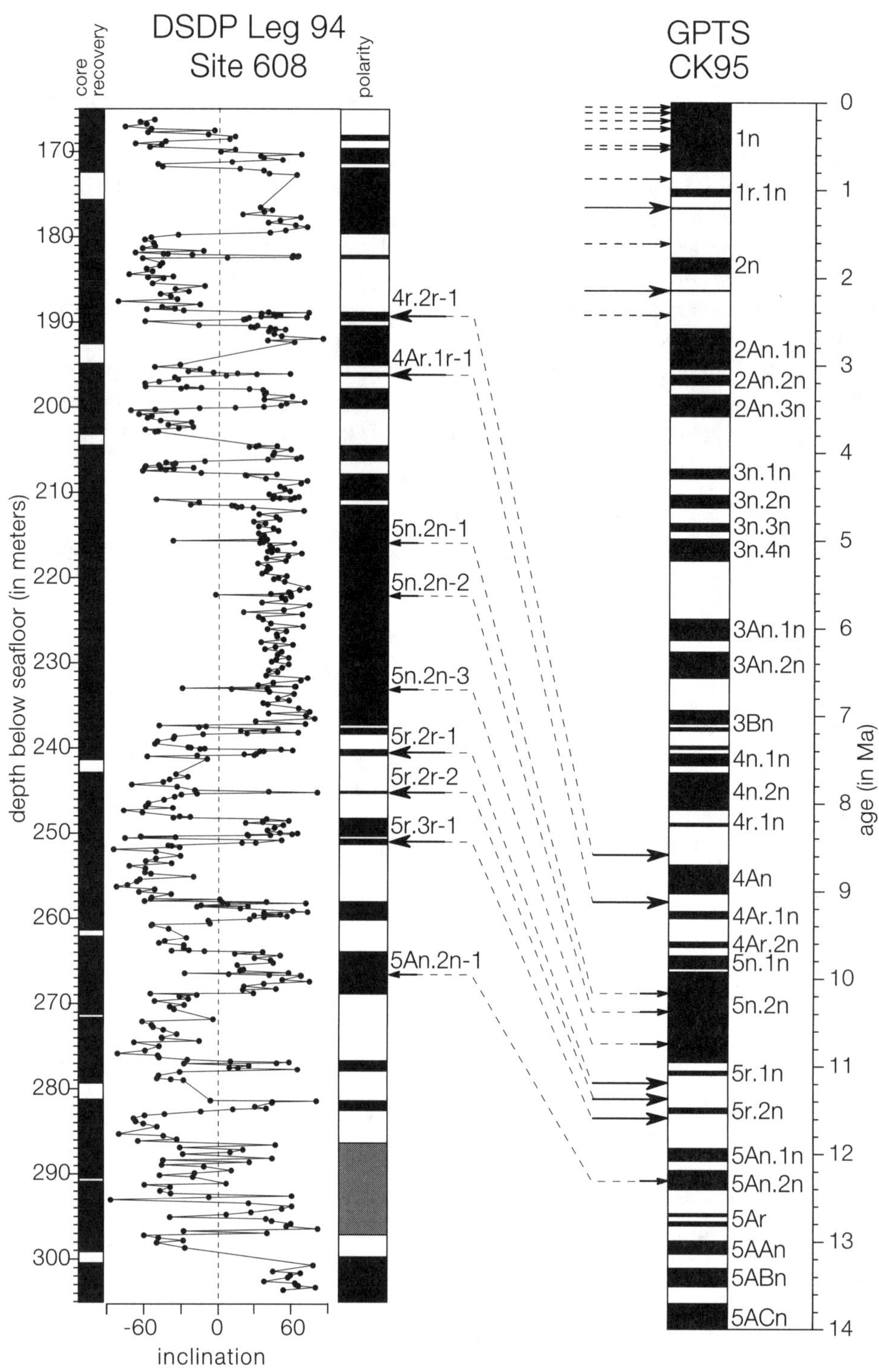

Figure 3. Correlation of the DSDP Site 608 magnetostratigraphy to the GPTS of *Cande and Kent* [1995]. Large (small) arrows indicate positions of short subchrons (excursions).

Table 2. Position in meters below seafloor (mbsf) of the short-term polarity fluctuations in DSDP Site 608 and their relative position within the corresponding subchron of the GPTS

Short-term fluctuation	Depth (mbsf) Site 608	Position Site 608	Duration (in kyr) 608	CK95	845	Orera	1092	Name
C4r.2r-1	188.76-189.81	C4r.2r(0.92-0.79)	61	16	44	.	24	C4r.2r-1**n**
C4Ar.1r-1	196.00-196.28	C4Ar.1r(0.45-0.38)	17	----	29	.	13	C4Ar.1r-1**n**
C5n.2n-1	215.65-215.68	C5n.2n(0.16)	1	8	----	.	3	DIP
C5n.2n-2	221.97-222.10	C5n.2n(0.41)	6	24	----	.	3	DIP
C5n.2n-3	232.89-233.02	C5n.2n(0.83)	6	16	----	.	5	DIP
C5r.2r-1	240.17-240.86	C5r.2r(0.25-0.18)	31	----	26	26	38	C5r.2r-1**n**
C5r.2r-2	245.02-245.27	C5r.2r(0.70-0.67)	15	----	----	15	5	C5r.2r-2**n**
C5r.3r-1	250.77-251.31	C5r.3r(0.13-0.06)	25	----	----	9	6	C5r.3r-1**n**
C5An.2n-1	266.44-266.50	C5An.2n(0.53-0.51)	3	----	<9	?	.	DIP

Values are in fractional position from the younger end of the subchron. Duration for each event is in kiloyears as calculated from sedimentation rates (Site 608, Site 1092), seafloor spreading rates (GPTS), and astronomical forcing (Orera). Fluctuations that have an overall duration greater than 9-15 kyr qualify as short subchrons and are labeled as such with a suffix designing polarity. Features of less than 10 kyr are interpreted to represent directional excursions associated in most cases with a decrease in paleointensity (DIP).

tion of the nine candidate subchrons that have been recorded as directional excursions can be determined from their average thickness and the average sedimentation rates of the above-mentioned intervals (Table 2).

3.3. Reliability of the Paleomagnetic Signal

The four directional excursions that occur within normal polarity intervals (C5n.2n-1, C5n.2n-2, C5n.2n-3, and C5An.2n-1) are all determined by a single specimen and hence are of relatively very short duration (<6 kyr; Table 2). The five new directional excursions that occur within reverse polarity intervals (C4r.2r-1, C4Ar.1r-1, C5r.2r-1, C5r.2r-2, and C5r.3r-1) are all determined by at least 3 samples but are inherently suspect as they may be related to a viscous component induced by the present day field. To validate the paleomagnetic signal, we have subjected selected samples covering all the individual directional excursions, as well as samples from the neighboring intervals of normal or reverse polarity, to isothermal remanent magnetization (IRM) component analyses. These experiments allow a diagnostic comparison between the magnetic carriers of the different intervals, which could be an indication for the reliability of the paleomagnetic signal. IRM acquisition curves were measured using a PM4 pulse magnetizer. The magnetic components are characterized by the saturation IRM (SIRM), the peak field at which half of the SIRM is reached ($B_{1/2}$), and the dispersion of the corresponding cumulative log-normal distribution according to the IRM component fitting method of [*Kruiver et al.*, 2001].

A total of 18 samples have been selected for the IRM component analysis. All samples are dominated by a relatively low-coercivity component with a $B_{1/2}$ value of approximately 40–50 mT, which is compatible with magnetite or partially

oxidized magnetite (Table 3). The results show that there is no significant difference in magnetic composition between samples from the directional excursions and from neighboring polarity zones (Table 3). This implies that the directional excursions are not simply explained by differences in lithology or lithology-related diagenetic processes, and that they thus probably reflect true variations of the geomagnetic field.

There is, however, a marked change in composition occurring in the section, which may be related to the change in sedimen-

Table 3. Results of IRM component analysis of selected samples of DSDP site 608 that straddle the short-term polarity fluctuations

Sample Code	Depth (mbsf)	Small-term fluctuation	Component 1 SIRM	$B_{1/2}$	%	DP
D21.3.085	189.15	C4r.2r-1	190	53	77	0.26
D21.4.014	189.94		134	50	79	0.28
D21.4.131	191.11		142	50	75	0.28
D22.1.034	195.24		260	53	85	0.28
D22.1.120	196.10	C4Ar.1r-1	190	56	86	0.27
D22.2.112	197.52		146	51	80	0.27
D24.2.005	215.65		770	43	94	0.28
D24.6.038	221.98	C5n.2n-1	1920	44	85	0.24
D25.7.027	232.97	C5n.2n-2	1570	42	91	0.27
D26.1.036	233.66		1470	41	92	0.27
D26.3.108	237.38		3000	44	92	0.25
D26.3.145	237.75	C5r.2r-1	9700	43	89	0.27
D27.1.077	243.67		2520	40	90	0.25
D27.2.079	245.19	C5r.2r-2	1890	43	78	0.23
D27.3.114	247.04		3220	41	83	0.25
D27.6.011	250.46		2200	40	86	0.24
D27.6.078	251.13	C5r.3r-1	2320	40	83	0.23
D27.6.109	251.44		2170	38	85	0.24

Values of the main component are given by Saturation IRM (SIRM) in 10^3 A/m, the peak field at which half of the SIRM is reached ($B_{1/2}$) in mT, percentage of the main component (%), and the dispersion parameter (DP).

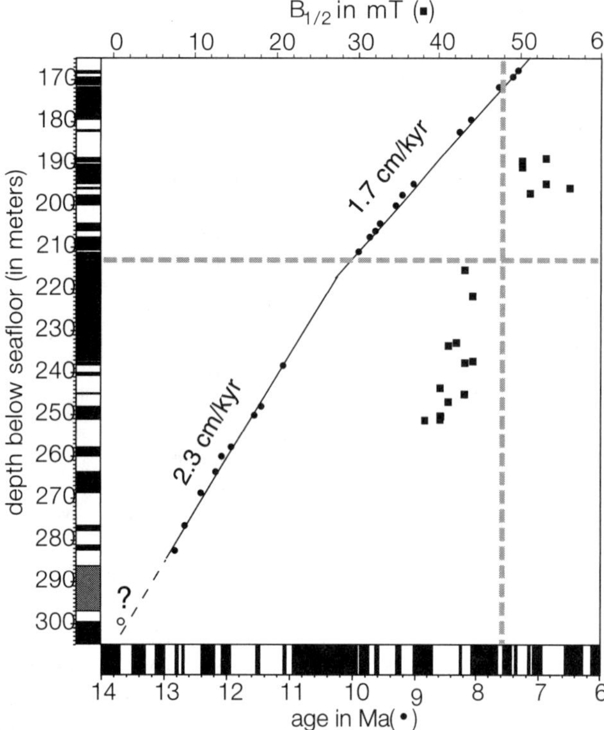

Figure 4. Polarity column for DSDP Site 608 as a function of depth (mbsf) versus age (Ma) in CK95, displayed by solid dots. Note the evident change in sedimentation rate from 2.3 cm/kyr in the lower part to 1.7 kyr in the upper part of the sequence. Squares represent $B_{1/2}$ values in (mT) of the dominant magnetic component in selected samples for IRM analysis. A change in $B_{1/2}$ values of about 40 mT to values higher than 50 mT roughly corresponds to the change in sedimentation rate.

tation rate at the top of C5n.2n at approximately 211 mbsf (Figure 4). Above this level, the $B_{1/2}$ values range between 50 and 56 mT, compatible with single domain magnetite. The dispersion parameters suggest a detrital origin for this component because they are generally too high for a biogenic origin [*Kruiver and Passier*, 2001]. In the lower part of the sequence, the $B_{1/2}$ values are around 40 mT whereas SIRM values are an order of magnitude higher. This implies that the magnetic component may suffer from viscous resetting as the lower part of the grain size distribution tails to the super-paramagnetic grain size range. The IRM results are in good agreement with the AF-demagnetization results, which show a very stable component in the upper part of the sequence and commonly a much larger secondary component in the lower part (Figure 2).

4. COMPARISON WITH OTHER RECORDS

Many of the short-term fluctuations in the geomagnetic field, either recorded as tiny wiggles in marine magnetic

anomalies or as excursions in direction in sedimentary geological records, have an uncertain origin. Consequently, the age, origin and nature of any short-term fluctuation needs to be confirmed by other records from different parts of the world, so that its status can be ascertained as simply a directional excursion (perhaps accompanied by a decrease in paleointensity) or a short polarity subchron. Although there are not yet many reliable paleomagnetic studies of sufficiently high-resolution from the Middle to Late Miocene time interval available, several studies have revealed the existence of directional excursions or subchrons in this time interval.

Detailed studies of the marine magnetic anomaly profiles of the ocean floor revealed several small-scale fluctuations in the Middle to Late Miocene time interval [*Blakely*, 1974; *Cande and LaBrecque*, 1974]. Four of these 'tiny wiggles' have been nominated as cryptochrons in CK92 (Table 2) and their duration has been calculated from sea floor spreading rates (C4r.2r-1 of 16 kyr, C5n.2n-1 of 8 kyr, C5n.2n-2 of 24 kyr and C5n.2n-3 of 16 kyr). The three cryptochrons of C5n had previously also been recognized from magnetic profiles of the northeast Pacific by Blakely [1974] who suggested that they correspond to short polarity subchrons and that they should be incorporated as such in the GPTS. In contrast to this, a thorough re-study of the C5n interval of the sea-surface and near-bottom profiles in the North and South Pacific suggested that the tiny wiggles in C5n most likely represent intensity fluctuations [*Bowers et al.*, 2001; *Bowles et al.*, 2003]. Anomaly data alone, however, cannot uniquely determine whether tiny wiggles in the anomaly profiles represent either short subchrons or paleointensity variations. Consequently, data from sedimentary sequences provide crucial information to determine the exact nature of each cryptochron.

The paleomagnetic record from the low latitude ODP Site 845 in the Equatorial Pacific (Figure 1) is perhaps one of the most valuable examples of a continuous mid-Miocene to mid-Pliocene magnetostratigraphic record (Figure 5; [*Schneider*, 1995; *Schneider et al.*, 1997]). The Miocene magnetostratigraphy of this site was generated partly from pass-through magnetometer measurements and partly by using discrete samples [*Schneider*, 1995]. Here, only the declination provided useful results, since inclination changes alone were rarely indicative of polarity because of the low site latitude (10°N). The combined magnetostratigraphic records from Holes 845A and 845B documented five short putative polarity intervals in the Middle to Late Miocene epoch (Figure 5). Only one of these events corresponds to a poorly defined tiny wiggle in the marine magnetic anomaly record, a cryptochron (C4r.2r-1) that is included in CK92. The three other short events (C4r.1r-1, C4Ar.1r-1, C5r.2r-1) were previously not identified. Correlation of the Site 845 magnetostratigraphy to the Site 608 record is straightforward and shows that three

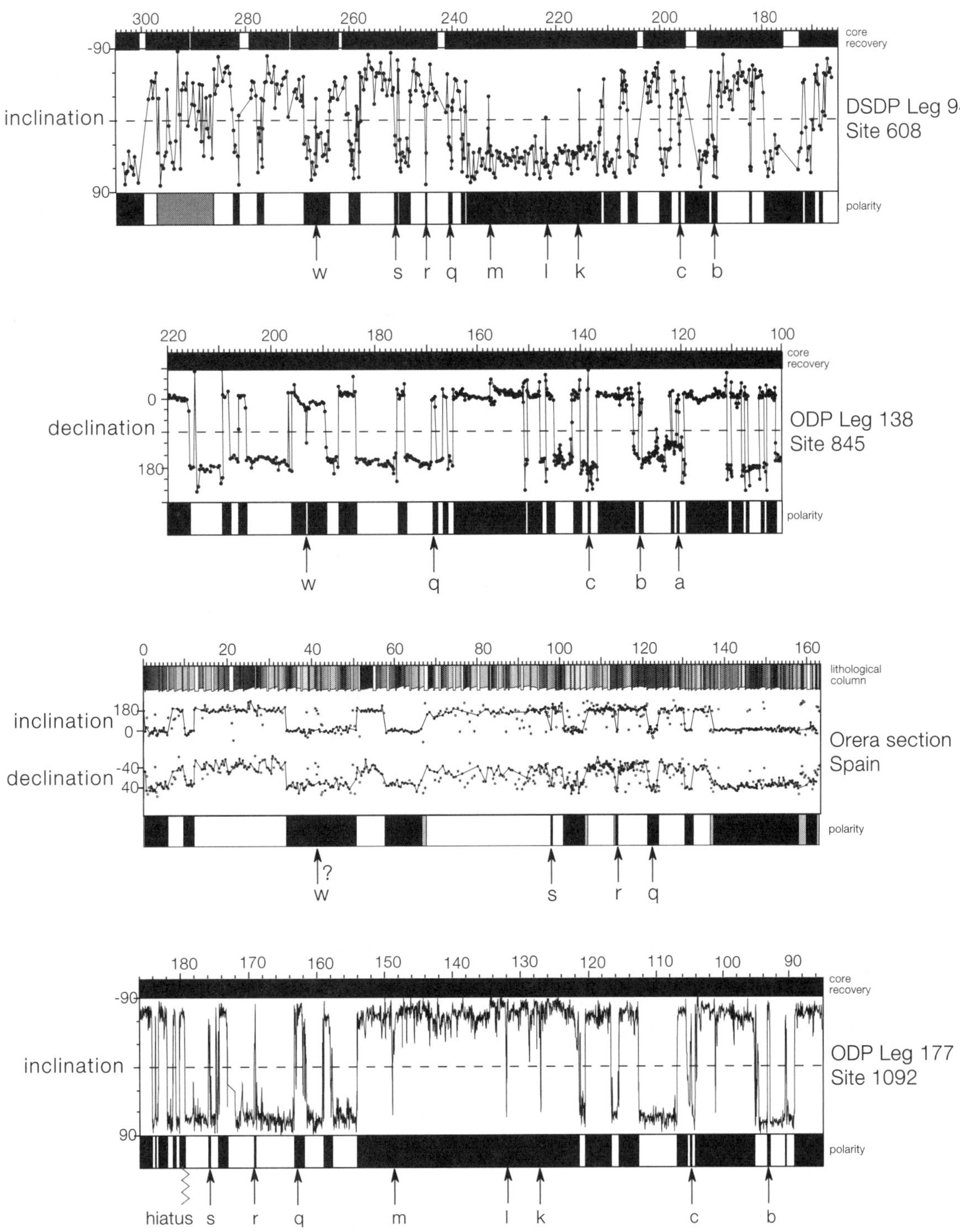

Figure 5. Comparison of the magnetostratigraphic records in local thickness units (meters) of DSDP Site 608, ODP Site 845 [*Schneider et al.*, 1997], the Orera Composite Section [*Abdul Aziz et al.*, 2000] and ODP Site 1092 (with adjustment of a revised composite depth section) [*Evans et al.*, 2004] showing that all short-term polarity fluctuations in Site 608 have been recorded elsewhere. a = C4r.1r-1, b = C4r.2r-1n, c = C4Ar.1r-1n, k = C5n.2n-1, l = C5n.2n-2, m = C5n.2n-3, q = C5r.2r-1, r = C5r.2r-2, s = C5r.3r-1, and w = C5An.2n-1. In the Orera Composite Section, the black signal line connects only the reliable directions, unreliable directions are represented by open circles and asterisks (for more detail see [*Abdul Aziz*, 2000].

of these short events (C4r.2r-1, C4Ar.1r-1 and C5r.2r-1) have been registered in both sites (Figure 5). The fourth event (C4r.1r-1) is not recognized in Site 608. By contrast, there are no indications in Site 845 of the three cryptochrons within C5n (C5n.2n-1, C5n.2n-2 and C5n.2n-3), neither of the two additional ones in C5r (C5r.2r-2 and C5r.3r-1). There is, however, a clear indication of a very short directional excursion within C5An.2n, which exactly correlates to our C5An.2n-1 event. This interval was not mentioned in the original interpretation by Schneider [1995]. Relative paleointensity studies of the intervals spanning the cryptochrons showed that C5r.2r-1 reflected a full normal polarity interval with distinct intensity recovery between the bounding pair of reversals, while intensity recovery was not seen for C4r.1r-1, C4r.2r-1, or C4Ar.1r-1 [Schneider, 1995]. Consequently, he only considered C5r.2r-1 as a 'true' subchron and calculated its duration (26 kyr) from thickness and sedimentation rate. We have, however, calculated the duration for the other short intervals as well to allow a more detailed comparison between the different records. Durations of 44 kyr for C4r.2r-1 (75 cm) and of 29 kyr for C4Ar.1r-1 (50 cm) have been derived by applying the ages of CK95 to the most evident magnetic reversals (Table 2). The interval corresponding to C5An.2n-1 is only determined by one sample (sampling resolution of 10-15 cm), which suggests that it should be less than 9 kyr in duration.

The continental sediments of the Orera Composite Section of the Calatayud Basin in Spain (Figure 1) provide another reliable and detailed record of the time interval ranging from 10.6 to 13 Ma (Figure 5; [Abdul Aziz et al., 2000]). This section was sampled with a resolution of approximately 5 kyr for the construction of an astronomical time scale for the continental realm, because of its very marked sedimentary cyclicity and its good paleomagnetic signal [Abdul Aziz et al., 2003]. The detailed magnetostratigraphic record of the Orera Composite Section showed three short normal polarity intervals that had not been registered in CK92. Detailed re-sampling and additional paleomagnetic, rock magnetic and geochemical analyses of these three intervals showed that they have a geomagnetic origin, and are not related to diagenetic or other sedimentary artifacts [Abdul Aziz, 2001; Abdul Aziz and Langereis, this volume]. The youngest of the three intervals corresponds to C5r.2r-1 of Schneider [1995], the older two were previously not determined and have been named C5r.2r-2 and C5r.3r-1 [Abdul Aziz, 2001; Abdul Aziz and Langereis, this volume]. Correlation to the Site 608 magnetostratigraphy shows that all these normal polarity intervals in C5r have been documented in the two records. The reverse polarity interval in C5An.2n was not noticed in the Orera Composite Section, although several levels showed reverse directions at its expected position (Figure 5). These directions were considered unreliable in the earlier interpretation because they correspond to components of very weak intensity [Abdul Aziz et al., 2000], but this interval certainly warrants a more detailed reinvestigation. The duration of the three new normal polarity intervals in C5r has been established from the number of sedimentary cycles, which were proven to be related to the astronomical cycle of precession with average periodicity of about 21.7 kyr [Abdul Aziz et al., 2003]. This resulted in durations of 26 kyr for C5r.2r-1, 15 kyr for C5r.2r-2 and 9 kyr for C5r.3r-1 (Table 2).

Another excellent Upper Miocene paleomagnetic record was published recently from ODP Site 1092 (Figure 1), located in the sub-Antarctic South Atlantic [Evans and Channell, 2003]. The magnetostratigraphy for this site was obtained from u-channel samples of the composite stratigraphic section (Figure 5). The complete stratigraphic record of Site 1092, after adjustment of a revised composite depth section [Evans et al., 2004], clearly revealed evidence for changes in directions corresponding to C4r.2r-1 and to the three cryptochrons of C5n.2n (Figure 5). In addition, the inclination record shows clear indications for a short normal interval at the stratigraphic position where C4Ar.1r-1 is expected, although not interpreted as such by Evans and Channell [2003]. The revised composite depth section of Site 1092 furthermore seems to favor a correlation of the lower part of the record different than presented by Evans and Channell [2003]. Biostratigraphic data from diatoms suggests the presence of a hiatus of approximately 800 kyr at a level of 178.83 meters composite depth [Censarek and Gersonde, 2002]. The interval between this hiatus and the base of C5n.2n matches the polarity pattern of the Orera Composite Section and the Site 608 record (Figure 5). This different correlation implies that the three additional directional excursions in C5r (C5r.2r-1, C5r.2r-2 and C5r.3r-1) have also been recorded at 1092, which have durations of 38 kyr (144 cm), 5 kyr (19 cm) and 6 kyr (21 cm), respectively (Table 2).

Sediments from ODP Site 887 in the North Pacific are also known for their excellent magnetostratigraphy, in which C5n could also be unambiguously recognized [Weeks et al., 1995]. Shipboard inclination data from Site 887 Hole C show that two short polarity intervals have been recorded that correspond to C4r.2r-1 and C5r.2r-1. There appears to be no indication for the other two normal polarity intervals in chron C5r seen in Site 608, Site 1092 and Orera, nor for the cryptochrons in C5n. A detailed study of the relative paleointensity record from this site, compared with the deep tow anomaly records of [Bowers et al., 2001], has led to the conclusions that the tiny wiggles in C5n are most likely paleointensity variations, and if reverse polarity intervals exist in C5n, that they are likely to be less than 5 kyr in duration [Bowles et al., 2003].

The paleomagnetic records recovered from ODP Site 884 in the North Pacific also comprise the entire C5n interval

[*Roberts and Lewin-Harris*, 2000; *Weeks et al.*, 1995]. Here, relative paleointensity data suggested that at two putative short polarity intervals the geomagnetic field collapsed at the polarity reversals and recovered to higher intensities within the short reverse polarity zones. This led to the conclusion that these two directional excursions within C5n represent real subchrons with estimated durations of 26 kyr and 28 kyr, respectively [*Roberts and Lewin-Harris,* 2000]. The magnetostratigraphic correlations of Site 884, however, must be qualified as somewhat ambiguous [*Bowles et al.*, 2003; *Evans and Channell*, 2003]. Indeed there appear to exist dramatic changes in sedimentation rate at this site, corresponding exactly to the C5n interval. Although Roberts and Lewin-Harris [2000] implied that the sedimentation rates were relatively uniform over long periods of time and averaged about 3.6 cm/kyr between 12.4 and 7 Ma, calculation of the sedimentation rates according to their magnetostratigraphic correlation shows instead that there were significant changes, from 2.0 cm/kyr in C5n.2n (21 m; 1029 kyr) to 10.7 cm/kyr in C5n.1n (15 m; 140 kyr). This suggests that the conclusions derived from the Site 884 record must be treated with caution.

5. DISCUSSION

Short-term polarity fluctuations in the geomagnetic field have been registered in many portions of the geological time scale, but in numerous cases serious debate is related to whether they correspond to short polarity subchrons or to directional excursions that are associated with a decrease in paleointensity (DIP). The classic definition of an excursion implies a short-term departure of the Virtual Geomagnetic Pole (VGP) that is considerably larger than that seen in secular variation (e.g., [*Langereis et al.*, 1997; *Merrill et al,* 1998]). Excursions are most likely events in which the geomagnetic field reverses in the liquid outer core but not in the solid inner core, which implies that expressions of excursions may not be equally significant around the globe [*Gubbins*, 1999]. By contrast, short polarity subchrons are characterized by an interval of opposite polarity, bounded by a pair of geomagnetic polarity reversals. Short-period polarity intervals presumably require the inner core to have reversed its polarity [*Gubbins*, 1999]. The magnetic field in the solid inner core can only change by diffusion, which would result in approximately 3–5 kyr for a full polarity reversal to take place [*Hollerbach and Jones, 1993*], and as such implying a minimum duration of 6–10 kyr for just the bounding reversals of a polarity subchron.

Duration may thus be a useful criterion to help distinguish a DIP-related excursion from a short polarity subchron. For example, Cande and Kent [1992a, 1992b] relegated 'tiny wiggles' whose apparent duration was less than 30 kyr as cryp-

tochrons because it was ambiguous whether the origin of these geomagnetic features recorded in the magnetic anomaly patterns was due to short intervals of opposite polarity or simply to DIPs of the geomagnetic field. If we follow the arguments of Gubbins [1999], only features having an overall duration greater than 9–15 kyr would thus qualify as short polarity subchrons whereas features less than about 10 kyr would more likely represent DIP-related excursions. On this basis, C4r.2r-1n, C4A.1r-1n, C5r.2r-1n, C5r.2r-2n, and C5r.3r-1n in Site 608, with supporting data described above from Site 845, Site 1092, and/or the Orera Composite Section, have been elevated to polarity subchrons (and are labeled as such with a suffix designating polarity) whereas C5n.2n-1, C5n.2n-2, C5n.2n-3 and C5An.2n-1, with apparent durations less than 10 kyr, would be better described as directional excursions (Table 2 and 4). Although all these polarity fluctuations have been observed in more than one record, the age and duration of most of them still awaits an accurate and reliable dating, preferably by astronomical calibration techniques. Apart from the three short polarity subchrons in the Orera Composite section, all other Middle to Late Miocene events have so far only been dated by assuming a uniform sedimentation rate. This may sometimes result in incorrect age and duration estimates, which can have significant consequences for global correlations and if we also accept duration as a useful criterion to help distinguish between excursions and short polarity subchrons.

The ongoing debate whether tiny wiggles represent short-period geomagnetic polarity intervals or relate to decreases in geomagnetic intensity has thus not yet been fully resolved, although we believe that the short duration of the features within C5n.2n at Site 608 and their inconsistent expression from site to site makes them more likely to reflect DIP-related excursions of the geomagnetic field. In addition, the numer-

Table 4. Position in meters below seafloor (mbsf) of the magnetic reversals of the short subchrons in DSDP site 608, with ages (Ma) according to the GPTS [*Cande and Kent*, 1995]

Short subchron	Site 608 depth (mbsf)	GPTS age
C4r.2r-1n (y)	188.84-188.67	8.606
C4r.2r-1n (o)	189.94-189.68	8.664
C4Ar.1r-1n (y)	196.05-195.94	9.097
C4Ar.1r-1n (o)	196.45-196.10	9.117
C5r.2r-1n (y)	240.23-240.11	11.167
C5r.2r-1n (o)	240.88-240.83	11.193
C5r.2r-2n (y)	245.08-244.95	11.352
C5r.2r-2n (o)	245.32-245.21	11.363
C5r.3r-1n (y)	250.83-250.71	11.555
C5r.3r-1n (o)	251.44-251.17	11.584

ous (18) cryptochrons in the Oligocene anomaly record have so far not been resolved in the sedimentary archives. Channell et al. [2003] recognize one new excursion in the Oligocene (C7Ar-1) and one in the early Miocene (C5Dr-1), but both of them do not correspond to a cryptochron of CK92. Other detailed paleomagnetic studies on Oligocene sediments did not reveal any excursions at all, which lead to the conclusion that the Oligocene tiny wiggles correspond to DIPs, which may have been accompanied by directional excursions [*Lowrie and Lanci,* 1994; *Lanci and Lowrie,* 1997; *Tauxe and Hartl,* 1997]. It is possible, however, that these Oligocene records suffer from diagenetic processes which might have obscured the geomagnetic signal, especially because the sedimentation rates of these sites are relatively low [*Roberts and Lewin-Harris,* 2000]. This confirms the need for detailed and independent paleomagnetic analyses corresponding to each tiny wiggle in the ocean floor anomaly pattern and for each short-term polarity fluctuation in the sedimentary record.

The paleomagnetic data of late Neogene sediments furthermore suggest that the occurrence of short-term polarity fluctuations is indeed non-uniformly distributed, with numerous excursions and two short subchrons (Cobb Mountain and Reunion-1) in the Brunhes-Matuyama interval, an interval with no evidence for excursions in the latest Miocene to early Pliocene, and at least the nine short events (of which five can be considered as short subchrons) in the early Late Miocene interval (Figure 3). It remains uncertain whether this apparent non-uniform occurrence of short-term polarity fluctuations is related to geomagnetic field behavior or to registration problems of the paleomagnetic signal. For instance, the CK92 time scale used magnetic anomaly profiles from the relatively slow-spreading NE Pacific for the interval corresponding to C5n and C5r [*Cande and Kent,* 1992a]. These profiles only have a marginal resolution in which short anomaly fluctuations are not easily recognized. Marine magnetic anomaly profiles from the central Pacific and Nazca plates have higher resolution and already revealed the existence of subchron C5r.2r-1n [*Abdul Aziz et al.,* 2003]. Registration problems in the magnetic anomaly patterns may thus be an explanation for the discovery of the new short polarity fluctuations in the Middle to Late Miocene time interval.

By contrast, the absence of short-term polarity fluctuations in the latest Miocene-early Pliocene is not easily explained by registration problems in the marine magnetic anomalies. Detailed anomaly data from various fast-spreading ridges were combined for the Pliocene, while reliable profiles from the relatively fast and uniformly spreading Chile Ridge and East Pacific Rise were used for the Late Miocene time scale [*Cande and Kent,* 1992a]. Many of the paleomagnetic records, however, show data of inferior quality during this time interval, especially during the Late Miocene. The Mediterranean

sections all suffered from the Messinian Salinity Crisis, which resulted in deposition of sediments that are less suitable for magnetostratigraphy [*Krijgsman et al.,* 1995, 1999]. This major paleoceanographic event may also have had its influence on the Atlantic Ocean sediments as the magnetic signal in the Site 608 record could not be properly recovered from the Messinian interval [*Clement and Robinson,* 1986]. In addition, the Messinian to early Pliocene of Site 1092 is represented by an interval of uninterpretable polarity stratigraphy [*Evans and Channell,* 2003], while also the Pliocene of Site 884 gives correlation problems. Consequently, this time interval still awaits more high-resolution, high-quality records, preferably accompanied by paleointensity data.

6. CONCLUSIONS

A detailed paleomagnetic resampling and remeasuring of DSDP Site 608 of the North Atlantic resulted in the registration of a complete magnetostratigraphic record from 13 to 7.5 Ma. In addition, nine short-term polarity fluctuations have been identified which do not correspond to subchrons in the GPTS. IRM component analysis show that there is no significant difference in magnetic composition between samples from the excursions and from neighboring polarity zones, implying that the polarity fluctuations reflect true variations of the past geomagnetic field. Comparison to other records show that all nine events have also been recognized in other sites from different part of the world.

If we follow the arguments of Gubbins [1999] and assume that short polarity subchrons should have a minimum overall duration of more than 9–15 kyr, five new subchrons are registered in the early Late Miocene record: C4r.2r-1n, C4Ar.1r-1n, C5r.2r-1n, C5r.2r-2n, and C5r.3r-1n. In addition, the three polarity fluctuations within C5n (C5n.2n-1, C5n.2n-2, C5n.2n-3) and the interval termed C5An.2n-1 in the paleomagnetic record of Site 608 qualify better as DIP-related excursions. The addition of just a few new subchrons makes the distribution of polarity interval lengths conform closer to a Poisson model wherein the probability of a reversal stays constant and does not depend on a previous occurrence [*Lowrie and Kent,* this volume].

Our results imply that the occurrence of short-term polarity fluctuations in the geomagnetic field is non-uniformly distributed through time, with several excursions and subchrons in the Brunhes-Matuyama (2.5–0 Myr) and early Late Miocene (12.5–8 Myr) intervals and none in the late Miocene to early Pliocene (8–2.5 Myr). Resolution problems in the marine magnetic anomaly profiles of the early Late Miocene time interval, however, may also help explain the discovery of new subchrons in magnetostratigraphic records in this time interval. In addition, poor quality paleomagnetic records could account for the apparent absence of short-term polarity fluctuations in the latest Miocene-Early Pliocene. Studies of the Oligocene, which is the next interval known for the common occurrence of cryptochrons, have not confirmed the existence of polarity subchrons. Instead, they suggested that these tiny wiggles most likely correspond to DIPs of the geomagnetic field. Consequently, it is uncertain if the observed non-uniform occurrence of short-term polarity fluctuations is truly related to specific behavior of the geodynamo. Nevertheless it remains interesting to notice that directional excursions have so far not been observed in the latest Miocene to early Pliocene records.

Acknowledgments. We thank Gar Esmay for his help with sampling of the Site 608 cores, Dave Schneider and Helen Evans for providing us with the data of ODP Site 845 and the adjusted record of ODP Site 1092, and Cor Langereis, Mark Dekkers and Hayfaa Abdul Aziz for fruitful discussions on an earlier version of the manuscript. We also acknowledge the constructive comments of Jim Channell and two anonymous reviewers. WK acknowledges financial support from the Netherlands Research Centre for Integrated Solid Earth Sciences (ISES) for his stay at LDEO. The work was carried out under the program of the Vening Meinesz Research School of Geodynamics. It is Lamont-Doherty Earth Observatory contribution #6553.

REFERENCES

Abdul Aziz, H., Astronomical forcing in continental sediments. An integrated stratigraphic study of Miocene deposits from the Calatayud and Teruel basins, NE Spain, *Geol.Ultraiectina*, *207*, 191 pp., 2001.

Abdul Aziz, H., F. Hilgen, W. Krijgsman, E. Sanz, and J. P. Calvo, Astronomical forcing of sedimentary cycles in the middle to late Miocene continental Calatayud Basin (NE Spain), *Earth Planet. Sci. Lett.*, *177*, 9–22, 2000.

Abdul Aziz, H., W. Krijgsman, F. J. Hilgen, D. S. Wilson, and J. P. Calvo, An astronomical polarity timescale for the late middle Miocene based on cyclic continental sequences, *J. Geophys. Res.*, *108*, EPM 5/1-16 doi: 10.1029/2002JB00181, 2003.

Abdul Aziz, H., and C. G. Langereis, Astronomical tuning and duration of three new subchrons (C5r.2r-1n, C5r.2r-2n and C5r.3r-1n) recorded in a Middle Miocene continental sequence from NE Spain, this volume, pp. 141–160

Blakely, R. J., Geomagnetic reversals and crustal spreading rates during the Miocene, *J. Geophys. Res.*, *79*, 2979–2985, 1974.

Bowers, N. E., S. C. Cande, J. S. Gee, J. A. Hildebrand, and R. L. Parker, Fluctuations of the paleomagnetic field during chron C5 as recorded in near-bottom marine magnetic anomaly data, *J. Geophys. Res.*, *106*, 26,379–26,396, 2001.

Bowles, J., L. Tauxe, J. Gee, D. McMillan, and S. C. Cande, The source of tiny wiggles in Chron C5: A comparison of sedimentary relative intensity and marine magnetic anomalies, *Geochem. Geophys. Geosyst.*, 4(6), 1049, doi:10.1029/2002 GC000489, 2003.

Cande, S. C., and D. V. Kent, A new geomagnetic polarity time scale for the Late Cretaceous and Cenozoic, *J. Geophys. Res.*, *97*, 13,917–13,951, 1992a.

Cande, S. C., and D. V. Kent, Ultrahigh resolution marine magnetic anomaly profiles: A record of continuous paleointensity variations?, *J. Geophys. Res.*, *97*, 15,075–15,083, 1992b.

Cande, S. C., and D. V. Kent, Revised calibration of the geomagnetic polarity time scale for the Late Cretaceous and Cenozoic, *J. Geophys. Res.*, *100*, 6093–6095, 1995.

Cande, S. C., and J. L. LaBrecque, Behaviour of the Earth's paleomagnetic field from small-scale marine magnetic anomalies, *Nature*, *247*, 26–28, 1974.

Celaya, M., and B. M. Clement, Inclination shallowing in deep-sea sediments from the North Atlantic, *Geophys. Res. Lett.*, *15*, 52–55, 1988.

Censarek, B., and R. Gersonde, Miocene diatom biostratigraphy at ODP sites 689, 690, 1088, 1092 (Atlantic sector of the Southern Ocean), *Marine Micropal.*, *45*, 309–356, 2002.

Channell, J. E. T., A. Mazaud, P. Sullivan, S. Turner, and M. E. Raymo, Geomagnetic excursions and paleointensities in the 0.9–2.15 Ma interval of the Matuyama Chron at ODP Site 983 and 984 (Iceland Basin), *J. Geophys. Res.*, *107 B(6)*, doi: 10.1029/2001JB000491, 2002.

Channell, J. E. T., S. Galeotti, E. E. Martin, K. Billups, H. D. Scher, and J. S. Stoner, Eocene to Miocene magnetostratigraphy, biostratigraphy, and chemostratigraphy at ODP Site 1090 (sub-Antarctic South Atlantic), *GSA Bull.*, *115*, 2003.

Clement, B. M., and D. V. Kent, Short polarity intervals within the Matuyama: transitional field records from hydraulic piston cored sediments from the North Atlantic, *Earth Planet. Sci. Lett.*, *81*, 253–264, 1987.

Clement, B. M., and F. Robinson, *The magnetostratigraphy of Leg 94 sediments*, 635–650 pp., U.S. Government Printing Office, Washington, D.C., 1986.

Evans, H. F., and J. E. T. Channell, Upper Miocene magnetic stratigraphy at ODP site 1092 (sub-Antarctic South Atlantic): recognition of 'cryptochrons' in C5n.2n, *Geophys. J. Int.*, *153*, 483–496, 2003.

Evans, H. F., Westerhold, T., and J .E. T. Channell, Revised composite depth section has implications for Upper Miocene "cryptochrons", *Geophys. J. Int.*, *156*, 195–199, 2004.

Garcés, M., J. Agustí, L. Cabrera, and J. M. Parés, Magnetostratigraphy of the Vallesian (late Miocene) in the Vallès-Penedès Basin (northeast Spain), *Earth Planet. Sci. Lett.*, *142*, 381–396, 1996.

Garcés, M., W. Krijgsman, and J. Agustí, Chronostratigraphic framework and evolution of the Fortuna Basin (Eastern Betics) since the Late Miocene, *Basin Research*, *13(2)*, 199–216, 2001.

Gartner, S., Miocene nannofossil chronology in the North Atlantic, DSDP Site 608, *Marine Micropal.*, *18*, 307–313, 1992.

Gubbins, D., The distinction between geomagnetic excursions and reversals, *Geophys. J. Int. -Fast Track*, *137*, F1–F3, 1999.

Hilgen, F. J., W. Krijgsman, C. G. Langereis, L. J. Lourens, A. Santarelli, and W. J. Zachariasse, Extending the astronomical (polarity) time scale into the Miocene, *Earth Planet. Sci. Lett.*, *136*, 495–510, 1995.

Hollerbach, R., and C. A. Jones, Influence of the Earth's inner core on geomagnetic fluctuations and reversals, *Nature*, *365*, 541–543, 1993.

Krijgsman, W., F. J. Hilgen, C. G. Langereis, A. Santarelli, and W. J. Zachariasse, Late Miocene magnetostratigraphy, biostratigraphy and cyclostratigraphy in the Mediterranean, *Earth Planet. Sci. Lett.*, *136*, 475–494, 1995.

Krijgsman, W., F. J. Hilgen, I. Raffi, F. J. Sierro, and D. S. Wilson, Chronology, causes and progression of the Messinian salinity crisis, *Nature*, *400*, 652–655, 1999.

Kruiver, P. P., M. J. Dekkers, and D. Heslop, Quantification of magnetic coercivity components by the analysis of acquisition curves of isothermal remanent magnetisation, *Earth Planet. Sci. Lett.*, *189*, 269–276, 2001.

Kruiver, P. P., and H. F. Passier, Coercivity analysis of magnetic phases in sapropel S1 related to variations in redox conditions, including an investigation of the S-ratio, *Geochem. Geophys. Geosyst.*, doi:*2001GC000181*, 2001.

Lanci, L., and W. Lowrie, Magnetostratigraphic evidence that 'tiny wiggles' in the oceanic magnetic anomaliy record represent geomagnetic paleointensity variations, *Earth Planet. Sci. Lett.*, *148*, 581–592, 1997.

Langereis, C. G., M. J. Dekkers, G. J. de Lange, M. Paterne, and P. J. M. van Santvoort, Magnetostratigraphy and astronomical calibration of the last 1.1 Myr from an eastern Mediterranean piston core and dating of short events in the Brunhes, *Geophys. J. Int.*, *129*, 75–94, 1997.

Langereis, C. G., and F. J. Hilgen, The Rosello composite: a Mediterranean and global reference section for the Early to early Late Pliocene, *Earth Planet. Sci. Lett.*, *104*, 211–225, 1991.

Li, J.-J., X.-M. Fang, R. Van der Voo, J.-J. Zhu, C. MacNiocaill, Y. Ono, B.-T. Pan, W. Zhong, J.-L. Wang, T. Sasaki, Y.-T. Zhang, J.-X. Cao, S.-C. Kang, and J.-M. Wang, Magnetostratigraphic dating of river terraces: Rapid and intermittent incision by the Yellow River of the northeastern margin of the Tibetan Plateau during the Quaternary, *J. Geophys. Res.*, *102*, 10,121–10,132, 1997a.

Li, J. J., X. M. Fang, R. V.d. Voo, J .J. Zhu, C. MacNiocaill, J. X. Cao, W. Zhong, H. L. Chen, J. L. Wang, J. M. Wang, and Y. C. Zhang, Late Cenozoic magnetostratigraphy (11–0 Ma) of the Dongshanding and Wangjiashan sections in the Longzhong Basin, western China, *Geologie en Mijnbouw*, *76*, 121–134, 1997b.

Lowrie, W., and D. V. Kent, Geomagnetic polarity timescale and reversal frequency regimes, this volume, pp. 117–129.

Lowrie, W., and L. Lanci, Magnetostratigraphy of Eocene-Oligocene boundary sections in Italy: No evidence for short subchrons within chrons 12R and 13R, *Earth Planet. Sci. Lett.*, *126*, 247–258, 1994.

Lund, S. P., G. Acton, B. Clement, M. Hastedt, M. Okada, T. Williams, and L.S. Party, Geomagnetic field excursions occurred often during the last million years, *EOS, Trans. AGU*, 178–179, 1998.

Merrill, R. T., M. W. McElhinny, and P. L. McFadden, *The Magnetic Field of the Earth: Paleomagnetism, the Core, and the Deep Mantle*, 531 pp., Academic Press, San Diego, 1996.

Miller, K. G., M. D. Feigenson, J. D. Wright, and B. M. Clement, Miocene isotope reference section, Deep Sea Drilling Project Site 608: an evaluation of isotope and biostratigraphic resolution, *Paleoceanography*, *6*, 33–52, 1991.

Roberts, A. P., and J. C. Lewin-Harris, Marine magnetic anomalies: evidence that 'tiny wiggles' represent short-period geomagnetic polarity intervals, *Earth Planet. Sci. Lett.*, *183*, 375–388, 2000.

Roperch, P., G. Herail, and M. Fornari, Magnetostratigraphy of the Miocene Corque basin, Bolivia: Implications for the geodynamic evolution of the Altiplano during the late Tertiary, *J. Geophys. Res.*, *104*, 20,415–20,429, 1999.

Rösler, W., and E. Appel, Fidelity and time resolution of the magnetostratigraphic record in Siwalic sediments: high-resolution study of a complete polarity transition and evidence for cryptochrons in a Miocene fluviatile section, *Geophys. J. Int.*, *135*, 861–875, 1998.

Ruddiman, W. F., R. B. Kidd, E. Thomas, and e. al, Site 608, in: *Init. Rep. DSDP*, edited by W. F. Ruddiman, R. B. Kidd, and E. Thomas, pp. 149–246, U.S. Government Printing Office, Washington, D.C., 1987.

Schneider, D. A., Paleomagnetism of some Leg 138 sediments: Detailing Miocene magnetostratigraphy, *Proc. ODP Sci. Res.*, *138*, 59–72, 1995.

Schneider, D. A., J. Backman, W. P. Chaisson, and I. Raffi, Miocene calibration for calcareous nannofossils from low-latitude Ocean Drilling Program sites and the Jamaican conundrum, *GSA Bull.*, *109*, 1073–1079, 1997.

Singer, B. S., K. A. Hoffman, A. Chauvin, R. S. Coe, and M. S. Pringle, Dating transitionally magnetized lavas of the late Matuyama Chron: Toward a new ^{40}Ar/^{39}Ar timescale of reversals and events, *J. Geophys. Res.*, *104*, 679–693, 1999.

Tauxe, L., and P. Hartl, 11 million years of Oligocene geomagnetic field behaviour, *Geophys. J. Int.*, *128*, 217–229, 1997.

Tauxe, L., and N. D. Opdyke, A time framework based on magnetostratigraphy for the Siwalik sediments of the Khaur Area, Northern Pakistan, *Paleogeogr., Paleoclimatol. Paleoecol.*, *37*, 43–61, 1982.

Van Vugt, N., J. Steenbrink, C. G. Langereis, F. J. Hilgen, and J. E. Meulenkamp, Sedimentary cycles in the early Pliocene lacustrine sediments of Ptolemais (NW Greece) correlated to insolation and to the marine Rossello section (S. Italy), *Earth Planet. Sci. Lett.*, *164*, 535–551, 1998.

Weeks, R. J., A. P. Roberts, K. L. Verosub, M. Okada, and G. J. Dubuisson, Magnetostratigraphy of upper Cenozoic sediments from leg 145, North Pacific Ocean,, *Proc. ODP Sci. Res.*, *145*, 491–521, 1995.

D. V. Kent, Department of Geological Sciences, Rutgers University, Piscataway, New Jersey 08854, and Lamont Doherty Geological Observatory, Palisades, New York 10964. (dvk@ldeo.columbia.edu)

W. Krijgsman, Paleomagnetic laboratory "Fort Hoofddijk", Utrecht University, Budapestlaan 17, 3584 CD Utrecht, Netherlands. (krijgsma@geo.uu.nl)

^{40}Ar/^{39}Ar Chronology of Late Pliocene and Early Pleistocene Geomagnetic and Glacial Events in Southern Argentina

Brad S. Singer[1], Laurie L. Brown[2], Jorge O. Rabassa[3], and Hervé Guillou[4]

K-Ar dating and paleomagnetic directions from the lava sequence atop Cerro del Fraile, Argentina, contributed to the nascent Geomagnetic Polarity Time Scale (GPTS), recording the Réunion event, and the Olduvai and Jaramillo subchrons [*Fleck et al.*, 1972]. New stratigraphy, paleomagnetic analyses, ^{40}Ar/^{39}Ar incremental heating ages, and unspiked K-Ar dating of 10 lava flows on Cerro del Fraile place these eruptions between 2.181±0.097 and 1.073±0.036 Ma and enhance this unique record, which includes seven tills interbedded with the lavas. The Réunion event is recorded by three lavas with transitional, normal, and reversed polarity that yielded identical ^{40}Ar/^{39}Ar isochron ages and a weighted mean age of 2.136±0.019 Ma. When combined with ^{40}Ar/^{39}Ar ages from lavas on Réunion Island and a normal tuff in the Massif Central, the age of the Réunion event is 2.137±0.016 Ma and is older by ~50 kyr than the 2.086±0.016 Ma Huckleberry Ridge event. The onset and termination of the Olduvai are similarly constrained to 1.922±0.066 Ma and 1.775±0.015 Ma, whereas the onset of the Jaramillo occurred 1.069±0.011 Ma. A discordant age spectrum from another transitional lava gave a total fusion age of 1.61 Ma and an unspiked K-Ar age of 1.43 Ma. It is uncertain whether this corresponds to the Gilsa, Gardar, Stage 54, or Sangiran events, or represents an unrecognized period of geomagnetic instability. Deposition of till on the piedmont surface prior to 2.186 Ma and six subsequent tills between 2.186 Ma and ~1.073 Ma mark frequent glaciations of southern South America during marine oxygen isotope stages 82 to 48.

1. INTRODUCTION

Astrochronologic dating of sediments coupled with precise ^{40}Ar/^{39}Ar ages of key lava and tuff sequences triggered a major revision of the Geomagnetic Polarity Time Scale (GPTS) over the past dozen years [*Shackleton et al.*, 1990*; Hilgen*, 1991*; Cande and Kent*, 1995*; Berggren et al.*, 1995]. In addition to improving global stratigraphic correlation and validating space geodesy measurements of plate motions [*e.g., Baksi*, 1994], an accurate and precise chronology of geomagnetic reversals and short-lived events or subchrons is critical to understanding how the Earth's magnetic field originates and is modulated within the core and lowermost mantle [*Gubbins*, 1999*; Glatzmaier et al.*, 1999*; Singer et al.*, 2002*; Hoffman and Singer*, this volume]. Yet, the existence and precise

[1]Department of Geology and Geophysics, University of Wisconsin-Madison, Wisconsin.

[2]Department of Geosciences, University of Massachusetts, Amherst, Massachusetts.

[3]Laboratorio de Geología del Cuaternario, CADIC-CONICET, Ushuaia, Tierra del Fuego, Argentina.

[4]Laboratoire des Sciences du Climat et de l'Environnement, Domaine du CNRS, Gif-sur-Yvette, France.

Timescales of the Paleomagnetic Field
Geophysical Monograph Series 145
Copyright 2004 by the American Geophysical Union
10.1029/145GM13

timing of several reversals or aborted reversal attempts, recorded as short-lived polarity events, including for example the Réunion event, are still disputed [*Kidane et al.*, 1999; *Baksi and Hoffman* 2000; *Baksi*, 2001; *Lanphere et al.*, 2002]. This reflects obstacles to magnetostratigraphic correlation among critical sedimentary sequences, difficulty in assigning uncertainties to astrochronologic ages, a limited number of salient volcanic materials available for study, and different K-Ar and ^{40}Ar/^{39}Ar approaches that have been used in various laboratories to date these volcanic rocks.

More than three decades ago, the Plio-Pleistocene lava sequence at Cerro del Fraile, Argentina yielded important confirmation of the fledgling GPTS, specifically revealing in a single, continuous, radioisotopically dated section a record of the three normal polarity events that occurred during the Matuyama reversed chron [*Fleck et al.*, 1972]. The information came from seven successive basaltic lava flows on which Fleck et al. [1972] obtained replicate K-Ar ages and paleomagnetic directions from hand specimens that were oriented in the field. We revisited this classic lava flow sequence that is interbedded with several glacial tills with goals of assessing in further detail its stratigraphy, and revising the chronology of geomagnetic and glacial events recorded through modern ^{40}Ar/^{39}Ar dating and paleomagnetic analysis.

We report the results of new stratigraphy, paleomagnetic analyses, ^{40}Ar/^{39}Ar incremental heating experiments, and unspiked K-Ar dating of a sequence of 10 lava flows from Cerro del Fraile. In light of these data, the timing of the Réunion event, the Olduvai subchron, and the Jaramillo subchron that are recorded in these lavas are critically examined and an initial Geomagnetic Instability Time Scale [GITS; *Singer et al.*, 2002] for the interval between 2.14 and 0.79 Ma is presented. Moreover, the temporal constraints provided by the combination of magnetic and radioisotopic data provide a unique opportunity to further quantify the number and timing of major glaciations of the southern Andes and to comment briefly on the pace of Pleistocene landscape evolution.

2. GEOLOGIC SETTING, STRATIGRAPHY, AND SAMPLING

Cerro del Fraile (50.5° S, 72.7° W) is an 8 km^2 mesa located 40 km east of the Andean Cordillera and 10 km south of Lago Argentino (Figure 1) that comprises Cretaceous sediment

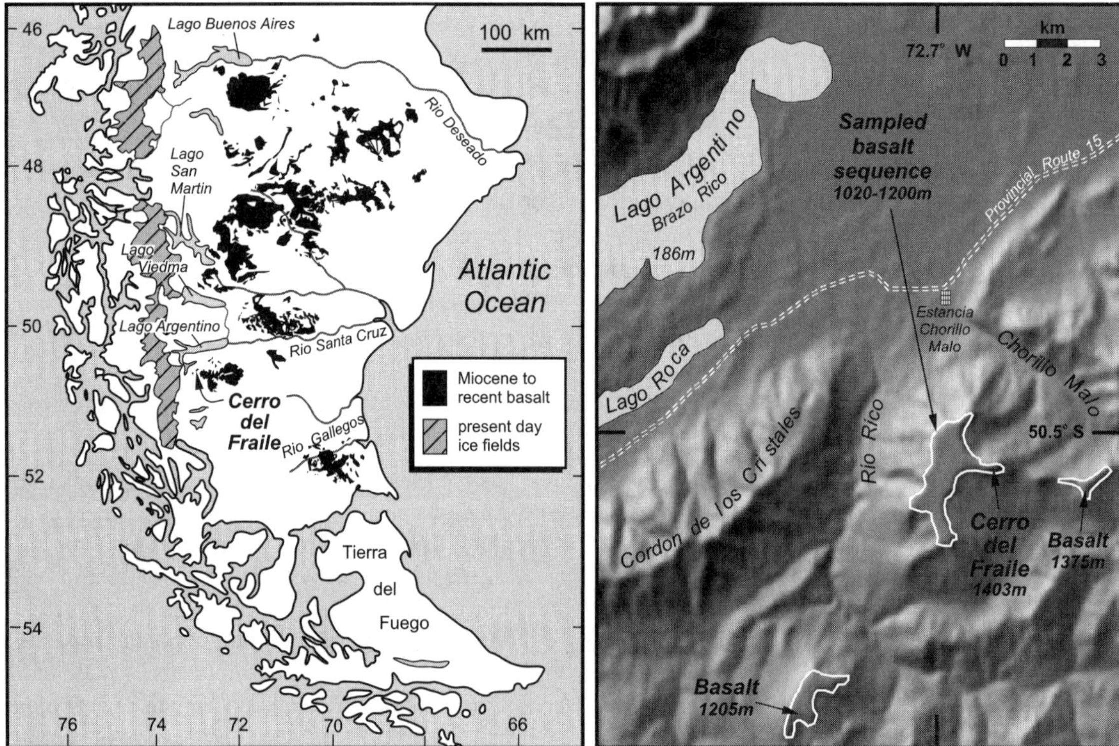

Figure 1. Location of Cerro del Fraile in southwestern Santa Cruz province, Argentina. The three mesas outlined in white on the shadowed digital elevation model comprise Plio-Pleistocene basalt flows or eroded volcanic necks that overlie Cretaceous marine sediment. Relief between Lago Argentino and the top of Cerro del Fraile is 1200 m. Access to the sampled section is by foot from Estancia Chorillo Malo on the gravel provincial route 15.

unconformably capped by Plio-Pleistocene basaltic lava flows [*Feruglio*, 1944; *Mercer*, 1969; *Fleck et al.*, 1972]. Where sampled by Mercer [1969] and Fleck et al. [1972] these lavas crop out between 1020 and 1200 masl and are interbedded with a series of glacial tills and fluvial sediments (Figure 2). Feruglio [1944] and Fleck et al. [1972] noted that the stratigraphy of the Plio-Pleistocene units varies along the northwestern exposure with some flows pinching out or thickening laterally. The section sampled by Mercer and described by Fleck et al. [1972] includes 8 lava flows which they numbered sequentially A–H and 6 interbedded glacial tills. Contact relations between the lava flows and tills, including baked soil atop the thickest tills, cobbles and fluvial sediment in some of the tills, and lack of ice-contact textures in any of the basalt flows, led Fleck et al. [1972] to conclude that glaciers advanced eastward at least six times across Cerro del Fraile. The ensuing tills, coarse outwash sediment, and soil were episodically buried and baked by the thick basaltic flows. We sampled an adjacent section through the exposure of Plio-Pleistocene units in 1996 and 1998 and found a vertical sequence of lava flows that we labeled 1 through 10 which are interbedded with at least 7 tills.

The stratigraphic sequence and relationship of the lavas analyzed by Fleck et al. [1972] to those which we have measured are illustrated in Figure 2. The key differences are that we identified and sampled a 2–3 m thick, discontinuous flow between Fleck et al.'s Flows E and F, and a 6–8 m thick, columnar-jointed and relatively continuous flow, overlain by a till between Fleck et al.'s Flows F and G (Figure 2). The latter till, which is 3–5 m thick, underlies our Flow 9 (Fleck et al.'s Flow G); however, we determined that the highest till, comprising a sparse lag of striated cobbles and granitic boulders up to 2 m in diameter, covers only the western portion of the mesa and our Flow 9 (Fleck et al.'s G; Figure 2). We did not find this highest (7th) till in the sequence underlying our Flow 10 (Fleck et al.'s H), which covers only the eastern third of the mesa (Figures 1 and 2).

Based on replicate analyses of whole-rock samples from 7 of the lavas, Fleck et al. [1972] obtained conventional K-Ar ages that range from 2.12±0.07 to 1.05±0.06 Ma when recalculated using the decay constants of Steiger and Jäger [1977] and weighting the ages by the inverse-variance [*Taylor*, 1982]. Excepting the lowermost flow, from which two of four experiments yielded a consistent age of 2.05 ± 0.04 Ma, the K-Ar ages agree with the stratigraphy (Figure 2). Progressive four-step alternating field (AF) demagnetization procedures used by Fleck et al. [1972] on 3–4 cores from oriented hand specimens of 5 of the same 7 flows yielded either normal or reversed magnetization directions (Figure 2). Flow B yielded an intermediate direction with a Virtual Geomagnetic Pole (VGP) of 32° N, 307°W [*Fleck et al.*, 1972; Figure 2]. For paleomagnetic meas-

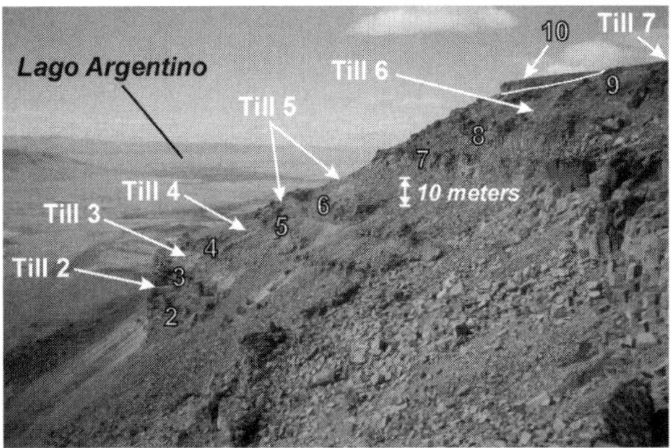

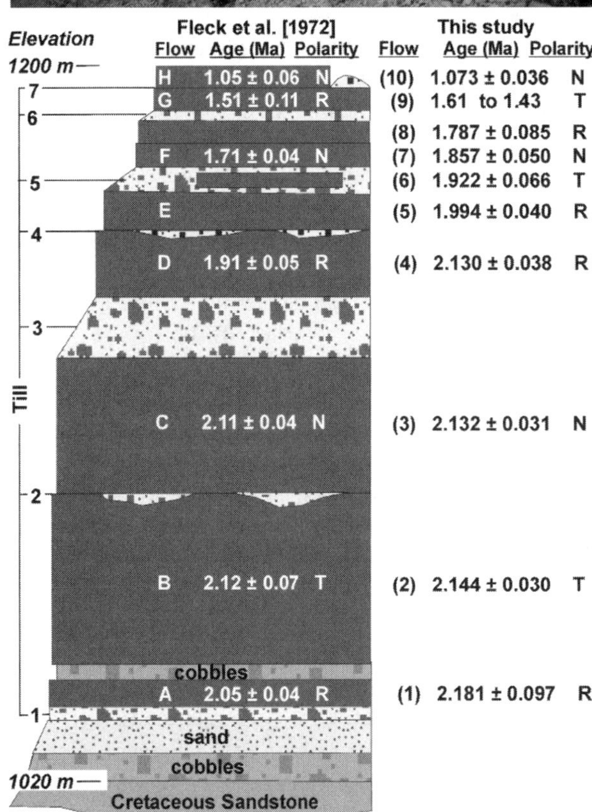

Figure 2. Stratigraphy of the upper 200 m on the western edge of Cerro del Fraile. Top: View north along the mesa. Basaltic lava flows are numbered 2–10. Six glacial tills are interbedded in this portion of the section. The oldest till and the overlying basalt flow number 1 are not visible. The large main body of Lago Argentino is visible 30 km in the distance to the northeast 1000 m below these lavas. Bottom: summary of the basalt and till sequence. Flow sequence A–H of Fleck et al. [1972] is shown with mean K-Ar ages updated using modern decay constants and with errors of replicate analyses weighted by the inverse of their variance, and the measured paleomagnetic directions. ^{40}Ar/^{39}Ar isochron and unspiked K-Ar ages for the 10 flows analyzed in this study are listed along with their paleomagnetic directions (N = normal polarity; R = reversed; T = transitional).

urements, we collected up to 10 cores over a several m^2 area on 8 of the 10 lava flows using a gasoline-powered, water-cooled diamond coring drill. These 2.5 cm diameter cores were oriented using magnetic and sun compasses. Because we ran out of water and sunlight at the end of a single day of drilling, cores from Flows 6 and 10 were drilled later from 50x20x30 cm oriented blocks. Samples for geochronology were chiseled from the outcrops at the same locations spanned by the drilling.

3. ^{40}Ar/^{39}Ar AND K-Ar WORK

3.1. Geochronologic Standards and the GPTS

The ^{40}Ar/^{39}Ar method requires that the age of a sample be calculated relative to a mineral standard which has been previously dated, usually by conventional K-Ar techniques. This has occasionally led to confusion when comparing ages determined in different laboratories and to erroneous conclusions regarding the timing of geomagnetic events [*e.g., Baksi,* 2001; *Lanphere et al.,* 2002]. The monitor minerals used here were sanidines from the Taylor Creek (TCs) and Alder Creek (ACs) rhyolites. The age of TCs was determined to be 27.92 Ma relative to the USGS primary standard SB-3 biotite at 162.9 Ma [*Duffield and Dalrymple,* 1990; *Lanphere and Dalrymple,* 2000]. *Turrin et al.* [1994] measured the age of ACs at 1.186 Ma relative to sanidine from the Fish Canyon tuff (FCs) with an age of 27.84 Ma taken from Cebula et al. [1986]. However, in light of recent intercalibration of these ^{40}Ar/^{39}Ar standards relative to 98.79±0.96 Ma GA-1550 biotite, we have adopted ages of 28.34±0.16 Ma for TCs and 1.194±0.007 Ma for ACs [*Renne et al.,* 1998], though a consensus regarding the age of the GA-1550 standard has not yet been reached [*Lanphere and Dalrymple,* 2000; *Lanphere et al.,* 2002].

Our decision to use the intercalibrated standard ages of Renne et al. [1998] reflects, in part, their consistency with astrochronologically determined ages of several magnetic chron boundaries. Adopting an age of 27.92 Ma for the TCs standard would shift the ages reported here 1.5% younger. Conversion of our ages to make them consistent with yet other values for standard ages [*e.g., Baksi et al.,* 1996; *Villeneuve et al.,* 2000] can be done simply using the equation in Dalrymple et al. [1993]. Via the intercalibration of Renne et al. [1998], the ages reported here correspond to an age of 28.02 Ma for the Fish Canyon sanidine (FCs) standard, that yields ages 0.6% older than the value of 27.84 Ma which was adopted by Cande and Kent [1995] and Berggren et al. [1995] to revise portions of the GPTS. The significance of the values chosen for standard minerals will become apparent below where our results are compared against published radioisotopic age determinations that bear on the timing of the Réunion event, Olduvai subchron, and other geomagnetic events.

In this paper, we calculate ^{40}Ar/^{39}Ar and K-Ar ages using the decay constants of Steiger and Jäger [1977] and report ±2F uncertainties that include analytical and inter-calibration terms. This is appropriate for comparing our results to ^{40}Ar/^{39}Ar ages from other studies, provided all ages are calculated relative to a common standard value as described above. An additional source of systematic error arises from uncertainty in the ^{40}K decay constant [*Renne et al.,* 1998], that contributes between 0.1 and 0.4% of uncertainty to each of our age determinations. Adding this latter error component is appropriate should our ages be com pared directly independent chronometers including U-Pb zircon or astrochronologic ages [*Renne et al.,* 1998].

3.2. The ^{40}Ar/^{39}Ar Methods

^{40}Ar/^{39}Ar dating of the basaltic lavas followed the procedures outlined in Singer et al. [2004]. To avoid xenocrystic contamination, 125–315 µm holocrystalline groundmass separates of 50 or 100 mg each were prepared by crushing, sieving, magnetic sorting, and hand-picking under a binocular microscope. These were wrapped into 99.99% Cu foil packets and along with several packets of TCs or ACs standard grains were loaded into 5 mm i.d. quartz vials that were evacuated and sealed. Samples were irradiated for 1 or 2 hours at the Oregon State University Triga reactor in the Cadmium-Lined In-Core Irradiation Tube (CLICIT). Corrections for undesirable nucleogenic reactions on ^{40}K and ^{40}Ca, based on previous measurements of Ca- and K- free salts [*Wijbrans et al.,* 1995], are [^{40}Ar/^{39}Ar]$_K$ = 0.00086; [^{36}Ar/^{37}Ar]$_{Ca}$ = 0.000264; [^{39}Ar/^{37}Ar]$_{Ca}$ = 0.000673. Incremental degassing using a resistance furnace, temperature measurement, gas clean-up, mass spectrometry, mass discrimination and blank corrections, and standard measurements using a CO$_2$ laser were similar to Singer et al. [2004].

For each analysis uncertainties include estimates of the analytical precision on peak signals, the system blank, and spectrometer mass discrimination. Inverse-variance weighted mean plateau ages and standard deviations were calculated according to Taylor [1982] and uncertainties multiplied by the square root of the MSWD where it is >1. The uncertainty in J, the neutron fluence parameter, was 0.5% to 0.6% (2σ); this uncertainty was also propagated into the final plateau and isochron ages for each analysis, yet it contributes less than 0.1% to the total uncertainty in these age determinations. For comparing to astro-chronologic ages, fully propagated uncertainties including the decay constant term are reported alongside the analytical uncertainties in Table 1.

Criteria used to determine whether an incremental heating experiment gave meaningful results were: (1) plateaus must be defined by at least three contiguous steps all concordant in age at the 95% confidence level and comprising >50% of the ^{39}Ar

Table 1. Summary of $^{40}Ar/^{39}Ar$ incremental heating experiments on basalt groundmass samples, Cerro del Fraile, Argentina

Flow #	Sample ID	Experiment	wt. mg	K/Ca total	Total fusion Age (Ma) ± 2σ	Increments used, °C	^{39}Ar %	Age Spectrum Age (Ma) ± 2σ	MSWD	N	SUMS (N-2)	$^{40}Ar/^{36}Ar_i$ ± 2σ	Isochron Analysis Age (Ma) ± 2σ[d]	± 2σ[e]	Mag
10	CF-09	98G1342[a]	50	1.000	1.021 ± 0.096	830-1080	48.2	1.078 ± 0.015	0.72	4/11	1.03	297.0 ± 10.0	1.073 ± 0.036	± 0.041	N
9	CF-08	97GE629[a]	50	1.313	1.612 ± 0.060			*Discordant spectrum, use total fusion and unspiked K-Ar age as limits*							T
8	CF-07	96GE615[a]	50	0.362	1.778 ± 0.018	750-920	58.4	1.775 ± 0.020	0.68	7/16	0.82	294.0 ± 12.0	1.787 ± 0.085	± 0.090	R
7	CF-06	96GE614[a]	50	0.298	1.817 ± 0.015	675-925	72.1	1.834 ± 0.010	1.19	9/14	1.20	291.5 ± 8.4	1.857 ± 0.050	± 0.059	N
6	CF-11	UW10E99[b]	50	0.278	1.946 ± 0.238	775-920	64.6	2.052 ± 0.044	0.63	5/12	0.45	303.7 ± 3.6	1.922 ± 0.066	± 0.074	T
5	CF-05	96GE613[a]	50	0.494	1.973 ± 0.015	600-950	96.1	1.993 ± 0.010	2.40[c]	13/16	2.50[c]	295.0 ± 11.0	1.994 ± 0.040	± 0.053	R
4	CF-04	96GE612[a]	50	0.761	2.134 ± 0.014	820-1030	59.9	2.108 ± 0.011	4.60[c]	9/18	4.40[c]	289.0 ± 10.0	2.130 ± 0.038	± 0.053	R
3	CF-03	UW01E55[b]	100	0.330	2.160 ± 0.030	820-1080	94.0	2.163 ± 0.021	3.10[c]	9/14	2.20	296.4 ± 4.4	2.150 ± 0.045	± 0.058	
		96GE611[a]	50	0.397	2.096 ± 0.026	800-1020	71.7	2.132 ± 0.016	1.80	8/18	2.20	300.0 ± 14.4	2.106 ± 0.091	± 0.098	
								Combined isochron:		17/32	3.70[c]	297.4 ± 3.4	2.132 ± 0.031	± 0.048	N
2	CF-02	UW01E54[b]	100	0.306	2.100 ± 0.030	865-1150	81.7	2.160 ± 0.028	4.20[c]	8/16	0.25	296.0 ± 12.0	2.155 ± 0.079	± 0.087	
		96GE610[a]	50	0.431	2.290 ± 0.096	525-1140	100.0	2.280 ± 0.180	0.24	22/22	5.10[c]	295.8 ± 7.1	2.280 ± 0.180	± 0.184	
								Combined isochron:		30/38	1.40	298.4 ± 3.9	2.144 ± 0.030	± 0.048	T
1	CF-01	96GE609[a]	50	0.237	2.420 ± 0.098	610-1100	57.3	2.270 ± 0.120	0.69	12/14	0.65	300.2 ± 3.4	2.181 ± 0.097	± 0.104	R

[a]Measured in Geneva, ages calculated relative to 28.34 Ma Taylor Creek Rhyolite sanidine [*Renne et al.*, 1998].
[b]Measured in Wisconsin, ages calculated relative to 1.194 Ma Alder Creek Rhyolite sanidine [*Renne et al.*, 1998].
[c]MSWD or SUMS/(N-2) larger than expected due to incorporating additional plateau increments (see text for discussion).
[d]Analytical and intercalibration uncertainties only.
[e]Fully propagated uncertainties, including ^{40}K decay constant (see text and *Renne et al.*, 1998).

released, and (2) a well-defined isochron exists for the plateau points as defined by the F-variate statistic *SUMS/(N-2)*. Because the isochron approach makes no assumption regarding the trapped component and combines estimates of analytical precision plus internal disturbance of the sample, the isochron ages (Table 1) are preferred over the weighted mean plateau ages. Many age spectra exhibit subtle discordances with small age differences (at the 95% confidence level) distinguishing ends of what otherwise would be statistically valid plateaus. Thus, for some experiments criteria 1, which is somewhat arbitrary, was relaxed to accommodate significantly larger fractions of gas into the age plateau and isochron calculations. Although this results in slightly excessive of MSWD and *SUMS/(N-2)* values, the reported isochron ages conservatively measure time since eruption.

3.3. The Unspiked K-Ar Method

Unspiked K-Ar dating differs from the conventional isotope dilution method in that argon extracted from the sample is measured in sequence with purified aliquots of an atmospheric argon standard at an identical total pressure during each measurement. This is done by changing the volume of the mass spectrometer via an adjustable bellows. Differences between the ^{40}Ar/^{36}Ar ratio of the atmospheric argon standard and the sample can reveal quantities of radiogenic ^{40}Ar* as small as 0.2% of the total ^{40}Ar [*e.g., Guillou et al.*, 1998]. The groundmass from Flow 9 yielded a strongly discordant age spectrum and no plateau. Thus, using methods completely described in Singer et al. [2004], replicate unspiked K-Ar age determinations (Table 2) were completed at Gif-sur-Yvette, France on sub-samples of the same material prepared for ^{40}Ar/^{39}Ar analysis. The volume of the spike-free inlet system is calibrated by measuring mineral standards of known molar ^{40}Ar* concentration, including GL-O glauconite, Mmhb-l horn-blende, LP-6 biotite, and HD-B1 biotite [see *Charbit et al.*, 1998, for details].

3.4. The ^{40}Ar/^{39}Ar and K-Ar Results

The dozen ^{40}Ar/^{39}Ar incremental heating experiments yielded nearly concordant to moderately discordant age spectra for all but Flow 9, which produced highly discordant results (Table 1, Figure 3). Plateau segments comprising 48–100% of the gas released characterize the other samples. As noted earlier, experiments on Flows 2, 3, 4, and 5 gave age spectra that met the rigorous criteria for defining an age plateau, but also additional gas steps that were only slightly discordant at the 95% confidence level (Figure 3). These few additional steps were included in the plateau and isochron age calculations, with resulting MSWD and *SUMS/(N-2)* values slightly higher than expected (Table 1; Figure 3). The relatively low apparent ages that characterize the low temperature increments from Flows 2, 3, and 4 suggest that minor argon loss, possibly associated with alteration, has affected these lavas. The age spectrum of Flow 4 suggests that in addition to minor argon loss, a small amount of ^{39}Ar recoil—leading to slightly decreasing apparent ages prior to the plateau (Figure 3)—may have occurred. Similarly, the strongly discordant spectrum from Flow 9 suggests that a combination of argon loss, accompanied by ^{39}Ar recoil—possibly reflecting petrographically undetectable alteration of matrix glass—have compromised this lava. The ^{40}Ar/^{36}Ar$_i$ values of 300.2±3.4 and 303.7±3.4 from Flows 1 and 6 indicate that small amounts of excess argon are present in these basalt flows, whereas the other lavas contain an initial trapped component indistinguishable from the atmosphere. The isochrons calculated from each lava flow, including the combination of replicate plateau analyses from Flows 2 and 3, give preferred ages between 2.181±0.097 and 1.073±0.036 Ma that agree with the stratigraphic succession (Figure 2).

The uncertainty estimates for the K-Ar ages reported by Fleck et al. [1972] were made by taking a simple standard deviation of the ages from the replicate experiments on each whole-rock sample without regard to the size of the associated analytical errors. This approach underestimates the uncertainty of these measurements. A more appropriate estimate of the uncertainty for each K-Ar age is to take the inverse-variance weighted mean age and uncertainty [*Taylor*, 1982], as is done to calculate the ^{40}Ar/^{39}Ar plateau ages. When uncertainties are estimated in this manner, the ^{40}Ar/^{39}Ar isochron ages for Flows 1, 2, 3, and 10 are indistinguishable at the 95% confidence level from the K-Ar ages determined by Fleck et al. [1972], but are more precise (Figure 2). In contrast, the isochron ages from Flows 4 and 7 are ~10% older than the K-Ar ages.

Table 2. Replicate unspiked K-Ar analyses, groundmass from 9th flow, Cerro del Fraile, Argentina[a]

Flow #	Sample	Weight Molten (g)	K (wt. %)	^{40}Ar* (%)	^{40}Ar* (10^{-12}mol/g)	Age (Ma) ± 2σ	Weighted mean age (Ma) ± 2σ
9	CF-08	1.42223	0.668 ± 0.007	6.331	1.651	1.424 ± 0.030	
	CF-08	1.12758	" "	6.284	1.654	1.427 ± 0.031	1.425 ± 0.021

[a]K by flame photometry. Herve Guillou analyst at Gif-sur-Yvette, France.

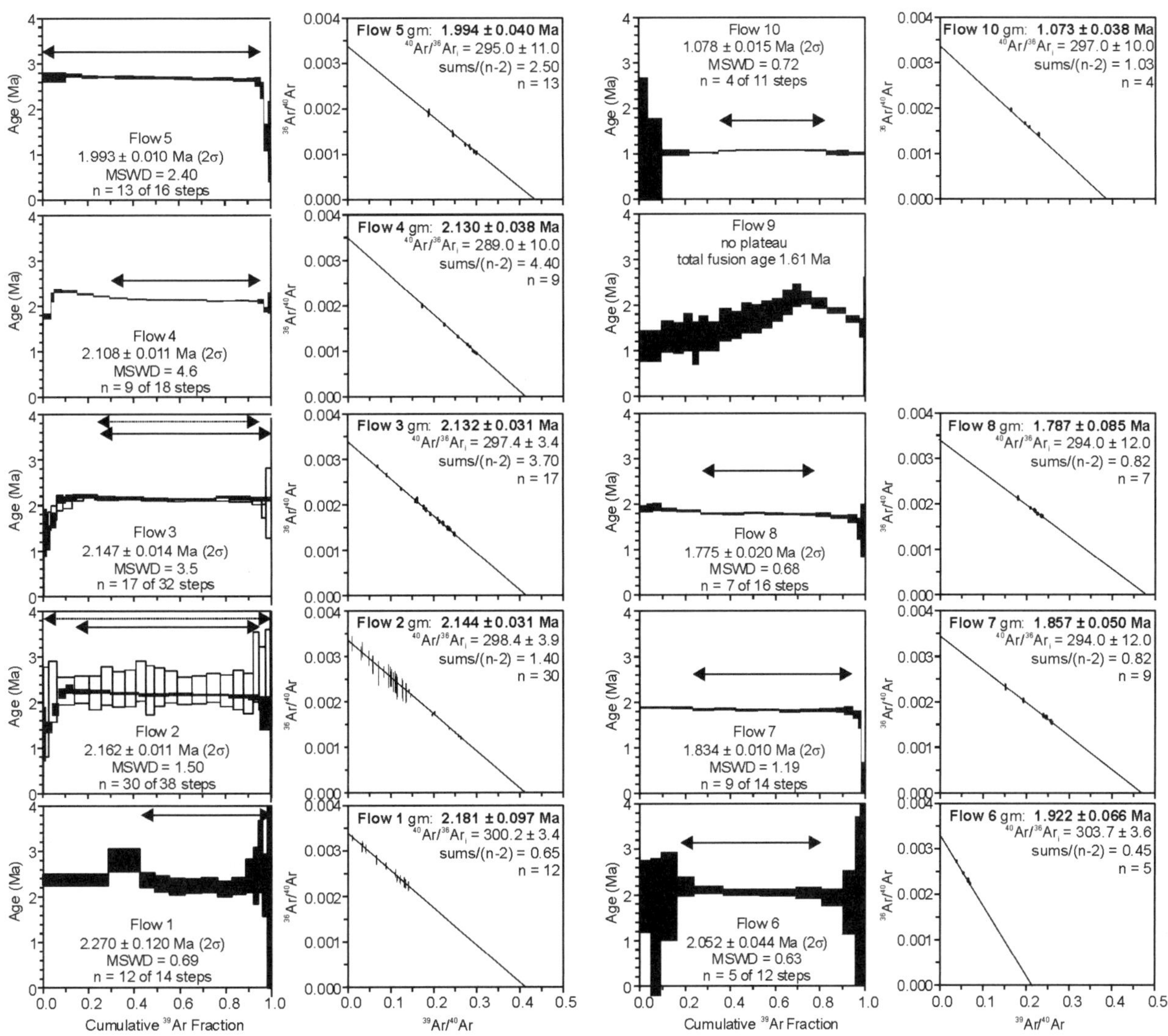

Figure 3. ^{40}Ar/^{39}Ar age spectra and inverse isochron diagrams for the 10 lava flows in Figure 2. Ages calculated relative to sanidines from the 28.34 Ma Taylor Creek Rhyolite sanidine or 1.194 ma Alder Creek Rhyolite (both equivalent to 28.02 Ma Fish Canyon Tuff sanidine).

The discordant ^{40}Ar/^{39}Ar result from Flow 9 (Figure 3) precluded generating an isochron for this lava. The total fusion age for this experiment, which should be equivalent to a K-Ar age, is 1.612±0.060 Ma; however, if the sample was affected by ^{39}Ar recoil to the extent that some ^{39}Ar leaked from it during irradiation, it is possible that unsupported radiogenic ^{40}Ar overestimates the time since eruption. Alternatively, argon loss during weathering may have lowered the apparent ages of the initial gas increments (Figure 3). The unspiked K-Ar age of Flow 9, 1.425±0.021 Ma (Table 2), is immune to ^{39}Ar recoil and significantly younger than the ^{40}Ar/^{39}Ar total fusion age. Its age is thus constrained to between 1.612±0.060 and 1.425±0.021 Ma and is identical to the K-Ar age of 1.51 ± 0.11 Ma determined by Fleck et al. [1972] (Figure 2).

4. PALEOMAGNETIC WORK

4.1. Methods

Magnetic data was processed in the Paleomagnetism Laboratory at the University of Massachusetts. Samples were measured on a 2G cryogenic magnetometer, model 755R, with alternating field and thermal demagnetization done using a Molspin AF demagnetizer and an ASC thermal demagnetizer, respectively. Susceptibility measurements were made using a Sapphire susceptibility meter.

Paleomagnetic samples were subjected to detailed demagnetization using both alternating field (AF) and thermal techniques. AF demagnetization was performed in steps from 0 to 100 mT, with 10 to 14 steps per sample. Thermal demagnetization, done at 10 or more temperature steps from room temperature to 600°C, used paired specimens with AF studies. Thermal demagnetization, as discussed below, turned out to be unsatisfactory and was not used further. Characteristic directions were determined for both demagnetization methods using line-fitting techniques [*Kirschvink*, 1980].

4.2. Results

Natural remanent magnetization (NRM) and demagnetization behavior was measured on a total of 73 samples from the 10 flows at Cerro del Fraile. Susceptibility measurements, made on each core and averaged for each site range, from 0.59 to 3.57 x 10^{-2} SI, with an average value for all flows of 1.54 x 10^{-2} SI. Mean natural remanent magnetization (NRM) prior to demagnetization is 2.73 A/m, but sites fall into two distinct groups (Table 3). Three flows (2, 7 and 9) have NRM values less than 1 A/m, with a mean of 0.75 A/m, while the remaining 7 flows have NRM values greater than 3 A/m, with a mean of 4.88 A/m.

Paired specimens from the same core for each site were demagnetized using both AF and thermal techniques with interesting results. AF demagnetization on both normal and reversed flows removed minor overprints at low levels (Figure 4A and C) with relatively straight-line decays to the origin. Median destructive fields are all greater than 20 mT and many are 50 to 60 mT. Cores yielding transitional directions were often an order of magnitude less intense, with overprints being removed by 20 mT (Figure 4D). Thermal demagnetization on companion samples showed removal of overprints up to 400°C (Figure 4B) and then slow decay towards the origin. The example shown in

Figure 4B is the best of the thermal data; other specimens gave much more erratic results. The majority of thermal demagnetization studies were unable to give line-fitting results with a maximum angle of deviation (MAD) <5°. Due to this questionable behavior under thermal demagnetization, AF demagnetization was used on remaining samples and final results all use data obtained in this way. All AF produced directions were obtained from line-fitting with MAD values <5°.

Inclinations and declinations for all the flows, along with associated statistics and virtual geomagnetic poles (VGP) are given in Table 3 and plotted in Figure 5. Most flows show excellent within site statistics, precision parameters (k) and 95% circles of confidence (σ_{95}) of Fisher [1953]. Using a VGP latitude of <45° as a determination of transitional directions, the sequence of flows from bottom to top give a pattern of R-T-N-R-R-T-N-R-T-N (Figure 2). Although this data set is not large enough to warrant time-averaged field observations, the mean of the 7 normal and reversed flows (I=-61.8°, D=355.5°, α_{95} = 6.7°) is similar to, but slightly shallower than, results from Meseta del Lago Buenos Aries, 400 km north of Cerro del Fraile [*Brown et al.*, 2004]. The mean inclination is 5.8° shallower than the expected inclination at this latitude of 67.6°

Of the 3 sites with low NRM values, two of them have transitional directions, flows 2 and 9. None of the transitional flows show randomly scattered NRM directions or high NRM intensities usually attributed to lightning strikes, such as observed in flows from the southwestern United States recently studied by Tauxe et al. [2003]. All 3 flows have α_{95} values of 8.0° or less, and appear to accurately record the magnetic field at the time of their emplacement. As will be discussed below, the geochronology on transitional flows 2 and 6 indicate correspondence to known polarity boundaries.

Table 3. Summary of site mean paleomagnetic data from lavas of Cerro del Fraile, Argentina[a]

Flow#	Site	N/No	Polarity	J (A/m)	X (10^{-2} SI)	INC	DEC	K	α_{95}	Pole Lat	Pole Long
10[b]	H	3/3	N	-	-	-63	343	29	23	77	145
9	PFC 14	6/6	T	0.815	1.514	-58.5	108.4	72	8.0	19.4	56.1
8	PFC 13	8/9	R	4.228	1.113	58.2	188.3	393	2.8	-76.9	137.1
7	PFC 12	6/7	N	0.884	1.884	-69.9	324.2	68	8.2	68.1	175.0
6	PFC 15	7/7	T	3.152	0.847	-59.4	152.2	181	4.5	4.0	86.4
5	PFC 11	8/8	R	5.527	2.189	56.4	182.1	119	5.1	-76.4	114.5
4	PFC 10	7/7	R	3.576	3.565	63.5	189.8	75	7.0	-81.5	161.3
3	PFC 9	9/9	N	3.125	1.864	-58.4	5.2	204	3.6	78.1	307.1
2	PFC 8	9/9	T	0.541	0.588	14.5	10.9	78	5.9	31.3	300.0
1	PFC 7	8/8	R	5.432	1.802	57.5	164.3	153	4.5	-73.3	59.5

[a]N/No, number used in calculations/number measured; J, magnetic intensity; X, magnetic susceptibility; INC, inclination; DEC, declination; K and α_{95}, precision parameter and radius of 95% confidence circle [*Fisher*, 1953].
[b]Directional data from Fleck et al. [1972].

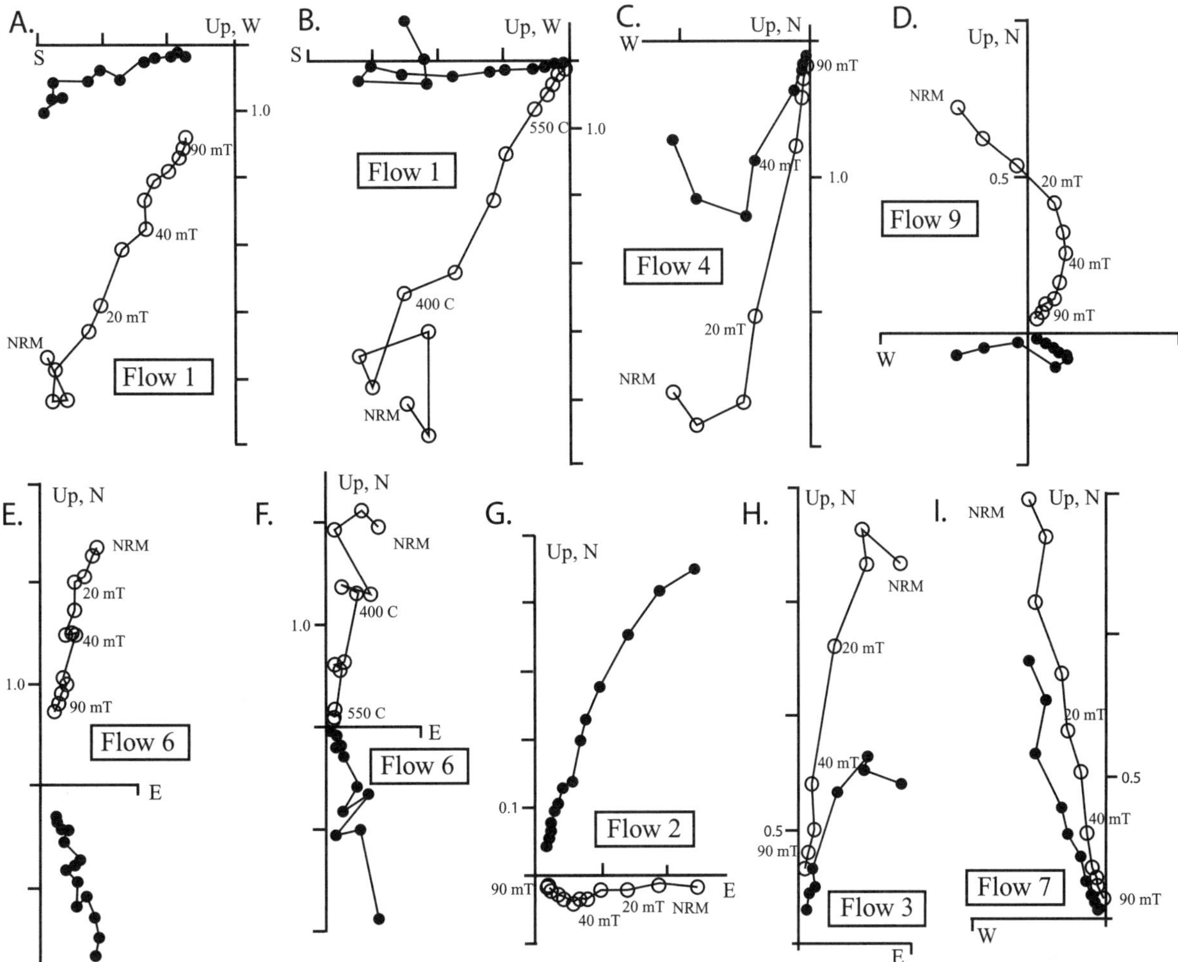

Figure 4. Vector end-point diagrams showing demagnetization behavior of normal, reversed, and transitional samples from Cerro del Fraile, Argentina. A, B, and C. Examples of demagnetization of samples with reversed directions. D, E, F, and G. Examples of transitional demagnetization behavior. H and I. Normal direction demagnetization examples. Open circles are projections onto the vertical plane; Solid circles are projections on to horizontal plane, Axes are labeled in A/m.

5. DISCUSSION

5.1. Toward a Geomagnetic Instability Time Scale for the Matuyama Chron

Refinements of the GPTS from 0–1.2 Ma have involved addition of several short-lived (<20 kyrs) geomagnetic "events" or cryptochrons, including excursions, aborted reversal attempts, and rapid back-to-back reversals, that have been calibrated using $^{40}Ar/^{39}Ar$ dating [*Singer et al.* 2002]. Although evidence from sediments and lava flows strongly suggests that at least 20 geomagnetic events occurred during the last 1.2 myr, the number and timing of events in the earlier part of the Matuyama reversed chron has remained more uncertain. Recently, however, sediment drifts in the North Atlantic Ocean that were deposited at high rates during the Matuyama Chron were cored at ODP sites 981, 983, and 984 and revealed a remarkably detailed record of reversals and excursions that are now astronomically dated [*Channell et al.*, 2002; 2003]. To the extent that these short-lived events can be temporally quantified, they will provide valuable tie points for global Plio-Pleistocene paleointensity stacks [*e.g., Guyodo and Valet*, 1999] as well as helping to refine models of the geodynamo that make explicit predictions as to the timing, frequency, and geometry of reversals and excursions [*e.g., Gubbins*, 1999; *Glatzmaier et al.*, 1999]. Here we review geomagnetic events during the Matuyama chron in light of the new radioisotopic ages from Cerro del Fraile and outline an initial Geomagnetic Instability Time Scale (GITS) [*Singer et al.*, 2002] for this interval (Figure 6).

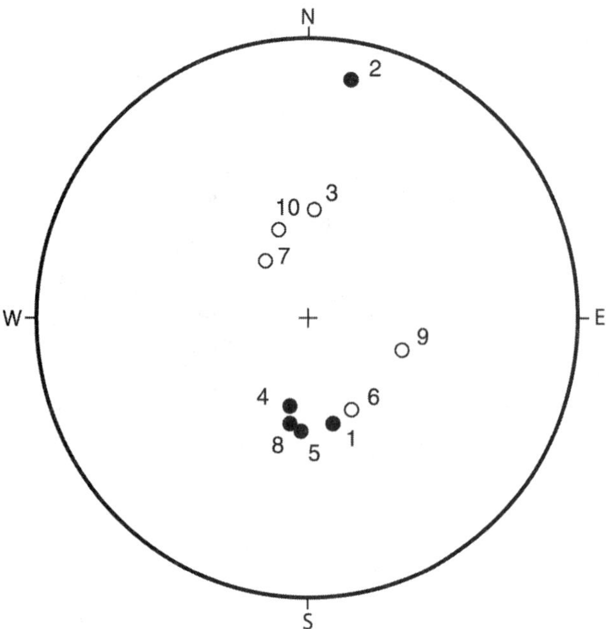

Figure 5. Equal area net mean inclination and declination, determined from principal component analysis, for each paleomagnetic site from Cerro del Fraile. Numbers correspond to flow number (Table 3). Open circles are upward pointing (negative) inclinations, solid circles are downward pointing (positive) inclinations.

5.1.1. The Réunion and Huckleberry Ridge events. The ^{40}Ar/^{39}Ar and unspiked K-Ar ages, together with the paleomagnetic directions, indicate that at least three intervals of normal polarity during the Matuyama reversed chron are recorded at Cerro del Fraile. The ^{40}Ar/^{39}Ar isochron ages suggest that the sequence of transitional—normal—reversed magnetic directions recorded in Flows 2, 3, and 4 correspond to the Réunion event. Because ages determined from each lava overlap one another at the 95% confidence level, the weighted mean of 2.136±0.019 Ma gives a very precise estimate of the time elapsed since the magnetic field transitioned from reverse to normal and back to reverse polarity.

The age of the Réunion event, and the possibility that it may comprise two separate intervals of normal polarity has been controversial [*Kidane et al.*, 1999; *Roger et al.*, 2000; *Baksi and Hoffman*, 2000; *Lanphere et al.*, 2002; *Baksi*, 2001]. Baksi and Hoffman [2000] and Lanphere et al., [2002] reviewed the geochronologic evidence, including unspiked K-Ar dated basalts from Gamarri, Ethiopia [*Kidane et al.*, 1999], and reached contrasting conclusions regarding whether the Réunion event comprises multiple periods of normal polarity. Baksi and Hoffman [2000] calculated the weighted mean ^{40}Ar/^{39}Ar plateau age of six normally or transitionally magnetized basalts from the Réunion Island at 2.139±0.034 (±2σ, normalized to standard values used here) and sug-

gested that the K-Ar ages from Gamarri are suspect due to high ^{36}Ar contents and alteration. Using ^{40}Ar/^{39}Ar laser fusion and furnace incremental-heating techniques, Lanphere et al. [2002] re-dated sanidine from the transitionally magnetized Huckleberry Ridge Tuff (HRT) and concluded that that at least two geomagnetic events occurred between 2.2 and 2.0 Ma. From sanidine separated out of pumice at three geographically distinct sites in the HRT, Lanphere et al. [2002] pooled together the total fusion, plateau, and isochron ages to arrive at an extraordinarily precise age of 2.090±0.008 Ma (±2σ, normalized to standard values used here) that overlaps some of the K-Ar ages from Gamarri, but is significantly younger than the Réunion Island basalts dated by Baksi and colleagues.

Because the total fusion, age plateau and isochron ages determined on each HRT sanidine separate by Lanphere et al. [2002] are not independent from one another, it is inappropriate to calculate such a precise weighted mean age for the three HRT samples by pooling all nine ages including the total fusion, plateau, and isochron ages of each sample. This is particularly true because a small amount of excess argon was detected in the incremental heating experiments [*Lanphere et al.*, 2002]. The most appropriate way to compare the age determined by Lanphere et al. [2002] to our age for the Réunion event is to calculate the weighted mean of the three isochron ages from the HRT sanidine; this yields an age for the HRT of 2.086±0.016 Ma. The weighted mean age of the six basalt flows from Réunion Island [*Baksi et al.*, 1993; *Baksi and Hoffman*, 2000] and the three flows from Cerro del Fraile is 2.137±0.016 Ma, which is older than Lanphere et al.'s [2002] age for the HRT by 51±23 kyr. Because the 2.130±0.030 Ma Flow 4 at Cerro del Fraile is reversely magnetized and significantly older than the mean age of the HRT, it is clear that there are at least two separate normal or transitional periods of magnetic polarity between 2.14 and 2.08 Ma (Figure 6).

The weighted mean of the five unspiked K-Ar ages for normally magnetized lavas at Gamarri is 2.06±0.05 Ma [±2σ; *Kidane et al.*, 1999]. This age overlaps with the 2.086 ±0.016 Ma derived from Lanphere et al.'s three isochrons for the HRT event. Thus, rather than corresponding to the Réunion event as claimed by Kidane et al. [1999], the Gamarri lavas are either: correlative with the younger HRT event, or, as discussed by Baksi and Hoffman [2000], the Gamarri lavas are too young due to alteration and these K-Ar ages cannot be relied upon to provide an accurate chronology. ^{40}Ar/^{39}Ar incremental heating analyses of groundmass separates, using the methods and standards employed here, will be essential to resolving the age of the Gamarri lavas and assigning them to either the Réunion or HRT events.

The only other ^{40}Ar/^{39}Ar age that closely constrains the age of the Réunion event comes from sanidine dated by Roger et

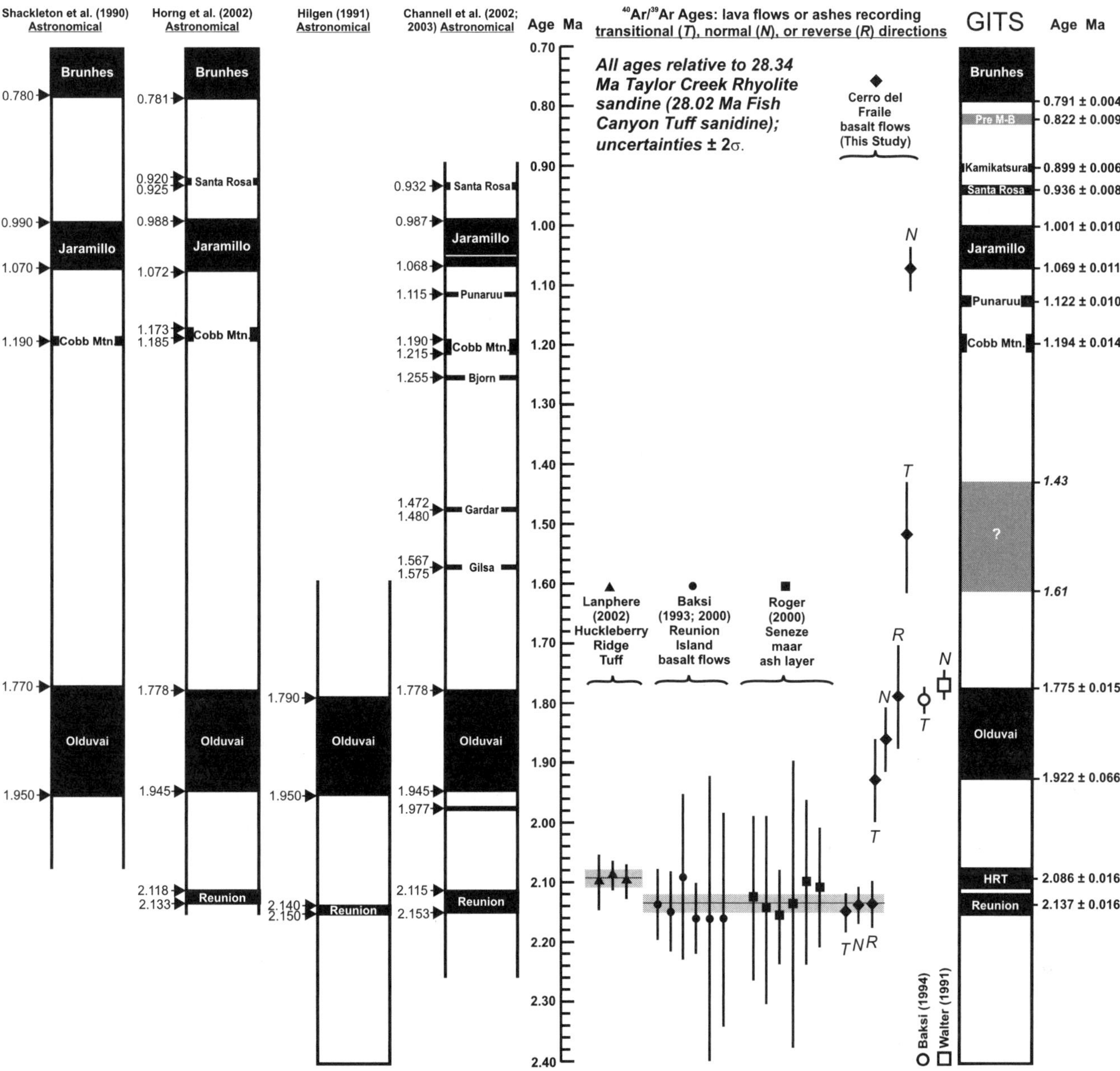

Figure 6. Comparison of ^{40}Ar/^{39}Ar radioisotopic ages from CerroFraile (♦) with other ^{40}Ar/^{39}Ar ages bearing on the age of the Réunion event [Baksi ● = *Baksi and Hoffman*, 2000 and *Baksi et al.,* 1993; Roger ■ = *Roger et al.* 2000; Lanphere ▲ = *Lanphere et al.*, 2002], the Olduvai subchron boundaries [Baksi ○ = *Baksi,* 1994; Walter □ = *Walter et al.*, 1991], and astrochronologic timescales. Ages at right for the upper part of the Matuyama chron are from Singer et al. [2002] and Singer and Brown [2002]. HRT = Huckleberry Ridge event of Lanphere et al. [2002]. All radioisotopic ages are calculated, or normalized to 28.02 Ma for the Fish Canyon Tuff sanidine standard and reported at ±2σ. The shaded horizontal bands are the weighted mean ages and uncertainties of the Réunion event and Huckleberry Ridge event discussed in the text. The Réunion and Cobb Mountain events are intervals of normal polarity bounded by reversals [*Baksi et al.*, 1993; *Clement and Kent*, 1986], thus the Geomagnetic Instability Timescale (GITS) for the Matuyama chron includes at least 14 ^{40}Ar/^{39}Ar-dated events.

al. [2000] at 2.10±0.02 Ma (±2σ, comparable directly to our ages via the 28.02 Ma value used for the FCs standard) that was recovered by drilling into a tephra within normally magnetized lacustrine sediment of the Senèze maar, France. Baksi [2001], however, pointed out several problems with the interpretation of Roger et al. [2000], including normalization of some the their ages to inconsistent value for one of two standards used, and the combination of five one-step total fusion ages and four two-step (partly degassed) fusion ages with the plateau ages from two laser incremental heating experiments to obtain a final weighted mean age. Because Roger et al.'s [2000] single-step total fusion ages were all younger than the plateau ages at the 95% confidence level, Baksi [2001] correctly stated that sanidine in this water-lain ash has been altered, and that the single-step total fusion ages should not be included in the mean age. The weighted mean of Roger et al.'s [2000] two plateau ages and the 5 fusion steps from the partly degassed crystals is 2.135±0.050 Ma, in excellent agreement with the ages from normally and transitionally magnetized basalt from Réunion Island [*Baksi and Hoffman*, 2000] and Cerro del Fraile (Figure 6). The acceptable single-crystal sanidine ages of Roger et al. [2000] have large uncertainties, thus when pooled together with the nine basalt flows the best estimate for the age of the Réunion event remains 2.137±0.016 Ma—that is to say, distinctly older than the Huckleberry Ridge event of Lanphere et al. [2002](Figure 6).

The radioisotopic age of 2.137±0.016 Ma for the Réunion event (±0.040 Ma with fully propagated errors) is consistent with a single short-lived normal polarity interval C2r,1n at 2.15–2.14 Ma in Cande and Kent's [1995] revised GPTS, as well as the astronomical ages estimated by Hilgen et al. [1991] from the Mediterranean sapropel record and Horng et al. [2002] using a giant piston core MD972143 from the Philippine Sea (Figure 6). Horng et al. [2002] detected transitional magnetic directions in a disturbed sandy layer below the normal polarity interval illustrated in Figure 6, however owing to the disturbances it is unclear whether this layer corresponds to an event earlier than 2.137 Ma.

It is notable that each of the three age determinations from Cerro Fraile, and the weighted mean of 2.137±0.016 Ma for all 14 radioisotopic ages, fall within the 38 kyr normal polarity interval between 2.153 and 2.115 Ma observed in North Atlantic drift sediments at ODP site 981 astronomically dated by Channell et al. [2003] (Figure 6). Evidence that the HRT event may have been recorded in ODP site 984 sediment includes a brief period of equator-crossing virtual geomagnetic poles associated with low paleointensity at about 2.06 Ma [*Channell et al.*, 2002]. Records to the southeast from site 981 also show shallow inclinations and low virtual geomagnetic poles at about 2.04 Ma which Channell et al. [2003] correlate to the HRT event.

5.1.2. The Olduvai Subchron. The ^{40}Ar/^{39}Ar ages of transitional, normal, and reversed polarity Flows 6, 7, and 8 help to tightly constrain the onset and termination of the Olduvai subchron. The only other salient ^{40}Ar/^{39}Ar data include the isochron age of 1.779±0.020 Ma (±2σ; normalized to standards used here) from a transitionally magnetized basalt on Moorea Island that was incrementally heated [*Baksi*, 1994] and the isochron age of 1.770±0.022 (±2σ; normalized to standards used here) obtained by totally fusing 36 aliquots of sanidine crystals from normally magnetized Tuff IF at Olduvai Gorge [*Walter et al.*, 1991]. The onset of the Olduvai subchron was recorded 1.922 ± 0.066 Ma by Flow 6 at Cerro del Fraile (Figure 6). The termination of the subchron must have occurred by the time Flow 8 erupted 1.787±0.085 Ma, because it is reversely magnetized. Since the age of Flow 8 overlaps the isochron ages for the transitionally magnetized basalt [*Baksi*, 1994] and the normal polarity tuff [*Walter et al.*, 1991], the weighted mean of these three ages, 1.775±0.015 precisely defines the termination of the Olduvai subchron (Figure 6). These new radioisotopic ages are consistent with GPTS of Cande and Kent [1995] and with the astrochronologic ages of Shackleton et al. [1990] and Hilgen [1991], but in detail best match the more recent astrochronologic estimates of Horng et al. [2002] and Channell et al., [2002] (Figure 6).

5.1.3. Event at 1.61–1.43 Ma. Flow 9 at Cerro Fraile records a transitional paleomagnetic direction that is imprecisely constrained between the ^{40}Ar/^{39}Ar total fusion and K-Ar ages of 1.61 and 1.43 Ma, thus it is unclear if a correlation exists with other potential records of transitional field behavior. For example, the Gilsa event has been controversial since is was originally defined by McDougall and Wensink [1966] as a brief period of normal polarity recorded in Icelandic lavas 1.61±0.05 Ma. Udagawa et al. [1999] undertook paleomagnetic measurements and reported two new conventional whole-rock K-Ar dates from the same Icelandic lavas that seemed to confirm an age of 1.62 Ma for the Gilsa event. However, Wijbrans and Langereis [2003] obtained ^{40}Ar/^{39}Ar laser incremental heating ages from the groundmass of basaltic lavas in these Icelandic sections; they concluded that no lavas are 1.61 Ma, that only the Réunion and Olduvai events are recorded, and the existence of a Gilsa event is questionable.

Clement and Kent [1986] discovered a short interval of steep positive inclinations in sediments from two cores at DSDP site 609 that, on the basis of extrapolating from bio- and magnetostratigraphic ages of known chron boundaries, and assuming average sedimentation rates, was estimated to have occurred 1.55 Ma. Clement and Kent [1986] correlated this excursional event with the Gilsa event in Iceland. Similarly, Biswas et al. [1999] discovered a prominent inclination anomaly in sediments of Osaka bay, Japan, that occurred between the Olduvai subchron and the Cobb Mountain event. Although an age was not assigned to this event, its position in the core suggests that it occurred during marine oxygen isotope stage 54, or about 1.60 Ma, which led Biswas et al. [1999] to suggest that this event may correspond to an early part of the Sangiran excursion recorded by sediment in Java [*Hyodo et al.*, 1992]. Unfortunately, the timing of the Sangiran excursion is very poorly constrained by fission-track ages from two tuffs above the main declination anomaly that are 1.51 and 1.48 Ma.

More recently, drift sediments from ODP sites 983 and 984 revealed clear evidence for two brief excursions, astronomically dated at 1.567–1.575 Ma and 1.472–1.480 Ma [*Channell et al.*, 2002], that fall within the period in question at Cerro del Fraile (Figure 6). Channell et al. [2002] correlated the older event with the Gilsa and named the younger the Gardar event after the drift deposit in which is was discovered (Figure 6). The imprecise age of Flow 9 at Cerro del Fraile limits the certainty of correlation with the Gilsa, Stage 54, Sangiran, or Gardar events, and does not rule out the possibility that it records a brief, previously unrecognized period of geomagnetic instability between the termination of the Olduvai subchron and the onset of the Cobb Mountain subchron (Figure 6).

5.1.4. Onset of the Jaramillo Subchron. The onset of the Jaramillo normal subchron is recorded by several transitionally magnetized basalt flows in the Punaruu Valley, Tahiti, one of which was incrementally heated to give an ^{40}Ar/^{39}Ar isochron age of 1.069±0.012 Ma (±2σ; normalized to standards used here) [*Singer et al.*, 1999]. The isochron age of normally magnetized Flow 10 at Cerro del Fraile, 1.073±0.036 Ma, is indistinguishable from that of the transitional basalt from Tahiti, thus it must record the very earliest Jaramillo time. The weighted mean age of these two lavas, 1.069±0.011 Ma gives the most precise radioisotopic constraint for the onset of this subchron and is identical to the estimated astronomical ages for this event (Figure 6).

5.2. Glacial History of the Southern Andes

Cerro del Fraile is one of several >1200 m high mesas located 30–40 km east of the Andean Cordillera that are capped by Plio-Pleistocene basaltic lavas interbedded with glacial till (Figure 1). From three of these mesas, including Cerro del Fraile, Mercer [1976; 1983] obtained K-Ar ages that revealed a fragmentary record of the earliest Cenozoic glaciations in South America. The others are Meseta del Lago Buenos Aires, 47° S, where till is preserved between basalt flows that were K-Ar dated at 7.01 and 4.58 Ma and ^{40}Ar/^{39}Ar dated at 7.49±0.10 and 5.12 ±0.08 Ma by us [*Ton That et al.*, 1999; ±2σ, normalized to standards used here], and Meseta Desocupada, 49.4°S, where till crops out between two reversely

magnetized basalt flows K-Ar dated at 3.57 and 3.64 Ma [*Mercer*, 1983]. Thus, the oldest glaciations in Patagonia occurred in the earliest Pliocene prior to 5.12 Ma and at ca. 3.6 Ma during the Gauss chron. It remains unclear whether this piecemeal record reflects poor preservation, the limited extent of glaciers, or lack of large ice caps during the period between 5.12 Ma and the base of the section at Cerro del Fraile.

Contact relations between the lava flows and underlying tills, together with the ^{40}Ar/^{39}Ar ages indicate that glaciers advanced eastward out of the Andean Cordillera and deposited till at least seven times from prior to 2.181±0.097 Ma to some time between 1.61 and 1.073±0.036 Ma. Cerro del Fraile preserves the only known record of frequent glaciation in southern South America during this Plio-Pleistocene interval [*Mercer*, 1976; 1983]. The marine δ^{18}O time series during this period—between stage 82 and 48—suggests that climate oscillations recorded in the oceans were dominated by orbital obliquity with a ~40 kyr periodicity [*Shackleton et al.*, 1990; *Zachos et al.*, 2002]. Deposition of at least four separate tills in the ca. 220 kyrs between Flows 2 and 6 (Figure 2) is consistent with a Patagonian ice cap that waxed and waned at frequency similar to that of the oceans. Carbon isotope variations in the South Atlantic Ocean during the last 2.9 Ma reveal that the most significant change in deep water circulation was an abrupt lowering of δ^{13}C values 1.55 Ma, possibly induced by the rapid expansion of Antarctic sea ice, and followed by increased coupling between North and South Atlantic water [*Venz and Hodell*, 2002]. The youngest till at Cerro del Fraile may have been deposited during or shortly after this 1.55 Ma transition, and may mark the change from relatively small piedmont glaciers to significantly more extensive oscillations of the Patagonian ice cap.

Glaciations continued to frequent the Pleistocene with ice reaching its farthest eastward extent during Mercer's [1976] Greatest Patagonian Glaciation (GPG), between 1.168±0.014 and 1.016±0.010 Ma—the ^{40}Ar/^{39}Ar ages of basaltic lava flows that under- and overlie till ~100 km east of the Andes in the Rio Gallegos and Lago Buenos Aires valleys (Figure 1) [*Singer et al.*, 2004]. Till of the GPG, and several terminal moraines, including those corresponding to the Last Glaciation Maximum ~25 ka were deposited on the floors of wide valleys occupied by lakes including Lago Argentino and Lago Buenos Aires at elevations of ca. 200 m (Figure 1). Fission-track cooling ages from rocks within the main Cordillera led Thomson et al. [2001] to conclude that east of the modern topographic divide of the Patagonian Andes denudation since the Late Cretaceous was less than 3 km, owing largely to the rain-shadow in effect since Oligocene-Miocene uplift. We find it remarkable that since deposition of the 7th till and eruption of Flow 10 onto the piedmont surface at Cerro del Fraile 1.073 Ma, valleys have been deepened by >1000 m of erosion

(Figure 1) at a rate of ~1 cm/yr. Evidently, nearly all of this denudation, including incision into the piedmont leaving isolated basalt-capped mesas, and a dramatic change in the landscape, occurred rapidly during the Quaternary, driven by vigorous, repeated glaciation that created vast trough-shaped valleys [*Singer et al.*, 2004].

6. CONCLUSIONS

Nine of ten basalt flows that cap Cerro del Fraile, Argentina, yielded precise ^{40}Ar/^{39}Ar isochron ages between 2.186 and 1.073 Ma that are consistent with the stratigraphy. The 9th flow in the sequence gave a discordant age spectrum suggesting that ^{39}Ar recoil has compromised this sample; its age is broadly constrained by its ^{40}Ar/^{39}Ar total fusion and unspiked K-Ar ages. Paleomagnetic directions of the ten flows, together with the new age determinations, indicate that three lavas record the onset and termination of the Réunion event, three others constrain the onset, middle, and termination of the Olduvai subchron, and the uppermost normally magnetized lava erupted during the earliest part of the Jaramillo subchron. A fourth geomagnetic event, recorded by the transitionally magnetized 9th flow, is imprecisely constrained between 1.61 and 1.43 Ma; thus, correlation of this event with other global records of geomagnetic instability between the Olduvai and Cobb Mountain events is uncertain. At least 14 geomagnetic reversals or excursional events occurred between 2.14 and 0.79 Ma. Moreover, if one considers the Laschamp, Albuquerque, and Big Lost events in the Brunhes chron, the Geomagnetic Instability Time Scale (GITS) includes 18 radioisotopically-dated events during the last 2.14 myr, as well as several others identified in sediments that remain to be radioisotopically dated [*Singer et al.*, 2002].

Specifically, the transitional, normal, and reverse polarity Flows 2, 3, and 4 that record the Réunion event at Cerro del Fraile have a weighted mean age of 2.136±0.019 Ma that is indistinguishable from ^{40}Ar/^{39}Ar ages obtained from both normally magnetized lavas on Réunion Island and ash in the Senèze maar, France. The weighted mean ^{40}Ar/^{39}Ar age of all these rocks, 2.137±0.016 Ma, gives a very precise radioisotopic date for the Réunion event that is ~50 kyr older than the Huckleberry Ridge event at 2.086±0.016 Ma. The ^{40}Ar/^{39}Ar ages of transitional, normal, and reverse polarity Flows 6, 7, and 8 at Cerro del Fraile, together with those from a transitional basalt on Moorea and a normal polarity tuff at Olduvai gorge, constrain the Olduvai subchron to between 1.922± 0.066 and 1.775±0.015 Ma, in agreement with astrochronologic estimates. The onset of the Jaramillo normal subchron, defined by transitionally magnetized lavas at Punaruu Valley Tahiti, and the youngest normal polarity flow at Cerro del Fraile, occurred 1.069±0.011 Ma.

Deposition of till on the piedmont surface prior to 2.186 Ma and six subsequent tills between 2.186 Ma and ~1.073 Ma record periodic growth of the Patagonian ice cap, but do not coincide with any known shift in Southern Ocean conditions. After ~1.5 to 1.2 Ma, the Patagonian icecap began to expand more dramatically. Repeated eastward glacial advances rapidly incised the piedmont adjacent to the mountains, deepened by >1000 m glacial troughs 10's of km wide at a rate of ~1 cm/yr, and profoundly altered the modern landscape.

Acknowledgments. We appreciate the assistance of L. Gualtieri, R. Ackert, D. Douglass, K. Brunstad, L. Powell, M. Relle, Y. Vincze, T. Ton-That, and C. Roig, and C. Kelly. This study was prompted by the pioneering use of K-Ar dating and meticulous field observations by John Mercer. Supported by U.S. NSF grants EAR-9909309, ATM-0212450, and EAR-0114055 (Singer) and EAR-9805424 (Brown).

REFERENCES

Ackert, R. P., B S. Singer, H. Guillou, M. R. Kaplan, and M. D. Kurz, Long-term cosmogenic ^{3}He production rates from ^{40}Ar/^{39}Ar and K-Ar dated Patagonian lava flows at 47°S, *Earth Planet. Sci. Lett., 210*, 119–136, 2003.

Baksi, A. K., Concordant sea-floor spreading rates obtained from geochronology, astrochronology, and space geodesy, *Geophys. Res. Lett., 21*, 133–136, 1994.

Baksi, A. K., Comment on: " ^{40}Ar/^{39}Ar dating of a tephra layer in the Pliocene Senèze maar lacustrine sequence (French Massif Central): constraint on the age of the Réunion-Matuyama transition and implications for paleoenvironmental archives" by Roger et al., *Earth Planet. Sci. Lett., 192*, 627–628, 2001.

Baksi, A. K., K. A. Hoffman, and M. McWilliams, Testing the accuracy of the geomagnetic polarity time-scale (GPTS) at 2–5 Ma, utilizing ^{40}Ar/^{39}Ar incremental heating data on whole-rock basalts, *Earth Planet. Sci. Lett., 118*, 135–144, 1993.

Baksi, A. K., D. A. Archibald, and E. Farrar, Intercalibration of ^{40}Ar/^{39}Ar dating standards, *Chem. Geol., 129*, 307–324, 1996.

Baksi, A. K., and K. A. Hoffman, On the age and morphology of the Réunion Event, *Geophys. Res. Lett., 27*, 2997–3000, 2000.

Berggren, W. A., F.J. Hilgen, C. G. Langereis, D. V. Kent, J. D. Obradovich, I. Raffi, M. E. Raymo, and N. J. Shackleton, Late Neogene chronology: New perspectives in high-resolution stratigraphy, *Geol. Soc. Am. Bull., 107*, 1272–1287, 1995.

Biswas, D. K., M. Hyodo, Y. Taniguchi, M. Kaneko, S. Katoh, H. Sato, Y. Kinugasa, and K. Mizuno, Magnetostratigraphy of Plio-Pleistocene sediments in a 1700-m core from Osaka Bay, southwestern Japan and short geomagnetic events in the middle Matuyama and early Brunhes chrons, *Palaeogeog.., Palaeoclimat., Palaeoecol., 148*, 233–248, 1999.

Brown. L. L., B. S. Singer, and M. L. Gorring, Paleomagnetism and ^{40}Ar/^{39}Ar chronology of lavas from Meseta del Lago Buenos Aires,

Patagonia,, *Geochem., Geophys., Geosys.* in press, 2004.

Cande, S. C., and D. V. Kent, Revised calibration of the geomagnetic polarity time scale for the Late Cretaceous and Cenozoic, *J. Geophys. Res., 100, 6093–6095, 1995.*

Cebula, G. T., M. J. Kunk, H. H. Mehnert, C. W. Naeser, J. D. Obradovich, and J. F. Sutter, The Fish Canyon Tuff, a potential standard for ^{40}Ar-^{39}Ar and fission-track dating methods. *Terra Cognita, 6,* 139–140, 1986.

Channell, J. E. T., J. Labs, and M. E. Raymo, The Réunion Subchronozone at OPD site 981 (Feni Drift, North Atlantic). *Earth and Planet Sci Lett., 215,* 1–12, 2003.

Channell, J. E. T., A. Mazaud, P. Sullivan, S. Turner, and M. E. Raymo, Geomagnetic excursions and paleointensities in the Matuyama Chron at Ocean Drilling Program sites 983 and 984 (Iceland Basin). *J. Geophys. Res., 107*(B6), doi: 10.1029/2001JB000491, 2002.

Charbit, S., H. Guillou, and L. Turpin, Cross calibration of K-Ar standard minerals using an unspiked Ar measurement technique, *Chem. Geol., 150,* 147–159, 1998.

Clement, B. M., and D. V. Kent, Short polarity intervals within the Matuyama" transitional field records from hydraulic piston cored sediments from the North Atlantic, *Earth Planet. Sci. Lett., 81,* 253–264, 1986/87.

Dalrymple, G. B., G. A. Izett, L. W. Snee, and J. D. Obradovich, ^{40}Ar/^{39}Ar age spectra and total-fusion ages of tektites from the Cretaceous-Tertiary boundary sedimentary rocks in the Beloc Formation, Haiti, *U.S. Geol. Surv. Bull., 2065,* 20 p., 1993.

Duffield, W. A., and G. B. Dalrymple, The Taylor Creek Rhyolite of New Mexico: a rapidly emplaced field of lava domes and flows: *Bull. Volcanol., 52,* 475–487, 1990.

Feruglio, E., Estudios geológicos and glaciológicos en la región del Lago Argentino (Patagonia). *Bol. Acad. Nacional Ciencias Córdoba, 37,* 3, 1944.

Fisher, R. A., Dispersion on a sphere, *Proc. R. Soc. Lond. A217,* 295–305, 1953.

Fleck, R. J., J. H Mercer, A. E M. Nairn, and D. M. Peterson, Chronology of late Pliocene and early Pleistocene glacial and magnetic events in southern Argentina: *Earth Planet. Sci. Lett., 16,* 15–22, 1972.

Glatzmaier, G. A., R. S. Coe, L. Hongre, and P. H. Roberts, The role of the Earth's mantle in controlling the frequency of geomagnetic reversals, *Nature, 401,* 885–890, 1999.

Gubbins, D., The distinction between geomagnetic excursions and reversals, *Geophys. J. Int., 137,* F1–F3, 1999.

Guillou, H., J. C. Carracedo, and S. J. Day, Dating of the Upper Pleistocene–Holocene volcanic activity of La Palma using the unspiked K–Ar technique: *J. Volcanol. Geotherm. Res., 86,* 137–149, 1998.

Guyodo, Y., and J.-P. Valet, Global changes in intensity of the Earth's magnetic field during the past 800 kyr, *Nature, 399,* 249–252, 1999.

Hilgen, F. J., Astronomical calibration of Gauss to Matuyama sapropels in the Mediterranean and implications for the geomagnetic polarity time scale, *Earth Planet. Sci. Lett., 104,* 226–244, 1991.

Hoffman, K. A., and Singer, B. S., Regionally recurrent paleo-magnetic transitional fields and mantle processes, this volume, pp 233–244.

Horng C.-S., M.-Y. Lee, H. Palike, K.-Y. Wei, W.-T. Liang,, Y. Iizuka, M. Torii, Astronomically calibrated ages for geomagnetic reversals within the Matuyama chron, *Earth, Planets, Space, 54,* 679–690, 2002.

Hyodo, M., W. Sunata, and E. E. Susanto, A long-term geomagnetic excursion from Plio-Pleistocene sediments in Java, *J. Geophys. Res., 97,* 9323–9355, 1992.

Kidane, T., J. Carlut, V. Courtillot, Y. Gallet, X. Quidelleur, P. Y. Gillot, and T. Haile, Paleomagnetic and geochronological identification of the Réunion subchron in Ethiopian Afar, *J. Geophys. Res., 104,* 10405–10419, 1999.

Kirschvink, J. L., The least squares line and plane and the analysis of paleomagnetic data, Geophys. *J. Roy. Astron. Soc., 62,* 699–718, 1980.

Lanphere, M. A., D. E. Champion, R. L. Christiansen, G. A. Izett, and J. D. Obradovich, Revised ages for tuffs of the Yellowstone Plateau volcanic field: Assignment of the Huckleberry Ridge Tuff to a new geomagnetic polarity event, *Geol. Soc. Am. Bull., 114,* 559–568, 2002.

Lanphere, M. A. and G. B. Dalrymple, First-principles calibration of ^{38}Ar tracers: implications for the ages of ^{40}Ar/^{39}Ar fluence monitors: *U.S. Geol. Survey Prof. Paper 1621,* 2000.

McDougall, I., and H. Wensink, Paleomagnetism and geochronology of the Plio-Pleistocene lavas in Iceland, *Earth Planet. Sci. Lett., 1,* 232–236, 1966.

Mercer, J. H., Glaciation in southern Argentina more than two million years ago, *Science, 164,* 823–825, 1969.

Mercer, J. H., Glacial history of southernmost South America: *Quat. Res., 6,* 125–166, 1976.

Mercer, J. H., Cenozoic glaciation in the southern hemisphere, *Ann. Rev. Earth Planet. Sci.,* 11, 99–132, 1983.

Renne, P. R., C. C. Swisher, A. L. Deino, D. B. Karner, T. L. Owens, and D. J. DePaolo, Intercalibration of standards, absolute ages and uncertainties in ^{40}Ar/^{39}Ar dating: *Chem. Geol., 145,* 117–152, 1998.

Roger, S., C, Coulon, N. Thouveny, G. Feraud, A. Van Velzen, S. Fauquette, J. J. Cocheme, M. Prevot, and K. L. Verosub, ^{40}Ar/^{39}Ar dating of a tephra layer in the Pliocene Senèze maar lacustrine sequence (French Massif Central): constraint on the age of the Reunion-Matuyama transition and implications for paleoenvironmental archives, *Earth Planet. Sci. Lett., 183,* 431–440, 2000.

Shackleton, N. J., A. Berger, and W. R. Peltier, An alternative astronomical calibration of the lower Pleistocene timescale based on ODP site 677, *Trans. Roy. Soc. Edinburgh, Earth Sci., 81,* 251–261, 1990.

Singer, B. S., K. A. Hoffman, A. Chauvin, R. S. Coe, and M. S. Pringle, Dating transitionally magnetized lavas of the late Matuyama Chron: Toward a new ^{40}Ar/^{39}Ar timescale of reversals and events, *J. Geophys. Res., 104,* 679–693, 1999.

Singer, B. S., M. R. Relle, K. A. Hoffman, A. Battle, H. Guillou, C. Laj, and J. C. Carracedo, Ar/Ar ages of transitionally magnetized lavas on La Palma, Canary Islands, and the Geomagnetic Instability

Timescale: *J. Geophys. Res., 107 (B11)*, 2307, doi: 10.1029/2001JB001613, 2002.

Singer, B. S., and L. L. Brown, The Santa Rosa event: ^{40}Ar/^{39}Ar and paleomagnetic results from the Valles rhyolite near Jaramillo Creek, Jemez Mountains, New Mexico, *Earth Planet. Sci. Lett., 197*, 51–64, 2002.

Singer, B. S., R. P. Ackert Jr., and H. Guillou, ^{40}Ar/^{39}Ar and K-Ar chronology of Pleistocene glaciations in Patagonia, *Geol. Soc. Am. Bull., 116,* in press, 2004.

Steiger, R. H., and E. Jäger, Subcommission on geochronology: convention on the use of decay constants in geo- and cosmochronology: *Earth Planet. Sci. Lett., 5,* 320–324, 1977.

Tauxe, L., C. Constable, C. L. Johnson, A. A. P. Koppers, W. R. Miller, and H. Staudigel, Paleomagnetism of the southwestern U.S.A. recorded by 0–5 Ma igneous rocks, *Geochem., Geophys., Geosys., 4(4)*, 8802, doi:10.1029/2002GC000343, 2003.

Taylor, J. R., *An Introduction to Error Analysis*, University Science Books, Mill Valley, Calif., 270 pp.,1982.

Thomson, S. N., F. Hervé, and B. Stöckhert, Mesozoic-Cenozoic denudation history of the Patagonian Andes (southern Chile) and its correlation to different subduction processes, *Tectonics, 20*, 693–711, 2001.

Ton That, T., B. S. Singer, N. A. Mörner, and J. Rabassa, Datación de lavas basalticas por ^{40}Ar/^{39}Ar geología glacial de la region del lago Buenos Aires, provincia de Santa Cruz, Argentina: *Revisita de la Asociación Geológica Argentina, 54*, 333–352, 1999.

Turrin, B. D., J. M. Donnelly-Nolan, and B. C. Hearn, ^{40}Ar/^{39}Ar ages from the rhyolite of Alder Creek, California: Age of the Cobb Mountain Normal-Polarity Subchron revisited, *Geology, 22*, 251–254, 1994.

Udagawa, S., H. Kitagawa, A, Gudmundsson, O. Hiroi, T. Koyaguchi, H. Tanaka, L. Kristjansson, and M. Kono, Age and Magnetism of lavas in the Jokuldalur area, Eastern Iceland: Gilsa event revisited, *Physics Earth Planet. Inter., 115*, 147–171, 1999.

Venz, K., and D. A. Hodell, New evidence for changes in Plio-Pleistocene deep water circulation from Southern Ocean ODP Leg 177 Site 1090, *Palaeogeog., Palaeoclimat., Palaeoecol., 182*, 197–200, 2002.

Villenueve, M., H. A. Sandeman, and W. J. Davis, A method for intercalibration of U-Th-Pb and ^{40}Ar/^{39}Ar ages in the Phanerozoic, *Geochim. Cosmochim. Acta, 64*, 4017–4030, 2000.

Walter, R C., P. C. Manega, R. L. Hay, R. E. Drake, and G. H. Curtis, Laser-fusion ^{40}Ar/^{39}Ar dating of Bed I, Olduvai Gorge, Tanzania, *Nature, 354*, 145–149, 1991.

Wijbrans J., and C. Langereis, Elusive Gilsa: Finally laid to rest in Northeast Iceland, *Geophys. Res. Abstracts, (2003 AGU-EUG Joint Assembly) 5*, 11595, 2003.

Wijbrans, J. R., M. S. Pringle, A. A. P. Koppers, and R. Scheveers, Argon geochronology of small samples using the Vulkaan argon laserprobe, *Proc Dutch Acad. Sci., 98, (2)*, 185–218, 1995.

Zachos, J., M. Pagani, L. Sloan, E. Thomas, K. Billups, Trends, rhythms, and aberrations in global climate 65 Ma to present, *Science, 292*, 685–693, 2001.

Laurie L. Brown, Department of Geosciences, University of Massachusetts, Amherst, Massachusetts.

Hervé Guillou, Laboratoire des Sciences du Climat et de l'Environnement, Domaine du CNRS, Bat.12, Avenue de la Terrasse, 91198, Gif-sur-Yvette, France.

Jorge O. Rabassa, Laboratorio de Geología del Cuaternario, CADIC-CONICET, C.C. 92, 9410 Ushuaia, Tierra del Fuego, Argentina.

Brad S. Singer, Department of Geology and Geophysics, University of Wisconsin-Madison, 1215 West Dayton Street, Madison, Wisconsin.

After the Dust Settles: Why Is the Blake Event Imperfectly Recorded in Chinese Loess?

Josep M. Parés, Rob Van der Voo, and Maodu Yan

Department of Geological Sciences, University of Michigan, Ann Arbor, Michigan

Xiaomin Fang

Department of Geography, Lanzhou University, Gansu, China

The aeolian deposits in China are known to provide a continuous and fairly complete magnetostratigraphy in sections with alternating horizons of loess and paleosols that correlate with cold-dry and warm-humid periods, respectively. These deposits have the potential, therefore, to provide a precise timing of the Blake Event relative to Pleistocene climatic shifts. However, paleomagnetic studies in loess/paleosol deposits in China have shown that the Blake Event is imperfectly recorded, appearing in different stratigraphic positions or being completely absent. We sampled four loess/paleosol successions in the western Chinese Loess Plateau to investigate the inconsistent recording of the Blake Event. Disappointingly, the reverse Blake Event is completely masked by the dominantly normal Brunhes Chron in these four sections. Rock-magnetic results indicate the enhanced presence of superparamagnetic magnetite, suggesting that new magnetic grain growth occurred long after deposition also in the loess. We argue that this caused the remanent magnetization acquisition process to have been rather protracted throughout the interval of 130-100 ka. Together, paleomagnetic evidence and rock-magnetic observations are compatible with a model of long-delayed magnetization acquisition, similar to that which earlier had already been surmised from the misplaced (too old) position of the Matuyama-Brunhes boundary in loess layer L8. Delayed magnetization acquisition is most likely the result of pedogenesis in paleosols as well as loess, as suggested by the absence of marked contrasts between the normally-distinct loess and paleosol horizons in our studied sections. Similarly to widespread remagnetizations in orogenic belts, the enhanced presence of secondary SP magnetite in loess seems to be a fingerprint for protracted magnetization.

1. INTRODUCTION

Geomagnetic reversals are one of the most intriguing features of the Earth Magnetic Field. Although on average about

two reversals occur per million years, the process is highly irregular so that the geomagnetic reversal time scale shows a broad spectrum of polarity-interval durations, ranging from a few thousand to a few million years, or even longer in the case of superchrons such as the Kiaman and Cretaceous Long Normal. The most recent polarity Chron C1n (Brunhes) is thought to contain about eleven or even fourteen geomagnetic events [e.g., *Langereis et al.*, 1997; *Lund et al.*, 1998

Timescales of the Paleomagnetic Field
Geophysical Monograph Series 145
Published in 2004 by the American Geophysical Union
10.1029/145GM14

respectively] of very short duration (a few thousand years); about half of these are not well established. Most events, even the ones that have been widely observed in magnetostratigraphic studies, are not seen in marine magnetic anomaly profiles. Of all the putative geomagnetic events within the Brunhes Chron, some of which may be artifacts or simple 'excursions', the Blake Event is one of the better established. It has been reported in both sedimentary and volcanic rocks, is observed globally and with full-scale (near-180°) reversal records. Because it occurs in or close to the Eemian interval of the Middle Pleistocene, the Blake Event has great potential in Quaternary stratigraphy as a time line and correlation tool and hence has received a lot of attention ever since Smith and Foster [1969] first documented it.

While it seems clear that the Blake Event (BE) is a truly short polarity interval and not an artifact or excursion, there is no consensus on its age, duration or structure. Stratigraphic links with marine isotope stages (MIS) suggest that the BE is best placed in MIS 5d–5e (~105–120 ka), based on cores in the Mediterranean [*Tucholka et al.*, 1987; *Dinarès-Turell et al.*, 2001] and studies in or near the Tibetan Plateau [*Chen et al.*, 1995; *Fang et al*, 1997; *Hu et al.*, 1999] (Figure 1). There are a number of sections in volcanic or volcanoclastic rocks that have yielded evidence for the BE [*Zanella and Laurenzi*, 1998, and references therein], but unfortunately the error margins are

rather large so that the ages do not really allow a precise age of the Blake Event to be established.

The structure of the BE is equally ambiguous, as multiple reversals are often, but not always, seen in the BE. Some authors have detected two reversed-polarity episodes separated by one of normal polarity [*Lund et al.*, 1998; *Creer et al.*, 1980; *Tric et al.*, 1991], while others document three negative-polarity episodes [*Herrero-Bervera et al.*, 1989; *Reinders and Hambach*, 1995]. In the Chinese Loess Plateau, the BE appears as a single [*Yue et al.*, 1991; *Chen et al.*, 1995], double [*Fang et al.*, 1997], or even triple event [*Zhu et al.*, 1994]. Furthermore, whereas near Xining in NE Tibet the BE occurs in loess L2-2 (=L2-b) [*Zhu et al.*, 1994], in Lanzhou and in the Zoige Basin in Tibet the BE is found in the bottom part of paleosol S1 below L2-2 [*Fang et al.*, 1997; *Chen et al.*, 1995]. Granted that pedostratigraphic correlations are correct, it follows that some sections in or near the Tibetan Plateau have shown the BE in sediments corresponding to a 'cold' episode (loess L2-2), whereas other sections contain the BE in a 'warm' episode (paleosol S1-c, also labeled S1-3 [*Fang et al.*, 1997]). This mismatch is alarming, given that the loess-paleosol stratigraphy is generally quite consistent in the Loess Plateau.

In order to better understand the causes of the variable occurrence of the BE in Chinese loess, we have undertaken a paleomagnetic study in the western Loess Plateau of four selected sections where pedostratigraphy allowed us to locate the expected position of the Blake Event with confidence.

2. GEOLOGICAL SETTING AND METHODS

The Chinese Loess Plateau covers an area of more than 300,000 km^2 [e.g., *Heller and Liu*, 1984], between latitudes 35° and 47° N in northern China (Figure 2). The loess varies in thickness, reaching 500 m in the northwestern part of the Loess Plateau along the Huang He (Yellow River).

Although typically referred to as simply "loess", this deposit consists of alternating layers of wind-blown dust (loess *sensu stricto*) and buried soils (paleosols). In this paper, we will refer to "loess", when we mean those parts of the sequence that are apparently less (or not at all) affected by pedogenesis, and use the term "loess-paleosol sequence" for the entire package. The paleosol-loess alternations in the sediments constitute a paleoclimate record [see, e.g., the review by *Evans and Heller*, 2001]; transport and settling of dust occurred during cool and dry stages, whereas weathering and soil formation took place during warm and wet stages. Through decades of studies it has become widely appreciated that these changing climatic conditions are reflected in the magnetic properties. The magnetic susceptibility signal contained in the records of the loess-paleosol sequences has been convincingly matched to deep-sea oxygen isotope records [*Heller and Liu*, 1986; *Kukla et al.*,

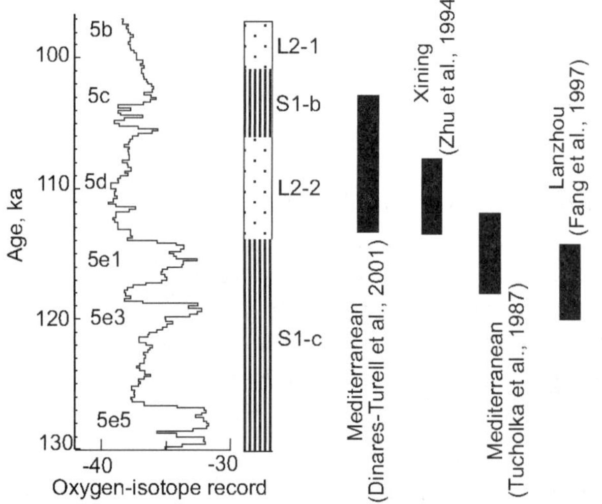

Figure 1. Position of the Blake Event in the Chinese Loess Plateau and in the Mediterranean within the marine isotope stage framework (5b through 5e5). Solid vertical bars represent the extent of the recorded events. Records are correlated to the MIS using susceptibility profiles (for Xining, Lanzhou) or oxygen curves (for the Mediterranean records). Oxygen-isotope climatic record is from the Greenland GRIP ice core [*Grootes et al.*, 1993].

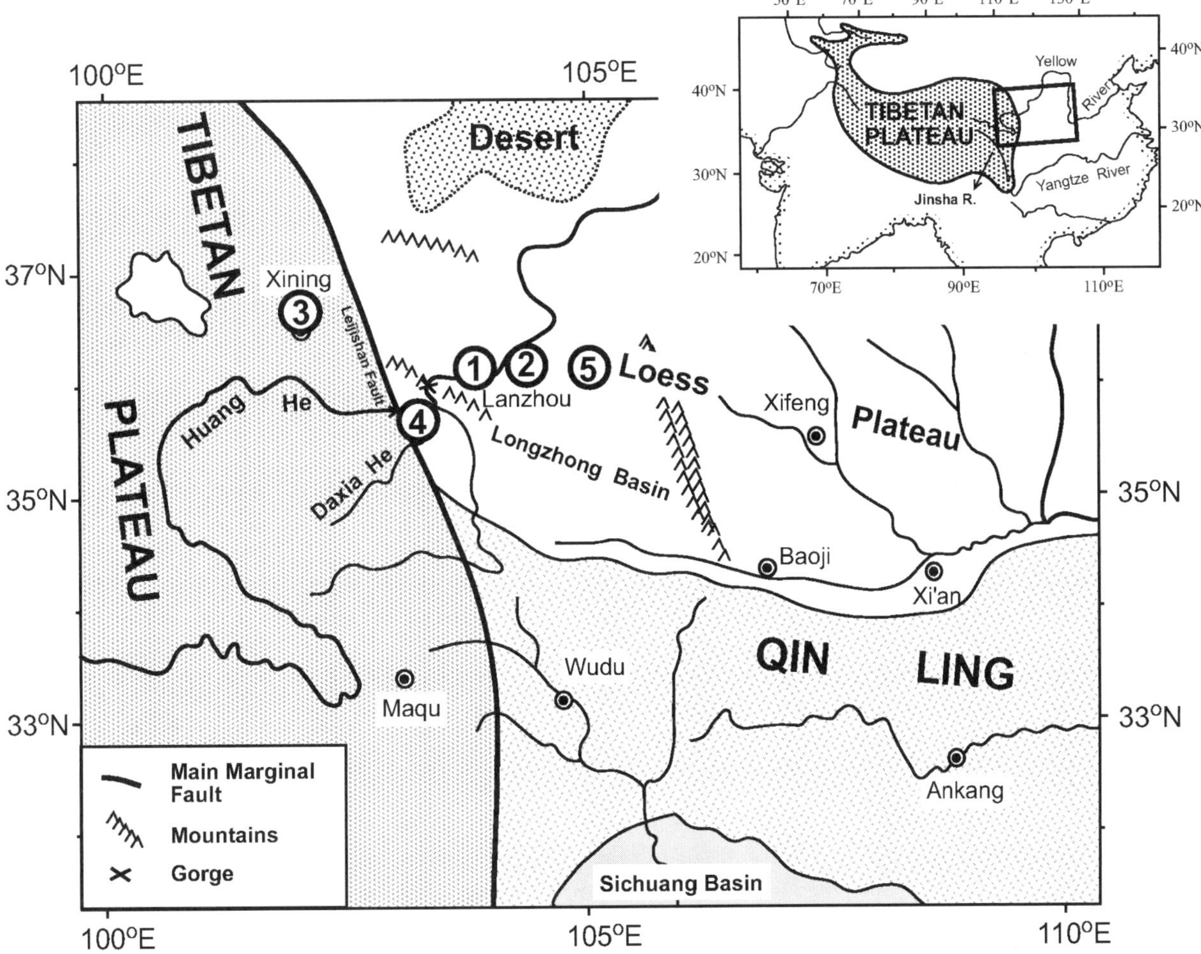

Figure 2. Location map of the studied loess–paleosol sections. 1- Lanzhou [*Fang et al.*, 1997], 2- Chaoxian (=Jingyuan), 3- Tuxiangdao, 4- Beiyuan, 5- Duanxian.

1988; *Evans and Heller*, 2001]. Magnetic susceptibility tends to reach higher values in the Chinese paleosols than in the loess horizons. Detailed mineral magnetic investigations have shown that the enhanced susceptibility in paleosols is due to a higher concentration of fine-grained ferromagnetic materials [*Evans and Heller*, 1994; *Eyre and Shaw*, 1994; *Forster et al.*, 1994]. Thus, the pedostratigraphy and distinctions between loess-paleosol couplets are very often based on measurements of the susceptibility variations within the sequence.

For this study, we have sampled four different loess-paleosol successions in the western Chinese Loess Plateau, which will be referred to as Chaoxian, Tuxiangdao, Beiyuan and Duanxian (Figure 2). Unbeknownst to us at the time of sampling and laboratory measurement, an unpublished record already existed for Tuxiangdao [a study by F.H. Chen, cited in *Evans and Heller*, 2001, Figure 5].

About one and a half meter deep and one meter wide trenches were dug at the studied localities to ensure fresh loess outcrop. Samples were obtained in the field by cutting 8 cm^3 cubes by hand, or with a cordless saw in the more indurated layers, and were then stored in plastic boxes. For each sampling layer we collected three specimens, at a horizontal spacing of ~2–3 cm. We typically sampled an interval comprised between the upper part of loess L2-3 to the upper part of L2-2, including the entire intervening paleosol S1-c (see Fang et al. [1997] for pedostratigraphic labels and divisions). Age estimates for the sampled intervals are derived from correlations with the marine isotope stages and span ~100–120 ka (Figure 1). At Chaoxian (referred to also as Jingyuan), where the thickest loess deposit in the Loess Plateau, and probably in the world, can be found (~530 m), we based our sampling on a preliminary study by Yue et al. [1991]. Their section contained the upper Jaramillo, upper Matuyama, and Brunhes Chrons. The average accumulation rate is about 30 cm / kyr [*Yue et al.*, 1991]. At around a depth of ~ 54 m from the sub-recent soil S$_0$, Yue et al. [1991] reported a horizon with southerly and upward magnetization directions, interpreted as the Blake Event (Figure 3). Our sampling was designed to complement this result by obtaining a higher resolution magnetostratigraphic record.

Remanent magnetization was measured on a three-axis 2G cryogenic magnetometer, housed in a shielded room with a residual field no greater than 200 nT. Both thermal and alternating field demagnetization procedures, using ASC and a SI-4 apparatus respectively, were applied to isolate the Characteristic Remanent Magnetization (ChRM) of the samples. Friable samples were impregnated with a 1:1 sodium silicate solution for consolidation prior to the thermal treatment. Standard orthogonal demagnetization plots [*Zijderveld*, 1967] were used to inspect the arrangement of magnetization components and principal component analysis [*Kirschvink*, 1980] was applied to calculate ChRM directions.

Measurements of the susceptibility and remanence at low temperatures were made by cooling the samples in liquid nitrogen and letting them simply warm up, using time as a (non-linear) proxy for increasing temperature. For remanence measurements, warming up started immediately in the cryogenic magnetometer, whereas for susceptibility measurements the samples were immersed in liquid nitrogen, so that this liquid needed to boil off before warming could begin. Bulk

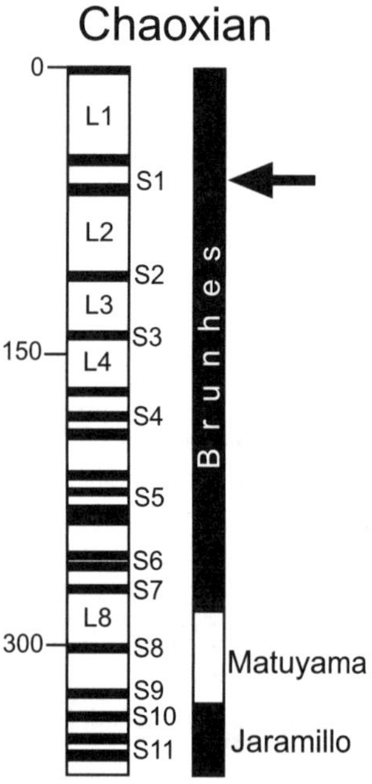

Figure 3. Magnetostratigraphic results of Yue et al. [1991] from the loess-paleosol sequence at Chaoxian (see Figure 2 for location). Arrow indicates position of reversely magnetized samples reported by Yue et al. [1991], which they interpreted as a record of the Blake Event. Our sampling was designed to examine that part of the stratigraphic section in greater detail.

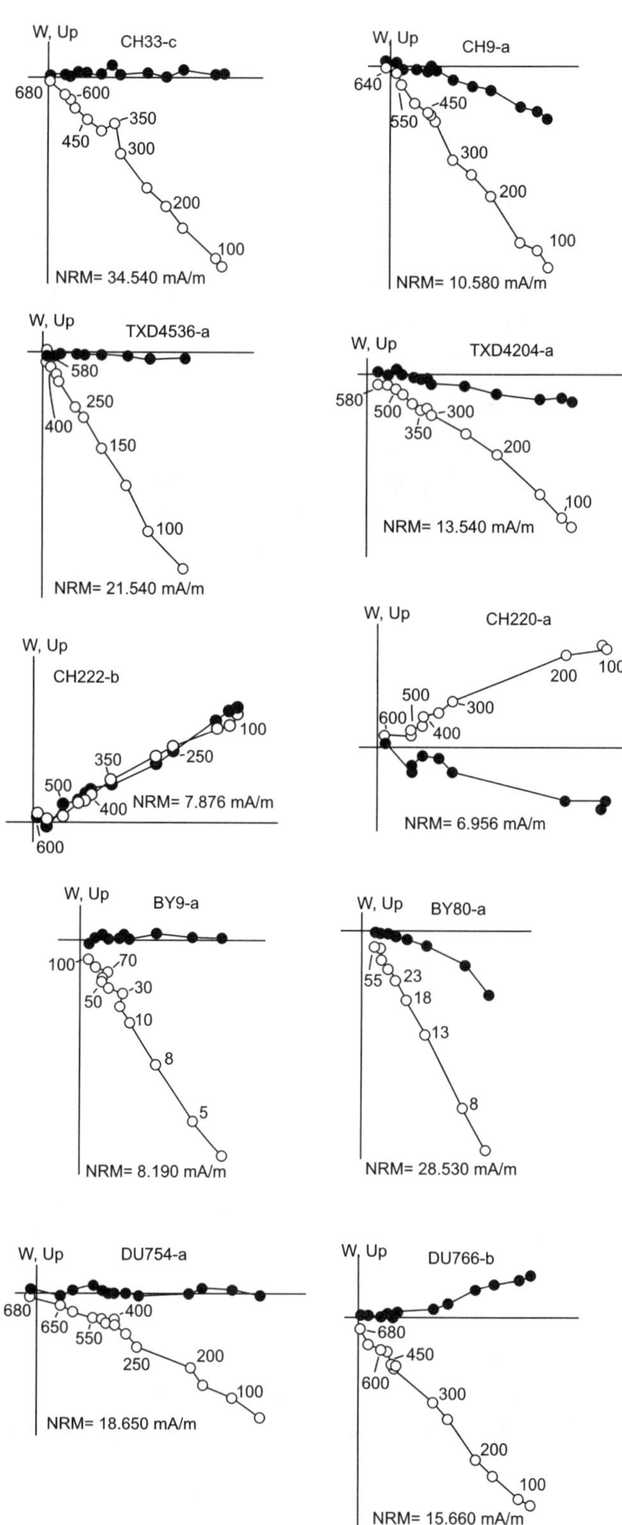

Figure 4. Orthogonal thermal and alternating-field demagnetization diagrams [*Zijderveld*, 1967] of representative samples from the studied sections. Open (closed) symbols represent the vertical (horizontal) component of the magnetization.

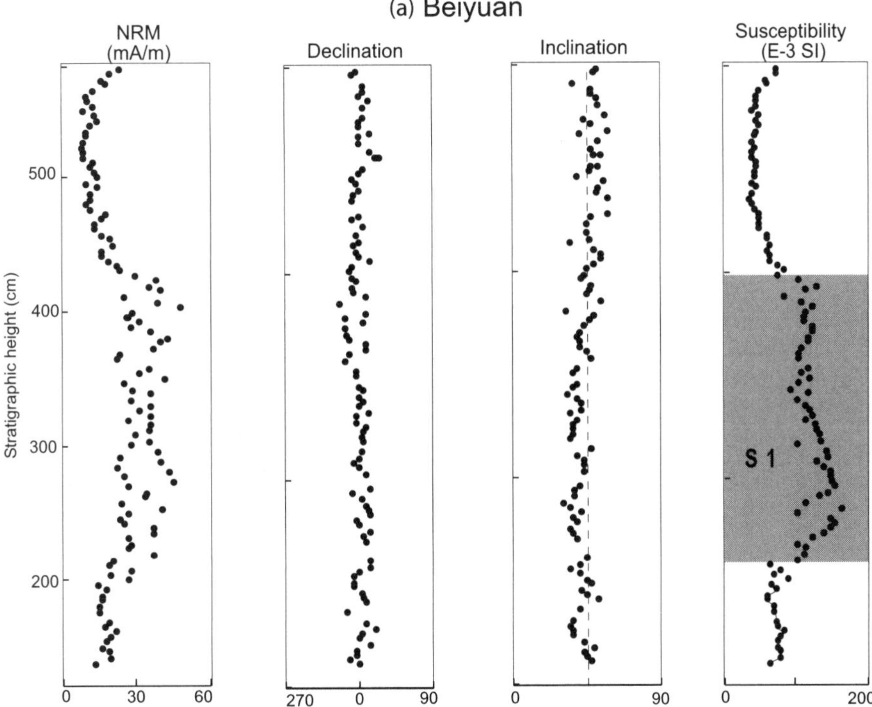

Figure 5. (a,b,c,d) Characteristic remanent magnetization direction plots for the studied sections. For each locality NRM intensity, declination, inclination and bulk susceptibility are shown. S1- Position of paleosol S1. In Figure 5b (Chaoxian) shaded intervals show the location of the S1 doublet in Figure 3.

magnetic susceptibility was measured with a Bartington MS2W jacket-insulated. Anisotropy of magnetic susceptibility (AMS) measurements were performed with a KLY-2 (Kappabridge) in order to determine the sediment fabric and possible relation with the paleomagnetic signature.

3. RESULTS

3.1 Paleomagnetism

Typical demagnetization diagrams show a progressive and linear decay of the magnetization towards the origin of the orthogonal plots (Figure 4). Thermal demagnetization reveals maximum unblocking temperatures typically around 580°C, although a few samples have magnetization persisting up to 680°C, which indicates a minor contribution of hematite to the remanence, in addition to the dominant magnetite (or perhaps maghemite). Alternating field demagnetization typically removes ~ 80% of the initial NRM at a peak field of 60 mT. Overall, the demagnetization results agree with those of previously published studies of other loess-paleosol sequences in the vicinity [e.g., *Zhu et al.*, 1994; *Fang et al.*, 1997].

Paleomagnetic directions obtained from principal component analysis are shown in Figure 5 versus depth for the four strati-

graphic sections. Changes in bulk susceptibility assist in locating the position of paleosol S1-c and, hence, the expected position of the BE. In none of the four sections have we been able to identify a clear southerly and upward magnetization component. It appears that the BE is escaping detection completely in these four sections. The record obtained by Chen for Tuxiangdao [*Evans and Heller*, 2001] similarly shows no evidence of the BE, despite the fact that this locality is close to Xining, where Zhu et al. [1994] found a triple BE.

Whereas the BE was found in Lanzhou within paleosol S1-c [*Fang et al.*, 1997], it was reported as occurring in loess between paleosols S1-c and S1-b, i.e., at a higher stratigraphic level, in Xining [*Zhu et al.*, 1994]. Given our conclusion that the BE may not be observable because of a delayed magnetization acquisition, it is logical to expect that the longer this delay, the deeper in the sequence any evidence of the BE would be found. In turn this implies that we should be looking for the BE at a higher stratigraphic level, if the magnetization is rather quickly acquired. This is the reason that our sections include much of the loess (L2-2) overlying paleosol S1-c or even the lower part of paleosol layer S1-b, where it could be identified in the field. Despite this strategy, our data show no traces of reversed magnetization across the entire couplet S1-c—L2-2. Rather, paleomagnetic directions from the

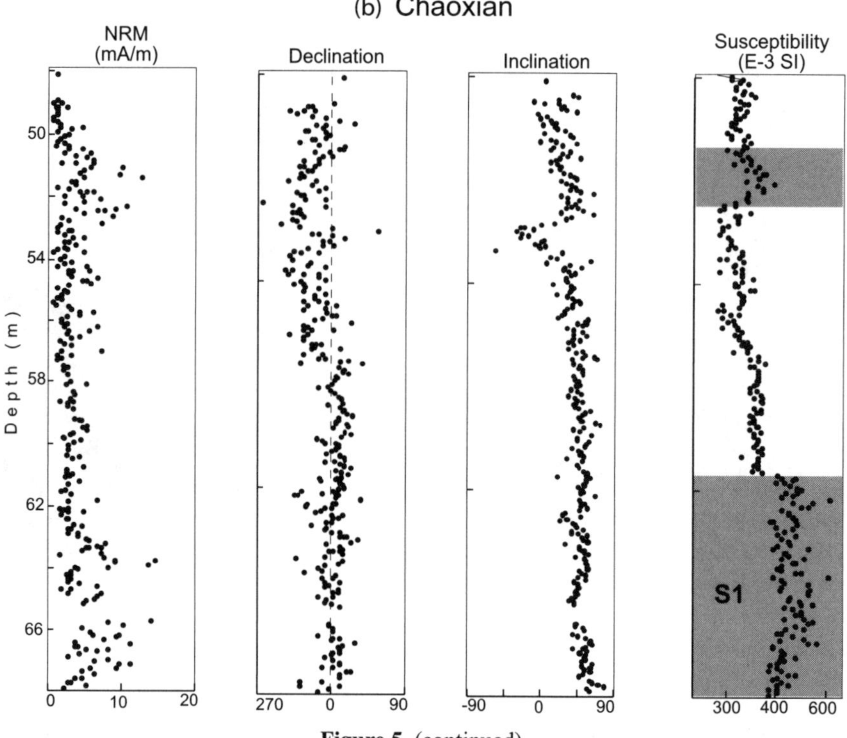

Figure 5. (continued)

sections of Beiyuan, Duanxian and Tuxiangdao cluster around the expected Quaternary normal-polarity reference direction. Paleomagnetic directions at Chaoxian show a relatively high dispersion, including a few northerly and upwards ChRM directions. The demagnetization diagrams for such magnetization directions show a stable behavior during progressive thermal cleaning (Figure 4, samples CH222-b and CH220-a). These negative directions appear at ~8 m above paleosol S1-c and could be the trace of a strongly masked reversed-magnetization interval. Alternatively they could simply be an artifact, due to some unrecognized disturbance of the sediments by solifluction, slumping or creep [e.g., *Hus et al.*, 1993, *Derbyshire et al.*, 1995]. It is interesting to point out that the stratigraphic position of these 'anomalous directions' is well above the location of the BE of Zhu et al. [1994] in Xining. It is safe to say that the studied loess-paleosol successions do not show any compelling evidence for the Blake Event in the interval comprised between L2-3 to L2-2.

3.2 Rock Magnetism

Conventional rock-magnetic analyses have been carried out in order to detect characteristics among loess and paleosol horizons that may indicate possible explanations as to why the BE has not been observed. The experiments include low-temperature susceptibility, AMS, low-temperature remanence measurements and determinations of magnetic viscosity. Previous studies have shown that some of the more discriminating magnetic parameters to determine the degree of pedogenesis of loess-paleosol deposits are obtained by low-temperature measurements of both susceptibility and remanence [e.g., *Maher and Thompson*, 1999], because they are very informative about the grain size and/or domain state. Most studies [*Liu et al.*, 1994; *Hunt et al.*, 1995; *Maher and Thompson*, 1999] reveal that upon warming from liquid nitrogen temperature (77 K) loess-paleosol samples show a distinctive behavior in magnetic susceptibility. First, the Verwey transition at ~110 K is usually clearly defined, showing the contribution of multi-domain magnetite. Above the transition, loess samples show a slight decrease in susceptibility with increasing temperature, reflecting minor paramagnetic contributions of clay minerals that diminish with increasing temperature. In contrast, paleosol samples typically show a progressive increase of susceptibility upon warming above 110 K, characteristic of superparamagnetic oxide grains. Our own susceptibility-versus-temperature results reveal entirely similar behavior (Figure 6). Plots for both loess and paleosol samples show an initial increase in susceptibility until ~110 K (estimated as half way in "time" between 77 K and 240 K in Figure 6), indicative of the Verwey transition. At higher temperatures the susceptibility remains constant or increases slightly in paleosol samples and decreases slightly in loess

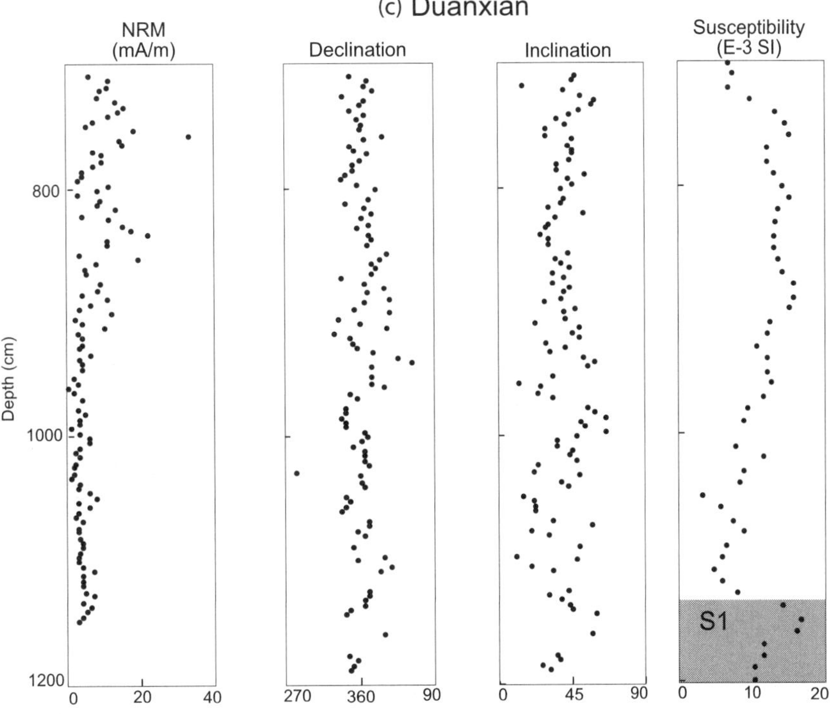

Figure 5. (continued)

samples. As already explained above, in loess this reveals the paramagnetic contribution of clays, whereas in paleosols we believe that we detect the presence of superparamagnetic oxides. However, we stress that the differences between loess and paleosol samples are not as large as has sometimes been observed in other Chinese sequences.

Low-temperature remanence data complement the susceptibility measurements. Specifically, analysis of the thermal decay of an isothermal remanent magnetization (IRM) imparted at low-temperatures provides an effective means of identifying magnetic minerals and can allow quantification of superparamagnetic (SP) grain concentrations [*Hunt et al.*, 1995; *Fang et al.*, 1999]. After cooling samples to 77 K, an IRM was given in a peak field of 1.3 T. Then the samples were taken out of the liquid-N bath and placed in zero field in the cryogenic magnetometer, where the changes in IRM are measured up to room temperature (Figure 7, where, again, time is a (non-linear) proxy for increasing temperature). The resulting remanence curves show sharp drops early on. We estimate that this occurs at around ~110 K and represents the Verwey transition, because the samples warm up very quickly initially and more slowly later on. This remanence drop at the transition is indicative of MD magnetite. Above ~110 K the quasi-exponential decay of magnetization is associated with thermal unblocking of progressively more and more particles that are SP at room temperature [*Banerjee et al.*, 1993; *Hunt*

et al., 1995]. Paleosol as well as loess samples from Beiyuan show SP contributions, whereas the loess sample from Tuxiangdao does not.

The presence of SP magnetite can also be determined by measuring magnetic susceptibility at two different frequencies [e.g., *Thompson and Oldfield*, 1986]. The measurement exploits the fact that SP particles have very short relaxation times and will therefore reach an equilibrium magnetization in a relatively short time period. The procedure involves comparing the susceptibility in an applied field at two different frequencies. Grains measured at a lower frequency, corresponding to a period that is longer than the relaxation time, will reach equilibrium. Conversely, a higher frequency will result in non-equilibrium magnetization and hence will produce a lower susceptibility value. We measured the frequency-dependent susceptibility at 0.46 kHz and 4.6 kHz for samples from the Beiyuan profile (Figure 8) and calculated the so-called frequency factor $F = (K_{0.46} - K_{4.6})/K_{0.46}) \times 100\%$. Our results show that the average of the F-factors is greater than ~6, and thus is higher than the typical average for loess-paleosol sequences (see Figure 6 of Evans and Heller [2001]). Importantly, our F-factor values at Beiyuan are rather high for many parts of the loess sections and show only a minor contrast between loess and paleosol S1, where bulk magnetic susceptibility has higher values (120×10^{-5} SI on average). In contrast, the Xifeng profile [*Evans and Heller*, 2001] shows *F*

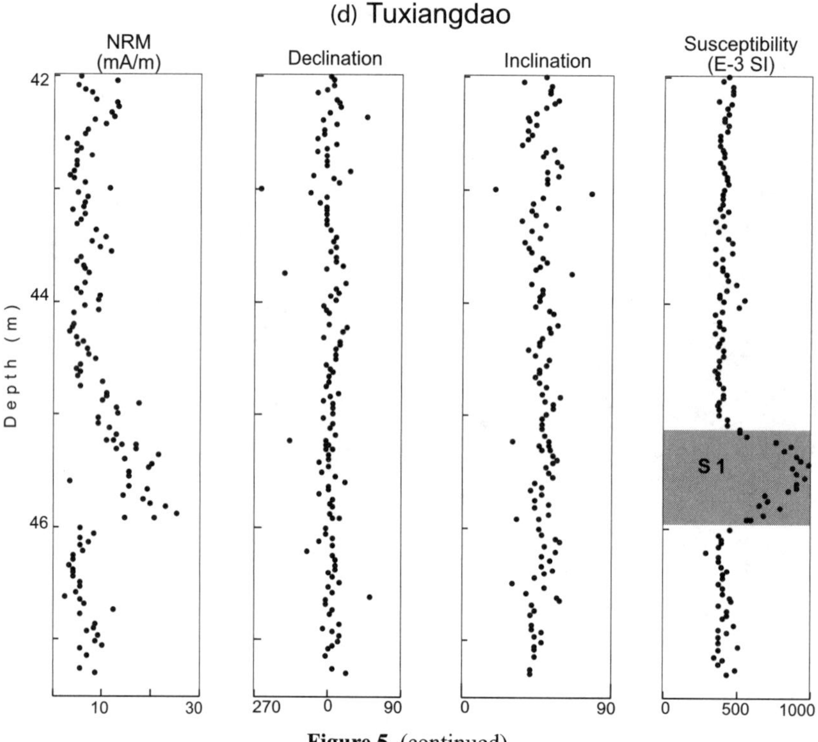

Figure 5. (continued)

varying between typical values of about 3 for loess and about 9 for paleosols. Lower (~6) as well as even higher F-factors (up to 15) have been proposed as diagnostic values for SP grains [*Stephenson*, 1971; *Forster et al.*, 1994; *Dearing et al.*

1996; *Eyre*, 1997]. If we take the value of F of about 6 as average, we find that the bulk of our Beiyuan section, including—interestingly—much of the loess, is higher than average. Thus, we conclude that it is likely that SP grains have grown in our loess as well as paleosol samples.

3.3 Anisotropy of Magnetic Susceptibility

Anisotropy of magnetic susceptibility (AMS) data can be examined to determine whether a depositional sedimentary fabric is preserved in relatively undeformed strata. The distribution of the principal axes of susceptibility ($K_{max} > K_{int} > K_{min}$) in sediments defines an ellipsoid that should mimic, if unperturbed, the orientation of grains in the depositional environment, being typically oblate and parallel to the paleohorizontal. In the Central Loess Plateau, a number of studies have reported such a sediment fabric in loess based on AMS [*Derbyshire et al.*, 1988; *Liu et al.*, 1988; *Thistlewood and Sun*, 1991; *Fang et al.*, 1997]. Together, these results show that the magnetic ellipsoid for loess-paleosol sequences for that area is typically characterized by a horizontal magnetic foliation that contains the K_{max} and K_{int} axes, consistent with a depositional (primary) fabric. In addition, it has occasionally been observed that the principal axes of maximum susceptibility are clustered, defining a magnetic lineation. The origin of such a lineation is still unclear. Some authors [*Matalucci et*

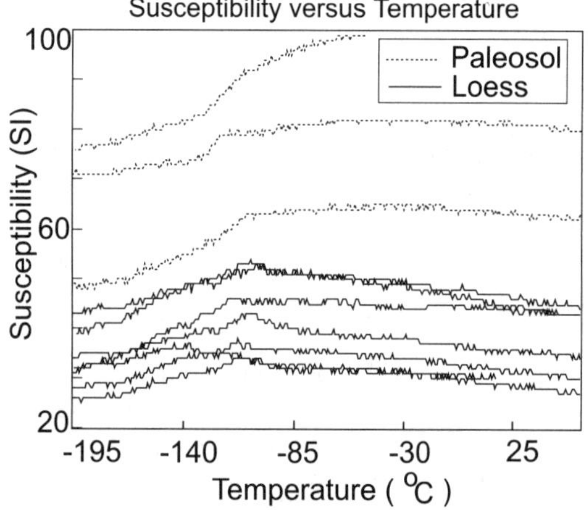

Figure 6. Plot of magnetic susceptibility versus time (= proxy for increasing temperature). Samples were immersed in liquid nitrogen at a temperature of -196°C and allowed to warm up to room temperature slowly after the liquid-N evaporated, while susceptibility was being measured with a Bartington sensor. The sharp increases midway between -196° and -30°C are thought to represent the Verwey transition at about -160°C.

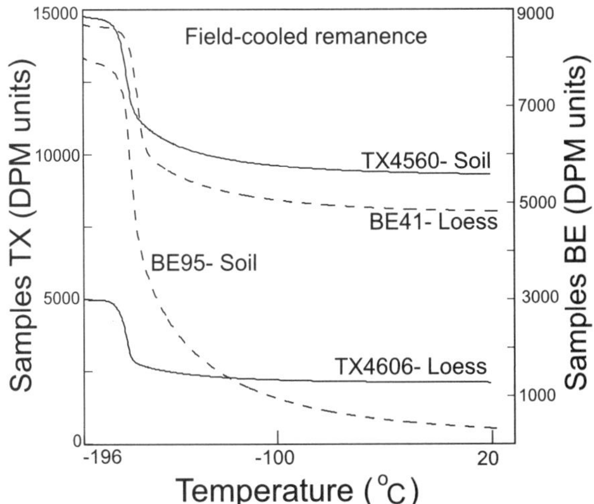

Figure 7. Remanence versus time (= proxy for increasing temperature). 8-cc samples were cooled in liquid nitrogen, whereupon an IRM was imparted in a field of 1.3T, and the samples were allowed to warm up quickly. Magnetization intensity changes were monitored with a cryogenic magnetometer until samples reached room temperature. The sharp decreases in magnetization soon after -196°C are thought to represent the Verwey transition at about -160°C. The subsequent progressive decay of magnetization up to room temperature is attributed to grains becoming superparamagnetic (SP). DPM- Digital panel meter voltage output of cryogenic magnetometer (reading is proportional to magnetization intensity and is normalized to sample volume, 8 cc). BE- Beiyuan, TX- Tuxiangdao.

al., 1969; *Lagroix and Banerjee*, 2002] have argued that mineral lineations correspond to the prevailing paleowind direction, but others maintain that they coincide with local paleoslope gradients [*Derbyshire et al.*, 1988].

Our AMS results contrast with those of the existing literature in that the principal axes of susceptibility are highly dispersed (Figure 9). Notably, K_{min} axes deviate often by more than 50° from the vertical (= pole to the depositional plane). Considering that most published AMS results from the Loess Plateau show typical depositional fabrics, the AMS data presented here suggest that the original fabric has been altered, leading to the conclusion that post-depositional grain-growth has modified the original sedimentary magnetic foliation.

Mathé et al. [1997] and Lagroix and Banerjee [2002] obtained similar results in soils elsewhere, confirming that secondary magnetic fabrics associated with soil formation can be characterized by scattered orientations of AMS principal axes.

3.4 Magnetic Viscosity

Our observations on the magnetic viscosity of samples are the only ones to suggest a really major difference between paleosols and loess, in that a large fraction of the remanence

of the paleosol horizons may be viscous. In contrast, our experiments do not show significant viscous behavior in loess samples.

A set of samples from both loess and paleosol layers were placed in the ambient Earth's magnetic field for a period of more than 2,500 hours. Special attention was paid in orienting the sample's NRM direction parallel to the Earth's magnetic field. Afterwards, samples were placed in zero-field space created by a μ-metal shield, whereupon their remanent magnetization was monitored over several thousands hours with a cryogenic magnetometer (Figure 10). There is a marked difference between loess and paleosol sediments, with loess samples barely changing their remanence intensity and paleosol samples losing 30% of their remanence in about 6000 hours (250 days). The capacity of paleosol sediments to acquire viscous remanent magnetization was previously noted by Heller and Liu [1984, 1986] who showed that viscosity coefficients could account for a strong Brunhes overprint. Also, Pan et al. (2001) recently reported that no less than 35% of the total NRM intensity in loess from the central Plateau is a VRM contribution.

It seems reasonable to conclude, given the large fraction of NRM that can be accounted for by viscous remanence

Beiyuan

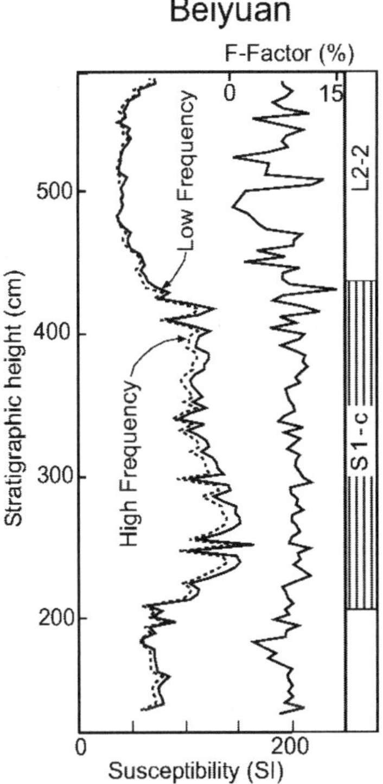

Figure 8. Dual frequency susceptibility for the Beiyuan section. The frequency factor (F) increases only slightly in paleosol S1, revealing only minor changes in SP concentration, on average, from loess (e.g., L2-2) to paleosol (S1-c).

(VRM) during a period of less than one year in the studied samples, that an equivalent or larger fraction of the remanence of the paleosols must be viscous [e.g., *Lowrie and Kent*, 1978]. VRM growth is likely due to MD magnetite and may accompany a protracted magnetization in paleosols, but this can probably not account for the complete masking of the Blake Event in the loess.

4. DISCUSSION

The Blake Event is a well-documented short polarity interval of the Middle Pleistocene and hence its absence in the studied sections is intriguing. Several causes and explanations for the absence of the event can be postulated. (1) A hiatus occurs at just the right intervals in the four sections through erosion or interrupted sedimentation. (2) The sampled sequences have missed a record of the event because of incorrect age assignments. (3) The Blake event is not observed because a complete secondary overprint has masked the original reversed-polarity record. (4) Imperfect recording of the geomagnetic field occurred as a result of delayed acquisition of the remanence or a significant lock-in depth representing multiple thousands of years.

Pedostratigraphy of loess-paleosol couplets in the Chinese Loess Plateau is typically accomplished by using magnetic susceptibility variations along the profiles. A plethora of rock-magnetic studies has documented that the magnetic susceptibility in loess-paleosol sequences arises mostly from the presence of a mixture of magnetite and maghemite grains, dominantly of multi-domain (MD) size, with very little contribution from hematite or paramagnetic minerals. It is generally assumed that these minerals are detrital and carry a

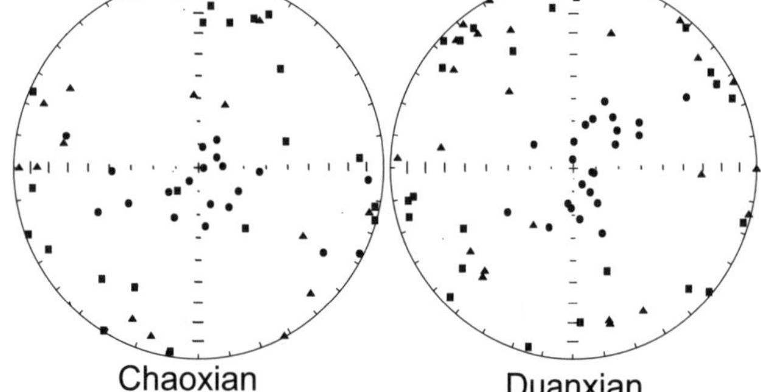

Figure 9. Stereographic projection of the principal axes of magnetic susceptibility for two of the sections. Squares, triangles and dots represent maximum, intermediate and minimum susceptibility axes, respectively. Note that the K_{min} axes often depart significantly from the vertical.

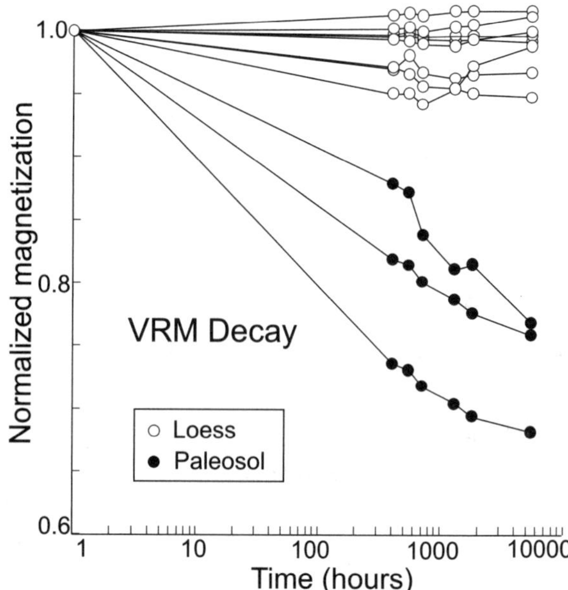

Figure 10. Viscosity test for representative samples. Note the different behavior between loess and paleosol samples. See text for discussion.

(post-) depositional remanence. In paleosols, additional magnetic minerals grow later as superparamagnetic (SP) or larger magnetite and maghemite grains so that the room-temperature susceptibility is usually higher than in the loess horizons [*Heller et al.*, 1991; *Evans and Heller*, 1994; *Eyre and Shaw*, 1994; *Foster et al.*, 1994; *Maher and Thompson*, 1999]. Thus, magnetic susceptibility in loess deposits has commonly been used to correlate stratigraphic sections among different localities and to identify the loess-paleosol couplets. Specifically, enhanced magnetic susceptibility has been used as a proxy for paleosol horizons. However, pedogenetic modification of iron minerals is common in loess-paleosols and could dramatically change the magnetic susceptibility, not just in paleosols but also in loess. The use of magnetic susceptibility as a regional pedostratigraphic tool has, in fact, been questioned by some authors [*Ding et al.*, 1993], because seasonal changes of the water table can induce remobilization of Fe and Mn through the sedimentary layers. In addition, other common processes may include leaching [*Anderson and Hallet*, 1995; *Derbyshire et al.*, 1997] and changing local redox conditions [*Feng and Johnson*, 1995; *Maher*, 1998; *Williams*, 1992] and these can modify the magnetic susceptibility. As a result, magnetic susceptibility may well be the net result from a combination of factors that include pedogenesis as one of many causes. If true, then pedostratigraphy and age assignments based on it could be in some cases problematic. As a consequence, post-depositional processes can alter the susceptibility signal and this can lead to an incorrect assignment of the loess-paleosol horizons.

Despite these caveats, we believe that in our studied profiles, the well-established pedostratigraphy allowed us to unequivocally identify horizons L1, L2 and S1. The loess-paleosol couplets can be unambiguously recognized in the field, by placing them in a sequence that starts at the very top with the youngest soil S_0 and can be followed downwards until at least S_3 or even S_{14} in Chaoxian [*Yue et al.*, 1991]. Age control is particularly well constrained at Chaoxian, where more than fourteen loess-paleosol couplets, combined with previous magnetostratigraphic studies [*Yue et al.*, 1991; *Heslop et al.*, 1999], allowed the Matuyama/Brunhes boundary and the Jaramillo Subchron to be determined. It is therefore unlikely that in our four sections we misidentified the paleosols or that we failed to recognize significant hiatuses in each of the four sections.

The third possibility is that the BE is not observed because of an overprint. There is a priori no unambiguous test to distinguish primary Brunhes-Chron directions from recent or present-day overprints in paleosol-loess deposits; a pervasive remagnetization of the loess-paleosol successions is therefore not detectable and may go unrecognized. Nevertheless, the fact that one can clearly observe reversed Matuyama, and older normal Jaramillo and Olduvai Chrons in loess-paleosol sequences below L8 argues against the ad-hoc proposition that young Brunhes normal magnetizations may have been replaced over large parts of sections by secondary chemical remagnetizations. Thus, we think that it is unlikely that the entire sections that we have studied have been completely remagnetized during recent times.

This brings us to the fourth possible explanation. An offset or delay in the acquisition of magnetization can explain the absence of short polarity intervals such as the Blake Event. The Matuyama/Brunhes boundary (MBB), for example, has widely been recorded in Chinese loess layer L8, equivalent to the cold glacial interval MIS20, whereas its marine occurrence is in a warm MIS interval (19) that would correspond to Chinese paleosol S7 [*Tauxe et al.*, 1996]. The lock-in depth or, alternatively, delay in magnetization acquisition, corresponds to a couple of tens of thousands of years [*Heller et al.*, 1987; *Evans and Heller*, 2001; *Spassov et al.*, in press], as estimated from the stratigraphic positions of microtektites [*Zhou and Shackleton*, 1999] found below the MBB in marine sediments, whereas in loess-paleosol deposits they are found above it.

Moreover, Heslop et al. [2000] developed a loess/paleosol timescale based on the combination of astronomically tuned monsoon records and the oxygen isotope record. They concluded that polarity-interval boundaries are shifted downwards in the loess-paleosol deposits, corresponding to time delays of ~25 kyr. Spassov et al. [in press] have recently developed a model to explain the "misplaced" MBB in loess-

paleosol deposits. The model is founded on the ideas of Bleil and von Dobeneck [1999] and presumes that the deposits carry two types of magnetization: a post-depositional remanent magnetization (pDRM) and a chemical remanent magnetization (CRM). pDRM is acquired by mechanical fixing of magnetic particles after dust settling. The CRM is proposed to have been acquired by paleosol horizons due to pedogenetic processes. The latter include oxidation of titanomagnetites [*Heller and Liu*, 1984], alteration of ferromagnesian minerals (e.g., biotite, augite [*Liu*, 1985]) and/or reprecipitation of iron in carbonate-rich environments [*Perel'man*, 1977]. Spassov and colleagues assume two different lock-in zones for the two remanence components, pDRM and CRM, and show that it is possible to explain the stratigraphic shift of the Matuyama-Brunhes boundary in loess sections with respect to the marine records.

In their original model, Bleil and von Dobenek [1999] showed that the simple concept of delayed remanence acquisition can also account for the appearance of artifacts that are simple interference phenomena. From such a model it follows that preservation of the magnetizations, acquired during short polarity intervals, is determined by both the lock-in range and the duration of the polarity interval. The lack of short polarity intervals in loess-paleosol deposits could then be caused by a large ratio of lock-in depth to sediment accumulation rate. The model of Bleil and von Dobeneck [1999] implies indeed that only polarity events lasting longer than the median lock-in depth divided by the sedimentation rate will be recorded in sediments.

In their model for loess-paleosol deposits, Spassov et al. [in press] show that small changes in the pDRM/CRM ratio near a polarity reversal can have a dramatic effect on the NRM polarity pattern. Variations in such a ratio will occur, for example, with loess-paleosol alternations. We next examine whether this can apply to the incomplete and/or missing record of the Blake Event in the Chinese Loess Plateau, where its expected position is just above paleosol S1-c.

Overall, our rock-magnetic experiments show evidence for MD and SP grains in paleosol horizons and for SP grains in many parts of the loess profile in Beiyuan (Figure 8). AMS results confirm the growth of new magnetic grains in paleosols as well as loess that perturb the original depositional pattern. These results contrast to some extent with those of the published literature [e.g., *Zhou et al.*, 1990; *Liu et al.*, 1994; *Hunt et al.*, 1995; *Maher and Thompson*, 1999] where it is emphasized that the presence of SP magnetite is particularly significant in the paleosol horizons. Together, our results are reminiscent of those by Katz et al. [2000] and Weil and Van der Voo [2002], who reported an increase in the relative abundance of SP magnetite grains associated with a profound chemical remanent magnetization.

The presence of SP grains in both loess and paleosol horizons suggests secondary growth of magnetic grains in the entire sampled sequence. If these newly grown oxides included single-domain grains, they may have become an important carrier of a secondary remanence of chemical nature (CRM). This CRM likely was acquired over an extended period of time, being particularly protracted during interglacial and interstadial events (=paleosol horizons), but probably also present during colder-drier intervals such as represented by loess L2-2. It is thus plausible that the process of remanence acquisition was sufficiently long that short polarity events such as the Blake have been completely masked by magnetizations acquired subsequently in the normal Brunhes Chron. A protracted magnetization process produces both a delay in acquisition, explaining the MBB offset, and the disappearance of short events, such as the Blake.

5. CONCLUSIONS

The Blake Event is inconsistently recorded in Chinese Loess. Its absence in the eastern part of the Loess Plateau was noted by Zhu et al. [1998], who thought that it could be explained by regional environmental variations; the then-known records of the Blake Event [*Zhu et al.*, 1994; *Chen et al.*, 1995; *Fang et al.*, 1997] were all from rocks in the western part of the region. We now know that the Blake Event is absent in sections in the western region as well, so that this is a less likely explanation.

It also appears that the stratigraphic position of the BE-record in the sections is confusing; in the vicinity of Xining, the BE appears as a triple feature within loess L2-2, whereas in Lanzhou it shows up as a doublet within paleosol S1-c. Here we reported on the results from four sections from the western Loess Plateau that do not show the Blake Event at all. Since pedostratigraphy seems to have been well enough constrained to locate the likely level of its occurrence, it is concluded that the absence of the Blake Event is due to the magnetization acquisition process in loess/paleosol sequences. Both CRM and pDRM are postulated to co-exist in loess/paleosol deposits [e.g., *Spassov et al.*, in press], the former being particularly important in paleosol horizons. Our study suggests that CRM may occur in loess as well. We argue that protracted magnetization acquisition masked and obscured the short-lived Blake Event. The preservation of short polarity events such as the BE depends mostly on the ratio of lock-in depth to sediment accumulation rate. Net downward water movement and the proportion of silt and clay in the sediment control the lock-in depth [*Assallay et al.*, 1998]. Local variations of such parameters can easily explain why in some areas within the Loess Plateau short polarity events are recorded, whereas in some other areas they are not. The process

by which short polarity events are obscured in loess/paleosols is similar to the well-established delay of magnetization acquisition that plagued the record of the Matuyama-Brunhes boundary in loess L8 and displaced it by a sedimentary thickness representing some 25 kyr.

Acknowledgments. We thank Friedrich Heller for providing us with a copy of the manuscript by Spassov et al. [in press]. We gratefully acknowledge the hard work of Jim E.T. Channell, Dennis V. Kent and Bill Lowrie, in organizing the successful *Chapman Conference on Timescales of the Geomagnetic Field* in Gainesville, Fl., in March 2003, and are pleased to dedicate this paper to birthday guy Neil D. Opdyke. Three anonymous referees provided helpful comments and suggestions. This study was supported by the National Science Foundation, Division of Earth Sciences (Grant EAR-9903074) and by the Scott Turner Fund, Department of Geological Sciences, University of Michigan.

REFERENCES

Anderson, R.S. and Hallet, B., imulating magnetic susceptibility profiles in loess as an aid in quantifying rates of dust deposition and pedogenic development, *Quat. Res.*, 45, 1016, 1995.

Assallay, A.M., Jefferson, I., Rogers, C.D.F. and Smalley, I.J., Fragipan formation in loess soils: development of the Bryant hydro-consolidation hypothesis, *Catena*, 83, 1–16, 1998.

Banerjee, S.K., Hunt, C.P. and Liu, X.M., Separation of local signals from the regional paleomonsoon record of the Chinese loess plateau: a rock-magnetic approach, *Geophys. Res. Lett.*, 20, 843–846, 1993.

Bleil, U. and von Dobeneck, T., Geomagnetic events and relative paleointensity records, *in Clues to high-resolution paleomagnetic chronostratigraphies of Late Quaternary marine sediments?, in Use of Proxies in Paleoceanography: Examples from the South Atlantic*, edited by G. Fisher and G. Wefer, pp. 635–654, Springer-Verlag, Berlin. 1999.

Chen, F., Wang, S., Li, J., Shi, Y.F., Li, J., Cao, J., Zhang, Y., Wang, Y. and Kelts, K., Study of paleomagnetic stratigraphy of a core from Zoige Lake on the Qinghai-Tibetan Plateau (in Chinese), *Sci. China* B25, 772–777, 1995.

Creer, K.M., Readman, P.W. and Jacobs, A.M., Palaeomagnetic and palaeontological dating of a section at Gioa Tauro, Italy: Identification of the Blake event, *Earth Planet. Sci. Lett.*, 50, 289–300, 1980.

Dearing, J.A., Dann, R.J.L., Hay, K., Lees, J.A., Loveland, P.J., Maher, B.A. and O'Grady, K., Frequency-dependent susceptibility measurements of environmental materials, *Geophys. J. Int.*, 124, 228–240, 1996.

Derbyshire, E., Billard, A., Van Vliet-Lanoe, B. and Cremaschi, M., Loess and palaeoenvironment: some results of a European joint programme of research, *Jour. Quat. Sci.*, 3, 147–169, 1988.

Derbyshire, E., Keen, D.H., Kemp, R.A., Rolph, T.A., Shaw, J. and Meng, X., Loess-palaeosol sequences as recorders of palaeoclimatic variations during the last glacial-interglacial cycle: some problems of correlation in North-Central China, *Quat. Proc.*, 4, 7–18, 1995.

Derbyshire, E., Kemp, R.A. and Meng, X., Climate change, loess and paleosols: proxy and resolution in North China, *Jour. Geol. Soc. London*, 154, 793–805, 1997.

Dinarès-Turell, J., Sagnotti, L. and Roberts, A.P., The Blake geomagnetic episode in cores from the Mediterranean Sea: Evidence for a "double" event, EGS XXV General Assembly, Nice (France), 2000.

Ding, Z.L., Rutter, N.W. and Liu, T.S., Pedostratigraphy of Chinese loess deposits and climatic cycles in the last 2.5 Ma, *Catena*, 20, 73–91, 1993.

Evans, M.E. and Heller, F., Magnetic enhancement and paleoclimate: study of a loess/paleosol couplet across the Loess Plateau of China, *Geophys. J. Int.,* 117, 257–264, 1994.

Evans, M.E. and Heller, F., Magnetism of loess/palaeosol sequences: recent developments, *Earth Sci. Rev.*, 54, 129–144, 2001.

Eyre, J.K., Frequency dependence of magnetic susceptibility for populations of single-domain grains, *Geophys. J. Int.*, 129, 209–211, 1997.

Eyre, J.K. and Shaw, J., Magnetic enhancement of Chinese loess- The role of ?Fe_2O_3?, *Geophys. J. Int.*, 117, 265–271, 1994.

Fang, X.M., Li, J.J., Van der Voo, R., McNiocaill, C., Dai, X.R., Kemp, R.A., Derbyshire, E., Cao, J.X., Wang, J.M. and Wang, G., A record of the Blake Event during the last interglacial paleosol in the western Loess Plateau of China, *Earth Planet. Sci. Lett.*, 146, 73–82, 1997.

Fang, X.M., Li, J.J., Banerjee, S.K., Jackson, M., Oches, E.A. and Van der Voo, R., Millennial-scale climatic change during the last interglacial period: Superparamagnetic sediment proxy from paleosol S1, western Chinese Loess Plateau, *Geophys. Res. Lett.*, 26, 2485–2488, 1999.

Feng, Z.D. and Johnson, W.C., Factors affecting the magnetic susceptibility of a loess-soil sequence, Barton County, Kansas, USA, *Catena*, 24, 25–37, 1995.

Forster, T., Evans, M.E. and Heller, F., The frequency dependence of low field susceptibility in loess sediments, *Geophys. J. Int.*, 118, 636–642, 1994.

Grootes, P.M., Stuiver, M., White, J.W.C., Johnsen, S. and Jouzel, J., Comparison of oxygen isotope records from the GISP2 and GRIP Greenland ice cores, *Nature*, 366, 552–554, 1993.

Heller, F. and Liu, T.S., Magnetostratigraphical dating of loess deposits in China, *Nature*, 300, 431–433, 1982.

Heller, F. and Liu, T.S., Magnetism of Chinese loess deposits, *Geophys. J.R. Astron. Soc.*, 77, 125–141, 1984.

Heller, F. and Liu, T.S., Palaeoclimatic and sedimentary history from magnetic susceptibility of loess in China, *Geophys. Res. Lett.*, 13, 1169–1172, 1986.

Heller, F., Meili, B., Wang, J.D., Li, H.M. and Liu, T.S., *Magnetization and sedimentary history of loess in the central loess plateau of China, in Aspects of Loess Research*, edited by Liu, T.S., China Ocean Press, Beijing, pp. 147–163, 1987.

Heller, F., Liu, X.M., Liu, T.S. and Xu, T.C., Magnetic susceptibility of loess in China, *Earth Planet Sci. Lett.*, 103, 301–310, 1991.

Herrero-Bervera, C.E., Helsley, C.E., Hammond, S.R. and Chitwood, L.A., A possible lacustrine palaeomagnetic record of the Blake episode from Pringle Falls, Oregon, USA, *Phys. Earth Planet. Int.*, 56, 112–123, 1989.

Heslop, D., Shaw, J., Bloemendal, J., Chen, F., Wang, J. and Parker, E., Sub-Millennial scale variations in East Asian monsoon systems recorded by dust deposits from the North-Western Chinese Loess Plateau, *Phys. Chem. Earth*, 24, 785–792, 1999.

Heslop, D., Langereis, C.G. and Dekkers, M.J., A new astronomical timescale for the loess deposits of Northern China, *Earth Planet. Sci. Lett.*, 184, 125–139, 2000.

Hu, S., Appel, E., Wang, S., Wu, J., Xue, B., Wang, Y., Qian, J. and Xiang, L., A preliminary magnetic study on lacustrine sediments from Zoige Basin, Eastern Plateau, China: Magnetostratigraphy and environmental implications, *Phys. Chem. Earth*, 24, 811–816, 1999.

Hunt, C.P., Banerjee, S.K., Han, J.M., Solheid, P.A., Oches, E., Sun, W.W. et al., Rock magnetic proxies of climate change in the loess-paleosol sequences of the western Loess Plateau of China, *Geophys. J. Int.*, 123, 232–244, 1995.

Hus, J.J., Paepe, R. and Geeraerts, R., The influence of periglacial activity on the remanent magnetization of sediments, *Geol. Mijnbouw*, 72, 225–235. 1993

Katz, B., Elmore, R.D., Cogoini, M., Michael, H.E. and Ferry, S., Associations between burial diagenesis of smectite, chemical remagnetization, and magnetite authigenesis in the Vocontian trough, SE France, *J. Geophys. Res.*, 105, 851–868, 2000.

Kirschvink, J.L., The least-squares line and plane and the analysis of paleomagnetic data. Geophys, *J. Roy. Astron. Soc.*, 62, 699–718, 1980.

Kukla, G., Heller, F., Liu, X.M., Xu, T., Liu, T.S. and An, Z.A., Pleistocene climates in China dated by magnetic susceptibility, *Geology*, 16, 811–814, 1988.

Lagroix, F. and Banerjee, S.K., Paleowind directions from the magnetic fabric of loess profiles in central Alaska, *Earth Planet. Sci. Lett.*, 195: 99–112, 2002.

Langereis, C.G., Dekkers, M.J., de Lange, G.J., Paterne, M. and van Santvoort, P.J.M., Magnetostratigraphic and astronomical calibration of the last 1.1 Myr from an eastern Mediterranean piston core and dating of short events in the Brunhes, *Geophys. J. Int.*, 129, 75–94, 1997.

Liu, T.S., *Loess and the Environment*, pp. 1–251, China Ocean Press, Beijing, 1995.

Liu, X.M., Xu, T. and Liu, T.S., The Chinese loess in Xifeng. II. A study of anisotropy of magnetic susceptibility of loess from Xifeng, *Geophys. J.* 92, 92, 349–353, 1988.

Liu, X.M., Rolph, T., Bloemendal, J., Shaw, J. and Liu, T.S., Remanence characteristics of different magnetic grain-size categories at Xifeng, Central Chinese Loess Plateau, *Quat. Res.*, 42, 162–165, 1994.

Lowrie, W. and Kent, D.V., Characteristics of VRM in oceanic basalts, *J. Geophys.*, 44, 297–315, 1978.

Lund, S.P., Acton, G., Clement, B., Hastedt, M., Okada, M., Williams, T. and the ODP Leg 172 Scientific Party, Geomagnetic Field Excursions occurred often during the last million years, *EOS Transactions*, AGU, 79, 178–179, 1998.

Maher, B.A., Magnetic properties of modern soils and Quaternary loessic paleosols: paleoclimatic implications, *Palaeogeog., Palaeoecol., Palaeoclimat.*, 137, 25–54. 1998.

Maher, B.A. and Thompson, R., Magnetism and palaeoclimate in the Chinese Loess, *in Quaternary Climates, Environments and Magnetism,* edited by B.A. Maher and R. Thompson, pp. 81–125, Cambridge University Press, 1999.

Matalucci, R.V., Shelton, J.W. and Abdel-Hady, M., Grain orientation in Vicksburg loess, *J. Sediment. Petrol.*, 39, 969–979, 1969.

Mathé, P.E., Rochette, P and Colin, F., The origin of magnetic susceptibility and its anisotropy in some weathered profiles, *Phys. Chem. Earth*, 22, 183–187, 1997.

Pan, Y., Zhu, R., Shaw, J., Liu, Q. and Guo, B., Can relative paleointensities be determined from the normalized magnetization of the wind-blown loess of China?, *Jour. Geophys. Res.*, 106, 19,221–19,232, 2001.

Perel'man, A.I., *Geochemistry of elements in the Supergene Zone*, pp. 1–266, Keter, Jerusalem, 1977.

Reinders, J. and Hambach, U., A geomagnetic event recorded in loess deposits of the Tönchesberg (Germany): identification of the Blake magnetic polarity episode, *Geophys. J. Int.*, 122, 407–418, 1995.

Smith, D.J. and Foster, J.H., Geomagnetic reversal in Brunhes normal polarity epoch, *Science*, 163, 565–567, 1969.

Spassov, S., Heller, F., Evans, M.E., Yue, L.P. and von Dobeneck, T., A lock-in model for the complex Matuyama-Brunhes boundary record of the loess/paleosol sequence at Lingtai (Central Chinese Loess Plateau), *Geophys. J. Int.*, in press.

Stephenson, A., Single domain grain distributions. 1. A method for the determination of single domain distributions, *Phys. Earth Planet. Int.*, 4, 353–360, 1971.

Tauxe, L., Herbert, T., Shackleton, N.J. and Kok, Y.S., Astronomical calibration of the Matuyama-Brunhes boundary: consequences for magnetic remanence acquisition in marine carbonates and Asian loess sequences, *Earth Planet. Sci. Lett.*, 140, 133–146, 1996.

Thislewood, L. and Sun, J., A paleomagnetic and mineral magnetic study of the loess sequence at Liujiapo, Xian, China, *J. Quat. Sci.*, 6, 13–26, 1991.

Thompson, R. and Oldfield, E., *Environmental Magnetism*, 227 pp., Allen and Unwin, Winchester, Mass, 1988.

Tric, E., Laj, C., Valet, J.P., Tucholka, P., Paterne, M. and Guichard, F., The Blake geomagnetic event: transition geometry, dynamical characteristics and geomagnetic significance, *Earth Planet. Sci. Lett.*, 102, 1–13, 1991.

Tucholka, P., Fontugne, M., Guichard, F. and Paterne, M., The Blake magnetic polarity episode in cores from the Mediterranean Sea, *Earth Planet. Sci. Lett.*, 86, 320–326, 1987.

Weil, A.B. and Van der Voo, R., Insights into the mechanism for orogen-related carbonate remagnetization from growth of authigenic Fe-oxide: A scanning electron microscopy and rock magnetic study of Devonian carbonates from northern Spain, *J. Geophys. Res.* 107 (B4), 2063, 10.1029/2001JB000200, 2002.

Williams, M., Evidence for the dissolution of magnetite in recent Scottish peat, *Quat. Res.*, 37, 171–182, 1992.

Yue, L., Lei, X. and Qu, H., A magnetostratigraphic study on the Jingyuan loess section, Gansu (in Chinese), *Quat. Sci.*, 4, 349–353. 1991.

Zanella, E. and Laurenzi, M.A., Evidence for the Blake event in volcanic rocks from Lipari (Aeolian Archipelago), *Geophys. J. Int.*, 132, 149–158, 1998.

Zhou, L.P. and Shackleton, N.J., Misleading positions of geomagnetic reversal boundaries in Eurasian loess and implications for correlation between continental and marine sediment sequences, *Earth Planet. Sci. Lett.*, 168, 117–130, 1999.

Zhou, L.P., Oldfield, F., Wintle, A.G., Robinson, S.G. and Wang, J.T., Partly pedogenic origin of magnetic variations in Chinese loess, *Nature*, 346, 737–739, 1990.

Zhu, R.X., Zhou, L.P., Laj, C., Mazaud, A. and Ding, Z.L., The Blake geomagnetic polarity episode recorded in Chinese loess, *Geophys. Res. Lett.*, 21: 697–700, 1994.

Zhu, R., Coe, R.S., Guo, B., Anderson, R. and Zhao, X., Inconsistent palaeomagnetic recording of the Blake event in Chinese loess related to sedimentary environment, *Geophys. J. Int.*, 134: 867–875, 1998.

Zijderveld, J. A.C. demagnetization or rocks: Analysis of results, *in Methods in Paleomagnetism*, edited by D. Collinson, K. Creer and S. Runcorn, pp. 254–286, Elsevier, New York, 1967.

X. Fang, Department of Geography, Lanzhou University, Gansu 730000, China.

J. M. Parés, R. Van der Voo, and M. Yan, Department of Geological Sciences, University of Michigan, 2534 C.C. Little Building, Ann Arbor, Michigan. (jmpares@umich.edu, voo@umich.edu, maoduyan@umich.edu)

The Matuyama Chronozone at ODP Site 982 (Rockall Bank): Evidence for Decimeter-Scale Magnetization Lock-in Depths

J. E. T. Channell

Department of Geological Sciences, University of Florida, Gainesville, Florida

Y. Guyodo[1]

Institute for Rock Magnetism, Department of Geology and Geophysics, University of Minnesota, Minneapolis, Minnesota

At ODP Site 982, located on the Rockall Bank, component magnetizations define a polarity stratigraphy from the middle part of the Brunhes Chronozone to the top of the Gauss Chronozone (0.3–2.7 Ma). The Cobb Mountain and Reunion sub-chronozones correlate to marine isotope stages (MIS) 35 and 81, respectively. The Gauss/Matuyama boundary correlates to the base of MIS 103. The slope of the natural remanence (NRM) versus the anhysteretic remanence (ARM) during stepwise alternating field demagnetization is close to linear (r > 0.9) for the 900–1700 ka interval indicating similarity of NRM and ARM coercivity spectra. The normalized remanence (NRM/ARM) can be matched to paleointensity records from ODP Sites 983/984 (Iceland Basin) and from sites in the Pacific Ocean. The age model based on δ^{18}O stratigraphy at ODP Site 982 indicates sedimentation rates in the 1–4 cm/kyr range. At ODP Site 980/981, located ~200 km SE of Site 982 on the eastern flank of the Rockall Plateau, sedimentation rates are 2–7 times greater. According to the δ^{18}O age models at Site 982 and Site 980/981, the boundaries of the Jaramillo, Cobb Mountain, Olduvai and Reunion subchronozones appear older at Site 982 by 10–15 kyr. We model the lock-in of magnetization using a sigmoidal magnetization lock-in function incorporating a surface mixed layer (base at depth M corresponding to 5% lock-in) and a lock-in depth (L) below M at which 50% lock-in is achieved. The observed age offsets can be simulated by values of DM (difference between values of M at Sites 982 and 980/981) in the 16–23 cm range, and values of L <10 cm. The low values of L imply that the magnetization is rapidly locked (within 10 cm) after sediment passes through the decimeter-scale surface mixed-layer.

[1] Now at Laboratoire des Sciences du Climat et de l'Environnement, Gif-sur-Yvette, France.

Timescales of the Paleomagnetic Field
Geophysical Monograph Series 145

1. INTRODUCTION

ODP Site 982 (at 57.52°N, 15.87°W, 1135 m water depth) is located in the Hatton-Rockall Basin, a small depression on the Rockall Plateau (Figure 1). The site has provided a coupled magnetic/δ^{18}O record for the 0.3–2.7 Ma interval. The record can be used to correlate polarity reversals to marine isotope

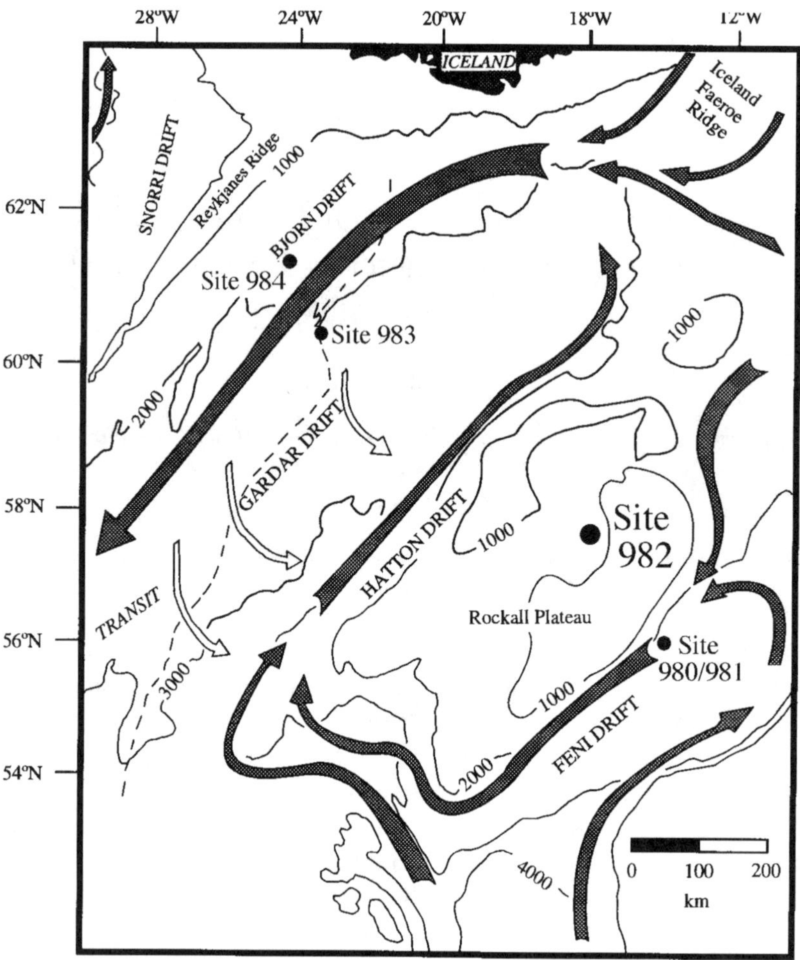

Figure 1. Location map for ODP Site 982. Bathymetry in meters. Dashed line indicates crest of Gardar Drift. Arrows indicate inferred bottom current flows [after *Manley and Caress*, 1994; *McCave et al.*, 1980].

stages (MIS), and provides a normalized remanence (paleointensity) record for comparison with other records spanning the same time interval.

The correlation of Plio-Pleistocene isotopic stages to polarity chrons and subchrons has been established using data from DSDP Sites 607/609 [*Raymo et al.*, 1989], ODP Site 659 [*Tiedemann et al.*, 1994], ODP Site 846/849 [*Shackleton et al.*, 1995a; *Schneider*, 1995], ODP Site 983/984 [*Channell and Kleiven*, 2000; *Channell et al.*, 2002] and Italian land sections [*Lourens et al.*, 1996]. The correlation is not always straightforward due to uncertainties in interpretation of either the MIS or the magnetic stratigraphy. For example, the boundaries of the Olduvai Subchronozone and the Gauss-Matuyama boundary at ODP Site 659 are poorly defined [*Tauxe et al.*, 1989], and the lack of foraminifera in the Matuyama Chronozone at ODP Site 983/984 requires that the susceptibility record be used as a proxy for $\delta^{18}O$ [*Channell et al.*, 2002].

An additional motive for this study is that the comparison of polarity chron/MIS correlations for different sedimentation rates offers the possibility of gaining insights into the nature of magnetization lock-in. ODP Site 980/981 located on the eastern slope of the Rockall Plateau (Figure 1) has sedimentation rates 2–7 times greater than at Site 982. High resolution magnetic and isotope records for the Matuyama Chronozone at Site 980/981 have been published elsewhere [*Channell and Raymo*, 2003; *Channell et al.*, 2003]. Here we report the magnetic and isotope records from ODP Site 982, and we compare these records with those from Site 980/981.

2. LITHOSTRATIGRAPHY AND AGE MODEL

The sedimentary sequence recovered at ODP Site 982 comprises nannofossil ooze with clay and silt [*Shipboard Scientific Party*, 1996]. The composite section, the optimal section recovered at the site, was compiled from the shipboard multi-sensor track (MST) as a splice from the four holes drilled at the site [*Shipboard Scientific Party*, 1996]. This paper deals with the 7–66 meters composite depth (mcd) interval, correlative to the 0.3– 2.7 Ma time interval. The 0–7 mcd interval was not sampled due to drilling disturbance.

Samples for stable isotope analyses were collected at 5 cm intervals from the composite section at Site 982. Stable isotope data were acquired from two benthic foraminifera, *Cibicidoides wuellerstorfi* and *Cibicidoides kullenbergi* (Figure 2). Oxygen isotopic ratios were measured on carbon dioxide gas released on treatment with orthophosphoric acid at 90°C using a VG Isogas mass spectrometer at the University of Florida. Values are reported relative to PDB. The analytical procedure, and the benthic and planktic isotope records for the last

1.2 Myrs from Site 982, have been reported elsewhere [*Venz et al.*, 1999; *Flower et al.*, 2000; *Kleiven et al.*, 2003].

Carbonate percentage, calculated from reflectance data and coulometric titration, varies widely in the 90–20% range [*Ortiz et al.*, 1999; *Venz et al.*, 1999]. Highs in susceptibility (lows in carbonate) are associated with glacial isotopic stages (Figure 2), reflecting enhanced surface water (carbonate) productivity during interglacial stages. Highs in susceptibility also correspond to highs in percentage ice rafted debris [*Venz et al.*, 1999]. Down-section from just above the Gauss-Matuyama boundary, located at 57.3 mcd, the percentage carbonate values are uniformly high and volume susceptibility values are uniformly low (Figure 2).

We construct the age model for Site 982 by matching the benthic δ^{18}O record to the chronology of Shackleton et al.

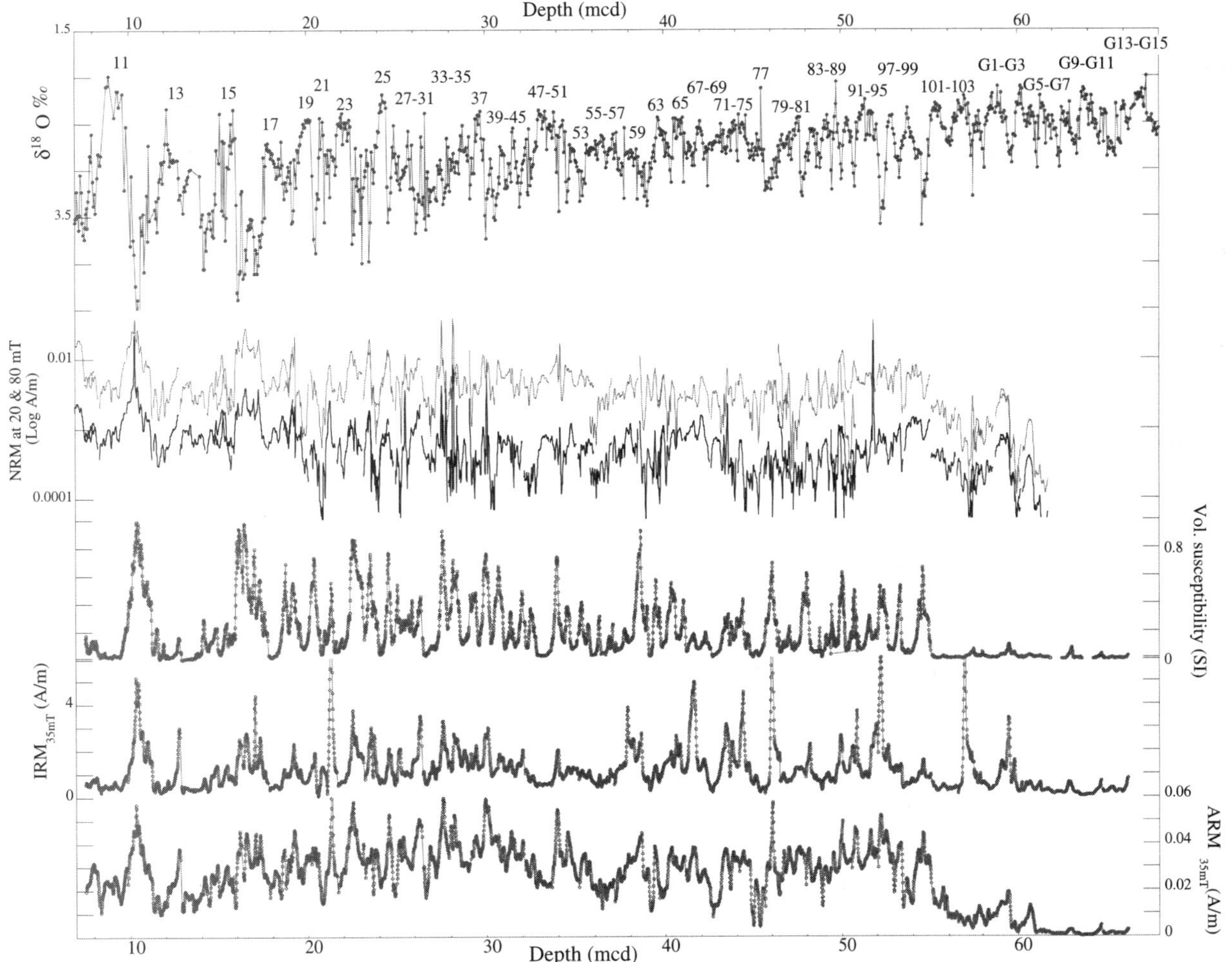

Figure 2. ODP Site 982: benthic δ^{18}O data [*Venz et al.*, 1999; *Flower et al.*, 2000; *Kleiven et al.*, 2003] with marine isotope stages numbered, natural remanent magnetization (NRM) after demagnetization at peak fields of 20 mT and 80 mT, volume magnetic susceptibility, isothermal remanence (IRM) and anhysteretic remanence (ARM), both remanences after demagnetization at peak fields of 35 mT. Magnetic measurements on u-channel samples.

Table 1. Depth/age tie-points for ODP Site 982

mcd	ka	mcd	ka	mcd	ka
6.87	336.0	30.05	1251.0	47.69	2153.0
8.39	370.0	30.55	1290.0	47.80	2162.1
9.84	408.0	31.90	1365.0	48.00	2179.6
14.19	531.0	34.10	1461.0	48.35	2192.0
16.15	630.0	35.25	1542.0	48.80	2243.0
17.78	678.3	37.73	1653.0	49.50	2282.0
19.18	750.0	38.48	1701.0	50.78	2363.0
20.53	798.0	39.03	1730.0	52.18	2441.0
22.58	873.0	40.40	1788.0	52.53	2465.0
24.58	966.0	41.60	1875.0	53.28	2486.0
26.13	1041.0	42.45	1916.0	53.73	2507.0
26.43	1050.0	44.15	1988.0	54.58	2525.0
26.68	1060.0	44.85	2045.0	55.28	2543.0
27.68	1098.0	45.80	2084.0	56.63	2579.0
28.35	1131.0	47.45	2137.9	57.33	2597.0
29.20	1197.0	47.65	2144.2	58.00	2642.0

[1990, 1995a,b] as defined by his TARGET curve (http://delphi.esc.cam.ac.uk/coredata/v677846.html) shown in Figure 3. Depth/age tie points are given in Table 1. The TARGET curve in the interval of interest comprises data from ODP Site 677 (0.34–1.81 Ma) and ODP Site 846 (1.81–2.70 Ma) [*Shackleton et al.*, 1990; 1995a,b]. In Figure 3, the match of the Site 982 δ^{18}O record with the TARGET curve is compared with the benthic δ^{18}O record from DSDP Site 607 [*Raymo et al.*, 1989] which was independently correlated to the Shackleton time scale used here. Note that the Site 982 age model of Venz et al. [1999], subsequently used by Flower et al. [2000] and Kleiven et al. [2003], utilizes a different target curve and hence differs slightly from the age model used here. The age model for ODP Site 980/981 [*Channell and Raymo*, 2003; *Channell et al.*, 2003] utilizes the same TARGET curve used here for Site 982.

3. MAGNETIC MEASUREMENTS

Magnetic remanence and susceptibility measurements were made on u-channel samples (with 2x2 cm square cross-sections) collected from the archive half of the Site 982 composite section in the 7–66 mcd interval. Magnetic remanence measurements were made using a pass-through 2G magnetometer designed for u-channel samples [see *Weeks et al.*, 1993]. Volume magnetic susceptibility measurements utilized a u-channel track incorporating a Sapphire Instruments 3.5 cm-square susceptibility loop [*Thomas et al.*, 2003]. Shipboard treatment of archive halves of the composite section involved AF demagnetization of natural remanent magnetization (NRM) at maximum peak fields of 25 mT [*Channell and Lehman*, 1999]. Subsequent treatment of NRM on u-channel samples constituted alternating field (AF) demag-

netization at peak fields of 20 to 60 mT (5 mT steps), 70 mT and 80 mT. The natural remanent magnetization (NRM) intensity falls by an order of magnitude between the 20 mT and 80 mT demagnetization steps (Figure 2) indicating that the NRM is carried by a low coercivity mineral such as magnetite. Thermal demagnetization of isothermal remanent magnetization and hysteresis ratios at neighboring sites (ODP Sites 980/981, 983 and 984) indicate pseudo-single domain (PSD) magnetite as the principal remanence carrier in the sediment drifts of the Iceland Basin [*Channell et al.*, 1997, 1998; *Channell*, 1999; *Channell and Raymo*, 2003].

Orthogonal projections of AF demagnetization data from ODP Site 982 indicate the presence of a well-defined, low coercivity, magnetization component (e.g. samples at 39.84 mcd and 40.43 mcd in Figure 4). In the immediate vicinity of polarity zone boundaries, multi-component magnetizations are occasionally observed. For example, at the top of the Olduvai Subchronozone in the 40.11–40.33 mcd interval (Figure 4), a lower coercivity post-reversal magnetization appears to be superimposed on a higher coercivity pre-reversal magnetization.

Component declination and inclination were computed from u-channel data by applying the standard least squares method [*Kirschvink*, 1980] to the 20–70 mT demagnetization interval. The demagnetizaton interval was modified for some samples at polarity zone boundaries (e.g. those at 40.19 and 40.24 mcd in Figure 4). Maximum angular deviation (MAD) values are less than 10° for the majority of the magnetization components (Figure 3). Component declination and inclination define a polarity stratigraphy from the mid-Brunhes Chronozone to the Gauss Chronozone (Figure 3).

Anhysteretic remanence (ARM) was applied to u-channel samples using 100 mT AF and a 0.05 mT DC field. The ARM was measured and then progressively demagnetized in the same steps applied to the NRM. ARM intensity, after AF demagnetization in peak fields of 35 mT, varies over a range of 0.1 to 0.6 A/m in the 7 to 55 mcd interval (Figure 2). ARM values are close to zero below 55 mcd where carbonate percentages reach 90% [*Ortiz et al.*, 1999]. A similar trend is observed for isothermal remanence (IRM), applied in a 500 mT DC field (Figure 2). Anhysteretic susceptibility (k_{arm}), the ARM normalized by the biasing DC field divided by volume susceptibility (k), can be used as a measure of magnetite grain size [*King et al.*, 1983]. For ODP Site 982, the k_{arm}/k ratios (Figure 5) imply magnetite grain sizes of <0.1 μm for low susceptibility intervals (full interglacial stages) and grain sizes of ~1–5 μm for high susceptibility intervals (full glacial stages).

Volume susceptibility, ARM and IRM fluctuate by more than an order of magnitude, particularly below 45 mcd (Figure 2). In order to test the coercivity match of NRM to ARM and IRM, and evaluate the potential for deriving paleointen-

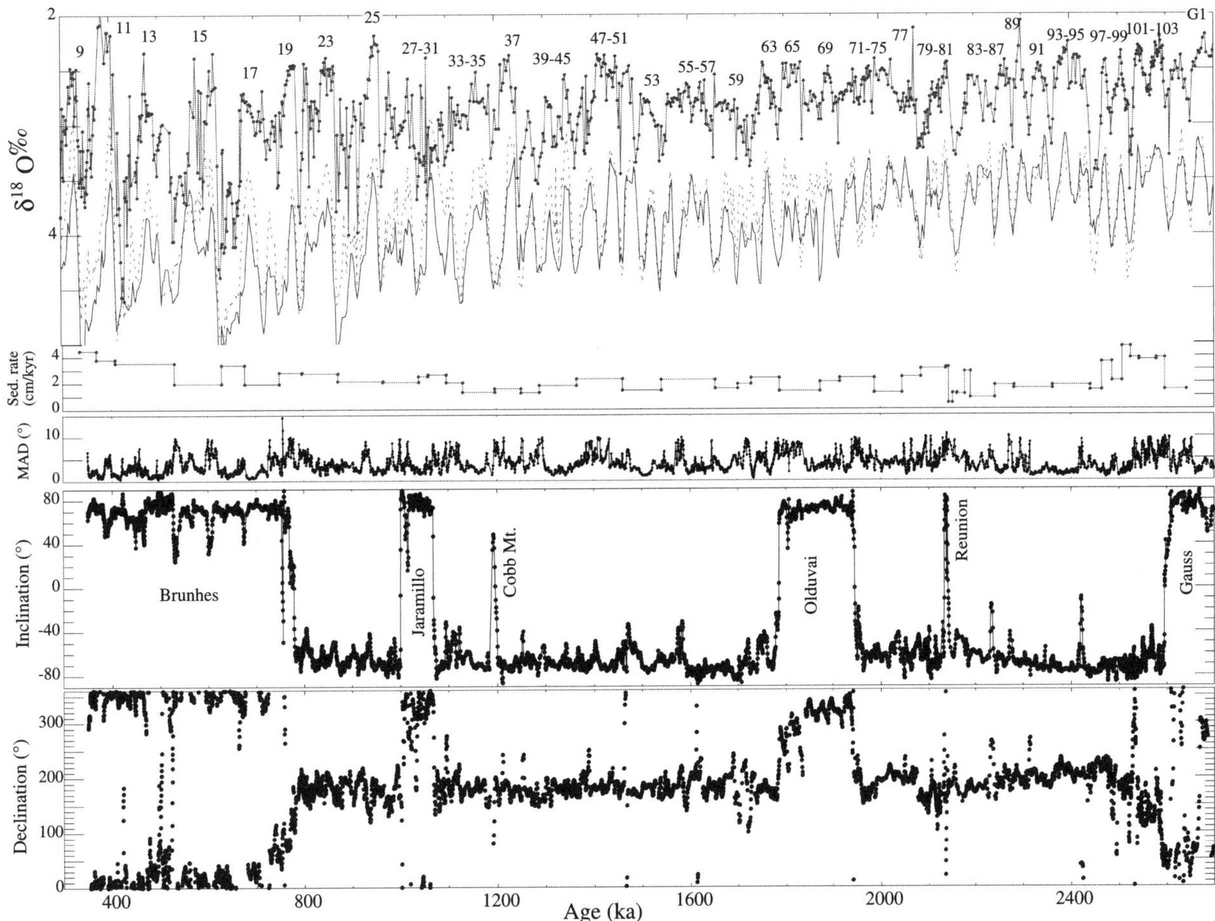

Figure 3. ODP Site 982: Benthic $\delta^{18}O$ data versus age from ODP Site 982 (fine line with points) compared to the chronology of Shackleton et al. [1990] as defined by his TARGET curve(http://delphi.esc.cam.ac.uk/coredata/v677846.html) (fine continuous line) and benthic $\delta^{18}O$ data from DSDP Site 607 [*Raymo et al.*, 1989] (dashed line) tuned to the same TARGET curve. The TARGET curve comprises data from ODP Sites 677 (0.34–1.81 Ma) and 846 (1.81–2.70 Ma) [*Shackleton et al.*, 1990; 1995a,b]. Interval sedimentation rates according to the age model, maximum angular deviation (MAD) values associated with component magnetization directions, component inclinations and declinations placed on the oxygen isotope age model.

sity proxies, we plot the slope of NRM versus ARM (or IRM) in the 25–45 mT demagnetization range (Figure 6b). In ideal circumstances, the slopes constitute paleointensity proxies and the linear correlation coefficients (r) associated with the slopes should be close to unity. In the vicinity of the Matuyama-Brunhes boundary (Figure 6), the linear correlation coefficients depart significantly from unity. This indicates that neither IRM nor ARM have coercivity that matches that of NRM in this interval. For the 900–1700 ka (22–38 mcd) interval, however, linear correlation coefficients are usually above 0.9 indicating that the coercivity of NRM matches the coercivity of IRM and ARM. This is also the interval over which the values of IRM and ARM are less variable (Figure 2).

The paleointensity proxy derived from the slope of NRM/ARM can be compared with other records for the 900–1700 ka interval. In Figure 7, we compare the ODP Site 982 normalized remanence record with the paleointensity records from the Ontong-Java Plateau [*Kok and Tauxe*, 1999], the California Margin [*Guyodo et al.*, 1999], the Equatorial Pacific [*Valet and Meynadier*, 1993; *Meynadier et al.*, 1995], and ODP Sites 983/984 [*Channell and Kleiven*, 2000; *Channell et al.*, 2002]. Mean sedimentation rates in the Matuyama Chronozone are ~2 cm/kyr or less for all records in Figure 7, apart from ODP Sites 983/984 where mean sedimentation rates are ~15 cm/kyr. To compensate for differing sedimentation rates, the ODP Site 983/984 records in Figure 7 have been smoothed using a 10 kyr running mean.

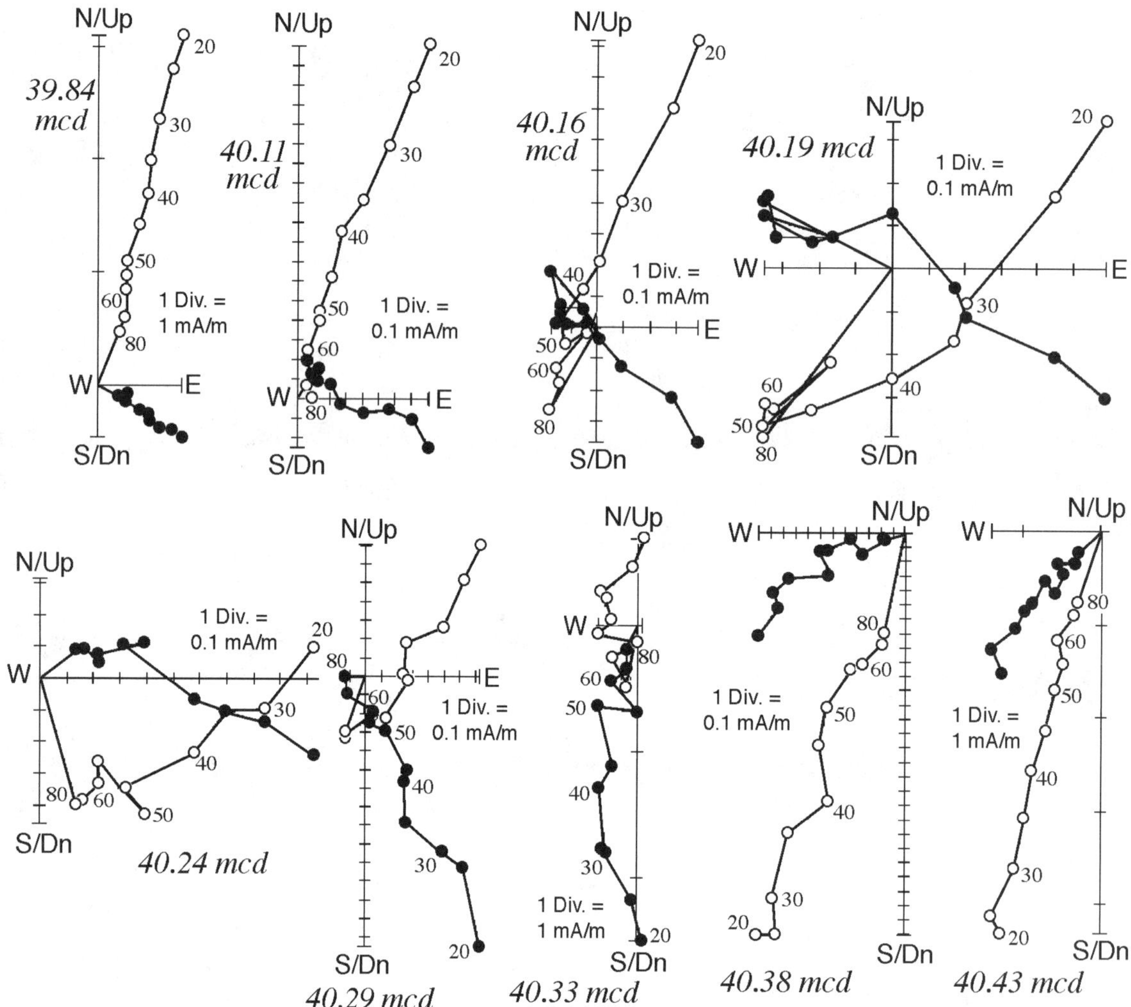

Figure 4. Orthogonal projection of alternating field demagnetization data for samples from the polarity reversal at the top of the Olduvai Subchronozone at ODP Site 982. Superimposed magnetization components are apparent in the 22 cm transition interval (40.11–40.33 mcd). Representative projections are shown. Data acquired at stratigraphic interval of 1 cm on u-channel samples with a magnetometer response function width of ~3.5 cm. Open and closed symbols indicate projection on the vertical and horizontal planes, respectively. The positions of the samples in meters composite depth (mcd) are indicated. The peak alternating fields range from 20 mT to 80 mT.

Some offsets in Figure 7 can be explained by discrepancies in chronology. The Ontong-Java stack was dated by linear interpolation between polarity reversals at the Matuyama-Brunhes boundary (MBB), the boundaries of the Jaramillo Subchronozone, and the Gauss-Matuyama boundary [*Kok and Tauxe*, 1999]. The age control at ODP Site 1021 (California Margin) for the Matuyama Chronozone was based on linear interpolation from the MBB to the boundaries of the Jaramillo Subchronozone [*Guyodo et al.*, 1999]. The age control for the Equatorial Pacific ODP Leg 138 record [*Valet and Meynadier*, 1993; *Meynadier et al.*, 1995] was based on orbital tuning of the GRAPE density stratigraphy [*Shackleton et al.*, 1995a]. The age models for ODP Sites 983/984 [*Channell and Kleiven*, 2000; *Channell et al.*, 2002] were derived by tuning to the Ice Volume Model of Imbrie and Imbrie [1980] for the 700–1100 ka interval, and by matching the $\delta^{18}O$ and susceptibility records to the ODP Site 677 $\delta^{18}O$ record [*Shackleton et al.*, 1990] for the older part of the record.

Paleointensity lows at ~990 ka, 1070 ka and 1200 ka in Figure 7 correspond to the boundaries of the Jaramillo Subchron (TJ/BJ) and to the Cobb Mountain Subchron (C). The normalized remanence record from ODP Site 982 for the 900–1600 ka interval can be matched to paleointensity records from the Iceland Basin (ODP Sites 983/984) and to records

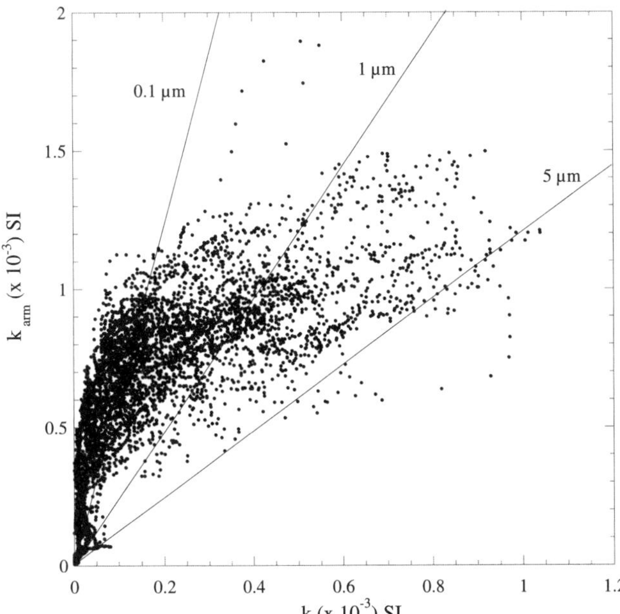

Figure 5. Anhysteretic susceptibility (k_{arm}), the ARM normalized by the biasing DC field, plotted against volume susceptibility (k) for ODP Site 982. Magnetite grain size estimates from King et al. [1983].

from the Pacific Ocean (Figure 7). The Equatorial Pacific record and the Site 983/984 records are better aligned with the Site 982 record than the California Margin and Ontong-Java records (Figure 7). This reflects the enhanced age control for the Equatorial Pacific record and for ODP Sites 982/983/984 as outlined above. Paleointensity lows correspond to the boundaries of the Jaramillo Subchronozone (TJ/BJ), the Santa Rosa excursion (SR), the Punaruu excursion (P) [*Singer et al.*, 1999], the Cobb Mountain excursion (C) [*Clement and Kent*, 1987], the Ontong-Java excursion (OJ) [*Valet and Meynadier*, 1993; *Gallet et al.*, 1993], the Gardar excursion (Ga) [*Channell et al.*, 2002] and the Gilsa excursion (Gi) [*Clement and Kent*, 1987; *Channell et al.*, 2002].

4. REVERSAL AGE OFFSETS AT SITES 980–982

The correlations of Pliocene-Pleistocene isotopic stages to polarity chrons and subchrons at ODP Site 982 are consistent with correlations established elsewhere (Table 2). The magnetic and $\delta^{18}O$ data from Site 982 and from neighboring Site 980/981 (Figure 1) are sufficiently detailed to allow an assessment of the precise position of reversals relative to isotopic stages. Mean sedimentation rates at Site 982 and Site 980/981 differ by factors of 2–7, and hence the precise correlation of reversals to isotopic stages at these sites may give insights into the magnetization lock-in function. Mean sedimentation rates in the Matuyama Chron at Site 982 vary in the 1–3

cm/kyr range (Figure 3). At Site 980/981, mean sedimentation rates in the Matuyama Chron vary in the 5–30 cm/kyr range [*Channell and Raymo*, 2003; *Channell et al.*, 2003].

The age models at Site 982 and Site 980/981 were derived by matching the benthic $\delta^{18}O$ records to the same TARGET curve (see above). For the two sites, the ages of polarity reversals at the boundaries of the Jaramillo Chronozone, the Cobb Mountain Subchronozone, the Olduvai Subchronozone and the Réunion Subchronozone are discrepant by 10–15 kyr (Figure 8). With the exception of the base of the Réunion Subchronozone (Figure 8d), the ages of polarity reversals appear older for Site 982, the site with lower mean sedimentation rates. This could be explained by incorrect correlation of the $\delta^{18}O$ records to the TARGET curve and inadequacy of the age model. For example, the Site 982 $\delta^{18}O$ records at the boundaries of the Jaramillo Chronozone lack distinctive features (Figure 8a), and may therefore be misaligned with the TARGET curve. The consistent age offset, however, with older ages for Site 982, tends to indicate that the offset is related to the magnetization lock-in function. If the physical lock-in functions were comparable for the two neighboring sites, the lock-in *time* would have been more extended for the site with the lower sedimentation rates. The modeling of this process, developed below, indicates that lock-in depth for the lower sedimentation rate site (Site 982) must have been greater to account for the observed age offsets.

Post-depositional remanent magnetization (pDRM), stemming from the work of Irving and Major [1964], has been modeled by an exponential function [*Hyodo*, 1984; *Meynadier and Valet*, 1996; *Mazaud*, 1996; *Teanby and Gubbins*, 2000; *Guyodo and Channell*, 2002]. Teanby and Gubbins [2000] added an 8 cm uniform mixing layer (magnetization = 0) at the top of the sedimentary column to simulate the bioturbated surface mixed layer. The assumption of a well-mixed layer, in which the sediment is consumed by organisms such as sipunculid or echiuran worms, is valid in deep-sea sediments where the mixing coefficient exceeds the product of mixed layer depth and sedimentation rate [*Guinasso and Schink*, 1975]. In deep-sea sediments, isotopic tracers indicate mean mixed layer thicknesses of about 10 cm (values vary by an order of magnitude from 3–30 cm) that is largely independent of sedimentation rate [*Boudreau*, 1994, 1998]. The main control on the mixed-layer thickness appears to be organic carbon flux derived from surface water productivity [*Trauth et al.*, 1997; *Smith and Rabouille*, 2002]. We would, therefore, expect the mixed layer thickness at Site 982, with its lower detrital component and higher organic carbon flux, to exceed that at Site 980/981. Using ^{14}C of the bulk carbonate fraction as the isotopic tracer, Thomson et al. [2000] estimated mixed-layer thicknesses of 10–20 cm in box cores collected close to the Rockall Plateau at water depths comparable to Sites 980–982. These values are greater than the 2–13 cm mixed-layer thicknesses obtained from further south in the North Atlantic using ^{14}C in foraminifera [*Trauth et al.*, 1997; *Smith and Rabouille*, 2002]. Estimates of mixed-layer thickness are grain-size sensitive [see *Bard*, 2001] and would be expected to be lower for the coarse fraction (foraminifera) than for the bulk carbonate (nannofossils). For this reason, mixed layer thickness estimates based on ^{14}C and other isotopic tracers should be considered as minimum estimates for the fine (PSD) grains that carry stable magnetization. In the studies of Thomson et al. [2000] and Trauth et al. [1997], uniform ^{14}C values in the mixed layer, and an abrupt change to

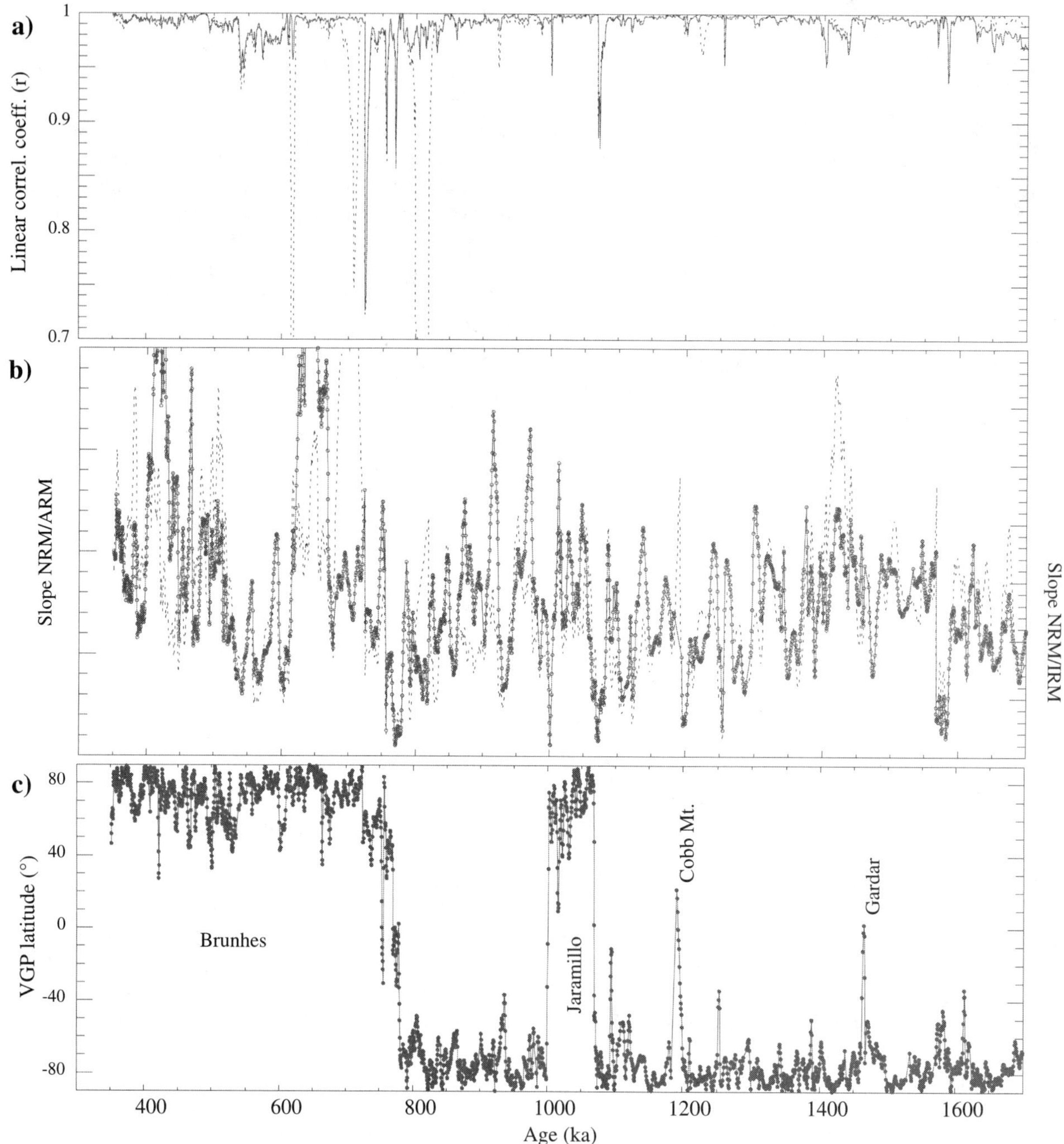

Figure 6. ODP Site 982: Linear correlation coefficients (r) and slopes of NRM versus ARM (solid line) and NRM versus IRM (dashed line) for the 25–45 mT peak field range in the 300–1700 ka interval. Virtual geomagnetic polar (VGP) latitudes derived from the component directions in Figure 4.

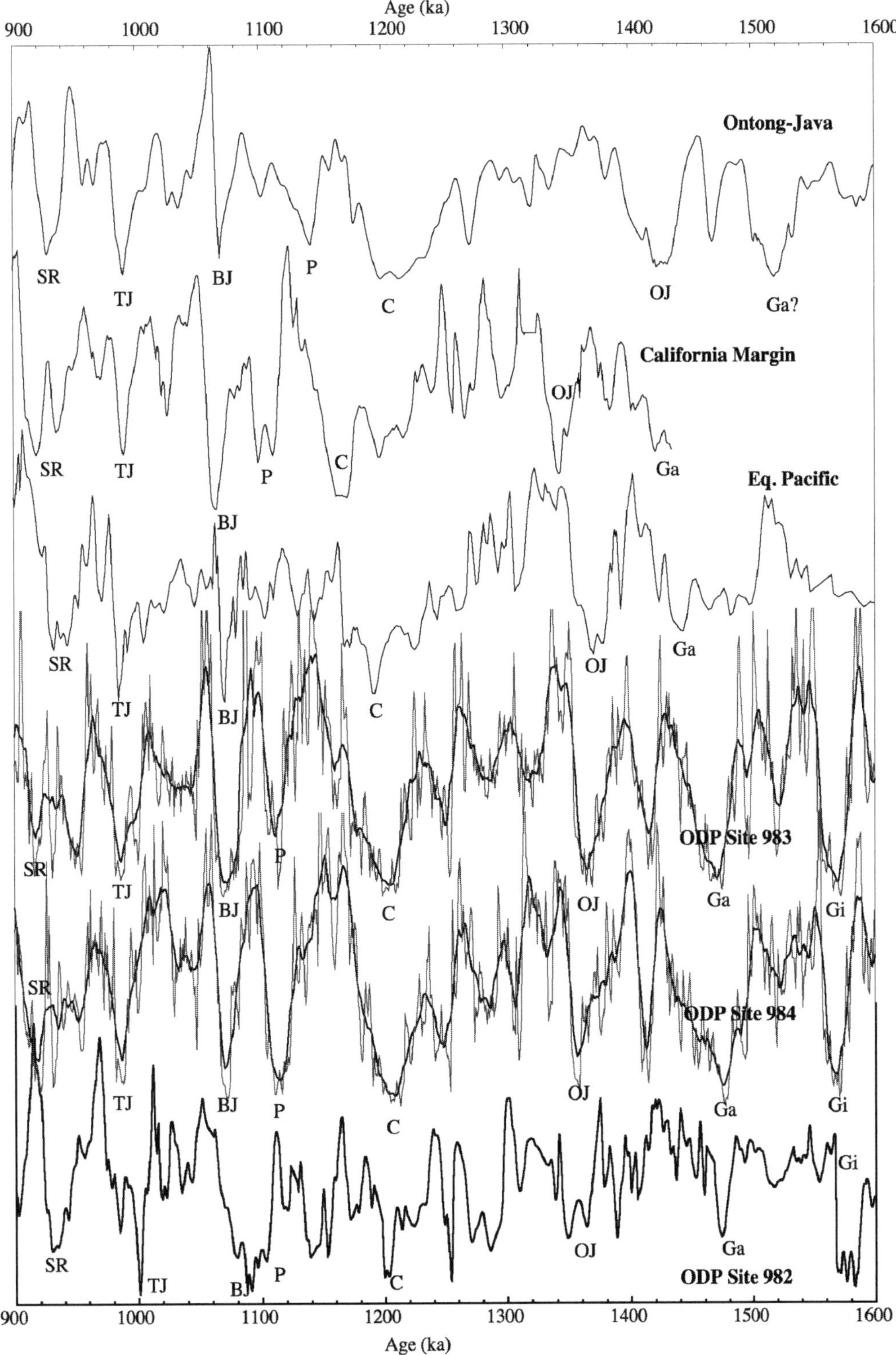

Figure 7. Comparison of NRM/ARM (slope) for ODP Site 982 with paleointensity data from Ontong-Java Plateau [*Kok and Tauxe*, 1999], California Margin [*Guyodo et al.*, 1999], Equatorial Pacific [*Valet and Meynadier*, 1993; *Meynadier et al.*, 1995], and from ODP Sites 983 and 984, before and after smoothing with a 10 kyr running mean [*Channell and Kleiven*, 2000; *Channell et al.*, 2002].

Table 2. Positions of reversals and polarity subchronozones relative to marine isotope stages (MIS)

Subchronozone	Subchron label*	MIS[1] Sites 607, 609, 677	MIS[2] Italy	MIS[3] Site 980/981	MIS[4] Site 983	MIS[4] Site 984	MIS Site 982 (this paper)
top Jaramillo	end 1r.1n	mid 27	27	mid 27	mid 27	mid 27	base 27
base Jaramillo	onset 1r.1n	mid 31	31	top 31	mid 31	mid 31	mid 31
Punaruu	1r.2r.1r				base 33	base 33	base 33
top Cobb. Mt.	end 1r.2r.2n	base 35		mid 35	mid 35	mid 35	base 35
base Cobb Mt.	onset 1r.2r.2n	base 35		base 35	mid 36	mid 36	base 35
top Olduvai	end 2n	base 63	64	63	base 63	base 63	base 63
base Olduvai	onset 2n	base 71	72	top 71		base 71	71
Réunion	2r.1n	79-80	81	79-81		base 81	81

MIS: marine isotope stage

* [nomenclature of *Cande and Kent*, 1992]

[1] [*Ruddiman et al.*, 1989; *Raymo et al.*, 1989; *Shackleton et al.*, 1990]

[2] [*Lourens et al.*, 1996]

[3] [*Channell and Raymo*, 2003; *Channell et al.*, 2003]

[4] [*Channell et al.*, 2002]

uniformly increasing ^{14}C ages below, imply that the mixing coefficient (efficiency) is high in the mixed layer on millennial (^{14}C) timescales.

A simple way of explaining the reversal age offsets observed between Site 982 and Site 980/981 would be to consider a larger surface mixed layer at Site 982. In view of the observed average age offset and the difference in average sedimentation rates between the two sites, an increased mixed layer thickness of about 25 cm would be required. In order to study this problem in a more comprehensive manner, we consider a number of parameters: (1) estimated interval sedimentation rates, (2) age control procedures, (3) offsets due to magnetization lock-in, and (4) sample measurement effects.

Rather than using a mixed layer (magnetization = 0) overlying an exponential lock-in function [*Teanby and Gubbins*, 2000], we have used a sigmoidal pDRM function based on tanh(x) (Figure 9a). This function has the attribute of describing a smooth transition between the mixing zone and the lock-in zone. The function can be defined by a surface mixing layer depth (M) at which 5% of the magnetization is acquired, and a lock-in depth (L) that is the depth below M at which 50% of the magnetization is acquired (Figure 9a). With this pDRM function, increasing M by x centimeters leads to an increase (by x cm) in the depth at which the magnetization is locked. An increase in L also leads to an increase in the depth at which the magnetization is locked but, more importantly, it leads to a marked increase in the depth interval over which the reversal transition is recorded (Figure 9a). Increasing L also reduces the time resolution of the record, as the lock-in function acts as a filter [*Teanby and Gubbins*, 2000; *Guyodo and Channell*, 2002].

The modeling procedure involved the following steps: (1) conversion from the time domain to the depth domain of a model virtual geomagnetic polar (VGP) latitude time series using sedimentation rate estimates from the age model, (2) simulation of the magnetization (pDRM) acquisition process, (3) simulation of the smoothing inherent in the u-channel measurement procedure, and (4) conversion back into the time domain.

The input VGP latitude time series was constructed as a step function representing a succession of reversals corresponding to the onset and termination of Jaramillo, Cobb Mountain, Olduvai, and Réunion subchrons. The initial duration of the reversals was set to 4 kyr, based on the distribution of reversal durations obtained from ODP Sites 983 and 984 [*Channell et al.*, 2002]. For each simulation, the age-scale of this initial time series was converted to a depth-scale using the sedimentation rates specific to each site investigated (Figure 8). The magnetization acquisition process was subsequently simulated as a convolution between this input signal and the pDRM function described above. In this function, both L and M can be varied systematically in order to define the curve that will best fit the real data (i.e., yield the appropriate mid-reversal position and transition duration). A more precise fit would require simulation of the details of each reversal record, including the (amplified) secular variation during reversal transitions. Note that the mixing depth necessary to match the real data depends on the initial age chosen for the reversals. Since the precise absolute age of reversals is not known with sufficient accuracy, the position of the reversals in the model signal was arbitrarily fixed at some time before the occurrence of the reversals in the real data. Using this method, only relative values of the mixing depth between sites (hereafter referred to as DM) could be obtained for a given input. Subsequently, another function was applied to simulate the smoothing introduced by the magnetometer during measurement of the u-channel. This function was chosen as the *z axis* response function of the 2-G Enterprises cryogenic magnetometer located at the University of Florida [*Guyodo et al.*, 2002]. The resulting signal was then converted back into the time domain using site-specific sedimentation rates.

Using this approach, the observed reversal offsets (Figure 8) can be simulated by values of DM ($M_{Site\ 982} - M_{Site\ 980/981}$) of ~20 cm for the boundaries of the Jaramillo and for the Cobb Mt. (Figure 9b), 16 cm for the base of the Olduvai and 36 cm for the top of the Réunion (Figure 9c). High values of

L do not generate the desired offset but extend the apparent duration of the polarity reversal (Figure 9a). Values of L < 10 cm appear to satisfy the observations (Figures 9 b,c). Evidently, this value depends to some extent on the duration chosen for the initial reversals (here 4 kyr), and therefore could vary by a few centimeters. Large fluctuations in sedimentation rates at the onset of the Olduvai Subchron and within the Réunion Subchron make the results of these simulations more problematic than the estimates for the boundaries of the Jaramillo and Cobb Mt. subchrons. Indeed, a slight change

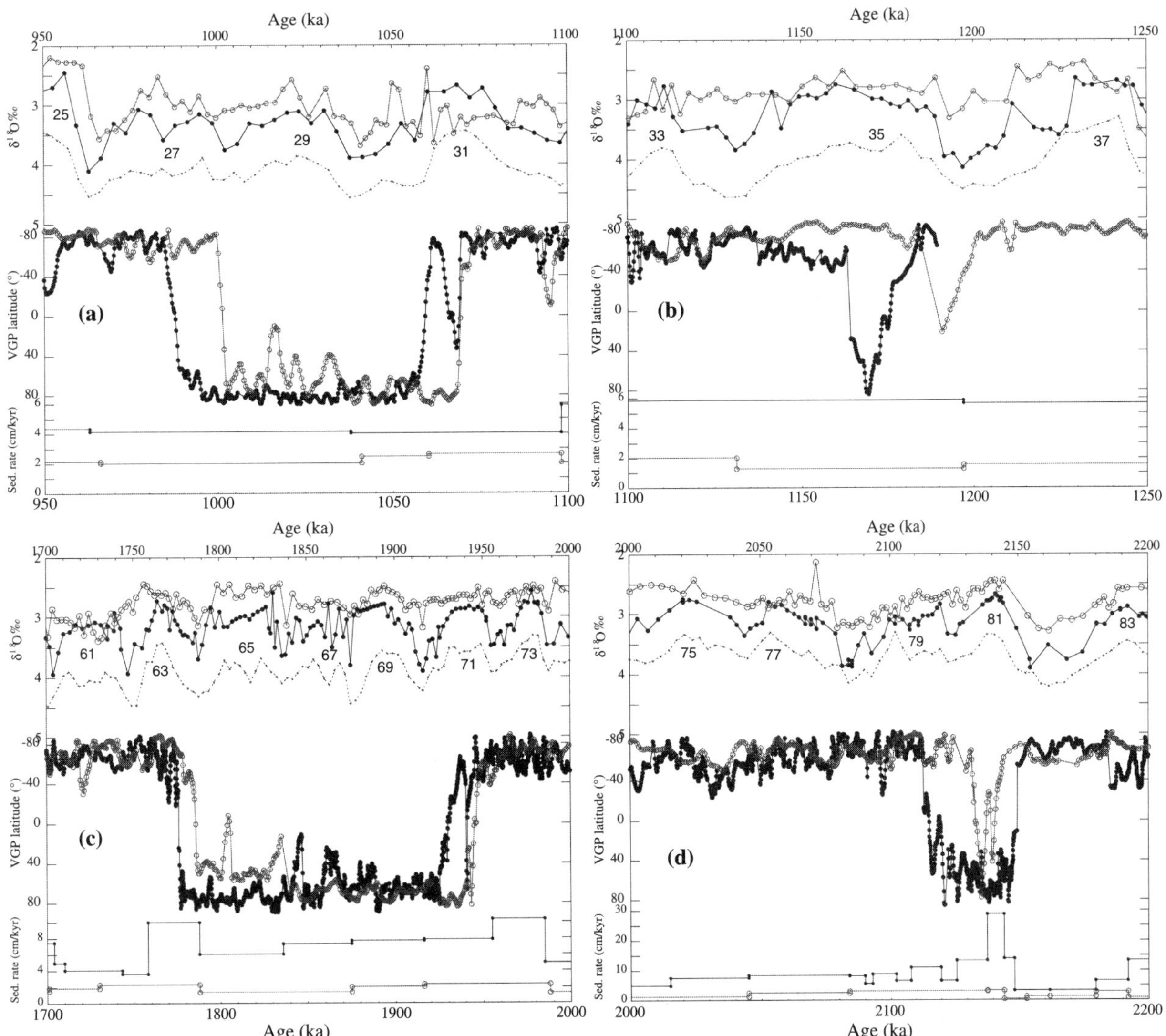

Figure 8. Comparison of Site 982 and Site 980/981. (a) Jaramillo Subchronozone, (b) Cobb Mountain Subchronozone, (c) Olduvai Subchronozone, (d) Reunion Subchronozone. Benthic $\delta^{18}O$ with marine isotope stage numbers for Site 982 (thin line, open symbols) and Site 980/981 (thick line, closed symbols) correlated to the TARGET curve (dashed line) from Shackleton et al. [1990, 1995a,b]. Virtual geomagnetic polar latitudes (VGPs) from Site 982 (thin line, open symbols) and Site 980/981 (thick line, closed symbols). Sedimentation rates for Site 982 (thin line, open symbols) and Site 980/981 (thick line, closed symbols).

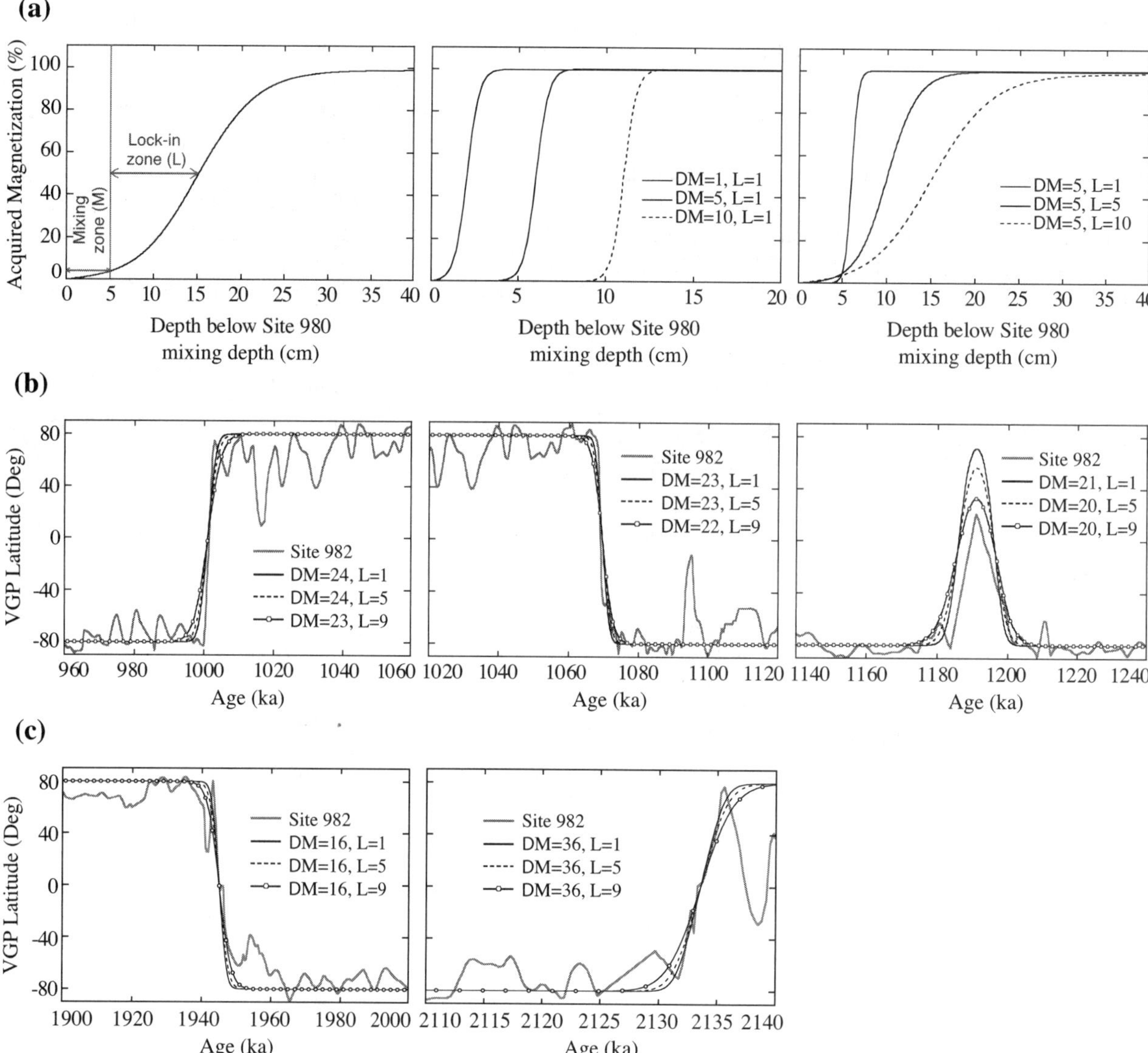

Figure 9. (a) The sigmoidal function based on tanh(x) used to model the mixing zone (thickness M) and the underlying lock-in zone (thickness L). The effect of varying the difference in mixing zone depth (DM) between Sites 982 and 980/981, and lock-in depth (L), is illustrated. The sigmoidal function can be used to model the reversal offsets between the two sites for (b) the boundaries of the Jaramillo and Cobb Mt. subchronozones and (c) the base of the Olduvai and top of the Réunion subchronozones.

in the positioning of a reversal (in the input signal) relative to the boundary between two different sedimentation rate intervals (i.e., an age model tie-point), may place this reversal either in a higher or a lower sedimentation rate interval, which will alter the model output.

5. DISCUSSION

A recent re-deposition study has suggested that, at least for some lithologies, inter-granular interaction could reduce sig-

nificantly the extent of pDRM, and bioturbation would not enhance, but rather disrupt, the remanent magnetization below the sediment/water interface [*Katari et al.*, 2000]. The conclusions of this paper are consistent with studies by Tauxe et al. [1996], based on a comparison of magnetic and oxygen isotope records at the MBB, that imply that magnetization lock-in depth is insignificant in these (low sedimentation rate) marine sediments. Other studies [*DeMenocal et al.*, 1990; *Lund and Keigwin*, 1994; *Kent and Schneider*, 1995] invoked pDRM, and hence a finite lock-in depth, to explain reversal/isotope offsets, and observed attenuation of secular variation records.

At Sites 982 and 980/981, the consistent sense of age offset of reversals between Sites 982 and 980/981 can, according to the model presented above, be explained by a mixing depth (zone) that is significantly thicker (by ~20 cm) at Site 982 than at Site 980/981. The simulations imply that magnetization is acquired very soon after the sediment passes through the mixed layer (M), the lock-in layer (L) beneath the mixed layer being just a few centimeters thick. Estimates of mixed layer thickness, based on ^{14}C, from Holocene pelagic sediments in the eastern North Atlantic give mixed layer thicknesses of ~2–20 cm [*Thomson et al.*, 2000; *Trauth et al.*, 1997]. Large variations in M are found for pelagic sediments, with no clear dependence on sedimentation rate [*Boudreau*, 1994, 1998]. Lower mean sedimentation rates at Site 982, relative to Site 980/981 (Figure 8), do not imply that the mixed layer depth (M) would be greater at Site 982 than at Site 980/981. Organic carbon flux is believed to be a control on mixed layer thickness [*Trauth et al.*, 1997; *Smith and Rabouille*, 2002]. Higher carbonate content (lower detrital concentration) and shallower water depth at Site 982, relative to Site 980/981, implies that organic carbon flux to the seafloor was greater for Site 982 than for Site 980/981. The higher organic carbon flux at ODP Site 982 may, therefore, account for the larger surface-sediment mixed-layer thickness at ODP Site 982.

The difference in mixed layer thickness between the two sites is proposed as an explanation for the apparent age offset of reversals in the Matuyama Chron. It is important to note that this difference in mixed layer thickness would affect the δ^{18}O record, and hence the age model. The small number of foraminifera (1–3) required by modern mass spectrometers for precise estimates of δ^{18}O has exacerbated the effect of reworking on δ^{18}O records in general. Reworking down-section is limited by the extent of the mixed layer, but no such limitation exists for reworking up-section [see *Guinasso and Schink*, 1975]. As a result of lower sedimentation rates and larger mixing layer thickness at Site 982, the definition of the isotope stratigraphy (and hence the age model) would be reduced at Site 982 relative to that at Site 980/981. Mixed layer effects on the isotopic data should be taken into account in future, more comprehensive simulations. Nevertheless, it is very unlikely that reworking and fortuitous picking of specimens would offset the Site 982 δ^{18}O record to account for the observed offsets in ages of polarity reversals.

Acknowledgments. Research supported by US National Science Foundation (EAR-98-04711). We are indebted to the staff of the Ocean Drilling Program (ODP) for facilitating this study, particularly the staff at the ODP Bremen Core Repository. We thank Chris Bayliss for careful laboratory measurements. The paper benefited from discussions with Kathy Venz-Curtis and John Jaeger, and from the comments of three anonymous reviewers.

REFERENCES

Bard, E., Paleoceanographic implications of the difference in deep-sea sediment mixing between large and fine particles, *Paleoceanography*, 16, 235–239, 2001.

Boudreau, B. P., Is burial velocity a master parameter for bioturbation?, *Geochim. Cosmochim. Acta*, 58, 1243–1249, 1994.

Boudreau, B. P., Mean mixing depth of sediments: The wherefore and the why, *Limnol. Oceanogr.*, 43, 524–526, 1998.

Cande, S. C., and D. V. Kent, A new geomagnetic polarity timescale for the late Cretaceous and Cenozoic, *J. Geophys. Res.* 97, 13917–13951, 1992.

Channell, J .E. T., D. A. Hodell and B. Lehman, Relative geomagnetic paleointensity and δ^{18}O at ODP Site 983 (Gardar Drift, North Atlantic) since 350 ka, *Earth Planet. Sci. Lett.*, 153, 103–118, 1997.

Channell, J. E. T., D. A. Hodell, J. McManus and B. Lehman, Orbital modulation of geomagnetic paleointensity, *Nature*, 394, 464–468, 1998.

Channell, J. E. T., and B. Lehman, Magnetic stratigraphy of Leg 162 North Atlantic Sites 980–984, In: Jansen, E., Raymo, M.E., Blum, P. and Herbert, T. (Eds.), *Proc. ODP, Sci. Results*, 162: College Station, TX (Ocean Drilling Program), 113–130, 1999.

Channell, J. E. T. and H. F. Kleiven, Geomagnetic palaeointensities and astrochronological ages for the Matuyama-Brunhes boundary and the boundaries of the Jaramillo Subchron: palaeomagnetic and oxygen isotope records from ODP Site 983, *Phil. Trans. R. Soc. Lond.* A, 358, 1027–1047, 2000.

Channell, J. E. T. and M. E. Raymo, Paleomagnetic record at ODP Site 980 (Feni Drift, Rockall) for the past 1.2 Myrs, *Geochem., Geophys. Geosyst.*, doi:10.1029/2002GC000440, 2003.

Channell, J. E. T., A. Mazaud, P. Sullivan, S. Turner, and M. E. Raymo, Geomagnetic excursions and paleointensities in the 0.9–2.15 Ma interval of the Matuyama Chron at ODP Site 983 and 984 (Iceland Basin), *J. Geophys. Res.*, 107 (B6), doi:10.1029/2001JB000491, 2002.

Channell, J.E.T., J. Labs, and M. E. Raymo, The Réunion Subchronozone at ODP Site 981 (Feni Drift, North Atlantic), *Earth Planet. Sci. Lett.*, 215, 1–12, 2003.

Clement, B. M., and D. V. Kent, Short polarity intervals within the Matuyama: transition field records from hydraulic piston cored sediments from the North Atlantic, *Earth Planet. Sci. Lett.*, 81, 253–264, 1987.

DeMenocal, P. B., W. F. Ruddiman, and D. V. Kent, Depth of post-depositional remanence acquisition in deep-sea sediments: a case study of the Brunhes-Matuyama reversal and oxygen isotopic stage 19.1, *Earth Planet. Sci. Lett.*, 99, 1–13, 1990.

Flower, B. P., D. W. Oppo, J. F. McManus, K. A. Venz, D. A. Hodell, and J. A. Cullen, North Atlantic intermediate to deep water circulation and chemical stratification during the past 1 Myr, *Paleoceanography*, 15, 388–403, 2000.

Gallet, Y., J. Gee, L. Tauxe and J. A. Tarduno, Paleomagnetic analyses of short normal polarity magnetic anomalies in the Matuyama chron, In: Berger, W.H., Kroenke, L.W. and Mayer, L.A. (Eds.). *Proc. ODP Sci. Results*, 130: 547–559, 1993.

Guinasso, N. L. and D. R. Schink, Quantitative estimates of biological mixing rates in abyssal sediments, *J. Geophys. Res.*, 80, 3032–3043, 1975.

Guyodo, Y. and J. E. T. Channell, Effects of variable sedimentation rates and age errors on the resolution of sedimentary paleointensity records, *Geochem., Geophys. Geosyst.*, doi:10.1029/2001GC000211, 2002.

Guyodo, Y., C. Richter and J-P. Valet, Paleointensity record from Pleistocene sediments (1.4–0 Ma) off the California Margin, *J. Geophys, Res.*, 104, 22,953–22,964, 1999.

Guyodo, Y., J. E. T. Channell, and R. G. Thomas, Deconvolution of u-channel paleomagnetic data near geomagnetic reversals and short events, *Geophys. Res. Lett.*, 29, 1845, doi:10.1029/2002GL014927, 2002.

Hyodo, M., Possibility of reconstruction of the past geomagnetic field from homogeneous sediments, *J. Geomag. Geoelectr.*, 36, 45–62, 1984.

Imbrie, J., and J. Z. Imbrie, Modeling the climatic response to orbital variations, *Science*, 207, 943–953, 1980.

Irving, E., and A. Major, Post-depositional detrital remanent magnetization in a synthetic sediment, *Sedimentology*, 3, 135–143, 1964.

Katari, K., L. Tauxe, and J. King, A reassessment of post-depositional remanent magnetism: preliminary experiments with natural sediments, *Earth Planet. Sci. Lett.*, 183, 147–160, 2000.

Kent, D., and D. Schneider, correlation of paleointensity records in the Brunhes/Matuyama polarity transition interval, *Earth Planet. Sci. Lett.*, 129, 135–144, 1995.

King, J. W., S. K. Banerjee, and J. Marvin, A new rock-magnetic approach to selecting sediments for geomagnetic paleointensity studies: application to paleointensity for the last 4000 years, *J. Geophys. Res.*, 88, 5911–5921, 1983.

Kirschvink, J. L., The least squares lines and plane analysis of paleomagnetic data, *Geophys. J.R. Astr. Soc.* 62, 699–718, 1980.

Kleiven, H. F., E. Jansen, W. B. Curry, D. A. Hodell and K. Venz, Atlantic ocean thermohaline circulation changes on orbital to suborbital timescales during the mid-Pleistocene, *Paleoceanography*, 18, 1008, doi:10.1029/2001PA000629, 2003.

Kok, Y. S. and L. Tauxe, A relative geomagnetic paleointensity stack from Ontong-Java Plateau sediments for the Matuyama, *J. Geophys. Res.*, 104, 25,401–25,413, 1999.

Lourens, L. J., A. Antonarakou, F. J. Hilgen, A. A. M. Van Hoof, C. Vergnaud-Grazzini and W. J. Zachariasse. Evaluation of the Plio-Pleistocene astronomical timescale, *Paleoceanography*, 11, 391–413, 1996.

Lund, S., and L. Keigwin, Measurement of the degree of smoothing in sediments paleomagnetic secular variation records: an example from late Quaternary deep-sea sediments of the Bermuda rise, western North Atlantic ocean, *Earth Planet. Sci. Lett.*, 122, 317–330, 1994.

Manley, P. L. and D. W. Caress, Mudwaves on the Gardar Sediment Drift, NE Atlantic, *Paleoceanography*, 9, 973–988, 1994.

Mazaud, A., Sawtooth variation in magnetic intensity profiles and delayed acquisition of magnetization in deep sea cores, *Earth Planet. Sci. Lett.*, 139, 379–386, 1996.

Meynadier, L., and J.-P. Valet, Post-depositional realignment of magnetic grains and asymetrical saw-toothed pattern of magnetization intensity, *Earth Planet. Sci. Lett.*, 140, 123–132, 1996.

Meynadier, L., J-P. Valet, and N. J. Shackleton, Relative geomagnetic intensity during the last 4 M.Y. from the equatorial Pacific, In: Pisias, N. G., Janacek, L. A., Palmer-Julson, A., and Van Andel, T. H. (Eds.). *Proc. ODP Sci. Results*, 138: 779–793, 1995.

McCave, I. N., P. F. Lonsdale, C. D. Hollister and W. D. Gardner, Sediment transport over the Hatton and Gardar contourite drifts, *J. Sediment. Pet.*, 50, 1049–1062, 1980.

Ortiz, J., A. Mix, S. Harris, and S. O'Connell, Diffuse spectral reflectance as a proxy for percent carbonate content in north Atlantic sediments, *Paleoceanography*, 14, 171–186, 1999.

Raymo, M. E., W. F. Ruddiman, J. Backman, B. M. Clement, and D. G. Martinson, Late Pliocene variation in Northern Hemisphere ice sheets and North Atlantic deep water circulation, *Paleoceanography*, 4, 413–446, 1989.

Ruddiman, W. F., M. E. Raymo, D. G. Martinson, B. M. Clement, and J. Backman, Pleistocene evolution: northern hemisphere ice sheet and north Atlantic ocean, *Paleoceanography*, 4, 353–412, 1989.

Schneider, D. A., Paleomagnetism of some Leg 138 sediments: detailing Miocene magnetostratigraphy, In: Pisias, N. G., Janacek, L. A., Palmer-Julson, A., and Van Andel, T. H. (Eds.). *Proc. ODP Sci. Results*, 138: 59–72, 1995.

Shackleton, N. J., A. Berger, and W. R. Peltier, An alternative astronomical calibration of the lower Pleistocene timescale based on ODP Site 677, *Trans. Roy. Soc. Edinburgh: Earth Sci.* 81, 251–261, 1990.

Shackleton, N. J., S. Crowhurst, T. Hagelberg, N. G. Pisias, and D. A. Schneider, A new Late Neogene time scale: application to Leg 138 Sites, In: Pisias, N. G., Janacek, L. A., Palmer-Julson, A., and Van Andel, T. H. (Eds.). *Proc. ODP Sci. Results*, 138: 73–101, 1995a.

Shackleton, N. J., M. A. Hall, and D. Pate, Pliocene stable isotope stratigraphy of Site 846, In: Pisias, N. G., Janacek, L. A., Palmer-Julson, A., and Van Andel, T. H. (Eds.), *Proc. ODP Sci. Results*, 138, 337–355, 1995b.

Shipboard Scientific Party, Site 982, In: Jansen, E., M. Raymo, P. Blum, et al., eds., *Proc. ODP Init. Repts.*, 162: College Station, TX (Ocean Drilling Program), 91–138, 1996.

Singer, B. S., K. A. Hoffman, A. Chauvin, R. S. Coe, and M. S. Pringle, Dating transitionally magnetized lavas of the late Matuyama Chron: toward a new ^{40}Ar/^{39}Ar timescale of reversals and events, *J. Geophys. Res.*, 104, 679–693, 1999.

Smith, C. R. and C. Rabouille, What controls the mixed-layer depth in deep-sea sediments? The importance of POC flux, *Limnol. Oceanogr.*, 47, 418–426, 2002.

Tauxe, L., Sedimentary records of relative paleointensity of the geomagnetic field: theory and practice, *Rev. Geophys.*, 31, 319–354, 1993.

Tauxe, L., J-P. Valet and J. Bloemendal, Magnetostratigraphy of Leg 108 advanced hydraulic piston cores, In: Ruddiman, W., Sarnthein et al. (Eds.). *Proc. ODP Sci. Results*, 108: 429–439, 1989.

Tauxe, L., T. Herbert, N. J. Shackleton, and Y. S. Kok, Astronomical calibration of the Matuyama-Brunhes boundary: consequences for the magnetic remanence acquisition in marine carbonates and Asian loess sequences, *Earth Planet. Sci. Lett.*, 140, 133–146, 1996.

Teanby, N., and D. Gubbins, The effect of aliasing and lock-in processes on paleosecular variation records from sediments, *Geophys. J. Int.*, 142, 563–570, 2000.

Thomas, R., Y. Guyodo and J.E.T. Channell, U-channel track for susceptibility measurements, *Geochem. Geophys. Geosyst.*, 1050, doi:10.1029/2002GC000454, 2003.

Thomson, J., L. Brown, S. Nixon, G.T. Cook and A.B. McKenzie, Bioturbation and Holocene sediment accumulation fluxes in the northeast Atlantic Ocean (Benthic Boundary Layer experiment sites), *Marine Geology*, 169, 21–39, 2000.

Tiedemann, R., M. Sarnthein and N.J. Shackleton, Astronomic timescale for the Pliocene Atlantic $\delta^{18}O$ and dust flux records of Ocean Drilling Program Site 659, *Paleoceanography*, 9, 619–638, 1994.

Trauth, M.H., M. Sarnthein and M. Arnold, Bioturbational mixing depth and carbon flux at the seafloor, *Paleoceanography*, 12, 517–526, 1997.

Valet, J.P., and L. Meynadier, Geomagnetic field intensity and reversals during the past four million years, *Nature*, 366, 234–238, 1993.

Venz, K.A., D.A. Hodell, C. Stanton and D.A. Warnke, A 1.0 Myr record of glacial North Atlantic Intermediate Water variability from ODP Site 982 in the northeast Atlantic, *Paleoceanography*, 14, 42–52, 1999.

Weeks, R., C. Laj, L. Endignoux, M. Fuller, A. Roberts, R. Manganne, E. Blanchard and W. Goree, Improvements in long-core measurement techniques: applications in palaeomagnetism and palaeoceanography. *Geophys. J. Int.* 114, 651–662, 1993.

J.E.T. Channell, Department of Geological Sciences, PO Box 112120, University of Florida, Gainesville, Florida 32611-2120.

Y. Guyodo, Laboratoire des Sciences du Climat et de l'Environnement, Avenue de la Terrasse, 91198 Gif-sur-Yvette, France.

The Complexity of Reversals

Robert S. Coe*

Earth Sciences Department, University of California, Santa Cruz, California

Jonathan M.G. Glen

U.S. Geological Survey, Menlo Park, California

Geomagnetic reversals could be far more complex than even the best, most detailed paleomagnetic record in hand. Sequences of lava flows can give accurate spot readings of the field but yield records that are necessarily shot full of holes, whereas sedimentary recordings smooth temporal variations of the field and may also be subject to hiatuses and delayed, non-uniform lock-in of remanence. Thus, paleomagnetic records are always incomplete and give only lower bounds on how rapidly changing and complex the behavior of the reversing field may have been. Nonetheless, the combined evidence from several high-deposition-rate sedimentary records, multiple lava-flow records of one reversal from the same region, and a reversal record from a dynamically self-consistent numerical geodynamo simulation suggest that at least some reversals are much more complex than typically portrayed, with episodes of oscillatory and very rapid field change.

1. INTRODUCTION

Reversals of the dipole field, surely the most dramatic attribute of the geodynamo, are known only because rocks can record the ancient magnetic field. Reliable paleomagnetic records showing how the field changed during a reversal of polarity, though difficult to obtain, are important for evaluating geodynamo models and the deep-earth parameters prescribed for their operation. For instance, many people working on polarity transitions have suggested that the field direction may move rapidly between points of greater stability, but how rapidly and the locations and significance of stable points remain important questions [*Bogue and Coe*, 1982; *Mankinen et al.*, 1985; *Hoffman*, 1992; *Hoffman*, 2000]. In the same vein, reversal records have been intensely scrutinized for longitudinal confinement and geographical preference of transitional virtual geomagnetic poles (VGPs) because they could carry information about lateral variation of temperature and other properties of the lowermost mantle [e.g., *Clement*, 1991; *Laj et al.*, 1991; *Merrill and McFadden*, 1999, and references therein].

Geologically speaking, reversals occur very quickly and thus involve rates of change in field direction that challenge the capability of natural magnetic recorders. Lava flows are generally considered to offer the most accurate spot readings of the field because, typically, they are almost isotropic in their bulk magnetic properties and become magnetized by a relatively simple mechanism in a time ranging from a few days to a few years. However, the discontinuous and episodic nature of eruption is a major drawback that must always be kept in mind, as well as possible distortion of the local field arising from the magnetization of the lava flows themselves. Sedimentary rocks offer the attractive advantage of quasi-continuous recording, and thus an estimate of duration, but how they get magnetized is more complex, more vari-

* To whom correspondence should be addressed.

Timescales of the Paleomagnetic Field
Geophysical Monograph Series 145
Copyright 2004 by the American Geophysical Union
10.1029/145GM16

able, and not as well understood as for lavas. How much a particular sedimentary record has smoothed the geomagnetic signal, shallowed the inclination, and suffered from other recording artifacts associated with variable lock-in depth and formation of magnetic minerals during diagenesis is difficult to assess. Thus, although we are not always so cautious, it is entirely justifiable to ask about any reversal record whether it is incomplete or inaccurate or both.

With such questions in mind, we consider selected reversal records and geodynamo simulations, ranging from simpler to more complex, concentrating more on changes in direction than intensity. We propose that, although the complexity of many sedimentary polarity transition records probably owes largely to recording artifacts, there are some complex sedimentary records that may well contain important information about the character of the reversing field. Support for our view may be found, paradoxically, in discrepant directions among multiple records of the same reversal in lava flow sections, and also in some geodynamo simulations.

2. SOME ILLUSTRATIVE POLARITY TRANSITION RECORDS

The conventional view of what happens as the field reverses comes from a collective, 'common-sense' evaluation of transition records. The earliest example is the classic study by van Zijl et al. [1962] of a reversal recorded by the Jurassic Stormberg flood basalts of Lesotho in southern Africa. Rather than a simple progression, they found a sequence of flow-mean directions after blanket AF demagnetization to 22 mT that exhibited significant structure through the transition zone. One feature, however, is clearly the result of episodic volcanism, because the large gap around the midpoint of the reversal is marked by a sandstone interbed. Comparison of AF demagnetization curves of NRM and total TRM produced in the laboratory suggested that the field strength during the transition dropped by a factor of four to five compared to its average value before and after. More recent studies of the same section, while confirming the general conclusions of the earlier study, showed that the demagnetization and paleointensity procedures that had been employed were adequate to obtain accurate directions from only some of the transitional flows and reliable paleointensities from almost none of them [*Prévot et al.*, 2003; *Kosterov et al.*, 1997].

Four years later Ninkovich et al. [1966] presented a contrasting picture with their paleomagnetic study of an azimuthally unoriented deep-sea sediment core from the north Pacific spanning the last three reversals. In this record the inclination progresses smoothly between full polarity states in just a few samples, while the intensity of remanence drops by 80–90% and recovers again over a somewhat greater inter-

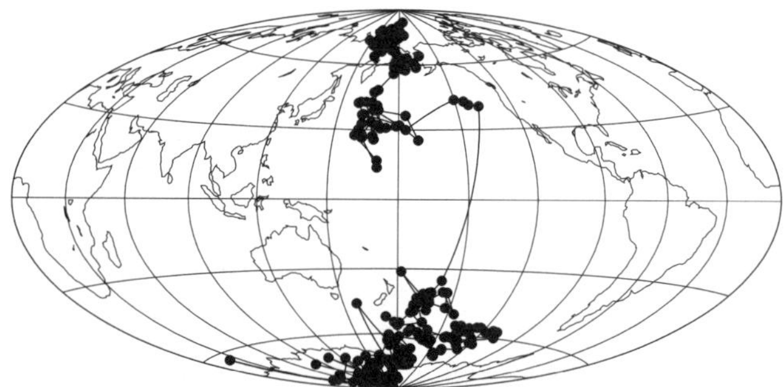

Figure 1. Detailed record of virtual geomagnetic poles (VGPs) spanning the lower Jaramillo reversal from a deep-sea sediment core from the southern Indian Ocean [*Clement and Kent*, 1984].

val. Based on the average sedimentation rate of around 1 cm/kyr between reversal boundaries, the duration of the transition zones is 1 to 2 kyr and the intensity lows about ten times longer. At the time it was an open question whether these simple sedimentary records or the more complicated lava-flow records discussed above better represent transitional field behavior. Today, however, we ascribe the simplicity of the directional change to smoothing of the geomagnetic signal because of the slow deposition rate.

Clement and Kent [1984] restudied the lower Jaramillo polarity transition in a deep-sea core from the southern Indian Ocean previously investigated by Opdyke et al. [1973]. The sediment carried a particularly strong and well-behaved remanence and was deposited at a higher than average rate (5.9 cm/kyr after downward revision to accord with the more recent estimate for duration of the Jaramillo subchron [*Cande and Kent*, 1995]). Their record (Figure 1) of VGPs, sampled at very close intervals of 0.5 cm through the transition zone, shows more structure than did the previous, lower-deposition-rate record. Most notably, the transition consists of three clusters of VGPs, with a 60° gap in latitude straddling the equator between two of them and considerable east-west VGP movement of 60 to 90° in longitude. Using the average deposition rate, the duration of the directional transition (VGP latitude between 60 S and 60 N) is calculated to be 9.0 kyr, whereas the intensity low (<50% of the full-polarity average) lasted about 14 kyr. Some of the details of this record, however, likely reflect artifacts of common sedimentological processes. For example, the signal has almost certainly been averaged over a longer time interval than the nominal 100 years of deposition spanned by a single sample—perhaps an order of magnitude longer—because post-depositional remanent magnetization (PDRM) is acquired over a non-zero depth range [*deMenocal et al.*, 1990; *Kent and Schneider*, 1995]. Indeed, the amplitude of full-polarity directional variation in this record appears to be significantly lower than is typical for secular variation on the 1000-year timescale observed in

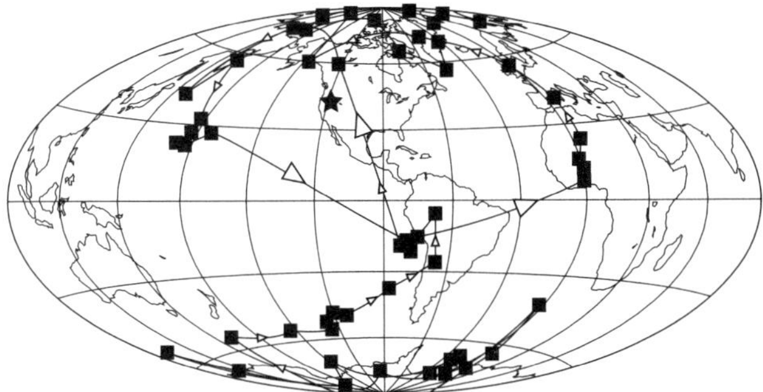

Figure 2. Detailed record of VGPs spanning a Miocene (~16 Ma) reversal from superposed basalt flows at Steens Mountain, Oregon, USA [*Mankinen et al.*, 1985].

archeomagnetic and lake-sediment studies. As a consequence, one can conclude that the absence of VGPs in the latitudinal gap between VGP clusters in Figure 1 marks a depositional hiatus rather than a rapid change in field direction, as suggested in the original paper on sedimentological grounds.

The most studied and still the most detailed volcanic polarity transition record is that found in the middle Miocene Steens Mountain basalt flows. This record consists of two phases (Figure 2): (i) an initial progression of VGPs from reversed-to-normal (R-N) followed by (ii) a rebound to transitional VGPs and then back to stable normal polarity [*Mankinen et al.*, 1985]. The VGP transition path is not confined longitudinally, spanning 90° during the first phase and 180° during the second phase with large swings to the northeast and southeast. Detailed paleointensity determinations by the method of Thellier and Thellier [1959] show that the field strength fell by 80–90% as the direction became intermediate, recovered to almost typical full-polarity values during the temporary normal, then plummeted again during the second phase of the transition until finally regaining strength as normal polarity became established [*Prévot et al.*, 1985]. The VGP revisits the same intermediate cluster near the coast of Peru during both phases of the transition, perhaps for a considerable time: in total, thirteen lava flows and one baked sediment zone record this direction. Moreover, it jumped to this cluster from another cluster of VGPs in the northwest Pacific recorded by sixteen superposed flows. It is particularly interesting that twice during the transition, once as the VGP arrived at the cluster in the northwest Pacific and again as it rejoined the cluster near the coast of Peru, the paleointensity as recorded in three superposed flows decreased progressively by a factor of three without any significant change in field direction. These observations are consistent with the notion that the transitional field may sometimes be dominated by a few flux patches that wax and wane in strength independently [*Hoffman*, 1992; *Gubbins and Coe*, 1993]. In the Steens record the three largest gaps, ranging from 80 to 100°, all occur as the VGP jumps to

or away from a cluster (Figure 2). These gaps suggest that the VGP may have moved rapidly as one of the flux patches turned on or off, a notion that is also consistent with the unusual streaking of remanence directions documented in particular individual lava flows emplaced at the beginning of two of the three large gaps [*Coe and Prévot*, 1989; *Coe et al.*, 1995]. Episodic volcanism coupled with selective partial overprinting of the interiors of these particular flows could be an alternative explanation for these observations, but it cannot account for the simultaneous decrease in field intensity and stationary field direction described above with regard to these VGP clusters.

Figure 3 shows a much more complex VGP path than the previous example, along with two simpler ones, for the Matuyama/Brunhes polarity transition [*Valet et al.*, 1989], although all three paths are more longitudinally confined than the Steens Mountain reversal path. The complex record, which features many large oscillatory swings in both inclination and declination during the transition, is from Lake Tecopa in the western U.S. [*Valet et al.*, 1988], whereas the two simple records are from marine cores in the northern [*Clement and Kent*, 1987] and equatorial [*Valet et al.*, 1989] Atlantic Ocean. Could the marine records have completely smoothed rapid field oscillations that the lake sediment record was able to resolve? For this to be true the remanence lock-in window would have had to be much narrower for the Lake Tecopa than for the marine sediments, the more so because the Tecopa transition zone is only 10–12 cm thick, half that of the marine transition zones. Moreover, Valet et al. [1988] noted that secondary components were often quite difficult to remove from Tecopa samples. Thus it seems likely that the directional oscillations reflect unremoved overprints, although some of these swings have a large east-west component that cannot be reconciled with simple normal-polarity overprinting.

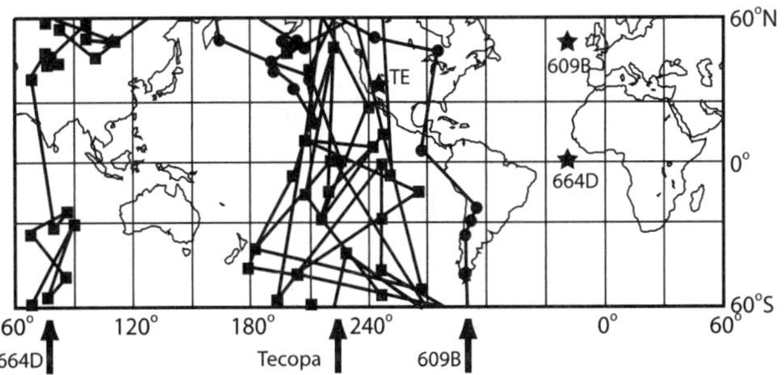

Figure 3. Comparison of three records of VGPs spanning the Matuyama/Brunhes transition from sediments [*Valet et al.*, 1989]. Localities marked by stars—TE: outcrops of ancestral Lake Tecopa sediment [*Valet et al.*, 1988]; 609B: DSDP core from the northeastern Atlantic [*Clement and Kent*, 1987]; 664D: DSDP core from the eastern equatorial Atlantic [*Valet et al.*, 1989].

Van Hoof and colleagues addressed the question of fidelity of sedimentary recording in a series of papers describing their detailed studies of many successive reversals in marls from Sicily [*van Hoof and Langereis*, 1991, 1992; *van Hoof et al.*, 1993a & b]. The deposition rate for these marls is typically 5 cm/kyr and, much like the previous example, these records have VGP paths that are confined to longitudinal bands near the Americas and contain many large oscillations (e.g., Figure 4a). These authors discovered evidence for delayed acquisition of remanence by authigenic magnetite, which they thought probably formed during reduction diagenesis. They noted two components of magnetization in many samples, both carried by magnetite: a lower temperature (LT) and higher temperature (HT) remanence. Because reversal boundaries and other identifiable magnetic features generally occur stratigraphically lower in the HT than in the LT record, the HT component must have formed after the LT component. They invoke a grain-growth, chemical-remanence mechanism to produce both components [*van Hoof et al.*, 1993b], although why the secondary component should have higher unblocking temperatures is not explained. Whatever the mechanism, acquisition of varying amounts of a delayed component of magnetization across a reversal boundary could produce multiple oscillations of remanence direction that might be mistaken

for real oscillations of the field during the reversal. Moreover, ordinary sedimentary averaging of the near-transitional or pre- and post-transitional fields can be called upon to explain the longitudinal confinement of the VGP paths [*Langereis et al.*, 1992].

This satisfyingly self-consistent explanation, that the Sicilian-marl reversal paths are rock-magnetic artifacts, is challenged by transition paths of slightly older reversals that strongly resemble the VGP paths of Figure 4a that were obtained from more rapidly deposited marl on Crete that did not reveal such LT and HT components. Furthermore, the extraordinarily similar paths of the same reversals reported by Clement et al. [1998] in Fiji (Figure 4b) offer an even more direct challenge. One of the transition zones spanned 1.8 m of section, much greater than any estimate of remanence lock-in interval, making it unlikely that the longitudinal confinement in that record could be the result of smoothing between pre- and post-transitional directions [*Valet et al.*, 1986, and references therein]. These Fijian VGP paths for the lower and upper Nunivak transitions are recorded by marl that is reported to be more uniform lithologically and deposited at twice the rate (10 cm/kyr) of the Sicilian marl. In two parallel sections 4 km apart the major features of these paths (though not their individual oscillations) are remarkably reproducible. The

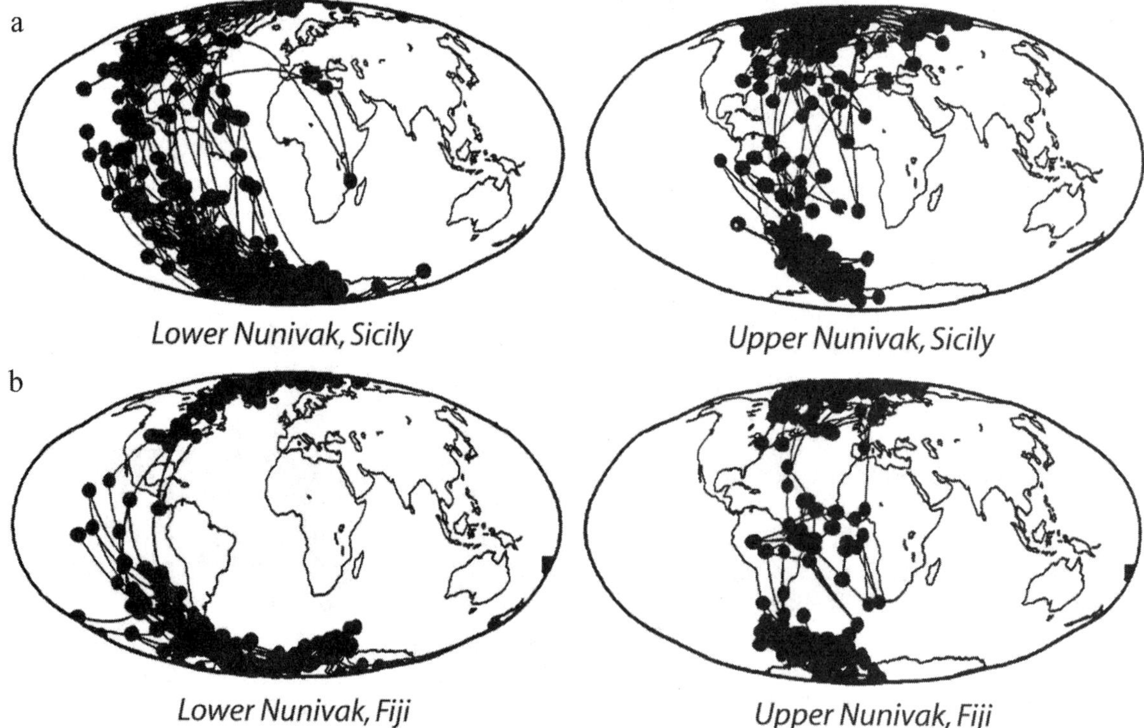

Figure 4. Comparison of four records of VGPs spanning back-to-back reversals from marl outcrop deposits in (a) Sicily [*van Hoof et al.*, 1993a] and (b) Fiji [*Clement et al.*, 1998].

authors attribute the high-frequency oscillations to variations in the timing of remanence acquisition but regard the similarity in longitude of the VGP bands recorded in Fiji and Sicily as evidence for a strong dipolar influence during these reversals. Of course, if the time-averaged, near-transition field were characterized by an equatorial-dipole component at the longitude of the Americas, then sedimentary averaging could account just as well for the coincidence of the VGP bands and also for their longitudinal confinement. Even so, the low-latitude cluster of VGPs in the upper Nunivak records would still imply a strong downward flux patch in that region for a significant period of time during the reversal. In addition, even though both Fiji and Sicily are roughly 90° from their VGP bands through the Americas, one cannot explain the coincidence of their paths with the proposition of Quidelleur and Valet [1994]. This explanation—that spuriously shallow transitional inclinations may be brought about by the inability of the greatly weakened field to maintain alignment of magnetic grains against gravity during deposition—fails because these marls almost certainly carry a post-depositional remanence.

A study by Channell and Lehman [1997] of the last two geomagnetic reversals recorded in high-deposition-rate (12–14 cm/kyr) sediment-drift cores from the North Atlantic offers support for the possible stop-and-go character of reversals, as suggested by the Steens Mountain basalt record presented above (Figure 2), but with many more oscillations. Nine transition records derived, from measurements at 1-cm intervals along U-channels passed through a cryogenic magnetometer and AF demagnetization apparatus, exhibit longitudinally unconfined VGP paths consisting of long tracks and large loops. Figure 5 shows three of them from three separate cores at one locality that record the Matuyama/Brunhes reversal. These paths are striking in that they visit almost the whole globe. Because the magnetometer pick-up coils have a characteristic response width of around 4 cm, these paths should be regarded as smoothed already over four measurements. Although the paths from the three cores are not identical, strong similarities are evident. The tracks oscillate between turning points around South America and in the western and central Pacific. The density of transitional VGPs tends to be greatest near the turning points, especially in the South American and in northeastern Asia/Pacific regions, indicating times of order one thousand years spent during each visit. Some of these 'hang-up' regions are in or not far from two of the flux patches identified by Hoffman [Hoffman, 1992] where today's geomagnetic field stripped of its axial dipole component is nearly vertical, whereas others are not. Between these places the VGP density is low, indicating much faster rates of change: along part of one of them four measurements span more than 180°! If the real field behavior during the reversal at this site

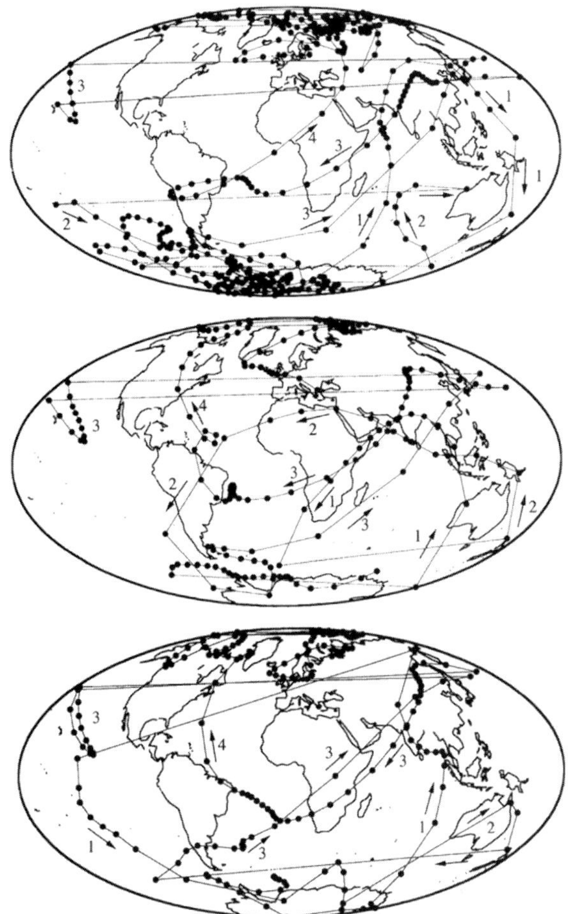

Figure 5. Three high-resolution records of VGPs spanning the Matuyama-Brunhes transition from ODP site 984 drift sediments from the northern Atlantic [*Channell and Lehman*, 1997].

was at least as complex as indicated by Figure 5, then the many much simpler sedimentary Matuyama/Brunhes records from around the globe [see *Jacobs*, 1994, for examples] must be very smoothed, very incomplete (or both), or affected by some other recording artifact.

Figure 6 shows another complex, high-resolution record of a different reversal, the Gauss/Matuyama transition recorded in a core of lacustrine sediment from eastern California [*Glen et al.*, 1999b]. Based on detailed magnetostratigraphy, the average deposition rate was about 22 cm/kyr during late-Gauss and early Matuyama time, almost twice that of the previous example, and the lithololgy is uniform [*Liddicoat*, 1982; *Glen et al.*, 1999a]. The core was sampled continuously at 1-cm spacing through a 2.1 m stratigraphic interval centered on the transition, with two or three 1-cm cubic specimens measured at each level. The quality of thermal demagnetization results are variable, with the lowest quality predictably during the transition when the field

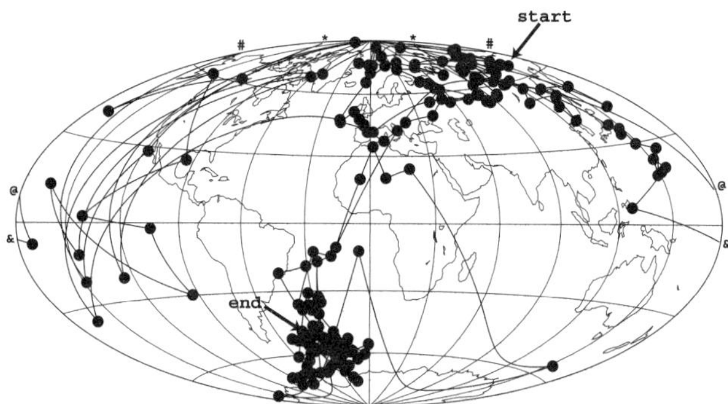

Figure 6. High-resolution record of VGPs spanning the Gauss/Matuyama transition from a long core of sediment from ancestral Searles Lake, eastern California [*Glen et al.*, 1999b].

intensity was weakest, but the broad features of the transition are well-defined even if individual points are sometimes not. From a swath of normal VGPs that extends from the north-polar region to northeastern Siberia, two excursions reach to transitional latitudes in the western Pacific and southern Europe. After a return to normal polarity, the VGPs undergo four large oscillations of 90 to 130° southwestward into the eastern equatorial Pacific and back to normal again. Three of these oscillations contain only one or two transitional VGPs each, but the first oscillation comprises seven. Following another spell near the pole, the VGPs march southeastward across Eurasia into the west-equatorial Pacific and eventually reach reversed polarity via a complex route. A final excursion from reversed polarity to the equatorial Atlantic and back ends the directional transition, which all together took an estimated 5300 years. Note that there are several VGP clusters: one comprising southern Europe and northern Africa, which is visited early, mid, and late in the transition; another in the Atlantic just south of the equator consisting of mid- and late-transitional VGPs; and a third, broad cluster comprising the far-reaching oscillatory VGPs in the east equatorial Pacific.

The large swings of direction accomplished in only a few centimeters of core in the last two, highest-resolution transition records raise some difficult questions. Do they signal an otherwise undetectable dramatic drop in deposition rate, or did the field actually change so rapidly? Could sediment even resolve changes as rapid as a literal interpretation of these records would demand? Is the complexity of reversals indicated by these transition records real, or just the product of faulty magnetic recording? If the oscillations in Figure 6 were caused by patchy remagnetization accompanying diagenesis down the core, then why is there no record higher in the transition zone of a later standstill of the field in the southern Pacific Ocean that could produce the remagnetization?

3. THE CASE FOR COMPLEXITY

Even though there are common features among the transition paths illustrated in Figures 4, 5 and 6 that argue for some degree of true geomagnetic content in these high-resolution sedimentary records, independent sources of evidence are needed to help answer the questions posed above. One completely different body of information we can turn to for insight comes from numerical geodynamo simulations. Glatzmaier and Roberts' [1996] geodynamo model is a dynamically self-consistent simulation and the first to exhibit spontaneous reversals. It solves the equations for the time-dependent thermal, compositional, velocity and magnetic fields in the spherical geometry appropriate for the core. 'Simulation time' (henceforth referred to simply as 'time' when discussing simulations) is scaled to give the appropriate free decay time for electrical currents in the earth's core. Spatial resolution is limited by the necessity of using a time step sufficiently long that the simulations span enough time for reversals to occur, requiring the representation of small-scale eddies by eddy diffusion, as is done in models of the oceans and atmosphere. How the simulations will change as greater computational power enables greater resolution and physical parameters closer to those of the earth is an interesting question. Quite possibly the behavior at larger scales will remain similar, whereas at smaller scales more rapid and complex field variations may emerge.

A suite of eight simulations, each with a different pattern of heat flux prescribed at the core-mantle boundary (CMB) but identical in all other regards including total heat flux out of the core, exhibited different styles of reversals [*Glatzmaier et al.*, 1999]. The six simulations with CMB heat flux that varied symmetrically about the rotation axis yielded fourteen full reversals, eleven of them with relatively simple transition paths and durations (3 to 10 kyr) typically estimated for actual reversals on earth [see *Coe et al.*, 2000, for examples]. Two of the exceptions were for an unstable case in which the dynamo 'died' (intensity dropped more than 95% and never recovered) as it first reversed, and the third was for the case in which the CMB heat flux varied monotonically with latitude from pole to pole. The two simulations with CMB heat flux that varied with longitude reversed three times, two of them with complex VGP paths and directional durations around 15 to 30 kyr. The most relevant of these for the earth is the second reversal of the tomographic simulation, so named because the heat-flux at the CMB was patterned after the large-scale variation of seismic velocity from tomographic studies of the lowermost mantle. The rationale for this case is that relatively more heat can conduct into overlying mantle that is cooler than average, i.e. where seismic velocity is presumed to be higher than average.

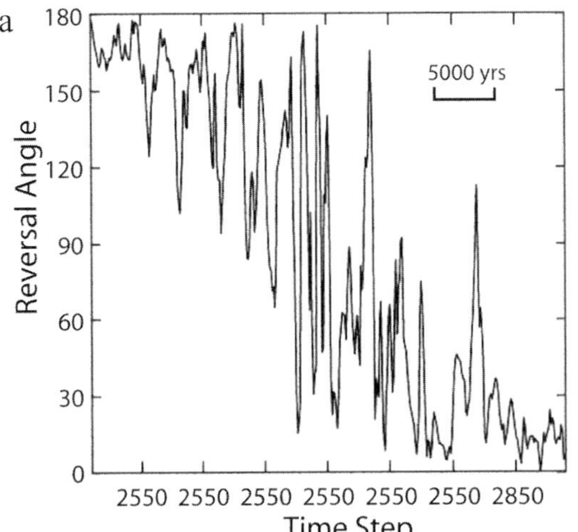
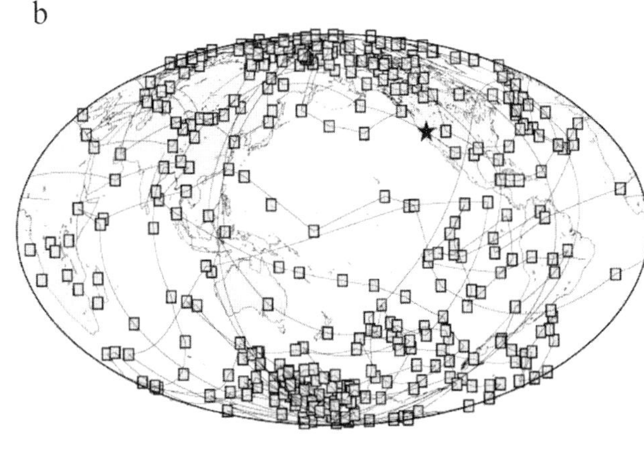

Figure 7. Record of a reversal from the tomographic geodynamo simulation of Glatzmaier et al. [1999] at 40°N, 240°E. (a) Angle between the simulated field and the centered axial dipole field direction versus simulation time step (95 yr). (b) corresponding VGP path spanning the reversal (star shows observation site). Redrawn from Coe et al. [2000].

Figure 7a shows the temporal behavior of the field direction during the complex reversal of the tomographic simulation at a site in the northern hemisphere [*Coe et al.*, 2000]. The record exhibits many oscillations that grow in amplitude and frequency to a maximum at the midpoint of the transition and then decrease as stable normal polarity is achieved. Measured from the first to the last large oscillation, the transition took 26 kyr. The definition of duration is debatable: one could consider the last oscillation to be a post-transitional excursion because it occurs about 4 kyr after the previous one. The VGP path (Figure 7b) is characterized by many long tracks, with VGPs closely spaced in some places and with gaps elsewhere of 90 to almost 180° between successive points (sampling interval 95 years). The density of VGPs is distinctly greater than average in a longitudinal band over the Americas, reminiscent of (but less confined than) the VGP bands first reported by Laj et al. (1991) and exhibited in the paleomagnetic records of Figure 4. A second, more diffuse band is discernible over eastern Asia, and a much lower density of VGPs occurs in the central Pacific and Atlantic regions. Notice as well the similarity between the simulation record and the Matuyama-Brunhes record discussed above. By sprinkling a number of hiatuses into (i.e., removing a number of VGP sequences from) the tomographic VGP path of Figure 7b, we could obtain a record quite similar in character to that of Figure 5. At the same time, it is important to recognize that the tomographic simulation departs very significantly from some time-averaged properties of the earth's field [*Coe et al.*, 2000], so the similarity to a real record must be taken as only suggestive.

The simulation also provides some insight into the degree to which records of the same reversal may vary from place to place over globe. Plate 1 shows the variation of VGP latitude as a function of time through the reversal around seven circles of latitude between 80°N to 80°S. Transitional VGPs first appear in the southern hemisphere around latitude 65°S, but not till about 4 kyr later does the VGP cross the equator. Focusing on these zero crossings for clarity, which are shown by the dashed lines, we see that the field instability spread quickly throughout the southern hemisphere, but much more slowly in the northern hemisphere. There is a tendency for the zero crossings to propagate westward at mid-southern latitudes, whereas at mid-northern latitudes they propagate eastward around the middle of the record and westward later on. Records of the reversal vary greatly with position on the globe. For instance, at 65°S/270°E the VGP crosses the equator 39 times during the transition, whereas at 65°N/180°E it crosses only 9 times. At a given moment the field direction may appear to be fully normal at one place on the globe, reversed at another, and transitional at another—even for points around one circle of latitude. Duration of the transition depends on the definition. Defined as the first and last equatorial crossing of VGPs, it varies greatly between extremes of about 3 kyr at 80°N to 30 kyr for western longitudes at 65°S.

Thus, geodynamo simulations make the case for complex reversals plausible. Can we find more direct evidence for complexity? To do so we return to reversal transitions recorded by lava flows because, even though a gap of unknown length exists at each flow boundary, the stable direction of each flow usually gives a reliable estimate of the true field direction at the time the flow cooled. We return to the 990-meter Steens Mountain section of flow-on-flow basalt that has yielded a detailed record of a Miocene rever-

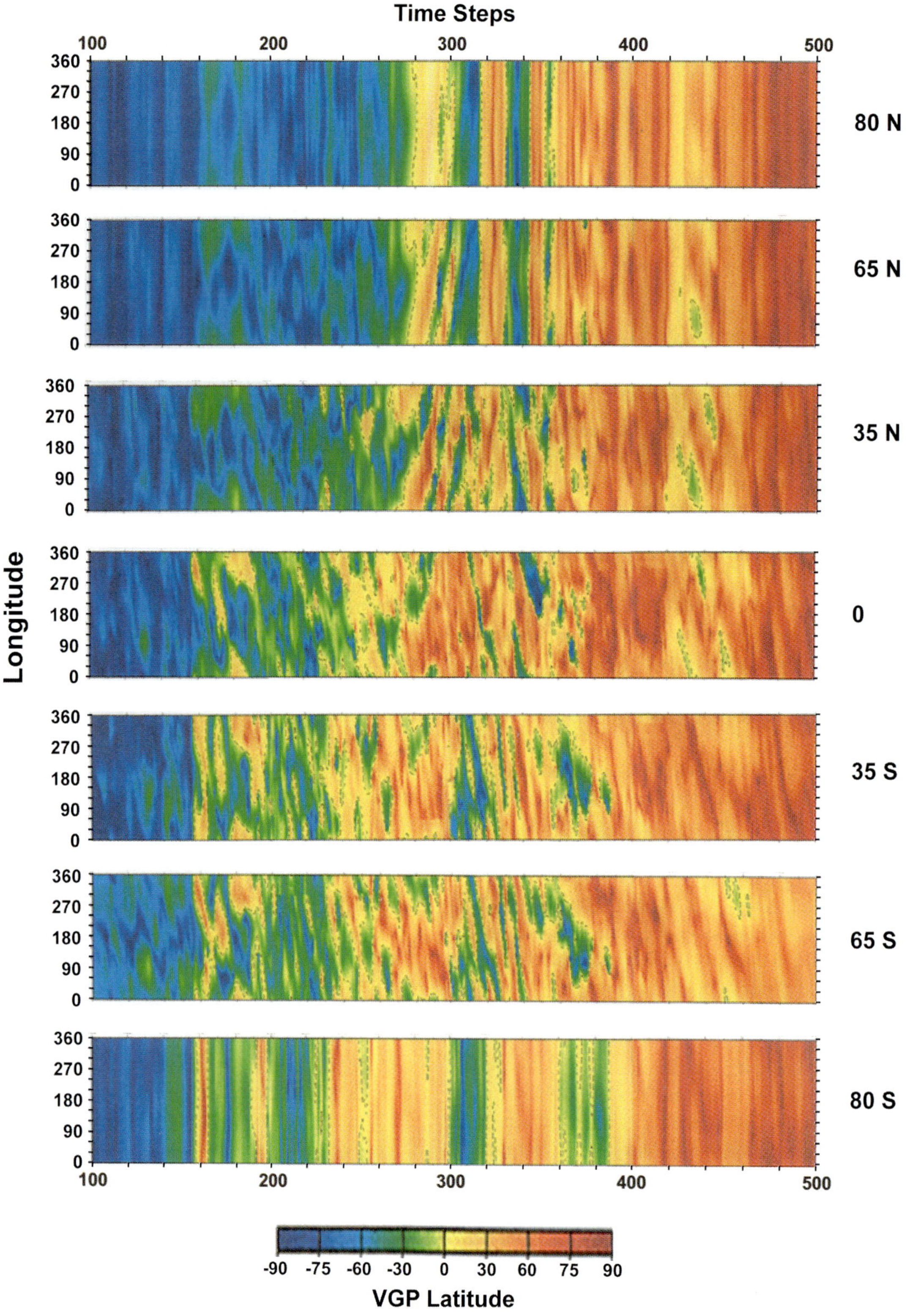

Plate 1. Space-time variation of VGP latitude through the reversal from the same geodynamo simulation as in Figure 7. Time steps along the abscissa are 95 yr (38 kyr total). See text for discussion.

sal with 26 distinct directional groups documenting the transition (Figure 4). Despite its detailed character, we now know that the actual behavior of the reversing field was more complex—perhaps much more complex—than this record shows. A recent study [*Camps et al.*, 1999] of a parallel section only 1.7 km from the original section uncovered a second unsuccessful swing of the field direction to normal polarity that is not represented in the original record (Figure 8). The new swing is definitely not the result of unrecognized sills or fault-repeated section. Rather, it is recorded by three flows that must pinch out before reaching the original section. Note that the rest of the new record matches the original directional path well, even though it is clear that there are additional instances of flows at one section that don't reach the other. It is interesting that the uppermost new direction (flow B33 in Figure 8) represents a swing back toward the cluster of west and down directions. Instead of jumping directly from the B33 to the B32 direction as shown in Figure 8, the transitional field may have completed the oscillation back to the cluster. Indeed, any number of more complicated scenarios are possible.

Other sections from the surrounding region containing the same reversal recorded by the Steens Basalt have been summarized by Mankinen et al. [1987]. The Steens Basalt is now recognized as the initial outpouring of the Yellowstone hot spot flood basalts [*Hooper et al.*, 2002], which explains how a pulse of flows that recorded the reversal can be widely distributed. Some of these records contain features quite different from the record at Steens (Figure 9). In an early record from Poker Jim Ridge, 80 km to the west, the field makes a large excursion during reversed polarity that is not seen at Steens, visits the same two cluster spots as at Steens and then moves directly to normal polarity, missing much of the rest of the Steens record [*Goldstein et al.*, 1969]. In two records from the Santa Rosa Range 120 km to the southeast, the reversal ends along the track by which the Steens reversal passes to stable normal polarity, but exhibits VGPs in western Australia, southern Siberia, the western Indian Ocean and the central Atlantic that are not found at Steens [*Larson et al.*, 1971; *Roberts and Fuller*, 1990].

There are other instances of volcanic reversal records that also suggest that reversals can be too complex and too rapidly

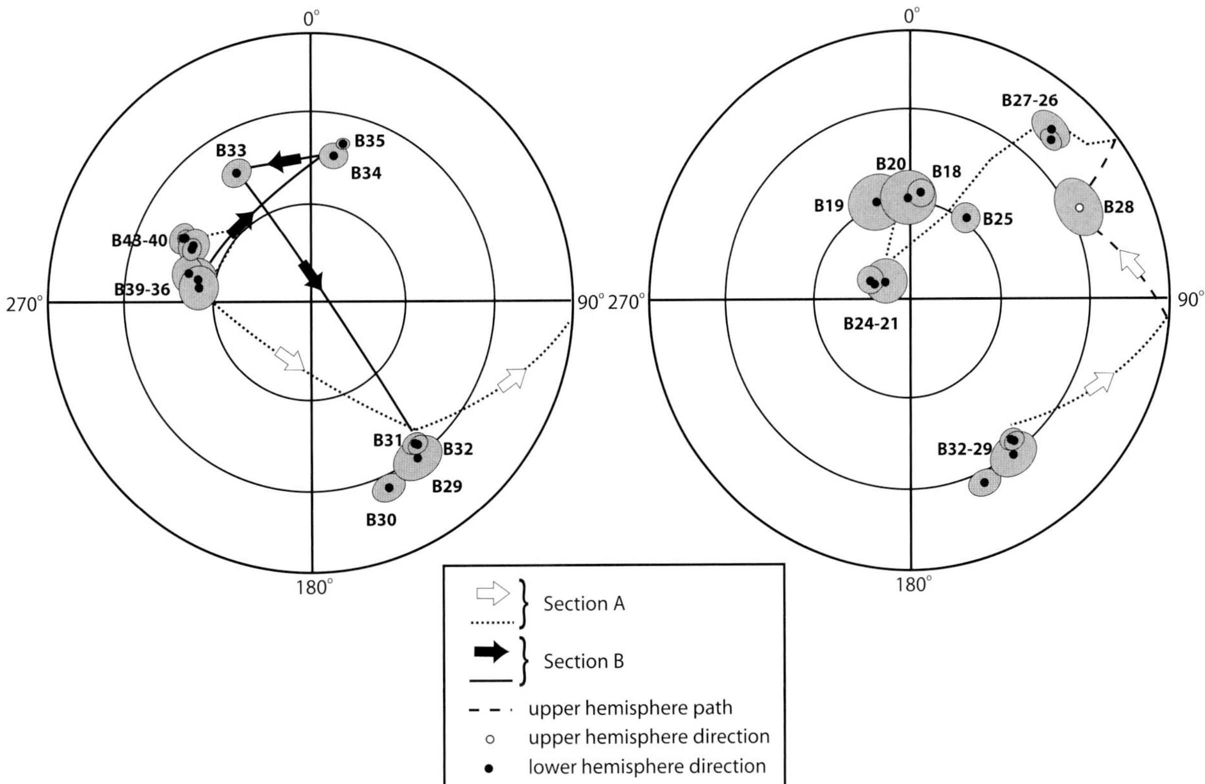

Figure 8. Flow-mean directions and confidence limits from lava flows spanning last half of the Steens Mountain reversal at a new Steens B section [*Camps et al.*, 1999], 1.7 km from original Steens A section. Dotted and dashed line shows progression of directions at Steens A from Mankinen et al. [1985]. They follow the flow means for Steens B quite well, except for the swings from flow B36 to flows B35/B34 and then back to B33 (see text). Modified from Camps [1999].

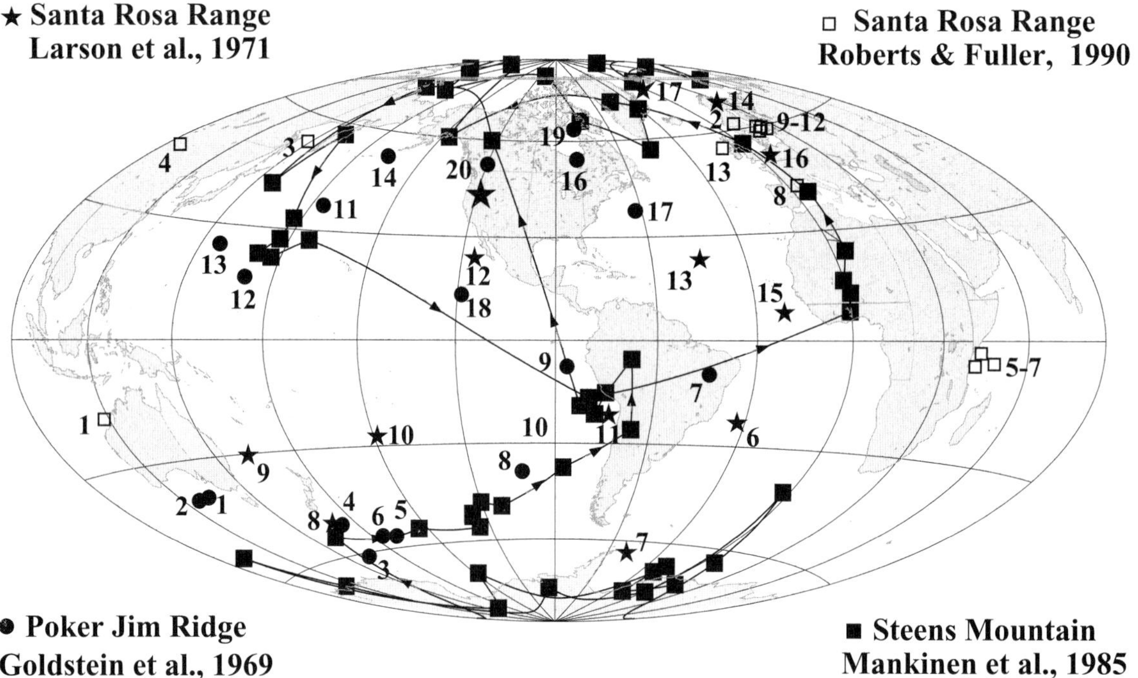

Figure 9. Reversal records showing VGPs of the polarity transition at Steens Mountain and from Steens-equivalent lava flows at Poker Jim Ridge and the Santa Rosa Range. These latter records include some VGPs not found in the Steens Mountain record (see text).

changing to be resolved by even thick sequences of rapidly emplaced flows. One example is the recently published study by Riisager et al. [2003] of the C27n–C26r transition recorded at ~61 Ma by West Greenland flood basalt. The stratigraphy and paleomagnetic directions in two thick sections, KI and NQ, comprising 35 and 45 sampled lava flows indicate that the former overlaps the first half of the latter. Even though the sections are only separated by 16 km and the lava production rate there is estimated to have been an order of magnitude greater than over the current Iceland hot spot, a large swing of the VGP path to Central America and rebound back to normal polarity recorded in the KI section is absent at NQ. Several common directions appear to tie the two sections together, but most of the flows in one section are not matched by flows in the other section. This result is in accord with helicopter photogrammetry of superb sea-cliff exposures, by which individual flows can be frequently observed to pinch out within 2 km. The necessary conclusion is that the records comprise independent paleomagnetic time series at irregular intervals, some of which are too great to track the course of the reversing field adequately.

4. CONCLUSIONS

Geologically speaking, reversals take so little time to happen that they challenge the ability of the best paleomagnetic recorders to resolve the behavior of the transitional field. There are many mechanisms that are capable of degrading paleomagnetic records in sediments. Doubtless many sedimentary reversal records are severely compromised by smoothing, by patchy remagnetization, and perhaps also by breakdown of recording fidelity in weak transitional fields. Even though many sedimentary records presumably do retain some information about the reversing field, it is difficult to know how to separate fact from artifact. Thus it is all too easy to dismiss sedimentary records that are unusually simple or extremely complicated as completely obscured by recording problems. Some recent detailed studies in high-deposition-rate sediments, however, collectively suggest that at least some reversals may be truly complex, with swings and oscillations of field direction that are so rapid that even these exceptional records may smooth or even miss many of them.

This hypothesis requires independent support, because sedimentary recording problems are always a possibility. Numerical geodynamo simulations, though still in their near-infancy in terms of the spatial resolution of fluid motions that can be computed for long-enough simulation times that spontaneous reversals occur, provide such support: namely, examples of complex (as well as simple) polarity transitions with tracks, standstills and oscillations like those envisaged above. It is interesting and perhaps significant that complex paths did not occur in simulations with highly symmetrical patterns of CMB heat flux (i.e., symmetrical about both the rotation axis and equator).

Independent evidence directly supporting the occurrence of complex reversals comes from transition-zone records in lava flows. No single record is as complex as we propose some reversals may be, but lack of agreement between multiple records through the same reversal implies that the underlying transitional field behavior was much more complicated than indicated by any one record. This inference is based on the assumption that lava flows, unlike sediments, almost always yield reasonably reliable estimates of the instantaneous field direction. When sampling and laboratory demagnetization are carried out correctly, we hold that this is the case.

In summary, sedimentary and volcanic paleomagnetic records are complementary, but both are always incomplete and give only lower bounds on how rapidly changing and complex the behavior of the reversing field may have been. Geodynamo simulations provide a theoretical approach to the question, but limitations in computer power have prevented them from operating near the parameter regime appropriate for the core, with concomitant loss of spatial and temporal resolution. Nonetheless, the combined evidence from all three approaches appears sufficient to conclude that at least some reversals are much more complex than typically portrayed, with episodes of oscillatory and very rapid field change.

Acknowledgments. We are grateful to three anonymous reviewers whose comments were helpful in improving the manuscript. The writing of this paper was supported by NSF grant EAR-0310316 to RSC and JMG.

REFERENCES

Bogue, S.W., and R.S. Coe, Successive paleomagnetic reversal records from Kauai, *Nature*, 295, 399–401, 1982.

Camps, P., R.S. Coe, and M. Prevot, Transitional geomagnetic impulse hypothesis: Geomagnetic fact or rock-magnetic artifact?, *J. Geophys. Res.*, 104, 17747–17758, 1999.

Cande, S.C., and D.V. Kent, Revised calibration of the geomagnetic polarity timescale for the Late Cretaceous and Cenozoic, *J. Geophys. Res.*, 100, 6093–6095, 1995.

Channell, J.E.T., and B. Lehman, The last two geomagnetic polarity reversals recorded in high-deposition-rate sediment drifts, *Nature*, 389, 712–715, 1997.

Clement, B.M., Geographical distribution of transitional VGPs—Evidence for non-zonal equatorial symmetry during the Matuyama-Brunhes geomagnetic reversal, *Earth Planet. Sci. Lett.*, 104, 48–58, 1991.

Clement, B.M., and D.V. Kent, A detailed record of the Lower Jaramillo polarity transition from a southern-hemisphere, deep-sea sediment core, *J. Geophys. Res.*, 89, 1049–1058, 1984.

Clement, B.M., and D.V. Kent, Geomagnetic polarity transition records from 5 hydraulic piston core sites in the North Atlantic, *Deep Sea Drilling Project, Initial Rep.*, 94, 831–852, 1987.

Clement, B.M., P. Rodda, L. Sierra, and E. Smith, Lower Pliocene polarity transitions from the Suva Marl, Fiji, *J. Geophys. Res.*, 103, 15483–15495, 1998.

Coe, R.S., L. Hongre, and G.A. Glatzmaier, An examination of simulated geomagnetic reversals from a palaeomagnetic perspective, *Phil. Trans. Roy. Soc. London*, Series A, 358, 1141–1170, 2000.

Coe, R.S., and M. Prévot, Evidence suggesting extremely rapid field variation during a geomagnetic reversal, *Earth Planet. Sci. Lett.*, 92, 292–298, 1989.

Coe, R.S., M. Prévot, and P. Camps, New evidence for extraordinarily rapid change of the geomagnetic field during a reversal, *Nature*, 374, 687–692, 1995.

deMenocal, P.B., W.F. Ruddiman, and D.V. Kent, Depth of post-depositional remanence acquisition in deep-sea sediments: a case study of the Brunhes-Matuyama reversal and oxygen isotopic Stage 19.1, *Earth Planet. Sci. Lett.*, 99, 1–13, 1990.

Glatzmaier, G.A., R.S. Coe, L. Hongre, and P.H. Roberts, The role of the Earth's mantle in controlling the frequency of geomagnetic reversals, *Nature*, 401, 885–890, 1999.

Glatzmaier, G.A., and P.H. Roberts, An anelastic evolutionary geodynamo simulation driven by compositional and thermal convection, *Physica*, D97, 81–94, 1996.

Glen, J.M.G., R.S. Coe, and J.C. Liddicoat, A detailed record of paleomagnetic field change from Searles Lake, California 2. The Gauss Matuyama polarity reversal, *J. Geophys. Res.*, 104, 12883–12894, 1999b.

Glen, J.M.G., J.C. Liddicoat, and R.S. Coe, A detailed record of paleomagnetic field change from Searles Lake, California 1. Long-term secular variation bounding the Gauss Matuyama polarity reversal, *J. Geophys. Res.*, 104, 12865–12882, 1999a.

Goldstein, M.A., D.W. Strangway, and E.E. Larson, Paleomagnetism of a Miocene transition zone in southeastern Oregon, *Earth Planet. Sci. Lett.*, 7, 231–239, 1969.

Gubbins, D., and R.S. Coe, Longitudinally confined geomagnetic reversal paths from non-dipolar transition fields, *Nature*, 362, 51–53, 1993.

Hoffman, K.A., Dipolar reversal states of the geomagnetic field and core mantle dynamics, *Nature*, 359, 789–794, 1992.

Hoffman, K.A., Temporal aspects of the last reversal of Earth's magnetic field, *Phil. Trans. Roy. Soc. London*, Series A-Mathematical Physical and Engineering Sciences, 358, 1181–1190, 2000.

Hooper, P.R., G.B. Binger, and K.R. Lees, Ages of the Steens and Columbia River flood basalts and their relationship to extension-related calc-alkalic volcanism in eastern Oregon, *Geol. Soc. Amer. Bull.*, 114, 43–50, 2002.

Jacobs, J.A., *Reversals of the Earth's Magnetic Field*, 346 pp., Cambridge University Press, Cambridge, 1994.

Kent, D.V., and D.A. Schneider, Correlation of paleointensity variation records in the Brunhes/Matuyama polarity transition interval, *Earth Planet. Sci. Lett.*, 129, 135–144, 1995.

Kosterov, A.A., M. Prévot, M. Perrin, and V.A. Shashkanov, Palaeointensity of the Earth's magnetic field in Jurassic: new results from a Thellier study of the Lesotho basalt, Southern Africa, *J. Geophys. Res.*, 102, 24859–24872, 1997.

Laj, C., A. Mazaud, R. Weeks, M. Fuller, and E. Herrero-Bervera, Geomagnetic reversal paths, *Nature*, 351, 447–447, 1991.

Langereis, C.G., A.A.M. van Hoof, and P. Rochette, Longitudinal confinement of geomagnetic reversal paths as a possible sedimentary artefact, *Nature*, 358, 226–230, 1992.

Larson, E.E., D.E. Watson, and W. Jennings, Regional comparison of a Miocene geomagnetic transition in Oregon and Nevada, *Earth Planet. Sci. Lett.*, 11, 391–400, 1971.

Liddicoat, J.C., Gauss-Matuyama polarity transition, *Phil. Trans. Roy. Soc. London*, A306, 121–128, 1982.

Mankinen, E.A., E.E. Larson, C.S. Grommé, M. Prevot, and R.S. Coe, The Steens Mountain (Oregon) Geomagnetic Polarity Transition. 3. Its Regional Significance, *J. Geophys. Res.*, 92, 8057–8076, 1987.

Mankinen, E.A., M. Prévot, C.S. Grommé, and R.S. Coe, The Steens Mountain (Oregon) geomagnetic polarity transition. 1. Directional history, duration of episodes, and rock magnetism, *J. Geophys. Res.*, 90, 393–416, 1985.

Merrill, R.T., and P.L. McFadden, Geomagnetic polarity transitions, *Rev. Geophys.*, 37, 201–226, 1999.

Ninkovich, D., N.D. Opdyke, N.D. Heezen, and J.H. Foster, Paleomagnetic stratigraphy, rates of deposition and tephrachronology in North Pacific deep-sea cores, *Earth Planet. Sci. Lett.*, 1, 476–492, 1966.

Opdyke, N.D., D.V. Kent, and W. Lowrie, Details of magnetic polarity transitions recorded in a high deposition rate deep-sea core, *Earth Planet. Sci. Lett.*, 20, 315–324, 1973.

Prévot, M., E.A. Mankinen, R.S. Coe, and C.S. Grommé, The Steens Mountain (Oregon) geomagnetic polarity transition. 2. Field intensity variations and discussion of reversal models, *J. Geophys. Res.*, 90, 417–448, 1985.

Prévot, M., N. Roberts, J. Thompson, L. Faynot, M. Perrin, and P. Camps, Revisiting the Jurassic geomagnetic reversal recorded in the Lesotho Basalt (Southern Africa), *Geophys. J. Int.*, 155, 367–378, 2003.

Quidelleur, X., and J.P. Valet, Paleomagnetic records of excursions and reversals—possible biases caused by magnetization artefacts, *Phys. Earth Planet. Interiors*, 82, 27–48, 1994.

Riisager, J., P. Riisager, and P. Pedersen, The C27n–C26r geomagnetic polarity reversal recorded in the west Greenland flood basalt province: How complex is the transitional field?, *J. Geophys. Res.*, 108, 2155, 2003.

Roberts, N., and M. Fuller, Similarity of new paleomagnetic data from the Santa Rosa Mountains with those from Steens Mountain gives wide regional evidence for a 2-stage process of geomagnetic-field reversal, *Geophys. J. Int.*, 100, 521–526, 1990.

Thellier, E., and O. Thellier, Sur l'intensité du champ magnétique terrestre dans le passé historique et géologique, *Annales de Géophysique*, 15, 285–376, 1959.

Valet, J.P., C. Laj, and P. Tucholka, High-resolution sedimentary record of a geomagnetic reversal, *Nature*, 322, 27–32, 1986.

Valet, J.P., L. Tauxe, and D.R. Clark, The Matuyama-Brunhes transition recorded from Lake Tecopa sediments (California), *Earth Planet. Sci. Lett.*, 87, 463–472, 1988.

Valet, J.P., L. Tauxe, and B. Clement, Equatorial and mid-latitude records of the last geomagnetic reversal from the Atlantic Ocean, *Earth Planet. Sci. Lett.*, 94, 371–384, 1989.

van Hoof, A.A.M., and C.G. Langereis, Reversal records in marine marls and delayed acquisition of remanent magnetisation, *Nature*, 351, 223–225, 1991.

van Hoof, A.A.M., and C.G. Langereis, The upper and lower Thvera sedimentary geomagnetic reversal records from southern Sicily, *Earth Planet. Sci. Lett.*, V114, 59–75, 1992.

van Hoof, A.A.M., B.J.H. Vanos, and C.G. Langereis, The upper and lower Nunivak sedimentary geomagnetic transitional records from southern Sicily, *Phys. Earth Planet. Interiors*, 77, 297–313, 1993a.

van Hoof, A.A.M., B.J.H. Vanos, J.G. Rademakers, C.G. Langereis, and G.J. Delange, A paleomagnetic and geochemical record of the upper Cochiti reversal and two subsequent precessional cycles from southern Sicily (Italy), *Earth Planet. Sci. Lett.*, 117, 235–250, 1993b.

van Zijl, J.S.V., K.W.T. Graham, and A.L. Hales, The paleomagnetism of the Stormberg lavas of South Africa II. The behaviour of the magnetic field during a reversal, *Geophys. J. Roy. Astron. Soc.*, 7, 169–182, 1962.

Robert S. Coe, Earth Sciences Department, University of California, Santa Cruz, California 95064. (rcoe@es.ucsc.edu)

Jonathan M. G. Glen, U.S. Geological Survey, 345 Middlefield Road, Menlo Park, California 94025. (jglen@usgs.gov)

Regionally Recurrent Paleomagnetic Transitional Fields and Mantle Processes

Kenneth A. Hoffman

Physics Department, Cal Poly State University, San Luis Obispo, California

Brad S. Singer

Department of Geology and Geophysics, University of Wisconsin-Madison, Wisconsin

Whether Earth's lower mantle observably influences the geodynamo as it attempts to reverse polarity has been hotly debated for more than a decade. Although several paleomagnetic records of transitional events support this contention, analyses of worldwide Cenozoic reversal data, when considered *en masse*, appear inconclusive. With this controversy in mind we present transitional field data associated with a sequence of lavas on Tahiti-Nui, $^{40}Ar/^{39}Ar$-dated at 578.6±8.8 ka, that record the Big Lost Event. The recorded field behavior is dominated by a cluster of virtual geomagnetic poles *(VGPs)* near the west coast of Australia, a common feature seen in previously published records of reversals and geomagnetic events obtained from lava sequences erupted at the Society Island hot spot over 2.3 Myr during the Plio-Pleistocene. This correspondence suggests that recurrent, or geographically "preferred," *VGP* behavior is most apparent given regionally-partitioned paleomagnetic data. Synthetic removal of the axial dipole term from modern-day and historic field models produces a field configuration at the hot spot site compatible with these transitional paleodirectional findings. Various analyses of the core-surface modern-day, historic, and time-averaged paleomagnetic fields suggest the existence of a long-lived concentration of magnetic flux beneath Australasia. At those times when the axial dipole is weak or nonexistent, such a patch of mantle-held flux apparently can possess sufficient strength to regionally dominate the magnetic field structure. Lower-mantle seismic studies show the region beneath Australasia to be moderately anomalous, offering a possible clue toward an understanding of the physical conditions that may produce such a magnetic flux gathering.

1. INTRODUCTION

The fact that Earth's magnetic field is capable of reversing its polarity and that transitional behavior can be recorded in rocks for later paleomagnetic investigation insures a unique source of data relevant to the problem of the geodynamo: observations of the field when the axial dipole is anomalously weak. Two particular questions that have been addressed by such paleomagnetic data are *1)* whether or not there exists observable systematics to the process of field reversal [see *Jacobs,* 1994], and if so, *2)* whether the lower mantle plays a significant role in influencing what is experienced at Earth's surface [see *Merrill et al.,* 1996]. In the early 1990's scenarios involving mantle control were proposed, based on analy-

Timescales of the Paleomagnetic Field
Geophysical Monograph Series 145
Copyright 2004 by the American Geophysical Union
10.1029/145GM17

ses of late-Cenozoic transitional field records. First, Clement [1991] and Laj et al. [1991] claimed that transitional *VGP* pathways followed preferred bands of longitude. Later, Hoffman [1992] argued for preferred locations where transitional *VGPs* cluster. In the years that followed a number of papers were published, some in opposition, others in support of geographically preferred *VGP* behaviors [see e.g. the review by *Merrill and McFadden,* 1999].

Of the supporting publications, the work of Constable [1992] was particularly intriguing as it appeared to tie elements of the modern-day field to the claim of preferred *VGP* bands. On the negative side of the issue Prévot and Camps [1993] tested the contention of preferential *VGP* behavior by amassing available transitional *VGP* data obtained from late-Cenozoic lava sequences and found no hint of such behavior, whether in the form of banded or clustered *VGPs*. Love [1998] conducted a similar analysis and concluded otherwise, arguing that preferred *VGP* pathways were in fact a statistically significant feature of recorded transitional field behavior. These seemingly contradictory findings underscore the complexity of the problem, and suggest that the addition to the dataset of yet undiscovered transitional field records is unlikely to definitively solve it.

Several of the most reliable and detailed paleomagnetic reversal records presently available contain one or more clusters of sequential transitional *VGPs,* many of which reside in either of the two claimed patches shown in Figure 1 *(upper row left).* For the case of the Brunhes-aged Blake Event recorded in Mediterranean sediments [*Tric et al.,* 1991], the *VGP* is seen (Figure 1, *upper row center*) to first loiter within one patch and then move with apparent rapidity to the other before returning to normal polarity. Several other examples of clustered *VGPs* found in either the region of the Southwest Atlantic or Australasia are displayed in Figure 1, records obtained from various rocktypes and which span the Cenozoic. However, since these data account for only a fraction of the Cenozoic transitional field dataset, such a compilation in itself cannot be considered a robust confirmation of the existence of preferred quasi-stationary states of the reversing field.

With the goal of shedding more light on the question of preferred transitional field configurations and, hence, the possible controlling presence of the lower-most mantle on the geodynamo, we now consider an approach in which paleomagnetic data is analyzed in a regional context. Specifically, we first revisit the case of the "last reversal," the Matuyama-

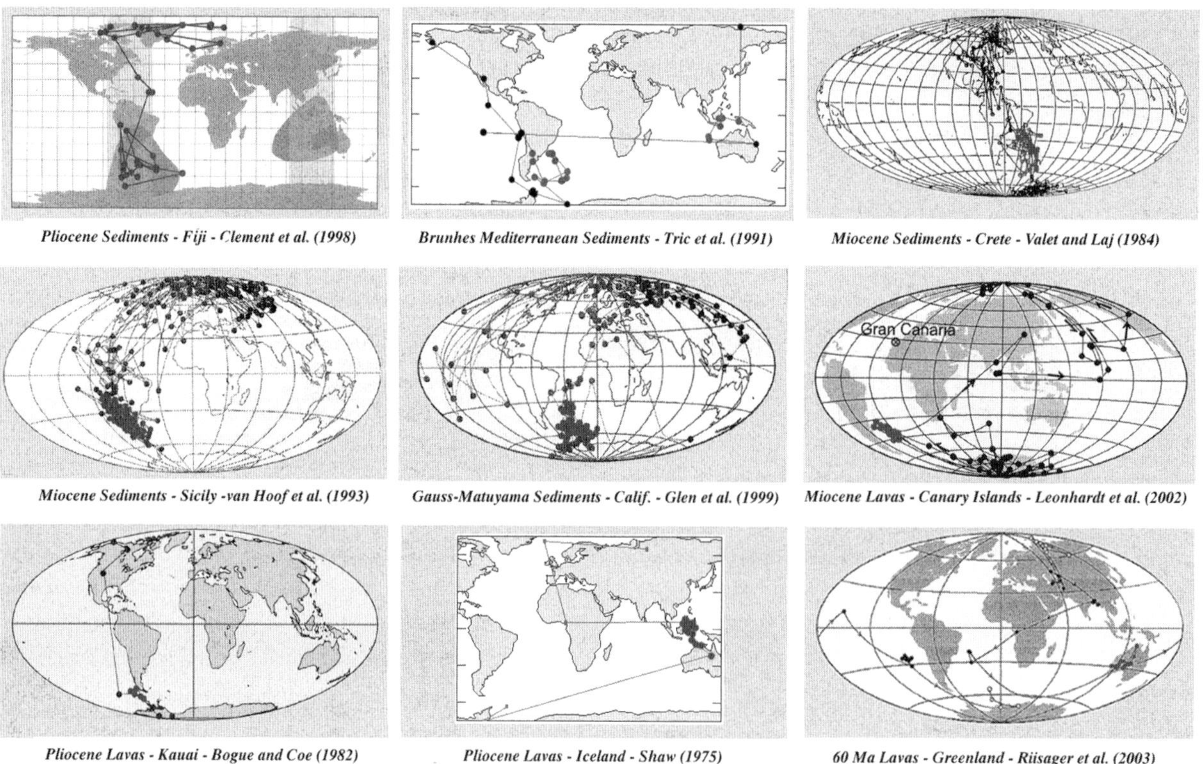

Figure 1. Examples of paleomagnetic transition records spanning the Cenozoic that contain a clustering of sequential virtual geomagnetic poles *(VGPs)* in the region of the South Atlantic and/or Australasia. Of the nine records shown, five and four were obtained from sediments and lavas, respectively. The claimed *VGP* preferred bands [*Laj et al.,* 1991] and *VGP* cluster patches [*Hoffman,* 1992] are shown in the upper left plot.

Brunhes polarity transition, for which by far the largest number of records are presently available. We then consider accounts of transitional field behavior obtained from lavas erupted at the Society Island hot spot over the past 3 Myr, and include in this dataset our newly-analyzed record of the Brunhes-aged Big Lost Event recorded in lavas from northern Tahiti-Nui. Later, we explore these results alongside aspects of the modern-day, historic, and paleomagnetic time-averaged fields as well as seismic studies of the lower-most mantle.

2. THE MATUYAMA-BRUNHES REVERSAL

Recordings of the last significant reversal of Earth's magnetic field, the Matuyama-Brunhes reverse-to-normal transition, have been obtained from various types of sedimentary materials as well as lavas. Hoffman [2000] analyzed records satisfying the set of reliability criteria imposed by Love and Mazaud [1997] and which these authors included in their select database. Love and Mazaud [1997] reported that the complete database demonstrated a statistically significant preference for transitional *VGPs* to lie within the two longitudinal bands claimed by Clement [1991] and Laj et al. [1991]. Following Hoffman's [2000] analysis, we separate those Matuyama-Brunhes data obtained from sites between latitudes 45°S and 45°N (region *A*) from those at higher latitudes (region *B*). Regardless of the overall complexity of transitional field behavior possessed by a given record, each of five examined region *A* recordings were found to contain a *VGP* cluster in Australasia (Plate 1a).

In contrast, records included in region *B* contain significant *VGP* behavior in the South Atlantic. Most striking in this regard is the complex transitional field behavior recorded in three North Atlantic marine sediment cores at Hole 984 (Plate 1b) reported by Channell and Lehman [1997] at the time of the database publication by Love and Mazaud [1997]. Possessing remarkable similarity, each of these records display a back and forth "diagonal" transit of the *VGP* between the South Atlantic and Central Asia. Plate 1b shows the 3-record composite *VGP* geographic histogram for that part of the recorded reversal behavior.

The regional dissimilarity in cluster location between the two Matuyama-Brunhes sub-datasets suggests, first, that during much of the last reversal the transitioning field was non-dipolar. More specifically, the correspondence of *VGP* cluster location associated with records from sites within region *A*—sites that encompass much of the globe—suggests that much of the power in the transitioning field resided in non-dipole terms of low-order. If so, this finding could imply that the spatial nature of the primary sources responsible for such apparently stationary transitional fields may be ultimately decipherable.

3. TRANSITIONAL EVENTS RECORDED AT THE SOCIETY ISLAND HOT SPOT

We now turn to the recordings at a particular region *A* site, the Society Island hot spot. There presently exist four reasonably detailed transitional field recordings (each containing a minimum of 5 transitional *VGPs*) obtained from lavas erupted at the site of the hot spot. Three of these come from a sequence of four Pleistocene reversal records obtained from Tahitian lavas exposed along the Punaruu Valley reported by Chauvin et al. [1990]: the Matuyama-Brunhes reversal (see Plate 1a), the upper and lower boundaries of the Jaramillo subchron, and what was assumed at the time of its publication to be the Cobb Mountain event. Each record was precisely dated by $^{40}Ar/^{39}Ar$ age determinations [see *Singer et al.,* 1999].

Of these records only the lower Jaramillo contains too few transitional *VGPs* (3 in total) to be considered further. (The transitional record then thought to be a recording of the Cobb Mountain event was later found by Singer et al. [1999] to be some 100 kyr younger. Singer et al. [1999] renamed this recorded event the Punaruu after the site at which it was first discovered.) Each of the three remaining records contain a cluster of transitional *VGPs* in essentially the same location off the southwest coast of Australia [*Chauvin et al.,* 1990]. In addition, Roperch and Duncan [1990] reported a 2.9 Ma transitional event obtained from lavas exposed on Huahine, another island in the Society Island hot spot chain. These transitionally magnetized Pliocene flows were found to be dominated by two dense concentrations of *VGPs* off the northwest coast of Australia.

Although these four Society Island hot spot records possess varying degrees of complexity, all contain a number of sequential *VGPs* found near western Australia not unlike the composite Matuyama-Brunhes field behavior displayed in Plate 1a. Of course, the possibility that one or more of these clusters represents a period of rapid-fire lava eruptions cannot be discounted [e.g. *Prévot and Camps,* 1993; *Riisager et al.,* 2003]. However, a number of examples of rather dense *VGP* clustering (see e.g. Figures 1 and 2a) is associated with sedimentary material. Hence, it is likely that such features often correspond to quasi-invariant states of the transitioning field spanning, perhaps, considerable time.

4. A NEW RECORD OF THE BIG LOST EVENT FROM TAHITIAN LAVAS

4.1. The Paleomagnetic Results

We attempted to expand the Tahitian Matuyama-Brunhes dataset by sampling near a flow previously K-Ar dated by Le Roy [1994] at about 794 ka at the end of the *Chemin des*

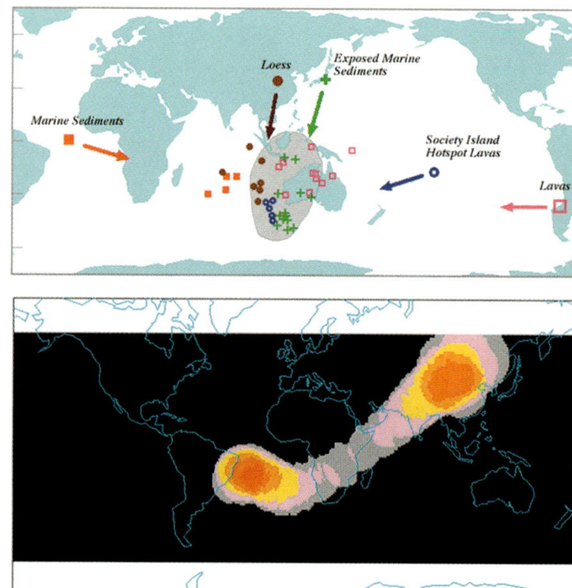

Plate 1. *VGP* systematics observed in records of the Matuyama-Brunhes. *(top)* Records containing a *VGP* cluster in the vicinity of Australia from the select database of Love and Mazaud [1997] associated with sites between 45°N and 45°S latitude [after *Hoffman*, 2000]. The corresponding sites and rocktypes are indicated. *(bottom)* Composite geographic histogram of *VGPs* (method as in *Hoffman* 2000) contained in three parallel records obtained from Hole 984 in the North Atlantic [*Channell and Lehman*, 1997] during back and forth polar movement between the South Atlantic and Central Asia.

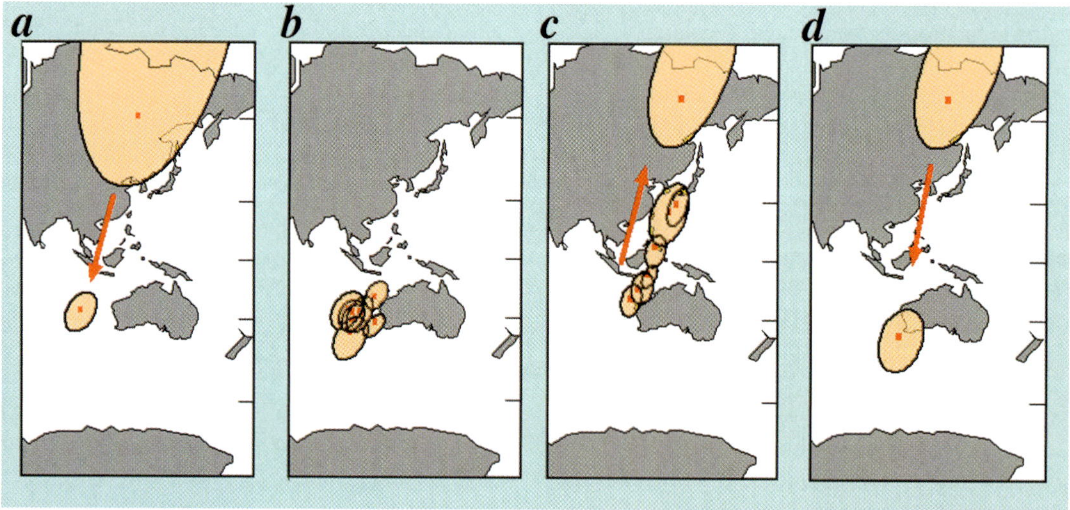

Plate 2. Path of the *VGP* (flow mean poles along with α_{95} ovals of confidence) as recorded in the lava sequence at the end of the *Chemin des Milles Sources*. The transitional field behavior begins in frame *a* with a Siberian *VGP* and ends in frame *d* with a west-Australian *VGP*. [Note: Although the recorded field behaviors of the lowermost and uppermost flows (frames *a* and *d*, respectively) appear to be quite similar, we found no evidence of faulting that could have caused flow replication within the gully section.]

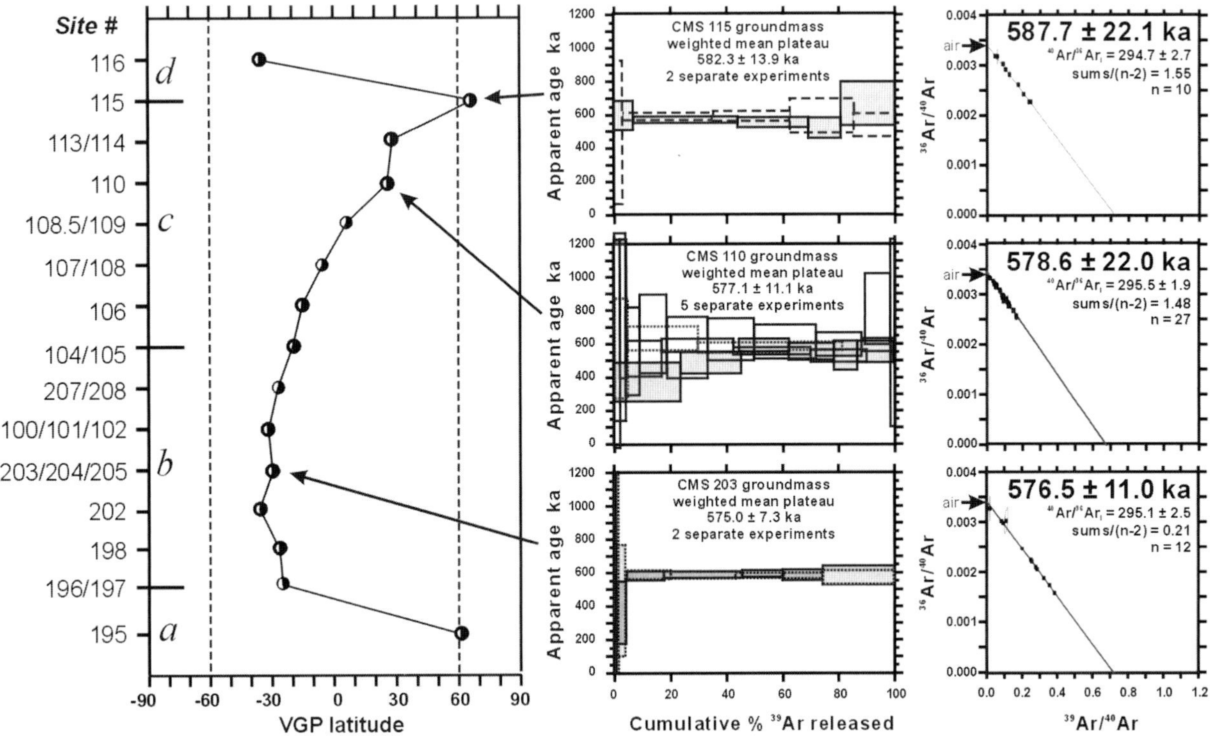

Figure 2. Apparent age spectra *(at center)* and inverse isochron diagrams *(at right)* for transitionally magnetized basaltic lava flow sites CMS-203, CMS-110, and CMS-115. The isochrons combine 12, 27, and 10 analyses, respectively, from the age plateaux, and give the preferred age of each lava flow. Virtual geomagnetic pole latitudes of lavas comprising the stratigraphic sequence are shown *(at left)*.

Milles Sources (CMS) in the northern part of the island some 14 km northeast from the Punaruu Valley exposure. Some 27 sites were sampled along the accessible part of this continuous spring-fed gully section. The number of cores drilled at a given site was dependent on the amount of exposure. Demagnetization was accomplished by both thermal and alternating field techniques. Principal component analysis was employed to identify the primary thermoremanent magnetization component. All but two of the sites rendered a reliable mean paleodirection.

It was sometimes difficult to determine in the field what constituted a distinct flow unit. However, in these cases the section contained sets of successive sites that possessed nearly identical paleodirections. These results were combined and the 25 successful site-mean analyses were reduced to 15 distinctive directional groups (Table 1). Plate 2 shows the path of the *VGP*, along with 95% confidence ovals, associated with the recorded paleodirections. As can be seen, the attainable record starts with a *VGP* at high northern latitudes in Siberia (frame *a*), transits across the equator to a location off the west coast of Australia, lingers there (frame *b*) for an unknown amount of time associated with the eruption of several lavas, returns along a longitudinally-confined path (frame *c*) to a

location near that at the onset, and then rebounds (frame *d*) to the southwest coast of Australia near the cluster of *VGPs* seen in frame *b*.

4.2. The Geochronologic Results

Given both the earlier-determined K-Ar date by Le Roy [1994] for a nearby flow at the end of the *CMS* as well as the fact that the grouping of transitional *VGPs* off the west coast of Australia is virtually identical to the findings for the Matuyama-Brunhes reversal recorded in the lavas exposed in the Punaruu Valley [*Chauvin et al.,* 1990], we were convinced that the *CMS* lavas also recorded the last reversal of Earth's magnetic field. However, nine ^{40}Ar/^{39}Ar incremental-heating experiments on groundmass separated from 3 lavas within the transitionally-magnetized flow sequence yielded isochron ages with 2σ uncertainties of 576.5±11.0 ka, 578.8±22.0 ka, and 587.7±22.1 ka, respectively (Figure 2).

The ^{40}Ar/^{39}Ar incremental heating experiments were carried out on ca. 100 mg aliquots of purified basaltic groundmass using a resistance furnace at the University of Wisconsin–Madison following analytical and data reduction procedures [*Singer et al.,* 2002]. The neutron fluence monitor was

Table 1. Paleodirectional data from *Chemin des Milles Sources* lavas, Tahiti-Nui

Site #	Group[a]	N[b]	INC	DEC	α_{95}	k	Frame[c]	VGP$_{Long}$	VGP$_{Lat}$
195/	1	3	-23.7	330.2	20.7	15	a	125.3	60.8
196/197	2	7	-18.7	238.9	5.8	83	a/b	98.6	-25.8
198/	3	7	-8.3	239.7	5.6	90	b	104.5	-27.2
202/	4	5	-7.8	229.4	4.3	216	b	99.9	-36.7
203/204/205	5	7	-12.4	235.8	3.5	222	b	100.6	-30.0
100/101/102	6	13	6.4	237.4	2.3	298	b	111.5	-32.0
207/208	7	8	-13.8	238.5	1.9	712	b	101.1	-27.2
104/105	8	7	-1.9	248.7	3.8	196	b/c	111.3	-20.0
106/	9	9	1.5	254.3	4.6	102	c	114.9	-15.2
107/108	10	13	4.9	263.6	2.0	385	c	119.4	-6.8
108.5/109	11	12	2.3	278.0	3.97	104	c	122.6	7.3
110/	12	4	2.1	297.3	12.5	31	c	129.0	25.5
113/114	13	5	5.6	301.6	4.3	210	c	132.6	28.9
115/	14	7	-18.9	335.7	10.5	25	c/d	134.5	65.1
116/	15	6	11.1	233.0	12.2		d	112.6	-36.9

[a] Directional group number
[b] Number of specimens
[c] Plot location in Plate 2

either 28.34 Ma Taylor Creek rhyolite sanidine or 1.194 Ma Alder Creek rhyolite sanidine that have been intercalibrated against one another [*Renne et al.*, 1998]. Because the resulting isochron ages are compared to ^{40}Ar/^{39}Ar ages from other lavas obtained using the same methods and standards [*Singer et al.*, 2002], the 2σ uncertainties reported with the ages include uncertainties in the analytical procedure, and where appropriate, the standard intercalibration. From samples CMS-203, −100, and −115, gas increments that gave concordant "plateau" ages comprise more than 98, 83, and 99% of the gas released. The age spectra imply that sample CMS-110 lost a small fraction of radiogenic argon, possibly due to petrographically undetectable post-eruption alteration. Notwithstanding, the isochron age of CMS-110 overlaps those of the over- and underlying lava flows.

The combination of concordant analyses from two to five experiments on each lava flow improved regression of the isochrons such that ages have uncertainties of 1.3 to 2.4%. It is standard to determine the best age from a group of determinations with variable analytical uncertainty by calculating the inverse-variance weighted mean age of the population. Thus, samples yielding the highest quantities of radiogenic argon and the most precise ages are proportionally more influential than those which give much more uncertain results. The weighted mean isochron age of the three lavas, 578.8±8.8 ka, gives our best estimate of time elapsed since the eruption of this flow sequence. This age is about 200 kyr younger than the widely accepted 780–790 ka age for the Matuyama-Brunhes reversal [*Singer and Pringle*, 1996; *Tauxe et al.*, 1996], but is identical to the 580.2±7.8 ka age [*Singer et al.*, 2002]

recently determined for the Big Lost Event [*Champion et al.*, 1988] using the same ^{40}Ar/^{39}Ar procedures on a lava flow sequence at La Palma, Canary Islands.

4.3. Discussion of Results

Given the ^{40}Ar/^{39}Ar age determinations, Australian-clustered *VGPs* are associated with each of five transitional field records obtained from Society Island hot spot lavas spanning 2.3 Myr from 2.9 Ma to about 0.6 Ma. If these correspondences are not simply fortuitous, they suggest the presence of similar structures of geomagnetic flux—at least as seen from the site of the Society Island hot spot—at times when the axial dipole field has been weak and/or when the geodynamo has been in a transitional state. Sources of such an enduring pattern of magnetic flux may reside either in close proximity to the hot spot volcanoes themselves or in the outer core. More specifically, the two most likely possible explanations are: *1)* the existence of an anomalous crustal field source which locally dominates during times when the core field is very weak, and *2)* long-standing control by the lowermost mantle of magnetic flux emanating from the outer core [e.g. *Bloxham and Gubbins*, 1987].

5. THE MODERN-DAY FIELD AT THE SOCIETY ISLAND HOT SPOT

If in fact a long-held pattern of flux emanating from the core-mantle boundary is responsible for the recurring transitional field behavior at the Society Island hot spot, one may attempt to find evidence of its presence in an analysis of the modern-day geomagnetic field. Since polarity reversal requires a change in sign of Earth's axial dipole, we explore the above possibility by initially removing the axial dipole term, g_1^0, from the *IGRF* reference field for the years 1900, 1950 and 2000 [data listed in *http://www.ngdc.noaa.gov/IAGA/wg8/table1.txt*], and proceed to determine the location of the respective *VGPs* at the Society Island hot spot if only this non-axial-dipole *(NAD)* field was present. Note that since the *IGRF* models have harmonic resolution to order and degree 10, they are not capable of resolving local crustal field sources. Hence, a finding in which modern-day *NAD*-field *VGPs* are compatible with the transitional paleomagnetic clusters from the hot spot site, would be supportive of far deeper, long-standing field sources that still exist.

Since the dynamo process is "blind" to the sign of magnetic flux [see e.g. *Merrill et al.,* 1996], it is equally valid to analyze either north or south *VGPs*—which are antipodal. Such a procedure has been referred to as the "geomagnetic convention" [see *Prévot and Camps,* 1993]. For the Society Island hot spot site the *north NAD*-field *VGPs* associated with the twentieth century geomagnetic field are found in the Caribbean. The *south NAD*-field *VGPs*, however, lie near the coast of western Australia in close proximity to the clustered *VGPs* from the, now, 5 available transitionally-magnetized lava records obtained from the Society Islands (Plate 3).

6. DISCUSSION

Although global databases of late-Cenozoic transitional paleodirections have not found the Australasian *VGP* feature to be clearly resolvable [*Prévot and Camps,* 1993; *Love,* 1998], the correspondences seen in Plate 3 suggest that the repeatable nature of the feature may become quite evident when regional data alone are considered. This apparent inconsistency is compatible with either or both of the contentions that *1)* the reversal process is, typically, complex [*Coe and Glen,* this volume] and *2)* transitional fields are often non-dipolar [e.g. *Prévot*

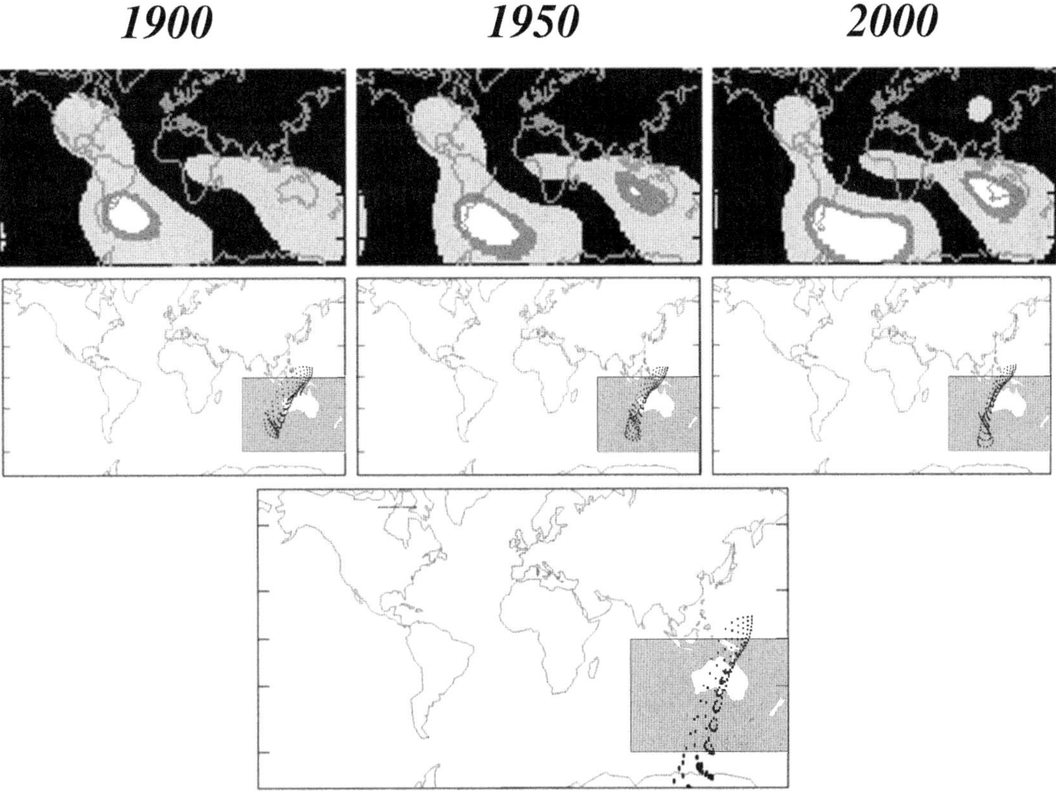

Figure 3. *(top)* Absolute value of the vertical component of the *NAD*-field for the 1900, 1950 and 2000 *IGRF*, each normalized to the strongest intensity present. Regions in white have 90–100% of the strongest value. Note: The two most intense regions have opposite sign. *(middle)* Geomagnetic convention *VGPs* associated with a grid of 247 sites within Australasia (shaded region), each corresponding to the *NAD*-field shown immediately above. *(bottom)* *VGPs* associated with the same Australasian sites corresponding to the *IGRF* 2000 *non-dipole* field.

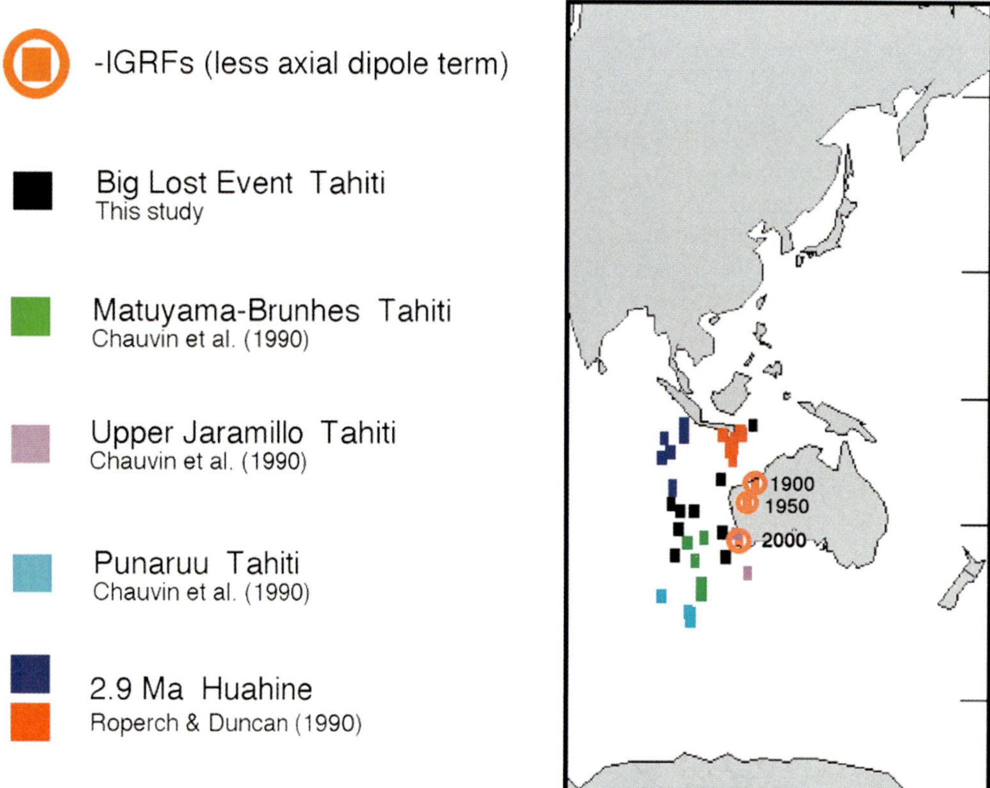

Plate 3. Composite plot of all Australasian *VGP* clusters from available Society Island hot spot lava records and the *south VGP* (geomagnetic convention) for the non-axial dipole *(NAD)* part of the *IGRF* for years 1900, 1950, and 2000, as indicated.

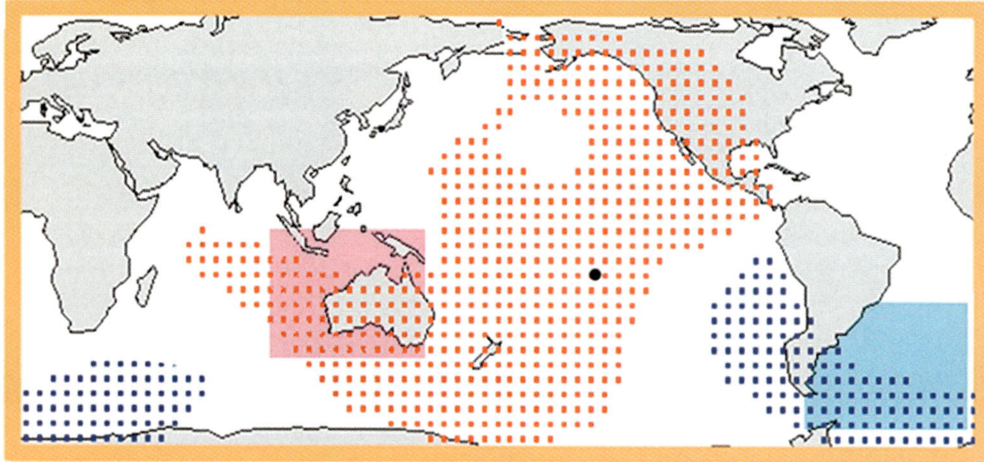

Plate 4. Those sites about the globe that, given the modern-day *NAD*-field (year 2000 *IGRF*), would be associated with magnetic directions having a *south VGP* in Australasia (region indicated in red) or a *north VGP* in the South Atlantic (region indicated in blue). The site of the Society Island hot spot is also shown.

and Camps, 1993; *Love*, 1998]. Such a situation could arise if midway in the process—at a time when the axial dipole field has nearly vanished—a small number of long-lived, localized flux concentrations at the core surface largely define the structure of the residual field. Recorded field behavior, then, would be strongly site-dependent: those sites nearest to the coordinates of a significant flux concentration would be most affected by it, the influence of the feature greatly diminishing with distance from it on Earth's surface. In this way, regional fields above mantle-held flux bundles may take on more "dipolar characteristics."

We now explore this idea: the three uppermost plots in Figure 3 show the absolute value of the radial component at Earth's surface of the 1900, 1950 and 2000 *IGRF NAD*-fields. Each plot is normalized to the strongest value present. The plots indicate two significant features, one in the South Atlantic, the other about Australasia. Although the chronology shows both vertical *NAD*-field features to be growing in strength, over the last century the intensity of the Australasia has been increasing in relative strength to the South Atlantic feature.

Figure 3 *(middle three plots)* shows the *VGPs* associated with the geomagnetic convention *NAD*-fields for the 1900, 1950 and 2000 *IGRF's*, respectively, for a 5° x 5° grid of 247 sites from within the indicated region in and surrounding Australasia (latitude range 60°S–0°; longitude range 90°E–180°E). As can be seen, regardless of the relative significance at Earth's surface of the Australian flux patch from 1900 to 2000, the *VGPs* for sites in this region are found in a rather confined north-south sweep through western Australia, displaying negligible westward drift.

Although the equatorial dipole has its pole at the equator in Australasia, the full explanation for the placement of the *NAD*-field *VGPs* appears to involve a more complex field source: Figure 3 *(bottom plot)* shows the *VGPs* for the same region of sites for the *IGRF* 2000 *non-dipole* field. As is seen, even with the removal of all dipole terms from the *NAD*-field, the *VGPs* still show a strong preference for the Australasian region. This positive correlation supports the contention that the principal field source causing this confinement of *VGPs* physically resides in the core beneath Australasia and has a significant influence on non-dipole as well as dipole spherical harmonic terms.

The observation of the low-to-mid latitude Austalasian *NAD*-field patch at Earth's surface is consistent with features of the core surface field modeled for the modern-day [*Jackson et al.*, 2000; *Johnson et al.*, 2003], historic [*Bloxham and Jackson*, 1992] and time-averaged paleomagnetic [*Kelly and, Gubbins*, 1997] fields (see e.g. Figure 4). Evidence for the existence of physically anomalous conditions beneath Australasia come from deep-Earth seismic investigations: Kendall and Shearer [1994] report that reflections by *S*-waves indi-

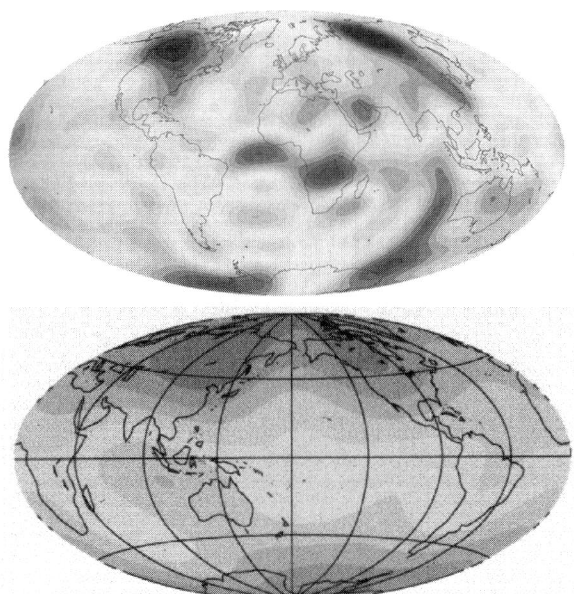

Figure 4. The radial component of the geomagnetic field at the surface of the core *(top)* for the year 1990 [from *Jackson et al.*, 2000], and *(bottom)* for the time-averaged paleomagnetic field over the last 5 Myr [from *Kelly and Gubbins*, 1997]. In both figures increased field intensity is displayed by an increased degree of darkness.

cate that the *D''*-layer beneath Australasia is highly variable in thickness. In addition, anomalously fast seismic wave velocities deep within the mantle beneath Australasia have been documented, first, through normal mode analysis [*Dziewonski and Woodhouse*, 1987], and, later, through shear wave tomography [e.g. *Li and Romanowicz*, 1996]. Thus, findings from *1)* transitionally magnetized lavas erupted at the Society Island hot spot since the Pliocene, *2)* analyses of the modern-day, historic geomagnetic and time-averaged paleomagnetic fields, and *3)* deep-mantle seismic tomographic studies, provide support for a model in which the observation of recurrent, regionally-controlled transitional field states are related to heterogeneity of the lowermost mantle and its effect on dynamo flux.

Hypothetically, in the limiting case in which only the flux concentration beneath Australasia is present during the reversal process, depending on sign, either *north VGPs* or *south VGPs* associated with sites over much of the globe would congregate in a region directly above it. As the significance of other flux concentrations may increase, however, the region that would be most influenced by the Australasian patch would clearly shrink. The Brunhes-aged event recorded in Mediterranean sediments (Figure 1: *upper row center*) may be an example whereby the South Atlantic flux feature first dominated the transitional field, only to be thoroughly overtaken by the Australasian feature.

Although groupings of *VGPs* located near western Australia are seen to recur from one reversal to the next at the site of the Society Island hot spot, transitional paths of the *VGP* are often observed to be far more complex [see *Coe and Glen*, this volume]. Hence, it is likely that the degree of dominance by mantle-controlled flux beneath mid-to-low latitude sites like Australasia, largely varies over the course of a complete reversal; perhaps, only on occasion during the dynamo process may sustained field behavior be recorded. Further, even though an extensive region above such a flux concentration may display field characteristics that appear to be dipole-dominated (see Figure 3), the global field may be far more complex (see Plate 1). In addition, recurrent clusters of *VGPs* have been found in successive Pliocene reversal records from Oahu, Hawaii [*Herrero-Bervera and Coe*, 1999]. Interestingly, these groupings lie within Africa, suggesting that other mantle-controlled localities exist at the core-surface that, on occasion, may regionally dominate transitional fields.

Finally, Plate 4 shows the pattern of sites (*shown in red*) that, for the year 2000 *NAD*-field, are associated with *south VGPs* within the indicated Australasian cluster patch. The extent of coverage seen clearly indicates the considerable influence this flux feature has over the global *NAD*-field. At the same time, the Southern Hemisphere high latitude sites (*shown in blue*) are associated with *north VGPs* within the indicated South Atlantic cluster patch. Given a continuing descent in strength of the modern-day axial dipole, these findings suggest that field changes associated with the onset of the *next* geomagnetic reversal may be more predictable than previously thought.

Acknowledgments. KAH wishes to acknowledge the several undergraduate physics majors—D. Soukup, M. Jock, A. Battle, B. Bose and A. Demogines—who assisted in the field and/or in the Cal Poly Paleomagnetism Laboratory. We also thank M. Relle for her ^{40}Ar/^{39}Ar age determinations conducted at the University of Wisconsin. This study was supported through National Science Foundation grants EAR-9418862, EAR-9627927, EAR-9805065, EAR-9909309, and EAR-014055.

REFERENCES

Bloxham, J., and D. Gubbins, Thermal core-mantle interactions, *Nature, 325*, 511–513, 1987.

Bloxham, J., and A. Jackson, Time-dependent mapping of the magnetic field at the core-mantle boundary, *J. Geophys. Res., 97*, 19537–19563, 1992.

Bogue S. W., and R. S. Coe, Successive paleomagnetic reversal records from Kauai, *Nature, 295*, 399–401, 1982.

Champion, D. E. and M. A. Lanphere, Evidence for a new geomagnetic reversal from lava flows in Idaho: discussion of short polarity reversals in the Brunhes and Late Matuyama polarity chrons, *J. Geophys. Res., 93*, 11,667–11,680, 1988.

Channell, J. E. T., and B. Lehman, The last two geomagnetic polarity reversals recorded in high-deposition-rate sediment drifts, *Nature, 389*, 712–715, 1997.

Chauvin, A., P. Roperch, and R. A. Duncan, Records of geomagnetic reversals from volcanic islands of French Polynesia, 2, paleomagnetic study of a flow sequence (1.2–0.6Ma) from the island of Tahiti and discussion of reversal models, *J. Geophys. Res., 95*, 2727–2752, 1990.

Clement, B. M., Geographical distribution of transitional VGPs: evidence for non-zonal equatorial symmetry during the Matuyama-Brunhes geomagnetic reversal, *Earth Planet. Sci. Lett., 104*, 48–58, 1991.

Clement, B. M., L. Sierra, E. Smith, and P. Rodda.. A record of the upper Cochiti polarity transition from the southern hemisphere Fiji). *Surv. Geophys., 17*: 189–196, 1996.

Coe, R. S., and J. M. G. Glen, The complexity of reversals, this volume.

Constable, C., Link between geomagnetic reversal paths and secular variation of the field over the last 5 Myr., *Nature, 358,* 230–233, 1992.

Dziewonski, A., and J. Woodhouse, Global images of the Earth's interior, *Science, 236,* 37–48, 1987.

Glen, J. M. G., R. S. Coe, and J. C. Liddicoat, A detailed record of paleomagnetic field change from Searles Lake, California: 2. The Gauss/Matuyama polarity reversal, *J. Geophys. Res., 104*, 12,883–12,894, 1999.

Herrero-Bervera, E., and R. S. Coe, Transitional field behavior during the Gilbert-Gauss and Lower Mammoth reversals recorded in lavas from the Waianae volcano, O'ahu, Hawaii, *J. Geophys. Res., 104*, 29,157–29,173, 1999.

Hoffman, K. A., Dipolar reversal states of the geomagnetic field and core-mantle dynamics, *Nature , 359,* 789–794, 1992.

Hoffman, K. A., Temporal aspects of the last reversal of Earth's magnetic field, *Phil. Trans. R. Soc. Lond. A, 358,* 1181–1190, 2000.

Jackson, A., A. R. T. Jonkers, and M. R. Walker, Four centuries of geomagnetic secular variation from historic records, *Phil. Trans. R. Soc. Lond. A, 358,* 957–990, 2000.

Jacobs, J. A., *Reversals of the Earth's Magnetic Field*, 2nd ed. (Cambridge University Press, New York), pp. 346, 1994.

Johnson, C. L., C. G. Constable, and L. Tauxe, Mapping Long-Term Changes in Earth's Magnetic Field, *Science, 300*, 2044–2045, 2003.

Kelly, P., and D. Gubbins, The geomagnetic field over the past 5 Myr, *Geophys. J. Int., 128*, 315–330, 1997.

Kendall, J-M., and P. M. Shearer, , Lateral variations in D" thickness from long-period shear wave data, *J. Geophy. Res., 99*, 11575–11590, 1994.

Laj, C., A. Mazaud, R. Weeks, M. Fuller, and E. Herrero-Bervera, Geomagnetic reversal paths, *Nature 351*, 447, 1991.

Leonhardt, R., J. Matzka, F. Hufenbecher, H. C. Soffel, and F. Heider, A reversal of the Earth's magnetic field recorded in mid-Miocene lava flows of Gran Canaria: Paleodirections, *J. Geophys. Res., 107(B1),* 10.1029/2001JB000322, 2002.

Le Roy, I., *Evolution des Volcans en Systeme de Point Chaud: Ile de Tahiti, Archipel de la Societe (Polynesie Francaise),* thesis, l'Universite Paris XI Orsay. 1994.

Li, X. D., and B. Romanowicz, Global mantle shear velocity model developed using nonlinear asymptotic coupling theory, *J. Geophys. Res., 101,* 22,245–22,273, 1996.

Love, J. J., Paleomagnetic volcanic data and geometric regularity of reversals and excursions, *J. Geophys. Res., 103,* 12,435–12,452, 1998.

Love, J. J. and A., Mazaud, A database for the Matuyama-Brunhes magnetic reversal, *Phys. Earth Planet. Int., 103,* 207–245, 1997.

Merrill, R. T., M. W.,McElhinny, and P. L McFadden, *The Magnetic Field of the Earth: Paleomagnetism, the Core and the Deep Mantle* (Academic Press, San Diego) pp. 531, 1996.

Merrill R.T., and P.L. McFadden, Geomagnetic polarity transitions. *Rev. Geophys., 37,* 201–226, 1999.

Prévot, M. and P. Camps, Absence of preferred longitudinal sectors for poles from volcanic records of geomagnetic reversals, *Nature, 366,* 53–57, 1993.

Renne, P. R., C. C. Swisher, A. L. Deino, D. B. Karner, T. Owens, and D. J. DePaolo, Intercalibration of standards, absolute ages and uncertainties in ^{40}Ar/^{39}Ar dating, *Chem. Geol., 145,* 117–152, 1998.

Riisager, J., P. Riisager, and A.K. Pedersen, The C27n–C26r geomagnetic polarity reversal recorded in the west Greenland flood basalt province: How complex is the transitional field?, *J. Geophys. Res. 108,* NO. B3, 2155, doi:10.1029/2002JB002124, 2003.

Roperch, P., and R. A. Duncan, Records of geomagnetic reversals from volcanic islands of French Polynesia 1. Paleomagnetic study of a polarity transition in a lava sequence from the island of Huahine. *J. Geophys. Res., 95,* 2713–27126, 1990.

Shaw, J., Strong geomagnetic fields during a single Icelandic polarity transition. *Geophys. J.R. Astron. Soc., 40,* 345–50, 1975.

Singer, B. S., K. A. Hoffman, A. Chauvin, R. S. Coe, and M. S. Pringle, Dating transitionally magnetized lavas of the Matuyama-Chron: Toward a new timescale of reversals and events, *J. Geophys. Res., 104,* 679–693, 1999.

Singer, B. S., and M. S. Pringle, The age and duration of the Matuyama-Brunhes geomagnetic polarity reversal from ^{40}Ar/^{39}Ar incremental heating analyses of lavas, *Earth Planet. Sci. Lett., 139,* 47–61, 1996.

Singer, B. S., M. K. Relle, K. A. Hoffman, A. Battle, C. Laj, H. Guillou, and J. C. Carracedo, Ar/Ar ages of transitionally magnetized lavas on La Palma, Canary Islands, and the Geomagnetic Instability Timescale, *J. Geophys. Res., 107 (B11),* 2307 doi:10.1029/2001JB001613, 2002.

Tauxe, L., T. Herbert, N. J. Shackleton, and Y. S Kok,, Astronomical calibration of the Matuyama-Brunhes boundary: consequences for magnetic remanence acquisition in marine carbonates and the Asian loess sequences, *Earth Planet. Sci. Lett., 140,* 133–146, 1996.

Tric, E., C. Laj, J. P. Valet, P. Tucholka, M. Paterne, and F. Guichard,. The Blake geomagnetic event: transition geometry, dynamical characteristics and geomagnetic significance, *Earth Planet. Sci. Lett., 102,* 1–13, 1991.

Valet, J.-P., and C. Laj, Invariant and changing transitional field configurations in a sequence of geomagnetic reversals, *Nature, 311 ,* 552–555, 1984.

van Hoof, A. A. M., van Os, B.J. H. and C. G. Langereis, The upper and lower Nunivak sedimentary geomagnetic transitional records from Southern Sicily. *Phys. Earth Planet. Inter., 77,* 297–313, 1993.

Kenneth A. Hoffman, Physics Department, Cal Poly State University, San Luis Obispo, California 93407.

Brad S. Singer, Department of Geology and Geophysics, University of Wisconsin-Madison, Madison, Wisconsin 53706.

Paleomagnetic Intensity Data as a Time Sequence: Opening a Window Into Dynamics of Earth's Fluid Core?

Ross Baker and Keith Aldridge

Graduate Program in Physics and Astronomy, York University, Toronto, Canada

Paleomagnetic intensity data from single (ODP 983) and composite (NAPIS-75) oceanic cores are analyzed for evidence of a rotating parametric instability (RPI) in Earth's fluid outer core. RPI, observed in laboratory experiments for over three decades, in Earth's core would be a resonant triad of two inertial modes coupled through straining of the core-mantle boundary. Small deformations in both amplitude, due to the semi-diurnal tide, and direction, from precession, produce growth of the instability on a geomagnetic time scale. Just as in laboratory observations of the RPI, alternating sequences of growth and decay continue at irregular intervals while straining is maintained and as long as the dissipation is sufficiently small. Presented here as e-folding times, decays show much less variance than growths while both are geophysically plausible. Combining growths and decays using linear stability theory yields e-folding times with reduced variance for the single and composite cores.

1. INTRODUCTION

Earth's magnetic field would disappear in approximately 10,000 years without the existence of some mechanism continuously providing energy to maintain the field against ohmic decay. Currently, it is believed that this mechanism is a convectionally driven dynamo in Earth's liquid outer core.

A significant property of the geomagnetic field is its variability over both long and short time scales, reversing over a few thousand years and having 21 reversals over the last 5.3 Ma. At the other extreme, geomagnetic jerks are seen on time scales of order years, suggesting that the geomagnetic field is of hydromagnetic origin. It is argued that gravitational instability of either thermal or compositional origin is responsible for buoyancy driven convection of the fluid core to maintain

the field. There can be, however, other types of fluid instability present in the core.

Rotational parametric instability (RPI) can exist in the core, a rapidly rotating fluid contained by the mantle, through tidal perturbation and precessional forcing. Elliptical strain from the tidally driven periodic forcing of the core-mantle boundary (CMB) and predominately shear strain from Earth's precession [*Kerswell*, 1993; *Aldridge*, 2003] produce fluid instabilities through coupling with the fluid core's otherwise stable rotational modes. Experimental work spanning more than three decades [*Aldridge*, 2003] has demonstrated the existence of RPIs in contained rotating fluids; accordingly, RPIs should exist within the liquid outer core at sufficiently small dissipation rates.

We use paleointensity as a proxy for the fluid motion in Earth's core. It has been demonstrated [*Moffatt*, 1978] that the spectra of turbulent velocity fields of conductive fluids with helicity and their resulting magnetic fields are proportional [*Davidson*, 2001] when the fluid is subject to low dissipation. For our investigations, this spectral equivalence between the core's velocity field and the geomagnetic field is

Timescales of the Paleomagnetic Field
Geophysical Monograph Series 145

10.1029/145GM18

assumed, as viscous and ohmic decays are considered to be sufficiently weak in the outer core.

Consequently, we have been seeking a signature of RPI activity within the paleomagnetic intensity records. This paper extends an earlier study [*Aldridge and Baker*, 2003] on the search for the paleomagnetic intensity signature of RPI in Earth's core. In that work, some 400 kyr of the relative paleomagnetic intensity time sequence recovered from sediments of the Ocean Drilling Program Site 983 (ODP 983) [*Channell et al.*, 1997] were analyzed. The search of the full ODP 983 record, and the relative paleointensity derived from the stacking [*Laj et al.*, 2000] of six North Atlantic sediment cores (NAPIS-75) is presented here.

In laboratory experiments, a characteristic of RPIs has been their exponential increase in amplitude followed by the RPI's collapse into small scale turbulent motion. While the rate at which its amplitude grows is predictable, the time between successive growths seems to be random [*Aldridge et al.*, 1997]. Additionally, the amount of time for which the amplitude in an RPI will continue to grow before collapse into turbulence is uncertain. This behaviour has been observed in experiments involving the generation of RPIs.

It is expected, therefore, that the evidence of an RPI within the paleomagnetic intensity data will show as an increase followed by the decay of the intensity and its inevitable collapse. This continual cycle of growth and decay of RPIs as seen in laboratory experiments is accordingly modeled in growth and decay of paleointensity.

2. CONDITIONS AFFECTING RPI GROWTH RATES

Tidal deformation and precession of Earth produce elliptical and shear strains in otherwise circular streamlines in the liquid outer core. Based on RPIs observed in the laboratory, we would expect these perturbations to determine the rates at which RPIs grow. Growth rates of RPIs have been shown [*Kerswell*, 1993] to be proportional to the product of the perturbation amplitude and to the rotation rate of the fluid.

For a fluid of kinematic viscosity v of dimension L, rotating at rate Ω, the initial growth rate is

$$\sigma = (\alpha\varepsilon - \beta E^{\frac{1}{2}})\Omega \tag{1}$$

where ε is the perturbation strain, $E = v/\Omega L^2$ is the Ekman number, and α and β, typically of order unity, are determined by the coupling of the boundary to the fluid core and the geometry of the perturbation.

It is important to note that the growth rate given in equation (1) is for the initial linear stage of the RPI and that the full development of the instability will go through a non-linear phase of growth not considered here. It is consistently observed

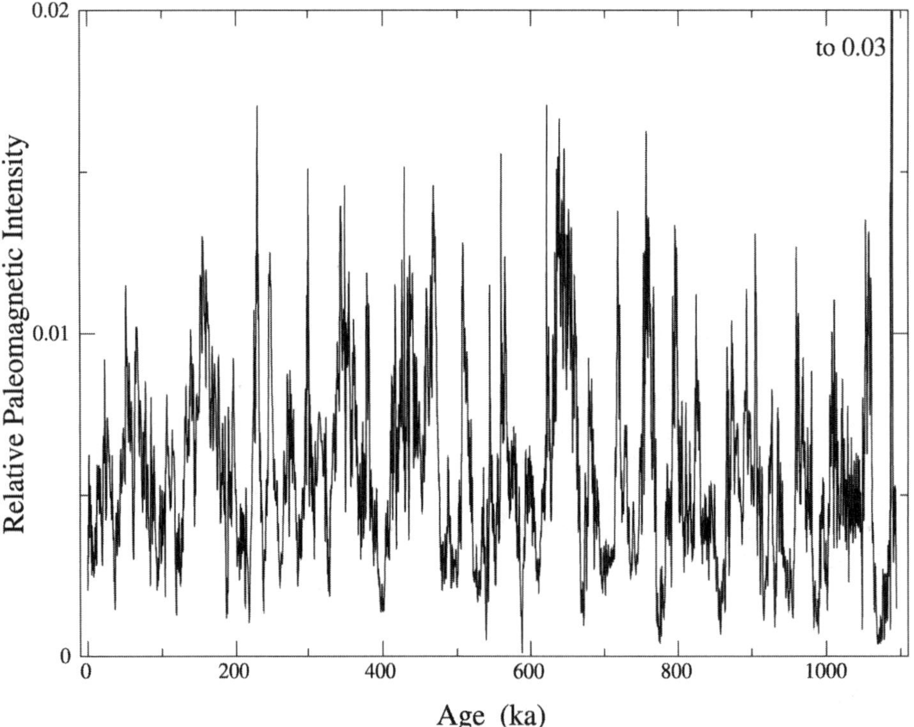

Figure 1. Relative paleointensity data from ODP 983. Note that by plotting against age, real time runs from right to left.

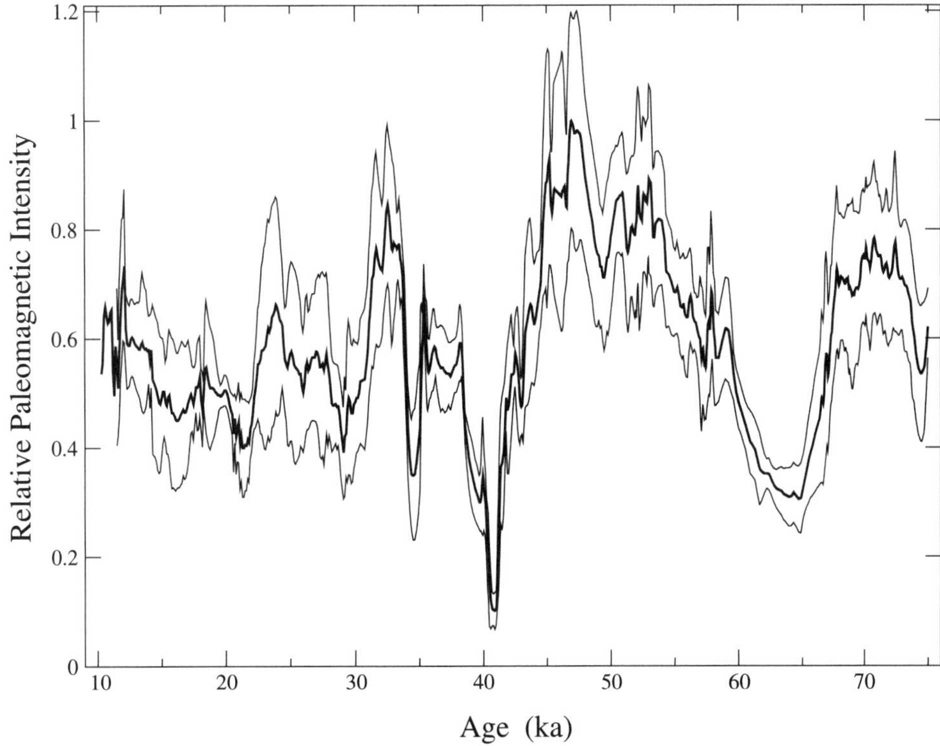

Figure 2. Relative paleointensity data from NAPIS-75, a stacked record of 6 North Atlantic cores including ODP 983. The Mono Lake and Laschamp geomagnetic excursions occur at 34 ka and 40 ka respectively in this record. The displayed error bounds were determined using bootstrap techniques.

in laboratory experiments [*Aldridge et al.*, 1997; *Aldridge*, 2003] that growth rates of RPIs are faster than predicted by the linear stability theory just described; however, analysis of the current observations on paleomagnetic intensity follows from linear stability theory.

3. RELATIVE PALEOMAGNETIC INTENSITY RECORDS

The relative paleomagnetic intensity data used in this study originate from ocean sediment cores. The ODP 983 record of 13,000 points shown in Figure 1 [*Channell et al.*, 1997] was derived from a single core located on the Gardar Drift south of Iceland. This 1100 kyr record contains the last geomagnetic pole reversal, the Brunhes-Matuyama, at about 780 ka. Mean sedimentation rate at the ODP 983 was 11.3 cm/kyr.

Measurement of normal remanent magnetization (NRM) of the sediment was normalized with anhysteretic remanent magnetization (ARM) to account for variation in the sediment's magnetic grain content. The NRM/ARM relative paleomagnetic intensity proxy was dated using nanofossil planktic and benthic oxygen isotope data. The $\delta^{18}O$ signal was tied to a combined SU90-08 [*Vidal et al.*, 1997], (0–130 ka) and SPECMAP reference curve [*Martinson et al.*, 1987]

with assumed linear sedimentation rates between age tie points, resulting in a non-equispaced sequence of relative paleomagnetic intensities.

The NAPIS-75 data stack of 3275 points displayed in Figure 2 was compiled from intensity data of six sediment cores including ODP 983. The records used in the stack were placed on the GISP2 age scale [*Grootes and Struiver*, 1997]. A discussion of the NAPIS-75 relative paleointensity data stack [*Laj et al.*, 2000] describes the individual records as well as their stacking methodology which included selective rejection of outliers across the records, and interpolation to generate an equispaced time series of intensities. Error bounds at the level of 2 standard deviations for the stacked data shown in Figure 2, were estimated using bootstrap techniques [*Tauxe et al.*, 1991].

Uncertainties in paleointensity records that arise from imperfect timing of the sediment's depth and recovery of remanences are discussed at greater length below.

4. OBSERVED GROWTH AND DECAY FROM PALEOMAGNETIC INTENSITY RECORDS

Segments in the paleomagnetic intensity records were identified as regions of net growth or decay of intensity. Criteria

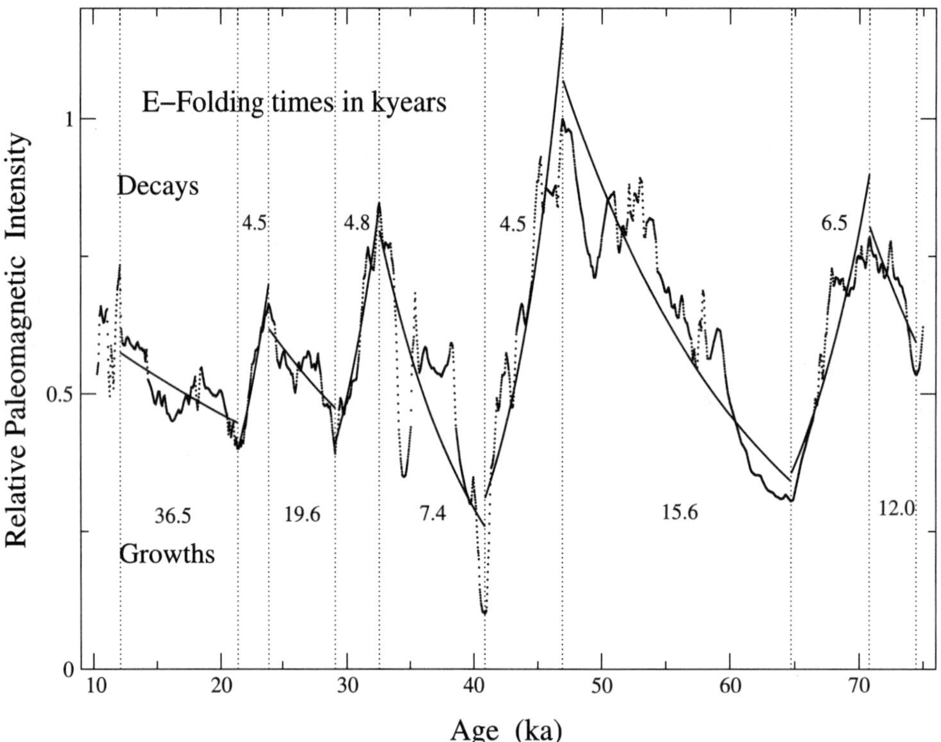

Figure 3. Demonstration of exponential fits to segments for NAPIS-75. The paleointensity data (y_i, t_i) is shown by the small circles; calculated paleointensities $\widehat{y}_i$ from equation (3) are the thin smooth lines. Recovered growth (decay) e-folding times are shown below (above) the curves.

of the segmenting algorithm are discussed below in relation to NAPIS-75 segments. These segments are modeled as exponentials as would be expected from the development of an RPI:

$$y_i = A \exp \sigma t_i + \varepsilon_i \quad i = 1, 2, ... N \quad (2)$$

where ε_i is the error between the model and the data. Positive (negative) values of σ correspond to decay (growth) as real time corresponds to decreasing age. y_i is the paleomagnetic amplitude at age t_i for the N points in a particular data segment.

Estimates of the parameters A, σ are determined from regression so that the calculated paleointensities are

$$\widehat{y}_i = \widehat{A} \exp \widehat{\sigma_{g,d}} t_i \quad i = 1, 2, ... N. \quad (3)$$

Recovered from the least squares fits are observed growth $\widehat{\sigma}_g < 0$ or decay $\widehat{\sigma}_d > 0$ with corresponding e-folding times, $\tau_g = 1/|\widehat{\sigma}_g|$ and $\tau_d = 1/\widehat{\sigma}_d$ which are reported below for ease of comparison.

Figure 3 shows the $\widehat{y}_i$ fits to the NAPIS-75 data segments. Vertical dotted lines divide the time series into individual segments. Growth (decay) e-folding times are listed below (above) each segment. These data subsets and their regression curves

show patterns in both the segments and the choice of segment age bounds. Segments exhibit an overall increase or decrease in intensity over their duration. For example, the Mono Lake excursion at 34 ka is rejected as a decay–growth pair as it is a brief digression from the overall 32.5–41 ka growth.

Decays have similar durations at 3–6 kyr compared to growths which have greater range up to 16 kyr. Decays are regular and tend to fall off smoothly at similar rates as seen in the scatter shown below. In comparison, growths tend to exhibit complicated oscillations during increase of intensity with greater variability in growth rate.

The fits to the NAPIS-75 segments also illustrate that there is clearly more than just simple exponential growth and decay occurring as specified by our model (2). For example, the growth from 47–64 ka initially exhibits exponential growth (up to 59 ka). Beyond this initial development of RPI, non-linear effects may influence the core's response to straining as well as other processes not modeled here.

The e-folding times of the recovered rates for the NAPIS-75 segments are displayed in Figure 4. The mean e-folding times are $\overline{\tau}_g = 13.8$ kyr for growth, and $\overline{\tau}_d = 5$ kyr for decay.

E-folding times are plotted for the 40 growth and 41 decay segments of ODP 983 in Figure 5. The open circles (squares)

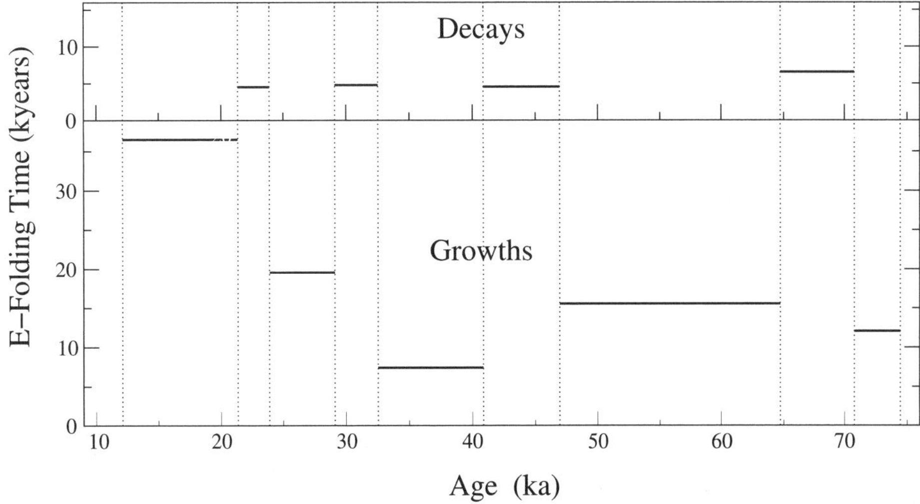

Figure 4. Summary of recovered growth and decay e-folding times for NAPIS-75. E-folding times are drawn across the duration of individual segments.

indicating the e-folding time of decays (growths) are plotted at each segment's younger age and connected by straight lines. Durations of segments for ODP 983 range from a few to several 10's of kyr with averages 15 kyr for growth and 11 kyr for decay. The mean e-folding times are $\overline{\tau}_g$ = 9.2 kyr for growth: and $\overline{\tau}_d$ = 6.2 kyr for decay. Generally, the scatter of recovered e-folding times is a few 10's of kyr, but ranging from a few to more than 100 kyr.

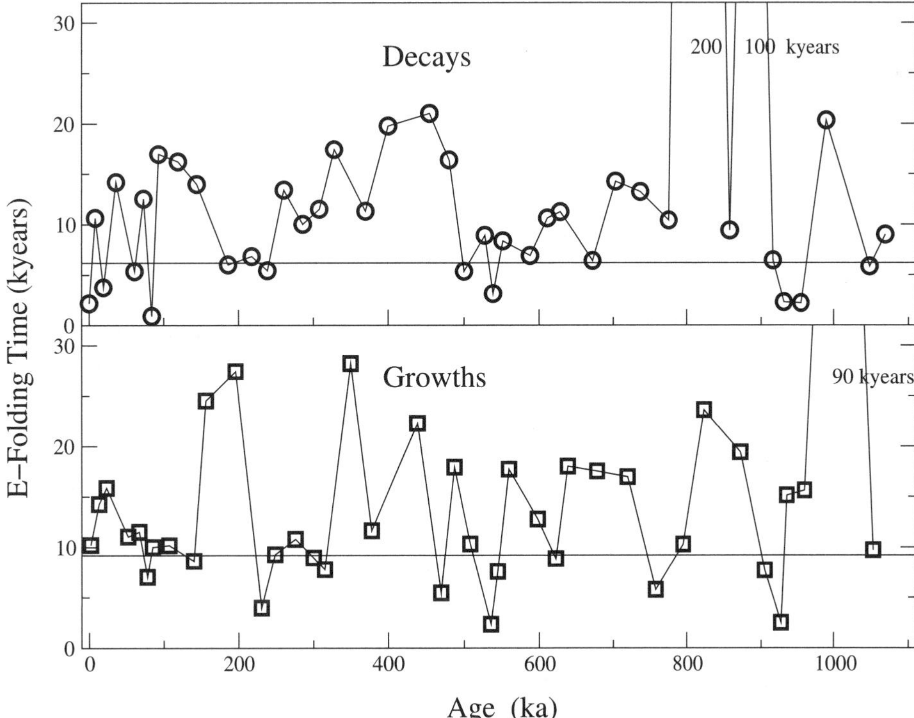

Figure 5. Summary of recovered growth and decay e-folding times for ODP 983. The line connects the plotted circles (squares) for decay (growth) of individual segments. Horizontal lines show the e-folding times for the averages of the growth and decay rates.

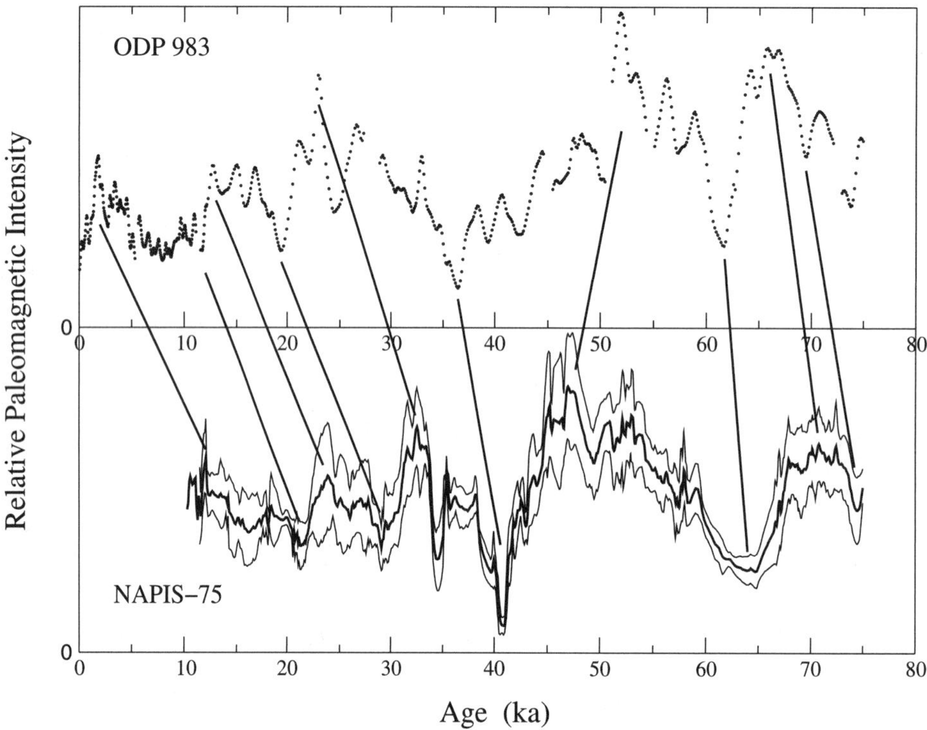

Figure 6. Comparison of relative durations and locations of corresponding segments of ODP 983 (top) and NAPIS-75 (bottom). The slanted lines connect corresponding age boundaries.

5. SOURCES OF ERROR IN INVERSIONS

The large scatter recovered from the single record ODP 983 is not unexpected. It would arise from irregular skewing in dating the ocean sediments, and error in measurement of magnetic remanences used in estimating the relative paleomagnetic intensity proxy. Local fluctuations of the geomagnetic field will tend to be smoothed out in a stack where the geomagnetic field is being sampled regionally (North Atlantic). Additionally, stacking will tend to suppress errors from individual records due to any acquisition of anomalous remanences.

Errors in recovered e-folding times will arise from uncertainty in the intensity itself as well as errors in timing due to age offsets. The recovered intensities from sediments will effectively be low–pass filtered during remanence measurements at sub-millennial scales (estimated at 4–5 cm separation or about 300 years for the ODP 983 record), and from overprinting of remanences during the sediment accumulation.

Error in e-folding time due to uncertain intensity can be quantified from the error bound estimates shown in Figure 2 of the NAPIS-75 stack of six independent records. The relative error in e-folding times can be as much as 44% for the decay at 22 ka or as small as 5% for the growth at 37 ka. The mean e-folding time error due to intensity is 17% for the 9 segments.

Uncertainty in age determination of the sediment's depth will also generate e-folding time error. This timing problem has been considered using synthetic data from geodynamo models [*McMillan et al.*, 2002] to investigate the errors from tie point displacement and non-linear sedimentation rates which can severely degrade the spectral content in inferred paleointensity. A model using mean rates of sedimentation up to 15 cm/kyr [*Guyodo and Channell*, 2002] found that stacked records should be favoured for high frequency resolution (4 kyr), and sparse sedimentation rates (1 cm/kyr) could result in up to a 50 kyr smoothing of extracted intensities.

Figure 6 shows the effect by mapping sediment core records onto different age models. The slanted lines in Figure 6 connecting the age boundaries of corresponding segments of ODP 983 and NAPIS-75 highlight the effect of age offsets. Where these lines are parallel the segment is displaced in time without change of duration and there will be no difference in the recovered e-folding time. The largest change in duration is for the segment containing 45 ka where the e-folding time of the NAPIS-75 segment would be a factor of 2.5 larger. The average relative difference from comparing timing of age models is 14%.

Also apparent in the comparison of the data sets in Figure 6 is skewing of data within segments resulting from assumed linear sedimentation rates between tie points. In the

Table 1. Expected pure growths derived from linear stability theory [*Kerswell*, 1993]

Type of perturbation	Deformation amplitude ε	Pure growth rate $\alpha\varepsilon\Omega$ $(\sec^{-1})$	Pure E-folding time $(\alpha\varepsilon\Omega)^{-1}$ (kyr)
Tidal (elliptical strain)	8.5×10^{-8}	3.1×10^{-12}	10
Precessional (shear strain)	4.0×10^{-8}	1.7×10^{-12}	18

NAPIS-75 record from 65–75 ka, the flat or stable section around 70 ka appears to be relatively compressed in ODP 983. While this random back and forth displacement of the data may lead to large errors in e-folding times for individual segments, the overall expectation is a null mean effect across a whole record; the change from expansion in one segment would be balanced out by a compression in a nearby segment as can be seen for NAPIS-75 decays at 21–24 and 29–32.5 ka in Figure 6.

In summary, the largest error in observed e-folding times may be as severe as a factor of 3, while there may be no error at all in particular segments. Thus corresponding fluctuations in our modeled growths and decays almost certainly mask any trends in these results. Improvements in recovered paleointensities and their age should eventually reveal Earth orbital elements and evolving internal structure that determine these rates as anticipated by the RPI model.

6. APPLICATIONS TO EARTH'S CORE

RPIs, whether due to elliptical or shear strain, will develop in Earth's fluid core as long as the dissipation does not overcome the perturbation driving the growth. At some finite intensity, RPI in the core can no longer build. The coupled inertial modes which underly the RPI development are modeled to decay through Ekman boundary layer dissipation. The second term of equation (1) represents this dissipation and is recovered in decays of paleointensity;

$$\widehat{\sigma}_{d} = \beta E^{\frac{1}{2}}\Omega. \qquad (4)$$

From the fits for decay (Figure 4, and Figure 5) with means of 5.0 and 6.2 kyr we can estimate the Ekman number [*Kerswell*, 1993],

$$E \sim 10^{-15}$$

which is in accordance with lower estimates of kinematic viscosity of the fluid core [*Lumb and Aldridge*, 1991].

Observed growths $\widehat{\sigma}_{g}$ and decays $\widehat{\sigma}_{d}$ may also be used with equation (1) to recover pure growth rates $\alpha\varepsilon\Omega$, which are cal-

culated for Earth's core in Table 1. Combining consecutive pairs of recovered rates, decay–growth and growth–decay about their common central age provide an estimate of the expected pure growth

$$\widetilde{\alpha\varepsilon\Omega} = \widehat{\sigma}_{d} - \widehat{\sigma}_{g}, \qquad (5)$$

which is a sum since the recovered values, $\widehat{\sigma}_{d}$ and $\widehat{\sigma}_{g}$ are of opposite sign in our model (3). The e-folding estimates $\widetilde{\alpha\varepsilon\Omega}$ are displayed in Figure 7 for the 80 pairs of ODP 983 and the 8 pairs of NAPIS-75. Mean values of estimated e-folding times for pure growth, 3.8 and 3.6 kyr for ODP 983 and NAPIS-75 respectively are shorter than calculated values of 10 and 18 kyr (Table 1) for elliptical and shear RPIs respectively. As described earlier, this is not an unexpected result as growth rates in laboratory experiments [*Aldridge et al.*, 1997] are larger than predictions from linear stability theory.

Recovered pure growths have limited range in comparison with the recovered growths and decays of ODP 983 that included values ranging from a few to more than 100 kyr. Scatter in the estimates, $\widetilde{\alpha\varepsilon\Omega}$ are minimal at 20 kyr.

Another treatment of the results is to predict what the observed growth rate should be from combining the calculated pure growths $\alpha\varepsilon\Omega$ shown in Table 1 and the mean of observed decays $\widehat{\sigma}_{d}$. Following equation (1) the expected e-folding times for observed growth for ODP 983 are 3.8 and 4.6 kyr for elliptical and shear RPIs respectively while the mean e-folding time recovered for growth is 9.2 kyr; for NAPIS-75 corresponding expected e-folding times are 3.3 and 3.9 kyr compared with 13.8 kyr.

7. DISCUSSION AND CONCLUSIONS

A search of relative paleomagnetic intensity as a proxy for fluid flow in the core has been applied to the 1100 kyr ODP 983 record and 75 kyr NAPIS-75 stack. Recovery of growth and decay using an exponential regression model is consistent with the existence of rotational parametric instabilities being present in Earth's fluid core.

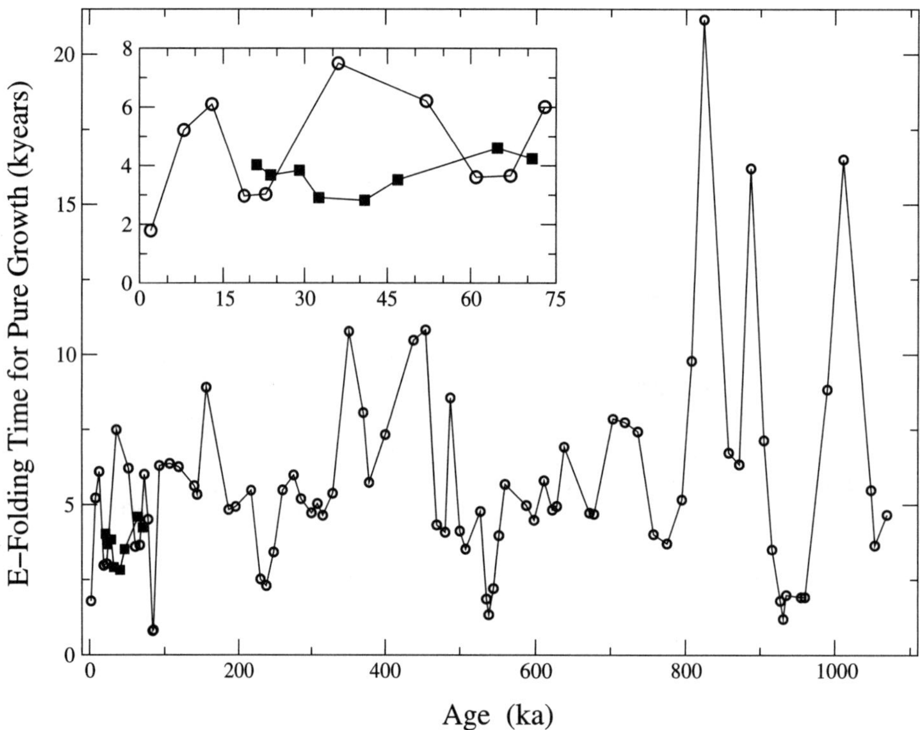

Figure 7. Pure e-folding times $\widetilde{\alpha\varepsilon\Omega}$ from exponential fits. The two sets of 80 (8) points are shown for ODP 983 (NAPIS-75) plotted as circles (squares) and connected by line segments.

Scatter of several 10's of kyr in the observed e-folding times of the ODP 983 paleomagnetic intensity record were discussed above. Uncertainties are not as prevalent in the NAPIS-75 data indicated by the reduction to a few kyr of scatter in the results for decay.

Laboratory experiments with contained rotating fluids have clearly demonstrated the existence of rotational parametric instabilities due to shear and elliptical strains. Earlier experimental observations [*Malkus, 1968*] of fluid flows generated in a precessing spheroidal contained fluid were not initially interpreted as a parametric instability [*Aldridge, 2003*]. Periodic perturbation, though small in amplitude, will generate large flows in a rotating contained fluid over time. Neglect of this input of energy [*Loper*, 1975; *Rochester et al.*, 1975] led to rejecting precession of Earth in maintaining the geodynamo [*Malkus*, 1968].

The power available from tidal origin alone is about 4 TW, most of which is thought to be dissipated in shallow seas and about 0.1 TW in the solid Earth. Since the rate of energy dissipation of the geodynamo is estimated at 0.2 TW, a few percent of the tidal dissipation rate, it appears that an RPI remains a candidate to maintain the geomagnetic field.

Evidence of Milankovitch orbital modulations of eccentricity (100 kyr) and obliquity (41 kyr) have been identified in paleomagnetic ocean sediment records. It is uncertain whether

orbital variations in paleomagnetism are effectively lithological variation in bulk magnetic properties of sediments due to climate forcing or if the geomagnetic variations are driven by the external torques acting on the fluid core of Earth. For example, variation in paleomagnetic intensity corresponding to Earth's obliquity have been initially reported for the Brunhes epoch from a mid-latitude core RC10-167 [*Kent and Opdyke*, 1977]. A reassessment of this record with improved age control [*Kent and Carlut*, 2001] reported no evidence for obliquity control on paleointensity or a correlation to pole reversals.

Spectral–coherency analysis of the ODP 983 intensity record shows modulation of intensity corresponding to orbital obliquity in the Brunhes epoch [*Channell et al.*, 1998] and variations of both eccentricity and obliquity in the Jaramillo Subchron (700–1100 ka) [*Channell and Kleiven*, 2000]. Conversely, a wavelet analysis to demarcate subsections of the record which contain orbital power [*Guyodo et al.*, 2000] concluded that orbital frequencies embedded in the paleointensity record are an expression of lithological variations. In another 2.2 Ma sediment record from the West Caroline Basin [*Yamazaki and Oda*, 2002], obliquity periodicity is reported in the inclination error with a phase shift at the Brunhes–Matuyama boundary.

We consider here what would be expected from our RPI model in paleomagnetic records as a result of orbital modulations. Such modulations would produce periodic components in the strain ε, given in equation (1), of the RPI growth rate. Thus there should be no expectation of orbital modulation in the intensity, but these would show as periodicities in the rate of growth of intensity over geological age. Additionally, pure growth on geological time spans could become stronger or weaker dependent on core parameters. As the inner core slowly grows a shift in eigenfrequencies may result in a jump to new resonant coupling of modes responsible for RPI. Thus, with pure growth being modulated, there exists a possibility of a balance between growth and decay, resulting in development of an RPI to a stable amplitude which will persist, as observed in laboratory experiments [*Eloy et al.*, 2000]. Closer examination of the paleointensity data itself reveals the possibility of these stable states. Prominent examples are apparent in the ODP 983 record of Figure 1 at 1010–1040 ka and near 70 ka in the NAPIS-75 data shown in Figure 2.

It is concluded from the evidence presented here for growth and decay of paleomagnetic intensity that the role of rotational parametric instabilities needs to be addressed in understanding the geodynamo.

Acknowledgments. We thank Yohan Guyodo and an anonymous reviewer for their valuable comments and suggestions that have improved the presentation of this work. Support for our research came from a grant awarded by the Natural Sciences and Engineering Research Council of Canada (KA) and an Ontario Graduate Scholarship (RB).

REFERENCES

Aldridge, K. D., Dynamics of the Core at Short Periods: Theory, Experiments and Observations in Earth's Core and Lower Mantle, in *The Fluid Mechanics of Astrophysics and Geophysics*, edited by Christopher Jones, Andrew Soward, and Keke Zhang, Taylor and Francis, London, U. K., 2003.

Aldridge, K. D. and R. E. Baker, Paleomagnetic intensity data: a window on the dynamics of Earth's fluid core?, *Phys. Earth Planet. Inter., 140*, 91–100, 2003.

Aldridge, K. D., B. Seyed-Mahmoud, G. A. Henderson, and W. van Wijngaarden, Elliptical instability of the Earth's fluid core, *Phys. Earth Planet. Inter., 103*, 365–374, 1997.

Channell, J. E. T., and H. F. Kleiven, Geomagnetic paleointensities and astrochronological ages for the Matuyama–Brunhes boundary and the boundaries of the Jaramillo Subchron: paleomagnetic and oxygen isotope records from ODP Site 983, *Phil. Trans. R Soc. Lond. A., 358*, 1027–1047, 2000.

Channell, J. E. T., D. A. Hodell, and B. Lehman, Relative geomagnetic paleointensity and at ODP Site 983 (Garder Drift, North Atlantic) since 350ka, *Earth Planet. Sci. Lett., 153*, 103–118, 1997.

Channell, J. E. T., D. A. Hodell, J. McManus, and B. Lehman, Orbital modulation of the Earth's magnetic field intensity, *Nature, 394*, 464–468, 1998.

Davidson, P. A., *An Introduction to Magnetohydrodynamics*, Cambridge Texts in Applied Mathematics, pp. 183–184, Cambridge Univ. Press, Cambridge, U. K., 2001.

Eloy, C., P. Le Gal, and S. Le Dizès, Experimental study of the multipolar vortex instability, *Phys. Rev. Lett., 85*, 3400–3403, 2000.

Grootes, P. M., and M. Stuiver, $^{18}O/^{16}O$ in Greenland snow and ice with 10^{-5} to 10^5 year time resolution, *J. Geophys. Res., 102*, 26,455–26,470, 1997.

Guyodo, Y., and J. E. T. Channell, Effects of variable sedimentation rates and age errors on the resolution of sedimentary paleointensity records, *Geochem. Geophys. Geosyst., 3*(8), 1048, doi:10.1029/2001GC000211, 2002.

Guyodo, Y., P. Gaillot, and J. E. T. Channell, Wavelet analysis of relative geomagnetic paleointensity at ODP Site 983, *Earth Planet. Sci. Lett., 184*, 109–123, 2000.

Kent, D. V., and J. Carlut, A negative test of orbital control of geomagnetic reversals and excursions, *Geophys. Res. Lett., 28*, 3561–3564, 2001.

Kent, D. V., and N. Opdyke, Paleomagnetic field intensity variations recorded in a Brunhes epoch deep-sea sediment core, *Nature, 266*, 156–159, 1977.

Kerswell, R. R., The instability of precessing flow, *Geophys. Astrophys. Fluid Dynam., 72*, 107–114, 1993.

Laj, C., C. Kissel, A. Mazaud, J. E. T. Channell, and J. Beer, North Atlantic paleointensity stack since 75 (NAPIS-75) and the duration of the Laschamp event, *Phil. Trans. R. Soc. Lond. A, 358*, 1009–1025, 2000.

Loper, D., Torque balance and energy budget for the precessionally driven dynamo, *Phys. Earth Planet. Inter., 11*, 43–60, 1975.

Lumb, L. I., and K. D. Aldridge, On viscosity estimates for the Earth's fluid core, *J. Geomag. Geoelectr., 43*, 93–110, 1991.

Malkus, W. V. R., Precession of the Earth as the Cause of Geomagnetism, *Science, 160*, 259–264, 1968.

Martinson, D. G., N. G. Pisias, J. D. Hays, J. Imbrie, T. C. Moore Jr., and N. J. Shackleton, Age dating and the orbital theory of Ice Ages: development of a high-resolution 0 to 300,000-year chronostratigraphy, *Quat. Res., 27*, 1–29, 1987.

McMillan, D., C. G. Constable, and R. L. Parker, Limitations on stratigraphic analyses due to incomplete age control and their relevance to sedimentary paleomagnetism, *Earth Planet. Sci. Lett., 201*, 509–523, 2002.

Moffatt, H. K., *Magnetic Field Generation in Electrically Conducting Fluids*, Cambridge Monographs on Mechanics and Applied Mathematics, p. 290, Cambridge Univ. Press, Cambridge, U. K., 1978.

Rochester, M. G., J. A. Jacobs, D. E. Smylie, and K. F. Chong, Can precession power the geomagnetic dynamo? *Geophys. J. R. Astron., 43*, 661–678, 1975.

Tauxe, L., N. Kylstra, and C. Constable, Bootstrap statistics for paleomagnetic data, *J. Geophys. Res., 96*, 11,732–11,740, 1991.

Vidal, L., L. Labeyrie, E. Cortijo, M. Arnold, C.J. Duplessy, E. Michel,

S. Becqué, and T. C. E. Van Weering, Evidence for changes in the North Atlantic deep water linked to meltwater surges during the Heinrich events, *Earth Planet. Sci. Lett., 146*, 13–27, 1997.

Yamazaki, T., and H. Oda, Orbital influence on Earth's magnetic field: 100,000 year periodicity in inclination, *Science, 295*, 2435–2438, 2002.

Keith Aldridge, Department of Earth and Atmospheric Science, York University, 4700 Keele St., Toronto, Ontario, Canada, M3J 1P3. (keith@yorku.ca)

Ross Baker, Box 111, Petrie Science and Engineering Building, York University, 4700 Keele St., Toronto, Ontario, Canada, M3J 1P3. (rossb@yorku.ca)

High Resolution Global Paleointensity Stack Since 75 kyr (GLOPIS-75) Calibrated to Absolute Values

Carlo Laj and Catherine Kissel

Laboratoire des Sciences du Climat et de l'Environnement, (CEA-CNRS), Gif-sur-Yvette, France

Juerg Beer

Department of Surface Waters, EAWAG, Duebendorf, Switzerland

We have obtained a global relative paleointensity stack using a selection of published records from marine cores with sedimentation rates in excess of 7 cm/kyr, from the North and South Atlantic Oceans, the Mediterranean and the Indian Ocean. After correlation of the cores, the results were stacked using a new approach to minimize possible local disturbances. This relative paleointensity stack extends from 75 kyr to about 10–12 kyr BP. This new stack presents a larger overlap than previously published records with the archeomagnetic/volcanic absolute paleointensity determinations that are all well constrained in age for the period 0–12 kyr. This allows calibration to absolute values, but additional work is still necessary on this point. Superimposed to the long-term features, rapid fluctuations are identified indicating that GLOPIS-75 has attained high global resolution. The duration of the Mono Lake and Laschamp excursions can be estimated to about 1500 years consistently with the suggestion that excursions are shorter than reversals.

1. INTRODUCTION

Records of geomagnetic field intensity in the past have proven very useful in discussing a variety of scientific issues, such as long-range correlation of sedimentary marine sequences [see e.g. *Stoner et al.*, 2002] or production of cosmogenic isotopes. More fundamentally, these records, which are the only archive for extending present-day geomagnetic measurements into the past, allow to explore the long-time behavior of the Earth's geomagnetic vector field and therefore the mechanisms of the geodynamo. As a very recent example, it has been suggested that records of geomagnetic paleointensity can be considered as time series related to fluctuations of the fluid velocity in the Earth's core: growth of the

Timescales of the Paleomagnetic Field
Geophysical Monograph Series 145
Copyright 2004 by the American Geophysical Union
10.1029/145GM19

inner core would alter the geometry determining instability properties that should be detectable in the observed growth and decay of paleointensity [*Baker and Aldridge*, this volume]. Records of geomagnetic field intensity better constrained in amplitude and with higher accuracy and resolution in time than the early records are therefore needed.

Records of relative field intensity from sediments have been rather scarce in the past in part because of the time-consuming aspect of the measurements on single samples. In recent years, however, the development of commercially available small-access, pass-through cryogenic magnetometer with high-resolution pick-up coils associated with u-channel samples [*Weeks et al*, 1993] has resulted in an increasing number of precise and reliable records. The large similarities present in these records from widely different regions worldwide have then progressively documented important characteristics of the geomagnetic field. For instance, the SINT-200 [*Guyodo and Valet*, 1996] and more recently the SINT-800 [*Guyodo and*

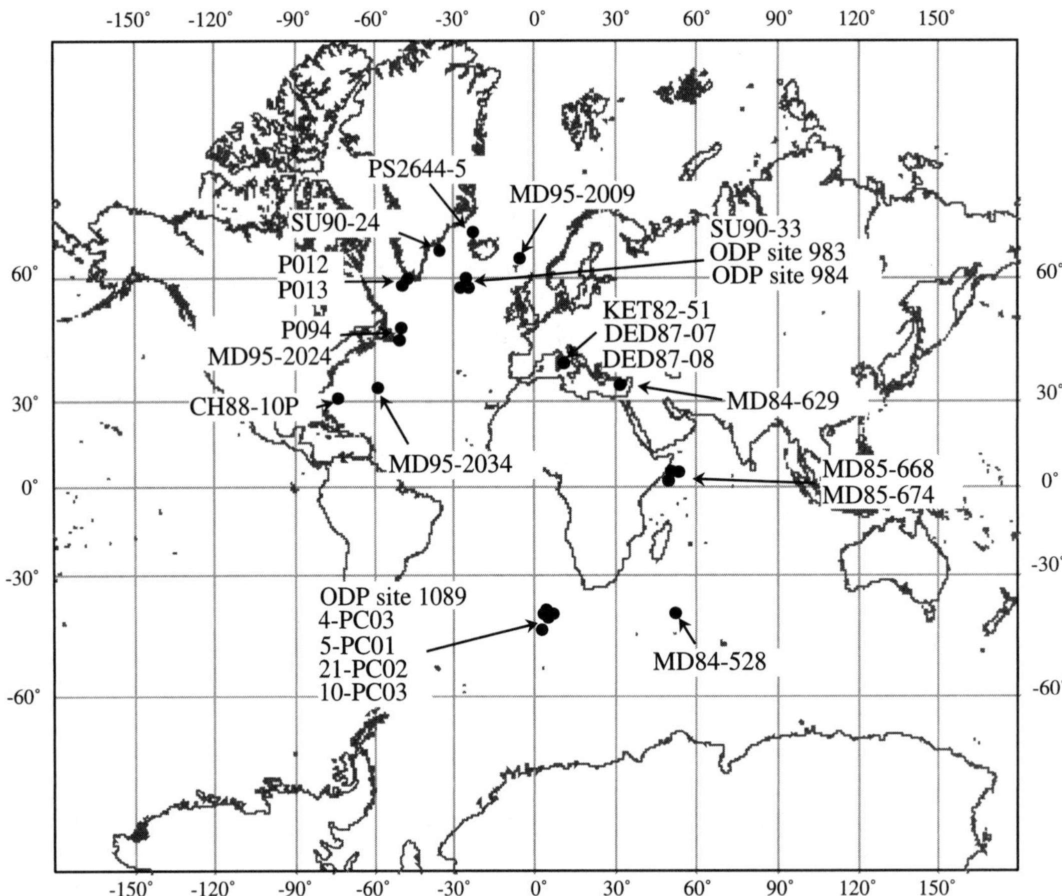

Figure 1. Schematic map showing the location of the sites.

Valet, 1999] paleointensity stacks, based respectively on 17 and 33 globally distributed marine paleointensity records and tied to the SPECMAP $\delta^{18}O$ record, have demonstrated that during the Brunhes chron long wavelength (= 10 kyr) intensity fluctuations are globally coherent and therefore that the intensity of the geomagnetic dipole field is largely fluctuating even during periods of stable polarity [*Guyodo and Valet,* 1999]. Shorter wavelength features are not apparent in these two stacks because of the low sedimentation rates of individual records, the smoothing effect of stacking individual records and of measuring with the u-channel technique.

In this article we use 24 published records, widely distributed worldwide, all but one characterized by average sedimentation rates in excess of 7 cm/kyr over the last 75 kyr. When studied with modern high-resolution cryogenic magnetometers, which have a spatial resolution of about 4 cm, submillenial time resolution may theoretically be attained. We stack these records after very careful correlation to the NAPIS-75 record, which is already placed on the GISP2 age model. We then use a new method for rejecting outliers to obtain a precise record of relative paleointensity

which is calibrated to absolute values using the archeomagnetic/ volcanic absolute paleointensity record for the last 10–20 kyr.

2. THE RECORDS, INTER-CORRELATION AND CHOICE OF AN AGE MODEL

The geographical distribution of the 24 records used in this article is shown in Figure 1 and their exact location is reported in Table 1. The sites are distributed in a longitudinal band between 73°W and 50°E and in latitude between 67°N and 47°S. All the records from these sites are characterized by average sedimentation rates ranging between 7 cm/kyr and 34 cm/kyr with the single exception of core MD85-668 from the Somali Basin, which has an average deposition rate of 5 cm/kyr. The cores have been studied using either the u-channel technique or discrete samples taken at regular intervals along the cores. The resulting time resolution of the magnetic analyses in this core varies between 500 and 110 years. One record fom the Sulu sea obtained from high deposition rate sediments (10 cm/kyr) [*Schneider and Mello*, 1996] has not been

Table 1. Location of the sites and characteristics of the records selected for GLOPIS-75

cores	Latitude	Longitude	sed rate (cm/kyr)	% of data	coeff correl	Reference
Sub-polar North Atlantic						
PS2644-5	67°52'N	21°46'W	12.4	77	0.885	[1]
MD95-2009	62°44'N	03°59'W	21.5	75	0.788	[1]
SU90-24	62°40'N	37°23'W	15.4	70	0.666	[1]
SU90-33	60°34'N	22°05'W	7.3	73	0.676	[1]
ODP site 983	60°24'N	23°38'W	10.8	60	0.832	[1, 2]
ODP site 984	61°24'N	24°06'W	18.7	45	0.611	[3]
Blake Outer Ridge and Bermuda Rise						
MD95-2034	33°41'N	57°34'W	34.5	74	0.884	[1]
CH88-10P	29°50'N	73°15'W	21.3	47	0.373	[6]
Labrador Sea						
P012	58°55'N	47°07'W	8	52	0.651	[4]
P013	58°13'N	48°22'W	12	64	0.718	[4]
P094	50°12'N	45°41'W	12	59	0.727	[4]
MD95-2024	50°12'N	45°41'W	24.0	70	0.715	[5]
Mediterranean						
KET82-51	39°29'N	14°10'E	12.5	77	0.701	[7]
DED87-07	39°25'N	13°21'E	8.6	67	0.752	[7]
DED87-08	39°25'N	13°20'E	17.5	56	0.663	[7]
MD84-629	36°02'N	33°05'E	14.4	34	0.761	[7]
South Indian Ocean						
MD84-528	42°06'S	53°02'E	7.7	55	0.641	[7]
South Atlantic						
ODP site 1089	40°56'S	09°53'E	18.1	72	0.716	[8]
4-PC03	40°56'S	09°53'E	25.1	60	0.763	[9]
5-PC01	41°00s	09°40'E	22.0	65	0.510	[9]
21-PC02	41°98'S	07°48'E	14.9	76	0.692	[9]
10-PC03	47°05'S	05°55'E	16.1	51	0.721	[9]
Somali Basin						
MD85-668	00°01'S	46°02'E	5.0	67	0.712	[10]
MD85-674	03°11'N	50°26'E	9.1	63	0.662	[10]

The sedimentation rate is the average one calculated over the time interval considered here. The percentage of data is the one remaning for each core after the 8th iteration (see text) and the correlation coefficient is the one obtained after correlating each individual record to NAPIS-75. References: [1] Laj et al., 2000; [2] Channell et al., 1997; [3] Channell, 1999; [4] Stoner et al., 1995, 1998; [5] Stoner et al., 2000; [6] Schwartz et al., 1998; [7] Tric et al., 1982; [8] Stoner et al., 2002; [9] Channell et al., 2000; [10] Meynadier et al., 1992.

considered because the sampling interval did not allowed to reach a time resolution better than 1 kyr.

Not all of the records cover the entire 0–75 kyr time interval. For example, core 4PC-03 from South Atlantic does not cover periods older than about 43 kyr while the top part of the record from core DED87-08 from the Mediterranean is at about 39 kyr. Combination of the records, however, allows to cover the last 75 kyr with a number of records varying from a minimum of 11 at 10 kyr, increasing to 23 at 21 kyr, reaching a maximum of 24 between 39 and 43 kyr and finally decreasing to 21 between 63 and 75 kyr (Figure 2).

Stacking of paleointensity records originating from different geographic regions has widely been used to assess the "true" geomagnetic character of the records, because spurious features of individual records are expected to be averaged out

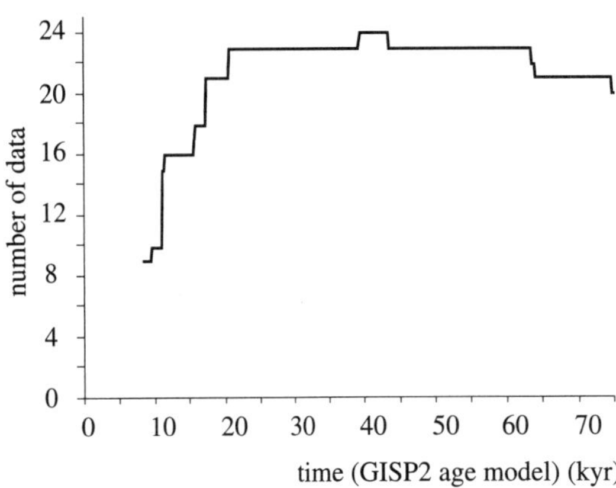

Figure 2. Diagram showing the number of data at each horizon between 10 and 75 kyr.

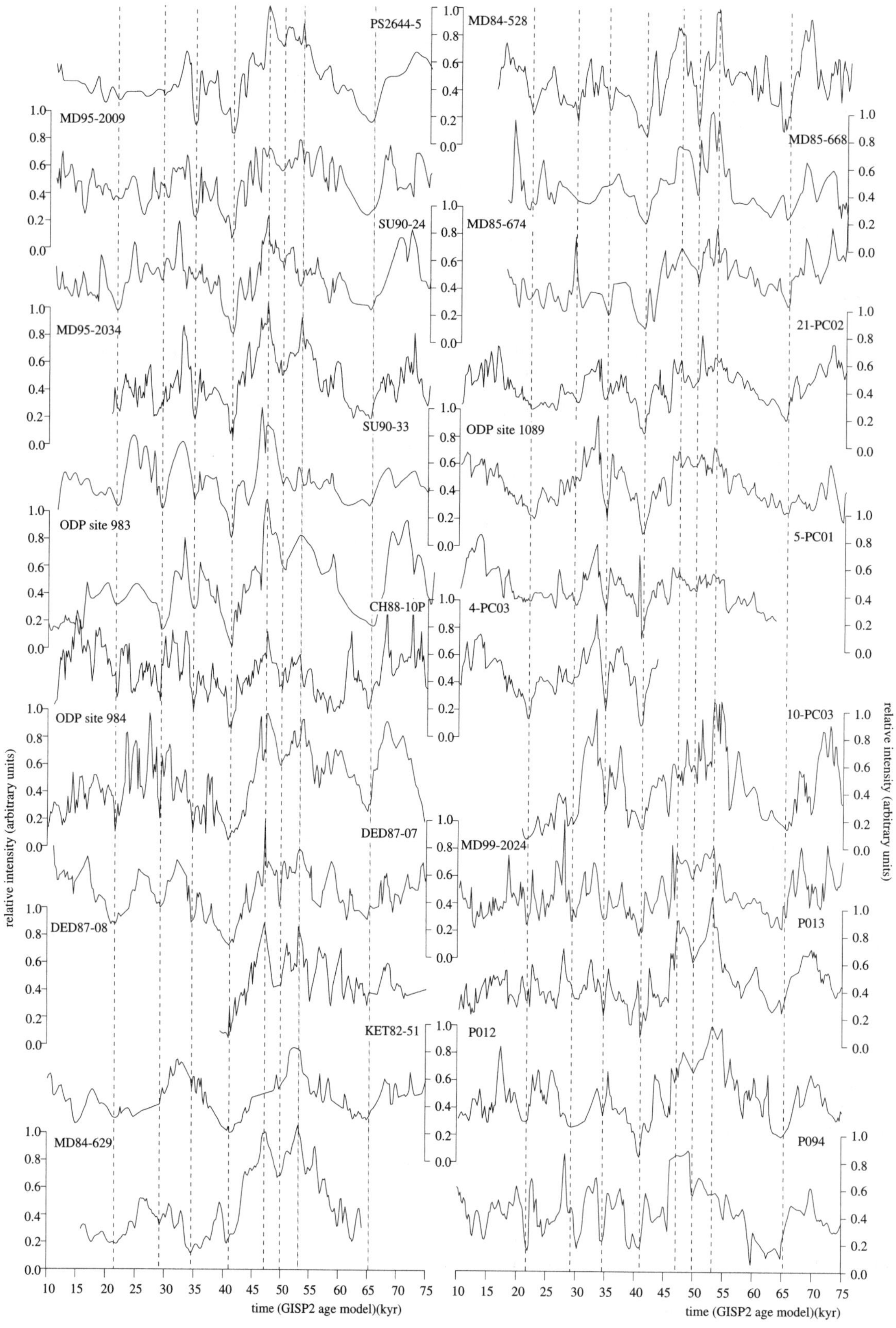

PS2644-5
MD84-528
MD95-2009
MD85-668
SU90-24
MD85-674
MD95-2034
21-PC02
SU90-33
ODP site 1089
ODP site 983
5-PC01
CH88-10P
4-PC03
ODP site 984
10-PC03
DED87-07
MD99-2024
DED87-08
P013
KET82-51
P012
MD84-629
P094
relative intensity (arbitrary units)
relative intensity (arbitrary units)
time (GISP2 age model)(kyr)
time (GISP2 age model)(kyr)

by the stacking process. Stacking implies that the individual records have been put on the same age model, because if the age models differ, even slightly, from one core to the other, the stack may be rather noisy. To put all the records on the same age model, we have chosen the NAPIS-75 record as a master curve, and then correlated all the other records to this master curve using the paleointensity records themselves. The rationale for this procedure is that it has been shown that NAPIS-75, through its independent correlation to the ice records [*Voelker et al.*, 1998; *Wagner et al.*, 2000], has a higher and more precise time resolution than other marine records. In addition, it has been shown that post-depositional phenomena of magnetization acquisition are negligible in this stack [*Kissel et al.*, 1999; *Laj et al.*, 2000]. Precise correlation of the individual records to NAPIS-75 was possible because peculiar features of the paleointensity records were unambiguously identified in every record and allow a much more precise intercorrelation in the studied interval than SPECMAP, used so far to intercorrelate marine records. Indeed, $\delta^{18}O$ records used to establish timescale in marine cores do not have any clear structure during stage 3 and when obtained from planktic foraminifera, they may, in addition, also be perturbed by local effects. Correlation was then obtained with the software Analyseries [*Paillard et al.*, 1996].

Figure 3 shows the different records after correlation to NAPIS-75. It also indicates the main correlation tie points. No constraints or stratigraphic ties are violated in doing this correlation. For a very large majority of the records the age adjustments that have been necessary to achieve maximum correlation, oscillate between ± 2000 years. Extreme values attain ± 4200 years for two specific horizons in only 2 cores. These values are well within the limits of accuracy of the individual age models. After correlation, the correlation coefficients of each record to NAPIS-75 are comprised between 0.6 and 0.8, with only one very low value (0.38) for core CH88-10P (Table 1).

3. CONSTRUCTION OF A STACK AND REJECTION OF OUTLIERS

After inter-correlation, each record was sampled with a spacing of 200 years, which corresponds for the fastest deposited cores to a slight smoothing of the record. Records were then normalized to the same average and stacked together arithmetically. The resulting stack is shown in Figure 4a. Notice that the most recent 10 kyr of this record have been cut, because this period corresponds to the upper part of the cores

(1–2 m), which is very disturbed by the coring and handling processes, due to the high water content.

It is apparent from Figure 4a that small, short-lived differences between records exist at a few discrete horizons distributed over the entire time interval, as evidenced from the significant scatter (calculated here as a simple arithmetic scatter) which tends to decrease in the oldest part of the record but is rather high in the 10–30 kyr interval.

Two types of factors may be at the origin of these differences. First, small, unavoidable imperfections in the inter-correlation of the cores may result in small inconsistencies. But, and most important, geomagnetic factors may be invoked: each individual sedimentary sequence will record the local dipole and non-dipole fields at the site. Although the mechanisms of sedimentary magnetization acquisition will tend to smooth out the high frequency component of the non-dipole field, there is no reason to expect the dipole field to be uniquely and perfectly recorded at each site even considering the high deposition rate of the selected sediments. As mentioned above, the rationale for stacking records from different sites is precisely to diminish local effects, including geomagnetic non-dipole components, to a minimum. Finally, non-geomagnetic factors, such as sedimentary effects (e.g. rapid local changes in accumulation rate or perturbations) may affect the quality of a record at a given site and for a given short time interval.

Therefore, a record affected by environmental factors in a given time interval, and consequently yielding in this interval a less accurate image of the field, may provide accurate recording in some other intervals. Rejection of the entire record from the stack, based for instance on low values of the correlation coefficient calculated over the entire interval, may therefore result in significant loss of reliable information. Rejection of short intervals has already been successfully done in cores where changes in magnetic mineralogy or grain size made these intervals unacceptable with respect to the criteria established for the reliability of relative paleointensity records [*Stoner et al.*, 1995; *Weeks et al.*, 1995]. However, there are (probably more numerous) cases where specific intervals in the core may have unfaithfully recorded the geomagnetic field changes and still meet the standard selection criteria for mineralogy, grain size and concentration changes [*Tauxe*, 1993]. We have, therefore, tried to devise a new method specifically designed for removing intervals of a record which yield results inconsistent with the other records, while keeping the reliable part of the record in other intervals

Two approaches were, in fact, attempted: in the first, at each horizon we discarded the points which lie farther than ±1 standard deviation from the average. This latter was then recalculated using the remaining data points and considered as the final record. We found, however, that this straightforward approach has some drawbacks: as a variable number

---

Figure 3. Individual paleointensity records reported on the GISP2 age model after correlation with NAPIS-75. The dashed lines are used to visualize the main tie points used for the correlation.

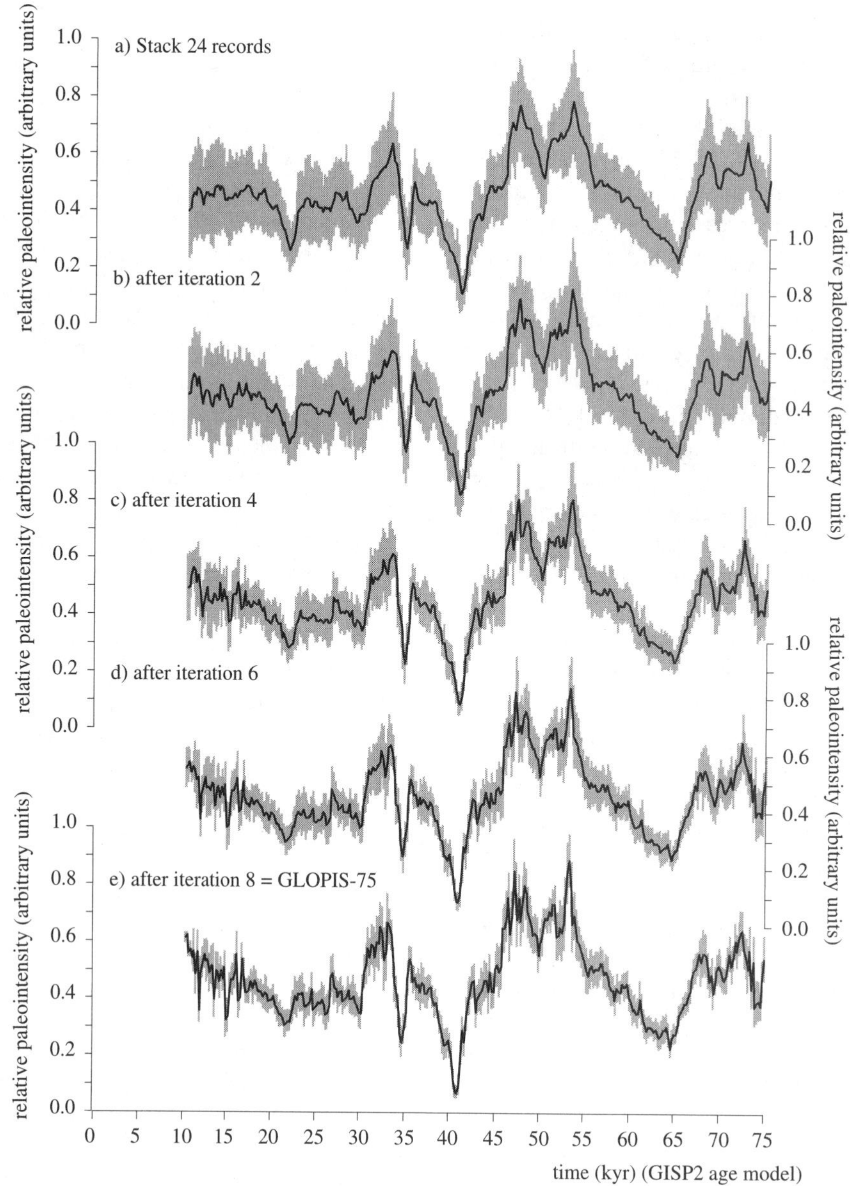

Figure 4. a) Stack obtained from the 24 individual records. b) to e) Records obtained after the 2nd, 4th, 6th and 8th iterations respectively. The iterations consists in calculating a new stack after rejecting in each time interval the datum lying the farthest from the average (see text). The 8th iteration reported in e) is considered as the most reliable one and it is called GLOPIS-75.

of points were rejected at each horizon, the final record is itself composed of a variable number of points (which made, in particular, the use of the incertitudes somewhat difficult). Furthermore, in this one-step approach the average with respect to which the data points were rejected or kept, has been calculated using all the data points, i.e. including those which were finally rejected. The average used as a criterion for acceptance/rejection is, consequently, different from the "true" average.

We have, therefore, devised a more progressive, multiple-steps approach. In the first step, after calculation of the average, we rejected at each horizon the datum lying the farthest from the average (in absolute value). A new average was then calculated for the (n-1) remaining data and the procedure repeated. The procedure was stopped when successive iterations gave results differing by less than 1.5% , as calculated by a least squares method at each horizon. It seems to us that this second approach has some advantages. First, the average, recalculated at each step, tends to move progressively and in fact rather rapidly toward the final, "true" average, making the "distance-from-the-average" criterion for rejection or acceptance more robust than in the one-step approach. Also, as the same number of points is rejected at each horizon, the final record is composed by the same number of points initially present, minus the number of iterations done. Different from usual stacks, however, the data points do not all necessarily originate from the same cores at the different horizons.

In Figure 4b to 4e we show the results obtained after 2, 4, 6 and 8 iterations. Beyond this limit the average difference between successive stacks, was smaller than 1.5%. We first verified that the global character of the stack has not been affected by the rejection process. As shown in Figure 5, the stack at each horizon is determined by values from geographically well widespread individual records, and therefore the stack is still global in any stratigraphic interval. Also, Table 1 gives the percentages of data points from each individual core used in the stack obtained after the 8th iteration. Globally, the records most affected by the rejection were MD84-629 from the Eastern Mediterranean (only 34% of the data were kept) ODP site 984 (45%), and the Blake Outer Ridge (47%).

As expected, the scatter is very significantly reduced at each step. In addition, there is a significant change in the trend of the record in the interval 20–10 kyr: in this interval, rejection of outliers results in a monotonic increase which

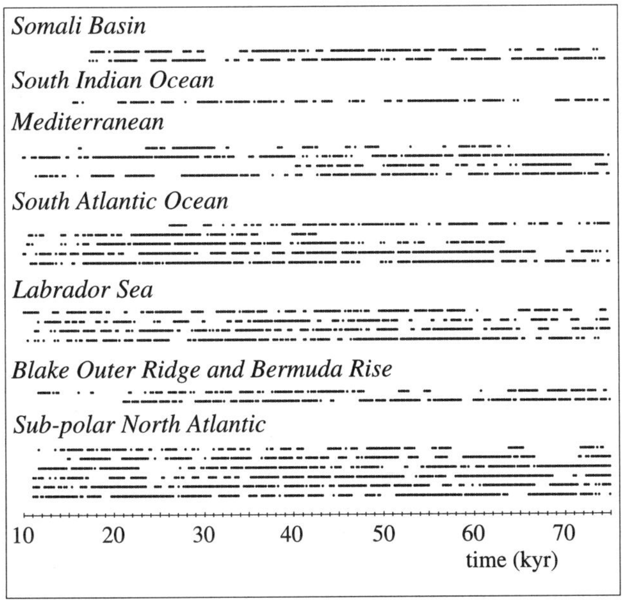

was not apparent in the original stack or in NAPIS-75. This trend is, on the contrary, present in the SINT-200 [*Guyodo and Valet*, 1996], SINT-800 [*Guyodo and Valet*, 1999] and in the compilation of archeomagnetic and volcanic records previously used [see e.g. *Valet* 2003].

4. CALIBRATION OF GLOPIS-75 TO ABSOLUTE VALUES (VADM)

GLOPIS-75 is based on marine sedimentary records and covers the 10–75 kyr time period. As mentioned above, usually the uppermost part (i.e. younger than ~10 kyr) of the cores does not reliably record the geomagnetic changes. Therefore, to extend GLOPIS-75 in the 0–10 kyr interval and to scale it to absolute values, we used the average of the archeomagnetic compilation of Yang et al. [2000] and of the available volcanic data for the last 20 kyr. Data were averaged over 500 years intervals between 0 and 19.5 kyr. The volcanic data can be found in the database Pint(00) at (<http://www.ngdc.noaa.gov/seg/potfld/paleo.html>). This database was completed with the most recent geochronological datings and paleointensity results [*Laj et al.*, 2002]. In calculating the average, each datum was given equal weight. Therefore, for the most recent period where the archeomagnetic data are much more numerous than the volcanic data, the average is very close to the archeomagnetic curve. Prior to about 8 kyr the number of archeomagnetic and volcanic data becomes comparable and finally no archeomagnetic data exist prior to 12 kyr and the record down to 19.5 kyr is entirely determined by the volcanic data. Around 41 kyr, we also used the paleointensity values obtained for transitional lava flows corresponding to the Laschamp event [*Marshall et al.*, 1988; *Roperch et al.*, 1988; *Chauvin et al.*, 1989; *Levi et al.*, 1990].

GLOPIS-75 reliably extends in time to about 10 kyr. This is an improvement with respect to NAPIS-75 which yielded reliable results only up to about 15–18 kyr. Moreover, as mentioned above, GLOPIS-75 exhibits a monotonic trend in the 20–10 kyr interval which was not apparent in NAPIS-75, but which is similar to that observed in the SINT-200 and in the compilation of volcanic records. These two features allow to calibrate GLOPIS-75 to absolute values with a better precision than in the case of NAPIS-75.

Figure 6 shows how the sedimentary and the absolute records were connected. First we used as a calibration point the intensity low at 41 kyr, where the absolute value (0.8×10^{22} Am^2) has been directly determined with the Thellier-Thellier technique from the volcanic outcrops at Laschamp in Central France and in Iceland. The reverse magnetization of these lavas unambiguosly shows that they were emplaced during the Laschamp event. We have therefore used the GISP2 age (41 kyr) for these data, which we believe to be more precise than

Figure 5. Diagram showing the distribution of the rejected data at each horizon in the different cores. Each core is reported as a line with one dot every 200 years. The dot has been taken out when the data were rejected. The lines have been grouped depending on the oceanic basin the cores are coming from.

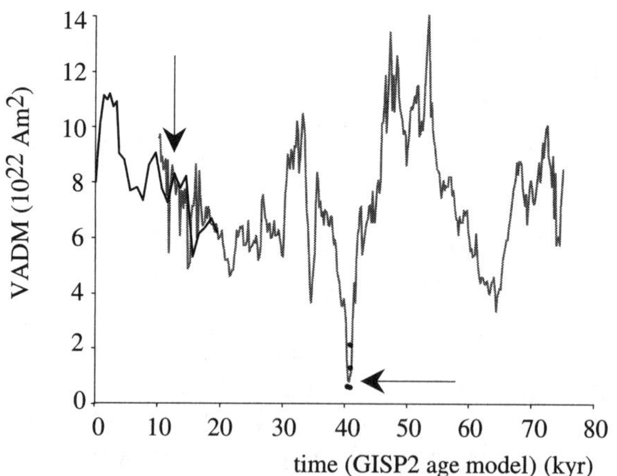

Figure 6. Transfer of GLOPIS-75 on the absolute scale. The black line is the combined archeomagnetic and volcanic curve constructed for the last 19.5 kyr (*Laj et al.*, 2002), the full dots are the virtual axial dipole moment (VADM) obtained from transitional lava of the Laschamp event (*Marshall et al.*, 1988; *Roperch et al.*, 1988; *Chauvin et al.*, 1989; *Levi et al.*, 1990). The grey line corresponds to GLOPIS-75. The large overlap of the two curves between 10 and 20 kyr is clear. The arrows indicate the two points which were used to connect GLOPIS-75 to absolute values.

the K/Ar or thermoluminescence ages reported in the original publications. The second calibration point was chosen at 12.5 kyr, where we imposed for the sedimentary record the value of the absolute archeomagnetic/volcanic curve. For ages older than 12.5 kyr the absolute intensity record is based on scarce volcanic data only and is therefore very scattered. When these two calibration points are used, the increasing slope of the sedimentary record is very similar to the increasing slope obtained from averaging the volcanic data in the 20–10 kyr interval. This is evidence toward the accuracy of the connection between the two records. Nevertheless, and despite this good overlap between GLOPIS-75 and the volcanic record, this calibration has still some uncertainties. These arise, to a large extent, from the scatter of the volcanic data in the 12–20 kyr interval (about 15%) in addition to that observed in GLOPIS-75. Moreover, dating of volcanic formation is delicate for such young ages, making the time correlation of the two records difficult.

The composite record obtained by combining archeomagnetic, volcanic and high resolution sedimentary data is reported in Figure 7. For comparison, we have also reported the calibrated SINT-200 record. The improved resolution of GLOPIS-75 clearly allows to better describe in absolute values the highs and lows of the geomagnetic paleointensity over the last 75 kyr than does SINT-200 even if the high-resolution (HR) version of this record [*Valet*, 2003] is used.

Although there is a clear improvement in accuracy with respect to previous estimates, it is still difficult to precisely

estimate the accuracy of the calibration of GLOPIS-75 to absolute values. We feel confident, however, that the calibrated records reflects absolute paleointensities to ±15%. This uncertainty arises mainly from the large scatter of the volcanic data used to calibrate. Improvements on this issue need either a much more complete volcanic database, or, more realistically, additional sedimentary records extending to much younger ages, so that it will be possible to calibrate the intensity directly with the much more extensive archeomagnetic database.

5. CHARACTERISTICS OF THE GEOMAGNETIC PALEOINTENSITY FOR THE LAST 75 KYR

5.1. Main Trends

The main trends of the records are very similar to those of NAPIS-75. However, at least for some specific intervals, it appears that the outliers rejection procedures has succeeded in attaining a better resolution. For instance, the maximum at about 50 kyr appears as a broad high in NAPIS-75, but has a well defined structure in GLOPIS-75, with two sharp highs at 53 and 47 kyr and significant lower strength between. There is then a very impressive drop of the field strength to the lowest values documented in the time interval explored here (~0.8x10^{22} Am^2) corresponding to the Laschamp event, also marked by directional changes in some of the records. After recovery of the field, another very marked low is observed at ca. 34.6 kyr. This corresponds in time to the most recent estimate for the Mono Lake event [*Benson et al*, 2003]. Although its resolution is significantly higher (see below), GLOPIS-75 documents the same

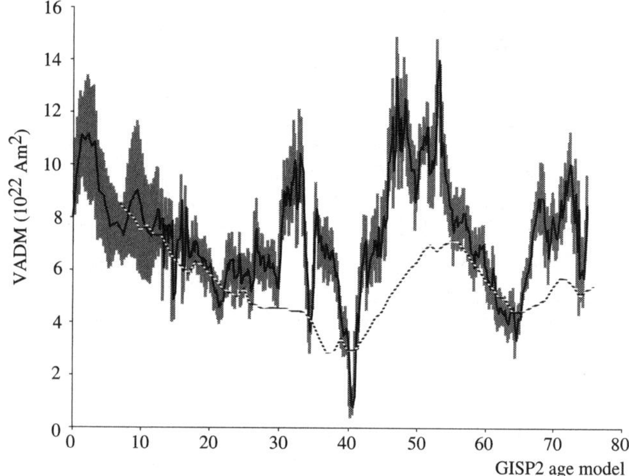

Figure 7. Combined archeomagnetic, volcanic and high resolution sedimentary curve reported with the ±1s standard deviation. The dashed line is the SINT-200 record reported for comparison.

general characteristics observed in NAPIS-75 in this interval. On the other hand, a significant long term difference with NAPIS appears in the 20–10 kyr interval. The observed trend documents an increase in dipole strength from about 4.5 to 8.3×10^{22} Am^2 and is therefore rather large. This trend was not documented in NAPIS, at least not as clearly. Its existence emerges during the iterative rejection of the outliers and is mostly due (specifically in the 20–10 kyr period) to rejection out of the 17 records covering entirely this time interval, of data from ODP sites 983 and 984 and cores P-013 and MD95-2024, and from core SU90-24 used in NAPIS-75.

5.2. High Frequency Features

Superimposed to the long-term changes, high frequency variations are observed in GLOPIS-75. Short-lived fluctuations are also visible in specific intervals of the individual records, but the stacking combined with the rejection of outliers has locally increased the signal/noise ratio. For example between the maximum around 47 kyr and the Laschamp event at 41 kyr, 5 to 6 of these fluctuations are well apparent (Figure 7). Assuming constant deposition rate, their duration is of the order of 600–700 years. Interestingly, this value appears to be very close to the shortest time constants associated with dipole field variations [*Hulot and LeMouël*, 1994]. Features with shorter time constants are associated with non-dipole terms, and do not necessarily correlate worldwide. In this interval GLOPIS-75 is determined by 15–16 records (after 8 iterations) well distributed worldwide. This would suggest that the fluctuations have a dipolar nature and in turn that GLOPIS-75 has attained a resolution comparable to the time constants of the fastest globally correlatable features, at least in some specific intervals. This is consistent with the theoretical resolution attainable, given the deposition rate of the sediments and the measuring techniques.

5.3. Duration of the Mono Lake and Laschamp Events

In GLOPIS-75 records the Mono Lake excursion (MLE) and the Laschamp event have been unambiguosly identified on the basis of their age and the corresponding ^{36}Cl and ^{10}Be records in the ice cores [*Laj et al.*, 2000; *Wagner et al.*, 2000; *Benson et al.*, 2003]. In 1999, Gubbins has proposed that excursions occur when the geomagnetic field reverses in the outer core, which has a typical overturn time of about 500 years, but not in the solid core, where diffusion of the field occurs on a typical timescale of 3000 years. On the other hand, a full reversal occurs when the field reverses in both the outer an inner cores, and requires a time significantly longer than an excursion. In addition, there has been some recent controversy about the nature of the Mono Lake excur-

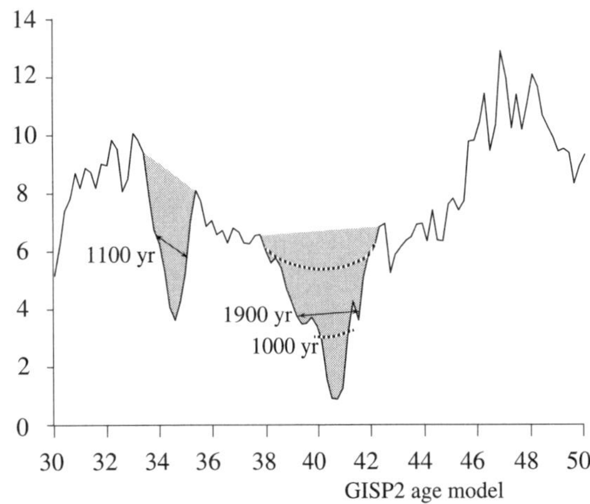

Figure 8. Enlarged view of the record over the period covering the Mono Lake and the Laschamp excursion. The grey areas are the ones taken into account to calculate the full width at half maximum. Two different baselines may be considered for the Laschamp event: the straight line limiting the grey area and resulting in a duration of 1900 years and the concave line resulting in a duration of 1000 yrs.

sion [*Kent et al.*, 2002; *Benson et al.*, 2003] which may also result in significantly different durations. It is therefore interesting to evaluate the duration of the two intensity lows recorded in GLOPIS-75.

It is usually hazardous to evaluate the duration of a short event from single geomagnetic records, because the sediment accumulation rate may undergo significant and undetected changes on short timescales. We believe, however, that we can reliably retrieve this duration from GLOPIS, obtained from 24 records from widely different and independent sedimentary environments which all yield very consistent results. It is therefore very unlikely that the duration of the events is significantly distorted in GLOPIS.

We have defined the duration as the full width at half maximum of the intensity lows. This was evaluated with respect to a baseline graphically estimated from the records. This was straightforward for the Mono Lake excursion (Figure 8), and yielded a value of 1100 years. On the other hand, it was more difficult for the Laschamp event, because of the "shoulders" apparent on both sides of the low. There is, therefore, some ambiguity in the determination of the duration of the Laschamp event which is between 1000 and 1900 years depending on the choice of the baseline (Figure 8). Both cases, therefore, provide supporting evidence for Gubbins' hypothesis for a short duration of excursions. On the other hand, GLOPIS-75 does not provide unambiguous evidence that the MLE is shorter than the Laschamp event, although a shorter duration for the MLE would be consistent with the fact that no directional changes associated with the MLE have been documented in our records.

6. CONCLUSIONS

GLOPIS-75 is a precise global record of geomagnetic paleointensity. Through its correlation to the GISP2 ice core oxygen isotope record, its time resolution is significantly higher than that in previous published global records of geomagnetic paleointensity. Moreover, the method of rejecting outliers used has allowed to identify short-lived fluctuations which appear globally coherent and therefore linked to fluctuations of the dipole field. The time constants associated with these fluctuations, of the order of 600–700 years, correspond to the shortest ones usually associated with dipole field fluctuations. This indicates that the resolution attained by GLOPIS-75, at least in certain intervals, is close to the maximum theoretical limit which can be attained on a global scale. Fluctuations of the field intensity on shorter time constants is indeed linked to non-dipole terms which fluctuate on a more restricted geographical scale. The duration of the Laschamp and Mono Lake event, of the order of 1500 years, determined from this record, is consistent with Gubbins' suggestion that geomagnetic excursions should have a shorter duration than reversals. Calibration to absolute values using published archeomagnetic and volcanic records is more precise than the calibration of the NAPIS-75 record. However, the large scatter of the volcanic data in the 20–10 kyr interval, does not allow calibration to better than about ±15%. Improvements can be expected from sedimentary records extending in time toward more recent periods.

Acknowledgments. We thank the CEA and the CNRS for the financial support as well as the European Program POP (EVK2-2000-22067). Very constructive reviews by J. Stoner and D. McMillan are gratefully acknowledged. This is LSCE contribution no. 1040.

REFERENCES

Baker R. and K. Aldridge, Paleomagnetic intensity data as a time sequence: opening a window into dynamics of Earth's fluid core?, this volume, pp. 245–254

Benson, L., J. Liddicoat, J. Smoot, R. Sarna-Wojcicki, S. Negrini, S. Lund, Age of the Mono Lake excursion and associated tephra. *Quaternary Sci. Reviews, 22*, 135–140, 2003.

Chauvin, A., R.A. Duncan, N. Bonhommet and S. Levi, Paleointensity of the Earth's magnetic field and K-Ar dating of the Louchardiere lava volcanic flow (central France) new evidence for the Laschamp excursion, *Geophys. Res. Lett. 16*, 189–1192, 1989.

Channell, J. E. T., D. A. Hodell, and B. Lehman, Relative geomagnetic paleointensity and δ18O at ODP Site 983 (Gardar Drift, North Atlantic) since 350 ka. *Earth Planet. Sci. Lett.* 153, 103–118, 1997.

Channell, J. E. T., Geomagnetic paleointensity and directional secular variation at ocean Drilling Program (ODP) Site 984 (Bjorn Drift) since 500 ka : Comparisons with ODP 983 (Gardar Drift), *J. Geophys. Res 104*, 22937–22951, 1999.

Channell, J.E.T., J. S. Stoner, D. A. Hodell and C. Charles, Geomagnetic paleointensity from late Brunhes-age piston cores from the subantarctic South Atlantic. *Earth Planet. Sci. Lett, 175*, 145–160, 2000.

Gubbins, D., The distinction between geomagnetic excursions and reversals, *Geophys. J. Int., 137*, F1–F3, 1999.

Guyodo, Y., J. P Valet, Relative variations in geomagnetic intensity from sedimentary records: the past 200 thousand years, *Earth Planet. Sci. Lett., 143*, 23–26, 1996.

Guyodo, Y., J. P Valet, Global changes in intensity of the Earth's magnetic field during the past 800 kyr. *Nature 399*, 249–252, 1999

Hulot G. and J.L. Le Mouël, A statistical approach to the Earth's main magnetic field, Phys. *Earth Planet. Inter. 82*, 167–183, 1994.

Kent, D. V., Hemming, S. R., Turrin, B. D., Laschamp excursion at Mono Lake? *Earth Planet. Sci. Lett., 197*, 151–164, 2002.

Kissel, C., C. Laj, L. Labeyrie, T. Dokken, A. Voelker, and D. Blamart, Rapid climatic variations during marine isotopic stage 3: magnetic analyses of sediments from nordic seas and North Atlantic, *Earth Planet. Sci. Lett. 171*, 489–502, 1999.

Laj, C., Kissel, C. , Mazaud, A. , Channell, J. E. T., Beer, J., North Atlantic Paleointensity Stack since 75 ka (NAPIS-75) and the duration of the Laschamp event. *Phil. Trans. Roy. Soc. Series A 358*, 1009–1025, 2000.

Laj, C., Kissel, C., Mazaud, A., Michel, E., Muscheler, R., Beer, J. Geomagnetic Field Intensity, North Atlantic Deep Water Circulation and Atmospheric Δ14C during the last 50 kyr. *Earth Planet. Sci. Lett., 200*, 179–192, 2002.

Levi, S., H. Audunsson, R.A. Duncan, L. Kristjansson, P.Y. Gillot, and S.P. Jakobsson, Late Pleistocene geomagnetic excursion in the icelandic lavas : confirmation of the Laschamp excursion, *Earth Planet. Sci. Lett. 96*, 443–457, 1990.

Marshall, M., A. Chauvin, and N. Bonhommet, preliminary paleointensity measurements and detailed geomagnetic analyses of basalts from the Skalamaelifell excursion, southwest Iceland, *J. Geophys. Res. 93*, 11681–11698, 1988.

Meynadier, L., J. P. Valet, R. Weeks, N. Shackleton, V. L. Hagee, Relative geomagnetic intensity of the field during the last 140 ka. *Earth Planet. Sci. Lett. 114*, 39–57, 1992.

Paillard, D., L. Labeyrie, P. Yiou, Macintosh program performs time-series analysis, *Eos Trans. AGU 77*, 379, 1996.

Roperch, P., N. Bonhommet, S. Levi, Paleointensity of the earth's magnetic field during the Laschamps excursion and it's geomagnetic implications, *Earth and Planet. Sci. Lett. 88*, 209–219, 1988.

Schneider, D. A., and G. A. Mello, A high-resolution marine sedimentary record of geomagnetic intensity during the brunhes Chron, *Earth Planet. Sci. Lett., 144*, 297–314, 1996.

Schwartz, M., S. P. Lund, T. C. Johnson, Geomagnetic field intensity from 71 to 12 ka as recorded in deep-sea sediments of the Blake Outer Ridge, North Atlantic Ocean. *J. Geophys. Res. 103*, 30,407–30,416, 1998.

Stoner, J.S., J.E.T. Channell and C. Hillaire-Marcel, Late Pleistocene relative geomagnetic paleointensity from the deep Labrador Sea—

Regional and global correlations, *Earth Planet. Sci. Lett., 134*, 237–252, 1995.

Stoner, J.S., J.E.T. Channell, C. Hillaire-Marcel, A 200 kyr geomagnetic chronostratigraphy for the Labrador Sea: indirect correlation of the sediment record to SPECMAP. *Earth Planet. Sci. Lett. 159*, 165–181, 1998.

Stoner, J. S., J. E. T. Channell, C. Hillaire-Marcel, C. Kissel, Geomagnetic paleointensity and environmental record from Labrador sea core MD95-2024 : global marine sediment and ice core chronostratigraphy for the last 110 kyr. *Earth Planet. Sci. Lett. 183*, 161–177, 2000.

Stoner, J. S. , C. Laj, C., J. E. T. Channell, C. Kissel, South Atlantic (SAPIS) and North Atlantic (NAPIS) geomagnetic Paleointensity Stacks (0–80ka): implications for inter-hemispheric correlation. *Quat. Sci. Reviews 21*, 1141–1151, 2002.

Tauxe, L., Sedimentary records of relative paleointensity of the geomagnetic field: theory and practice, *Rev. Geophys., 31*, 319–354, 1993.

Tric, E., J. P. Valet, P. Tucholka, M. Paterne, L. Labeyrie, F. Guichard, L. Tauxe, M. Fontugne, Paleointensity of the geomagnetic field for the last 80,000 years. *J. Geophys. Res. 97*, 9337–9351, 1992.

Valet, J. P.. Time variations in geomagnetic intensity, *Rev. of Geophysics, 41*(1), 1004, doi:10.1029/2001RG000104, 2003.

Voelker, A., M. Sarnthein, P. M. Grootes, H. Erlenkeuser, C. Laj, A. Mazaud, M. J. Nadeau, M. Schleicher, Correlation of marine 14C ages from the Nordic sea with GISP2 isotope record: implication for 14C calibration beyond 25ka BP. *Radiocarbon 40*, 517–534, 1998.

Wagner G., J. Beer, C. Laj, C. Kissel, J. Mazarik, R. Muscheler, H-.A. Synal,. Chlorine-36 evidence for the Mono Lake event in the Summit GRIP ice core. *Earth Planet. Sci. Lett. 181*, 1–6, 2000.

Weeks, R.J., C. Laj, L. Endignoux, M.D. Fuller, A.P. Roberts, R. Manganne, E. Blanchard, and W. Goree, Improvements in long core measurement techniques: applications in palaeomagnetism and palaeoceanography, *Geophys. J. Int., 114*, 651–662, 1993.

Weeks, R.J., C. Laj, L. Endignoux, A. Mazaud, L. Labeyrie, A. P. Roberts, C. Kissel, E. Blanchard, Normalised natural remanent magnetisation intensity during the last 240000 years in piston cores from the Central North Atlantic Ocean : geomagnetic field intensity of environmental signal? *Phys. Earth Planet. Inter., 87,* 213–229, 1995.

Yang, S.., H. Odah, and J. Shaw, Variations in the geomagnetic dipole moment over the last 12,000 years, *Geophys. J. Int. 140*, 158–162, 2000.

J. Beer, Department of Surface Waters, EAWAG, CH-8600 Duebendorf, Switzerland.

C. Laj and C. Kissel, Laboratoire des Sciences du Climat et de l'Environnement, CEA-CNRS, Avenue de la Terrasse, Bat 12, 91198 Gif-sur-Yvette Cedex, France.

Historic Archaeomagnetic Results From the Eastern U.S.,
and Comparison With Secular Variation Models

Stacey Lengyel

Department of Anthropology, University of Arizona, Tucson, Arizona

Rob Sternberg

Department of Earth and Environment, Franklin & Marshall College, Lancaster, Pennsylvania

Archaeomagnetic data can supplement direct observations of the geomagnetic field in the development of high-resolution records of regional secular variation. Archaeomagnetic data from historic archaeological features are particularly well suited for this because of the highly accurate and precise dating afforded through historic documents and artifacts. Here, archaeomagnetic directional data from 20 well-dated historic archaeological features in the eastern U.S. are presented and compared to existing models of secular variation and historic field measurements to demonstrate the compatibility between archaeomagnetic and direct measurements of geomagnetic field directions. Then, a previously undated historic brick kiln from Virginia is archaeomagnetically dated to the late 17th-early 18th centuries by statistically comparing the kiln's archaeomagnetic data to existing secular variation models.

1. INTRODUCTION

The geomagnetic field changes in various ways over many timescales and at different spatial scales. Prior to the last hundred years or so, the best records of regional secular variation came from land- and ship-based direct observations of the magnetic field [*Malin and Bullard*, 1981; *Barraclough*, 1982; *Alexandrescu et al.*, 1996; *Alexandrescu et al.*, 1997; *Jonkers et al.*, 2003]. Studies such as these have provided global field models at the surface [*Yukutake and Tachinaka*, 1969; *Barraclough*, 1974; *Braginskiy*, 1972; *Matsushima and Honkura*, 1988; *Bondar et al.*, 2002] and at the core-mantle boundary [*Bloxham*, 1986; *Bloxham et al.*, 1989; *Bloxham and Jackson*, 1992; *Jackson et al.*, 2003]. Ultimately such data can help us better understand the nature of the core and geomagnetic dynamo [*Courtillot and Le Mouël*, 1988; *Courtillot and Valet*, 1995; *Merrill et al.*, 1996].

Timescales of the Paleomagnetic Field
Geophysical Monograph Series 145
Copyright 2004 by the American Geophysical Union
10.1029/145GM20

While direct measurements of the geomagnetic field are the preferred source of secular variation data, they are temporally limited to the past 400 years for declination, 300 years for inclination, and less than 200 years for intensity [*Jonkers et al.*, 2003]. Pre-satellite data are also geographically constrained to areas monitored by repeat magnetic stations and observatories or by marine and field expeditions that usually measured only declination. For earlier epochs, direct measurements can be supplemented by archaeomagnetic data to yield more complete records of secular variation that have greater temporal and geographic coverage [*Thompson*, 1982; *Thompson and Barraclough*, 1982; *Gubbins*, 1986; *Hutcheson and Gubbins*, 1990; *Barraclough et al.*, 2000]. Archaeomagnetic data recovered from historic archaeological features (ca. post-AD 1600 in North America) are particularly well suited for this since they often can be accurately and precisely dated through historic records. Complete vector directions and paleointensities can be recovered from the thermoremanent magnetization (TRM) of archaeological features such as kilns, hearths, and fired house floors or walls. Principles and applications of archaeomagnetism are discussed by Eighmy et al. [1980], Tarling [1983], Eighmy and Sternberg [1990], Sternberg [1997], Kovacheva et al. [1998], Hedley [2000], and Sternberg [2001].

In North America, archaeomagnetic studies have focused on prehistory, with emphases on both secular variation and archaeomagnetic dating [*DuBois*, 1989; *Sternberg*, 1989a, 1989b; *Wolfman*, 1990; *Kean et al.*, 1997; *Lengyel and Eighmy*, 2002]. Only Dunlop and Zinn [1980], Aitken et al. [1987], and Gose et al. [1994] have considered historic archaeomagnetic features. In this study, we report archaeomagnetic directions from historic archaeological features in the eastern U.S. Comparison of these directions to direct observations of the magnetic field and existing secular variation models can be used to assess the reliability of the archaeomagnetic results. Additionally, archaeomagnetic dating of one historic brick kiln will be illustrated.

2. MATERIALS AND METHODS

2.1. HUSDIR Data Set

We will refer to the two data sets reported here as HUSDIR and HUSDAT. The HUSDIR data come from a series of archaeological sites in the eastern U.S. (Table 1). These data were collected from several types of cultural features, all historic in age, such as Native American (NA) or Euroamerican (EA) cooking hearths or basins, clay or brick-lined floors of burned structures, and Euroamerican brick-lined kilns or chimneys. In some cases, such as the features from Fort James or the lime kiln at the Tellico Blockhouse, historic documents were used to determine the date, or date range, of magnetization. For instance, the Fort James warehouse (feature JR-2) is known, through historic documentation, to have been destroyed by fire in the year 1644 [Eric Deetz, personal communication, 1999]. Likewise, it is known that the east lime kiln at Tellico (feature 1270) was constructed sometime between 1860 and 1866 and was most likely used only once during that period [*Polhemus*, 1977]. However, in this case we know only the age range during which the kiln could have been in operation, rather than the actual date of firing.

Other features, such as the fire basins from the James White Site, were dated through a combination of historic documentation and direct association with diagnostic archaeological artifacts. In these cases, the beginning of the date ranges was determined by the presence of particular diagnostic artifacts, such as medicine bottle shards that post date the mid-19th century (feature 1959) and a white pressed milk glass button (feature 1960). The end of these two date ranges, however, was determined through historic documents that report the year in which that part of the site was no longer used [*Faulkner*, 1984].

Table 1. Archaeomagnetic features included in the HUSDIR data set. Features 90-9, 90-14, 90-15, 90-16, 91-2, JR-1 and JR-2 were collected and measured by Lengyel; JR002 was measured by Sternberg; the remainder were collected and measured by Robert DuBois and students, but are not previously published.

Feature number	Site name	Location	Feature type	Date (all AD)	Dating method[a]
90-9	40BT90	35.7°N, 83.8°W	NA hearth	1600-1740	CO
90-14	40BT90	35.7°N, 83.8°W	NA hearth	1600-1740	CO
90-15	40BT90	35.7°N, 83.8°W	NA hearth	1600-1740	CO
90-16	40BT90	35.7°N, 83.8°W	NA pottery kiln	1600-1740	CO
91-2	40BT91	35.7°N, 83.8°W	NA hearth	1600-1740	CO
990	Chota	35.6°N, 84.2°W	NA hearth	1700-1760	HD
991	Chota	35.6°N, 84.2°W	NA hearth	1720-1760	HD
1263	Toqua	35.6°N, 84.2°W	NA hearth	1675-1725	DA
1270	Tellico Blockhouse	35.6°N, 84.2°W	EA lime kiln	1860-1866	HD
1334	Toqua	35.6°N, 84.2°W	NA hearth	1675-1725	DA
1385	Tomotley	35.6°N, 84.2°W	NA hearth	1750-1776	HD &DA
1386	Tomotley	35.6°N, 84.2°W	NA hearth	1750-1776	HD & DA
1394	Hermitage	36.2°N, 86.6°W	EA kiln	1817-1890	HD
1395	Hermitage	36.2°N, 86.6°W	EA kiln	1817-1890	HD
1959	James White Site	36.0°N, 83.9°W	EA fire basin	1850-1887	HD &DA
1960	James White Site	36.0°N, 83.9°W	EA fire basin	1840-1887	HD &DA
1961	James White Site	36.0°N, 83.9°W	EA chimney base	1852-1887	HD
JR-1	Fort James	37.2°N, 76.7°W	EA furnace base	1663-1698	HD
JR-2	Fort James	37.2°N, 76.7°W	EA burned warehouse subsoil	1644	HD
JR002	Jamestown	37.2°N, 76.7°W	NA root cellar floor	1800	DA

[a] HD=historic documents; DA=diagnostic artifacts; CO=cultural occupation

Table 2. Laboratory results for the HUSDIR data set. N - number of samples; *Demag* - the level of demagnetization in mT; I_s - mean archaeomagnetic inclination; D_s - mean archaeomagnetic declination; α_{95} - the 95% confidence limit; k - the Fisher precision parameter; λ_p - the VGP latitude; ϕ_p - the VGP longitude; dm and dp - the semi-major and minor axes of the VGP confidence oval.

Feature	N	Demag (mT)	I_s	D_s	α_{95}	k	λ_p	ϕ_p	dm	dp
90-9	11	5	62.9	355.5	2.9	244	80.7	255.7	4.6	3.6
90-14	12	NRM	60.9	356.5	2.1	411	83.2	253.5	3.3	2.5
90-15	12	5	61.2	354.5	2.9	232	82.1	245.0	4.4	3.4
90-16	10	10	56.3	356.6	3.6	177	87.0	210.3	5.3	3.8
91-2	9	5	62.2	356.8	2.5	415	81.9	259.7	3.9	3.1
990	8	NRM	64.1	2.3	1.6	1217	79.6	284.7	2.5	2.0
991	8	10	63.2	4.2	1.2	2107	80.4	294.0	1.9	1.5
1263	8	NRM	61.0	356.6	1.6	1243	83.1	254.2	2.4	1.8
1270	8	10	65.0	7.6	1.4	1548	77.3	300.2	2.3	1.8
1334	9	10	57.3	356.7	3.9	171	86.5	228.2	5.8	4.2
1385	8	10	66.3	0.7	2.7	414	76.8	277.8	4.5	3.7
1386	9	NRM	66.7	6.8	2.4	474	75.4	293.7	3.9	3.2
1394	6	10	66.1	5.2	0.6	13943	77.2	289.1	0.9	0.8
1395	8	NRM	65.3	8.8	2.2	628	77.0	300.8	3.6	2.9
1959	7	40	66.2	18.3	5.8	110	71.5	317.1	9.5	7.8
1960	7	10	68.2	13.4	3.3	331	71.9	303.9	5.6	4.7
1961	8	10	66.0	9.7	5.8	93	75.7	303.2	9.5	7.7
JR-1	9	NRM	63.5	7.6	2.0	692	80.3	316.8	3.1	2.5
JR-2	7	NRM	62.0	347.4	5.5	121	78.7	229.4	8.6	6.7
JR002	5	15	73.7	8.4	3.3	537	67.0	88.6	5.9	5.3

Finally, some of the features could only be dated to the cultural occupation of the site. This was particularly true for the Cherokee hearths from sites 40BT90 and 40BT91 in eastern Tennessee, for which no historic documents or diagnostic artifacts were available. Style of house construction allowed the features to be dated to the historic Cherokee period of the region, but it was not possible to determine a more precise date for these features.

Although roughly half the features in data set HUSDIR were collected by previous researchers, standard archaeomagnetic collecting procedures [*Eighmy*, 1980, 1990] were used in all cases [DuBois, personal communication, 2003]. Between 8 and 12 samples were cut into the feature; each sample was then leveled, encased in gypsum plaster, and the azimuth of one side of the horizontal cube recorded before removal to the laboratory for analysis. The only real difference in collecting procedures is that DuBois used a 1.7-inch mold while Lengyel used a 1-inch mold to collect samples.

In the laboratory, Lengyel and Eighmy measured NRMs (natural remanent magnetism) using a Schonstedt spinner magnetometer. The samples were then magnetically cleaned using alternating field (AF) demagnetization. The samples from 40BT90 and 91 and from Fort James were demagnetized and remeasured at 5 mT intervals, typically up to 15–20 mT. It was assumed that the primary component was isolated at the level with the least directional dispersion. DuBois's samples were also AF demagnetized; however, he generally demagnetized stepwise one pilot sample to 60 mT in order to identify the optimal field, and then demagnetized the rest of the samples directly at that blanket level [DuBois, unpublished data, 1995].

2.2. HUSDAT Data Set

The HUSDAT sample came from the Homeland Road brick clamp (i.e., a temporary kiln, probably used to make bricks for a single household; *Gurcke*, 1987) at site 18CH664 (latitude 38.54°N, 76.78°W) in Hughesville, Maryland, as part of a Phase II site evaluation by John Milner Associates for the Maryland State Highway Administration. The brick clamp was thought to date to the late 1700s, but no independent dating of the clamp was available, and the structures built from this batch of bricks could not be found [Cheek, personal communication, 2003]. Eight samples of baked clay from the floor of the clamp were collected and designated as feature HB001, samples I-P. Eight oriented bricks (samples A–H) from the base of the clamp were also collected but have not been cut or measured. Each sample was demagnetized stepwise to 100 mT on a Molspin alternating field demagnetizer and measured on a Molspin Minispin magnetometer. Principal component analysis, or PCA [*Kirschvink*, 1980], was done using the Paleomag software [*Jones*, 2002]. Room temperature susceptibilities were measured for each sample at low (0.46 kHz) and high (4.6 kHz) frequencies with a Bartington MS2 system and MS2B sensor. Two samples were used for isothermal remanent magnetization (IRM) acquisition, using an ASC Scientific Model IM-10 Impulse Magnetizer. Sternberg collected these samples and supervised the laboratory work.

3. RESULTS

3.1. HUSDIR Data Set

The directional data for the HUSDIR data set are summarized in Table 2. Because these data were collected from two distinct geographic areas, eastern Tennessee and Jamestown, Virginia, they are each analyzed separately. To this end, secular variation curves were developed using the Geomagix software program [http://geomag.usgs.gov/ Freeware/geomagix.htm] for a reference site in each region (36°N, 84°W for Tennessee; 37.2°N, 76.7°W for Jamestown). A number of geomagnetic field models were used via Geomagix to calculate the mean VGP at 10-year intervals for each reference site, including British Geologic Survey (BGS) historic models [*Barraclough,* 1974] for 1600, 1650, 1700, 1750, 1800, 1850; IGRF models for 1900, 1910, 1920, 1930, 1940; DGRF models for 1950, 1960, 1970, 1980; and USGS models for 1990,

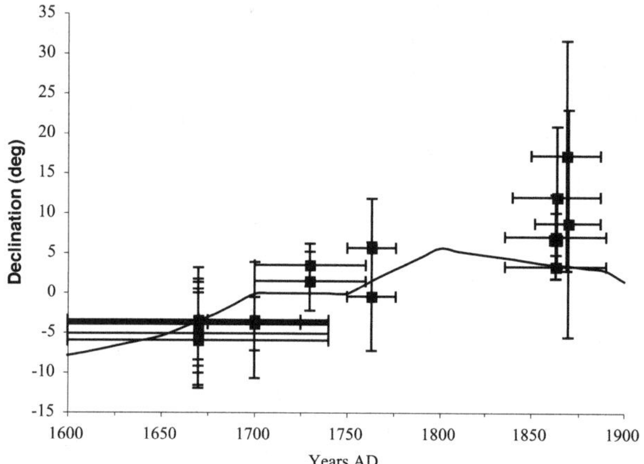

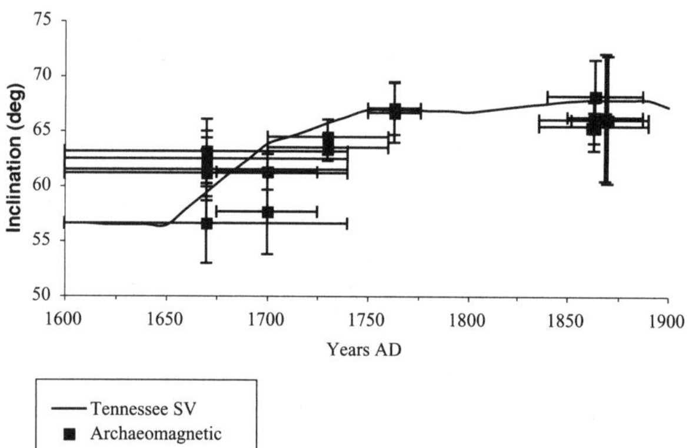

Figure 1. The Tennessee archaeomagnetic declination (top) and inclination (bottom) data compared with secular variation curves for eastern Tennessee. The curves were derived for the reference site 36°N, 84°W from Geomagix freeware [http://geomag.usgs.gov/Freeware/geomagix.htm] and geomagnetic field models BGS 1600, 1650, 1700, 1750, 1800, 1850; IGRF 00, 10, 20, 30, 40; DGRF 50, 60, 70, 80; US 1990, 2000. Values were calculated at 10-year intervals. Horizontal error bars on the archaeomagnetic data represent the date range associated with each result. Vertical error bars represent 95% confidence limits on the magnetic directions. The archaeomagnetic data are corrected to the reference site.

Table 3. Summary of the results of statistically comparing the Tennessee archaeomagnetic directions with the Geomagix derived secular variation models for eastern Tennessee.

Feature	Independent Date	Statistical Date	Peak Probability
	Years AD	Years AD	%
90-9	1600-1740	1680-1720	25
90-14	1600-1740	1670-1690	68
90-15	1600-1740	1670-1690	29
90-16	1600-1740	1600-1670	71.
91-2	1600-1740	1680-1700	52
990	1700-1760	1700-1730	53
991	1720-1760	none	-
1263	1675-1725	1700-1720	75
1270	1860-1866	none	-
1334	1675-1725	1600-1680	91
1385[a]	1750-1776	1710-1800; 1800-1900+	96 65
1386	1750-1776	1750-1900+	97
1394	1817-1890	none	-
1395	1817-1890	1780-1830	25
1959	1850-1887	1780-1880	12
1960	1840-1887	1790-1870	9
1961	1852-1887	1690-1900+	80

[a] there are two dating options for sample 1385

2000. After developing the regional secular variation curves, the directional data were converted via virtual geomagnetic poles [*Noël and Batt*, 1990] to their respective reference sites.

Figure 1 depicts the corrected declination and inclination data for the Tennessee archaeomagnetic samples against the Tennessee inclination and declination curves. A modified version of Sternberg and McGuire's [1990] statistical method was then used to quantitatively compare the independently dated archaeomagnetic directions with the modeled secular variation curves at 10-year intervals between AD 1600 and 1900. This test dates an archaeomagnetic sample by determining those intervals of time for which the archaeomagnetic direction and the secular variation curve do not significantly differ at the 5% significance level. The results of this test are summarized in Table 3.

Figure 2 depicts the corrected declination and inclination data for the Jamestown archaeomagnetic features against the Jamestown secular variation curves. As with the Tennessee

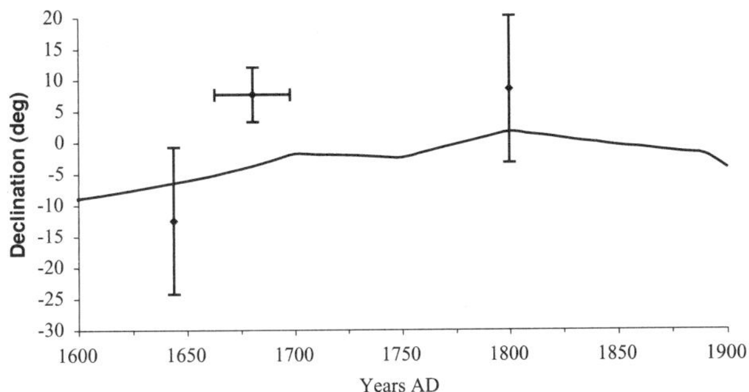

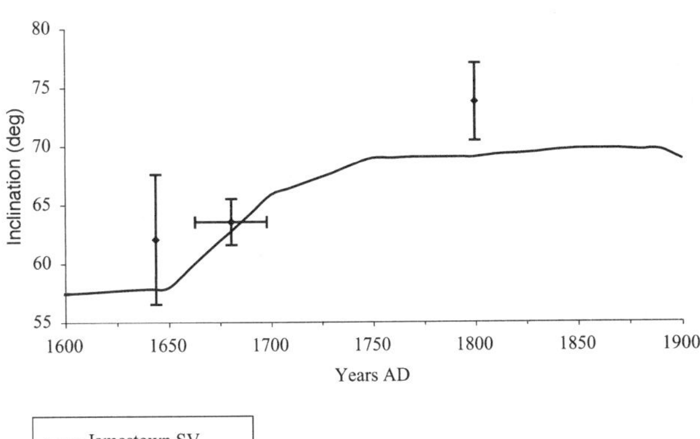

Figure 2. The Jamestown declination (top) and inclination (bottom) data compared with secular variation curves for Jamestown. The curves were derived for the reference site 37.2°N, 76.7°W as in Figure 1. Values were calculated at 10-year intervals. Horizontal error bars on the archaeomagnetic data represent the date range associated with each result. Vertical error bars represent 95% confidence limits on the magnetic directions. The archaeomagnetic data are corrected to the reference site.

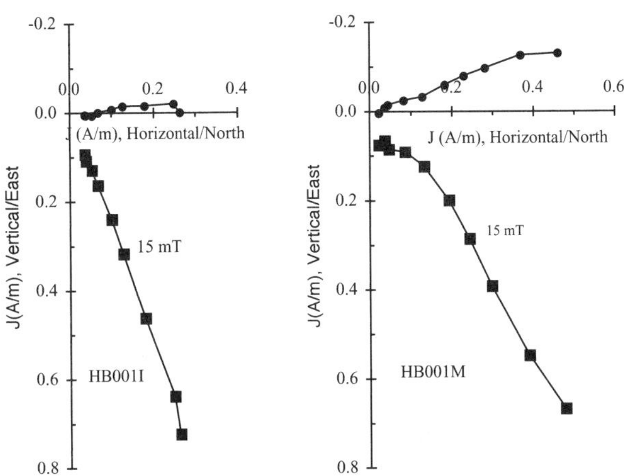

Figure 3. Vector endpoint diagrams for samples HB001I (left) and HB001M (right), interpreted with linear and planar fits, respectively. Circles represent declinations, and square are inclinations.

data, these samples were statistically compared to the Jamestown secular variation curves in order to identify archaeomagnetic dating options. No date determinations were obtained for features JR-1 or JR002, while JR-2 generated two dating options: AD 1600–1610 and AD 1660–1680. The peak probabilities for the two date ranges were 7.5% and 11%, respectively.

3.2. HUSDAT Data Set

Table 4 summarizes the magnetic properties of these samples. Samples are assumed to have volumes of 1×10^{-6} m^3, which is approximately correct, although variable and inexact because the irregular clay samples are encased in the standard sized plaster cubes. The Koenigsberger ratios suggest the remanence is a TRM or a partial TRM [*Jordanova et al.*, 2003; *Schnepp et al.*, 2003]. The shapes of the AF demagnetization curves as characterized by the median destructive fields (MDF) are consistent with the remanence being carried by predominantly single domain titanomagnetite [*Evans and Jiang*, 1996; *Jordanova et al.*, 2003]. The relatively slow saturation of the two IRM acquisitions suggest more bulk hematite than inferred from the softer IRM curves for archaeomagnetic baked clays seen by Jordanova et al. [1997] and Schnepp et al. [2003]. The values for the frequency dependence of susceptibility, or percent frequency effect,

$$\text{PFE} = \frac{k_{lf} - k_{hf}}{k_{lf}} \times 100\%$$

indicate a relatively high superparamagnetic component [*Jordanova et al.*, 2003].

Table 4. Summary of magnetic properties for HUSDAT samples.   Sample; k_{LF} and k_{HF} - susceptibilities at low and high frequencies; PFE - percent frequency effect for susceptibility; NRM - strength of NRM, Q - Koenigsberger ratio; MDF - median destructive field for AF demagnetization; SIRM - strength of IRM at 1 T; IRM(.1)/SIRM - ratio of IRM at 0.1 T to SIRM.

Sample	k_{LF}	k_{HF}	PFE	NRM	Q	MDF	SIRM	IRM(.1)/SIRM
	x10⁻⁵, SI	x10⁻⁵, SI	%	A m⁻¹		mT	A m⁻¹	
HB001I	370	336	9.2	0.768	4.9	13.6		
HB001J	390	350	10.3	0.975	5.9	14.7		
HB001K	640	585	8.5	0.773	2.9	11.3	94	0.67
HB001L	595	535	10.1	0.927	3.7	13.9		
HB001M	775	710	8.3	0.820	2.5	13.5		
HB001N	290	280	3.5	0.642	5.3	15.4	43	0.58
HB001O	610	570	6.6	0.839	3.3	13.0		
HB001P	295	280	5.0	0.535	4.3	14.9		
median	493	442	8.4	0.797	4.0	13.8		

Directions for samples I, J, K, and O were determined from least squares line fits, with n > 5 and the maximum angular deviation (MAD) < 6° in all cases. Least squares plane fits were used for samples L, M, N, and P, with n > 5 and the MAD < 25° in all cases; the average direction combined lines and planes. Directions for sample M converged upon the mean with demagnetization, while directions for P diverged from the mean. Differences in behavior were not clearly related to sample position or other magnetic properties, although M had the lowest Koenigsberger ratio, and P had the weakest remanence. Vector endpoint plots for samples I and M are shown in Figure 3.

The average direction for the eight combined line and plane fits yielded a declination of 354.9°, an inclination of 65.8°, and an α_{95} of 3.8°. This direction is plotted on a stereonet in Figure 4, along with the secular variation curve derived from Geomagix software for Washington, D.C., at a great arc distance of 0.4° from Hughesville.

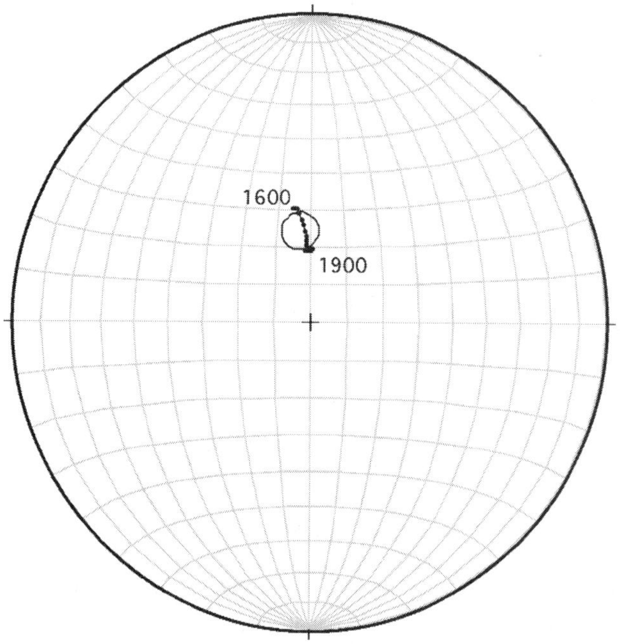

4. DISCUSSION

4.1. HUSDIR Data Set

The objective of the HUSDIR analysis was to assess the agreement between independently dated archaeomagnetic directions, secular variation models, and direct observations of the magnetic field, while additionally publishing new archaeomagnetic data for historic archaeological sites in the eastern U.S. This analysis found fairly good agreement between the archaeomagnetic directions and the secular variation models, as shown by Table 3, Figure 1, and Figure 2. Fourteen of the 20 features included in this data set were statistically dated to curve intervals that overlapped their independent date ranges, and for seven of the features the statistical date ranges were completely contained within the independent ranges. There is a tendency for the Tennessee archaeomagnetic declination values to trend easterly, and inclination values less steeply relative to the models. The Jamestown declination values also tend to trend easterly, but the inclination values trend more steeply than the models.

To further test the agreement between the archaeomagnetic directions and the secular variation models, we compared the larger Tennessee data set to direct observations of the magnetic field that were recorded between AD 1820 and 1920 at five

Figure 4. Equal area stereographic projection showing mean direction for HB001 and secular variation of the geomagnetic field for Washington, D.C., calculated at 10-year intervals from epoch AD 1600-1900, using Geomagix freeware [http://geomag. usgs.gov/Freeware/geomagix.htm] and geomagnetic field models BGS and IGRF models for AD 1600(10)1900.

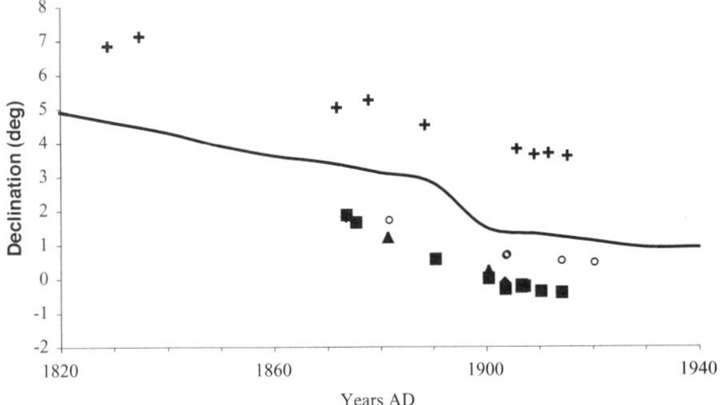

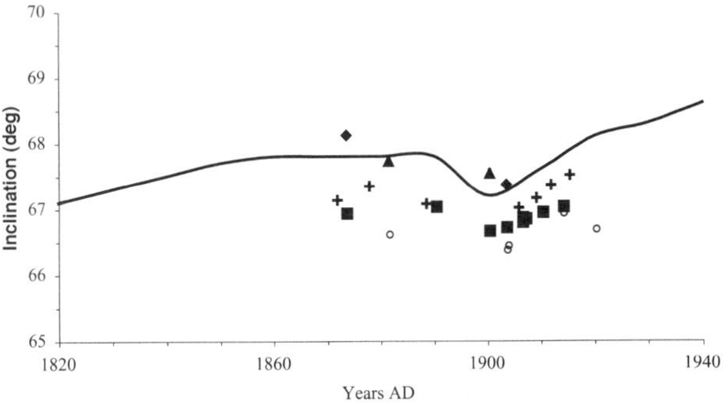

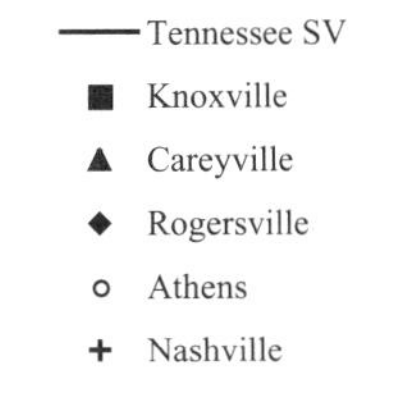

Figure 5. Historic measurements of declination (top) and inclination (bottom) compared with the Geomagix derived secular variation curves for eastern Tennessee. The historic observations are tabulated in Hazard [1917], and the observed inclination values are corrected to the reference site.

repeat stations in eastern Tennessee [Figure 5; *Hazard*, 1917]. All of the repeat stations but Nashville are located less than 1° angular distance from the reference site; Nashville is 2.3° from the reference site and was included because it provided two older declination measurements. All of the historic inclination measurements are associated with declination results, so the inclination values are reduced through the VGP to the Tennessee reference site. However, because some declination measurements do not have corresponding inclination measurements, the observed declination values shown in Figures 5 and 6 are the raw measurements.

As Figure 5 illustrates, the historic observations follow the general trend of the calculated secular variation curves, although the historic inclination measurements are slightly lower. Sim-

ilarly, while the uncorrected declination values depicted in Figure 5 are fairly symmetric around the curve, when these measurements are reduced to the reference site, they tend to trend about 1.8° below the curve. These systematic offsets suggest that the BGS and IGRF geomagnetic models used to calculate these secular variation curves could be improved for this area. Criss [2003] came to a similar conclusion when comparing direct declination measurements made by Lewis and Clark and several ship captains at Cape Disappointment, Washington, in the early 1800s with the BGS model calculations for the same time period. It is notable that the magnitude of the discrepancy between the observed and calculated declination values was similar to that found in this study.

Finally, Figure 6 compares the observed data, modeled data, and those archaeomagnetic data that have been independently dated to after AD 1820. When the modified statistical dating

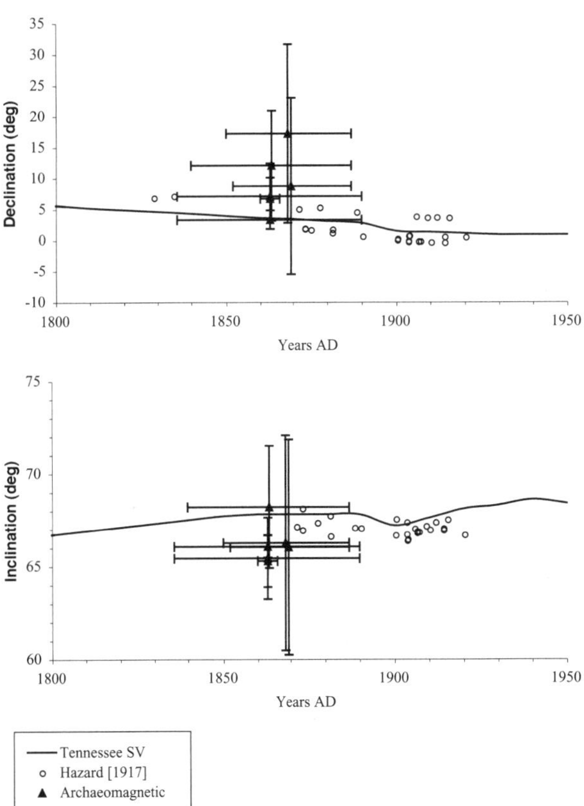

Figure 6. Archaeomagnetic and historic measurements declination (top) and inclination (bottom) compared with the Geomagix derived secular variation curves for eastern Tennessee. The historic observations are as in Figure 5. The archaeomagnetic data are the AD 1820-1920 subset of the Tennessee data depicted in Figure 1. Archaeomagnetic and observed inclinations are corrected to the reference site; archaeomagnetic declinations are corrected to the reference site.

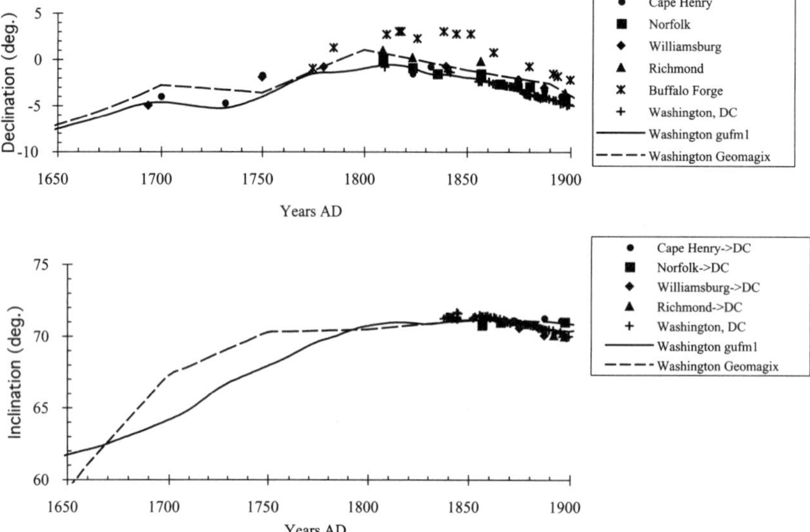

Figure 7. Secular variation models of Geomagix and gufm1 compared with direct measurements of the geomagnetic field for several Mid-Atlantic cities, tabulated in Hazard [1917]. Upper: declination. Lower: inclination.

method is employed, only one archaeomagnetic sample (feature 1961) is not statistically different from the historic data. As illustrated in Figure 6, the archaeomagnetic inclination data mirror the observed inclination values in that they tend to be less steep than the inclination curve; however, unlike the observed declination values, the archaeomagnetic declinations trend east of the declination curve. This is why only one archaeomagnetic sample matches the observed directions, despite the fact that many of the inclinations overlap. Efforts will be made to collect additional archaeomagnetic samples that date to this time period in order to further assess this situation.

Overall, there is good agreement between the HUSDIR archaeomagnetic inclinations and the directly measured geomagnetic inclinations, and both suggest that the models incorporated into the Geomagix software do not yield low enough inclinations for this area. Given this agreement, the difference in the declination values suggests that the archaeomagnetic data should be further scrutinized.

4.2. HUSDAT Data Set

The motivation for studying this feature was the derivation of an archaeomagnetic date. In making the interpretation, we have considered secular variation curves for the Washington, D.C., location computed by models Geomagix and gufm1 [*Jackson et al.*, 2003], the latter based on an enhanced database of nautical magnetic directions. Figure 7 compares these models with nearby direct measurements of the magnetic field on land tabulated by Hazard [1917]. Only data from sites with

older measurements, and with some continuity, are included. Declination measurements go back to AD 1700. Declinations shown are the raw measurements, although reducing these to a common location through the virtual geomagnetic pole has minimal effect. After AD 1750, the measured declinations are mostly contained within the envelope of the two secular variation curves. Model gufm1 gives a better fit to the Washington, D.C., data; otherwise the two models match the data equally well, with the exception of Buffalo Forge, which appears anomalous. The three measurements before 1750 are less well matched by the models. Inclination measurements cover a shorter period of time, only back to the 1850s. The inclinations in Figure 7 are adjusted to the common latitude of Washington, D.C. and are in good agreement with both models; the Washington, D.C., measurements are in better agreement with gufm1. Model gufm1 is based on few land-based measurements prior to AD 1850 [*Jackson et al.*, 2003; Jackson, personal communication]. A better regional model would include older historical data such as those reported here, augmented by earlier archaeomagnetic results, especially inclinations.

The statistical method for archaeomagnetic dating [*Sternberg and McGuire*, 1990] was used with both Washington, D.C., secular variation models calculated at 10-year increments of time, from AD 1600–1900. These curves are assumed to have no error, so the appropriate statistical test of Watson [1956] is used. Figure 8 shows a plot of the probability vs. time derived from each curve, as suggested by Le Goff et al. [2002]. If the dating range is taken as the age range where the probability exceeds 5%, then the archaeo-

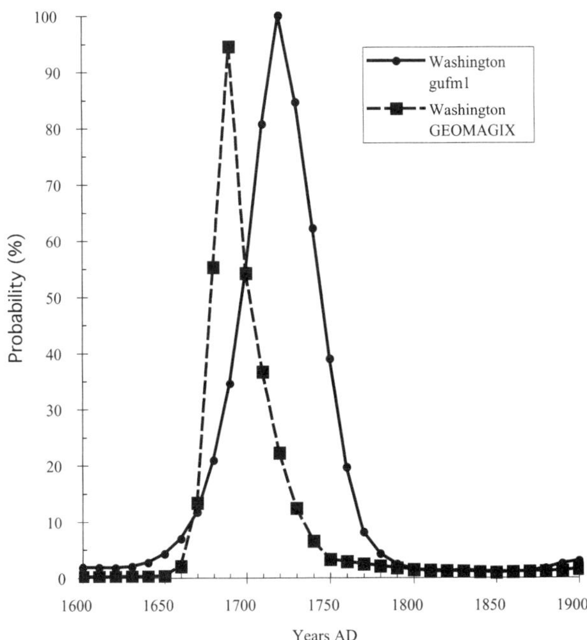

Figure 8. Probability at which directions for HB001 and the secular variation curves for Washington, D.C., are not significantly different, for both gufm1 (solid line) and Geomagix (dashed). The usual archaeomagnetic date would correspond to that interval of time when the probability is greater than 5%.

magnetic dates are AD 1665–1745 based on Geomagix, and AD 1655–1755 based on gufm1. The maximum probabilities occur at AD 1690 for Geomagix, and at AD 1720 for gufm1. There was originally no archaeometric or archaeological dating information for this brick clamp. Subsequent to the archaeomagnetic dating report, it was noted that water table bricks of the type made in this brick clamp were seen in Annapolis, Maryland, for two houses made in the 1760s [Cheek, personal communication, 2003]. This age is consistent with the latter portion of the archaeomagnetic date ranges. Thus it appears that archaeomagnetic dating can be useful for historic features when independent dating information is lacking. The utility of historic archaeomagnetic dating will be limited by the rate of secular variation. The slow secular variation in the 19th century (Figure 7) suggests archaeomagnetic dating would be problematic in this time period, while the more rapid changes of inclination around AD 1700 would tend to yield more precise archaeomagnetic dates.

In conclusion, the use of historic archaeomagnetic samples complements the use of historic igneous rocks in understanding the fidelity of the paleomagnetic recording process [*Doell and Cox*, 1963; *Holcomb et al.*, 1986; *Tanguy*, 1990]. Historic archaeomagnetic data are an important supplement to the historic record, especially for inclinations and intensities, which are more sparse than declination observations. During time periods when secular variation has been more rapid, archaeomagnetic dating of historic features can be useful when other chronological information is lacking.

Acknowledgments. We would like to thank Robert DuBois for allowing us to use his unpublished data in this study. Jeff Eighmy (Colorado State University Archaeometric Lab) provided the funding and equipment to process Lengyel's HUSDIR archaeomagnetic features and assisted in the interpretation of the results. Lynne Sullivan and Charlie Faulkner of the Department of Anthropology, University of Tennessee, provided assistance in dating the Tennessee archaeomagnetic features; Eric Deetz, from the Association for the Preservation of Virginia Antiquities, assisted in dating the two Fort James archaeomagnetic features. Collection and measurement of the HUSDAT samples was supported by John Milner Associates, Inc., Alexandria, Virginia; Charles Cheek advised on the archaeology of this feature. Thanks for lab measurements to Kaushik Katari for HUSDIR feature JR002-and Isaac Weaver for the HUSDAT feature. RS benefited from a discussion with Craig Jones on the use of principal component analysis and the PaleoMag software package. We would also like to thank Ted Evans, an anonymous reviewer, and co-editor Dennis Kent for comments on an earlier draft of this paper. This research was partially funded by NSF dissertation improvement grant #0308798. Measurements by Kaushik Katari were funded by the Franklin & Marshall College Hackman Scholars Program.

REFERENCES

Aitken, M. J., A. L. Allsop, G. D. Bussel, and M. Winter, Archaeomagnetic intensity determination: a nineteenth century pottery kiln near Jordan, Ontario, *Can. J. Earth Sci.*, *24*, 2392–2395, 1987.

Alexandrescu, M., V. Courtillot, and J. L. Le Mouël, Geomagnetic field direction in Paris since the mid-sixteenth century, *Phys. Earth Planet. Inter.*, *98*, 321–360, 1996.

Alexandrescu, M., V. Courtillot, and J. L. Le Mouël, High-resolution secular variation of the geomagnetic field in western Europe over the last 4 centuries: Comparison and integration of historical data from Paris and London, *J. Geophys. Res.*, *102* (B9), 20,245–20,258, 1997.

Barraclough, D. R., Spherical harmonic analyses of the geomagnetic field for eight epochs between 1600–1910, *Geophys. J. R. Astron. Soc.*, *36*, 497–513, 1974.

Barraclough, D. R., Historical observations of the geomagnetic field, *Philos. Trans. Roy. Soc. London*, *306*, 71–78, 1982.

Barraclough, D. R., J. G. Carrigan, and S. R. C. Malin, Observed geomagnetic field intensity in London since 1820, *Geophys. J. Int.*, *141*, 83–99, 2000.

Bloxham, J., Models of the magnetic field at the core-mantle boundary for 1715, 1777, and 1842, *J. Geophys. Res.*, *91* (B14), 13,954–13,966, 1986.

Bloxham, J., D. Gubbins, and A. Jackson, Geomagnetic secular variation, *Philos. Trans. R. Soc. London, Ser. A*, *329*, 415–502, 1989.

Bloxham, J., and A. Jackson, Time-dependent mapping of the magnetic field at the core-mantle boundary, *J. Geophys. Res.*, *97* (B13), 19,537–19,563, 1992.

Bondar, T. N., V. P. Golovkov, and S. V. Yakovleva, Spatiotemporal model of the secular variation of the geomagnetic field in the time interval from 1500 through 2000, *Geomagn. Aeron.*, *42*, 793–800, 2002.

Braginskiy, S. I., Spherical analyses of the main geomagnetic field in 1550–1800, *Geomagnetism and Aeronomy*, *12*, 464–468, 1972.

Courtillot, V., and J. L. Le Mouël, Time variations of the earth's magnetic field: from daily to secular, *Annu. Rev. Earth Planet. Sci.*, *16*, 389–476, 1988.

Courtillot, V., and J. P. Valet, Secular variation of the earth's magnetic field: from jerks to reversals, *Comptes Rendus de l'Academie des Sciences, Serie II. Sciences de la Terre et des Planetes*, *320*, 903–922, 1995.

Criss, R. E., Mid-continental magnetic declination: A 200-year record starting with Lewis and Clark, *GSA Today*, *13*, 4–11, 2003.

Doell, R. R., and A. Cox, The accuracy of the paleomagnetic method as evaluated from historic Hawaiian lava flows, *J. Geophys. Res.*, *68* (7), 1997–2009, 1963.

DuBois, R., Archaeomagnetic results from south-west United States and Mesoamerica, and comparison with some other areas, *Phys. Earth Planet. Inter.*, *56*, 18–33, 1989.

Dunlop, D. J., and M. B. Zinn, Archeomagnetism of a 19th century pottery kiln near Jordan, Ontario, *Can. J. Earth Sci.*, *17*, 1275–1285, 1980.

Eighmy, J. L., *Archaeomagnetism: A handbook for the archaeologist*, Heritage Conservation and Recreation Service Publication no. 58, Interagency Archaeological Services, U.S. Department of the Interior, Washington, D.C., 1980.

Eighmy, J. L., Archaeomagnetic dating: Practical problems for the archaeologist, in *Archaeomagnetic Dating*, edited by J. L. Eighmy and R. S. Sternberg, pp. 33–64, University of Arizona Press, Tucson, 1990.

Eighmy, J. L., and R. S. Sternberg (Eds.), *Archaeomagnetic Dating*, 446 pp., University of Arizona Press, Tucson, 1990.

Eighmy, J. L., R. S. Sternberg, and R. F. Butler, Archaeomagnetic dating in the American Southwest, *Antiquity*, *45*, 507–517, 1980.

Evans, M. E., and L. Jiang, Magnetomineralogy of archaeomagnetic materials, *J. Geomagn. Geoelectr.*, *48*, 1531–1540, 1996.

Faulkner, C. H, *An Archaeological and Historical Study of the James White Second Home Site*, Department of Anthropology Report of Investigations no. 28, University of Tennessee, Knoxville, 1984.

Gose, W. A., K. S. Collins, and M. B. Collins, Paleomagnetic studies from the Moore-Hancock Farmstead, Austin, Texas, *J. Field Archaeol.*, *21*, 125–129, 1994.

Gubbins, D., Global models of the magnetic field in historical times: Augmenting declination observations with archeo- and paleomagnetic data, *J. Geomagn. Geoelectr.*, *38*, 715–720, 1986.

Gurcke, K., *Bricks and Brickmaking: A Handbook for Historical Archaeology*, 326 pp., University of Idaho Press, Moscow, Idaho, 1987.

Hazard, D. L., *United States Magnetic Tables and Magnetic Charts for 1915*, Department of Commerce, U.S. Coast and Geodetic Survey, Washington, D.C., 1917.

Hedley, I. G., New directions in archaeomagnetism, *J. Radioanal. Nucl. Chem.*, *247*, 663–672, 2000.

Holcomb, R., D. Champion, and M. McWilliams, Dating recent Hawaiian lava flows using paleomagnetic secular variation, *Geol. Soc. Am. Bull.*, *97*, 829–939, 1986.

Hutcheson, K. A., and D. Gubbins, Earth's magnetic field in the seventeenth century, *J. Geophys. Res.*, *95* (7), 10,769–10,781, 1990.

Jackson, A., A. R. T. Jonkers, and M. R. Walker, Four centuries of geomagnetic data from historical records, *Philos. Trans. R. Soc. London, Series A*, *358*, 957–990, 2003.

Jones, C. H., User-driven integrated software lives: "Paleomag" paleomagnetics analysis on the Macintosh, *Comput. Geosci.*, *28*, 1145–1151, 2002.

Jonkers, A. R. T., A. Jackson, and A. Murray, Four centuries of geomagnetic data from historical records, *Rev. Geophys.*, *41*, 1–36, 2003.

Jordanova, N., E. Petrovsky, and M. Kovacheva, Preliminary rock magnetic study of archaeomagnetic samples from Bulgarian prehistoric sites, *J. Geomagn. Geoelectr.*, *49*, 543–566, 1997.

Jordanova, N., M. Kovacheva, I. Hedley, and M. Kostadinova, On the suitability of baked clay for archaeomagnetic studies as deduced from detailed rock-magnetic studies, *Geophys. J. Int.*, *153*, 146–158, 2003.

Kean, W. F., S. Ahler, F. M, and D. Wolfman, Archaeomagnetic record from Modoc Rock Shelter, Illinois, for the time range of 6200–8900 B.P., *Geoarchaeology*, *12*, 93–115, 1997.

Kirschvink, J. L., The least-squares line and plane and the analysis of paleomagnetic data, *Geophys. J. Roy. Astron. Soc.*, *62*, 699–718, 1980.

Kovacheva,, M., N. Jordanova, and V. Karloukovski, Geomagnetic field variations as determined from Bulgarian archaeomagnetic data. Part II: the last 8000 years, *Surveys in Geophysics 19*(5), 413–460, 1998.

Le Goff, M., Y. Gallet, A. Genevey, and N. Warmé, On archaeomagnetic secular variation curves and archaeomagnetic dating, *Phys. Earth Planet. Inter.*, *134*, 203–211, 2002.

Lengyel, S. N., and J. L. Eighmy, A revision to the U.S. Southwest archaeomagnetic master curve, *J. Archaeol. Sci.*, *29*, 1423–1433, 2002.

Malin, S. R. C., and S. E. Bullard, The direction of the earth's magnetic field at London, 1570–1975, *Philos. Trans. R. Soc. London*, *299*, 357–423, 1981.

Matushima, M., and Y. Honkura, Fluctuation of the standing and the drifting parts of the earth's magnetic field, *Geophys. J.*, *94*, 35–50, 1988.

Merrill, R. T., M. W. McElhinny, and P. L. McFadden, *The Magnetic Field of the Earth*, 531 pp., Academic Press, San Diego, 1996.

Noël, M. and C. M. Batt, A method for correcting geographically separated remanence directions for the purpose of archaeomagnetic dating, *Geophys. J. Int.*, *102*, 753–756, 1990.

Polhemus, R, *Archaeological Investigations of the Tellico Blockhouse Site*, Publications in Anthropology No. 16, Tennessee Valley Authority, Knoxville, 1977.

Schnepp, E., R. Pucher, C. Goedicke, A. Manzano, U. Müller, and P. Lanos, Paleomagnetic directions and thermoluminescence dating from a bread oven-floor sequence in Lübeck (Germany): A record of 450 years of geomagnetic secular variation, *J. Geophys. Res.*, *108* (B2), 1–13, doi:10.1029/2002JB001975, 2003.

Sternberg, R. S., Archaeomagnetic paleointensity in the American Southwest during the past 2000 years, *Phys. Earth Planet. Inter.*, *56*, 1–17, 1989a.

Sternberg, R. S., Secular variation of archaeomagnetic direction in the American Southwest, A.D. 750–1425, *J. Geophys. Res.*, *94* (B1), 527–546, 1989b.

Sternberg, R. S., Archaeomagnetic dating, in *Chronometric Dating in Archaeology*, edited by M. J. Aitken and R. E. Taylor, pp. 323–356, Plenum Press, New York, 1997.

Sternberg, R. S., Magnetic properties and archaeomagnetism, in *Handbook of Archaeological Sciences*, edited by D. R. Brothwell and A. M. Pollard, pp. 73–79, John Wiley & Sons, Chichester, 2001.

Sternberg, R. S., and R. H. McGuire, Techniques for constructing secular variation curves and for interpreting archaeomagnetic dates, in *Archaeomagnetic Dating*, edited by J. L. Eighmy and R. S. Sternberg, pp. 109–134, University of Arizona Press, Tucson, 1990.

Tanguy, J.-C., Abnormal shallow palaeomagnetic inclinations from the 1950 and 1972 lava flows, Hawaii, *Geophys. J. Int.*, *103*, 281–283, 1990.

Tarling, D. H., *Palaeomagnetism: Principles and Applications in Geology, Geophysics, and Archaeology*, 379 pp., Chapman and Hall, London, 1983.

Thompson, R., A comparison of geomagnetic secular variation as recorded by historical, archaeomagnetic and palaeomagnetic measurements, *Philos. Trans. R. Soc. London*, *306*, 103–112, 1982.

Thompson, R., and D. R. Barraclough, Geomagnetic secular variation based on spherical harmonic and cross validation analyses of historical and archaeomagnetic data, *J. Geomagn. Geoelectr.*, *34*, 245–263, 1982.

Watson, G. S., Analysis of dispersion on a sphere, *Mon. Not. Roy. Astron. Soc., Geophys. Suppl.*, *7*, 153–161, 1956.

Wolfman, D, Archaeomagnetic dating in Arkansas and border areas of adjacent states II, in *Archaeomagnetic Dating*, edited by J. L. Eighmy and R. S. Sternberg, pp. 237–260, University of Arizona Press, Tucson, 1990.

Yukutake, T., and H. Tachinaka, Spectrum of the earth's magnetic field into the drifting and the standing parts, *Bull. Earthquake Res. Inst.*, *47*, 65–97, 1969.

Stacey Lengyel, Department of Anthropology, University of Arizona, Tucson, Arizona 85721. (slengyel@u.arizona.edu)

Rob Sternberg, Department of Earth and Environment, Franklin & Marshall College, Lancaster, Pennsylvania 17604-3003. (Rob.Sternberg@FandM.edu)

Low Pacific Secular Variation

David Gubbins

School of Earth Sciences, University of Leeds, UK

Steven J. Gibbons

*School of Earth Sciences, University of Leeds, UK,
and Institutt for geologi, Universitet i Oslo, Norway*

The historical record shows that secular variation at Hawaii is limited to a few degrees in the last 400 years, whereas in the Atlantic hemisphere it often exceeds 30°. Paleomagnetic measurements from Hawaii show virtually no change in declination during the last 5 kyr and only a slow, millenium-scale inclination change of less than 20°. The usual directional scatter analysis of paleomagnetic data cannot discriminate between the two time scales. The disparity of time scales and difference in activity suggest different physical mechanisms for secular variation in the two hemispheres. This could arise from thermal core-mantle interaction. Seismic models of the lower mantle give a pattern of lateral variations that is nearly symmetric about the Pacific rim: two slow regions centered beneath the Pacific and Atlantic separated by a fast ring below the Pacific rim. These seismic anomalies are thought to be caused mostly by temperature variations in the bottom 200 km of the mantle. It is difficult to see how such a symmetric pattern could lead to long-term hemispheric differences. We show here a convection solution in a rapidly rotating sphere with heat flux on the outer boundary determined from a seismic model. The convection is suppressed beneath the Pacific but the usual drifting convection (Busse) rolls remain beneath the Atlantic. This mode of periodic convection arises because the Pacific hot region extends east-west and is much larger than a single convection roll, whereas the Atlantic hot region is elongated north-south and is about the same east-west size as a convection roll. This could explain the absence of normal, century-long secular variation in the Pacific: hot mantle suppresses short wavelength phenomena at the century time scale but not the longer wavelengths at the millenium time scale.

1. INTRODUCTION

There is growing evidence for influence of temperature anomalies in the overlying mantle on the Earth's magnetic field [*Bloxham*, 2000; *Bloxham and Gubbins*, 1987; *Olson*

Timescales of the Paleomagnetic Field
Geophysical Monograph Series 145
Copyright 2004 by the American Geophysical Union
10.1029/145GM21

and Glatzmaier, 1996]. Magnetic flux on the core-mantle boundary (CMB) tends to concentrate near regions of fast lower mantle seismic velocity, leading to departures from axial symmetry in both the present field [*Bloxham and Gubbins*, 1987] and the long-term time average [*Gubbins and Kelly*, 1993; *Johnson and Constable*, 1995; *Kelly and Gubbins*, 1997]; virtual geomagnetic poles tend to concentrate on a great circle enclosing the Pacific during reversals [*Laj et al.*, 1991]; and the Pacific region appears to have low secular

"

variation (SV) [*Merrill et al.*, 1998; *Walker and Backus*, 1996]. While all this evidence is the subject of current debate, the first two can be explained by a single theory in which cold mantle induces downwelling in the core and concentrates surface magnetic field. This, together with the repeated occurence of the same pair of longitudes around the Pacific where the seismic velocity is fast, is compelling evidence that the lower mantle does influence the geomagnetic field.

Rapid convection in the core maintains the temperature very close to the adiabat. The CMB may be considered isothermal for the purposes of mantle convection, which is much slower and allows departures from the adiabat of several hundred degrees. The appropriate boundary condition for core convection is fixed heat flux, including the lateral variations imposed by the convecting mantle. We may estimate the pattern, and to some extent the amplitude, of CMB heat flux from seismic tomography. If S-wave speed in the lowermost mantle is assumed to be caused entirely by temperature variations in a thermal boundary layer, the heat flow across the CMB is simply proportional to the variations in S-wave velocity. The constant of proportionality is difficult to estimate: it depends on the thermal conductivity of the lower mantle and the thickness of the boundary layer. A recent tomographic model [*Masters et al.*, 1996] of lower mantle S-wave velocity is shown in Figure 1. This is used as the pattern of boundary heat flux in the calculations of this paper. Other studies have used similar patterns. The dominant features are the ring of high velocity around the Pacific, taken to be cold mantle and high heat flow from the core, and low velocity in the central Pacific and Atlantic, taken to be low heat flow. The dominant

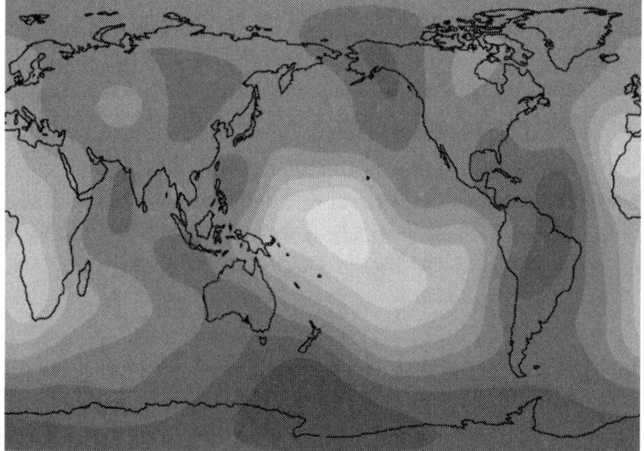

Figure 1. S-wave velocity anomalies for the lowermost 250 km of the mantle [*Masters et al.*, 1996], with spherical harmonic degree and order up to 10. Dark (light) regions represent higher (lower) velocities within a range of approximately ± 2.5%, corresponding to possible temperature variations of several hundred degrees in the lower mantle thermal boundary layer.

spherical harmonic in the expansion of the S-wave velocity has degree and order 2. Given such a pattern it is difficult to see how the Pacific hemisphere could behave any differently from the Atlantic hemisphere for a significant length of time. It is possible that the cold regions could produce a blocking effect, perhaps inhibiting SV in the Pacific for a few overturn times of core convection (1000 years), but it is very unlikely to do so for longer periods.

Zhang and Gubbins [1992] studied the effects of laterally varying heat flow on rotating convection in a sphere by calculating the steady thermal wind in the absence of heating from within and for stable stratification. Zhang and Gubbins [1993] explored the effects of simple patterns of boundary temperature on non-magnetic, rapidly rotating, infinite Prandtl number convection at low and moderate Rayleigh numbers. They found three effects: spatial resonance, where convection is dominated by the same azimuthal wavenumber m as the boundary condition; locking, where the drifting convection rolls become stationary; and secondary resonance, where convection was preferred with m a simple multiple of that of the applied boundary condition. Locking and resonance could, therefore, lead to a pattern of core convection that reflects the pattern of lower mantle heat flux.

Sun et al. [1994] studied some solutions at high Rayleigh number and found less evidence of locking or boundary influence. Zhang and Gubbins [1996] found locking to be less prevalent at finite Prantl number because of inertia. Olson and Glatzmaier [1996] were first to study magnetoconvection with boundary variations, finite Prandtl number, a highly supercritical Rayleigh number, and very strong lateral variations. Like Sun et al. [1994], they found no locking in either of the two combinations of parameters they examined. Sarson et al. [1997] were first to include inhomogeneous boundary conditions in a dynamo calculation. They found flux concentrated by downwellings but downwellings did not always coincide with high heat flux.

Inhomogeneous thermal boundary conditions have also been applied to dynamo models. Glatzmaier et al. [1999] examined a number of different boundary conditions and their effect on reversal behaviour, and found the most "realistic" reversal with homogeneous boundary conditions. Bloxham [2000] (see also *Bloxham* [2002]) studied the time-averaged field for a dynamo model with a spherical harmonic degree and order 2 pattern of heat flow and found strong similarity between the time average of his dynamo and the paleomagnetic time average of Kelly and Gubbins [1997]. He also speculated that low SV in the Pacific could arise from eastward convective flow, associated with displacement of the main convection rolls into that hemisphere, cancelling the general westward drift of the core as a whole. His model rarely generated a field like the present-day geomagnetic field and he concluded that the resemblance between the present-day field and the paleomagnetic time average is largely fortuitous. The most complete study to date of SV with a dynamo model is by Christensen and Olson [2003] [see also *Olson and Christensen*, 2002], who use a boundary heating based on the seismic tomography truncated back to spherical harmonic degree 4, 27% of the vertical heat flux, and three vertical heating strengths. Their model has rather weak SV for their main choice of heating and a highly variable dipole moment for stronger heating. They find strongest SV where the boundary heat flux is highest and variations in VGP scatter in both latitude and longitude. Westward drift is low in the east Pacific but not the west.

While some of these models hint at low SV in the hot Pacific region, none of them reproduce as clear a hemispheric division as we see in the historical record of Jackson et al. [2000]. The main question remains: is the present situation merely an occasional pattern that contributes to the time-average, as Bloxham suggests, or is it a semi-permanent feature of the geomagnetic field? If it is a long-term feature, how could it arise from a lower mantle heat flow pattern that is predominantly symmetric about the two hemispheres?

2. OBSERVATIONS

Hawaii offers an ideal dataset to examine SV. It lies close to the centre of the Pacific Basin, so that magnetic measurements made there are predominantly influenced by the CMB within the ring of high seismic velocity. It has a magnetic observatory and has been well sampled in both declination D and inclination I since James Cook's arrival in 1778. Earlier measurements of D were made nearby by navigators crossing the Pacific, the most notable voyage being that of Jacques L'Hermite [*Hutcheson*, 1990]. The global historical record dates from AD 1550; I measurements are rare anywhere before AD 1700 and non-existent in the Pacific; and intensity measurements are nonexistent before 1839. The time-dependent global field model of Jackson et al. [2000] gives reliable D in the Pacific back to at least 1600, and I can be relied upon back to 1700. Claims of a D anomaly in the Pacific in the 17th century ([*Yukutake*, 1993], discussed in *Merrill et al.* [1998]) are based on uncorrected catalogue data; using original sources [*Hutcheson and Gubbins*, 1990] and a vastly enlarged database [*Jackson et al.*, 2000] shows no strong SV in the Pacific in the last 400 years.

Determining SV from paleomagnetism in the Pacific has a long history, starting with Doell and Cox [1963]. A controversy centres on whether the lava flows sampled a sufficiently long time interval to provide a representative record. The time average in question is one that removes normal SV, namely fluctuations in the field that do not encompass tran-

sitions of polarity during excursions and reversals. Excursions occur every 20–50 kyr, so a shorter time is needed for the average to have its intended meaning. Bloxham found 5 kyr adequate for his dynamo, and 10–20 kyr should be plenty for the Earth. The time interval 1–50 kyr lies within the range of C^{14} dating and many Hawaiian lava flows have been dated. These dates, plus Hawaii's unique position in the central north Pacific, make the recent lava flows ideal for extending the historical record.

We first compare SV on Hawaii with that in the Atlantic hemisphere by plotting variations in D and I from model gufm [*Jackson et al.*, 2000] for a location on Kilauea volcano and a point on the same latitude on the Greenwich Meridian (Figure 2). Choosing the same latitude makes D and I directly comparable. Both D and I change by 30° in 150 years in the Atlantic, whereas the variation in Hawaii is less than 5°. This illustrates quite dramatically the lack of SV in Hawaii in historical times. The point on the Greenwich Meridian does not have particularly strong SV: the largest variations are in southern Africa. 400 years is sufficiently long to show the kind of variation seen in the Atlantic but it is too short a time to establish the Hawaiian low as a permanent feature: for that we need paleomagnetism.

The historical record also provides an opportunity to check the accuracy of published paleomagnetic data using lava flows that have erupted in historical times[1]. These are also plotted in Figure 2. They are consistent with model gufm. The systematic low paleomagnetic I has not been noted before: it could be caused by magnetisation of the seamount. Even with this systematic error paleomagnetism can detect Altantic-type SV: 30° is a huge variation and 150 years a short time. The problem is not accuracy of measurement but resolution in time.

Many recent lava flows have been dated by C^{14} [*Lockwood*, 1995]. While there are known problems with C^{14} dating in volcanic environments, this is still the best dating series available for SV studies anywhere. Paleomagnetism has been carried out on most of the younger lavas [*Hagstrum and Champion*, 1995]. Their compilation of D and I are plotted in Figure 3 for the last 5 kyr. If we drew a smooth curve through each of these sets of points, allowing 10° scatter to account for the paleomagnetic errors as suggested by the historical flows and for errors in the dating, we would conclude that D is effectively constant and I has suffered a dip of about 20° at age 1–2 ka. We could draw a much rougher line, with the 150-year time variation of Atlantic SV, and fit the points more accurately. Such a curve would be physically plausible because we see similar rapid SV in the Atlantic region today, but it would require over-fit-

---

[1]Historical flows include all those recorded since Cook's arrival in the Islands, including the AD 1750 and 1775 flows that were within living memory on Cook's arrival.

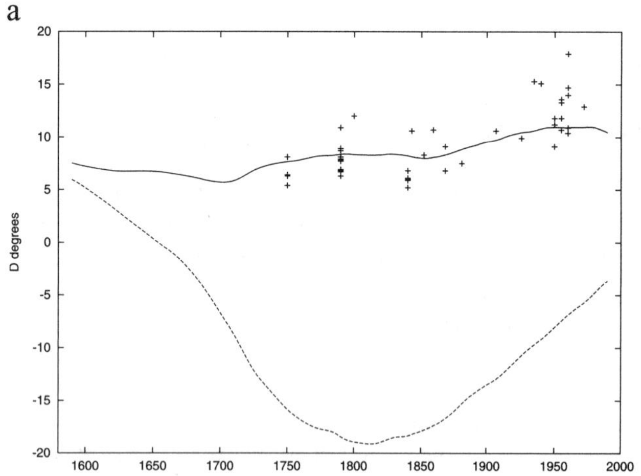

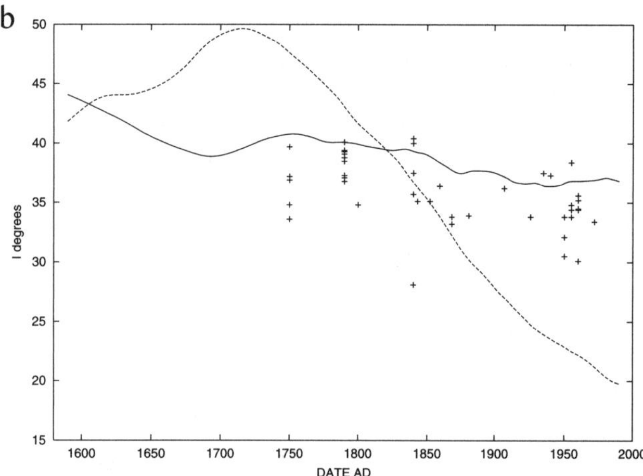

Figure 2. Comparison of magnetic field components on Hawaii (solid line) and representative point on same latitude in the Atlantic hemisphere (dashed). (a) *D* (b) *I*. Points give paleomagnetic measurements on historical lava flows.

ting of both the paleomagnetic and the age data. The only reasonable interpretation of Figure 3 is a slow change in *I* with a possible century-timescale SV limited in amplitude to about ± 10° in both *D* and *I*. The larger change in *I* (20°) takes place over a much longer time scale of about 5 kyr and is significantly smaller in amplitude than than is observed in the Atlantic hemisphere in a mere 150 years (>30°).

If we were to abandon the time series and examine VGP scatter statistics we would lose the difference in time scales between Atlantic and Pacific SV. Instead, we would find Hawaii registering a VGP scatter somewhat smaller than the global mean, which is the case for most of the Brunhes [*Love and Constable*, 2003]. This is not a very strong result, but the difference in timescales demands a different physical mechanism for the short term SV in the Atlantic, which defines normal "geomagnetic" SV, and the long term variations in

the Pacific, which can only be detected by paleomagnetism and is usually called "paleomagnetic" SV.

3. CORE CONVECTION

We have explored the influence of lateral variations in boundary heat flux on non-magnetic, rotating, convection in a spherical shell with the same dimensions as the Earth's outer core. A fixed heat flux is imposed on the outer boundary; fixed temperature on the inner boundary. We studied the thermal winds driven by the outer boundary condition in an earlier paper [*Gibbons and Gubbins*, 2000], which gives further details of the model formulation; in this paper we also drive convection from below. The solutions are time-dependent; they are obtained using the *GJZ* code from the convection and dynamo benchmark study [*Christensen et al.*, 2001] . The only modification necessary was the implementation of lat-

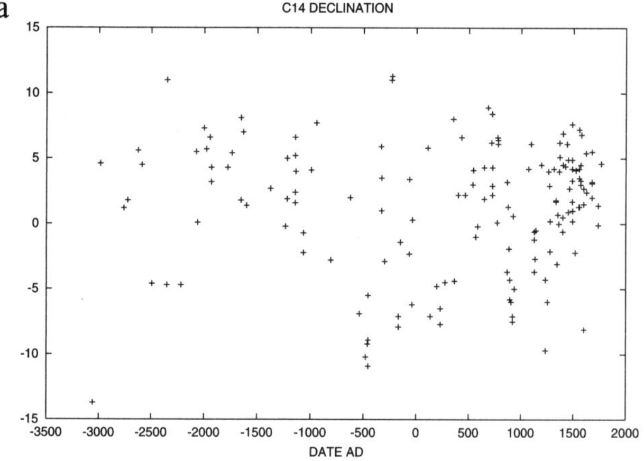

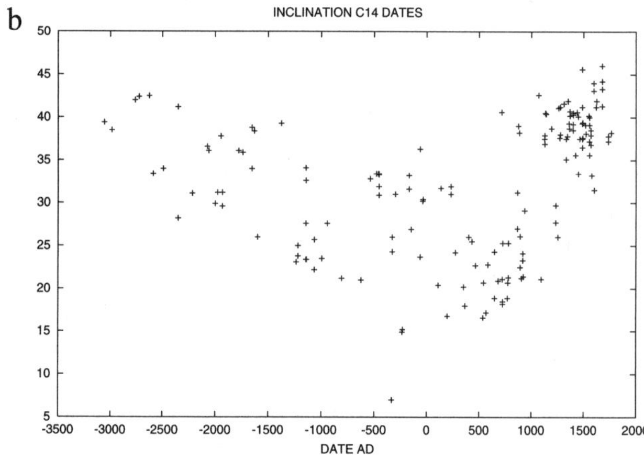

Figure 3. *D* and *I* from Hawaiian lavas with C[14] ages less than 5 ka. Note the scale on the *y*-axis: *D* varies by only ± 5° over 5 kyr compared with over 30° change in Figure 3(a). *I* dips by just 15° from 0–1000 AD, compared with a 30° change in 300 years in Figure 3(b).

erally heterogeneous heating at the boundary, which was tested by reproducing the time-dependent results of Zhang and Gubbins [1996].

The solution depends on four parameters: the Ekman number E, which measures the importance of rotation; the Prandtl number P_r, the ratio of diffusivities; the vertical Rayleigh number Ra^V, which measures the strength of vertical heating; and the horizontal Rayleigh number Ra^H, which measures the strength of the boundary variations:

$$E = \frac{\nu}{2\Omega d^2}, \quad P_r = \frac{\nu}{\kappa}, \quad Ra^V = \frac{\alpha\beta\gamma d^6}{\nu\kappa},$$

where ν is the kinematic viscosity, κ the thermal diffusivity, Ω the rotation rate, d the shell (outer core) thickness, α the coefficient of thermal expansion, β the mean temperature gradient on the boundary, and γ gravitational acceleration divided by radius. Ra^H is defined in the same wave as Ra^V with β replaced by the variation in temperature gradient around the boundary.

We illustrate the result relevant to this paper with three simple calculations. All have $E = 2 \times 10^{-4}$, $P_r = 1$, and $Ra^V = 1.15 \times 10^5$; they differ by $Ra^H = 0.1Ra^V, 0.3Ra^V, 0.7Ra^V$. The solution with homogeneous boundary conditions ($Ra^H = 0$) has critical $Ra^V_C = 1.04 \times 10^5$, so our choice of Rayleigh number (Ra^V) is $1.1\ Ra^V_C$. The solution at onset takes the usual form of prograde- (eastward)-drifting convection ("Busse") rolls aligned with the spin axis and stationary in a co-rotating frame of reference. The critical number of convection rolls is $m = 7$.

The three solutions with inhomogeneous boundary conditions are shown in Figure 4. All solutions are periodic in time, but they do not drift steadily. They cannot be stationary in a co-rotating frame because the boundary conditions are time-dependent in such a frame. The solution with weakest boundary variation, $Ra^H = 0.1Ra^V$, is very similar to the homogeneous case. The convection rolls are only slightly distorted by the boundary condition and the drift rate is nearly uniform. The second solution is dramatically different and exhibits convection rolls only in one half of the sphere, away from the Pacific region. The third solution is dominated by $m = 2$ convection and is almost stationary.

These results can be interpreted in terms of the resonance between the applied boundary heat flux pattern and the natural length scale of convection with homogeneous boundary conditions, as established in the study by Zhang and Gubbins [1993]. Weak boundary heating simply alters the uniform drift of the convection rolls so that it slows when convection brings heat up beneath regions of high boundary heat flux, and speeds up when convection brings heat up beneath regions of low boundary heat flux. There is no locking because the predominant azimuthal wavenumber ($m = 8$, Figure 4a) is so different from that of the boundary heating ($m = 2$). Strong boundary heating causes the resonance with $m = 2$ convection, and almost locks it (Figure 4c).

The intermediate case, $Ra^H = 0.3Ra^V$, is the most interesting. The drifting roll structure is maintained for about half the sphere. Rolls initiate to the east of the Pacific hot region and drift east, then dissipate at the western end of the hot region. An animation shows that, in between, they drift fairly uniformly, even beneath the Atlantic hot region. The Pacific hot region covers a far wider range of longitude than the Atlantic hot region and the resulting stable stratification penetrates far deeper into the core beneath the Pacific than beneath the Atlantic. The length scale of the underlying convection, which has $m = 8$, is comparable to the longitudinal extent of the Atlantic hot region but is considerably smaller than that of the Pacific hot region, with the result that rolls on this scale are completely suppressed beneath the Pacific, as could be expected with 30% lateral variation in heating and only 10% supercritical heating.

The appearance of rolls at one side of the Pacific and their dissipation at the other is, apart from the sense of drift, reminiscent of the way in which features in the geomagnetic field at the CMB drift westwards from the longitude of Australia and Indonesia towards that of the west coast of the Americas but no further. This very simple convection result can explain why there is a difference between the Atlantic and Pacific hemispheres despite a predominantly symmetric lower mantle structure: the Pacific anomaly is sufficiently extensive in longitude to suppress convection whereas the Atlantic anomaly is not.

Eastward rather than westward drift is an obvious limitation of this model. It is well known that, under the majority of conditions, convection rolls drift prograde (eastward) in the absence of a magnetic field (see, for example, *Zhang and Busse* [1987]). The presence of a magnetic field complicates things greatly; Zhang and Gubbins [2002] demonstrate how convection propagates either prograde or retrograde depending upon the exact form of the magnetic field applied. Most of the current numerical dynamo simulations (see, for example, *Christensen et al.*, [1999]) display retrograde or westward drifiting rolls. A companion study to this one, including magnetic fields, is underway and will be reported on elsewhere.

4. DISCUSSION

The historical record shows that ordinary geomagnetic ("Atlantic") SV is almost entirely absent from Hawaii. Paleomagnetism cannot contradict this result because it lacks the necessary time resolution to detect rapid SV. Paleomagnetism has recorded changes in direction in Hawaii, but they are smaller and ten times slower than those in the Atlantic region. A different physical mechanism is needed for the two hemi-

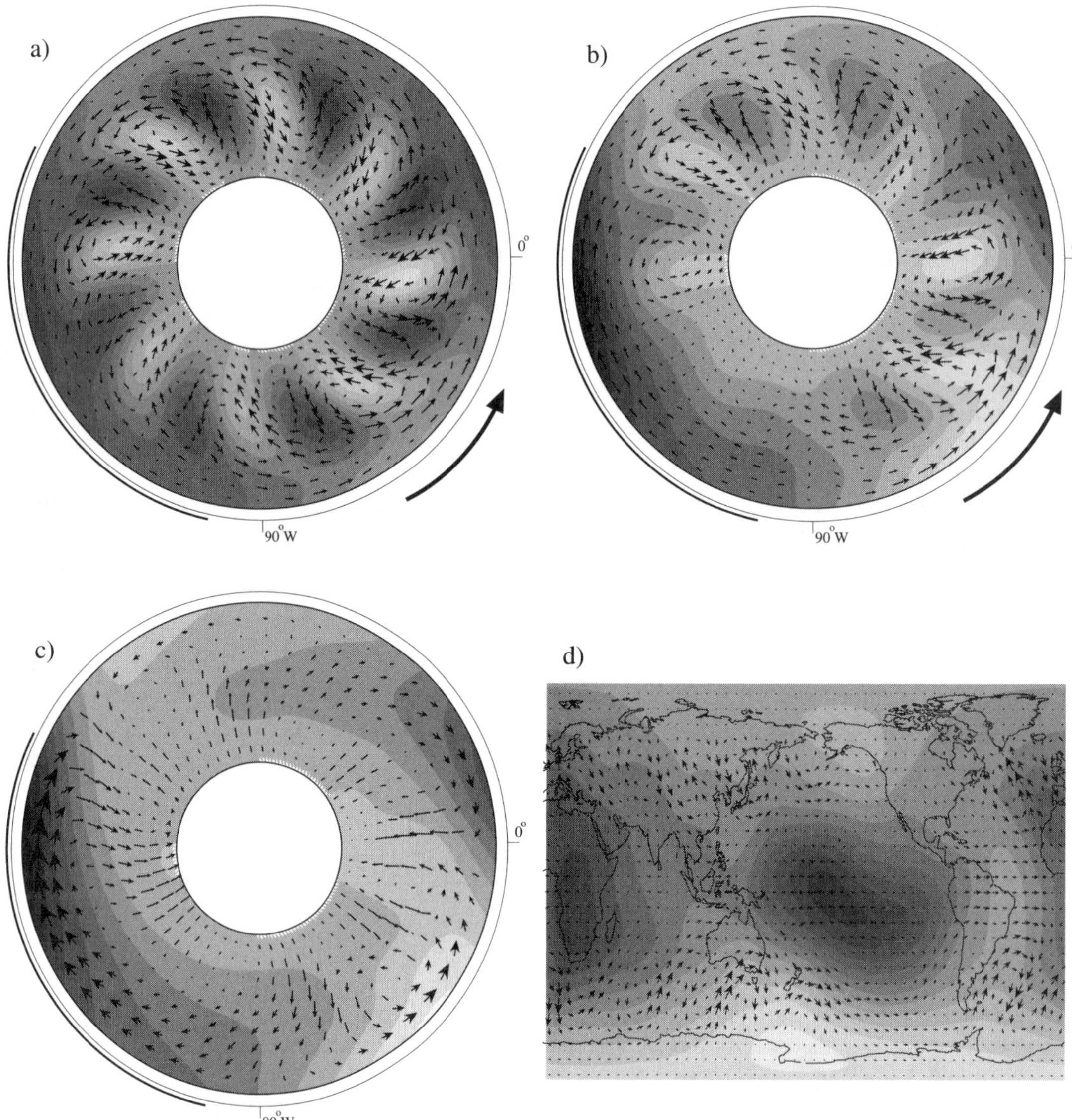

Figure 4. Snapshots of convective solutions subject to lateral heat-flux variations at the outer surface. Plots (a), (b), and (c) show contours of temperature perturbations and arrows of flow in an equatorial section for $Ra^H = 0.1Ra^V$, $Ra^H = 0.3Ra^V$, and $Ra^H = 0.7Ra^V$, respectively. Darker shades represent higher temperatures. The large arrows indicate the direction of propagation of the convection rolls and the thick arc indicates the approximate extent of the Pacific Ocean. Plot (d) shows temperature and flow at the outer surface corresponding to the solution in (b).

spheres; the simple convection model described here suggests that the effects of convection, at least in the upper reaches of the core, is suppressed beneath the Pacific. Hawaiian paleomagnetism could then register only deep and distant convective effects that involve global changes in the entire geomagnetic field.

Suppression of convection beneath the Pacific but not the Atlantic regions in our model arises because of the subtle difference in shape of these two main hot regions at the bottom of the mantle. The Pacific region is elongated east-west whereas the Atlantic region is elongated north-south. The Atlantic region is about as wide as a single convection roll in the calculation, which has $m = 8$. The Pacific region is much wider, occupying nearly half the core's circumference, and completely suppresses the smaller rolls.

None of the geodynamo simulations to date reproduce a hemispheric difference with anything like the clarity of either the Earth or the simple convection solution presented here.

It may be that we have not yet found the appropriate dynamo regime. The Rayleigh number may be too high in all the dynamo simulations. We know that high Ra^V ultimately removes the Taylor constraint and destroys the usual roll structure. This probably explains the lack of boundary effects in the high Rayleigh number studies of Sun et al. [1994] and Olson and Glatzmaier [1996]. When Christensen and Olson [2003] apply a sufficiently high Ra^V to obtain the correct time scale for SV the main dipole becomes weak and highly variable. Lowering Ra^V at these values of E results in dynamo failure, probably because the flow lacks sufficient chaotic time dependence to prevent flux concentration [*Gubbins et al.*, 2000].

The Earth's core has a much smaller E than the simulations. Heat flux constraints limit Ra^V to reasonable values. Jones [2000] and Gubbins [2001], using independent methods, estimate Ra^V to be less than 1,000 times the critical value for magnetoconvection, or about equal to the critical value for non-magnetic convection assuming turbulent values for the diffusivities. The numerical simulations are necessarily run at $Ra^V/Ra^V_c \approx 20$ or greater in order to get dynamo action, but such high supercritical values may not be needed at lower E. Keeping Ra^V/Ra^V_c constant and small while reducing E might ultimately yield dynamo action in the right regime, but such low Ekman numbers have yet to be achieved.

5. CONCLUSIONS

The historical record from Hawaii lacks any significant century-scale SV. The paleomagnetic record does show millenium-scale variations that are rather smaller than typical century-scale SV. The disparity in time scale demands different physical processes. Lateral variations in shear wave velocity have been taken as representative of temperature in the lower mantle boundary layer and hence heat flux through the CMB. The pattern is dominated by the spherical harmonic Y_2^2 and would not, at first sight, be expected to produce a long-term difference in core dynamics between the Pacific hemisphere and the Atlantic side. Recent geodynamo simulations using the seismic velocity for a thermal boundary condition do exhibit some hemispheric differences in the time average but differences in snapshots of the magnetic field at instants of time are difficult to discern. We have found a simple, periodic, example of non-magnetic convection showing permanent hemispheric differences arising from subtle departures of the boundary conditions from the Y_2^2 harmonic. The conditions for this solution to hold are that the underlying convection contain significant energy in azimuthal wavenumbers around 8, which allows convection rolls that can drift past the narrow Atlantic low velocity anomaly but not the broader Pacific anomaly. We suggest no similar dynamo example has yet been found because the Rayleigh number required for dynamo action at these high Ekman num-

bers is too large to produce energy in the right wavenumber band for resonance with the boundary conditions.

Acknowledgments. The work was supported in part by Natural Environment Research Council grant GR3/12825. We are grateful to Guy Masters and Peter Shearer for providing their shear wave velocity model.

REFERENCES

Bloxham, J., The effect of thermal core-mantle interactions on the paleomagnetic secular variation, *Philos. Trans. R. Soc. London Ser. A, 358,* 1171–1179, 2000.

Bloxham, J., Time-independent and time-dependent behaviour of high-latitude flux bundles at the core-mantle boundary, *Geophys. Res. Lett., 29,* 1854, 2002.

Bloxham, J. and D. Gubbins, Thermal core-mantle interactions, *Nature, 325,* 511–513, 1987

Christensen, U., J. Aubert, P. Cardin, E. Dormy, S. Gibbons, G. A. Glatzmaier, E. Grote, Y. Honkura, C. Jones, M. Kono, M. Matsushima, A. Sakuraba, F. Takahashi, A. Tilgner, J. Wicht and K. Zhang, A numerical dynamo benchmark, *Phys. Earth Planet. Int., 128,* 25–34, 2001.

Christensen, U. R. and P. Olson, Secular variation in numerical geodynamo models with lateral variations of boundary heat flow, *Phys. Earth Planet. Int., 138,* 39–54, 2003.

Christensen, U., P. Olson and G. A. Glatzmaier, Numerical modelling of the geodynamo: a systematic parameter study, *Geophys. J. Int., 138,* 393–409, 1999.

Doell, R. R. and A. Cox, The accuracy of the paleomagnetic method as evaluated from historic Hawaiian lava flows, *J. Geophys. Res., 68,* 1997–2009, 1963.

Gibbons, S. and D. Gubbins, Convection in the Earth's core driven by lateral variations in the core-mantle boundary heat flux, *Geophys. J. Int., 142,* 631–642, 2000.

Glatzmaier, G. A., R. S. Coe, L. Hongre, and P. H. Roberts, The role of the Earth's mantle in controlling the frequency of geomagnetic reversals, *Nature, 401,* 885–890, 1999.

Gubbins, D., The Rayleigh number for convection in the Earth's core, *Phys. Earth Planet. Int., 128,* 3–12, 2001.

Gubbins, D., C. N. Barber, S. Gibbons, and J. J. Love, Kinematic dynamo action in a sphere: I Effects of differential rotation and meridional circulation on solutions with axial dipole symmetry, *Proc. R. Soc., 456,* 1333–1353, 2000.

Gubbins, D. and P. Kelly, Persistent patterns in the geomagnetic field during the last 2.5 Myr, *Nature, 365,* 829–832, 1993.

Hagstrum, J. T. and D. E. Champion, Late Quaternary geomagnetic secular variation from historical and C-14-dated lava flows on Hawaii, *J. Geophys. Res., 100,* 24393–24403, 1995.

Hutcheson, K., Geomagnetic Field Modelling, Cambridge Univ., PhD Thesis, Cambridge, 1990.

Hutcheson, K. and D. Gubbins, A model of the geomagnetic field for the 17th century, *J. Geophys. Res., 95,* 10,769–10,781, 1990.

Jackson, A., A. R. T. Jonkers and M. R. Walker, Four centuries of geomagnetic secular variation from historical records, *Philos. Trans. R. Soc. London Ser. A, 358,* 957–990, 2000.

Johnson, C. and C. Constable, The time-averaged geomagnetic field as recorded by lava flows over the past 5 Myr, *Geophys. J. Int., 122*, 489–519, 1995.

Jones, C. A. , Convection-driven geodynamo models, *Proc. R. Soc., 873*, 873–897, 2000.

Kelly, P. and D. Gubbins, The geomagnetic field over the past 5 million years, *Geophys. J. Int., 128*, 315–330, 1997.

Laj, C. A., A. Mazaud, M. Weeks, M. Fuller, and E. Herrero-Bervera, Geomagnetic reversal paths, *Nature, 351*, 447, 1991.

Lockwood, J. P., Mauna Loa Eruptive History—the preliminary radiocarbon record, in *Mauna Loa Revealed: struture, composition, history and hazards,* pp. 81–94, AGU Geophysical Monograph 92, 1995.

Love, J. J. and C. G. Constable, Gaussian Statistics for Paleomagnetic Vectors, *Geophys. J. Int., 152*, 515–565, 2003.

Masters, G., S. Johnson, G. Laske, and H. Bolton, A shear-velocity model of the mantle, *Phil. Trans. R. Soc. Lond. A, 354*, 1385–1411, 1996.

Merrill, R. T., M. W. McElhinny and P. L. McFadden, *The Magnetic Field of the Earth (Paleomagnetism, the core, and the deep mantle)*, Academic, San Diego, Calif., 1998.

Olson, P. and U. R. Christensen, The time-averaged magnetic field in numerical dynamos with non-uniform boundary heat-flow, *Geophys. J. Int., 151*, 809–823, 2002.

Olson, P. and G. Glatzmaier, Magnetoconvection and thermal coupling of the Earth's core and mantle, *Philos. Trans. R. Soc. London Ser. A, 354*, 1–12, 1996.

Sarson, G. R., C. A. Jones, and A. W. Longbottom, The influence of boundary region heterogeneities on the geodynamo, *Phys. Earth Planet. Int., 101*, 13–32, 1997.

Sun, Z. P. , G. Schubert, and G. A. Glatzmaier, Numerical simulations of thermal convection in a rapidly rotating spherical shell cooled inhomogeneously from above, *Geophys. Astrophys. Fluid Dyn., 75*, 199–226, 1994.

Walker, A. D. and G. E. Backus, On the difference between the average values of B_r^2 in the Atlantic and Pacific hemispheres, *Geophys. Res. Lett., 23*, 1965–1968, 1996.

Yukutake, T. , The geomagnetic non-dipole field in the Pacific, *J. Geomagn. Geoelectr., 45*, 1441–1453, 1993.

Zhang, K. and F. H. Busse, On the onset of convection in rotating spherical shells, *Geophys. Astrophys. Fluid Dyn., 39*, 119–147, 1987.

Zhang, K. and D. Gubbins, On convection in the earth's core forced by lateral temperature variations in the lower mantle, *Geophys. J. Int., 108*, 247–255, 1992.

Zhang, K. and D. Gubbins, Convection in a rotating spherical fluid shell with an inhomogeneous temperature boundary condition at infinite Prandtl number, *J. Fluid Mech., 250*, 209–232, 1993.

Zhang, K. and D. Gubbins, Convection in a rotating spherical fluid shell with an inhomogeneous temperature boundary condition at finite Prandtl number, *Phys. Fluids, 8*, 1141–1148, 1996.

Zhang, K. and D. Gubbins, Convection-driven hydromagnetic waves in planetary fluid cores, *Math. Comp. Modelling, 36*, 389–401, 2002.

S. J. Gibbons, NORSAR, Box 53, 2027 Kjeller, Norway.

D. Gubbins, School of Earth Sciences, University of Leeds, Leeds LS2 9JT, UK.

Intensity-Inclination Correlation for Long-Term Secular Variation of the Geomagnetic Field and Its Relevance to Persistent Non-Dipole Components

Toshitsugu Yamazaki and Hirokuni Oda

Institute for Marine Resources and Environment, Geological Survey of Japan, AIST, Tsukuba, Japan

Yamazaki and Oda [2002] reported long-term secular variation in inclination with ~100 kyr periodicity and coherency between paleointensity and inclination variations using a sediment core from the western equatorial Pacific. In this paper, we first conducted a paleomagnetic study of another core taken nearby (~120 km away) to examine the reproducibility of the records. The new core covers the last 1.1 Myr. Long-term inclination variations from the new core are almost identical to those of the previous core after conversion to a common age scale by inter-core correlation based on relative geomagnetic paleointensity. Mutually consistent long-term inclination records were also obtained from two cores of Brunhes age in the north central Pacific after inter-core correlation using relative paleointensity. Next, we compared intensity-inclination correlations of cores in the western equatorial Pacific (WEP) with those in the north central Pacific (NCP). A cross-correlation analysis showed that the coherency between intensity and inclination is strong in the WEP but weak in the NCP. Another remarkable difference in inclinations of the two regions is that the cores in the WEP have a large negative inclination anomaly (ΔI), whereas the cores in the NCP have a small ΔI. These observations support the model of Yamazaki and Oda [2002] that long-term secular variations in inclination are controlled by changes in the relative strength of the geocentric axial dipole and persistent non-dipole components. This model predicts that the intensity-inclination correlation should be obvious in regions where ΔI is large, but not where ΔI is small.

1. INTRODUCTION

Secular variations of the geomagnetic field during the Holocene have been intensively studied for more than 30 years. Recent progress in relative paleointensity estimation from marine sediment cores has revealed long-term (10 to 100 kyr) secular changes of paleointensity during the Pleistocene [e.g., *Valet and Meynadier*, 1993; *Guyodo and Valet*, 1999; *Kok and Tauxe*, 1999; *Horng et al.*, 2003]. However,

Timescales of the Paleomagnetic Field
Geophysical Monograph Series 145
Copyright 2004 by the American Geophysical Union
10.1029/145GM22

directional secular variation records older than the last ca. 10 kyr are still scarce, although possible occurrences of long-term directional changes in the late Pleistocene have been suggested by several authors [*Negrini et al.*, 1988; *Lund et al.*, 1988; *Levi and Karlin*, 1989; *Creer et al.*, 1990; *Yamazaki and Ioka*, 1994; *Holt et al.*, 1996]. Recently, Yamazaki and Oda [2002] reported a 100-kyr periodicity in an inclination record during the last 2.3 Myr derived from a long piston core (MD982185) obtained from the West Caroline Basin (WCB) in the western equatorial Pacific Ocean, although the presence or absence of such orbital frequencies in inclination is still a matter of debate [*Yamazaki*, 2002; *Horng et al.*, 2003; *Roberts et al.*, 2003]. The amplitude of long-term inclination variations, even if they occur, is considered to be small (several degrees at most). Spurious changes from various error sources, including deformation of sediments and magnetic overprinting of viscous remanent magnetization (VRM) origin can easily obscure geomagnetic signals. Furthermore, global variations may also be obscured by local changes. It is, thus, necessary to accumulate inclination records with precise age control from geographically widespread areas to deduce global geomagnetic features.

Yamazaki and Oda [2002] reported an intriguing correlation between inclination and paleointensity variations in the core from WCB. A cross-correlation analysis showed that inclination and intensity are in-phase during the Brunhes chron: inclination tends to become more negative when paleointensity is low, whereas the two are anti-phase during the Matuyama chron between the Jaramillo and Olduvai subchrons. To explain this relationship, they proposed a model that long-term secular variations in inclination are controlled by changes in relative strength of the geocentric axial dipole (GAD) and persistent non-dipole components. Studies on the time-averaged geomagnetic field during the last 5 Myr revealed a relatively large contribution of a quadrupole component (g_2^0), ca. 5% of GAD, and that it flips sign with a reversal of GAD [Johnson and Constable, 1997; Hatakeyama and Kono, 2002]. The effect of the g_2^0 component is observed as the inclination anomaly (ΔI) in marine sediment cores in low latitudes [Schneider and Kent, 1990]. The inclination anomaly is defined as the observed inclination minus that expected from a GAD field. Because ΔI is negative in low latitudes during the Brunhes Chron, the model of Yamazaki and Oda [2002] predicts more negative inclinations in the WCB when GAD is weaker. Yamazaki [2002] examined the intensity-inclination correlation during the Brunhes Chron on the Ontong-Java Plateau using three short cores, which are also located near the equator but about 2800 km east of the coring site in the WCB, and obtained similar results. However, the inclination-intensity correlation has not yet been examined in middle to high latitudes, where ΔI is smaller than in low latitudes.

The Sint-800 paleointensity curve was established by integration of 33 globally distributed relative paleointensity records during the last 800 kyr [Guyodo and Valet, 1999]. The paleointensity variation represented by the Sint-800 curve is globally synchronous, therefore it can be used as a dating tool. Furthermore, recent studies using high-deposition-rate cores developed regional paleointensity master curves in the North Atlantic (NAPIS) and South Atlantic (SAPIS), and suggested that shorter-wavelength features, possibly up to millennial scale, are also globally synchronous or are at least coherent on a regional scale [Laj et al., 2000; Channell et al., 2000; Stoner et al., 2000, 2002]. This demonstrates that relative geomagnetic

paleointensity can be used as a high-resolution global and regional correlation tool. Relative paleointensity can also provide a common age scale for studying long-term directional secular variations [Yamazaki, 2002].

In this paper, we first present results of a paleomagnetic study from another long piston core from the WCB (core MD982183) to show its reproducibility with the record from core MD982185 [Yamazaki and Oda, 2002]. We utilize inter-core correlation based on relative paleointensity to confirm regional coevality of inclination variations and establish the occurrence of long-term inclination changes. Then, we discuss coherency between inclination and paleointensity variations during the Brunhes Chron based on records from the western equatorial Pacific and the north central Pacific.

2. CORE MD982183

2.1. Sampling and Measurements

A long piston core MD982183 of about 37m length was obtained from the WCB at 2°00.8′N, 135°01.3′E during the IMAGES IV cruise (Figure 1). The site is located about 120 km south of core MD982185. The water depth of the site, 4388m, is close to the carbonate compensation depth (CCD), which is at about 4600m at present [Berger et al., 1976]. The

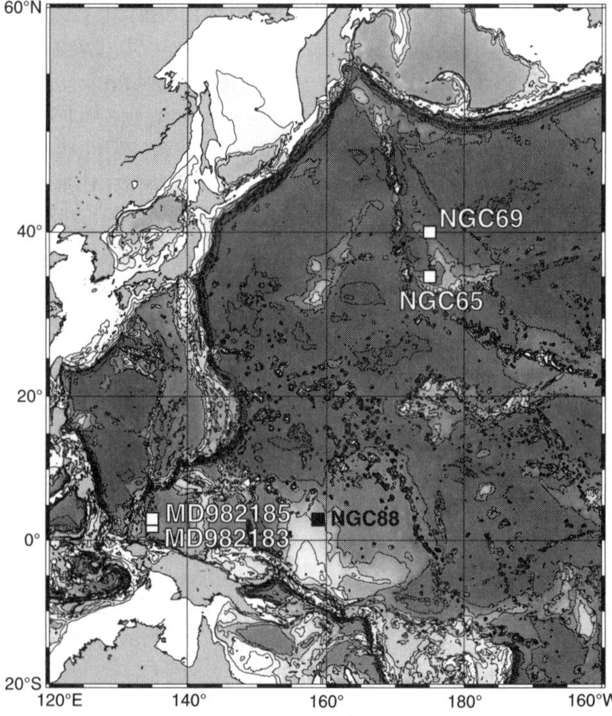

Figure 1. Location of cores used in this study superimposed on a simplified bathymetric map.

core consists mainly of hemipelagic clay with calcareous and siliceous microfossils, which is similar to that of core MD982185. The sediments show cyclic variations in carbonate content, which reflect glacial-interglacial changes of carbonate production and/or preservation, with higher carbonate content in glacial periods. Samples for paleomagnetic measurements were taken soon after the cruise from split core sections using u-channels of 1.5 m length with a cross-section of 2×1.8 cm^2. The uppermost 1.5 m section was the exception, and discrete cubic samples of 10 cm^3 were taken from it.

Magnetic susceptibility was measured at 1 cm intervals using a Bartington Instruments MS2 susceptometer with a loop sensor of 4.5 cm diameter. Remanent magnetization was measured at 1 cm intervals with a cryogenic magnetometer system (2G Enterprises model 760R) with in-line alternating-field (AF) demagnetizer. The bore diameter of the magnetometer is 7.6 cm. Stepwise AF demagnetization up to 80 mT was performed on the natural remanent magnetization (NRM). Anhysteretic remanent magnetizations (ARMs) were imparted by superimposing a DC biasing field of 0.1 mT on a smoothly decreasing AF with a peak field of 60 mT. To increase the resolution of the records, the deconvolution procedure of Oda and Shibuya [1996] was applied to the u-channel data: NRM data for each demagnetization step and the ARM data. For magnetic susceptibility data, the deconvolution procedure of Oda and Shibuya [1994] was applied. An example of the performance of the deconvolution is shown in Figure 2. The spatial resolution after the deconvolution is considered to be about 2 cm, which is close to that for discrete samples.

Stepwise AF demagnetization showed that the remanent magnetization of most samples consists of a single component except for the first few demagnetization steps. A magnetic overprint of probable VRM origin was gradually removed by AF up to 10 mT. Exceptionally, an AF of 20 to 30 mT was required to erase overprints in horizons where the NRM intensity is low. Examples shown are from the lower part of section 10 (Figures 2 and 3), which may correspond to the Iceland Basin event at about 190 ka [*Channell*, 1999; *Oda et al.*, 2002] as shown later.

2.2. Results

Paleomagnetic directions after AF demagnetization at 30 mT are shown in Figure 4. The Brunhes-Matuyama boundary is recognized at 30.3 m on the basis of the declination, and the Jaramillo Subchron is recognized from 34.0 to 35.3 m (Figure 4c). The age of the bottom of the core is estimated to be about 1.15 Ma from the magnetostratigraphy. The average sedimentation rate is about 32 m/Myr. It is about 70% higher than that of core MD982185 (2.25 Ma at 42 m in depth).

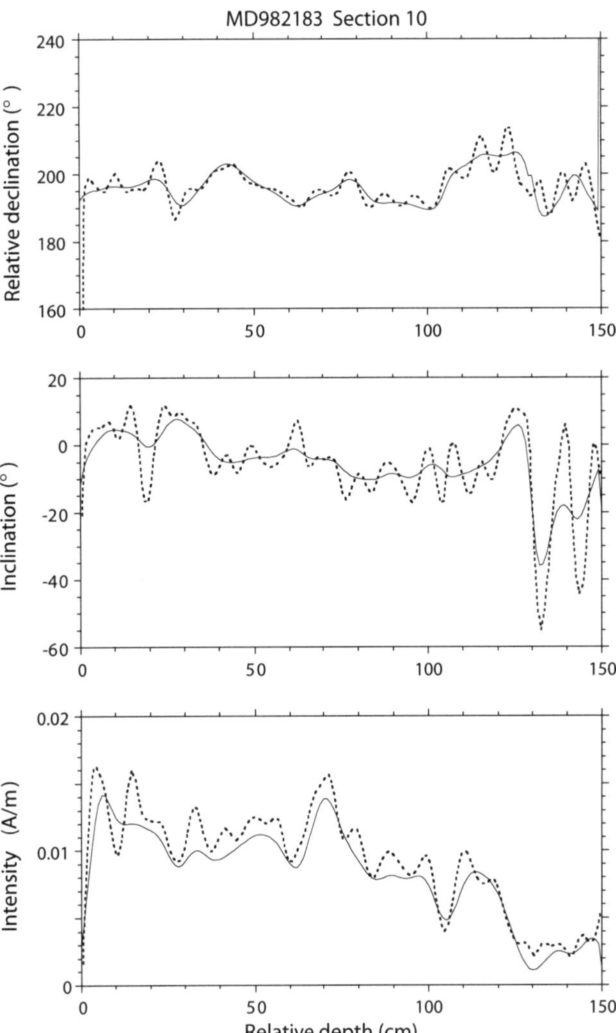

Figure 2. Natural remanent magnetization (NRM) of section 10 of core MD982183 sampled using u-channels after alternating-field (AF) demagnetization of 30 mT. The broken curves represent data after applying the deconvolution procedure of Oda and Shibuya [1996], and solid curves are the data before deconvolution. The top of the section is at 12.42 m from the top of the core.

Abrupt shifts in declination occur at several horizons, in particular in the upper part of the core (Figure 4c). Most of them correspond to core gaps or section boundaries, and are considered to be caused by core twisting. The average inclination is close to zero, and quasiperiodic variations with wavelengths on the order of several meters can be recognized (Figure 4d).

The magnetic susceptibility (k) fluctuates with carbonate content because of varying dilution, and is lower in glacial periods. Variations in k therefore mimic the oxygen-isotope (δ^{18}O) curve (Figure 4a). This style of k fluctuation is commonly observed in sediments from the WCB [*Yamazaki and*

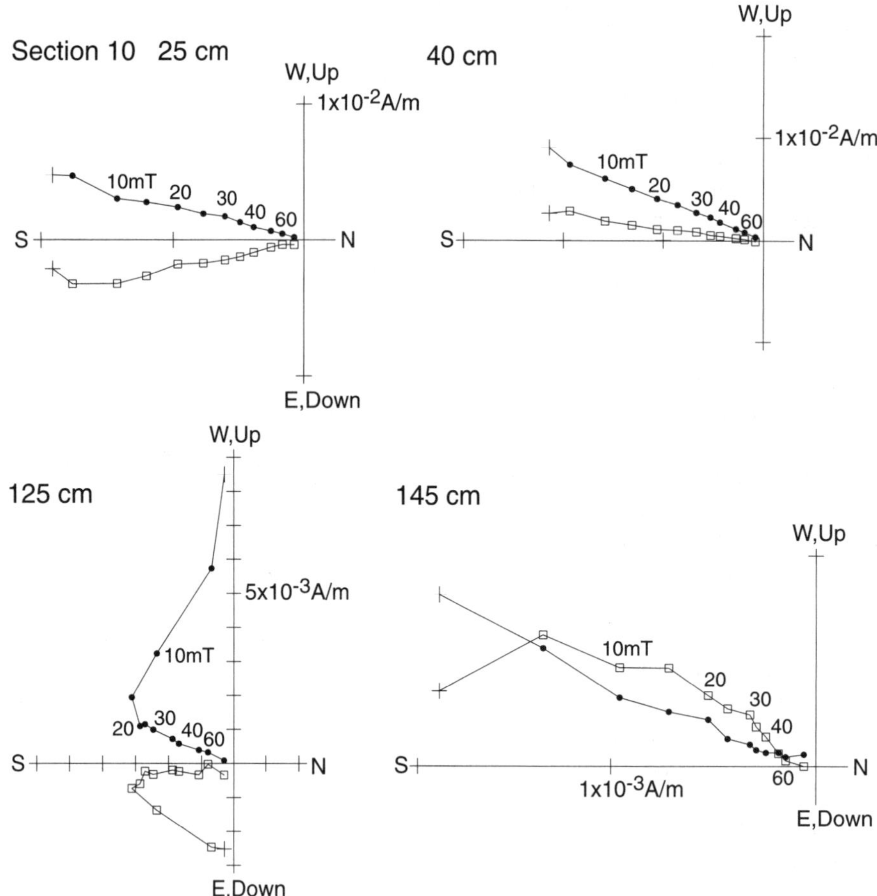

Figure 3. Stepwise AF demagnetization data after deconvolution. Examples are from section 10 shown in Figure 2. Open (solid) circles are projections of vector end-points onto the vertical (horizontal) plane. The core was not azimuthally oriented.

Ioka, 1994; *Yamazaki and Oda*, 2002]. The ratios of ARM susceptibility (k_{ARM}) to k do not vary significantly, except in an interval near 10 m (Figure 4b), which indicates uniformity of magnetic grain size [*Banerjee et al.*, 1981] and strength of magnetic interaction among grains [*Yamazaki and Ioka*, 1997]. The minimum in k_{ARM}/k at about 10m, which implies an increase in average grain size, occurs within δ^{18}O stage 6, and is coeval to that observed in core MD982185 [*Yamazaki and Oda*, 2002]. This could be caused by a change in the source of magnetic minerals and/or diagenetic dissolution of magnetic minerals [*Yamazaki et al.*, 2003] associated with a regional paleoenvironmental change.

Sediments in the WCB are known to be rock-magnetically homogeneous and suitable for relative paleointensity estimation [*Yamazaki and Ioka*, 1994; *Yamazaki and Oda*, 2002]. In the present core, variations of magnetic mineral concentration are less than a factor of ~7, as can be seen in k, and magnetic grain size is generally uniform, as shown by the k_{ARM}/k proxy. Relative geomagnetic paleointensity was obtained from the NRM intensity normalized by ARM (Figure 4e). Both NRM and ARM were AF demagnetized at 30 mT. The normalized intensity does not depend on the selection of the peak AF: almost identical results were obtained by normalization after AF demagnetization at 20 mT.

3. INTER-CORE CORRELATION

3.1. West Caroline Basin

Inter-core correlation between cores MD982183 and MD982185 was performed using relative paleointensity variations. First, the depths of core MD982185 were converted to those of core MD982183 by correlating paleointensity peaks and troughs. Then, the common depths were converted to ages by tying the relative paleointensity record of core MD982183 to the Sint-800 global paleointensity stack [*Guyodo and Valet*, 1999] (Figure 5). The relative paleointensity records of the two cores closely coincide with each other: even short-wavelength features of 10 to 20 kyr are commonly observed and are correlative. General variation patterns of the two records agree well with the Sint-800 stack. However, the records from the WCB have higher resolution: fine features that do not appear in the Sint-800 stack are present in the WCB records. Furthermore, the amplitude of the variations of the WCB records is larger than that of the Sint-800 stack. These differences are most probably caused by a loss of finer-scale features during

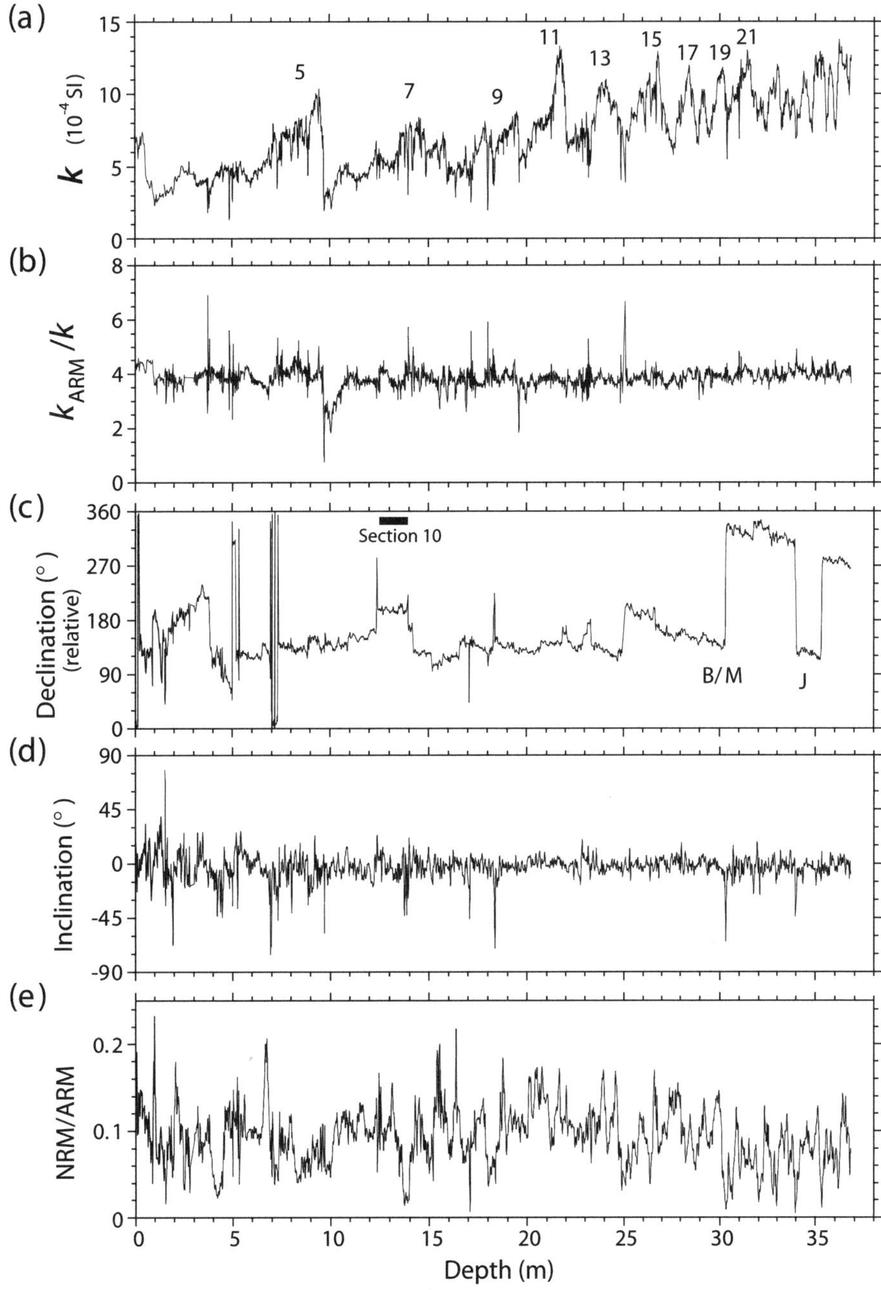

Figure 4. (a) Volume magnetic susceptibility of core MD982183. Numbers above the curve represent $\delta^{18}O$ interglacial stages. (b) Ratio of ARM susceptibility to magnetic susceptibility, which is a proxy of magnetic grain size and magnetic interactions. (c) Relative declination (i.e., without azimuthal orientation). B/M: Brunhes-Matuyama polarity boundary, J: Jaramillo Subchron. The depth interval of section 10, as shown in Figures 2 and 3, is marked. (d) Inclination. (e) Relative paleointensity derived from NRM intensity normalized by ARM. Both NRM and ARM were AF demagnetized at 30 mT.

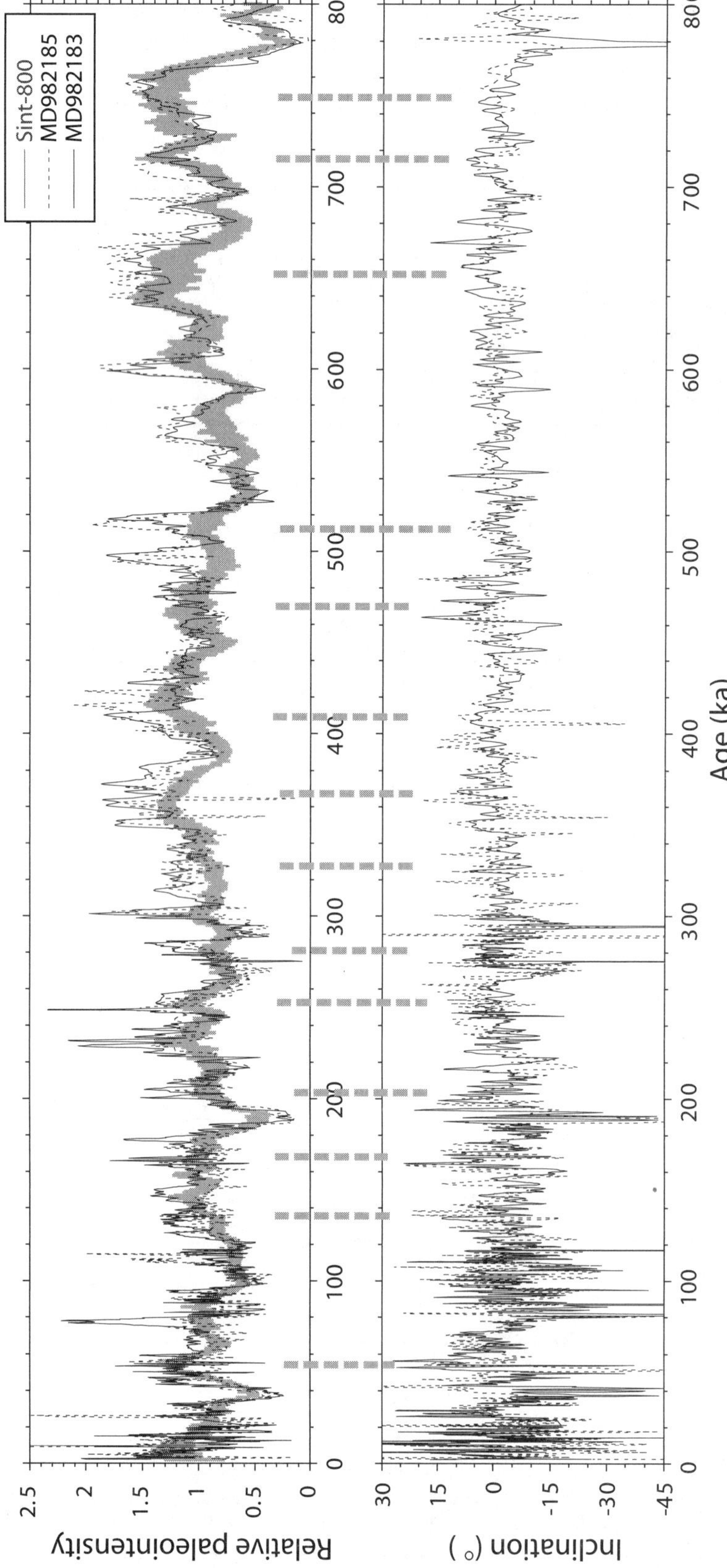

Figure 5. (Top) Inter-core correlation using relative paleointensity. Depths of the two cores from the West Caroline Basin were converted to a common depth scale by matching relative paleointensity curves, and then converted to ages by correlating to the Sint-800 paleointensity stack (gray) [*Guyodo and Valet*, 1999]. (Bottom) Inclination variations after converting to a common age scale using relative paleointensity. Correspondence of positive inclinations to paleointensity maxima is visible, as noted by thick gray vertical dashed lines.

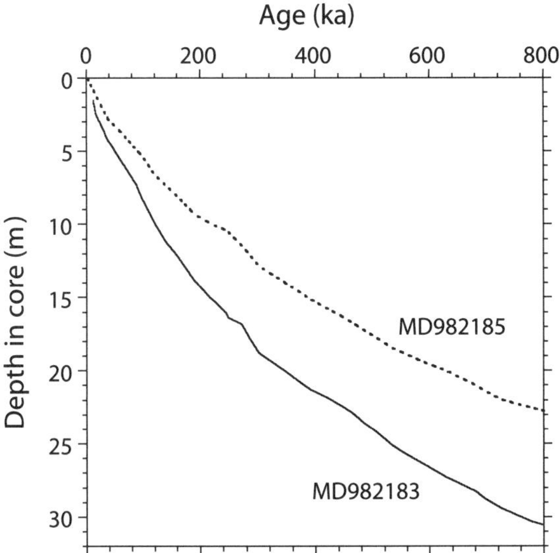

Figure 6. Age-depth curves for cores MD982183 and MD982185 using ages tied to the Sint-800 paleointensity stack.

the stacking procedure used to create the Sint-800 curve due to the uncertainty in ages and the inclusion of cores with low sedimentation rates [*Guyodo and Channell*, 2002]. The most prominent minimum in relative paleointensity, except for the Brunhes-Matuyama (B/M) boundary, is a low at about 190 ka. This minimum accompanies a larger inclination fluctuation. This fluctuation is correlative to the Iceland Basin event [*Channell*, 1999; *Oda et al.*, 2002]. Age-depth curves for the two cores from the WCB are based on correlation with the Sint-800 curve (Figure 6). The sedimentation rates of both cores significantly increase upward. The average of core MD982183 during the last 780 kyr is 38 m/Myr, and it is about 70 m/Myr for the last 200 kyr. Alternatively, apparent upward increases of the sedimentation rates may have been caused by oversampling in their upper part of the cores due to

suction of the piston during coring [*Skinner and McCave*, 2003]. Deposition of core MD982183 was 30 to 40% faster than core MD982185 during the Brunhes Chron.

Inter-core correlation using the relative paleointensity enables a detailed comparison of the inclination data [*Yamazaki*, 2002]. The inclination variations closely resemble each other (Figure 5). The inter-core consistency confirms the presence of long-term secular variations in inclination on the order of 10 to 100 kyr. Cyclic lithological changes may induce cyclic changes in the magnitude of inclination shallowing, and then inclination. However, this cannot explain the inclination variations observed here. In equatorial regions like this site, inclinations are close to zero, and thus inclination shallowing should be negligibly small. Variances of the inclination variations for the two cores are similar in magnitude despite differences in sampling and data processing: core MD982183 was sampled by u-channels and deconvolution was applied, while discrete 7 cm^3 samples were used throughout core MD982185. This implies that the deconvolution procedure worked well considering that the resolution of core MD982183 after deconvolution is expected to be about 2 cm. Discrete samples show somewhat larger scatter in some intervals, for example from 300 to 400 ka. A possible larger error in sampling for discrete samples may be the cause of the larger scatter. The amplitude of the inclination variations increases upward with the increase in sedimentation rate (Figure 5). It is probably due to a reduced low-pass filtering effect on post-depositional remanent magnetization (pDRM) acquisition processes caused by the increased sedimentation rate. Alternatively, the inclinations may have been affected by physical deformation of the sediments, which could be more or less stretched in their upper part.

Magnetic susceptibility records of the two cores using the age model based on relative paleointensity coincide with each other in the time range of glacial-interglacial variations (Figure 7.) There are, however, considerable differences in the

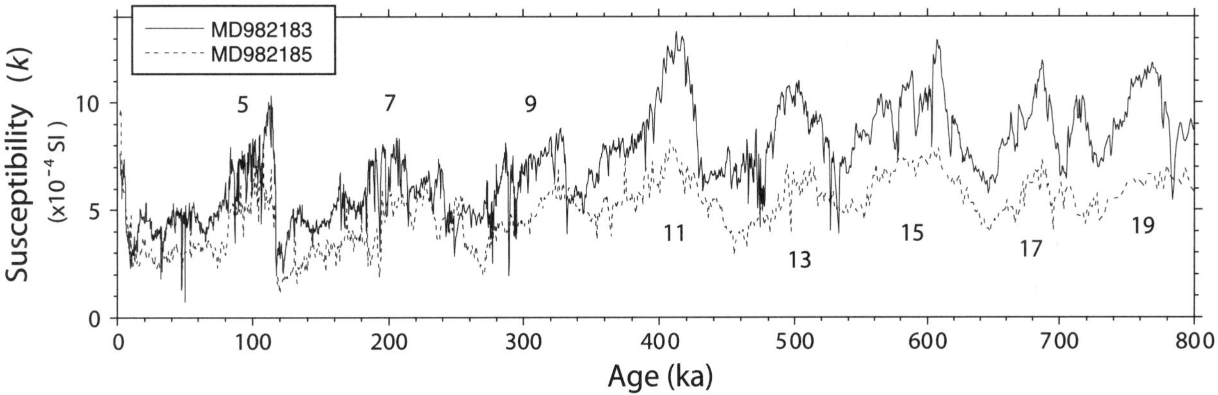

Figure 7. Comparison of magnetic susceptibility for cores MD982183 and MD982185 based on inter-core correlation using relative paleointensity.

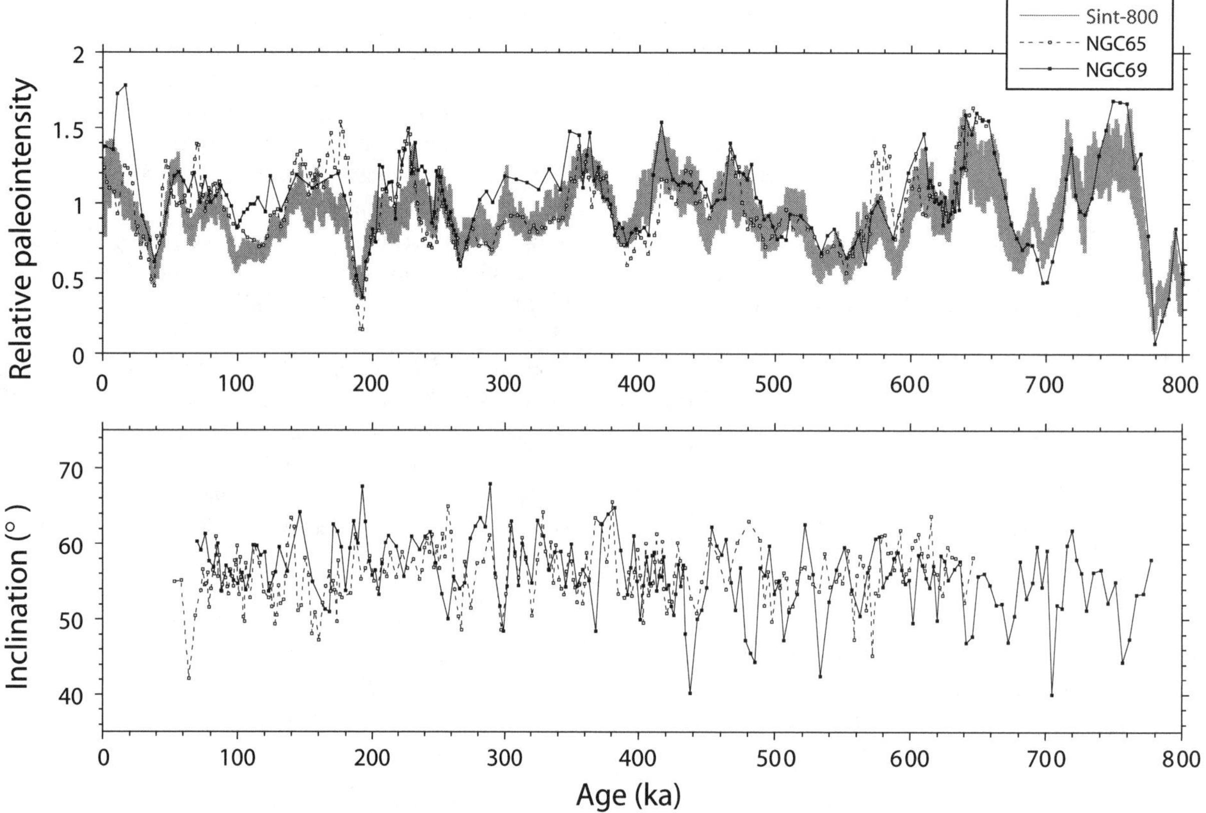

Figure 8. (Top) Inter-core correlation for two cores from the north central Pacific using relative paleointensity. The procedure is the same as that used to construct Figure 5. (Bottom) Inclination variations after converting to a common age scale using relative paleointensity.

variation patterns of shorter time periods: for example, changes at around 430 ka, which are correlative to the transition from $\delta^{18}O$ stage 12 to 11, are sharp in core MD982183, whereas they are gradual in core MD982185. Ages of minor peaks and troughs in k sometimes do not agree well. These discrepancies could reflect local environmental variations. This suggests that relative paleointensity could be more suitable for global and regional inter-core correlation than environmental proxies such as $\delta^{18}O$.

3.2. North Central Pacific

Two gravity cores, each of about 7m length, NGC65 and NGC69, were taken from the north central Pacific (NCP) Ocean along a longitude of 175°E (Figure 1). Relative paleointensity estimations for these cores were published by Yamazaki [1999], but paleomagnetic directions were not presented. The water depths of the sites, about 4900 m, are deeper than the CCD, and the cores consist of siliceous clay with little carbonate. Ages for the cores were estimated from magnetostratigraphy, radiolarian biostratigraphy, and S-ratio

changes associated with glacial-interglacial variations of eolian input. Sedimentation rates were relatively low, 10 m/Myr or less, and the cores cover the intervals from 650 to 900 ka. Paleomagnetic measurements were performed on discrete samples (10 cm³) taken continuously from the core sections. Relative paleointensity estimations were obtained from the NRM intensity normalized by saturation isothermal remanent magnetization (SIRM).

For inter-core correlation of cores NGC65 and NGC69, the same procedure as for the cores in the WCB was applied: a common depth scale was first established using the relative paleointensity results of Yamazaki [1999], and then they were converted to ages by tying to the Sint-800 paleointensity stack (Figure 8). The paleointensity records have temporal resolutions much lower than those of the WCB, but they are similar to that of the Sint-800 stack. The amplitude of the variations is also smaller than for the records from the WCB, but they are similar to those of Sint-800. These observations are due to the much lower sedimentation rates of the NCP cores. The lowest paleointensity, apart from the B/M boundary, occurs at about 190 ka, as is the case in the cores from the WCB. Inclination records for the two cores, based on the relative paleointensity timescale, generally coincide well with each other: features on the order of a few tens of thousands of years are observed in common. This again demonstrates the usefulness of relative paleointensity for inter-core correlation, and indicates the presence of long-term secular variations in inclination.

4. DISCUSSION

We have shown that a new paleomagnetic record from core MD982183 for the Brunhes Chron is almost identical to that of the nearby core MD982185 in the WCB. This indicates excellent inter-core consistency, and that the long-term secular variations in inclination observed in the WCB represent real geomagnetic field behavior. We have also confirmed the occurrence of long-term inclination variations using the records of two cores from the NCP. It has been demonstrated that inter-core correlation based on relative paleointensity is quite useful for detecting long-term inclination variations with small amplitudes.

We can recognize a correspondence of positive (negative) inclinations associated with relative paleointensity maxima (minima) in the records from the WCB (Figure 5), as pointed out by Yamazaki and Oda [2002]. On the contrary, no such correspondence is visible in the records from the NCP (Figure 8). We have conducted a cross-correlation analysis between relative paleointensity and inclination to more quantitatively evaluate this relationship. In the western equatorial Pacific (WEP), intensity and inclination have relatively strong coherence, and phase angles are generally near $0°$ where coherence is strong. That is, the two are in-phase for periods longer than a few tens of thousands of years (Figure 9). In this figure, the results for core NGC88 (see Figure 1 for the site location), which have the highest quality of the three short cores

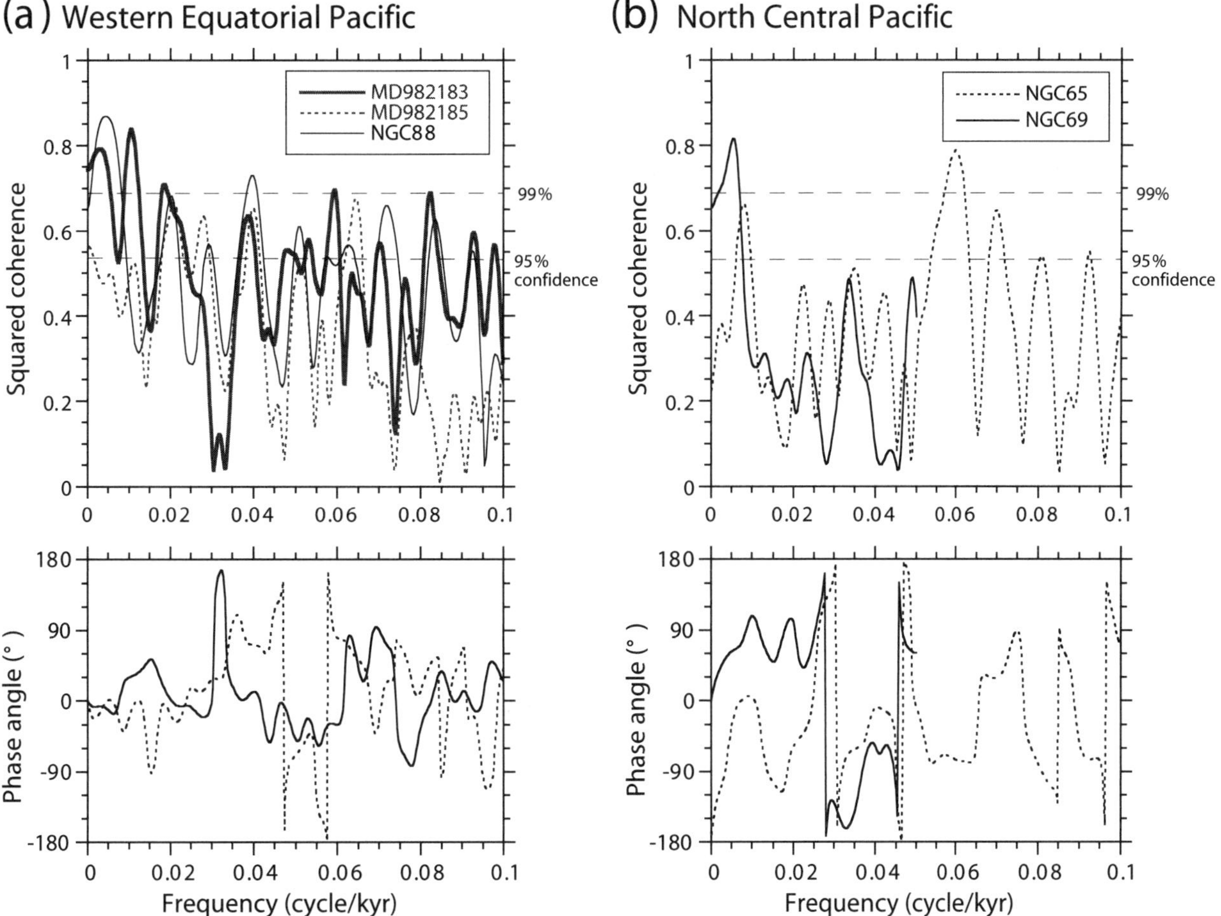

Figure 9. Coherence and phase angle between relative paleointensity and inclination derived from a cross-correlation analysis using the Blackman-Tukey method with a Bartlett window. The AnalySeries software [*Paillard et al.*, 1996] was used. (a) Cores from the western equatorial Pacific. Results from core NGC88 from the Ontong-Java Plateau are included: paleointensity and inclination records were presented by Yamazaki [2002]. (b) Cores from the north central Pacific.

Table 1. Summary of observed inclination anomaly (ΔI) and predictions from time-averaged field models

Cores	Position		ΔI TAF models			ΔI Observed
	Latitude	Longitude	HK02	JC97	KG97	
Western Equatorial Pacific						
MD982183	2°01'N	135°01'E	-4.14	-5.36	-8.81	-6.9
MD982185	3°05'N	135°00'E	-4.06	-5.32	-8.70	-7.5
NGC88	3°08'N	159°11'E	-1.15	-4.24	-4.17	-8.2
North Central Pacific						
NGC65	35°15'N	175°00'E	-2.11	-3.18	-4.98	1.3
NGC69	40°00'N	175°00'E	-1.87	-2.88	-5.40	-3.2

HK02: Hatakeyama and Kono [2002], JC97: Johnson and Constable [1997]; KG97: Kelly and Gubbins [1997]

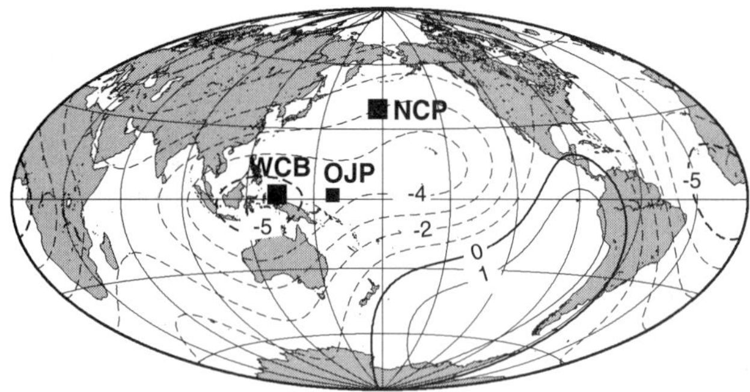

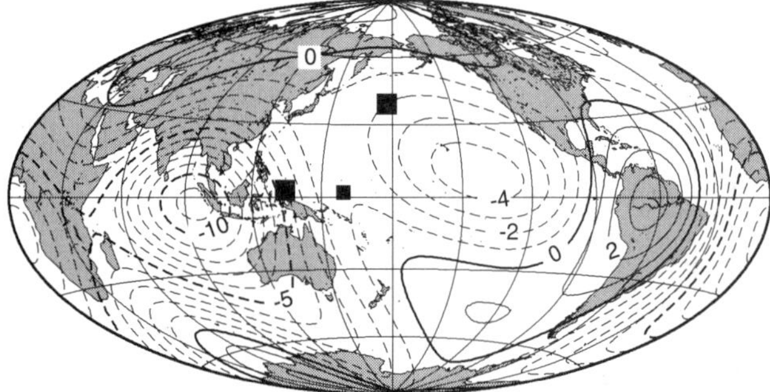

Figure 10. Inclination anomaly (ΔI) for the normal polarity predicted from time-averaged field models for the last 5 Myr (a) Johnson and Constable [1997] and (b) Hatakeyama and Kono [2002]. Contours are at 1° intervals. Solid curves are positive ΔI, and dashed curves are negative ΔI. Squares represent the sites of the studied cores, and observed ΔI values are summarized in Table 1. WCB: West Caroline Basin, OJP: Ontong-Java Plateau, NCP: north central Pacific.

taken from the Ontong-Java Plateau (OJP) [*Yamazaki*, 2002], are also shown with those of two cores from the WCB. In the NCP, on the other hand, coherence is much lower than that of the WEP, and the phase angle is not stable. This might be partly because of the lower time resolution of the NCP records, but this cannot explain the high coherence of the core NGC88 data from the OJP, whose sedimentation rate (8 m/Myr, Yamazaki [2002]) is as low as that of the NCP cores.

In the model of Yamazaki and Oda [2002], long-term secular variations in inclination are controlled by changes in the relative strength of the GAD and by persistent non-dipole components. This model predicts that long-term inclination variations should be obvious in the regions where ΔI is large, but not where ΔI is small. The average ΔI of the two NCP cores is $-1°$, and that of the three cores in the WEP is $-7.5°$ (Table 1). The observation that coherence between intensity and inclination is high in the WEP but low in the NCP supports the model. To further test the model, it is anticipated to accumulate data on intensity-inclination correlation from other regions, in particular from the southwest Pacific, where the sign of ΔI predicted from the time-averaged field (TAF) models for the last 5 Myr [*Johnson and Constable*, 1997; *Hatakeyama and Kono*, 2002] is opposite to that of other areas (Figure 10).

The observed contrast in ΔI between the WEP and NCP is larger than that of the TAF models (Table 1 and Figure 10). Other cores of Brunhes age in the WEP also show ΔI values similar to those mentioned above: $-5.0°$ for the mean of five cores in WCB [*Yamazaki and Ioka*, 1994], and $-7.2°$ for the mean of three cores from the OJP, including core NGC88 [*Yamazaki*, 2002]. Furthermore, strong negative ΔI values of $-10°$ or more were reported from the southwest Pacific (the North Fiji Basin and the Woodlark Basin; between 5°S and 20°S, 155°E and 175°E) [*Elmaleh et al.*, 2001]. Thus, the equatorial regions of large ΔI values in the TAF models centered at 120°E [*Johnson and Constable*, 1997] or 100°E [*Hatakeyama and Kono*, 2002] may extend to the central

Pacific. On the other hand, ΔI values of the TAF models in the NCP might be dominated by the large negative ΔI of Hawaiian data [*Herrero-Bervera and Valet*, 2003], and might be overestimated due to insufficient spatial coverage.

5. CONCLUSION

(1) New relative paleointensity and inclination records from the Brunhes Chron for core MD982183 from the West Caroline Basin (WCB) are almost identical to the records of the nearby (~120 km away) core MD982185, as reported by Yamazaki and Oda [2002]. The excellent inter-core consistency indicates that the long-term secular variations in inclination observed in the WCB represent real geomagnetic field behavior.

(2) Inter-core correlation based on relative paleointensity is quite useful for detecting long-term inclination variations with small amplitudes.

(3) In the western equatorial Pacific (WEP), long-term secular variations in paleointensity and inclination significantly correlate during the Brunhes Chron: positive (negative) inclinations correspond to paleointensity maxima (minima). In the north central Pacific (NCP), on the other hand, coherency between paleointensity and inclination variations is low.

(4) Cores from the WEP have a large negative inclination anomaly (ΔI): the mean of three cores is $-7.5°$. In the NCP, on the contrary, ΔI is small: the mean of two cores is $-1.0°$.

(5) The above observations support the model of Yamazaki and Oda [2002] that long-term secular variations in inclination are controlled by changes in relative strength of the GAD and persistent non-dipole components. This model predicts that the intensity-inclination correlation should be obvious in the regions where ΔI is large, but not where ΔI is small.

(6) The contrast of ΔI between the WEP and NCP is larger than the predictions of the time-averaged field models [*Johnson and Constable*, 1997, *Kelly and Gubbins*, 1997; *Hatakeyama and Kono*, 2002].

Acknowledgments. We thank Tadahiro Hatakeyama for providing the software for calculating inclination anomalies from TAF models, and Etsuko Usuda for help with the measurements. Part of this work was done when H.O. stayed at Paleomagnetic Laboratory Fort Hoofddijk, Utrecht University. We also thank Yuhji Yamamoto for reading the earlier version of the manuscript, and Andrew Roberts, Joe Kirschvink, and an anonymous reviewer for constructive reviews. The GMT software [*Wessel and Smith*, 1995] was used for Figures 1 and 10, and PMAG software [*Tauxe*, 1998] was used for Figure 3. This study was partly supported by the Grant-in-Aid for Scientific Research ((C)(2) No.14540402) of the Japan Society for the Promotion of Science.

REFERENCES

Banerjee, S. K., J. King,, and J. A. Marvin, Rapid method for magnetic granulometry with applications to environmental studies. *Geophys. Res. Lett.*, 8, 333–336, 1981.

Berger, W .H., C. G. Adelseck, Jr, and L. A. Mayer, Distribution of carbonate in surface sediments of the Pacific Ocean. *J. Geophys. Res.*, 81, 2617–2627, 1976.

Channell, J. E. T., Geomagnetic paleointensity and directional secular variation at Ocean Drilling Program (ODP) Site 984 (Bjorn Drift) since 500 ka: Comparison with ODP Site 983 (Gardar Drift). *J. Geophys. Res.*, 104, 22,937–22,951, 1999.

Channell, J. E. T., J. S. Stoner, D. A. Hodell, and C. D. Charles, Geomagnetic paleointensity for the last 100 kyr from the sub-Antarctic South Atlantic: A tool for inter-hemispheric correlation. *Earth Planet. Sci. Lett.*, 175, 145–160, 2000.

Creer, K., N. Thouveny, and I. Blunk, Climatic and geomagnetic influences on the Lac du Bouchet palaeomagnetic SV record through the last 110000 years. *Phys. Earth Planet. Inter.*, 64, 314–341, 1990.

Elmaleh, A., J.-P. Valet, and E. Herrero-Bervera, A map of the Pacific geomagnetic anomaly during the Brunhes chron. *Earth Planet. Sci. Lett.*, 193, 315–332, 2001.

Guyodo, Y., and J.-P. Valet, Global changes in intensity of the Earth's magnetic field during the past 800 kyr. *Nature*, 399, 249–252, 1999.

Guyodo, Y., and J. E. T. Channell, Effect of variable sedimentation rates and age errors on the resolution of sedimentary paleointensity records. *Geochem. Geophys. Geosyst.*, 3, doi: 10.1029/2001GC000211, 2002.

Hatakeyama, T., and M. Kono, Geomagnetic field model for the last 5 My: time-averaged field and secular variation. *Phys. Earth Planet. Inter.*, 133, 181–215, 2002.

Herrero-Bervera, E., and J.-P. Valet, Persistent anomalous inclinations recorded in the Koolau volcanic series on the island of Oahu (Hawaii, USA) between 1.8 and 2.6 Ma. *Earth Planet. Sci. Lett.*, 212, 443–456, 2003.

Holt, J. W., J. L. Kirschvink, and F. Garnier, Geomagnetic field inclinations for the past 400 kyr from the 1-km core of the Hawaii Scientific Drilling Project. *J. Geophys. Res.*, 101, 11,655–11,663, 1996.

Horng, C.-S., A. P. Roberts, and W.-T. Liang, A 2.14-Myr astronomically tuned record of relative geomagnetic paleointensity from the western Philippine Sea. *J. Geophys. Res.*, 108, 2059, doi:10.129/2001JB001698, 2003.

Johnson, C. L., and C. G. Constable, The time-averaged field: Global and regional biases for 0–5 Ma. *Geophys. J. Int.*, 131, 643–666, 1997.

Kelly, P., and D. Gubbins, The geomagnetic field over the past 5 million years. *Geophys. J. Int.*, 128, 315–330, 1997.

Kok, Y., and L. Tauxe, A relative geomagnetic paleointensity stack from Ontong-Java Plateau sediments for the Matuyama. *J. Geophys. Res.*, 104, 25,401–25,413, 1999.

Laj, C., C. Kissel, A. Mazaud, J. E. T Channell, and J. Beer, North Atlantic palaeointensity stack since 75 ka (NAPIS-75) and the duration of the Laschamp event. *Phil. Trans. R. Soc. Lond.* A, 358, 1009–1025, 2000.

Levi, S., and R. Karlin, A sixty thousand year paleomagnetic record from Gulf of California sediments: Secular variation, late Quaternary excursions and geomagnetic implications. *Earth Planet. Sci. Lett.*, 92, 219–233, 1989.

Lund, S. P., J. C. Liddicoat, K. R. Lajoie, T. L. Henyey, and S. W. Robinson, Paleomagnetic evidence for long-term (10^4 year) memory and periodic behavior in the Earth's core dynamo process. *Geophys. Res. Lett.*, 15, 1101–1104, 1988.

Oda, H., and H. Shibuya, Deconvolution of whole-core magnetic remanence data by ABIC minimization. *J. Geomag. Geoelectr.*, 46, 613–628, 1994.

Oda, H., and H. Shibuya, Deconvolution of long-core paleomagnetic data of Ocean Drilling Program by Akaike's Bayesian Information Criterion minimization. *J. Geophys. Res.*, 101, 2815–2834, 1996.

Oda, H., K. Nakamura, K. Ikehara, T. Nakano, M. Nishimura, and O. Khlystov, Paleomagnetic record from Academician Ridge, Lake Baikal: A reversal excursion at the base of marine oxygen isotope stage 6. *Earth Planet. Sci. Lett.*, 202, 117–132, 2002.

Paillard, D., L. Labeyrie, and P. Yiou, Macintosh program performs time-series analysis. *EOS Trans.*, 77, 379, 1996.

Roberts, A. P., M. Winklhofer, W.-T. Liang, and C.-H. Horng, Testing the hypothesis of orbital (eccentricity) influence on Earth's magnetic field. *Earth Planet. Sci. Lett.*, 216, 187–192, 2003.

Schneider, D. A., and D. V. Kent, The time-averaged paleomagnetic field. *Rev. Geophys.*, 28, 71–96, 1990.

Skinner, L. C., and I. N. McCave, Analysis and modelling of gravity- and piston coring based on soil mechanics. *Mar. Geol.*, 199, 181–204, 2003.

Stoner, J. S., J. E. T. Channell, C. Hillaire-Marcel, and C. Kissel, Geomagnetic paleointensity and environmental record from Labrador Sea core MD95-2024: Global marine sediment and ice core chronostratigraphy for the last 110 kyr. *Earth Planet. Sci. Lett.*, 183, 161–177, 2000.

Stoner, J. S., C. Laj, J. E. T. Channell, and C. Kissel, South Atlantic and North Atlantic geomagnetic paleointensity stacks (0–80 ka): Implications for inter-hemispheric correlation. *Quat. Sci. Rev.*, 21, 1141–1151, 2002.

Tauxe, L., Sedimentary records of relative paleointensity of the geomagnetic field: Theory and practice. *Rev. Geophys.*, 31, 319–354, 1993.

Tauxe, L., *Paleomagnetic Principles and Practice.* 299 pp., Kluwer, Dordrecht, 1998.

Valet, J.-P., and L. Meynadier, Geomagnetic field intensity and reversals during the past four million years. *Nature*, 366, 234–238, 1993.

Wessel, P., and W. H. F. Smith, New version of the generic mapping tools released, *EOS Trans.*, 76, 329, 1995.

Yamazaki, T., and N. Ioka, Long-term secular variation of the geomagnetic field during the last 200 kyr recorded in sediment cores from the western equatorial Pacific. *Earth Planet. Sci. Lett.*, 128, 527–544,1994.

Yamazaki, T., and N. Ioka, Cautionary note on magnetic grain-size estimation using the ratio of ARM to magnetic susceptibility. *Geophys. Res. Lett.*, 24, 751–754, 1997.

Yamazaki, T., Relative paleointensity of the geomagnetic field during Brunhes Chron recorded in North Pacific deep-sea sediment cores: Orbital influence? *Earth Planet. Sci. Lett.*, 169, 23–35, 1999.

Yamazaki, T., and H. Oda, Orbital influence on the Earth's magnetic field: 100-kyr periodicity in inclination. *Science*, 295, 2435–2438, 2002.

Yamazaki, T., Long-term secular variation in geomagnetic field inclination during Brunhes Chron recorded in sediment cores from Ontong-Java Plateau. *Phys. Earth Planet. Inter.*, 133, 57–72, 2002.

Yamazaki, T., A. L. Abdeldayem, and K. Ikehara, Rock-magnetic changes with reduction diagenesis in Japan Sea sediments and preservation of geomagnetic secular variation in inclination during the last 30,000 years. *Earth Planets Space*, 55, 327–340, 2003.

T. Yamazaki and H. Oda, Institute for Marine Resources and Environment, Geological Survey of Japan, AIST, 1-1-1 Higashi, Tsukuba, Ibaraki 305-8567, Japan. (toshi-yamazaki@aist.go.jp)

An Equivalent Source Model for the Geomagnetic Field

C. G. A. Harrison

Rosenstiel School of Marine and Atmospheric Science, University of Miami, Miami, Florida

A new model for the geomagnetic field is presented. It does not rely heavily on the present-day field, which has some ephemeral characteristics, but mostly on paleomagnetic data from lava flows, in particular the large Icelandic data set. The model consists of an axial dipole and vertical off centered dipoles representing the non-axial-dipole field. These off centered dipoles are drawn from a zero mean Gaussian distribution whose standard deviation is chosen to give model results agreeing with observations. These off centered dipoles are placed within the core so as to give the correct slope to the Lowes-Mauersberger function, and are concentrated toward the poles so as to produce the latitudinal variation of VGP angular standard deviation (ASD) seen in paleomagnetic observations. The central dipole varies around its mean value with a Gaussian distribution with a small standard deviation chosen to fit observations. The model is checked by comparing average field strengths as a function of VGP latitude with those obtained from intensities of magnetization in Icelandic lava flows. Because the off centered dipoles contribute to the first degree Gauss coefficients, the g_1^0 term can become small, sometimes resulting in low latitude VGPs. If the model is constrained to go through the Iceland ASD point, then it does a reasonable job of explaining the less well-defined lower latitude ASD results. To explain the results recently compiled by McElhinny and McFadden [1997] the standard deviation of the dipole intensities has to be reduced from 1060 nT to 740 nT.

1. INTRODUCTION

Over the years there have been many models produced for the secular variation of the Earth's magnetic field as well as its non-dipole component. These have been discussed and summarized by Harrison [1995]. There are two characteristics of most of these models, which are briefly, a considerable dependence on the characteristics of the present-day geomagnetic field, and the omission of low latitude VGP results when paleomagnetic fields are used to constrain the model. I believe that both of these characteristics have impeded our understanding of the geomagnetic field in all of its complex-

ity, and the model presented here does not have these particular characteristics.

Any particular representation of the field is usually given in terms of spherical harmonics for which the following equation shows some important features.

$$V(r, \theta, \phi) = a \sum_{n=1}^{N} \left(\frac{a}{r}\right)^{n+1} \sum_{m=0}^{n} \left\{g_n^m \cos m\phi + h_n^m \sin m\phi\right\}$$
$$\cdot P_n^m(\cos\theta) \tag{1}$$

V is the magnetic scalar potential at location radius r, colatitude θ, longitude ϕ, a is a reference radius normally taken to be the radius of Earth, n is the degree and m is the order of the harmonic, P_n^m are associated Legendre polynomials in the Schmidt quasi-normalized form and g_n^m, h_n^m are the Gauss coefficients that define the value of the potential at the location chosen. The degree of harmonic controls the wavelength

Timescales of the Paleomagnetic Field
Geophysical Monograph Series 145

of features that are produced by the harmonic. Differentiation of this equation along Cartesian axes will produce field components.

An example of a recent model which has these characteristics is model G of McFadden et al. [1988]. This was developed using the present-day field and the scatter of VGPs as a function of observation latitude. It was found that when the present-day field is rotated around the geographic pole, the resulting scatter in direction, when observed at various latitudinally varying points on Earth's surface, would give VGP scatters which agree with VGP scatters obtained from paleomagnetic analyses of rocks less than 5 Ma old, provided that low latitude VGPs were omitted from the paleomagnetic data set and an average of the present-day field over both hemispheres was taken. The paleomagnetic data were the same as those used by McFadden and McElhinny [1984] who employed a low latitude cutoff angle of 45°, in common with several other studies. The model is composed of contributions from the dipole family of spherical harmonics, in which the sum of degree and order is odd, and the quadrupole family of spherical harmonics, in which the sum of degree and order is even. Thus the equatorial dipole components (Gauss coefficients of g_1^1 and h_1^1) are in the quadrupole family. For the present-day field, the quadrupole family produces almost no latitudinal variation of VGP angular standard deviation, while the dipole family produces a large latitudinal variation. Paleomagnetic VGP scatters agree very well with the present-day field, but not very well with the 1840 field, and very poorly with the 1770 field [Harrison, 1995; Hulot and Gallet, 1996]. In addition, the present-day field gives exceedingly different results between the two hemispheres. It seems only feasible to discard any one instantaneous representation of the field as a major controller of a model, but deal with paleomagnetic fields where the whole range of magnetic field behavior is better represented.

This was the approach used by Camps and Prévot [1996], called CP in this paper. Because the model described in CP has many similarities to the model described in this paper, it will be discussed at some length. CP analyzed directions rather than VGPs. CP started with an axial dipole field exhibiting a normal distribution with a non-zero mean. A small quadrupole field was also included in the model. The standard deviation of the dipole field was made 0.45 of the mean for comparison with Icelandic data. For low latitude data, the standard deviation was 0.5. The non-dipole field was modeled by a Maxwell-Boltzmann probability density function (in which each Cartesian coordinate is independently modeled by a zero-mean Gaussian probability, all with the same standard deviation). The standard deviation for the Maxwell-Boltzmann distribution was made 0.2 for the high latitude Iceland data and for the low latitude data the value lay between 0.41 and 0.66.

These standard deviations are relative to the mean of the axial dipole field. CP showed that for the latitude of Iceland, their model produced a uniform distribution of azimuths, similar to the data from Iceland, fit the deviations from the axial field direction to the data, which showed a non-Fisherian distribution (it could be a Bingham distribution but this was not checked), and also fit the distribution of intensity as a function of angular deviation from the mean field direction. The variation in intensity showed a reduction to a factor of 0.23 of the mean field direction value when the field deviates by 90° from the mean.

There is one problem with the CP model which assumes that the central dipole field can be disassociated from the non-dipole field. Any internal field source with a dipole component will contribute to the three degree one Gauss coefficients. These coefficients represent the vector sum of all internal dipole sources, not just the central dipole source which is only represented by the degree 1 coefficients [Hurwitz, 1960; Chapman and Bartels, 1962]. So unless the non-central sources are quadrupolar or higher multipoles, it is impossible to separate the two types of field. By dealing with field sources rather than with fields themselves, this problem has been avoided in this paper.

In the model to be described below, use is made of Icelandic VGP data from all latitudes, as well as magnetic intensity data. In other words, no low latitude cutoff is used to develop the model. However, in order to compare with the most recent SV data evaluated by McElhinny and McFadden [1997] a cutoff of the model data was used. The specific cutoff angle employed by McElhinny and McFadden was originally proposed by Vandamme [1994] and is of the form

$$\Theta = 1.8 S_p + 5°\qquad(2)$$

where Θ (absolute colatitude of cutoff) and S_p (ASD of VGPs) are in degrees. This cutoff solves one problem with standard cutoffs in which VGPs are rejected if their latitudes fall closer to the equator than some standard value (often 45°). In this case more VGPs are discarded for high latitude observations than for low latitude observations because the ASD is higher for high latitude observations. The Vandamme cutoff deals with this problem.

In contrast to CP, I have used a distribution of vertical (radial) dipoles placed below the core surface to model the non-dipole part of the field. The model is schematically illustrated in Figure 1. This type of model has two advantages, in my opinion. First, it produces a model in which the sources are directly identified, allowing other calculations to be made. Second, it allows calculations to be made at any observation latitude, because the model is global. The CP model was developed using two data sets, one for low latitudes and the other for

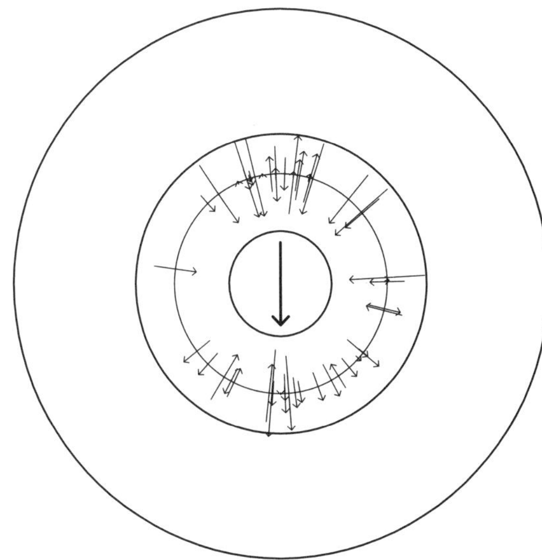

Figure 1. Schematic (2-D) illustration of model. This shows 50 vertical dipoles chosen from a zero mean Gaussian distribution. The length of the arrows shows the strength of the off centered dipoles. The standard deviation of the distribution is equivalent to 0.16. The four circles show Earth's surface (r = 1), core-mantle boundary (r = 0.5462), location of dipoles (r = 0.4) and inner core boundary (r = 0.1917). Dipoles are placed randomly in either hemisphere, and are preferentially located toward the poles using a Fisher distribution with a precision parameter of 3. This means that the chance of locating a dipole in a unit area at the equator is about a factor of ten less than locating a dipole in a unit area at a pole. The final model has about 90 dipoles rather than the 50 shown here (90 would make the figure too complicated). No attempt has been made to illustrate the longitudinal distribution of dipoles. The density of dipoles per unit latitude interval in this 2-D rendition of a 3-D situation is the same as the density of dipoles per unit area in the three-dimensional model. The axial central dipole is also shown, not to scale.

high latitude Icelandic data but the model parameters for the two latitudinal bands were different because CP were modeling local (i.e. latitudinally dependent) fields, and it is not clear how to merge the two results into a more global representation.

The work to be described in this paper owes much to the pioneering paper of Dodson [1980], who used vertical dipoles located at the core-mantle boundary to model the non-dipole field of Earth, for comparison with data from Iceland and the Canary Islands. Dodson used twenty vertical dipoles and a Fisher precision parameter of $\kappa = 5$ or 7 (see below for a definition of κ) to control the polar concentration of dipoles. The standard deviation of his zero-mean Gaussian distribution of dipole strengths was close to 3% of the central dipole intensity (or roughly 900 nT). The vertical dipole intensities used in this paper are about the same as those used by Dodson, which is at first sight surprising, because the number of dipoles used here is almost a factor of 5 greater. But since the dipoles used

in my model are placed well within the core, i.e. deeper than those used by Dodson, they will have in general a smaller effect on surface fields, and this may explain why they were larger in number. In addition, in order to obtain the correct latitudinal variation of VGP scatter, I had to use a precision parameter considerably smaller than the one used by Dodson. But despite these differences, the main points of the models are the same, radial dipoles with a Fisher distribution to describe the statistical concentration toward the geographic poles, random placement in either hemisphere and in longitude, and with a zero-mean Gaussian distribution of intensities. However, I did not use any dipole wobble, on the basis that it is so small that it has virtually no effect on the VGP scatter, compared with the other causes of scatter. The smallest precision parameter used by Dodson [1980] to describe dipole wobble was $\kappa = 32$. This gives an angular standard deviation of pole position of about 2°. Compared with observed angular standard deviations of 11° or greater, this is a trivial amount. Cox [1968] also used radial off centered dipoles to model the non-dipole field of the Earth. Lowes and Runcorn [1951] also used vertical off centered time varying dipoles to model the secular variation field and the dipoles were placed at a radius of between 0.4 and 0.5 R_e. Kono [1972] also used dipoles within the core in his mathematical model of the magnetic field.

In common with most other workers (including CP and *McFadden et al.* [1988]), I have used VGP positions rather than field directions to describe the field. This follows from the tendency of VGP positions to be more Fisherian in their distribution than field directions, which, especially for low latitude observations, tend to be elongated [e.g. *Creer et al.*, 1959]. The model presented here shows VGPs that are roughly symmetric about the mean pole. This model assumes that there is no preferred longitudinal concentration of low latitude VGPs. Although Love [2000] has demonstrated that there is a statistically significant concentration, the percentage of the concentration is not great, and other studies of lava flows have found no such concentration.

There has been a tendency recently to base magnetic field models solely on spherical harmonic coefficients. For instance, Constable and Parker [1988] used a giant Gaussian distribution of Gauss coefficients to describe the field, but their original model did not adequately allow for the latitudinal variation of VGP scatter. This caused Johnson and Constable [1996] to modify the model to produce a VGP scatter that increases as a function of observation latitude. Modifications to this basic model have also been suggested by Kono [1997a] so as to produce the desired latitudinal variation of VGP scatter. These models use only high latitude VGP data are so are not comparable with the model presented here. The model described here predicts that there should be some variation of strength of Gauss coefficients as a function of order, within each har-

Table 1. Methods of determining model parameter values

Item	Parameter	Method of determination
1	Mean central axial dipole intensity.	Free parameter, made equal to present-day dipole.
2	Central axial dipole intensity variation (standard deviation of central dipole).	Governed by comparison with latitudinal variation of Icelandic VGPs (Figure 9). Different standard deviations are each tried a number of times & compared with observed distribution of VGP latitudes using random numbers to generate specific models for a given SD. Results show that 0 to 10% variation is permissible, but that higher variation does not give as good a fit.
3	Off centered dipole depth (constant for a specific model).	Chosen to fit the Lowes-Mauersberger function for geomagnetic field over past 300 years (Figures 2 & 3). Depth is at 0.4 r_e.
4	Number of off centered dipoles.	Chosen to fit the latitudinal distribution of Icelandic VGPs (Figures 4-5). For each number of dipoles their intensity, as given by the standard deviation of a zero-mean Guassian distribution, was varied so as to give the best fit to the Icelandic data. The number of dipoles which gave the best fit was 90, the number used in later calculations.
5	Polar concentration of off centered dipoles.	Chosen by comparison with VGP scatter calculated for different latitudinal observations (Figure 7). Random numbers used to choose hemisphere (equal probability), absolute latitude (according to a Fisher distribution). The Fisher precision parameter, κ, is 3.
6	Longitude of off centered dipoles.	Chosen by random number so that probability of any longitude is uniform (Figure 11).
7	Variation of off centered dipoles.	Zero mean Gaussian distribution with standard deviation fixed by comparison with Icelandic VGP data (Figure 5). See item 4. With 90 dipoles, the standard deviation is 1040 nT. To fit more restrictive data sets the standard deviation has to be reduced to 740 nT (see Figure 7).

monic degree. Some models have been based on the variation of inclination and declination as a function of observation latitude [*Kono*, 1997a].

2. MODEL CHARACTERISTICS

Table 1 gives a summary of the model.

2.1. The Central Dipole

The model proposed here consists of a central axial dipole whose intensity is chosen randomly from a Gaussian distribution whose mean value is equivalent to the present-day g_1^0 term of intensity 30 μT. Any choice for the central dipole mean value could have been made, because for directional data, it is the ratios of the various spherical harmonic coefficients which matter, and for magnetization intensities, the constant governing the intensity of magnetization with the intensity of the magnetizing field is a free parameter. Therefore it is convenient to choose the value of the present-day axial dipole field. 30,000 nT is the average axial dipole term between 1960 and 2000 as determined from the IGRF, and this is what is used for the mean value of g_1^0. The standard deviation of the Gaussian distribution for the central axial dipole is chosen to agree with observations to be described later.

2.2. The Depth of Dipoles

The non-dipole field is modeled by off centered vertically oriented dipoles whose depth is fixed so as to give the correct slope of the Lowes-Mauersberger function [*Lowes*, 1966; *Mauersberger*, 1956] calculated for the present-day field plotted against degree of harmonic, the only time that the present-day field is used as a major constraint on the model. The Lowes-Mauersberger function gives the power of the field (mean square field) over Earth surface for each degree of harmonic and is shown below.

$$R_n = (n+1) \sum_{m=0}^{n} ((g_n^m)^2 + (h_n^m)^2) \tag{3}$$

This depth turns out to be 0.399 r_e. *Lowes* [1974] found a depth of 0.35 R_e for a much older spectrum. Figure 2 shows a series of Lowes-Mauersberger functions for ten different model representations of the field. The variation in the degree 1 component is due almost entirely to the vector sum of the off centered dipoles either adding to or subtracting from the central dipole, which was made almost constant (total variation less than 10%) for these calculations. The average slope of the functions, using degrees 2 through 10 is −0.55, which is almost exactly that for the present-day field [*Langel and Estes*, 1982].

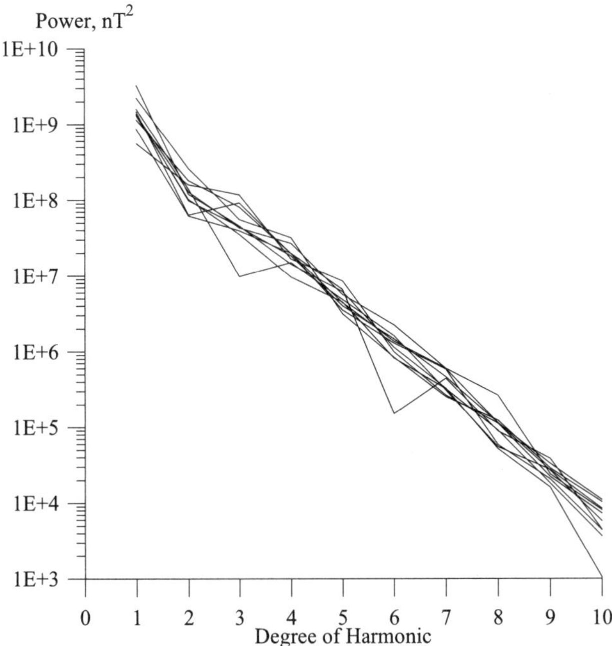

Figure 2. Lowes-Mauersberger functions for ten different representations of the field [*Mauersberger*, 1956; *Lowes*, 1966]. Note that several of the degree 2 harmonics are lower in intensity than a simple straight line projection would predict. This is similar to what is seen in the present-day field. The average slope of the function between degrees 2 and 10 is about -0.55, very similar to that for the present-day field. This slope is controlled by the depth at which the off centered dipoles are located, which in this case was at $0.399R_e$.

If the dipole sources are placed at the core-mantle boundary, the slope of the Lowes-Mauersberger function is greatly increased to -0.27. On inspection of Figure 2 and others generated using different sets of random numbers, it appeared as though the degree 2 harmonic was lower than might be expected. In the present-day field, the degree 2 harmonic lies below the best fitting straight line through the degrees 2 to 12 harmonics by about a factor of 1.6. In order to investigate this, 100 representations of the field were created (Figure 3) and the factor (represented by a logarithm) by which the degree 2 harmonic fell below the best fitting line to the degree 2–10 harmonics was calculated. This turned out to be 0.281 (i.e. a factor of 1.91) with a standard deviation of 0.216, indicating that the degree 2 harmonic was lower than the straight line prediction about 90% of the time. This is a small confirmation of the utility of the model. The spherical harmonic coefficients for each off centered dipole are calculated using the formulae in Hurwitz [1960]. The depth of $0.399\ R_e$ is close to the center of gravity of a narrow cone of the outer core, which is at a radius of $0.422\ R_e$.

In Figure 3 the geometric mean values of power are plotted along with the geometric standard deviation. The standard

error of each mean will be a factor of ten smaller than the standard deviation. The straight line is the best fitting line to degrees 3–10. The plus sign plotted at degree 1 is when the central dipole field is added to the harmonics produced by the non-dipole sources, and it can be seen that this is now close to the best fitting line through the degrees 3 to 10 harmonics.

Because the Lowes-Mauersberger function is made similar to that of the present-day field with a slope of -0.55, a downward continuation to the core-mantle boundary will not reveal the actual locations of the off centered dipoles. If the dipoles had been placed at the CM boundary, then a downward continuation to this boundary would reveal the positions of any of the dipoles with a significant dipole moment because a singularity would be produced in the field at the locations of each off centered dipole [*Hurwitz*, 1960].

The slope of the Lowes-Mauersberger function is a relatively stable characteristic of the geomagnetic field. During the 300-year period covered by the spherical harmonic models of Bloxham and Jackson [1992] the absolute value of the slope decreases by only a small amount, maybe because the

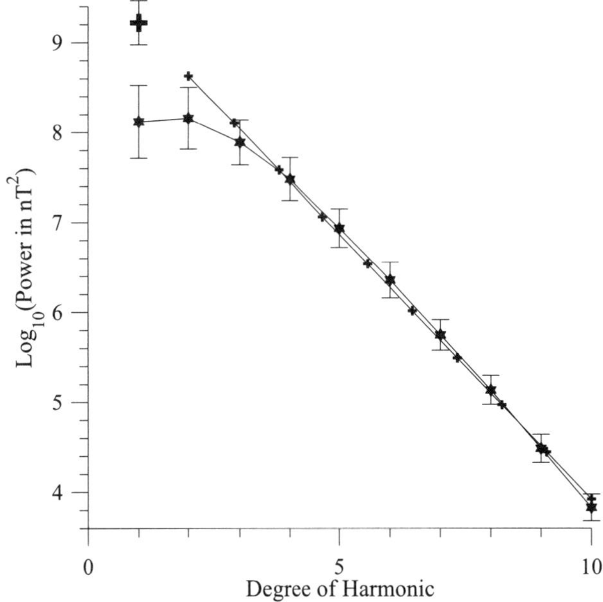

Figure 3. Average values of the log of the Lowes-Mauersberger function for 100 different representations of the field. The Fisher precision parameter for the polar concentration of the off centered dipoles is 3, the number of off centered dipoles is 90, and their standard deviation is 1060 nT. The vertical bars are the standard deviations for each degree of harmonic. The standard errors of the means are a factor of ten less than these standard deviations. The straight line is the best fitting line through degrees 3 through 10. The bold plus sign at degree 1 is the value of the power when the central dipole is included in the calculations. The mean value of the central dipole is 30,000 nT.

higher degree harmonics are not as well represented in the older field observations. Many of the other characteristics of the field change greatly, especially the latitudinal variation of VGP scatter produced if the fields are rotated about the geographic pole [*Harrison*, 1994]. During the same time the latitudinal variation of VGP scatter changes greatly. In 1990 the latitudinal variation was 9° whereas in 1770 the latitudinal variation was only 1.5°.

2.3. The Number and Strength of Dipoles

The number of vertical dipoles used in the model, as well as other factors, has a significant effect on the pattern of VGPs which is obtained. The number that gives the best fit to the distribution of VGPs from Iceland as a function of latitude is about 90. Using this value, it is possible to get a very good fit to the observed latitudinal distribution of Icelandic VGPs, as is shown in Figure 4. Only by plotting the difference between observed and model curves on an expanded scale (middle

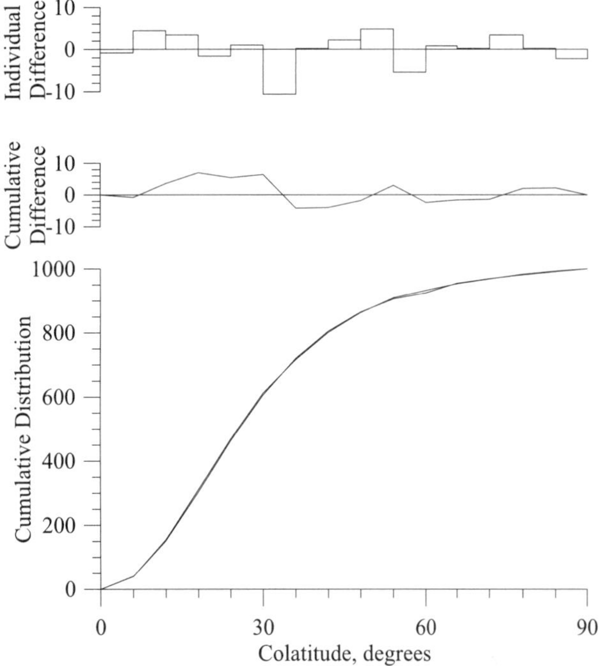

Figure 4. VGP distribution as a function of latitude for an observation latitude of 65° N. The lower curve shows the cumulative distribution of VGPs as a function of colatitude (inverting southern hemisphere VGPs into the northern hemisphere) for 1000 simulations of the field. Also shown is the observed cumulative distribution for Icelandic data, which is almost indistinguishable from the model distribution. The middle curve shows the difference in the cumulative distributions (model minus observed) on an expanded ordinate scale. The maximum difference is 7 observations. The top curve shows the individual differences, and is the differential of the middle curve.

figure) is it possible to see clearly the deviations. The method of comparison was a trial and error method whereby the agreement with the observed Icelandic data set from Kristjánsson [1985] was compared with the data set generated by the model, using a χ^2 test, the appropriate test for binned data [*Press et al.*, 1986]. This Icelandic data set was used because many more lava flows (2163) were used to generate it than other data sets such as the one presented by McElhinny and McFadden [1997] (781 high latitude VGPs). In order to determine the number and strength of radial dipoles it is necessary to use the distribution of VGPs as a function of VGP latitude, not just the ASD of the VGP distribution. In order to obtain a smooth curve of VGP distribution as a function of latitude (as shown in Figure 4) it is then necessary to have a distribution with a large number of data, so that the statistical variation of VGPs within each 6° bin is as small as possible. The age of the lavas was between 2 and 14 Ma. This wide spread in ages is not ideal because it has probably increased the VGP scatter, as can be seen by comparison with other data sets [*McElhinny and McFadden*, 1997]

This best fitting model was using 90 vertical dipoles with a standard deviation of 1060 nT, or 3.53% of the mean dipole field. In order to obtain the correct variation of angular standard deviation of VGPs as a function of observation latitude, it was found necessary to concentrate the vertical dipoles toward the geographic poles using a Fisher distribution. This will be discussed below, but the distribution used to obtain the results shown in Figure 4 was generated with a precision parameter κ of 3.0.

If different numbers of dipoles were used, the resulting pattern did not agree as well with the Icelandic data. The values of χ^2 for different models is shown in Figure 5. For each choice of the number of vertical dipoles, the standard deviation of the vertical dipole strength is varied so as to obtain the best fit. Parabolic curves have been drawn through each set of results, to help illustrate how the fit varies with standard deviation and with the number of dipoles. Although the number of dipoles estimated in this way seems large, it should be pointed out that Kono [1972] also used a large number of dipoles for his model (60 to 80). It has been suggested that the number of dipoles should be infinite, but a moment's thought tells us that this would produce zero non-dipole field for any finite value of the standard deviation of the vertical dipole strength. Any small patch of the core would have an infinite number of dipoles, drawn from a zero-mean Gaussian distribution. The observed mean would diminish as the number increases, and can be made arbitrarily small by increasing the number. On the other hand, too small a number of dipoles would make the non-dipole field very patchy, because each source would have to be larger (see next paragraph).

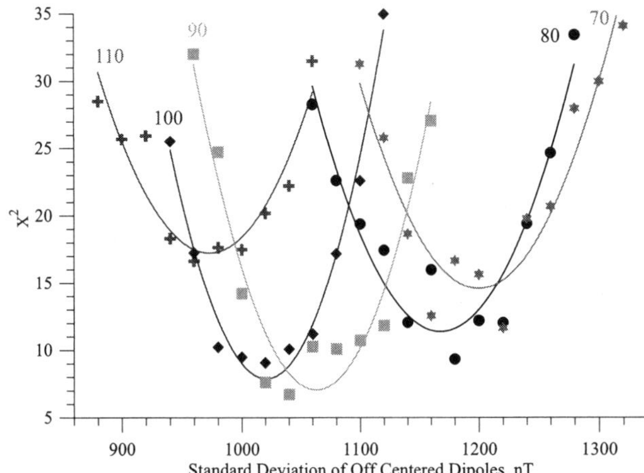

Figure 5. Values of χ^2 for comparisons of models with Icelandic VGP latitude data distributions. Five different sets of results are shown. The 110 dipole model (DM) is shown by the plusses, the 100 DM by the diamonds, the 90 DM by the squares, the 80 DM by the circles, and the 70 DM by the stars. For each choice of dipole number the standard deviation of dipole strength is varied so that a minimum value of χ^2 is reached. Each set of results has had a parabola fitted to it. The curves for 100 and 90 dipoles are narrower than the other three, indicating that the minima for these curves are better defined than those for the other curves.

For the five sets of results shown in Figure 5 there is a simple relationship between the number of off centered dipoles and the variance of the distribution from which the dipole magnitudes are calculated in order to get the best fit (i.e. at the minimum of the curves). This is that their product is about constant (100.8, 107.6, 101.1, 104.0 and 103.5 μT^2, respectively).

2.4. The Polar Concentration of Dipoles

In order to achieve a variation of scatter of VGPs which shows a dependence on the latitude of observation, it is necessary to concentrate the vertical dipoles toward the geographic poles. This is done by using a Fisher distribution of positions for the vertical dipoles. The same method was used by Dodson [1980]. The Fisher distribution [*Fisher*, 1953] was developed to deal with dispersions on a sphere. It is given by the following equation,

$$P_\theta d\theta = \frac{\kappa}{2 \sinh \kappa} e^{\kappa \cos\theta} \sin\theta \, d\theta \qquad (4)$$

where $P_\theta d\theta$ is the probability of finding a direction or pole in a belt between θ and $\theta + d\theta$ from the mean direction or pole. When integrated over the sphere this equation gives unit probability. The azimuthal distribution around the mean is uniform. The quantity κ is known as Fisher's precision parameter,

and has the characteristics of the invariance (1/variance) of a normal distribution. Large values of κ show tight groupings whereas small values show dispersed groupings. Irving [1964] has described the use of this statistical method in some detail.

It was also noted by Harrison [1994] that the spherical harmonic models of the current field have zeroth order (m = 0) harmonics which are relatively large, compared with other harmonics of the same degree, whereas the high order harmonics (with m > n/2) tend to be smaller. It was thought that this was connected with the fact that the ASD of VGPs as a function of observation latitude became larger at high latitudes, as high orders of harmonic are concentrated toward the equator because of a term $\sin^m\theta$ in the associated Legendre polynomials, where m is the order of harmonic, and θ is the colatitude.

Therefore an experiment was done in which 90 dipoles were placed on the sphere so that their distribution was random with respect to hemisphere and longitude, and randomly Fisherian with respect to absolute colatitude, with varying precision parameters. One hundred different arrangements were generated, and Figure 6 shows the results of this experiment. For each representation, the spherical harmonic coefficients in each degree above one were normalized according to the following procedure.

$$L_n = R_n /(n+1)^2 \qquad (5)$$

$$(G_n^m)^2 = \frac{(g_n^m)^2}{L_n} \qquad (6)$$

$$(H_n^m)^2 = \frac{(h_n^m)^2}{L_n} \qquad (7)$$

This procedure makes the mean squared value of each reduced Gauss harmonic order (i.e. $(G_n^m)^2 + (H_n^m)^2$) unity, since there are n + 1 orders per degree n. Then average values were calculated for zeroth order harmonics (top curve), harmonics in which the order was greater than zero and less than or equal to half the degree (middle curve) and the remaining high order harmonics (bottom curve). Calculations were done to degree 10. It is clear that as the Fisher precision parameter becomes greater, indicating a greater concentration of sources close to the geographic poles, the zeroth and low order harmonics increase in size, at the expense of the high order harmonics, which decrease in size. Based on the results of Dodson [1980] and Harrison [1994] a κ value of between 4 and 7 would seem to give the right concentration of sources close to the geographic poles, but in fact the best fit is obtained with a smaller value of κ of 3. But even this relatively small value of Fisher's precision parameter produces polar concentrations that are a factor of ten greater than equatorial concentrations.

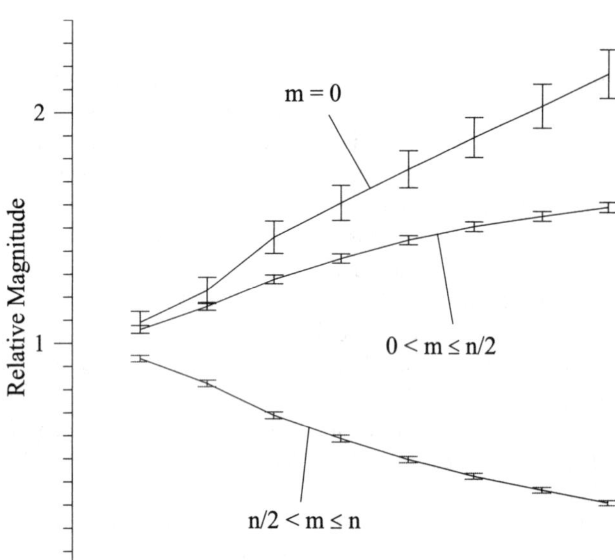

Figure 6. Values of SH coefficients as a function of Fisher precision parameter indicating the concentration of off center magnetic field sources close to the geographic poles. For each value of the precision parameter 100 representations of the field were generated. Within each degree of harmonic for each representation the Gauss coefficients were normalised so that their mean square value was unity (i.e. $\sum_{m=0}^{n}([G_n^m]^2 + [H_n^m]^2) = n + 1$. See text for definitions of G and H). Then averages were taken for harmonics between degree 2 and 10. The top curve represents the mean square value of the 900 zeroth order harmonics. The middle curve represents the mean square value of the 2500 harmonics where $0 < m \le n/2$ and the bottom curve represents the mean square value of the 2900 harmonics where $n \ge m > n/2$, where m, n are the order, degree of harmonic. The vertical bars are the standard error bars around the quantities. Compare these results with those in Harrison [1994] for the recent field.

By using a Fisher precision parameter of 3, 90 off centered dipoles and a standard deviation for the off centered dipole strength of 1060 nT, the variation of ASD as a function of observation latitude was as shown in Figure 7 (top curve). The ASD varies from 32° at the latitude of Iceland to 19.5° at zero latitude. These results can be compared with observations that are listed in Table 2. References to the original studies can be found in Shibuya et al. [1995] and Harrison [1995]. The 95% error for the ASD was obtained using the results in Cox [1969b].

Results 1 and 20–24 from Table 2 are plotted in Figure 7. The upper curve on this figure passes roughly through the non-Icelandic points (it was constrained to pass through the upper Icelandic point). There are many problems with the data set shown in Table 2. For instance, the low ASD for the Hawaiian data (which dominate result #23) is probably because so

many samples were collected from very young lava flows, erupted during the current normal polarity epoch and thus not sampling reversals or excursions. Likewise, the high ASD for Tahiti (which dominates result #24) may be caused by the researchers actively seeking out reversals to study, resulting in an oversampling of low latitude VGPs. Consequently, in Figure 7 arrows have been drawn from these two results, showing the direction that the values would move if these suppositions are correct. Yamamoto et al. [2002] have recently produced some more paleomagnetic results from the Society Islands. They calculate an ASD for all 140 results of 22.5°. In order to

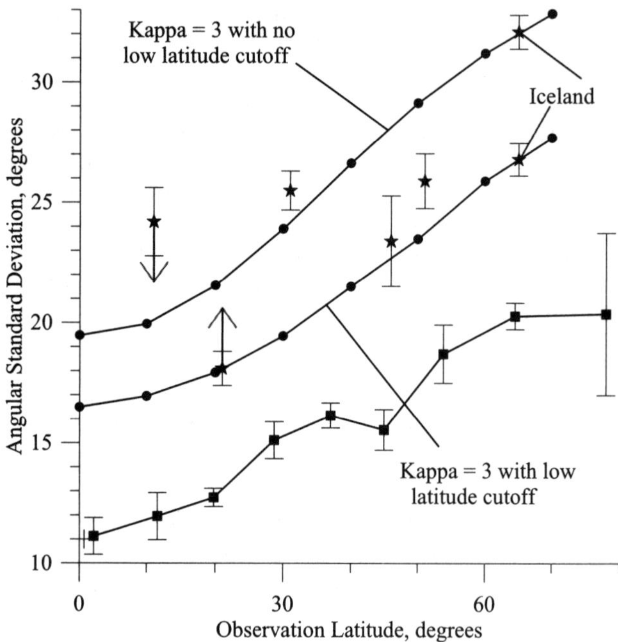

Figure 7. Angular standard deviation of VGPs plotted as a function of observation latitude. The number of off centered vertical dipoles is 90, the standard deviation of their intensity is 1060 nT and the Fisher precision parameter controlling the concentration of dipoles toward the geographic poles is 3. The top curve is where all VGPs are considered. The middle curve is where a low latitude cutoff [*Vandamme*, 1994] is applied to the VGPs, resulting in a distribution that has less scatter. Both of these curves agree completely with data from Iceland. The bottom curve is the result from McElhinny and McFadden [1997], which also has the same low latitude cutoff. The stars are the Shibuya et al. [1995] data grouped into 10° latitude bands (except for the 0–20° band) and plotted at the average latitude for each band, as given in Table 2. The star for the lowest latitude is dominated by results from Tahiti, and in these investigations reversals were actively sought so that it is likely that this result has too large an ASD. In contrast the star from the next lowest result is dominated by lava flows from the Hawaiian Islands, and these have many results from Hawaii, which has no reversals because of its young age, so that this result may have too small an ASD. Possible corrections to these results are shown by the arrows.

Table 2. Observed angular standard deviations

Item number	Location	No. of samples	Absolute latitude °	ASD °	95% CI for ASD °
1	Iceland	2163	65	32.8	0.70
2	Aleutians	92	53	21.5	
3	Brit. Tertiary	409	50	26.8	
4	Marion Is.	23	47	42.0	
5	France	42	46	17.7	
6	Crozet Is.	81	46	18.1	
7	Turkey	28	39	25.3	
8	Amsterdam Is.	45	38	30.4	
9	N. Zealand	104	37	33.0	
10	Canaries	292	30	23.8	
11	Deccan Traps	596	30	24.3	
12	Easter Is.	63	27	13.3	
13	Reunion	96	21	20.7	
14	Hawaii	484	21	17.5	
15	Mexico	40	20	24.6	
16	Pagan Is.	24	18	6.3	
17	Tahiti	123	18	34.5	
18	Comores Is.	39	12	11.6	
19	G. of Guinea	132	3	16.1	
20	Items 2-3	501	51*	26.9	0.70
21	Items 4-6	146	46*	23.4	1.88
22	Items 7-11	1065	31*	25.5	0.81
23	Items 12-15	683	21*	18.1	0.71
24	Items 16-19	318	11*	24.2	1.42

* Weighted mean latitude

see what would be obtained if the Vandamme [1994] cutoff criterion is used, a second curve was generated using the same model results, in which the ASD fell from 26.5° at the latitude of Iceland to 16.5° at zero latitude. Compared with the data obtained by McElhinny and McFadden [1997], also shown in Figure 7 as the bottom curve, the second model curve shows considerably greater ASD. But this is entirely due to having chosen a different observational data set with which to compare the model. The model fits the total Icelandic data set exceedingly well (see Figure 4). Applying the low latitude cutoff to the Icelandic data set shown in this figure and to the model results will obviously produce an agreement between the two, as is shown in the middle curve in Figure 7. The fact is that McElhinny and McFadden [1997] used a different Icelandic data set, which dominates their second highest latitude result. This Icelandic data set has a considerably smaller ASD than the data set obtained from Kristjánsson [1985]. This difference between ASDs may be due to the fact that the data used here span a time longer than the 5 Myr time span used by McElhinny and McFadden [1997]. Kristjánsson and Johannesson [1989] have shown that the ASD for Icelandic lava flows increases as the age of the collection becomes greater, from about 29° for lavas less than 5 Ma old, to about 34° for lavas between 10 and 15 Ma old. This may be due to tectonic effects.

It is however possible to obtain a very good agreement with the McElhinny and McFadden result by reducing the strength of the non-dipole field. This was done by lowering the standard deviation of the Gaussian distribution from which the off centered dipoles were chosen from 1060 nT to 740 nT. Results are shown in Figure 8.

The fact that the sources of the non-dipole field are concentrated toward the geographic poles, resulting in spherical harmonics in which low orders are of greater intensity than high orders, indicates that the giant Gaussian distribution of Constable and Parker [1988] is not ideal (see also Hulot and Le Mouël [1994]), because this model assumes that all harmonic coefficients in each degree are drawn from the same zero-mean Gaussian distribution. This has also been pointed out by Harrison [1994], by Quidelleur and Courtillot [1996], and by Kono and Tanaka [1995], who showed that the giant Gaussian distribution gives almost constant ASD of VGPs as a function of observation latitude, in contrast to observation. Kono [1997a] has suggested modifications to the Constable/Parker model, including (1) omitting the equatorial dipole terms g_1^1 and h_1^1, (2) adding a zonal quadrupole mean value different from zero $((\gamma_2)^2 > 0)$, (3) increasing the variance of the degree 2 order 1 harmonics (see also *Kono* [1997b]). By making

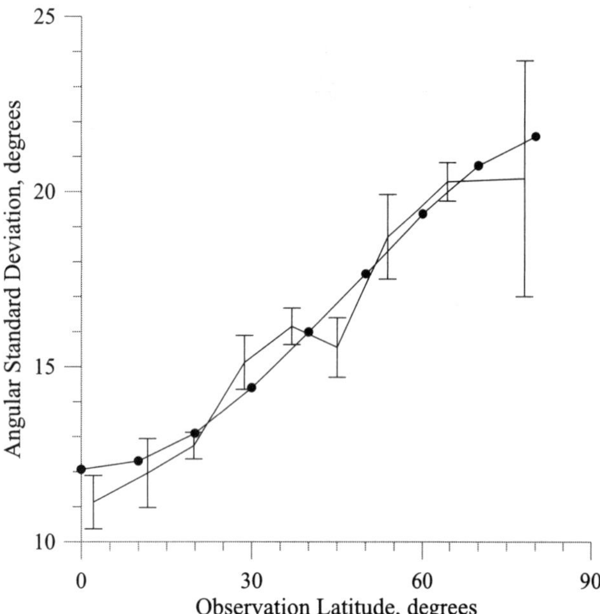

Figure 8. The angular standard deviation of VGPs plotted as a function of observation latitude for the model (filled circles). The model is generated using 90 vertical dipoles, with a precision parameter of 3 and with a standard deviation of dipole intensity 740 nT, in contrast with Figure 7. The data (shown with their standard errors) are from McElhinny and McFadden [1997]. Both data and model results have had a low latitude cutoff [*Vandamme*, 1994] applied. The agreement between model and data is good.

these adjustments, a spherical harmonic model can be made to fit the observed declination and inclination data. All of these models work with data from which results from low latitude VGPs have been winnowed, and so are not directly comparable with the model proposed here, which attempts to deal with VGPs from all latitudes. Johnson and Constable [1996] have proposed a modification of the giant Gaussian distribution that produces a VGP scatter that increases with observation latitude and agrees with observations. It consists of a giant Gaussian distribution of reduced intensity to which a random field is added chosen from a Fisherian distribution about the mean direction [*Constable and Johnson*, 1999].

Even when there is a uniform distribution of off centered dipoles over the core surface (i.e. $\kappa = 0$) equatorial regions still show a smaller scatter than high latitude regions. For instance, with 90 dipoles with an average moment of 740, the equatorially observed ASD is 18.0° whereas the ASD at the latitude of Iceland is 19.5°.

The relatively large number of non-axial dipole sources at high latitudes could be a result of the presence of the inner core. From studying figures of recent geodynamo simulations [e.g. *Glatzmaier et al.*, 1999; *Olson et al.*, 2002] it is seen that

much of the toroidal field within the core is located within the tangent cylinder (the cylinder tangent to the inner core whose axis is parallel to the rotation axis), thus facilitating the production of high latitude sources.

2.5. The Amount of Central Dipole Variation

One of the parameters which was allowed to vary was the amount of central dipole variation about the mean value of 30,000 nT. However, it was found that very little variation was necessary. To illustrate this, Figure 9 compares the latitudinal distribution of VGPs for the model and for Iceland. It shows the value of χ^2 when the standard deviation of the central dipole around its nominal mean value of 30,000 nT is allowed to vary. For each standard deviation chosen, the standard deviation of the off centered dipoles is allowed to vary so that the value of χ^2 passes through a minimum. It can be seen that there is little to choose if the axial dipole varies by up to 3200 nT, but at 6400 nT and especially at 12800 nT the fit to the observed Icelandic curve is degraded.

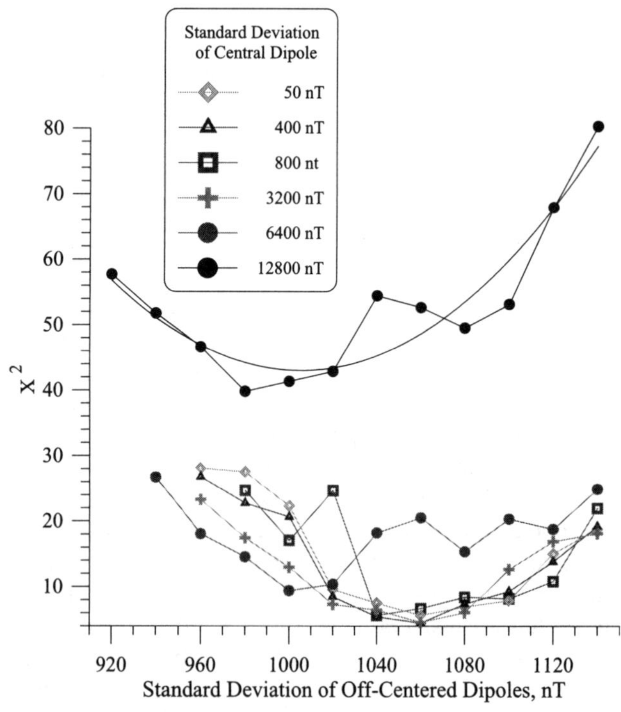

Figure 9. Values of χ^2 fit between model and Icelandic data when the standard deviation of the central axial dipole is varied. The abscissa is the standard deviation of the 90 off centered vertical dipoles that represent the non-dipole field. Six experiments were done with the indicated values of central dipole standard deviation. Central dipole mean intensity was 30,000 nT.

3. DISCUSSION

It was determined that the longitudinal distribution of VGPs for 1000 representations of the field observed at 0° latitude had a uniform distribution, (checked by χ^2 and Kolmogorov-Smirnov statistics) as is found in the observed data. Exactly the same result is obtained for an observation latitude of 65°. The azimuthal distribution of model-produced field directions around the mean direction for an observation point at the same latitude as Iceland has a uniform distribution which was checked both by a χ^2 test and by a Kolmogorov-Smirnov test. But for the field directions at zero latitude, the distribution is highly non-uniform as is seen in Figure 10. Since this is a cumulative distribution, excess numbers of directions are found where the slope is greater than normal, i.e. between an azimuth of −45° and 45° and also between 135° and 225°, whereas in the other directional segments, the number is less than average. This pattern shows that the model produces field directions at low geomagnetic latitudes that are not distributed according to Fisher statistics, just like observations of field directions in paleomagnetism [e.g. *Kono*, 1997b; *Quidelleur et al.*, 1994]. Tanaka [1999] has shown that the Hawaiian paleomagnetic data give an elongated pattern of directions, whereas the VGPs show a much more uniform distribution This is one major reason why paleomagnetists have

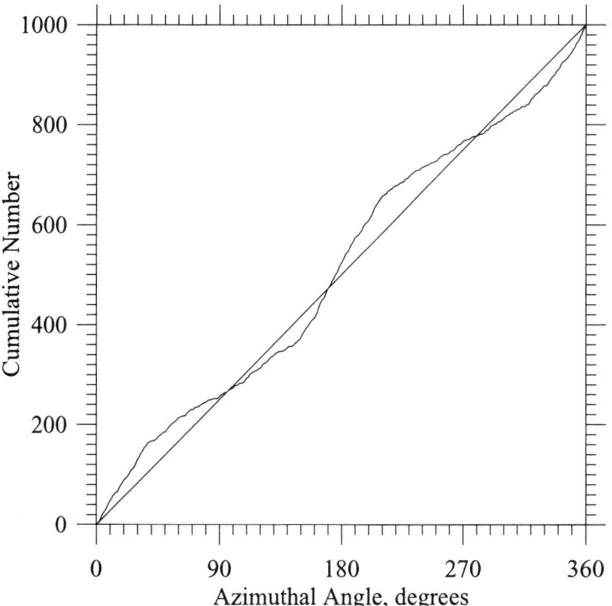

Figure 10. A plot of magnetic declinations derived from 1000 different realizations of the model and for 0° latitude, after having rotated all directions so that the mean direction is vertical. This shows a highly non-uniform distribution, in agreement with low latitude observations. Magnetic declinations for high latitude observations are uniformly distributed.

usually preferred to use VGP distributions rather than field distributions, because VGP distributions are azimuthally symmetric no matter where the point of observation. Observations obeying Fisher statistics can be described by three parameters (mean direction by inclination and declination and scatter by Fisher precision parameter) whereas non-azimuthally symmetric distributions need at least four parameters to describe them. Since the model was developed using high latitude (Iceland) distributions of VGPs but did not assume azimuthal symmetry of VGPs or directions, and did not use low latitude VGP distributions at all, this finding is an independent confirmation of the model's applicability.

I also calculated the field magnitude as a function of VGP latitude for an observation point on Iceland, using 8000 independent field estimates. This was to allow comparison with the magnetization intensities that have been compiled as a function of VGP latitude for the lava flows studied from this island. Magnetization intensities are used as a proxy for field intensities. A first comparison showed moderate agreement, but a comparison with a later data set [*Kristjánsson*, 1999, personal communication and compiled from data in the following papers. *Doell*, 1972; *Kristjánsson*, 1995; *Kristjánsson and Jóhannesson*, 1999; *Kristjánsson et al.*, 1980; *Kristjánsson et al.*, 1991; *Kristjánsson et al.*, 1992; *Kristjánsson et al.*, 1995; *Kristjánsson et al.*, 1998; *McDougall et al.*, 1976; *McDougall et al.*, 1977; *McDougall et al.*, 1984; *Watkins and Walker*, 1977; *Watkins et al.*, 1977] showed excellent agreement (Figure 11). These magnetization intensities are not as good as measurements designed to measure field intensity, but since there are many more magnetization intensity results, other factors besides field intensity that control magnetization intensity will tend to be averaged out in this large data set. Comparisons between model intensities and measured intensities using paleomagnetic techniques are discussed below. Because the Icelandic results used here are for magnetization intensity and not field intensity, and because the model field intensity results have an arbitrary scale factor in them, the linear ordinate scales have been adjusted to give the best fit. A total of 3514 lava flows was used to determine the experimental curve in this figure. The agreement is almost perfect, and gives excellent confirmation of one of the model characteristics.

On further analysis of the 8000 models calculated for the above exercise, it was found that in addition to the expected variation of field intensity at the position of Iceland as a function of VGP latitude there was also a variation as a function of VGP longitude. If the VGP calculated for Iceland happened to be close to the longitude of Iceland, decided by the VGP being in the longitudinal quadrant centered on the longitude of Iceland, the field intensity tended to be a little larger than if the VGP longitude were in one of the other quadrants. For VGPs in the southern hemisphere the quadrant used was that

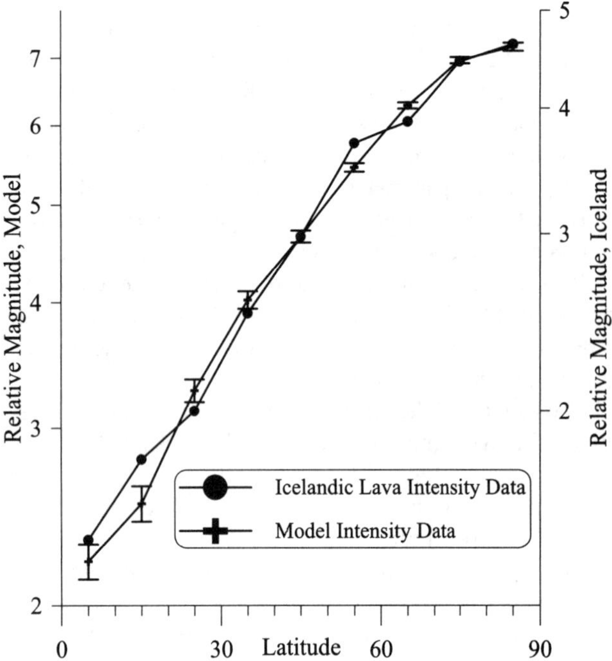

Figure 11. A plot showing the agreement between average field strength as a function of VGP latitude for the Icelandic data and for 8000 model results. The Icelandic data, shown as dots, are from Kristjánsson [1999], and are average magnetization values for each 10° latitude band. The model results are average field intensities for the same VGP latitude bands and are shown with standard error bars. The standard deviation for the centered dipole was 400 nT and 90 off centered dipoles with standard deviation of 1060 nT were used in the model calculations. The ordinate scales for each set of data have been chosen so that the two sets of data lie on top of each other. The agreement between the two curves is remarkable. Both ordinate scales are logarithmic.

centered on the antipode to Iceland. This is illustrated in Figures 12a and 12b, which show that at high latitudes, both intensities are very similar (differing by less than 1% at the pole), whereas at the equator, the difference between the two groups, as judged by the best fitting lines through the data, is 42%. The two best fitting lines are shown in Figure 12c. Cor-

---

Figure 12. Model data of field intensity. (a) For poles that lie in the longitudinal quadrant centered on Iceland, (b) for poles that lie in the three longitudinal quadrants centered opposite Iceland. These results (a, b) and those in Figure 14 were generated using the same 8000 realizations of the field. (c) The two best fitting lines from Figures 12a,b plotted together showing that if VGPs are at low latitudes, the field intensity is higher if the VGP longitude lies close to Iceland than if it lies further away. Upper line, best fitting line for poles (1912) located in longitudinal quadrant centered on Iceland (or its antipode in S. hemisphere). Lower line, best fitting line for other poles (6088). See caption to Figure 13a.

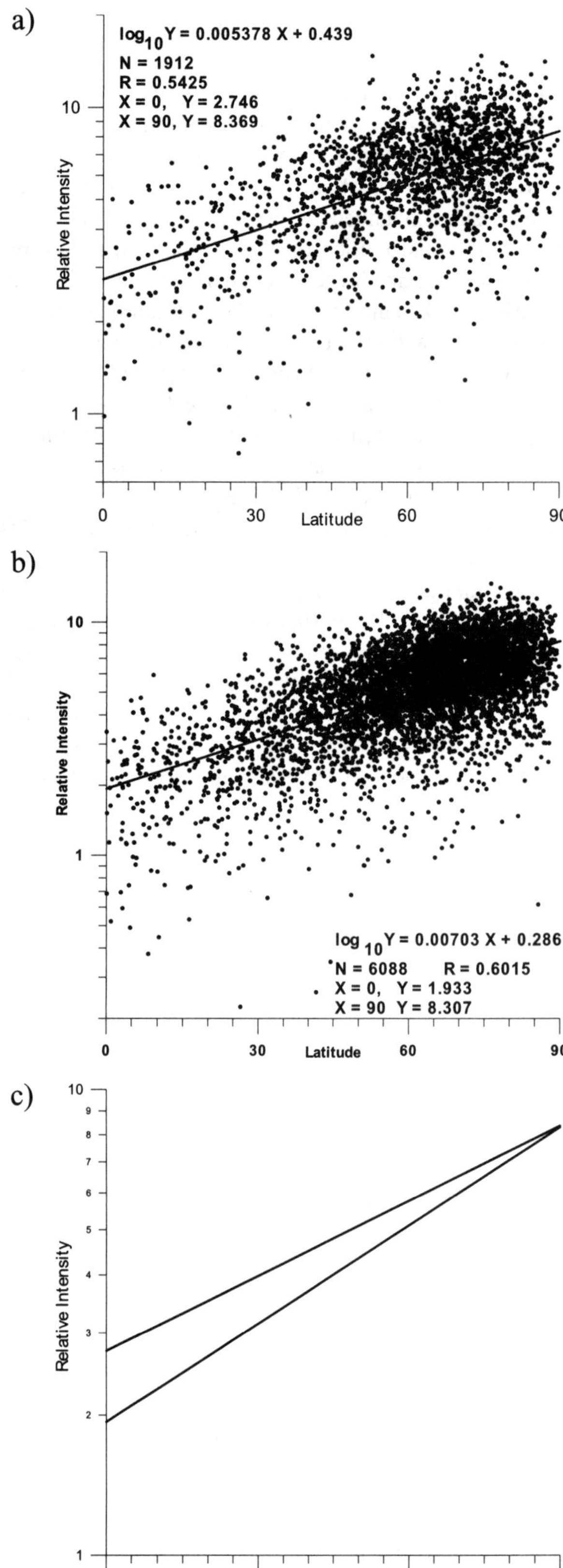

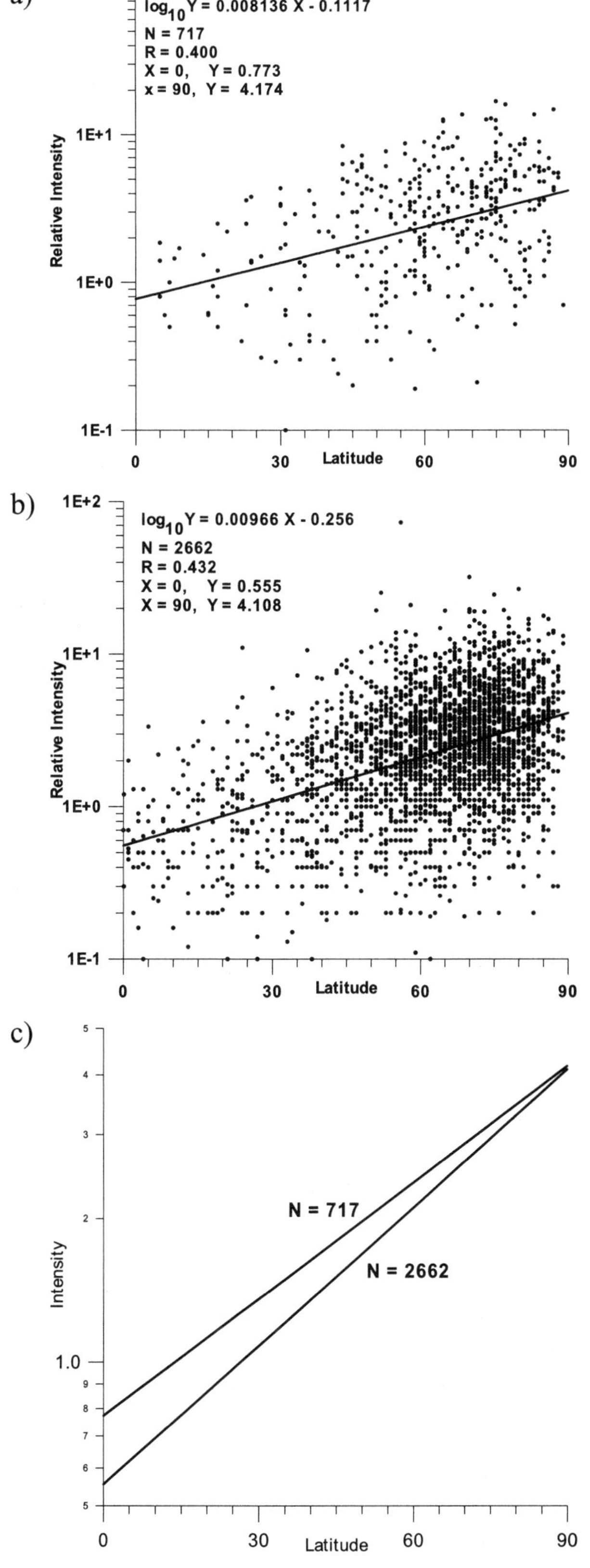

relation coefficients for these two lines are 0.543 and 0.602, and are larger for these log/linear relationships than for the linear relationship.

Data for the observed results were then examined to see if they gave the same results for intensity as a function of longitude. These results are shown in Figures 13a, b and c and they indeed show the same pattern. The intensity at the pole is again almost exactly the same for the two data sets. The difference at the equator is a little less than the model results (38%) but the overall pattern is quite similar. This phenomenon of the intensity of the field varying with the quadrant of the VGP is a consequence of the field intensity of a dipole being dependent on the colatitude, which will be less for VGPs in the longitudinal quadrant centered on Iceland's longitude than for other VGP positions.

One surprising result is that the variation of the strength of the central dipole is quite small. This can be understood in light of the following fact. The three first-degree Gauss coefficients (g_1^0, g_1^1 and h_1^1) represent the vector sum of all magnetic dipole sources within the sphere over which the spherical harmonics are calculated [*Chapman and Bartels*, 1962; *Hurwitz*, 1960]. Thus, although they can be represented by a central dipole, they can also be represented by off centered dipoles. Therefore the central dipole field experienced at the surface of Earth is caused by the central dipole source plus the vector sum of all off centered dipole sources of the geomagnetic field. So when these non-dipole sources add up in a direction opposite to the central dipole, the total dipole field at the surface will be low in magnitude, allowing there to be low latitude VGPs at some such times. This recalls the suggestion made by Cox [1969a] that the rapidly varying non-dipole field triggers a reversal when it acts strongly in the opposite direction to the central dipole field.

The sum of ninety off centered dipoles, with a precision parameter of 3 and with a standard deviation of 1060 nT, very rarely added up to the dipole strength. Out of a trial of 2000 representations, only 20 produced vector sums which were greater than 20,000 nT, and none were above 30,000 nT, This

Figure 13. Intensities of lava flows from Iceland for flows whose VGPs are (a) or are not (b) within the longitudinal quadrant centered on the longitude of Iceland (or its antipode if in the southern hemisphere). The best fitting lines are shown. The difference in slopes between these lines is significant at the 95% confidence level for a one sided tails test. Correlation coefficients are less than for Figure 12a,b because many different things affect magnetization intensity in addition to field intensity. (c) Best fitting lines from Figures 13a,b plotted together. Upper line, northern hemisphere poles located in quadrant centered around longitude of Iceland (or its antipode in the southern hemisphere). Lower line, all the rest of the poles.

means that the higher degree harmonics are sufficiently strong that they can produce low latitude VGPs often enough for the observed distribution (Figure 4).

A comparison was made between the model and field intensity results for Icelandic rocks. These are genuine measurements of the geomagnetic field strength, and not just proxy estimates of field strength made from magnetization intensities (Figures 14 and 15). In Figure 14, the observed magnetization intensity measurements on Icelandic rocks have been taken from CP and plotted as a cumulative curve as well as the best fitting Gaussian distribution. Results from the model described here have been plotted by adjusting the mean value to agree with the mean value from the curve of observed results. Differences between the observed results and the other three distributions have been plotted in Figure 15. An equatorial field of 4×10^{-5} T is equivalent to a Virtual Dipole Moment (VDM) of 10^{23} A m^2. Compared with the Gaussian distribution, both the observed curve and the model curve generated here have significant positive skewness (showing a tail toward the high side of the distribution). The model curve generated here fits the observed distribution somewhat better than the Gaussian fits the observed distribution, and much better than

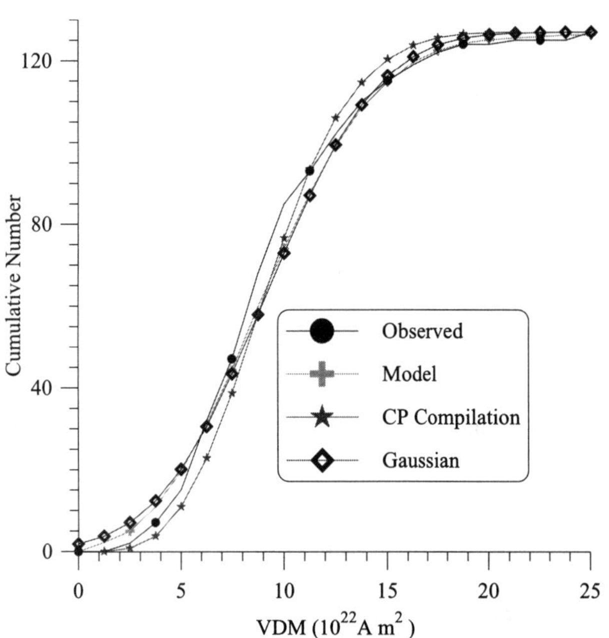

Figure 14. Cumulative distribution of VDM results. Observed results are from Icelandic lava flows compiled by Prévot et al. [1990]. These are genuine paleointensity measurements made on lava flows and are much better than the magnetization intensity results shown in Figures 11-13. Model results are from the model presented in this paper. CP compilation comes from the model presented by CP. Gaussian results come from a Gaussian model. All these results are limited to data that give VGPs within 45° of the poles.

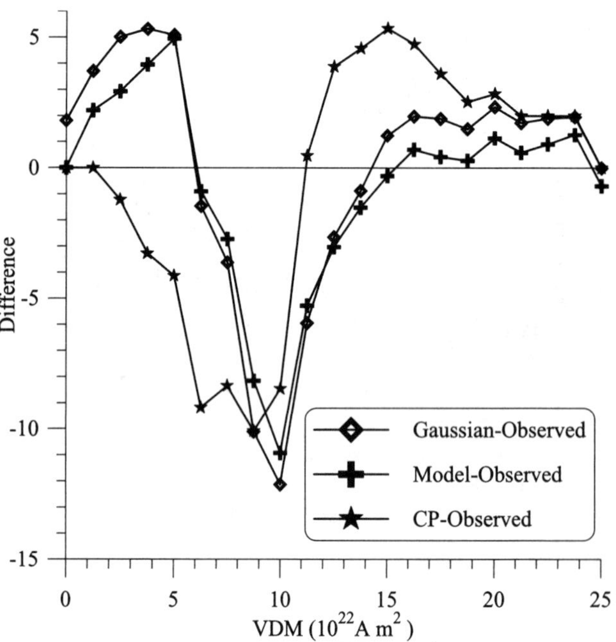

Figure 15. Difference between pairs of cumulative curves shown in Figure 14. The best agreement is between the model results generated here and the observed VDM magnitudes, which is slightly better than the Gaussian/Observed comparison.

the CP model fits the observed distribution. The intensity data and the model results come from data that give VGPs within 45° of the poles.

Because a Fisherian distribution of dipoles concentrated toward the poles and away from the equator has been shown to give the observed latitudinal variation of VGP scatter, physical models of the Earth's magnetic field must produce a similar latitudinal variation of sources. In order to determine the magnitude of this effect, we calculate the change in probability per unit solid angle for the Fisher probability distribution ongoing from poles to equator, remembering that we need to take both halves of the distribution into account.

$$C_{90/0} = \frac{e^{\kappa} + e^{-\kappa}}{2e^0} \qquad (8)$$

This ratio is close to 10 for $\kappa = 3$, the experimental result found in the model calculations (see Figures 2 and 3).

This model is called the H-GF model of the geomagnetic field, H because it comes after model G [*McFadden et al.*, 1988] and GF for Gaussian/Fisher because of the importance of these two distributions in determining the nature of the sources of the field.

In this model there is no particular sharp boundary between fields which give high latitude VGPs (so called "normal" or "reverse fields") and those fields which give low latitude

VGPs (intermediate fields or fields occurring during a reversal or excursion). This suggests that all fields should be regarded as coming from a continuum of behavior, which most of the time produces high latitude VGPs and occasionally produces low latitude VGPs. The actual number of low latitude VGPs with absolute latitudes less than $45°$ varies tremendously as a function of observation latitude, being only 2.1% at the equator but 15.6% at the latitude of Iceland for model H-FG.

4. CONCLUSIONS

A source model has been proposed for Earth's magnetic field. It consists of a central axial dipole which has very small variation in magnitude but can change polarity. The non-dipole field is represented by about ninety radial dipoles within the outer core at a depth so as to give the correct slope to the Lowes-Mauersberger function. These dipoles have a zero mean Gaussian distribution with standard deviation 3.53% of the mean central axial dipole. They are preferentially concentrated toward the poles so as to give the correct latitudinal variation of VGP angular standard deviation. The concentration is done with a double Fisher distribution in which the polar likelihood is about ten times the equatorial likelihood.

This model fits the distribution of VGP longitudes at both high & low latitudes & field azimuths at high latitudes & low latitudes. It fits the observations of average magnetization intensity as a function of VGP latitude derived from Icelandic lava flow magnetization intensities. It also fits the measurements of field intensity derived from Icelandic lava flows using Thellier-Thellier type measurements.

Icelandic data are available on an Excel spread sheet from the author.

Acknowledgments. Research supported by the US National Aeronautics and Space Administration. I appreciate discussions with Leo Kristjánsson Pierre Camps, Michel Prévot, Mike McElhinny and Phil McFadden, and reviews of an earlier draft by Dennis Kent, Mike Fuller and Bill Lowrie. I acknowledge useful reviews by Lisa Tauxe, Masaru Kono and Gauthier Hulot.

REFERENCES

Bingham, C. *Distributions on the sphere and the projective plane*, Ph. D. thesis, Yale University.

Bloxham, J, and A. Jackson, Time-dependent mapping of the magnetic field at the core-mantle boundary, *J. Geophys. Res.*, **97**, 19537–19563, 1992.

Camps, P., and M. Prévot, A statistical model of the fluctuation in the geomagnetic field from paleosecular variation to reversal, *Science*, **273**, 776–779, 1996.

Chapman, S., and J. Bartels, *Geomagnetism*, second edition, third impression, Clarendon Press, Oxford, 1049 pp., 1962.

Creer, K. M., E. Irving, and A. E. M. Nairn, Palaeomagnetism of the Great Whin Sill, *Geophys. J. Roy. Astron. Soc.*, **2**, 306–323, 1959.

Constable, C. G., and R. L. Parker, Statistics of the geomagnetic secular variation for the past 5 million years, *J. Geophys. Res.*, **93**, 11569–11581, 1988.

Constable, C. G., and C. L. Johnson, Anisotropic paleosecular variation models: implications for geomagnetic field observables, *Phys. Earth Planet. Int.*, **115**, 35–51, 1999.

Cox, A., Lengths of geomagnetic polarity intervals, *J. Geophys. Res.*, **73**, 3247–323260, 1968.

Cox, A., Geomagnetic reversals, *Science*, **163**, 237–245, 1969a.

Cox, A., Confidence limits for the precision parameter κ, *Geophys J. Roy. Astron. Soc.*, **18** 545–549, 1969b.

Doell, R. R., Palaeomagnetism studies of Icelandic lava flows, *Geophys. J. Roy. Astron. Soc.*, **26**, 459–479, 1972.

Dodson, R. E., Late Tertiary secular variation of the geomagnetic field in the North Atlantic, *J. Geophys. Res.*, **85**, 3606–3622, 1980.

Fisher, R. A., Dispersion on a sphere, *Proc. Roy. Soc. London*, **A217**, 295–305.

Glatzmaier, G. A., R. S. Coe, L. Hongre, and P. H. Roberts, The role of the Earth's mantle in controlling the frequency of geomagnetic reversals, *Nature*, **401**, 885–890, 1999.

Harrison, C. G. A., An alternative picture of the geomagnetic field, *J. Geomag. Geoelect.*, **46**, 127–142, 1994.

Harrison, C. G. A., Secular variation of the Earth's magnetic field, *J. Geomag. Geoelect.*, **47**, 131–147, 1995.

Hulot, G., and Y. Gallet, On the interpretation of virtual geomagnetic pole (VGP) scatter curves, *Phys. Earth Planet. Int.*, **95**, 37–53, 1996.

Hulot, G., and J. L. LeMouël, A statistical approach to the Earth's main magnetic field, *Phys. Earth Planet. Int.*, **82**, 167–183, 1994.

Hurwitz, L., Eccentric dipoles and spherical harmonic analysis, *J. Geophys. Res.*, **65**, 2555–2556, 1960.

Irving, E., *Paleomagnetism and Its Application to Geological and Geophysical Problems*, Wiley, NY, 399 pp, 1964.

Johnson, C. I., and C. G. Constable, Palaeosecular variation recorded by lava flows over the past five million years, *Phil. Trans. Roy. Soc. A*, **354**, 89–141, 1996.

Kono, M., Mathematical models of the Earth's magnetic field, *Phys. Earth Planet. Int.*, **5**, 140–150, 1972.

Kono, M., Paleosecular variation in field directions due to randomly varying Gauss coefficients, *J. Geomag. Geoelect.*, **49**, 615–631, 1997a.

Kono, M., Distributions of paleomagnetic directions and poles, *Phys. Earth & Planet. Int.*, **103**, 313–327, 1997b.

Kono, M., and H. Tanaka, Mapping the Gauss coefficients to the pole and the models of paleosecular variation, *J. Geomag. Geoelect.*, **47**, 117–130, 1995.

Kristjánsson, L., Some statistical properties of palaeomagnetic directions in Icelandic lava flows, *Geophys. J. Roy. Astron. Soc.*, **80**, 57–71, 1985.

Kristjánsson, L., New palaeomagnetic results from Iceland Neogene lavas, *Geophys. J. Int.*, **121**, 435–443, 1995.

Kristjánsson, L., R. A. Duncan, and A. Gudmundsson, Stratigraphy, palaeomagnetism and age of volcanics in the upper regions of Þjorsárdalur valley, central Iceland, *Boreas*, **27**, 1–13, 1998.

Kristjánsson, L., I. B. Fridleifsson, and N. D. Watkins, Stratigraphy and Paleomagnetism of the Esja, Eyrarfjall and Akrafjall mountains, SW-Iceland, *J. Geophys.*, **47**, 31–42, 1980.

Kristjánsson, L., A. Gudmundsson, and H. Haraldsson, Stratigraohy and Paleomagnetism of a 3-km-thick Miocene lava pile in the Mjoifjordur area, eastern Iceland, *Geol. Rundsch.*, **84**, 813–830, 1995.

Kristjánsson, L., and H. Johannesson, Variable dispersion of Negoene geomagnetic field directions in Iceland, *Phys. Earth Planet. Int.*, **56**, 124–132, 1989.

Kristjánsson, L., and H. Jóhannesson, Stratigraphy and Paleomagnetism of the lava pile south of Ísafjarðardjúp, NW-Iceland, *Jökull*, **44**, 3–16, 1996.

Kristjánsson, L., and H. Jóhannesson, Secular variation and reversals in a composite 2.5 km thick lava section in central Western Iceland, *Earth Planets Space*, **51**, 1–16, 1999.

Kristjánsson, L., H. Jóhannesson, and I. B. Fridleifsson, Paleomagnetic stratigraphy of the Mosfellsveit area, SW-Iceland: a pilot study, *Jökull*, **41**, 47–60, 1991.

Kristjánsson, L., H. Jóhannesson, and I. McDougall, Stratigraphy, age and Paleomagnetism of Langidalur, northern Iceland, *Jökull*, **42**, 31–44, 1992.

Langel, R. A., and R. H. Estes, A geomagnetic field spectrum, *Geophys. Res. Lett.*, **9**, 250–253, 1982.

Love, J. J., Statistical assessment of preferred transitional VGP longitudes based on palaeomagnetic lava data, *Geophys. J. Int.*, **140**, 211–221, 2000.

Lowes, F. J., Mean square values on the sphere of spherical harmonic vector fields, *J. Geophys. Res.*, **71**, 2179, 1966.

Lowes, F. J., Spatial power spectrum of the main geomagnetic field, and extrapolation to the core, *Geophys. J. Roy. Astron. Soc.*, **63**, 717–730, 1974.

Lowes, F. J., and S. K. Runcorn, The analysis of the geomagnetic secular variation, *Phil. Trans Roy. Soc. A.*, **243**, 525–546, 1951.

Mauersberger, P., Das Mittel der Energiedichte des Geomagnetischen Hauptfeldes an der Erdoberfläche und seine Säkulare Änderung, *Gerlands Beitr. Geophys.*, **65**, 207–215, 1956.

McDougall, I., L. Kristjánsson, and K. Saemundsson, Magnetostratigraphy and geochronology of northwest Iceland, *J. Geophys. Res.*, **89**, 7029–7060, 1984

McDougall, I, K. Saemundsson, H. Johannesson, N. D. Watkins, and L. Kristjánsson, Extension of the geomagnetic polarity time scale to 6.5 m.y.: K-Ar dating, geological and paleomagnetic study of a 3,500-m lava succession in western Iceland, *Geol. Soc. Amer. Bull.*, **88**, 1–15, 1977.

McDougall, I., N. D. Watkins, and L. Kristjánsson, Geochronology and Paleomagnetism of a Miocene-Pliocene lava sequence at Bessastadaa, eastern Iceland, *Am. J. Sci.*, **276**, 1078–1095, 1976.

McElhinny, M. W., and P. L. McFadden, Palaeosecular variation over the past 5Myr based on a new generalized database, *Geophys. J. Int.*, **131**, 240–252, 1997.

McFadden, P. L., and M. W. McElhinny, A physical model for palaeosecular variation, *Geophys. J. R. Astron. Soc.*, **78**, 809–830, 1984.

McFadden, P. L., R. T. Merrill, and M. W. McElhinny, Dipole/quadrupole family modeling of paleosecular variation, *J. Geophys. Res.*, **93**, 11583–11588, 1988.

Olson, P., I. Sumita, and J. Aurnou, Diffusive magnetic images of upwelling patterns in the core, *J. Geophys. Res.*, 107(B12), 2348, doi:10.1029/2001JB000384, 2002.

Press, W. H., B. P. Flannery, S. A. Teukolsky, and W. T. Vetterling, *Numerical Recipes The Art of Scienctific Computing*, Cambridge University Press, 818 pp, New York, 1986.

Prévot, M., M. El-Messaoud Derder, M. McWilliams, and J. Thompson, Intensity of the Earth's magnetic field: evidence for a Mesozoic dipole low, *Earth Planet. Sci. Lett*, **97**, 129–139, 1990.

Quidelleur, X., and V. Courtillot, On low-degree spherical harmonic models of paleosecular variation, *Phys. Earth Planet. Int.*, **95**, 55–77, 1996.

Quidelleur, X., J.-P. Valet, V. Courtillot, and G. Hulot, Long-term geometry of the geomagnetic field for the last five million years: an updated secular variation database, *Geophys. Res. Lett.*, **21**, 1639–1642, 1994.

Saemundsson, K., l. Kristjánsson, I. McDougall, and N. D. Watkins, K-Ar dating, geological and paleomagnetic study of a 5-km lava succession in northern Iceland, *J. Geophys. Res.*, **85**, 3628–3646, 1980.

Shibuya, H., J. Cassidy, I. E. M. Smith, and T. Itaya, Paleomagnetism of young New Zealand basalts and loongitudinal distribution of paleosecular variation, *J. Geomag. Geoelect.*, **47**, 1011–1022, 1995.

Tanaka, H., Circular symmetry of the paleomagnetic directions observed at low latitude volcanic sites, *Earth Planets Space*, **51**, 1279–1286, 1999.

Vandamme, D., A new method to determine paleosecular variation, *Phys. Earth, Planet. Int.*, **85**, 131–142, 1994.

Watkins, N. D., I. McDougall, and L. Kristjánsson, Upper Miocene and Pliocene geomagnetic secular variation in the Borgarfjordur area of western Iceland, *Geophys. J. R. Astron. Soc.*, **49**, 609–632, 1977.

Watkins, N. D., and G. P. L. Walker, Magnetostratigraphy of eastern Iceland, *Am. J. Sci.*, **277**, 513–584, 1977.

Yamamoto, Y., K. Shimura, H. Tsunakawa, T. Kogiso, K.Uto, H. G. Barsczus, H.Oda, T. Yamazaki, and E. Kikawa, Geomagnetic paleosecular variation for the past 5 Ma in the Society Islands, French Polynesia, *Earth Planets Space*, **54**, 797–802, 2002.

C.G.A. Harrison, Rosenstiel School of Marine & Atmospheric Science, University of Miami, 4600 Rickenbacker Causeway, Miami, Florida 33149. (charrison@rsmas.miami.edu)

Earth's Magnetic Field

Neil D. Opdyke and Victoria Mejia

Department of Geological Sciences, University of Florida, Gainesville, Florida

Recent studies of the time averaged field (TAF) shows that the paleofield of Earth is much less variable than the present magnetic field, which is asymmetric between the northern and southern hemispheres. The VGP dispersion in the southern hemisphere is much larger than in the northern hemisphere. If the present Earth's magnetic field were a good template for the field in the past then a larger VGP dispersion would be expected than is observed. The present field may therefore be leading to an excursion or a reversal.

1. INTRODUCTION

Earth's magnetic field has been of great interest to humans since the invention of the magnetic compass and a good record of Earth's field is known since the age of exploration. It was pointed out in the 19th century that Earth's magnetic field is dominantly dipolar and is of internal origin [*Gauss*, 1839]. It is also known that the declination and inclination of the field changes with time at a relatively rapid rate, i.e. minutes of degrees per year. This change is called secular variation. Secular variation is known from instrumental records for the last 400 years and the rate of change is variable over the surface of the Earth. This record can be carried back in time by using dated archeological and geological materials studied using paleomagnetic techniques. The study of Earth's magnetic field in recent decades has been greatly aided by the use of orbiting satellites.

Earth's magnetic field is thought to arise from thermal convection in the Earth's core. The magnetic field at present is asymmetric between the northern and southern hemispheres. Figure 1 shows the dispersion of inclination at randomly selected points on the Earth's surface in the two hemispheres plotted against the inclination predicted by the geocentric axial dipole [*Tauxe*, 1998]. It can be seen that the dispersion is greatest at high southern latitudes. A recent study by Hulot

et al. [2002] which compares satellite magnetometer results 20 years apart, shows the rapid growth of flux patches in both hemispheres which the authors believe may lead to a reversal of Earth's magnetic field. The intensity of the magnetic field has been steadily decreasing over the last 100 years and is part of a trend that began 2000 years ago. The intensity of Earth's field is still at a historically high level and the satellite measurements cited above may signal an increase in the rate of decay of Earth's magnetic field.

Studies of secular variation over timescales of hundreds of thousands to millions of years involve the determination of the scatter of directions recorded in lavas. These studies consist of a series of paleomagnetic sites that sample individual cooling units and consist of five to ten individually oriented cores. These cooling units preserve the direction of magnetization at the time of extrusion and cooling through their blocking temperature. The process takes usually under a year for most lava flows, and the resulting acquired magnetization represents a spot reading of the direction of the magnetic field at the time and geographic location of cooling of the flow. The individual studies usually consist of data from thirty or more flows from a restricted area spread over several millions of years. Figure 2 shows, as an example, the site mean directions and the resulting virtual geomagnetic poles (VGPs) from a recent study of volcanics in Patagonia [*Mejia et al.*, in press]. Data like this represents hopefully a random sampling of the magnetic field through time that can be used to determine a time averaged field (TAF): however, volcanicity is often episodic which may tend to bias the results when many flow units are erupted in a short interval of time.

Timescales of the Paleomagnetic Field
Geophysical Monograph Series 145
Copyright 2004 by the American Geophysical Union
10.1029/145GM24

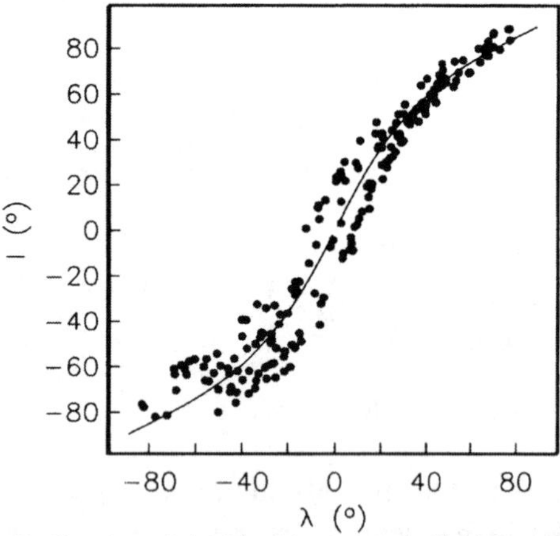

Figure 1. Inclination of the Earth's magnetic field from randomly selected sites on Earth's surface derived from values taken from the 1995 IGRF (dots) and inclination of the geocentric axial dipole (GAD). From Tauxe [1998].

The locality mean direction of magnetization is usually close to the direction of the geocentric axial dipole. Table 1 shows results of recent studies in which the number of sites is greater than 17 and where the data quality is high, i.e. the number of samples at each site is greater than five, the site mean α_{95} is small (generally <10°) and the directions have been determined by line fitting techniques. In some databases compiled for the study of paleosecular variation larger data sets exist but they are often of indifferent quality [*McElhinny and McFadden*, 1997]. Figure 3 shows that the mean inclination from these studies plot very close to the inclination expected

from the geocentric axial dipole field. It can be seen that several results from the reversed directions are shallower than the expected dipole field directions than are the normal directions; however the normal data sets are usually larger. The dashed line in Figure 3 gives the expected inclination of the field if a 5% quadrupole field is added. The precision of the data does not allow us to ascertain if a quadrupole field of this magnitude is present.

The directions from the studies can also be expressed as VGPs and the paleo-dispersion of the field is usually measured by determining angular standard deviation of the VGPs from the axis of rotation [*Merrill et al.*, 1996]. The dispersions are shown in Figure 4 plotted against latitude and compared with model "G" [*McFadden et al.*, 1991]. The data indicate that the effect of secular variation increases with increase in latitude from values less than 12° at lower latitudes to 18° or more at higher latitudes. In most studies poles that are more than 45° or so from the axis of rotation are regarded as directions of magnetization that were acquired during the time that the field was reversing polarity or during an excursion of the field. These directions are not all that common even in data sets that are quite large (Table 1). For instance in a study of lava flows from British Columbia by Mejia et al. [2002], consisting of 52 sites of Brunhes age, no excursional directions were observed. The same is true for a study of the Indian Heaven Volcanic field in Washington state by Mitchell et al. [1989] which consisted of 57 flows as well as for new results from Patagonia by Mejia et al. [in press] with both normal and reverse directions of magnetization and consisting of 46 sites

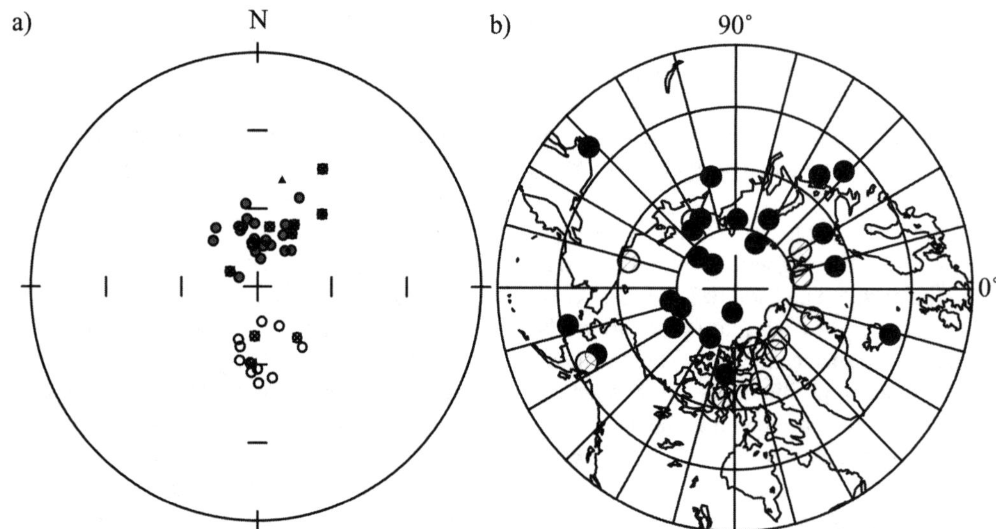

Figure 2. Paleomagnetic data from Southern Patagonia [*Mejia et al.*, in press]. Filled circles are normal sites and empty circles are reverse sites. (a) Site mean directions. Crossed sites indicate sites that did not pass the selection criteria. The triangle indicates the direction of the 2000 IGRF in the area. (b) Virtual pole positions of selected data.

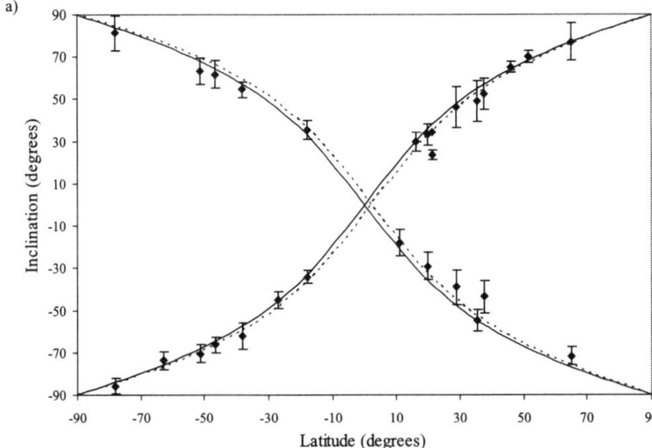

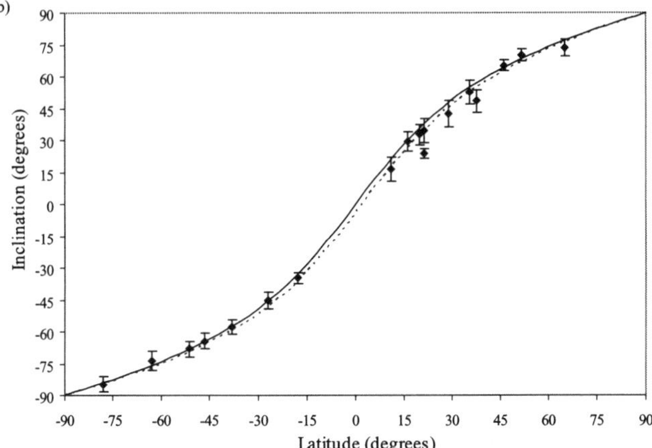

Figure 3. Magnetic inclination plotted against latitude. The figures show the mean inclination obtained from paleomagnetic studies listed in Table 1 (filled diamonds), inclination expected from GAD (solid lines) and inclination expected from a magnetic field composed of 95% GAD and 5% quadrupole components (dashed lines). (a) Normal and reverse data. (b) Normal and reverse data combined.

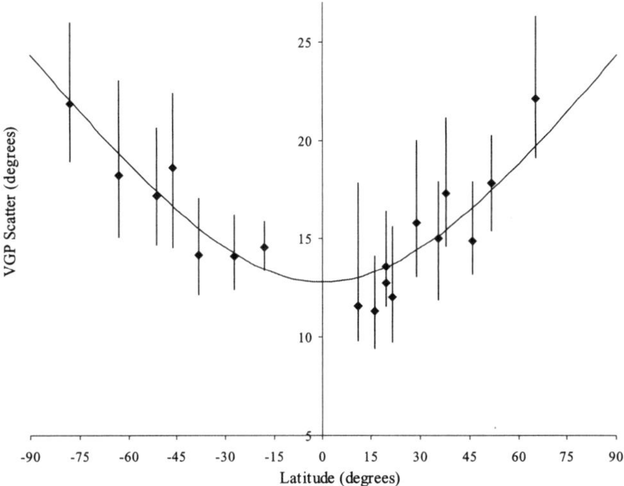

Figure 4. VGP scatter for studies from Table 1 compared to model "G".

between the northern and southern hemispheres. The result is a good fit to model "G". The way in which the paleosecular variation studies are done is very different from this averaging process and it would be expected that if the dispersion of the field were similar in the past to what it is today then larger dispersions would be expected. The true dispersion would be detected if the results are reasonably distributed in time and space. The magnetic field looks to be remarkably quiet when averaged over long periods of time.

The question that must be asked is how representative of the magnetic field in the past is the present field? Since it is the only one that we really know we have a tendency to believe that the present field is what we can expect Earth's field to be on

of Plio-Pleistocene age (Figure 2). However, a study of the Newer Volcanics of Australia [*Opdyke and Musgrave*, in press] yielded two intermediate directions of magnetization which are interpreted as transitional and both normal and reverse sites are represented in the data set. If the direction of the present magnetic field is determined at ten degrees longitudinal intervals along the small circle of the Earth corresponding to the latitude of the Newer Volcanics and VGPs determined from these values, the VGP scatter turns out to be 21 degrees (Figure 5), much greater than the value determined for the Newer Volcanics (14.2°). If however the sites that are believed to be transitional are included, then the VGP scatter is similar to that of the present field, i.e. 20.1°. The VGP dispersion plotted in Figure 4 is in agreement with model "G" of McFadden et al. [1991]. These authors fitted the present magnetic field to model "G" by averaging the magnetic field values

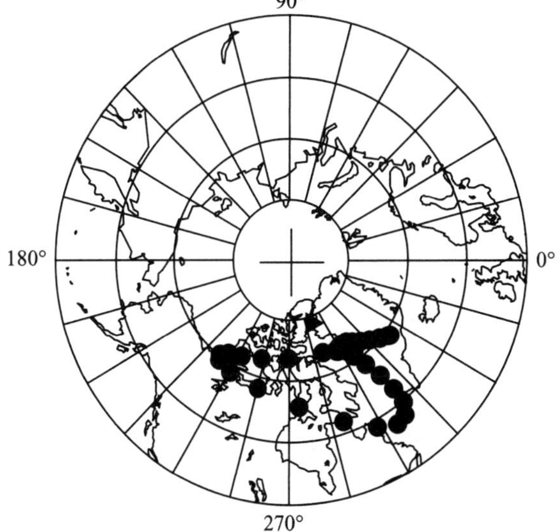

Figure 5. VGPs calculated every 10 degrees of longitude (from 2002 IGRF) around 37.5°S small circle of the Earth. VGP scatter is 21.2°.

Table 1. Summary of recent secular variation studies[a]

Reference	Location	Lat.	Long.	F.D.	Age	G	Dec	Inc	α_{95}	N	St/Sb	Sl	Su	Max disp.	ND1	Min VGP	ND2
Mankinen and Cox [1988]	Mc Murdo V.F.	-78.1	165.4	39	0-5	C	4.6	-84.7	3.5	38	21.9	18.9	25.9	6.2	0	48.7	1
						N	358.5	-85.8	3.9	29	---	---	---	---	---	---	---
						R	194.3	81.2	8.4	9	---	---	---	---	---	---	---
Baraldo et al. [2003]	Deception Island	-63.0	-60.6	27	Brunhes	N	348.8	-73.70	4.4	21	18.2	15.1	23.0	13.4	0	54.8	0
Mejia et al. [in press]	Patagonia	-51.2	-70.6	46	0 - 4.1	C	359.0	-68.2	3.5	33	17.2	14.7	20.6	≤5	13	56.4	0
						N	359.4	-70.6	4.3	22	---	---	---	---	---	---	---
						R	178.4	63.5	6.2	11	---	---	---	---	---	---	---
Camps et al. [2001][b]	Possession Is.	-46.4	51.8	45	0.5 - 5	C	5.2	-64.4	3.7	41	18.6	14.6	22.4	16	1	≥45	3
						N	359.4	-66.4	3.8	21	---	---	---	---	---	---	---
						R	190.7	61.8	6.4	20	---	---	---	---	---	---	---
Opdyke and Musgrave [2004]	Newer V.	-38.0	144.0	38	0 - 4	C	356.3	-57.7	3.5	33	14.2	12.1	17.0	≤6	3	≥40	2
						N	2.1	-62.3	6.3	13	---	---	---	---	---	---	---
						R	173.4	54.6	3.8	20	---	---	---	---	---	---	---
Brown [2002]	Easter Island	-27.1	-109.2	62	Brunhes	N	357.3	-45.2	4	55	14.1	12.4	16.2	≤10	7	53.7	0
Yamamoto et al. [2002]	Society Islands	-18.0	-148	154	0.62 - 3.45	C	2.1	-34.7	2.6	130	14.6	13.4	15.9	≤15	14	≥50	10
						N	3.8	-34.1	3.2	82	---	---	---	---	---	---	---
						R	179.7	35.5	4.5	48	---	---	---	---	---	---	---
Kidane et al. [2003][c]	Afar Depression	11.0	42.0	35	1.4 - 3.3	C	359.7	16.4	5.7	26	11.6	9.8	17.8	≤15	2	---	2
						R	179.6	-17.9	6	21	---	---	---	---	---	---	---
Carlut et al. [2000]	Guadalupe Island	16.0	-61.7	26	1 - 0.05	N	0.0	29.6	4.5	23	11.3	9.4	14.1	7.7	0	65.5	2
Mejia et al. [in prep.]	Mexico	19.6	-88.0	35	Pleistocene	C	356.2	32.6	4.6	32	13.6	11.6	16.4	≤5	---	---	---
						N	355.5	32.8	6.5	13	13.9	11.1	18.8	≤5	---	---	---
						R	176.7	-32.5	6.6	19	13.8	11.3	17.7	≤5	---	---	---
Laj et al. [1999][d]	Hawaii	21.4	157.5	105	3.22 - 3.11	N	2.8	23.7	2.3	103	12.7	11.6	14.1	13.6	1	53.4	1
Herrero and Valet [2002]	Hawaii	21.4	157.5	18	0.033 - 0.7	N	358.6	34.5	5.8	17	12.0	9.7	15.6	6.6	0	62.2	1
Tauxe et al. [2000]	La Palma Is.	28.8	-17.8	29	2 - 0.566	C	358.6	42.4	6.1	21	15.8	13.1	20.0	≤5	8	63.2	0
						N	355.1	46.1	9.6	9	---	---	---	---	---	---	---
						R	180.9	-39.1	8.4	12	---	---	---	---	---	---	---
Tauxe et al. [2003]	San Francisco V.	35.4	-111.8	47	0-5	C	351.9	52.6	5.5	22	15.0	11.9	17.9	k > 100	23	63.6	0
Tauxe et al. [2003]						N	349.9	49.1	9.4	12	---	---	---	---	---	---	---
						R	175.6	-54.7	5.2	10	---	---	---	---	---	---	---

Table 1. Continued.

Reference	Location	Lat.	Long.	F.D.	Age	G	Dec	Inc	α_{95}	N	St/Sb	Sl	Su	Max Disp.	ND1	Min VGP	ND2
Johnson et al. [1998]	Sao Miguel Is.	37.6	-26.4	35	0.78-0.88	N	357.4	48.5	5.3	28	17.3	14.7	21.2	k > 20	4	45.2	0
Mitchell et al. [1989]	Washington	46.0	-121.8	57	Brunhes	N	2.6	65.2	2.5	56	14.9	13.2	17.9	7.9	1	59.7	0
Mejia et al. [2002]	British Columbia	51.5	-122.4	52	Brunhes	N	356.9	70.2	2.8	45	17.8	15.4	20.2	≤5	7	52.5	0
Udagawa et al. [1999]	Iceland	65.1	-15.0	38	1.8 - 0.5	N	359.0	73.6	4	37	22.1	19.1	26.3	≤7.4	0	≥45	1

[a]Lat. and Long. are the approximate average latitude and longitude of the study areas, F.D. is the number of drilled flows, Age is given in Ma (for numerical values), G refers to the way the sites were grouped: N for normal, R for reverse, and C for combined normal and reversed sites. Dec. is declination, Inc. is inclination, St/Sb is the VGP scatter (either uncorrected or corrected for within site scatter). Sl and Su are the lower and upper 95% confidence limits of the scatter [*Cox*, 1969]. Max Disp. is the maximum α_{95} cut-off value used to consider a site for calculations (when preceded by "≤") or the maximum α_{95} value in the data set used for calculations (value not preceded by "≤"), except when the minimum k (dispersion parameter) is stated. ND1 is the number of sites excluded because of high scatter. Min VGP is the minimum VGP cut-off value used to consider a site (if value preceded by " ≥") or the minimum VGP value among the sites used for calculations (value not preceded by " ≥"). ND2 is the number of sites excluded because of low VGP.
[b]Confidence limits of scatter calculated using the method of Efron and Tibshirani [1986].
[c]Only the sites presented as tectonically stable were considered.
[d]Scatter of the VGP was recalculated in this study using the present position of the hot-spot to correct for plate movement.

average. However, when it is compared to the averaged field as determined from TAF studies it is clearly noisier than is expected. If Earth's field in the past is similar to the present field then a larger dispersion of the field would be expected in TAF studies. Therefore it is logical to presume that the present field is somehow anomalously noisy and represents Earth going into an excursion or a reversal. This is supported by the fact that the intensity of Earth's magnetic field is steadily decreasing. A similar conclusion was reached by Hulot et al. [2002] based on the change of the field over only 20 years.

2. CONCLUSIONS

Earth's magnetic field is much more variable at present than is expected from the study of the time averaged field. The instability of the present magnetic field and its decrease in intensity may indicate that the field is in the process of becoming unstable leading to an excursion or reversal of the field.

REFERENCES

Baraldo, A., A. E. Rapalini, H. Boehnel, and M. Mena, Paleomagnetic study of Deception Island, South Shetland Islands, Antarctica, *Geophys. J. Int.*, 153, 333–343, 2003.

Brown, L., Paleosecular variation from Easter Island revisited; modern demagnetization of a 1970s data set, *Phys. Earth Planet. Int.*, 133, 73–81, 2002.

Camps, P., B. Henry, M. Prevot, and L. Faynot, Geomagnetic paleosecular variation recorded in Plio-Pleistocene volcanic rocks from Possession Island (Crozet Archipelago, southern Indian Ocean), *J. Geophys. Res.*, 106, 1961–1971, 2001.

Carlut, J., X. Quidelleur, V. Courtillot, and G. Boudon, Paleomagnetic directions and K/Ar dating of 0 to 1 Ma lava flows from La Guadalupe Island (French West Indies): Implications for time-averaged field models, *J. Geophys. Res.*, 105, 835–849, 2000.

Cox, A., Confidence limits for the precision parameter k, *Geophys. J. R. Astron. Soc.*, 18, 545–549, 1969.

Efron, B., and R. Tibshirani, Bootstrap methods for standard errors, confidence intervals, and other measures of statistical accuracy, *Stat. Sci.*, 1, 54–77, 1986.

Gauss, C. F. Allegemeine Theorie des Erdmagnetismus, In *Resultate aus den Beobachtungen des magnetischen Vereins im Jahre 1838*, 57 pp., 1839. (Reprinted in *Werke*, 5, 121–193, 1877; Translated by Sabine, E., in Taylor, R., *Scientific Memoirs* Vol. 2, R. & J.E. Taylor, London, 1841).

Herrero-Bervera, E., and J. P. Valet, Paleomagnetic secular variation of the Honolulu volcanic series (33–700 ka), O'ahu (Hawaii), *Phys. Earth Planet. Int.*, 133, 83–97, 2002.

Hulot, G., C. Eymin, B. Langlais, M. Mandea, and N. Olsen, Small-scale structure of the geodynamo inferred from Oersted and Magsat satellite data, *Nature*, 6881, 620–623, 2002.

Johnson, C. L., J. R. Wijbrans, C. G. Constable, J. Gee, H. Staudigel, L. Tauxe, V. H. Forjaz, and M. Salgueiro, ^{40}Ar/^{39}Ar ages and paleomagnetism of Sao Miguel lavas, Azores, *Earth Planet. Sci. Lett.*, 160, 637–649, 1998.

Kidane, T., V. Courtillot, I. Manighetti, L. Audin, P. Lahitte, X. Quidelleur, P. Y. Gillot, Y. Gallet, J. Carlut, and T. Haile, New paleomagnetic and geochronologic results from Ethiopian Afar: Block rotations linked to rift overlap and propagation and determination of a ≈ 2 Ma reference pole for stable Africa, *J. Geophys. Res.*, 108, DOI 10.1029/2001JB000645, 2003.

Laj, C., H. Guillou, N. Szeremeta, and R. Coe, Geomagnetic paleosecular variation at Hawaii around 3 Ma from a sequence of 107 lava flows at Kaena Point (Oahu), *Earth Planet. Sci. Lett.*, 170, 365–376, 1999.

Mankinen, E. and A. Cox, Paleomagnetic investigation of some volcanic rocks from the McMurdo volcanic province, Antarctica, *J. Geophys. Res.*, 93, 11599–11612, 1988

McElhinny, M. W. and P. L. McFadden, Paleosecular variation over the past 5 Myr based on a new generalized database, *Geophys. J. Int.*, 131, 240–252, 1997.

McFadden, P. L., R. T. Merrill, M. W. McElhinny, and S. Lee, Reversals of the Earth's magnetic field and temporal variations of the dynamo families, *J. Geophys. Res.*, 96, 3923–3933, 1991.

Mejia, V., R. W. Barendregt, and N. D. Opdyke, Paleosecular variation of Brunhes age lava flows from British Columbia, Canada. *Geochem. Geophys. Geosys.*, 3, doi: 10.1029/2002GC000353, 2002.

Mejia, V., N. D. Opdyke, J. F. Vilas, B. S. Singer, and J. S. Stoner, Plio-Pleistocene time averaged field in Southern Patagonia recorded in lava flows. *Geochem. Geophys. Geosys.*, In press, 2004.

Merrill, R. T., M. W. McElhinny, and P. L. McFadden, *The magnetic field of the Earth*, 507 pp., Academic Press, San Diego, Ca, 1996.

Mitchell, R. J., D. J. Jeager, J. F. Diehl, and P. E. Hammond, Paleomagnetic results from the Indian Heaven volcanic field, south-central Washington, *Geophys. J. R. Astr. Soc.*, 97, 381–390, 1989.

Opdyke, N. D., and R. Musgrave, Paleomagnetic Results from the Newer Volcanics of Victoria: Contribution to the Time Averaged Field Initiative, *Geochem. Geophys. Geosys.*, In press, 2004.

Tauxe, L., C. G. Constable, C. L. Johnson, A. A. P. Koppers, W. R. Miller, and H. Staudigel, Paleomagnetism of the southwestern U.S.A. recorded by 0–5 Ma igneous rocks. Geochem. Geophys. Geosys., 5, doi: 10.1029/2002GC000343, 2003.

Tauxe, L., H. Staudigel, and J. R. Wijbrans, Paleomagnetism and $^{40}Ar/^{39}Ar$ ages from La Palma in the Canary Islands, Geochem. Geophys. Geosys., 1, 2000GC000063, 2000.

Tauxe, L., *Paleomagnetic principles and practices*, 299 pp., Kluwer Academic Publishers, Dordrecht, Netherlands, 1998.

Udagawa, S., H. Kitagawa, A. Gudmundsson, O. Hiroi, T. Koyaguchi, H. Tanaka, L. Kirstjansson, and M. Kono, Age and magnetism of lavas in Jokuldalur area, eastern Iceland; Gilsa event revisited, *Phys. Earth Planet. Int.*, 115, 147–171, 1999.

Yamamoto, Y., K. Shimura, H. Tsunakawa, T. Kogiso, K. Uto, H. G. Barsczus, H. Oda, T. Yamazaki, and E. Kikawa, Geomagnetic paleosecular variation for the past 5 Ma in the Society Islands, French Polynesia, *Earth, Planets Space*, 54, 797–802, 2002.

Victoria Mejia and Neil D. Opdyke, University of Florida, 241 Williamson Hall, P.O. Box 112120, Gainesville, Florida 32611-2120.